Carbonate Sequence Stratigraphy

Recent Developments and Applications

Edited by

Robert G. Loucks

and

J. Frederick Sarg

AAPG Memoir 57

Published by
The American Association of Petroleum Geologists
Tulsa, Oklahoma, U.S.A.
Printed in the U.S.A.

ISBN: 0-89181-336-5

Association Editor: Kevin T. Biddle
Science Director: Gary D. Howell
Publications Manager: Cathleen P. Williams
Special Projects Editor: Anne H. Thomas
Production: Custom Editorial Productions, Inc., Cincinnati, Ohio

Dust jacket photo: View of the Sassolungo (Langkofel) Group of peaks, western Dolomites, northern Italy. The high-standing light-colored peaks are formed by Triassic dolomite. The darker rocks in the foreground are basinal volcanic and carbonate turbidites and mudstones of the Triassic La Valle (Wengen) and San Cassiano (Cassian) formations. The photograph was taken from Col Rodella looking south. (Photograph by K. Biddle).

Related Titles from AAPG:
- **Cretaceous Carbonate Platforms** (AAPG Memoir 56), edited by J.A. Toni Simo, R.W. Scott, and J.-P. Masse
- **Carbonate Concepts from the Maldives, Indian Ocean** (AAPG Studies in Geology #34), by E.G. Purdy and G.T. Bertram
- **Carbonate Depositional Environments** (AAPG Memoir 33), edited by P.A. Scholle, D.G. Bebout, and C.H. Moore
- **Sedimentology and Sequence Stratigraphy of Reefs and Carbonate Platforms** (AAPG Continuing Education Course Note #34), by Wolfgang Schlager

These publications, and all other AAPG titles, are available from:
The AAPG Bookstore
P.O. Box 979
Tulsa, OK 74101-0979
Telephone: (918) 584-2555; (800) 364-AAPG (USA—book orders only)
FAX: (918) 584-0469; (800) 898-2274 (USA—book orders only)

AAPG
Wishes to thank the following
for their generous contributions
to

Carbonate Sequence Stratigraphy

Amoco Production Company

ARCO Exploration and Production Technology

The Donors of the Petroleum Research Fund,
Administered by the American Chemical Society

Kansas Geological Survey

Mobil Exploration and Producing Technical Center

*Contributions are applied against the production costs
of the publication, thus directly reducing the book's
purchase price and increasing its availability to a
greater audience.*

iii

About the Editors

◆

Robert G. Loucks received his B.A. degree from the University of New York at Binghamton in 1967 and his Ph.D. degree from the University of Texas at Austin in 1976. Before joining ARCO Exploration and Production Technology in 1983, he had already gained 15 years of research and exploration experience with Texaco, Texas Bureau of Economic Geology, Mobil, and Cities Service. He is presently a Research Advisor in ARCO's Reservoir Geology Group in Plano, Texas. His overall research interests include carbonate sequence stratigraphy, carbonate depositional systems, and carbonate and clastic diagenesis. Presently his research is concentrating on applying sequence stratigraphy to reservoir development, understanding rock fabric control on porosity and permeability, and reservoir heterogeneity in paleocave reservoir systems.

J. Frederick "Rick" Sarg has accumulated 18 years of petroleum exploration and production experience in research and operational assignments with major oil companies and as an Independent Consultant. Rick earned B.S. and M.S. degrees in geology from the University of Pittsburgh, Pittsburgh, Pennsylvania, and in 1976 he obtained his Ph.D. in geology from the University of Wisconsin, Madison. He was employed by Mobil in 1976, and then joined Exxon that same year. At Exxon, he became a member of the exploration research group that developed seismic/sequence stratigraphic concepts and techniques, and from 1985 to 1988 he was Supervisor of the Carbonate Group at Exxon Production Research Company. From 1990 to 1992 Rick operated as an Independent Consultant. Since 1992, he has been with Mobil Technology Center.

Rick Sarg has worldwide experience in integrated seismic-well–outcrop interpretation of siliciclastic and carbonate sequences. His emphasis has been on seismic stratigraphy, carbonate sequences, outcrop documentation of seismic-scale sequences, and Permian basin geology. Rick has authored and co-authored 15 papers on carbonate sedimentology and stratigraphy. He has prepared, coordinated, and presented seismic and sequence stratigraphy seminars to industry for 13 years. Rick is a member of AAPG, SEPM, GSA, PBS-SEPM, and WTGS. From 1985 to 1993 he served as an Associate Editor of the *AAPG Bulletin*. He was an AAPG Distinguished Lecturer in 1988–89. Currently he is a member of the Ocean Drilling Program, Sediments and Geochemical Processes Panel.

Table of Contents

◆

FOREWORD

This volume derives from the Research Symposium on Carbonate Sequence Stratigraphy presented at the 1991 national convention of the American Association of Petroleum Geologists. Selected papers from other technical meetings also are included. In formulating and compiling the papers for this volume, the coeditors recognized a need to bring together in one volume a representative cross section of current research in carbonate sequence stratigraphy. This volume is meant to be a "snapshot in time" of one of the fastest growing areas of stratigraphic research and application. We hope it will provide a stimulus for future research and for more successful application. Critical field observations, improvements in seismic resolution, and imaginative computer modeling are likely to make important future contributions to carbonate sequence stratigraphy. If this publication begins to generate new ideas and interdisciplinary discussion and debate, we will consider it to have accomplished its purpose.

The chapters in this volume clearly show the diverse response of carbonate platforms to subsidence, relative changes in sea level, and the related changes in environment and sediment supply. Some illustrate the similarities in geometry and accommodation response with siliciclastic sequences, and many demonstrate the unique responses carbonates show to tectonic subsidence, sea-level changes, and environmental fluctuations. The present volume covers a number of important topics in the area of carbonate sequence stratigraphy. These include: (1) conceptual models for interpretation; (2) the sedimentologic process-response of carbonate platforms to changing sea levels; (3) large-scale stratal patterns of second- and third-order sequences; and (4) small-scale, high-frequency cycle stacking patterns.

The chapters are grouped into three sections. The first describes sequence concepts and sedimentologic principles that permit and constrain interpretation. Handford and Loucks present conceptual models for interpreting carbonate sequences in a variety of depositional settings. Greenlee and Lehmann apply large-scale second-order sequence ideas to placing hydrocarbon-productive carbonate buildups in a stratigraphic framework, and Zempolich documents the platform response to long-term sea-level rises. Grammer et al. and Brown and Loucks describe the process/response of high-angle platforms to sea-level fluctuations.

The second section includes six chapters that describe seismic sequence-scale case studies involving both seismic and outcrop interpretations. The chapters of this section are grouped according to geologic age, from oldest to youngest, beginning with two chapters on the Devonian of the Canning basin (Southgate et al. and Holmes and Christie-Blick). These are followed by chapters interpreting: (1) the Upper Pennsylvanian stratigraphy of the Permian basin (Waite); (2) the drowning successions of a Mesozoic and a Tertiary platform (Erlich et al.); and (3) the Neogene reefs of Indonesia (Saller et al.) and the East Java Sea (Cucci and Clark).

The final section presents eight examples of high-frequency, meter-scale cycle deposition and stacking patterns. Montañez and Osleger describe parasequence stacking patterns from Cambrian platforms of the Great Basin, Nevada. Reid and Dorobek interpret parasequence sets in the context of larger scale second- and third-order sequences in the Mississippian of the western interior of the United States. Goldhammer et al. describe the high-frequency stratigraphy of an isolated, Triassic age platform in the Dolomites of northern Italy. Pomar describes the high-resolution stratigraphy of the Upper Miocene reefs on Mallorca, Spain, and compares outcrop geometry's to seismic progradational patterns. Franseen et al. qualify relative changes in sea level in similar Miocene cycles of southeast Spain. Four chapters from the Permian basin document the high-resolution cycle architecture of Upper Permian strata (Sonnenfeld and Cross), compare the internal characteristics of small-scale cycles to a Holocene shoal complex (Harris et al.), and interpret the diagenetic history related to cycle development (Hovorka et al. and Mutti and Simo).

The coeditors of this volume are very grateful for the authors' enterprise in the undertaking and for their patience in reworking their papers. We also appreciate the considerable interest by the AAPG–SEPM membership, whose large attendance at the 1991 Symposium and poster sessions indicates the importance the whole area of sequence stratigraphy has to petroleum exploration and development.

Robert G. Loucks
J. Frederick "Rick" Sarg

Carbonate Depositional Sequences and Systems Tracts—Responses of Carbonate Platforms to Relative Sea-Level Changes

C. Robertson Handford
Robert G. Loucks
ARCO Exploration and Production Technology
Plano, Texas, U.S.A.

ABSTRACT

Standard carbonate facies models are widely used to interpret paleoenvironments, but they do not address how carbonate platforms are affected by relative changes in sea level. An understanding of how the subtidal carbonate "factory" responds to relative sea-level changes and the role played by other environmental factors towards influencing the formation of carbonate platforms allows one to differentiate platform types and it helps establish a basis for constructing depositional sequence and systems tract models. The combination of in-situ production of carbonate sediment, which is also subject to transport, and local variations in depositional processes result in the formation of a wide variety of stratal patterns, some of which are unique to carbonate systems.

Fundamental carbonate-depositional principles and geologic-based observations were used to construct depositional sequence and systems tract models for a variety of rimmed shelves and ramps. The models show how, for example, depositional sequences made up of (1) carbonate, (2) carbonate-siliciclastic, or (3) carbonate-evaporite-siliciclastic facies are produced by depositional systems responding to lowstand, transgressive, and highstand conditions. *Lowstand:* Carbonate sediment production is reduced on rimmed shelves because a relatively small area of shallow seafloor is in contact with the carbonate "factory." Reduced sedimentation and subaerial exposure foster the retreat of shelf edges and slopes by erosion and slope failure during lowstands. As a result, thick debris-flow deposits may form. Karst development is important in humid climates and can affect large areas of a subaerially exposed platform. If siliciclastic sediments are available, they are delivered to the shelf edge and slope by fluvial-deltaic systems or, in arid climates, by wadis and advancing ergs. Under arid conditions, lowstand evaporites may fill an isolated or completely silled basin. *Transgression:* Carbonate sedimentation initiates in restricted environments and later as more open conditions develop, open marine facies, including patch reefs, may locally develop atop flooded platforms and ramps. Retrogradational parasequences comprising shallow-water carbonates form and subsequently drown, and shelf edges tend to aggrade, backstep, and drown if the rate of sea-level rise is high. *Highstand:* Seaward-prograding carbonate or siliciclastic coastal sediments and landward-prograding carbonate rimmed shelf edges may partially fill inner to outer shelf seas. Under arid conditions, evaporites and red beds commonly fill wide and shallow salinas. These strata onlap subaerially exposed rimmed shelf edges and prograding grainstone islands in ramps. Shelf edges and shorelines tend to prograde under the influence of high rates of carbonate sedimentation across the shelf and shelf edge. Slope and basinal environments receive excess shelf- and shelf-edge-derived sediment.

Factors listed above must be integrated with established facies models in order to arrive at comprehensive sequence and systems tract models. As should be the case with all models, however, they are not meant to serve as rigid templates within which all carbonate sequences must fit. Modification may be needed to accommodate each case. Once they are deemed applicable to a specific case, they function as working hypotheses to help geologists visualize how and why carbonate strata were laid down and fit together as they do. As a general predictor of facies, carbonate depositional sequence and systems tracts models may be used in conjunction with seismic records to identify depositional systems and to locate reservoir-, seal-, and source-prone facies.

INTRODUCTION

Over the past several decades, carbonate facies models of ramps (Ahr, 1973; Read, 1985), shelves (Wilson, 1975; Read, 1985), and craton settings (Irwin, 1965; Shaw, 1964) have been routinely used for describing and interpreting lateral facies relationships in ancient carbonate platforms (Figure 1). They offer a static representation of a carbonate platform by depicting an idealized distribution pattern of facies and paleoenvironments, usually during an instant of time and in the absence of relative sea-level changes. As pointed out by Irwin (1965), "Nature is never static. In the geologic past, epeiric seas transgressed and regressed over the continents during numerous periods." Environments migrate accordingly, but predicting the occurrence of a facies poses a difficult challenge to the explorationist and stratigrapher. During the history of a carbonate platform, paleoenvironments behave like "moving targets." They appear, migrate, disappear, and reappear to a large extent in response to depositional and erosional processes associated with marine transgressions and regressions imposed by relative changes in sea level. Thus, the predictive capacity of the facies models listed above is limited by their static view of time and relative sea-level changes.

Sequence stratigraphy integrates time and relative sea-level changes to track the migration of facies. Rooted mainly in seismic sequence analysis, the strength of sequence stratigraphy lies in its potential to predict facies within a chronostratigraphically constrained framework of unconformity-bound depositional sequences (Haq et al., 1987; Posamentier and Vail, 1988; Vail et al., 1977; Vail, 1987; Van Wagoner et al., 1990). Using the methodology developed for seismic sequence stratigraphy by Vail et al. (1977), interpreters analyze seismic reflections to describe stratal geometry and delineate the systematic patterns of lap-out and truncation of strata against chronostratigraphically constrained surfaces. In this manner, they establish the presence of unconformity-bound depositional sequences, deduce relative sea-level changes, and describe the depositional and erosional history of an area.

The methodology and practice of seismic sequence analysis led to the development of the Exxon sequence stratigraphic model and accompanying systems tract diagrams (Haq et al., 1987; Posamentier and Vail, 1988; Vail, 1987). This model summarizes the expected stratal patterns and geometries of siliciclastic depositional sequences that form in passive-margin depositional basins and in response to relative sea-level changes (Figure 2).

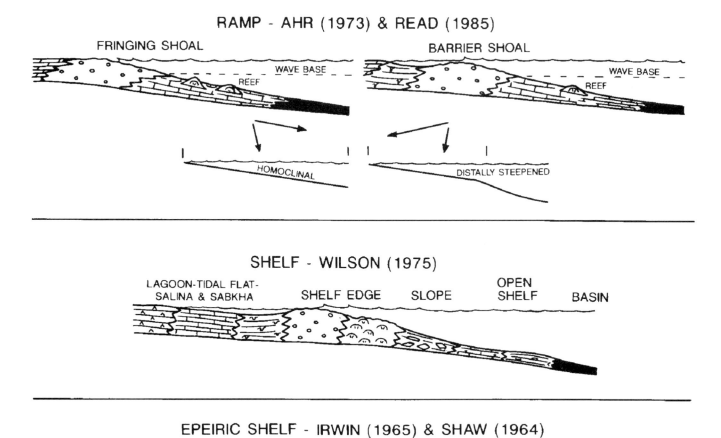

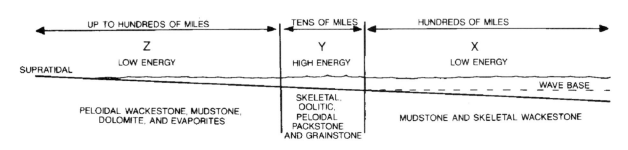

Figure 1. Standard facies models of ramps, rimmed shelves, and epeiric seas are static representations of carbonate platforms.

Sequence stratigraphy is now considered to be a practical methodology for analyzing the development and history of carbonate platforms (Eberli and Ginsburg, 1989; Handford and Loucks, 1990, 1991; Hardie et al., 1991; Hunt and Tucker, 1991; Jacquin et al., 1991; Rudolph and Lehmann, 1989; Sarg, 1988). Much of the existing work utilized or adapted the Exxon model of siliciclastic depositional sequence stratigraphy to help explain the evolution of carbonate depositional sequences. However, this siliciclastic-based model operates upon a different set of principles than those guiding carbonate deposition. The Exxon model assumes all sediment is extrabasinal in origin and is delivered to a marine basin largely by fluvial and deltaic systems which drain areas lying landward and updip of the basin (see figure 3 of

Vail, 1987). No such assumption, however, can be made for carbonate depositional sequences because carbonate sediments are not delivered to a basin, but rather are produced in the marine basin by organic and inorganic processes (James, 1979). This results in the unique ability of carbonate sediments to construct depositional topography and morphologically diverse platforms, such as rimmed shelves, isolated or detached platforms, and ramps. Siliciclastic systems can construct sequences with distinct shelf breaks but they lack rimmed margins. In addition, siliciclastic slope angles are generally lower than those of carbonates (Schlager and Camber, 1986). Pronounced depositional topography, due to organic buildups, is a hallmark of many carbonate sequences, and it alone guarantees a dissimilarity between car-

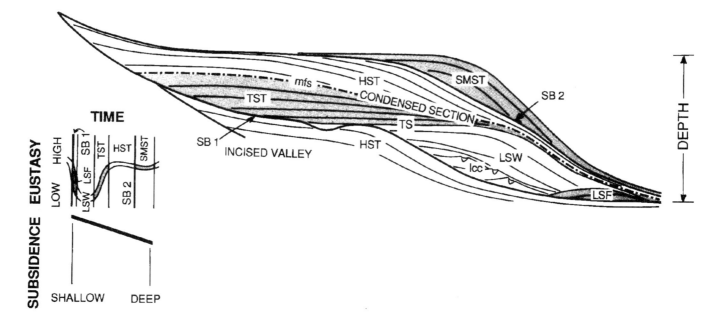

Figure 2. The depositional sequence model of Haq et al. (1987) and Vail (1987) was generally developed from studies of siliciclastic sequences. Abbreviations are as follows. Surfaces: SB1 = type 1 sequence boundary; SB2 = type 2 sequence boundary; mfs = maximum flooding surface; TS = transgressive surface. Systems Tracts: HST = highstand systems tract; TST = transgressive systems tract; LSW = lowstand wedge systems tract; LSF = lowstand fan systems tract; SMST = shelf-margin systems tract; lcc = leveed channel complex. Modified from Haq et al. (1987) and Vail (1987).

bonate and siliciclastic stratal geometries. Where carbonate and siliciclastic sequences do share similar geometries, it is only within platforms dominated by clastic carbonate sedimentation, such as ramps, or low-angle rimmed shelves (Sarg, 1988). This is because clastic carbonate and siliciclastic sediments respond similarly to wave- and current-transport and deposition.

Owing to the polygenetic origin of carbonate sediments, the strong environmental controls over carbonate sedimentation (Davies et al., 1989; Hopley, 1982; Schlager, 1991; Wilson, 1975), and the diverse ways that carbonates respond to relative changes in sea level, all of which differ from siliciclastics, we question the universal application of the siliciclastic-based Exxon depositional sequence model (Haq et al., 1987; Posamentier and Vail, 1988; Vail, 1987) to the interpretation of carbonate platform stratigraphy and relative sea-level history. Although the fundamental aspects of the Exxon siliciclastic sequence model are reasonably applicable to many carbonate sequences, the important differences between carbonates and siliciclastics mandate the development of separate carbonate sequence and systems tract models.

Sequence stratigraphic analysis is an evolving discipline whose underlying assumptions require a critical evaluation (Schlager, 1991). As a means of contributing to that evaluation, we have drawn upon geologic and geophysical data of modern and ancient carbonate systems, and utilized principles of carbonate sedimentation, existing carbonate facies models,

and the methodology of sequence stratigraphic analysis to:

1. identify carbonate stratal patterns, which differentiate them from siliciclastic stratal patterns, and
2. develop depositional sequence and systems tract models for various types of rimmed shelves and ramps.

INFRASTRUCTURE OF CARBONATE PLATFORMS

Carbonate platforms are similar to siliciclastic shelves to the extent that they are constructed and modified by depositional and erosional processes acting under the controls exerted by eustasy, tectonic subsidence, sedimentation rate, and climate (Sarg, 1988; Vail, 1987). The first three controls interact to regulate relative sea level. Climate mainly governs sediment type. Together, their interaction produces many of the variations in stratal patterns and facies in depositional sequences (Haq et al., 1987; Jervey, 1988; Kendall and Schlager, 1981; Sarg, 1988; Vail, 1987) and variations in gross morphology, or size and geometry, of carbonate platforms (Bosellini, 1989). The interaction of these and other factors are viewed as an infrastructure that describes the physical nature of carbonate platforms (Table 1). Assessing these factors can lead to a more thorough understanding of carbonate platforms and their evolution. We have used them to characterize some modern and ancient platforms (Table 2) and as a first step in performing

Table 1. Carbonate platform infrastructure.

INFRASTRUCTURAL ELEMENTS							
Eustasy	Lowstand		Transgression		Highstand		Stillstand
Subsidence/ Tectonic Setting	Thermal Cooling			Crustal Thinning		Crustal Loading	
	Intraplate			Plate Margin			
	Continental (craton)		Oceanic	Passive Margin		Convergent Margin	Transform Margin
Climate	Humid			Semiarid		Arid	
Carbonate Factory	Tropical (Low Latitude)					Temperate-Polar (Mid-High Latitude)	
	Biogenic Production by Phototrophic (Chlorozoan-Chloralgal) Organisms		Biogenic Production by Nonphototrophic (Foramol) Organisms & Red Algae		Nonbiogenic Production (Ooids, Peloids)	Nonphototrophic (Foramol) Organisms Plus Coralline Red Algae	
Continental Linkage	Attached			Detached			
				Solitary		Coalesced	
Morphology	Ramp			Rimmed Shelf		Flat-Topped Shelf	
	Homoclinal	Distally Steepened					
Width	Narrow (<10 km)			Wide (>10 km)			
Circulation	Open			Restricted			
	Tides	Waves		Currents	Storms		Upwelling
Type of Margin	Accretionary		Bypass		Erosional		
	Reef			Grain Shoal			
Orientation of Margin	Windward			Leeward			
Depositional Systems	Coastal & Nearshore		Offshore Shelf		Shelf-Edge	Slope and Basin Floor	
Lithology	Carbonate	Carbonate/Siliciclastic		Carbonate/Evaporite		Carbonate/Evaporite/Siliciclastic	

sequence stratigraphic analysis. The following summarizes many of these factors.

Carbonate Factory

Sedimentation rates across carbonate platforms are largely dependent on the productivity of the marine-subtidal carbonate factory. The ability of the carbonate factory to generate carbonate sediment hinges on the interaction of latitude, temperature, salinity, water depth, sunlight intensity, turbidity, water circulation, PCO_2, and nutrient supply (Hallock and Schlager, 1986; Lees, 1975; Lees and Buller, 1972; Milliman, 1974; Wilson, 1975). Where these fall within the range conducive to the production of biogenic and nonbiogenic carbonate sediment and interact properly, a vigorous carbonate factory can result.

The tropical marine factory operates in warm shallow seas between 0–30° north and south latitude. The factory exists as a shroud encompassing the surface waters of tropical, clear marine seas down to a depth of about 100 m but with the bulk of production occurring within the upper 10 m (Wilson, 1975). The sediment producers are chiefly phototrophic, or chlorozoan and chloralgal, organisms, and nonphototrophic organisms. The former includes hermatypic corals and calcareous red and green algae and the latter consists mainly of foraminifers, mollusks, bryozoans, and echinoderms. For healthy development of corals and calcareous green algae, minimum and maximum annual-water temperatures cannot fall below 15°C and 26°C, respectively (Lees and Buller, 1972). Nonskeletal grains, or ooids and peloids, and carbonate mud, derived from disintegration of calcareous algae and direct precipitation, are also produced almost exclusively in warm seas. Formation of modern ooids is further constrained by physical energy and salinity. Ooids tend to form only in shallow

Table 2. Some important carbonate platforms and their distinguishing characteristics.

Platform	Age	Tectonic Setting	Climate	Carbonate Factory	Linkage	Morphology	Width	Circulation	Type of Margin*	Orientation	Sediment Type
Bahamas	Q	Passive Continental Margin	Humid-Semiarid	Tropical	Detached	Rimmed Flat-Topped Shelf	Wide	Open-Restricted	A, B, E Reef-Grain Shoal	Windward-Leeward	Carbonate
Belize	Q	Transform Margin	Humid	Tropical	Attached	Rimmed Shelf	Narrow-Wide	Open-Restricted	A, B, E Reef	Windward	Carbonate-Siliciclastic
Great Barrier Reef, Australia	Q	Passive Continental Margin	Humid	Tropical	Attached	Rimmed Shelf	Wide	Open	A, ? Reef	Windward	Carbonate-Siliciclastic
Bermuda	Q	Oceanic Intraplate Hot Spot	Humid	Tropical	Detached	Rimmed Shelf	Narrow-Wide	Open	? Reef	Windward-Leeward	Carbonate
NE Yucatan, Mexico	Q	Transform Continental Margin	Humid	Tropical	Attached	Rimmed Shelf	Narrow-Wide	Open-Restricted	? Reef-Grain Shoal	Windward	Carbonate
Campeche Bank, Mexico	Q	Passive Continental Margin	Humid	Tropical	Attached	Ramp (Distally Steepened)	Wide	Open	? Grain Shoal/Reef	Leeward	Carbonate
West Florida	Q	Passive Continental Margin	Humid	Tropical	Attached	Ramp (Distally Steepened)	Wide	Open	A, E, Shoal	Leeward	Carbonate-Siliciclastic
Florida Keys	Q	Passive Continental Margin	Humid	Tropical	Attached	Rimmed Shelf	Wide	Open-Restricted	A, ?, Reef	Windward	Carbonate
Red Sea	Q	Early Rifted Continental Margin	Arid	Tropical	Attached	Rimmed Shelf	Narrow	Restricted	A, ? Reef	Windward-Leeward	Carbonate-Siliciclastic
South Australia	Q	Passive Continental Margin	Semiarid	Temperate	Attached	Ramp (Distally Steepened)	Wide	Open	A Reef	Windward	Carbonate
Maldives Islands, Indian Ocean	Q	Passive Oceanic Margin Hot Spot	Humid	Tropical	Detached	Rimmed Shelf	Wide	Open	A, B, E Reef	Windward-Leeward	Carbonate
Persian Gulf	Q	Convergent Continental Margin	Arid	Tropical	Attached	Ramp (Homoclinal)	Wide	Restricted	Reef/Grain Shoal	Windward	Carbonate-Evaporite-Siliciclastic

Table 2. Continued.

Platform	Age	Tectonic Setting	Climate	Carbonate Factory	Linkage	Morphology	Width	Circulation	Type of Margin*	Orientation	Sediment Type
Pulau Seribu, Indonesia	Q	Oceanic Intraplate Backarc	Humid	Tropical	Detached	Rimmed Shelf	Narrow	Restricted	A Reef	Windward-Leeward	Carbonate-Siliciclastic
Stuart City, Texas	K	Passive Continental Margin	Arid	Tropical	Attached	Rimmed Shelf	Wide	Open-Restricted	A, ? Reef/Grain Shoal	Leeward?	Carbonate-Evaporite-Siliciclastic
Permian Basin, Texas, New Mexico	Perm.	Convergent Continental Margin	Arid	Tropical	Attached	Rimmed Shelf	Wide	Restricted	A, B, E, Reef/Grain Shoal	Windward Leeward?	Carbonate-Evaporite-Siliciclastic
Smackover, Gulf Coast	Jur.	Passive Continental Margin	Arid	Tropical	Attached	Ramp/Rimmed Shelf	Wide	Open-Restricted	Grain Shoal/Reef	?	Carbonate-Evaporite-Siliciclastic
Golden Lane, Mexico	K	Passive Continental Margin	Semiarid?	Tropical	Detached	Rimmed Shelf	Wide	Open-Restricted	Reef/Grain Shoal	Windward-Leeward	Carbonate
Canning Basin, Western Australia	Dev.	Passive Craton Margin	Semiarid?	Tropical	Attached-Detached	Rimmed Flat-Topped Shelf?	Wide	Open?	A, B, E Reef/Grain Shoal	?	Carbonate-Siliciclastic
Dolomites, Italy	Tr	Passive Margin	Semiarid?	Tropical	Detached	Rimmed Flat-Topped Shelf	Narrow	Open Restricted?	A, B, E Reef/Grain Shoal	Windward-Leeward	Carbonate
Boone Formation, Arkansas	Miss.	Passive Continental Margin	Humid	Tropical	Attached	Ramp	Wide	Open	Grain Shoal	Windward	Carbonate
Natuna L, S. China Sea	Mio.	Rifted Continental Borderland	Humid	Tropical	Detached	Rimmed Shelf	Wide	Open?	A Reef/Grain Shoal	Windward-Leeward?	Carbonate

* A = Accretionary; B = Bypass; E = Erosional

(<2 m), agitated waters where daily wave and current activity is high (Bathurst, 1971; Loreau and Purser, 1973; Newell et al., 1960), and where salinity exceeds 35.8% (Lees, 1975).

In temperate and polar seas, the carbonate factory produces skeletal grains from calcareous red algae, and nonphototrophic organisms such as foraminifers, mollusks, bryozoans, echinoderms, and barnacles (also known as the foramol assemblage). Lacking calcareous green algae, temperate seas produce relatively minor amounts of carbonate mud. Where mud is produced, it is due mainly to mechanical abrasion and bioerosion of skeletal grains and the accumulation of coccoliths (Blom and Alsop, 1988).

Sediment production and accumulation rates are greater in tropical platforms than temperate settings. The following is summarized from prior work (Glaser and Droxler, 1991; James and Bone, 1991; James and MacIntyre, 1985; Sarg, 1988; Schlager, 1981; Wilson, 1975).

1. Holocene carbonate accumulation rates can match or exceed glacio-eustatic Holocene rise of sea level (30–700 cm kyr) (Schlager, 1981).

2. The long-term accumulation rates of ancient carbonate platforms (<1–50 cm kyr) are much less than Holocene rates (up to 1500 cm kyr). Though the differences are impressive, they do not take into account that Holocene rates are spot or local values whereas ancient rates record the entire platform. Furthermore, ancient rates do not take into account burial compaction, minor hiatuses, and long-term stillstands (Sarg, 1988).

3. High Holocene accumulation rates reflect the large amount of accommodation space available during the last rapid rise in sea level MacIntyre (1983) determined that Galeta Reef, Panama, accumulated at 390 cm kyr from 7000–4000 yr ago when the rate of sea-level rise was high. Subsequently, reef growth dropped to 60–70 cm kyr for the last 3000–4000 yr when the rate of rise slowed markedly. Schofield et al. (1983) recognized a similar pattern at Elizabeth Reef in the Tasman Sea.

4. Platforms often produce more sediment than can be accommodated by their tops. As a result, excess sediment is shed to adjacent slopes, which can result in progradation. Thus, relatively low accumulation rates do not always reflect low productivity and low growth potential. Measured low accumulation rates on platform tops can be offset by high rates of progradation.

5. While the average growth potential for reef-bearing carbonate platforms is 100 cm kyr, there are significant variations between environments. Shallow reef margins can accumulate at rates of 2000 cm kyr while ooid-sand margins have accumulation rates of ~50–200 cm kyr (Schlager, 1981). Much lower accu-

mulation rates characterize muddy platform-interior settings. With an average thickness of 0.65 m, a Holocene succession of carbonate mud has accumulated in the interior of Little Bahama Bank at an average net rate of 12 cm kyr (Neumann and Land, 1975). Where sediment thickness reaches 2 m, net accumulation rate is 36 cm kyr. If there were no offbank loss of algal sediments in Little Bahama Bank, accumulation rates would be 1.5 times the present net accumulation rate (Neumann and Land, 1975). Locally high accumulation rates are documented from mud banks in coastal lagoons of Yucatan. Brady (1971) recorded 11.5 m of muddy sediment in Nichupte Lagoon, which translates to an accumulation rate of 192 cm kyr. Very little or no sediment is transported off the platform in Yucatan because the coastal lagoons are protected from open seas.

6. Reef accumulation rates decrease with water depth— ~100–2000 cm kyr at <5 m water depth to 50–200 cm kyr in depths of 10–20 m (Schlager, 1981).

7. Highstand shedding of carbonate mud has led to accumulation rates of 200–270 cm kyr in water depths of 245–300 m on the flanks of Pedro Bank and Great Bahama Bank (Glaser and Droxler, 1991).

8. Carbonate sediment accumulation rates average 2.5 cm kyr in temperate seas and the normal range is 1–20 cm kyr (James and Bone, 1991). Calcite mud is accumulating in Bass Basin, southeastern Australia at a rate of <12 cm kyr (Blom and Alsop, 1988).

9. Locally high accumulation rates of carbonate sediment are possible in temperate platforms. Extensive seagrass banks in the protected central gulfs of the south Australian continental shelf provide a habitat for organisms that accumulate as skeletal sand. Rates reach 100 cm kyr (Gostin et al., 1988), which is equal to the average platform growth potential calculated by Schlager (1981).

Based upon estimates of biogenic carbonate productivity rates, Wilson (1975) and Schlager (1981) graphed profiles of total carbonate productivity versus depth. They clearly showed the influence of water depth, and consequently light intensity, over carbonate sediment production. In tropical seas, the threshold of abundant biogenic carbonate sediment production lies at depths less than 10–15 m. Another view, which complements their illustrations, schematically shows how productivity and accumulation rates can vary across tropical carbonate platforms (Figure 3).

Climate

Climate, which is a measure of air temperature, precipitation, atmospheric humidity, and wind, helps

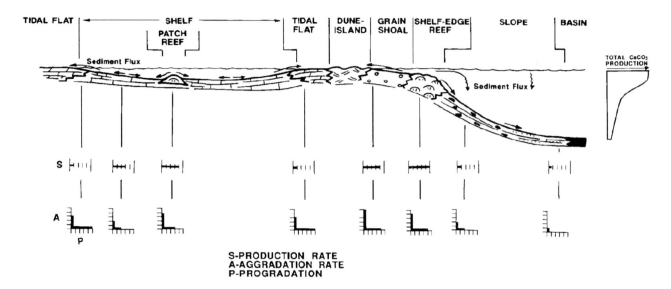

Figure 3. Carbonate sediment productivity and accumulation (aggradation and progradation) vary with depth and depositional setting. These graphic scales schematically show productivity and accumulation variations across an idealized platform. Note that supratidal environments produce little or no sediment, but accumulation rates can be high there. Arrows show sediment flux, or net general direction of sediment movement. Total CaCO$_3$ production after Schlager (1981) and Wilson (1975).

determine water conditions (salinity, water temperature, and circulation) and, hence, the nature of the carbonate factory (tropical or temperate) and the types of carbonate sediments produced. Shallow tropical-marine waters have a higher degree of CaCO$_3$ supersaturation than the temperate seas of the mid-latitudes. This difference affects the production, stability, and early lithification potential of carbonate sediments (Scoffin, 1987). Climate helps determine the types of sediments, besides carbonates, that will be deposited within a depositional sequence. Under arid conditions and restricted circulation, evaporite deposition may occur. If terrigenous sediment sources lie adjacent to a carbonate platform, differences in climate will affect the style of siliciclastic sediment delivery. Humid climates favor fluvial-deltaic deposition of siliciclastic sediments, and arid climates foster eolian siliciclastic deposition. The presence of these types of sediments in a carbonate-dominated stratigraphic succession is a clue not only to climatic conditions, but oftentimes, they signal relative sea-level changes. For example, the presence of thin eolian- and wadi-deposited sandstones in the Guadalupian carbonate-platform succession of New Mexico (Mazzullo et al., 1991) imply relative lowstand conditions. Shoreface and eolian sandstones, which disconformably overlie dissected karst and peritidal carbonates in the Middle Ordovician succession in north Arkansas, are directly linked to lowstand conditions.

Subsidence and Tectonic Setting

Without subsidence, long-term deposition, accumulation, and preservation of carbonate sediments will not take place. Subsidence, which takes place by thermal cooling, crustal thinning, and loading (Allen and Allen, 1990), plays a major role, with eustasy, in creating the space available for marine sedimentation to take place (Jervey, 1988). Rate of subsidence depends on the type of crust that is subsiding (oceanic or continental), its age, the type of stress causing the subsidence, lithospheric rheology, and position within a lithospheric plate, or tectonic setting. Subsidence may be driven by lithospheric thinning, cooling, creep of ductile lower crust towards the new oceans, and phase changes (gabbro to eclogite) in the lower crust or mantle. In addition, sediment loading may enhance tectonically driven subsidence. Shallow marine carbonate platforms form near and along convergent, divergent, and transform plate margins, as well as within plates made up of either oceanic or continental crust. The chance of preservation is improved by intraplate and passive margin settings. Tectonic setting helps to establish the following:

1. the location, elevation, and areal extent of surrounding terrain that may provide part, or all, of the detrital sediments contributed to the overall basin fill;
2. the basement geometry of the subsiding basin;
3. the initial geometry of shallow-marine carbonate sedimentation sites; and
4. the extent and style of marine influence.

Continental Linkage

Platforms may be isolated (detached) from or they may be linked (attached) to a large landmass, such as a continent or large island. Attached platforms are commonly long, linear features that face open seas, usually along passive continental margins. Large detached, or isolated, carbonate platforms develop on horst blocks along newly rifted continental margins and failed rifts. They may also form in oceanic plates by nucleating above hot spots around volcanoes and seamounts. Some detached platforms are solitary throughout their entire existence, but where several lie close to one another, they may coalesce into one larger platform, as in the case of the Great Bahama Bank (Eberli and Ginsburg, 1989).

Morphology

Platforms can assume three different morphologic profiles (Figure 4): (1) ramps have gentle homoclinal or distally steepened slopes; (2) rimmed shelves; and (3) unrimmed, flat-topped shelves that have a sharp break in slope along their seaward margins. Rimmed shelves form exclusively in tropical seas because they are made up chiefly of reef-building chlorozoan and chloralgal organisms, which require warm-water conditions. Unrimmed shelves are present in both tropical and temperate seas. Ramps comprise mainly clastic carbonate grains and mud and, thus, they occur in tropical and temperate seas. Drowned platforms are often recognized as a separate type of platform (Read, 1985; Tucker and Wright, 1990). However, they are not morphologically distinct from those listed above, but in fact are simply drowned ramps, rimmed shelves, or flat-topped shelves. Thus, they record a phase of development, and specifically a relative rise of sea level.

Width and Circulation

Carbonate platforms have highly variable widths, ranging from a few to >100 km. An arbitrary width of 10 km was chosen to differentiate between narrow (<10 km) and wide (>10 km) platforms. Circulation, which is responsible for delivering clear open-marine water to carbonate platforms is either dominated or

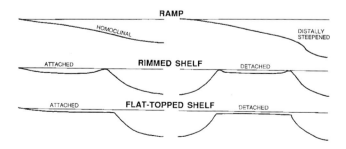

Figure 4. Morphologic profiles of carbonate platforms include ramps (homoclinal and distally steepened), rimmed shelves (attached and detached), and unrimmed, flat-topped shelves (attached and detached).

strongly influenced by wave- and wind-driven currents in shallow shelves (Johnson, 1978). Tides and waves, which are among the most important physical processes, are dependent on the size, shape, and bathymetry of the marine basin (Elliott, 1986). In general, tidal amplitude increases with shelf width over most of the world's continental shelves (Cram, 1979) and it decreases in restricted seas, with a small connection to open ocean where tidal waves originate. Thus, tidal range is minimal along narrow shelves or in restricted seas such as the Persian Gulf or Red Sea. Relative influence and the interactions between tide-, wind-, and wave-driven currents, in concert with nutrient levels, govern the nature of carbonate sedimentation from the shallow shelf edge, across the shelf, and into the supratidal environments of a carbonate platform.

Type of Margin and Orientation

Other factors controlling carbonate platform development are type of margin, its orientation relative to the wind, and sediment type. Shelf margins typically consist of grain shoals, reefs, or mixtures of the two. Depending on their length, continuity, and depth over their crests, they help control circulation of seawater within a platform. Continuous or nearly continuous rims hamper circulation and exchange of open seawater with shelves. Rimmed shelves with deep or less continuous rims and rimless flat-topped shelves are characterized by cross-bank circulation. Seaward slopes of shelf edges are marked by depositional accretion, sediment bypass, or erosion (McIlreath and James, 1979; Read, 1985). Bahamian workers (Eberli and Ginsburg, 1989; Hine et al., 1981a, b; Wilber et al., 1990) have pointed out that downwind, offbank transport and deposition of platform sediments have led to significant progradation of the leeward margins of modern and ancient platforms comprising the Great Bahama Bank. However, windward margins tend to be erosional, or bypass, and grow vertically (Eberli and Ginsburg, 1989).

Depositional Systems and Lithology

Carbonate platforms comprise a myriad of depositional systems, all of which are constrained by factors listed above, such as climate, morphology, width, circulation, and orientation of margin. Chief among carbonate-platform depositional systems are the following:

Coastal and nearshore systems
- Beaches, coastal dunes, tidal inlets, tidal deltas, and lagoons associated with mainland shorelines and barrier islands
- Fringing reefs
- Tidal flats, channels, and sabkhas
- Evaporite salinas

Offshore shelf systems
- Pinnacle reefs, patch reefs, mud banks, and sand shoals

- Open, shallow to deep, storm-dominated shelves
- Evaporite salinas (once open marine but subsequently isolated to become evaporitic)

Shelf-edge systems
- Reefs
- Grain shoals such as tidal-bar belts and marine sand belts
- Tidal channels and deltas
- Islands and eolian dunes

Slope and basin-floor systems
- Translational and rotational slides
- Channelized and nonchannelized sediment gravity-flow deposits
- Toe-of-slope aprons
- Submarine fans
- Submarine canyons and gullies
- Pelagic and hemipelagic basin-floors

Appraisal of the lithologic makeup of a platform is critically important to performing sequence stratigraphic analysis. Platforms are mostly lithologically mixed, especially attached platforms. Although some platforms are made up almost entirely of carbonate sediments, most comprise varying amounts of carbonate, siliciclastic, and evaporite sediments. The lithologic variation expressed by modern and ancient platforms directly records the depositional history and is one important indicator of how a platform responds to relative sea-level changes. For example, thin but widespread siliciclastic strata, which interrupt platform carbonates, often signify a relative drop or stillstand of sea level and they overlie sequence boundaries. In contrast, on clastic-dominated shelves, carbonate strata may signify transgressive conditions (Brown, 1989). Evaporites comprise a major part of many carbonate platforms, but their role in the sequence stratigraphic evolution of carbonate platforms has only recently been addressed (Goldhammer et al., 1991; Sarg, 1988; Tucker, 1991). The deposition of evaporites requires a more selective coincidence of specific eustatic, tectonic, geochemical, and climatic conditions than carbonate deposition. Widespread deposition of bedded subaqueous evaporites across a platform probably requires that the depositional site be almost completely isolated from open sea by some barrier (Lucia, 1972). The emergence of a barrier can be due to tectonic, depositional, and eustatic processes. A widespread (thousands of km^2) evaporite unit located behind a shelf-margin rim would suggest a minor relative lowstand during an overall highstand of sea level. If, however, the evaporite deposit is present only locally, the emergence of a barrier is probably due to depositional processes, such as the aggradation of storm deposits above sea level, or perhaps to local tectonic processes.

STRATAL GEOMETRY VARIATIONS

In most sedimentary basins, seismically derived descriptions of stratal geometry can be integrated with available well control to distinguish between carbonate and siliciclastic strata. In frontier basins lacking well control, geologists and geophysicists are confronted with the task of interpreting lithology solely from seismic records. In such basins, the task may be approached from a geophysical stance through an analysis of amplitude, frequency, and interval velocity (Bubb and Hatlelid, 1977). Another approach is to compare and contrast stratal geometries of these two rock types, which form in response to deposition and erosion. The tendency of carbonate sedimentation to create depositional topography and for solutional erosion of carbonates to create karst topography enhances the possibility for seismic stratigraphic differentiation and interpretation. In many cases, however, depositional and erosional topography may be too subtle for seismic resolution.

Carbonate and siliciclastic strata possess the same concordant and discordant stratal relationships to sequence boundaries. Where stratal discordance is recognized at sequence boundaries, carbonate and siliciclastic strata terminate by lap-out (onlap, downlap, and toplap) and truncation (erosional and structural). However, the capacity of carbonate sediments to be (1) generated and accumulate in situ as buildups, (2) transported and deposited as clastic particles with a wide array of textures, and (3) to erode subaerially, chiefly by dissolution, commonly results in peculiar to unique stratal-pattern associations. These include the following, which are keyed to circled numbers in Figure 5.

1. Karst-related dissolution or collapse leads to the formation of closed topographic depressions ranging from a few meters to kilometers in width. These include dolines, or sinkholes, blind valleys, and poljes. In plan view, karst depressions often have circular to oval outlines; however, blind valleys and poljes are commonly linear, but closed at both ends. Such views would be necessary to distinguish them from fluvially incised valleys. In cross section, karst strata are truncated at the margins of solution depressions. Onlapping depression-fill strata consist of lowstand lacustrine and fluvial deposits, terra rossa, and transgressive marine deposits. Residual hills, such as cone and tower karst, may also be present and, if preserved, would be onlapped by lowstand and transgressive deposits. Some karst areas lack depressions and residual hills. Such a lack of topography often implies long-term denudation by solution corrosion plains (Ford and Williams, 1989).
2. Construction of steep-sided, hummocky and lenticular reef mounds and reefs in shelf settings that downlap at their bases and are onlapped at their margins.
3. Leeward margins of rimmed shelf edges may prograde in a shelfward direction, and construct clinoforms that downlap lagoonal or

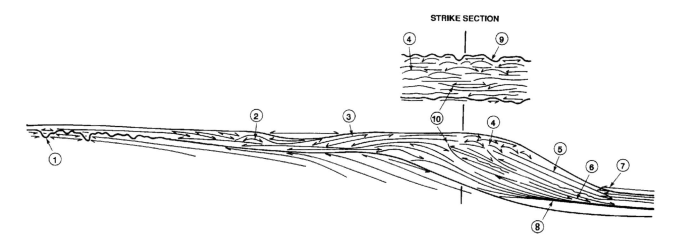

Figure 5. Idealized representation of common types of large-scale stratal patterns across carbonate platforms, many of which are seismically resolvable. Refer to text for discussion. (1) Karst-related truncation and onlap fill; (2) shelf mounds that downlap at their bases and are onlapped/downlapped at their margins; (3) shelfward prograding carbonate clinoforms along leeward margins of rimmed shelf edges and associated onlap of flat-lying shelf strata; (4) steep-sided hummocky to lenticular mounds (bioherms) along rimmed shelf edges and in downslope areas; (5) depositional slopes up to and exceeding the angle of repose; (6) downlapping clinoforms at the toe-of-slope; (7) alternating downlap and onlap, or onlap; (8) clinoforms may simply converge by thinning; (9) incisement of shelf edges at sequence boundaries or within sequences (10).

shelf strata. Flat-lying shelf strata may also onlap leeward-margin clinoforms.

4.   Steep-sided, hummocky and lenticular mounds (bioherms or reefs) are present along rimmed shelf edges and in downslope areas.

5.   Depositional foreslope angles off carbonate platforms generally range up to 29–30° (Schlager and Camber, 1986), and in some cases, they reach 45° (Kenter, 1990). However, vertical and even overhanging slopes (Grammer and Ginsburg, 1992; James and Ginsburg, 1979) are locally present. Such slopes may be depositional or erosional. Organic-binding of sediments commonly leads to vertical depositional slopes in modern reefs. Steep erosional slopes indicate that the sediments are lithified or they have a high shear stress, due to interlocking particle shape and packing arrangements.

Carbonate toe-of-slope deposits may show downlap (6), alternating onlap and downlap, or just onlap (7). Additionally, these deposits may converge by thinning (8). Siliciclastic strata also possess these characteristics. As demonstrated by (9), karstification, fluvial incisement, or submarine erosion can incise shelf edges at sequence boundaries. In (10), scoop-shaped incisement features formed by mass wasting can occur at sequence boundaries or within sequences.

The ability of carbonate slopes to build beyond the angle of repose is due to organic binding, early cementation, and the deposition of nonspherical, interlocking skeletal grains. Even muddy carbonate slopes attain higher declivities than siliciclastic muddy slopes (Schlager and Camber, 1986) because

of early, intensified lithification and the high shear strength of fine-grained carbonate sediments (Kenter and Schlager, 1989).

Some carbonate sequences consist of steeply dipping, downlapping clinoforms that alternate with flat-lying, onlapping basin-floor strata (Bosellini, 1984; Loucks and Brown, 1991). In those cases, a wide range of particle sizes was available for transport and deposition in slope environments. Steeply dipping clinoforms, which downlap bottomset strata, commonly comprise carbonate sand- to boulder-size sediment that was episodically deposited. As coarse-grained sedimentation diminished and the formation of downlapping clinoforms ceased, suspension deposition of fine-grained sediment became dominant and formed onlapping bottomset strata.

CARBONATE PLATFORM RESPONSES TO RELATIVE CHANGES IN SEA LEVEL

Depositional sequence and systems tract models were constructed for tropical rimmed shelves and ramps (Figures 6–11) based on geologic and geophysical data, carbonate depositional and erosional processes, and environmental factors governing carbonate sedimentation. Four major tenets form the basis of the models.

1.   Carbonate sediments are produced locally by biogenic and nonbiogenic processes.

2.   Climatic factors are fundamentally important to distinguishing platforms.

3.   Depending on climatic conditions, tectonic setting, and linkage, platforms may be com-

posed of any mixture of carbonate, siliciclastic, and evaporite sediments.

4. The stratigraphic positions of these sediment types within a sequence are not random and they offer important clues to relative sea-level history.

Depositional sequence and systems tract models are generally predictive because they show the pattern of change or growth and demise of the major types of carbonate platforms and associated depositional systems as they respond to relative sea-level changes. However, the models are idealized representations of the expected and known response of carbonate systems to various stands of sea level. Stratal patterns were drawn sequentially with their relative thicknesses and geometries conforming to expected and observed rates of carbonate productivity and accumulation. Although constructed to demonstrate the dynamic nature of carbonate platforms, the models should complement, rather than replace, the static models of Ahr (1973), Irwin (1965), Read (1985) and Wilson (1975). Furthermore, they are intended to supplement Sarg's (1988) carbonate sequence stratigraphic model and they can accompany the siliciclastic depositional sequence and systems tract models of Haq et al. (1987), Vail (1987), Posamentier and Vail (1988), and Van Wagoner et al. (1990).

As idealized representations, the models do not purport to address exceptions and anomalies. Thus, they are not rigid templates within which stratigraphers can forcefully fit a carbonate sequence. While not intended to represent final solutions to carbonate sequence stratigraphy, they are meant, however, to function as working hypotheses to help geologists visualize how and why carbonate strata were laid down and fit together as they do.

The following types of rimmed shelves and ramps were modeled:

1. humid carbonate rimmed-shelf and ramp (Figures 6, 7),
2. humid carbonate-siliciclastic rimmed shelf (Figure 8),
3. arid carbonate-evaporite-siliciclastic rimmed shelf and ramp (Figures 9, 10), and
4. humid carbonate, detached rimmed shelf (Figure 11).

Examples of these types of modern and ancient platforms are listed in Table 2. The following discussions focus on inferred and observed responses of these types of platforms to relative sea-level changes, including stratal patterns and facies characteristics.

Lowstand Conditions

Shorelines are dynamic environments, rarely remaining stationary for long periods of time, and will migrate depending upon eustasy, tectonic subsidence, and rate of sediment delivery. Seaward migration of a shoreline occurs with a relative fall of sea level, or lowstand conditions, and leaves in its wake an exposed shelf. In siliciclastic provinces, fluvial channels incise the exposed shelf, erode the inter-

fluves and transport sediment across the shelf to the new lowstand shoreline. Thus, the subaerially exposed shelf generally becomes a site for erosion and sediment bypass. If sea level drops to or below the shelf edge, a significant volume of sand may be delivered to point sources along the shelf edge by fluvial-deltaic systems and subsequently deposited in the slope and basin environments by sediment gravity-flow and mass-wasting processes. These depositional processes commonly lead to the accumulation of submarine fans and thick wedges of strata below the shelf margin that make up the lowstand systems tract (Haq et al., 1987; Posamentier and Vail, 1988; Vail, 1987).

Although the *rate* of siliciclastic sediment delivery to a coast fluctuates according to many factors operating outside the basin (tectonic uplift, climate, nature of sediment source, drainage characteristics, etc.), it is not greatly affected by relative sea-level changes. However, sea-level changes can govern the *dispersal* of siliciclastic sediment across shelves (Vail et al., 1977). A far different relationship exists in carbonate platforms when comparing sediment production/availability and relative sea level.

In carbonate environments, relative sea-level changes exert a strong control over sediment production and dispersal. If a wide, open, tropical marine shelf is covered by 10 m of seawater, for example, a healthy and productive carbonate factory should produce abundant amounts of carbonate sediment. However, during lowstand conditions, when sea level has dropped below the shelf margin, the once-flooded platform is now subaerially exposed and unable to produce sediment. The only productive part of the platform is the slope immediately seaward of the shelf margin, and its width is dependent on the slope gradient. Steep slopes correlate to narrow bands of sediment production and gentle slopes form wider bands (Figure 12). The width of the sediment-production zone for a homoclinal ramp, however, should not change substantially with a fall or rise in relative sea level (Figure 12).

As carbonate sediment production on rimmed shelves varies with relative sea level, so does the amount and type of sediment shed from the shelves onto the adjacent deep-water slopes. During highstands, shallow-water platforms produce large quantities of fine-grained sediment, and they shed a large portion of this sediment to the adjacent slopes and basins (Droxler and Schlager, 1985; Hine et al., 1981a; Kendall and Schlager, 1981; Mullins, 1983; Neumann and Land, 1975; Wilber et al., 1990). However, during lowstands (type 1 unconformity), sediment production is terminated on platform tops and is geographically constrained to shelf margins and upper slopes. Furthermore, since ooids are produced only when bank-tops are flooded, any loose sediments shed during a lowstand should be relatively free of ooids (Schlager, 1991).

Since a ramp has no marked break in slope, or shelf edge, the bathymetric profile is practically identical regardless of sea-level position. Thus, a humid carbonate ramp, which is dominated by ooid grain-

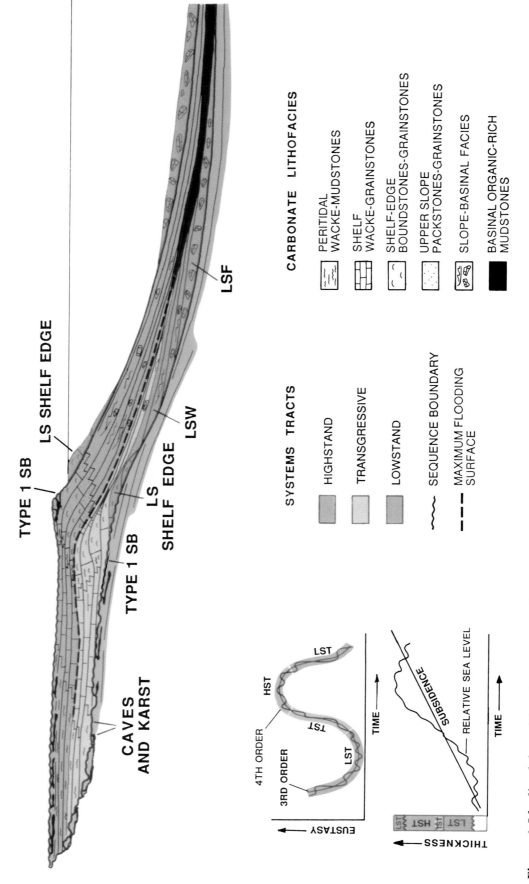

Figure 6. Idealized depositional sequence and systems tract models for a humid, carbonate rimmed shelf, associated with type 1 sequence boundaries. Abbreviations in this and subsequent depositional sequence diagrams are: SB = sequence boundary; LS = lowstand; LST = lowstand systems tract; LSW = lowstand wedge; LSF = lowstand fan; LCC = leveed channel complex; TST = transgressive systems tract; HST = highstand systems tract; SMST = shelf-margin systems tract. Figure 6 continues on page 17.

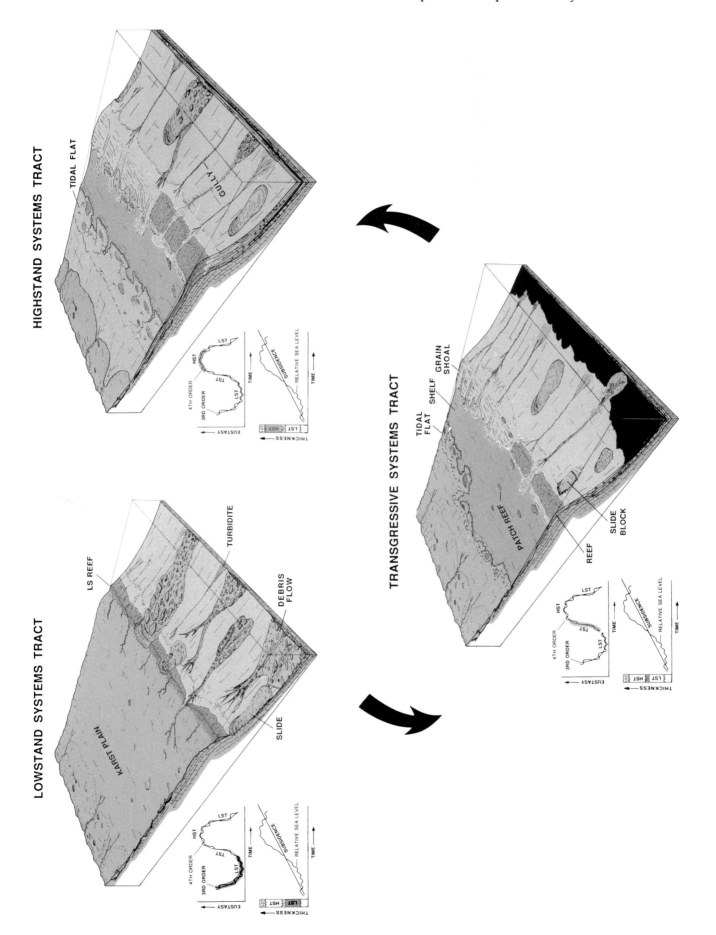

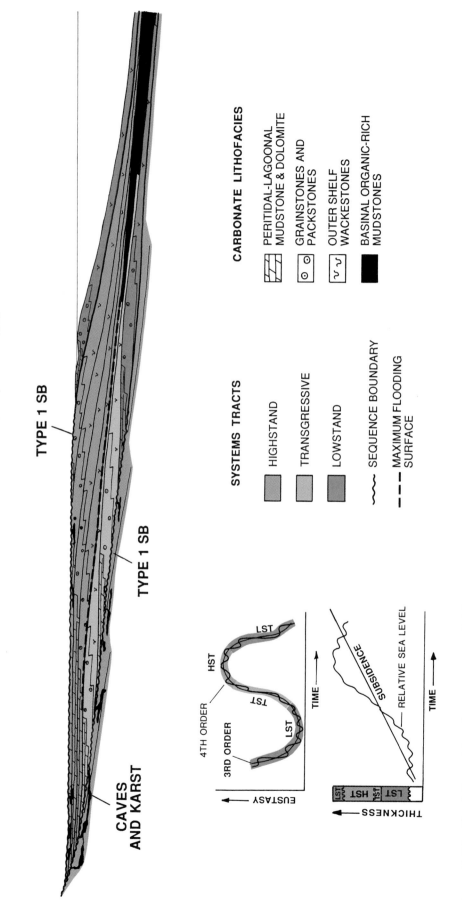

Figure 7. Idealized depositional sequence model for a humid carbonate ramp, associated with type 1 sequence boundaries, showing systems tracts, facies, and stratal patterns.

stone deposition during highstands of sea level, is likely to deposit a similar proportion of ooid grainstone deposits during lowstands, following a type 1 and type 2 sequence boundary, unless major environmental factors governing carbonate deposition have changed.

Fluvial-deltaic or eolian processes can spread siliciclastic sediments far and wide across subaerially exposed platforms during lowstands. Siliciclastic and carbonate environments may coexist but the influx of large amounts of silt and clay and large volumes of fresh water via river systems will terminate carbonate sedimentation along shelf edges. Episodic and infrequent influx may not seriously affect carbonate sedimentation or its ability to recover. For example, reefs are present ~4 km seaward of the prodelta slope on the inactive northern half of the Mahakam delta, Indonesia (Gerard and Oesterle, 1973; Magnier et al., 1975), and they are also present along the margins of alluvial fan-deltas in the Gulf of Elat, Red Sea, which are infrequently flash flooded (Friedman, 1988; Roberts and Murray, 1988).

What happens to the carbonate shelf edge during a relative sea-level fall is largely a function of accommodation and reduced sediment production as the areal extent of inundation on top of a platform decreases. Where sea-level falls below a shelf edge, carbonate sediment production is diminished and the potential for significant shelf-edge progradation by depositional means is reduced greatly. Shelf edges may advance laterally during a sea-level fall but the shelf break advances more as a result of sea-level lowering (reducing accommodation space) than to deposition (Posamentier et al., 1990). This leads to a shelf edge which advances in a downward shifting, offlapping manner (Figure 13A). This type of offlap is referred to as a *forced regression* (Posamentier et al., 1990). Carbonate examples of a forced regression are illustrated in Miocene shelf-edge reefs of southeastern Spain and Mallorca (Dabrio et al., 1981; Esteban and Giner, 1977; Franseen and Mankiewicz, 1991; Pomar, 1991, 1993). Despite the inferred subaerial exposure of the platform top, and hence, reduction of the carbonate factory, these lowstand shelf edges continued to prograde because accommodation space was reduced in front of the reefs (Dabrio et al., 1981; Esteban and Giner, 1977; Franseen and Mankiewicz, 1991; Pomar, 1991) and ample sediment was generated and deposited on the shelf edge and slope. An extreme example of a forced regression is present in Bonaire (Figure 13B), where Pleistocene reefs form steplike terraces such that the oldest terrace lies inland and ~100 m higher than the youngest terrace, which lies along the coast and ~10 m above sea level (De Buisonje, 1974; Kobluk and Lysenko, 1984).

Lowstand Karstification

When relative sea level falls below the shelf edge, subaerially exposed carbonate shelves commonly undergo a dramatic, geomorphic metamorphosis due to weathering by dissolution. The interaction of soluble carbonate minerals in limestone and chemically aggressive water (H_2CO_3, which dissociates into H^+ and HCO_3^-, created by charging rainwater with atmospheric and soil-derived CO_2 gas) penetrating fissures and voids creates unique dissolutionally modified landforms known collectively as karst. Modification of a carbonate terrain into a karst landscape along sequence boundaries routinely leads to a variety of unique landforms. They include collapse and solution sinkholes of various sizes, vertical shafts, blind valleys, dry valleys, and of course vadose and phreatic solution caves. Intense solutional weathering of limestone terrains forms terra rossa soils that mantle the karst surface. The formation of karst landforms and the development of a subterranean drainage network of caves are hallmarks of the karstification process. No other process is so singularly important in modifying limestone terrains, and virtually every major weathering and erosional product of a subaerially exposed carbonate terrain in a humid environment can be attributed to it.

A key to the development of sequence boundary karst is climate, and in particular the availability of water and CO_2, and temperature (White, 1988). Karst formation and the richness and diversity of karst landforms diminish with a decrease in rainfall (Ford and Williams, 1989; Jennings, 1971). In addition, matrix porosity and permeability of the host carbonate rock are important. Surface karst and cave development favors relatively dense carbonate rocks with faults, joints, and bedding plane breaks that focus the infiltration of carbonic acid–charged groundwater (Jennings, 1971). Focused groundwater flow through carbonate rocks with high matrix porosity, yet lacking fractures, is less apt to occur, and karst development may be hindered or result only in spongework-type caves (Palmer, 1991). These effects, however, may be mitigated by high rates of rainfall.

Surface drainage over karst terrain is liable to be intermittent, disrupted, widely spaced, or absent (Jennings, 1971). Rainwater flows only short distances across karst terrain before infiltrating rock matrix or spills into open joints, fractures, vugs, or other conduits. Even in semiarid and arid karst regions, rivers are absent or scarce. Thus, in those situations, incised valleys are not well developed. More frequent and elaborate river patterns are present in humid karst regions, but the drainage density is still generally less than on other types of rocks in the same region (Jennings, 1971). Most surface water drains into dissolutionally enlarged conduits (caves) beneath the surface and travels down gradient before resurging at the surface as springs. Long incised valleys are cut by through-flowing allogenic rivers with headwaters that rise on impervious rocks. Autogenic rivers, which begin their courses on karst terrain, are commonly initiated as large spring resurgences. The ability of a stream to cross a karst terrain is dependent on the capacity of the karst to absorb water (Ford and Williams, 1989) and the nature of the river alluvium which seals off the permeable carbonate beneath. This alluvial material may represent (1) terra rossa soils eroded from the karst surface, (2) detritus weathered

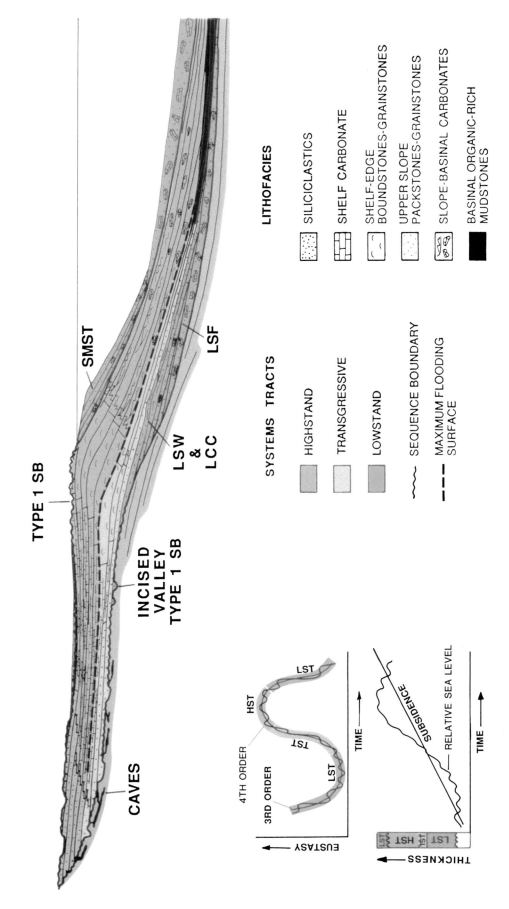

DEPOSITIONAL SEQUENCE MODEL
HUMID CARBONATE-SILICICLASTIC RIMMED SHELF

TYPE 1 SB

SMST

LSF

LSW
&
LCC

INCISED
VALLEY
TYPE 1 SB

CAVES

LITHOFACIES

- SILICICLASTICS
- SHELF CARBONATE
- SHELF-EDGE BOUNDSTONES-GRAINSTONES
- UPPER SLOPE PACKSTONES-GRAINSTONES
- SLOPE-BASINAL CARBONATES
- BASINAL ORGANIC-RICH MUDSTONES

SYSTEMS TRACTS

- HIGHSTAND
- TRANSGRESSIVE
- LOWSTAND
- SEQUENCE BOUNDARY
- MAXIMUM FLOODING SURFACE

4TH ORDER

3RD ORDER

LST
HST
TST
LST

EUSTASY

TIME

SUBSIDENCE

RELATIVE SEA LEVEL

TIME

LST HST LST

THICKNESS

Figure 8. Idealized depositional sequence and systems tract models for a humid, carbonate-siliciclastic rimmed shelf associated with type 1 sequence boundaries. Figure 8 continues on page 21.

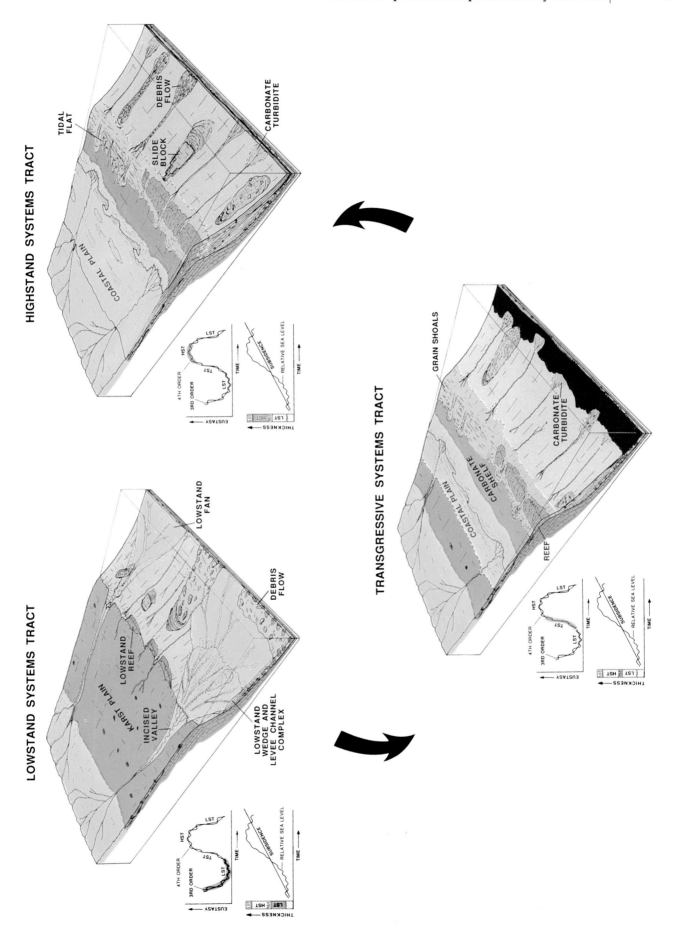

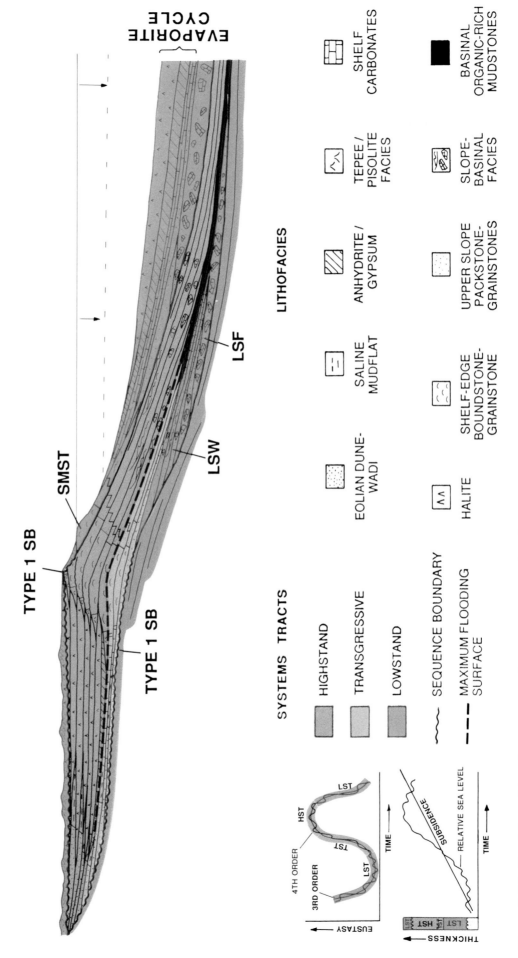

DEPOSITIONAL SEQUENCE MODEL
ARID CARBONATE-EVAPORITE-SILICICLASTIC RIMMED SHELF

EVAPORITE CYCLE

TYPE 1 SB

SMST

TYPE 1 SB

LSF

LSW

SYSTEMS TRACTS

HIGHSTAND

TRANSGRESSIVE

LOWSTAND

SEQUENCE BOUNDARY

MAXIMUM FLOODING SURFACE

LITHOFACIES

SHELF CARBONATES

BASINAL ORGANIC-RICH MUDSTONES

TEPEE / PISOLITE FACIES

SLOPE-BASINAL FACIES

ANHYDRITE / GYPSUM

UPPER SLOPE PACKSTONE-GRAINSTONES

SALINE MUDFLAT

SHELF-EDGE BOUNDSTONE-GRAINSTONE

EOLIAN DUNE-WADI

HALITE

LST
HST
TST
LST
4TH ORDER
3RD ORDER

EUSTASY
TIME

SUBSIDENCE
RELATIVE SEA LEVEL
TIME

THICKNESS
LST
HST
TST
HST
LST

Figure 9. Idealized depositional sequence and systems tract models for an arid carbonate-evaporite-siliciclastic rimmed shelf associated with type 1 sequence boundaries. Figure 9 continues on page 23.

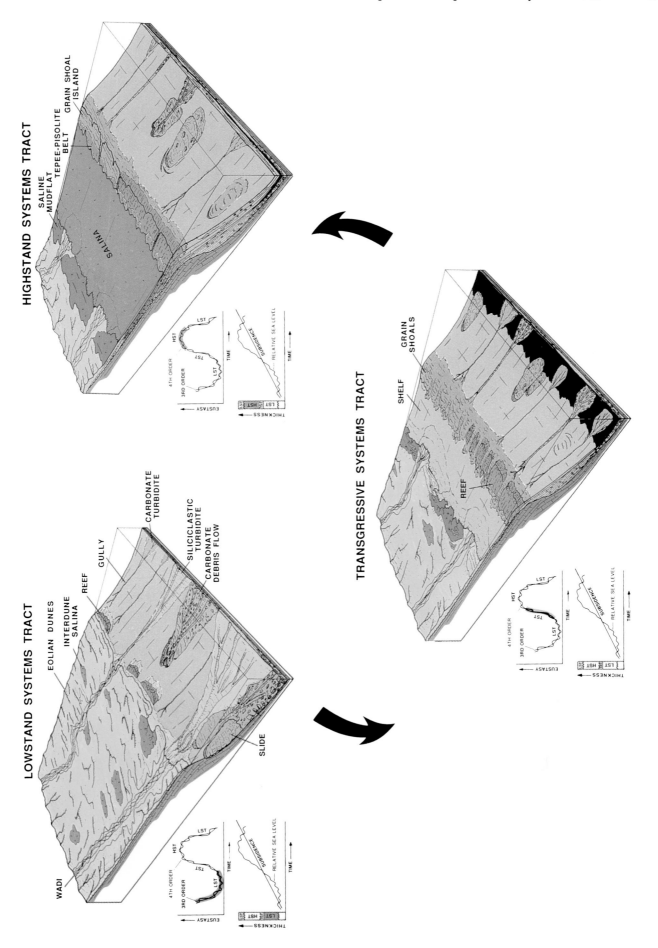

DEPOSITIONAL SEQUENCE MODEL
ARID CARBONATE-EVAPORITE-SILICICLASTIC RAMP

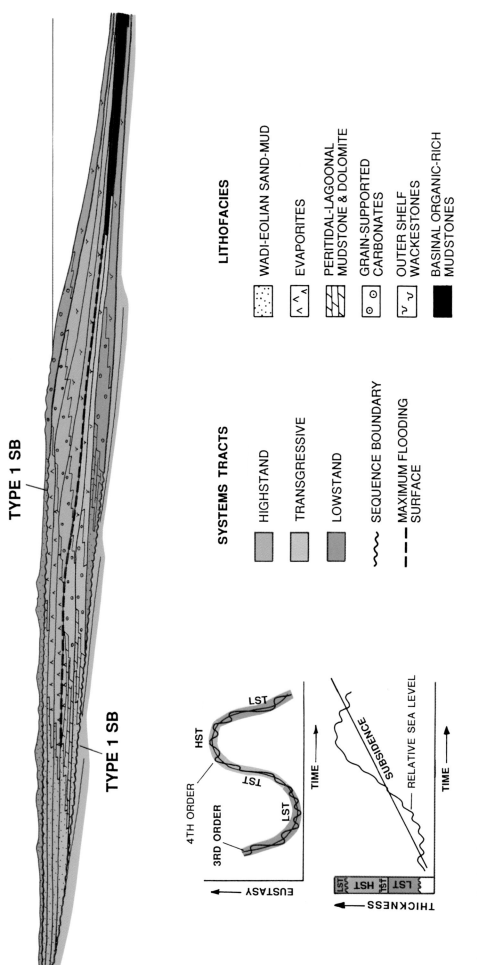

Figure 10. Idealized depositional sequence and systems tract models for an arid carbonate-evaporite-siliciclastic ramp associated with type 1 sequence boundaries. Figure 10 continues on page 25.

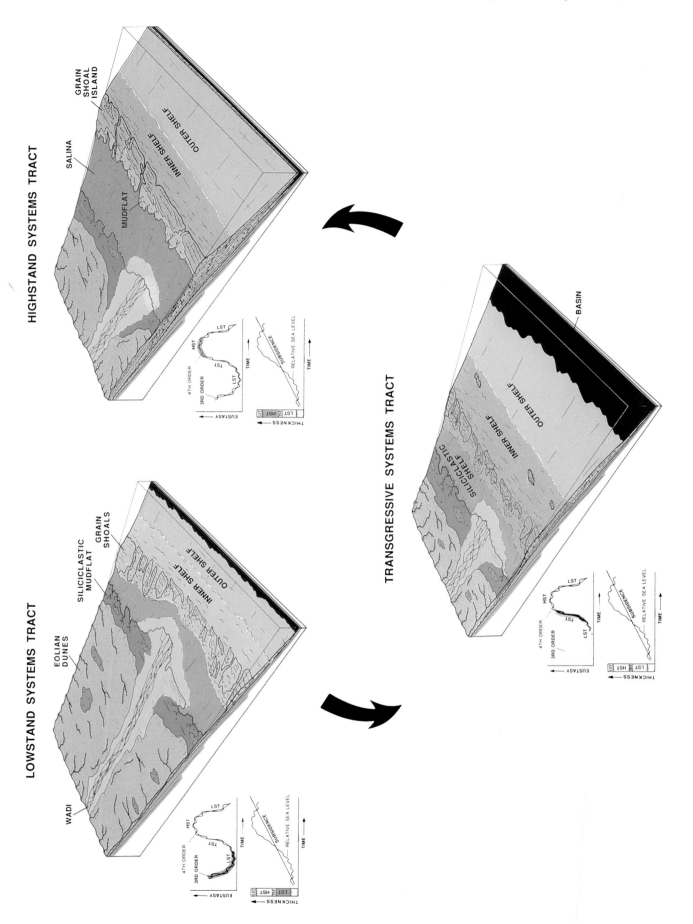

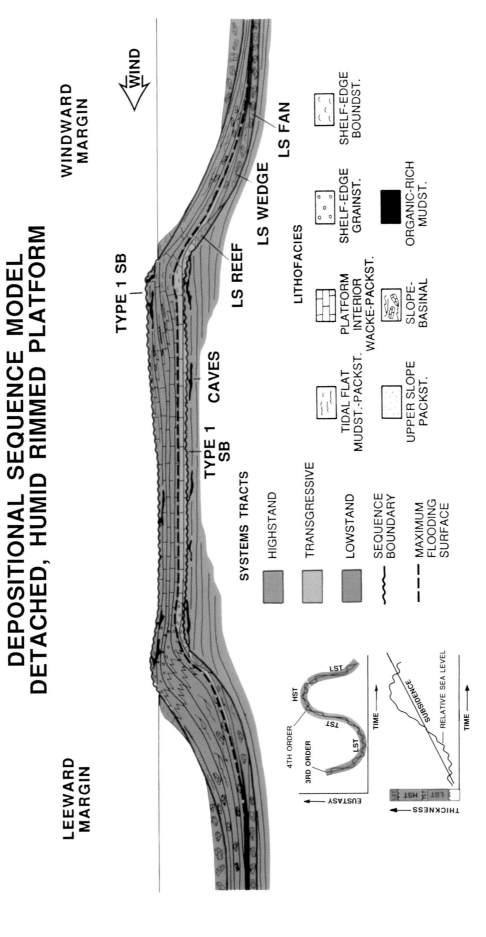

DEPOSITIONAL SEQUENCE MODEL
DETACHED, HUMID RIMMED PLATFORM

WINDWARD MARGIN

WIND

TYPE 1 SB

LS REEF

LS WEDGE

LS FAN

CAVES

TYPE 1 SB

LEEWARD MARGIN

LITHOFACIES

SHELF-EDGE GRAINST.	SHELF-EDGE BOUNDST.
PLATFORM INTERIOR WACKE-PACKST.	ORGANIC-RICH MUDST.
TIDAL FLAT MUDST.-PACKST.	SLOPE-BASINAL
UPPER SLOPE PACKST.	

SYSTEMS TRACTS

HIGHSTAND

TRANSGRESSIVE

LOWSTAND

SEQUENCE BOUNDARY

MAXIMUM FLOODING SURFACE

3RD ORDER

4TH ORDER

HST

TST

LST

LST

EUSTASY

TIME

SUBSIDENCE

RELATIVE SEA LEVEL

TIME

THICKNESS

Figure 11. Idealized depositional sequence and systems tract models for a detached, humid carbonate rimmed shelf associated with type 1 sequence boundaries. Figure 11 continues on page 27.

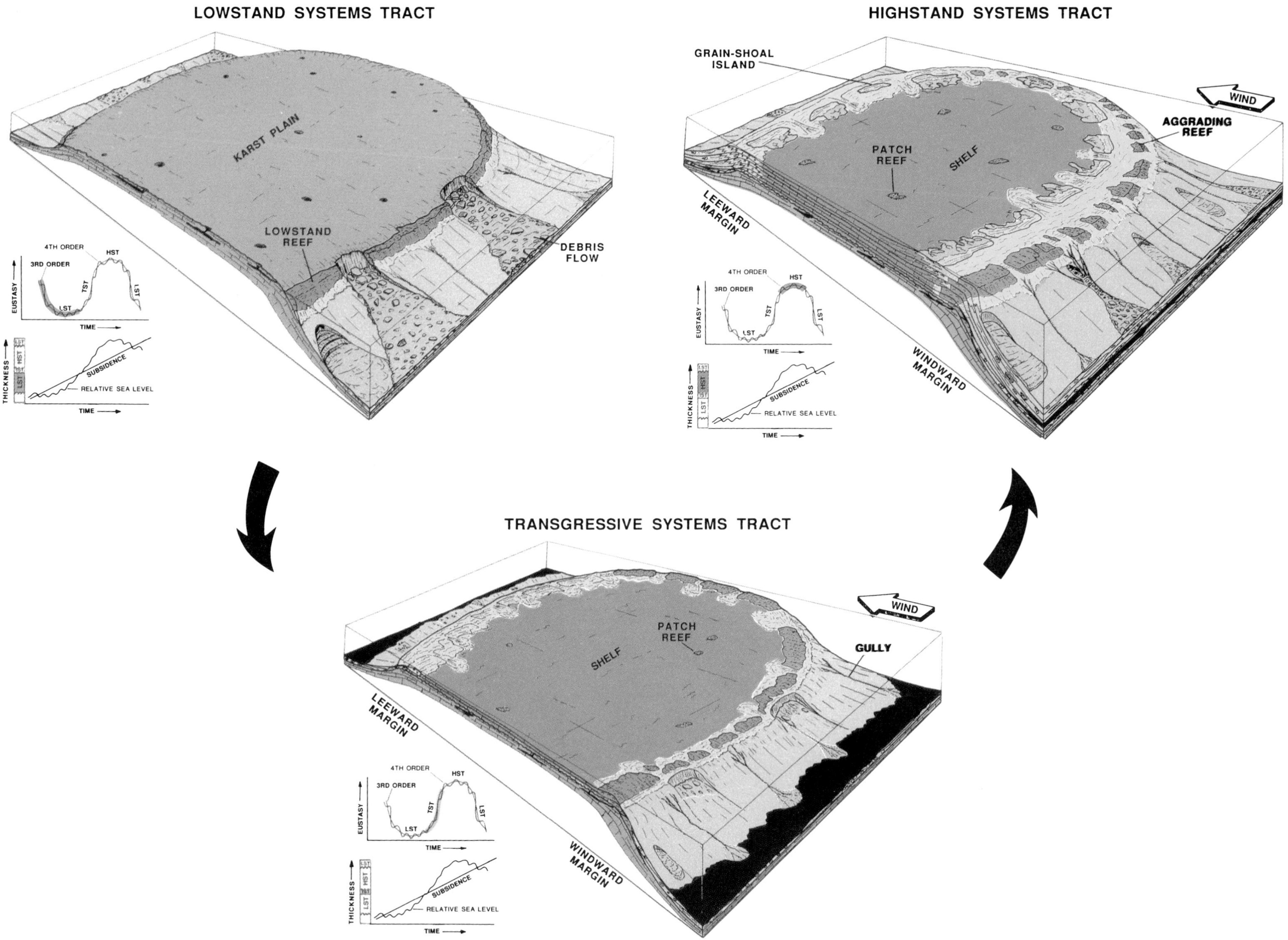

HIGHSTAND SYSTEMS TRACT

WIND

AGGRADING REEF

SHELF

PATCH REEF

GRAIN-SHOAL ISLAND

LEEWARD MARGIN

WINDWARD MARGIN

LST
HST
4TH ORDER
TST
LST
3RD ORDER
EUSTASY
TIME

SUBSIDENCE
RELATIVE SEA LEVEL
THICKNESS
TIME
LST HST LST

LOWSTAND SYSTEMS TRACT

KARST PLAIN

DEBRIS FLOW

LOWSTAND REEF

LST
HST
4TH ORDER
TST
LST
3RD ORDER
EUSTASY
TIME

SUBSIDENCE
RELATIVE SEA LEVEL
THICKNESS
TIME
LST HST LST

TRANSGRESSIVE SYSTEMS TRACT

WIND

GULLY

PATCH REEF

SHELF

LEEWARD MARGIN

WINDWARD MARGIN

LST
HST
4TH ORDER
TST
LST
3RD ORDER
EUSTASY
TIME

SUBSIDENCE
RELATIVE SEA LEVEL
THICKNESS
TIME
LST HST LST

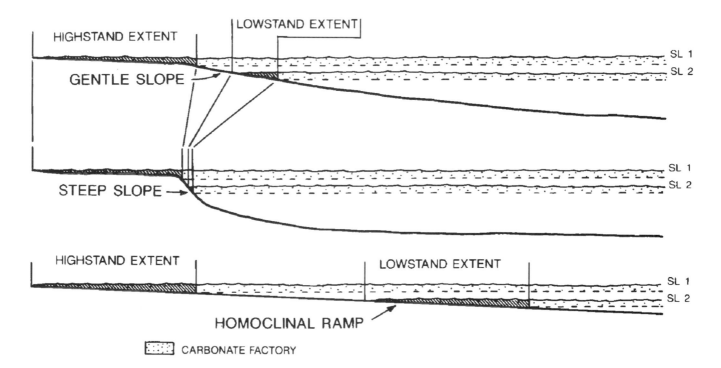

Figure 12. The width of a shelf bathed by the highly productive 10-m-thick portion of the tropical marine carbonate factory varies significantly depending on eustasy and the slope. At highstands of sea level, platform tops are inundated and the extent of the carbonate factory is great. During lowstands of sea level, the width of the shallow seafloor in contact with the 10-m-thick factory is directly proportional to the slope. The width is greater for gentle slopes than for steep slopes. For homoclinal ramps, the extents of the highstand and lowstand factories are equal.

and eroded from interbedded terrigenous rocks, or (3) detritus transported into the karst area by allogenic streams that drain uplands underlain by noncarbonate rocks. In general, allogenic streams carry a greater amount of terrigenous sediment than autogenic streams over carbonate terrain. If alluvium is lacking in a karst surface stream, incision is achieved mainly through corrosion. Where coarse detritus is present, streams may incise through corrosion and corrasion (Jennings, 1971). Thus, although there is a general perception that fluvial incision is unimportant on subaerially exposed carbonate platforms, incised valleys are present.

Given enough time and sufficient water, most carbonate terrains will develop a karst landscape. However, questions remain regarding whether carbonate platforms generally are exposed to subaerial conditions sufficiently long to develop karst topography, prior to another sea-level rise. Time may be an unimportant factor where precipitation rates are high, because rates of solutional denudation vary linearly with precipitation (White, 1988). The greatest limestone solution in the world occurs where it is wettest (Ford and Williams, 1989). For example, solutional denudation rates in karst terrain of Papua New Guinea range from 270–760 mm/1000 yr (Maire, 1981). Ford and Williams (1989) showed that for the last 240,000 yr, sea level at a tectonically stable area was −20 to −50 m below present sea level for about

46% of the time, or 110,400 yr. Thus, when subjected to a solutional denudation rate of 500 mm/1000 yr, a carbonate platform could be denuded by 55 m in 110,000 yr. Caves and karst can form quickly, as proven by their presence in many Pleistocene limestones, less than 100,000 yr old, of the Caribbean region (White, 1988). In fact, many karst landforms have formed during the last 10,000 yr (Ford and Williams, 1989). Thus, high-frequency sea-level cycles with durations as short as tens of thousands of years can result in the formation of karst features along sequence boundaries in humid climates.

Although a relative sea-level fall precludes the production of marine carbonate sediment on exposed shelves, the formation of karst features at and below sequence boundaries during lowstands of sea level attaches a distinctive facies overprint to the previously deposited highstand strata. Karst-formed features, which are recognizable in cores and outcrops, include terra rossa paleosols, cave fill (breakdown, clastic sediment, carbonate precipitates), and brecciated cave roofs (Loucks and Handford, 1992) (Figure 14). Where adequate subsurface control is available, karst topography may be identified in structure maps as closed depressions (dolines).

Karst features are not particularly facies- or site-specific. Karstification can affect all carbonate facies but may be better developed in some than in others. All portions of a platform which has been subaerially

A

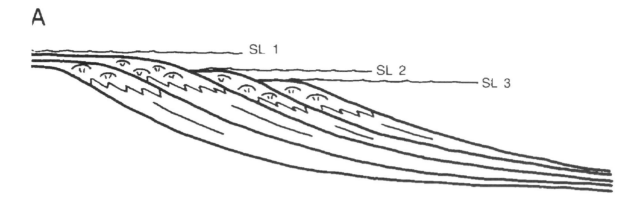

B

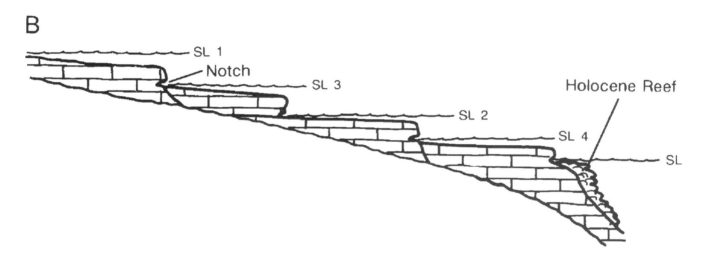

Figure 13. Two examples of a forced regression in carbonates. (A) Sea-level lowering (SL 1 to SL 3) and a reduction of accommodation space lead to continued progradation, but in a downstepping, offlapping manner. This style mimics the patterns recognized in the Miocene of Spain (Dabrio et al., 1981; Franseen and Mankiewicz, 1991). (B) A Pleistocene example from Bonaire, Netherlands Antilles illustrates the general downstepping through approximately 100 m of sea-level fall (Kobluk and Lysenko, 1984). Note, however, that the terrace formed at SL 3 overlies the older SL 2 terrace.

exposed and penetrated by meteoric water can be affected. If a platform lies within an arid climate, karst features may be less common and give way to caliches. Some subaerially exposed carbonates may show little evidence of karstification or calichification. In those cases (1) the length of exposure may have been too brief for the features to form; (2) surface-formed karst and caliche features formed but were subsequently stripped off by erosion; or (3) high intergranular permeabilities and porosities may have hindered the formation of conventional karst (Jennings, 1971; Meyers, 1988).

Lowstand Shelf-Edge to Slope Failure and Sedimentation

Slope failure is common in the marine environment, especially on muddy slopes. It can be a signifi-

cant erosional process on carbonate slopes and result in downslope resedimentation of large volumes of slope and shelf edge sediment. Failure is not constrained by relative sea-level position. It may happen during any sea-level stand, provided that sediments are poised for gravity failure and triggering mechanisms are available. Tectonic activity and seismicity are probably important, as suggested by Hine and Hallock (1991). They documented megabreccias, faulted and foundered carbonate platforms, scalloped bank margins, and large, displaced blocks in the Nicaraguan Rise, which lies near the transform fault-bounded margin of the Caribbean Plate. Earthquake shocks probably triggered the catastrophic collapse of many platform margins. This mechanism was invoked to explain the widespread, synchronous deposition of extraordinarily thick, graded carbonate

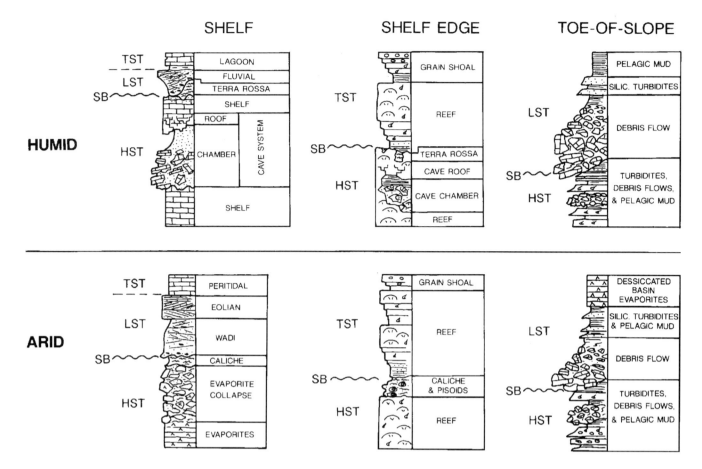

Figure 14. Idealized vertical succession of sequence bounding facies in shelf, shelf-edge, and basinal environments under humid and arid climates. No scale intended.

megaturbidites in the Pyrenees of Spain (Labaume et al., 1987). Whatever the triggering mechanism, it must either increase the stress on the sediment mass to the point of failure, or decrease the sediment strength so that the existing stress is sufficient to cause failure (Coleman and Prior, 1988).

Some triggering mechanisms may favor lowstand conditions. These include the following:

1. loss of buoying effect of seawater on marine sediments when subaerially exposed (Schwarz, 1982);
2. cyclic wave-loading and associated, pulsating pore pressures due to lowering of storm wave base into muddy sediments;
3. erosion and undercutting of a shelf edge shoreline during a sea-level fall by waves and currents;
4. depositional oversteepening and sediment loading during a previous highstand;
5. submarine erosion by deep-sea currents which migrate laterally in response to a sea-level drop (Pinet and Popenoe, 1985); and
6. dissolution of shelf edge strata by mixing-zone corrosion (Back et al., 1986).

The presence of scalloped shelf edges, which were recently documented by Mullins and Hine (1989) in the west Florida Platform, Pedro Bank, and in various parts of the Bahamas, suggests catastrophic failure of platform margins. Although the Bahamian features probably reflect tectonic failure (Mullins et al., 1991), some ancient examples may have formed during low-stands of sea level. For example, the Apulia Platform of southern Italy has scalloped margins, which Bosellini (1989) claimed was formed by catastrophic collapse during pronounced lowstands of the Early Turonian and Early Eocene times. Retreat of platform margins and the subsequent formation of scalloped margins (Mullins and Hine, 1989) truncate shelf and shelf-edge strata. Eroded sediments are resedimented as onlapping strata against the slope. Given adequate subsurface well control, seismic coverage, and stratigraphic resolution, structure and isochron maps may help delineate scalloped margins.

Timing of slope erosion relative to sea level is commonly difficult to determine in ancient platforms. Lowstands of sea level are often cited to be opportune times for slope erosion and large-scale slope failures (Mutti, 1985; Posamentier and Vail, 1988). However they are immediately followed by sea-level rises, which could also be opportune for slope erosion, due to landward migration of the carbonate factory and deposition of fine-grained sediment with low shear stress on the slope. In the Guadalupe Mountains, Texas, mixed siliciclastic-carbonate slope deposits of the Permian Cutoff Formation rest unconformably on the carbonate platform and muddy slope strata of the Bone Spring–Victorio Peak formations. The unconformity represents a sequence boundary along which erosion removed 250 m of Bone Spring–Victorio Peak strata over a lateral distance of 3 km prior to deposition of the Cutoff Formation (Rossen et al., 1988). Kirkby and Pray (in Pray, 1988) interpreted the sequence boundary as a submarine erosion surface which formed during a relative rise in sea level on the basis of an apparently deepening-upward sedimentological record in the upper Victorio Peak Formation. The origin of this sequence boundary is disputed, however. Rossen et al. (1988) believe that the sequence boundary formed during a relative fall in sea level. Key to the latter interpretation is the contention that the submarine erosion surface can be traced updip into a correlative, subaerially exposed erosional surface.

Slope failure leads to the formation of sediment slides, slide scars, and resedimented deposits at the toe-of-slope and into the basin. Large blocks of sediment detach along glide planes, which show up as inclined, scoop-shaped surfaces truncating underlying strata, and they move downslope by rotation or translation (Cook and Mullins, 1983; Nardin et al., 1979). Rotational slide blocks usually move short distances whereas translational slide blocks move longer distances prior to redeposition as coherent to chaotic masses at the toe-of-slope. Unless translational slide blocks break up during downslope movement, they may display concordant stratal relationships, which would render them difficult to recognize seismically. Where disruption does occur, translational slides typically show hummocky to chaotic stratal patterns. Chaotic patterns may reflect nearly complete disruption of a slide block due to soft-sediment deformation or transformation of the slide into a debris flow.

The volume of affected strata varies greatly, depending both on the frequency and scale of failure. Slope failure often forms a steep scarp which becomes a focal point for subsequent retrograding failures in an attempt to reach a lower, more stable inclination. Some slumps and slide blocks are believed to break up during downslope movement and spawn large-scale sediment-gravity flows. Flow volume increases with scale of catastrophic slope failure. One large-scale failure event or several progressive events will lead to an eroded slope and shelf edge facing a lowstand wedge made up of resedimented shelf-edge and slope sediments.

In the absence of slope and shelf-edge failure, lowstand slope-basinal deposits should be relatively thin. However, slope and shelf-edge failure is common and can lead to the deposition of thick, lowstand carbonate deposits (Jacquin et al., 1991; Sarg, 1988). Furthermore, although deep-water carbonate sediments may be introduced along some point sources to form fan deposits, there is a greater tendency in carbonate systems for line sources of sediment to operate and form toe-of-slope carbonate aprons (Mullins and Cook, 1986). Deposition of carbonate lowstand deposits leads to the formation of basinward thinning wedges that onlap slopes and are locally thick near the toe-of-slope. Lowstand fans are important in some areas (Jacquin et al., 1991). More complicated lowstand geometry will result where siliciclastic sediments are introduced to rimmed shelves and adjacent slopes, especially if there is any lag time between erosion of the shelf edge and slope and the input of siliciclastic sediments during lowstand conditions.

Lowstand slope and toe-of-slope carbonates are dominated by sediment gravity-flow deposits (Figure 14), some of which are thick (megaturbidite of Labaume et al., 1987) and contain abundant and exceptionally large clasts (megabreccia of Cook et al., 1972). Thick lowstand deposits imply catastrophic large-scale failure of slopes or shelf edges. Thinner deposits may imply small-scale failures and, if stacked one on top of the other, suggest recurrence over a period of time. The presence of abundant large clasts is fostered by the tendency for carbonate sediments to lithify in submarine and subaerial environments. Clast composition analyses are useful for determining which environment suffered more from collapse.

The composition of lowstand sediment gravity-flow deposits differs from highstand deposits and can be used cautiously as a key to interpreting relative sea level position. Lowstand carbonate turbidites may contain skeletal grains and clasts shed penecontemporaneously from lowstand shelf-edge environments and clasts derived from the older, subaerially exposed shelf edge. Ooids and peloids are scarce in lowstand turbidites because their formation requires flooded well-circulated platform tops, whereas the production of skeletal sands can be produced at shelf edges, regardless of sea-level stand (Schlager, 1991).

While the presence of a debris-flow deposit in a carbonate depositional sequence might raise suspicions of a lowstand origin, it should not be used without other supporting evidence to infer such an origin. For example, megabreccias making up Triassic slope deposits in the Dolomites of northern Italy were cited as a possible example of an allochthonous lowstand wedge in which the carbonate clasts were derived from a subaerially exposed platform (Sarg, 1988, based on Bosellini, 1984). Recent findings indicate that most carbonate blocks lack features which can be attributed to subaerial exposure (Yose, 1991). In addition, the presence of ooids and peloids, interstratified with the blocks, argue for sediment production on a flooded bank top, or highstand conditions. Yose concluded that sequence models in which carbonate megabreccias are viewed as lowstand deposits are not applicable to the Triassic Sciliar-Catinaccio buildup in northern Italy. It is possible, however, that the displaced blocks originated from submarine environments that were too deep to have been subaerially exposed, despite a sea-level fall. Furthermore, if the platform highstand ooids were made of calcite (ooid microfabric is preserved) rather than aragonite, subaerial cementation may have been limited so that the ooids could have been eroded as loose grains from the platform top during lowstand and redeposited downslope.

Since the extent of the carbonate factory is reduced along rimmed shelf edges during sea-level lowstands, finite amounts of contemporaneous carbonate sediment are available for resedimentation in slope and basin environments. However, if siliciclastic sediments are introduced by fluvial-deltaic or eolian processes during lowstands to form a shallow, prograding lowstand wedge which stands below the previous shelf edge, a ready-made platform is available for carbonate sedimentation during the earliest part of the following sea-level rise. This can happen only if siliciclastic sediment delivery declines during the early sea-level rise. This scenario would lead to siliciclastic lowstand wedge capped by progressively carbonate-rich slope and lowstand shelf-edge sediments, which onlap the older, subaerially exposed shelf edge (Figure 15).

Some carbonate platforms and adjacent basins that lie within cratons or along newly rifted continental margins are periodically cut off from world seas by relative lowstands of sea level. This helps establish an opportunity for the deposition of basin-center evaporites (Figure 9), such as the Permian Castile Formation of the Delaware basin, Texas and New Mexico, the Upper Permian Zechstein evaporites of western Europe (Tucker, 1991), and the Messinian evaporites of the sub-Mediterranean Sea (Schreiber et al., 1976). Water depth and the style of evaporite deposition varies, depending on the amount of evaporative drawdown. Laminated evaporites composed mostly of finely crystalline cumulates and lacking dissolution surfaces are probably deposited from relatively deep brines, whereas layers of bottom-precipitated evaporites record shallow brine deposition (Arthurton, 1973; Handford, 1990; Lowenstein and Hardie, 1985; Lowenstein et al., 1989). Basin-center evaporites onlap the adjacent slopes of carbonate platforms and commonly consist of one or more cycles of dark, fine-grained carbonate overlain by anhydrite, or gypsum, halite, and potash evaporites in some cases.

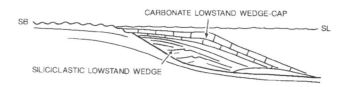

Figure 15. A siliciclastic lowstand wedge can form a shallow platform suitable for carbonate sedimentation to initiate upon if siliciclastic sedimentation rates are reduced sufficiently during the early sea-level rise.

The passage from highstand to lowstand conditions and the formation of a type 1 sequence boundary are not instantaneous events. Although the accuracy of published eustatic curves (Haq et al., 1987) is debatable (Miall, 1991), the curves do show that the falling limb of a eustatic drop in sea level can take as much as 1–2 m.y. to complete in third-order depositional sequences terminated by a type 1 unconformity. Sea-level lowering and reduced accommodation on the shelf over this amount of time lead to progressive erosion of shelf, shelf-edge, and slope environments, but marine carbonate sedimentation may continue. Where erosion outpaces the ability of the slope and shelf edge to heal by sedimentation during a drop in sea level, a distinctly discernible sequence boundary, which records the cumulative effects of erosion, often results. If, however, slopes are able to self-heal between periods of erosion during a fall, the approach into a type 1 lowstand succession may show up as a series of truncation and onlap surfaces within late highstand, or "falling limb," deposits (Figure 16). Seismic resolution of individual surfaces such as these may not be attainable if the interval of affected sediment is relatively thin. However, since seismic reflections usually are the interference composites of many subreflections (Sheriff, 1980), a single, distinctly discernible surface may paradoxically result.

Transgressive Conditions

In marine siliciclastic systems, a rapid, relative rise of sea level can force the loci of terrigenous sediment deposition to retreat with the shoreline, leading to the deposition of laterally extensive nearshore deposits, called the transgressive systems tract (Haq et al., 1987; Loutit et al., 1988; Posamentier and Vail, 1988; Vail, 1987). These deposits can form an aggradational to backstepping, or retrogradational, succession of parasequences (Posamentier and Vail, 1988; Van Wagoner et al., 1990) under progressively deeper water. With deepening conditions and greater isolation from siliciclastic sediment input, terrigenous sediment content in the transgressive systems tract decreases upward (Loutit et al., 1988). Maximum transgression commonly leads to sediment starvation and the deposition of hemipelagic and pelagic sediments over a large area of the shelf, thus forming the condensed section.

In some cases, a relative rise of sea level over a carbonate platform also will lead to sediment starvation and platform drowning. Under most conditions, however, the response often proceeds through three phases: (1) start-up phase, when carbonate accumulation lags behind the relative rise; (2) catch-up phase, when accumulation exceeds the rate of sea level rise and the platform builds to sea level; and (3) keep-up phase, when accumulation closely matches the rate of rise and the platform remains at or very close to sea level (Kendall and Schlager, 1981). Start-up of the carbonate factory lags behind initial transgression.

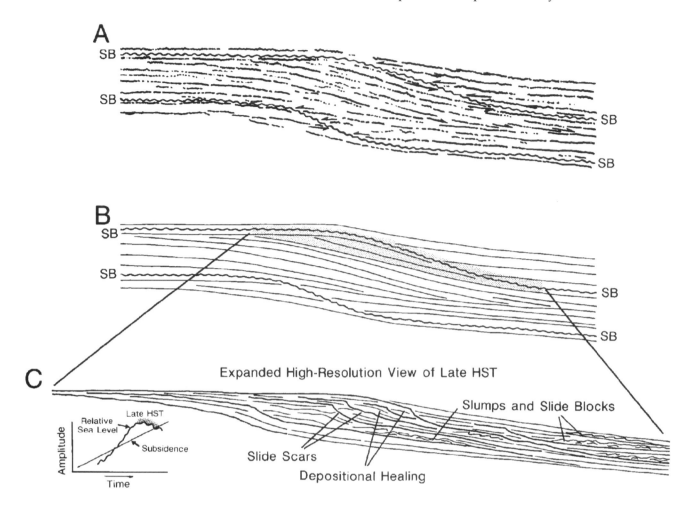

Figure 16. (A) A hypothetical seismic line across a carbonate platform shows reflections which have been interpreted to terminate by onlap, toplap, and erosional truncation at sequence boundaries. In (B), seismic reflections have been traced and smoothed to show inferred stratal geometry. Thus, interpretation suggests the presence of a discrete sequence boundary marked by truncation and onlap. If, however, greater seismic resolution were possible, perhaps a different interpretation would result. Formation of the late highstand systems tract (shaded area in B), may not be marked by a continuous sea-level fall. Instead the relative fall is interrupted by brief rises or stillstands, which can be manifested in the late highstand shelf-edge and slope stratigraphic record (C) as alternating episodes of mass wasting and partial depositional healing. Mass wasting creates slide scars, which show up as local truncation surfaces, and subsequent depositional healing results in the formation of onlapping stratal units above truncation surfaces.

Ginsburg (in Hardie, 1986) contends that the carbonate "factory" cannot go into full production and deliver sediment until sea level has risen enough to allow efficient circulation. Sedimentation does not track sea-level rise but initially lags behind, no matter how fast or slow the rate of sea-level rise. However, once water depth is great enough for adequate circulation, sediment production often catches up with sea-level rise to form aggradational or progradational shallowing-upward successions of organic buildups, grain shoals, and tidal flats along shelf edges and shorelines. Progradation may eventually stall as a result of building out into a deeper shelf during a continued relative rise of sea level. Subsequent flooding and catch-up sedimentation follow and when

repeated, the result is a transgressive systems tract made up of an aggrading or retrogradational parasequence set similar to siliciclastic examples documented by Van Wagoner et al. (1990). A progradational parasequence set is more likely to occur when sea-level highstand is reached.

The formation of carbonate transgressive systems-tract deposits across humid carbonate rimmed shelves and ramps (Figures 6, 7, 11) begins with the flooding of an eroded lowstand surface, commonly of karst origin, which is mantled by soil or caliche. Transgression usually reworks the surficial detritus into a lag deposit while the expanding carbonate factory produces new carbonate sediment. Open and partially collapsed caves and sinkholes may fill with marine carbonate sediments (Figure 14). If siliciclastic sediments were transported across the exposed shelf during the previous lowstand and fill the caves, they would reflect both lowstand and transgressive deposition. On the shelf, inherited topographic highs are onlapped and they may serve as nucleation sites for shoals and reefs. Once open-marine conditions develop and the carbonate factory is productive, subtidal carbonate sediments are accreted to shorelines as onlapping, locally prograding units. This is well illustrated in south Florida where Holocene sediment isopach maps of Enos (1977) illustrate: (1) sediment onlap and thinning against the Pleistocene limestone-cored Florida key islands; (2) local thickening of Holocene sediment toward offshore banks and tidal deltas; and (3) some offshore patches of sediment-bare seafloor. Given time, these sediments eventually will form nearly blanketlike accumulations over shallow shelves. However, thin and even condensed intervals may occur in the deeply flooded, interior shelves of some prominently rimmed platforms.

In mixed siliciclastic-carbonate sequences, carbonate sedimentation usually lags behind the transgression across platform tops. At first, landward encroachment of a high-energy shoreface and ravinement surface across a retreating coastal plain may rework part or all of the lowstand siliciclastics prior to start-up of the carbonate factory. Where siliciclastic facies are relatively minor sequence components, the transgressive shelf deposits may comprise a thin succession of clastics that grade upward into both transgressive and highstand carbonates. If siliciclastic input is great, such as in the Virgilian and Wolfcampian series of the Eastern Shelf and Midland basin, Texas, transgressive shelf deposits may consist of carbonate parasequences overlain by condensed marine shales and a progradational highstand fluvial-deltaic strata (Brown, 1989).

Transgression of an arid rimmed shelf or ramp covered by eolian-wadi sands also leads to coastal plain retreat. At first, low-lying interdune depressions are flooded and transformed into restricted ponds and lagoons. An advancing shoreface, however, subsequently planes off the tops of most eolian dunes and redeposits the siliciclastics into a transgressive marine sheet. Erosion may be less severe where a shoreface advances across a coastal plain made up of low-relief eolian sheet sands and wadi deposits. For example, Tucker and Chalcraft (1991) identified gradational contacts separating lowstand eolian-fluvial sandstones from overlying transgressive peritidal dolomites in the Queen Formation of the Central Basin Platform, Texas. In contrast, Mazzullo et al. (1991) documented a sharp contact between lowstand fluvial-eolian sandstones of the Shattuck Sandstone member and overlying transgressive lagoonal dolomite and anhydrite.

Formation of open-lagoon to shelf or restricted lagoonal environments during early transgression will depend in part on the bathymetry of the rimmed shelf margin. A relatively high rimmed shelf margin, which remains subaerially exposed during early transgression, impedes circulation and would force restricted conditions to develop in the lagoonal environments. A low-lying or discontinuous rim would not impede circulation during flooding and open-marine conditions would ensue.

Platform-interior transgressive systems tracts comprise stacked parasequences of shallowing-upward facies. Sequence boundaries at the bases of transgressive systems tracts often contain caliche or karst features (Figure 14). Lag conglomerates with bored, worn, encrusted, or mineral-stained clasts derived from underlying material are common and these may be succeeded by freshwater pond or marine facies (Enos and Perkins, 1979). Transgressive parasequences are cyclic and they may shallow upward into intertidal or supratidal environments with muddy or grainy caps. Similar to siliciclastic examples (Van Wagoner et al., 1990), carbonate parasequences are bounded by marine-flooding surfaces or their correlative surfaces. These flooding surfaces may form the upper boundary to subtidal, intertidal, and supratidal portions of parasequences.

Although Van Wagoner et al. (1990) have observed few transgressive-lag deposits above marine-flooding surfaces in siliciclastic parasequences, they do occur in both muddy and grainy carbonate parasequences. Clasts are often present and come from the transgressive reworking of both subtidal and subaerially exposed portions of parasequences. In the former case, carbonate clasts may originate from reworking of hardgrounds, cemented burrows, and reef-crest and -flat boundstone clasts. In the latter case, clasts are reworked from caliche crusts, supratidal and intertidal mudflats, intertidal beachrock layers, and water-table hardgrounds. Formation of these subaerial features does not require a change in base level because autocyclic mechanisms commonly lead to aggradation and accretion of carbonate sediments above sea level (e.g., tidal flats, islands, coastal dunes, and beaches), and their subsequent erosion by shoreline retreat.

Condensed deposits may occur atop platforms during maximum transgression (Loutit et al., 1988; Wendt, 1988). These sediments are typically thin when compared to coeval strata elsewhere. Thinning is due to very low sedimentation rates or nondeposition, long submarine exposure, erosion, and reworking. They are often composed of pelagic or hemipelagic sediments bearing planktonic and nektonic fauna and flora. Vagile benthonic organisms are generally rare, but there may be a diverse benthic microfauna. Skeletal remains are often fragmented and corroded. Condensed deposits, which have formed below the photic zone, lack calcareous algae and micrite envelopes. Authigenic minerals, such as glauconite, phosphorite, siderite, and organic matter are often present, and there commonly is an increase in $\delta^{13}C$ (Loutit et al., 1988).

In rimmed platforms, topographically high, back-stepping, and aggrading shelf edges are quickly reestablished during sea-level rise. If they are able to keep up with sea-level rise, rims will form thicker accumulations than the adjacent lagoons and shelves (Enos, 1977; Harris, 1979; Purdy, 1974). Backstepping of shelf edges during a relative rise of sea level tends to isolate slopes from sedimentation so that they become sediment bypass zones. Grammer and Ginsburg (1992) showed that slope deposition around the Tongue of the Ocean, Bahamas, ceased ~10,500 y.b.p. as sea level rose and began to flood the top of the vertical platform escarpment. Until about 7000 y.b.p., carbonate sand and mud produced on the shallow platform bypassed the steep upper slopes.

Early Holocene lowstand shelf edges around the Bahamas commonly were dominated by reefs, but some gave way to grain shoals during the subsequent eustatic rise. This transition may have been environmentally induced by the rise. Partially flooded lagoons that developed during sea-level rise were probably of variable salinity and temperature. If these waters periodically spilled off the platform and bathed lowstand reefs, coral death or reduced growth may have followed (Hine et al., 1981b; Schlager, 1981). In the Bahamas, lowstand shelf-edge reefs backstepped as much as 30 m vertically during 3000–4000 yr of the Holocene sea-level rise prior to giving up, being submerged more deeply, and finally being buried by shelf-edge carbonate sands that were transported offbank (Hine and Neumann, 1977). During the Holocene flooding of the Little Bahama Bank, some transgressive sand bodies were drowned and transformed into vegetated relicts because they were not able to keep pace with rising sea level (Hine, 1977). We have not seen any documentation of the lowstand and transgressive succession of lithofacies in ancient rimmed shelf margins, but a hypothetical example is offered in Figure 14.

Highstand Conditions

Highstand systems tract deposition occurs during the late part of a eustatic rise, a stillstand, and the early part of a eustatic fall (Van Wagoner et al., 1988). During this interval of time, shallow-marine sedimentation rates commonly exceed subsidence and the eustatic rise, thus leading to deposition of aggradational to progradational shelf, shelf-edge, and slope strata (Sarg, 1988). Although dependent on accommodation space and local water conditions, carbonate sedimentation is usually greatest during highstands of sea level because the extent of platform flooding and, therefore, the carbonate factory are greatest then. As the rate of accommodation increase begins to decline (Jervey, 1988) and overall sediment production rates remain high, platform shallowing occurs. Shallowing is due mainly to aggradation of the seafloor and progradation of islands, shoals, buildups, and the shoreline. Before aggradation or progradation can transpire in siliciclastic environments, the sediments must be delivered to a deposi-

tional site. However, aggradation and progradation in carbonate environments takes place as a result of in situ carbonate sediment generation and accumulation, as well as sediment delivery.

Progradation rates vary depending on water depth, energy, depositional processes, and sediment production and accumulation rates. The western margin of the Great Bahama Bank has been prograding into water depths of more than 400 m for the past 5.6 m.y. (Ginsburg et al., 1991) at an average rate of 1.3 m/1000 yr. On top of the platform, the southwest Andros Island tidal flats have advanced at a rate of 5–20 km/1000 yr since the Holocene transgression. The Persian Gulf tidal flats have prograded at a rate of 0.5–2 km/1000 yr (Kinsman, 1969). These progradation rates are comparable to those of Holocene siliciclastic shorelines (1.5–4.5 km/1000 yr) and deltas (2–22 km/1000 yr) (Evans, 1979). Although progradation rates of ooid shoreface deposits are considerably less [~100 m/1000 yr at West Caicos Island according to Lloyd et al. (1987), a much larger volume of sediment is deposited above sea level as eolian dunes up to 18 m high] than in the Andros Island tidal flats. When this difference is accounted for, the volume of sediment produced and, thus, the actual sedimentation rates at West Caicos Island and the Persian Gulf tidal flats are comparable.

Sediment production rates along rimmed shelf edges are greater than surrounding environments. The accommodation space is quickly filled along shelf edges at first by aggradation and later by progradation. Sediment wedges usually prograde seaward because sediment production rates and energy levels are highest there. In some cases, however, shelf edges prograde into shelf lagoons because of a strong bankward energy flux and sediment transport from seaward margins or because the lagoonal margins are open enough to promote high sediment production rates (Figure 5).

According to siliciclastic sequence stratigraphic concepts (Haq et al., 1987; Jervey, 1988; Posamentier and Vail, 1988; Vail, 1987; Van Wagoner et al., 1988), a downlap surface records maximum flooding conditions, and it forms just prior to a decline in the rate of accommodation development. As the rate of accommodation increase diminishes, regressive conditions develop and promote the formation of prograding clinoforms, which downlap a surface, or interval, of maximum flooding. This maximum flooding surface, or interval, separates the transgressive and highstand systems tracts. Similar features are present in carbonate sequences (Eberli and Ginsburg, 1989; Rudolph and Lehmann, 1989; Sarg, 1988) and record the regional progradation of platform margins into deep water. These are commonly recognizable on seismic lines. However, when examining outcrops, one must be cautious in identifying downlap, or so-called maximum flooding, surfaces. This is because even large-scale outcrops may contain local downlap surfaces that are not related to maximum flooding. Local downlap surfaces can form within any systems tract where beds, bedsets, parasequences, and parasequence sets (Van Wagoner et al., 1990) prograde into deeper water. All that is required is for the rate of deposition to exceed the rate at which accommodation space is created. These conditions are commonly met in carbonate environments where autochthonous sediment production and accumulation rates are high or where allochthonous sediment accumulation rates are high. These include muddy and sandy shorefaces, platform-interior patch reefs and grain shoals, shelf-edge reefs and shoals, and basin-margin toes-of-slopes. Careful examination, however, will usually show that the downlap surfaces are local in extent.

Highstands of sea level lead to higher rates of resedimentation in slopes and basins (Mullins, 1983). Large amounts of platform-derived fine-grained sediment are transported offbank by storm waves and currents. Slowly settling from suspension, they accumulate on slope and basin floors as periplatform oozes (Schlager and James, 1978). Prograding highstand shelf edges and slopes commonly oversteepen and fail. Collapse of shelf edges and slopes by way of rockfalls, sediment slides, and sediment gravity flows add resedimented debris to highstand systems tract deposits in the slope and basin floor.

Deposition and preservation of platform evaporites generally favor highstand conditions. Prograding rimmed shelf edges and ramp beach-dune complexes can depositionally isolate shelf environments. If these topographically high features can form effective barriers to marine circulation and disconnect the shelf from open-marine environments, evaporite sedimentation may follow. For platform evaporites, the prerequisite barrier could comprise shelf-edge ridges made up of eolian dunes and barrier bars, storm-deposited reef rubble, or reef buildups. As platform seawater evaporates and forms a brine, the brine level lowers and initiates a hydrodynamically driven seepage of seawater from the open-marine sea through the barrier to replenish that which is evaporated. Although a barrier can easily form during a sea-level fall, sea level must not fall below the top of the shelf. Otherwise, there would be no hydrodynamic head to drive seawater through the barrier into the shelf. Since disconnection from the open sea is required, platform evaporites probably onlap barriers. This scenario may help explain many of the Permian platform evaporites in west Texas and New Mexico, the Jurassic Buckner, and Cretaceous Ferry Lake evaporites of the Gulf Coast.

Where sufficiently arid climatic conditions prevail, the proper disconnection from open sea is established (Lucia, 1972), and sufficiently large volumes of seawater are available for evaporation, evaporite deposition (chiefly aggradation in subaqueous settings) can commence at a significantly rapid rate. Schreiber and Hsü (1980) determined that subaqueous evaporite depositional rates range from 1–100 m kyr^{-1}. Accumulation rates of the 1-km-thick upper Miocene evaporites in the Mediterranean Basin and the

Holocene Texada Halite in Lake MacLeod, Western Australia were 4–5 m kyr (Logan, 1987; Warren, 1989). For comparison, recall that the average growth potential of reef-bearing carbonate platforms is 100 cm kyr (Schlager, 1981). Evaporites have a low preservation potential, however. Low-lying evaporite environments are subject to flooding by low-salinity waters (continental or marine) that often results in the dissolution of evaporites. Thus, the net accumulation of a platform evaporite succession will not exceed the growth rate of the carbonate platform.

CONCLUSIONS

Sequence stratigraphic analysis is increasingly viewed as an essential methodology for studying carbonate platforms. The methodology pioneered by Vail et al. (1977) offers the potential for documenting the depositional and erosional history of a platform relative to the tectonic setting, eustasy, and rate of sedimentation. The analysis of stratal patterns is important to a sequence stratigraphic analysis of carbonates because of the unusual and unique patterns that often develop as a result of erosion and deposition in carbonate systems. A sequence stratigraphic approach is strengthened, however, by paying special attention to the myriad of environmental factors controlling the development of carbonate platforms. Fundamental differences between (1) the depositional origins of carbonate and siliciclastic sediments and (2) depositional and erosional responses of carbonate and siliciclastic systems to relative sea-level changes mandate development of independent depositional sequence and systems tract models. Thus, our intent was to develop a general hypothesis of carbonate depositional and erosional responses to relative sea-level changes. Models presented here honor the physical and biologic processes that help control carbonate platform development. As a result, they augment established carbonate-facies models and they complement sequence and systems tract models presented by Vail (1987), Posamentier and Vail (1988), and Sarg (1988).

While the focus of this study has been on concepts and stratigraphic responses to relative sea-level changes, we believe that these models can be used to explore for hydrocarbons. When used in conjunction with seismic records, well logs, and core data, the models may be applied carefully to specific case studies to help delineate reservoir-, seal-, and source-prone facies. Caution must be exercised, however, because the models are not templates that can be applied rigidly to any platform, modern or ancient. Similar to the Exxon models of Vail (1987), Posamentier and Vail (1988), and Sarg (1988), certain assumptions were made and simplified for the sake of presentation. However, these assumptions are not universally applicable and they must be modified to fit local needs.

Future work should attempt to test these and other models by conducting detailed stratigraphic studies of carbonate platforms, especially where adequate well-log, seismic, and rock control can constrain interpretations. High-resolution sequence stratigraphic studies of reservoirs will help determine the detailed makeup of sequences and parasequences, and they will lead to better and more sophisticated models in the future. Most studies have not adequately addressed tectonic influences on platform development. Thus, we lack information on how stratal geometries differ between quiescent and tectonically active areas, and we have insufficient criteria to distinguish relative influences of tectonic processes and eustasy in the creation of sequence boundaries.

ACKNOWLEDGMENTS

We gratefully acknowledge ARCO Exploration and Production Technology for supporting carbonate sequence stratigraphy research and permitting publication of this work. We appreciate the constructive reviews of H. W. Posamentier, J. R. Markello, J. F. Sarg, R. K. Suchecki, and J. Weber. Discussions with ARCO colleagues, such as A. A. Brown, J. A. Thorne, and M. G. Justice in the laboratory and in the field helped us solidify our concepts and ideas. We also appreciate the graphics support provided by Gary Garrett, Karen Ward, James Aston, and James Duke of ARCO.

REFERENCES CITED

Ahr, W. M., 1973, The carbonate ramp: An alternative to the shelf model: Transactions Gulf Coast Association of Geological Societies, v. 23, p. 221-225.

Allen, P. A., and J. R. Allen, 1990, Basin Analysis, Principles and Applications: Oxford, Blackwell Scientific Publications, 451 p.

Arthurton, R. S., 1973, Experimentally produced halite compared to Triassic layered halite-rocks from Cheshire, England: Sedimentology, v. 20, p. 145-160.

Back, W., B. B. Hanshaw, J. S. Herman, and J. N. van Driel, 1986, Differential dissolution of a Pleistocene reef in the ground-water mixing zone of coastal Yucatan, Mexico: Geology, v. 14, p. 137-140.

Bathurst, R. G. C., 1971, Carbonate Sediments and their Diagenesis. Developments in Sedimentology 12: Elsevier Publishing Company, Amsterdam, 620 p.

Blom, W. M., and D. B. Alsop, 1988, Carbonate mud sedimentation on a temperate shelf: Bass Basin, southeastern Australia, in C. S. Nelson, ed., Non-Tropical Shelf Carbonates—Modern and Ancient: Sedimentary Geology, v. 60, p. 269-280.

Bosellini, A., 1984, Progradation geometries of carbonate platforms: Examples from the Triassic of the Dolomites, northern Italy: Sedimentology, v. 31, p. 1-24.

Bosellini, A., 1989, Dynamics of Tethyan carbonate platforms, in P. D. Crevello, J. L. Wilson, J. F. Sarg,

and J. F. Read, eds., Controls on Carbonate Platform and Basin Development: SEPM Special Publication No. 44, p. 3-13.

Brady, M. J., 1971, Sedimentology and diagenesis of carbonate muds in coastal lagoons of NE Yucatan: unpublished Ph.D. dissertation, Rice University, 288 p.

Brown, L. F., 1989, A sequence stratigraphic and systems-tract model for the Virgilian and Wolfcampian Series, Eastern Shelf and adjacent Midland Basin, Texas: Texas Tech University Studies in Geology 2, Part II, Contributed Papers, p. 35-62.

Bubb, J. N., and W. G. Hatlelid, 1977, Seismic stratigraphy and global changes of sea level, Part 10: Seismic recognition of carbonate buildups, in C. E. Payton, ed., Seismic Stratigraphy—Applications to Hydrocarbon Exploration: AAPG Memoir 26, p. 185-204.

Coleman, J. M., and D. B. Prior, 1988, Mass wasting on continental margins, in Annual Review of Earth and Planetary Sciences, v. 16, p. 101-119.

Cook, H. E., P. N. McDaniel, E. W. Mountjoy, and L. C. Pray, 1972, Allochthonous carbonate debris flows at Devonian bank ('reef') margins, Alberta, Canada: Bulletin Canadian Petroleum Geology, v. 20, p. 439-497.

Cook, H.E., and H. T. Mullins, 1983, Basin margin environment, in P. A. Scholle, D. G. Bebout, and C. H. Moore, eds., Carbonate Depositional Environments: AAPG Memoir 33, p. 539-617.

Cram, J. M., 1979, The influence of continental shelf width on tidal range: Paleooceanographic implications: Journal of Geology, v. 87, p. 441-447.

Dabrio, C. J., M. Esteban, and J. M. Martin, 1981, The coral reef of Níjar, Messinian (Uppermost Miocene), Almería Province, S. E. Spain: Journal of Sedimentary Petrology, v. 51, p. 521-539.

Davies, P. J., P. A. Symonds, D. A. Feary, and C. J. Pigram, 1989, The evolution of the carbonate platforms of northeast Australia, in P. D. Crevello, J. L. Wilson, J. F. Sarg, and J. F. Read, eds., Controls on Carbonate Platform and Basin Development: SEPM Special Publication No. 44, p. 233-258.

De Buisonje, P. H., 1974, Neogene and Quaternary Geology of Aruba, Curacao and Bonaire: Ph.D. Thesis, University of Utrecht, Publication De Boer-Cuperus B. V., Utrecht, 293 p.

Droxler, A. W., and W. Schlager, 1985, Glacial versus interglacial sedimentation rates and turbidite frequency in the Bahamas: Geology, v. 13, p. 799-802.

Eberli, G. P., and R. N. Ginsburg, 1989, Cenozoic progradation of northwestern Great Bahama Bank, a record of lateral platform growth and sea level fluctuations, in P. D. Crevello, J. L. Wilson, J. F. Sarg, and J. F. Read, eds., Controls on Carbonate Platform and Basin Development: SEPM Special Publication No. 44, p. 339-351.

Elliott, T., 1986, Siliciclastic shorelines, in H. G. Reading, ed., Sedimentary Environments and Facies, 2nd Ed.: Boston, Blackwell Scientific Publications, p. 155-188.

Enos, P., 1977, Holocene sediment accumulations of the south Florida shelf margin, in P. Enos and R. D. Perkins, eds., Quaternary Sedimentation in South Florida: Geological Society of America Memoir 147, part I, p. 1-130.

Enos, P., and R. D. Perkins, 1979, Evolution of Florida Bay from island stratigraphy: Geological Society of America Bulletin, v. 90, p. 59-83.

Esteban, M., and J. Giner, 1977, Messinian coral reefs and erosion surfaces in Cabo de Gata (Almería, SE Spain): Acta Geológica Hispánica, v. 14, p. 97-104.

Evans, G., 1979, Quaternary transgressions and regressions: Journal of the Geological Society, London, v. 136, p. 125-132.

Ford, D. C., and P. W. Williams, 1989, Karst Geomorphology and Hydrology: London, Unwin Hyman, 601 p.

Franseen, E. K., and C. Mankiewicz, 1991, Depositional sequences and correlation of middle (?) to late Miocene carbonate complexes, Las Negras and Níjar areas, southeastern Spain: Sedimentology, v. 38, p. 871-898.

Friedman, G. M., 1988, Case histories of coexisting reefs and terrigenous sediments: The Gulf of Elat (Red Sea), Java Sea, and Neogene basin of the Negev, Israel, in L. J. Doyle, and H. H. Roberts, eds., Carbonate-Clastic Transitions: Developments in Sedimentology 42, Amsterdam, Elsevier Science Publishers B. V., p. 77-97.

Gerard, J., and H. Oesterle, 1973, Facies study of the offshore Mahakam area: Proceedings of Indonesian Petroleum Association, p. 187-194.

Ginsburg, R. N., D. F. McNeill, G. P. Eberli, P. K. Swart, and J. A. Kenter, 1991, Transformation of morphology and facies of Great Bahama Bank by Plio-Pleistocene progradation (abstract): Dolomieu Conference on Carbonate Platforms and Dolomitization, Abstracts, Ortisei, Italy, September 16-21, 1991, p. 88-89.

Glaser, K. S., and A. W. Droxler, 1991, High production and highstand shedding from deeply submerged carbonate banks, northern Nicaragua Rise: Journal of Sedimentary Petrology, v. 61, p. 128-142.

Goldhammer, R. K., E. J. Oswald, and P. A. Dunn, 1991, High frequency glacio-eustatic cyclicity in the middle Pennsylvanian of the Paradox Basin: An evaluation of Milankovitch forcing (abstract): Dolomieu Conference on Carbonate Platforms and Dolomitization, Abstracts, Ortisei, Italy, September 16-21, 1991, p. 91.

Gostin, V. A., A. P. Belperio, and J. H. Cann, 1988, The Holocene non-tropical coastal and shelf carbonate province of southern Australia, in C. S. Nelson, ed., Non-Tropical Shelf Carbonates - Modern and Ancient: Sedimentary Geology, v. 60, p. 51-70.

Grammer, G. M., and R. N. Ginsburg, 1992, Highstand versus lowstand deposition on carbonate platform margins: insight from Quaternary foreslopes in the Bahamas: Marine Geology, v. 103, p. 125-136.

Hallock, P., and Schlager, W., 1986, Nutrient excess and the demise of coral reefs and carbonate platforms: Palaios, v. 1, p. 389-398.

Handford, C. R., 1990, Halite depositional facies in a solar salt pond: A key to interpreting physical energy and water depth in ancient deposits? Geology, v. 18, p. 691-694.

Handford, C. R., and R. G. Loucks, 1990, Dynamic response of carbonate systems tracts to relative sea level changes and the development of carbonate depositional sequences in platforms and ramps (abstract): AAPG Bulletin, v. 74, p. 669.

Handford, C. R., and R. G. Loucks, 1991, Unique signature of carbonate strata and the development of depositional sequence and systems tract models for ramps, rimmed shelves, and detached platforms: AAPG Bulletin, v. 75, p. 588.

Haq, B. U., J. Hardenbol, and P. R. Vail, 1987, Chronology of fluctuating sea levels since the Triassic: Science, v. 235, p. 1156-1166.

Hardie, L. A., 1986, Stratigraphic models for carbonate tidal-flat deposition: Colorado School of Mines Quarterly, v. 81, p. 59-74.

Hardie, L. A., E. Newton-Wilson, and R. K. Goldhammer, 1991, Cyclostratigraphy and dolomitization of the Middle Triassic Latemar buildup, the Dolomites, northern Italy: Guidebook Excursion F, Dolomieu Conference on Carbonate Platforms and Dolomitization, Ortisei, Italy, 56 p.

Harris, P. M., 1979, Facies Anatomy and Diagenesis of a Bahamian Ooid Shoal: Sedimenta 7, Comparative Sedimentology Laboratory, University of Miami, 163 p.

Hine, A. C., 1977, Lily Bank, Bahamas: History of an active oolite sand shoal: Journal of Sedimentary Petrology, v. 47, p. 1554-1581.

Hine, A. C., and P. Hallock, 1991, Tectonic controls on carbonate platforms of the Nicaraguan Rise, Caribbean Sea (abstract): Dolomieu Conference on Carbonate Platforms and Dolomitization, Abstracts, Ortisei, Italy, September 16-21, 1991, p. 110.

Hine, A. C., and A. C. Neumann, 1977, Shallow carbonate bank margin growth and structure, Little Bahama Bank: AAPG Bulletin, v. 61, p. 376-406.

Hine, A. C., R. J. Wilber, J. M. Bane, A. C. Neumann, and K. R. Lorenson, 1981a, Offbank transport of carbonate sands along open, leeward bank margins, northern Bahamas: Marine Geology, v. 42, p. 323-348.

Hine, A. C., R. J. Wilber, and A. C. Neumann, 1981b, Carbonate sand bodies along contrasting shallow bank margins facing open seaways in northern Bahamas: AAPG Bulletin, v. 65, p. 261-290.

Hopley, D., 1982, The Geomorphology of the Great Barrier Reef: New York, John Wiley and Sons, 453 p.

Hunt, D., and M. Tucker, 1991, Sequence stratigraphic models for carbonate platforms (abstract): Dolomieu Conference on Carbonate Platforms and Dolomitization, Abstracts, Ortisei, Italy, September 16-21, 1991, p. 113.

Irwin, M. L., 1965, General theory of epeiric clear water sedimentation: AAPG Bulletin, v. 49, p. 445-459.

Jacquin, T., A. Arnaud-Vanneau, H. Arnaud, C. Ravenne, and P. R. Vail, 1991, Systems tracts and depositional sequences in a carbonate setting: a study of continuous outcrops from platform to basin at the scale of seismic lines: Marine and Petroleum Geology, v. 8, p. 122-139.

James, N. P., 1979, Facies models 9. Introduction to carbonate facies models, *in* R. G. Walker, ed., Facies Models: Geoscience Canada Reprint Series 1, p. 105-107.

James, N. P., and Y. Bone, 1991, Origin of a cool-water Oligo-Miocene deep shelf limestone, Eucla Platform, southern Australia: Sedimentology, v. 38, p. 323-341.

James, N. P., and R. N. Ginsburg, 1979, The Seaward Margin of Belize Barrier and Atoll Reefs, Morphology, Sedimentology, Organism Distribution and Late Quaternary History: International Association of Sedimentologists Special Publication No. 3, 191 p.

James, N. P., and I. G. MacIntyre, 1985, Carbonate depositional environments. I. Reefs. Quarterly Colorado School of Mines, 80, 70 p.

Jennings, J. N., 1971, Karst: Cambridge, Mass., The M. I. T. Press, 252 p.

Jervey, M. T., 1988, Quantitative geological modeling of siliciclastic rock sequences and their seismic expression, *in* C. K. Wilgus, B. S. Hastings, C. G. St. C. Kendall, H. W. Posamentier, C. A. Ross, and J. C. Van Wagoner, eds., Sea Level Changes: An Integrated Approach: SEPM Special Publication No. 42, p. 47-69.

Johnson, H. D., 1978, Shallow siliciclastic seas, *in* H. G. Reading, ed., Sedimentary Environments and Facies: New York, Elsevier, p. 207-258.

Kendall, C. G. St. C., and W. Schlager, 1981, Carbonates and relative changes in sea-level: Marine Geology, v. 44, p. 181-212.

Kenter, J. A. M., 1990, Carbonate platform flank: Slope angle and sediment fabric: Sedimentology, v. 37, p. 777-794.

Kenter, J. A. M., and W. Schlager, 1989, Comparison of shear strength in calcareous and siliciclastic marine sediments: Marine Geology, v. 88, p. 145-152.

Kinsman, D. J. J., 1969, Interpretation of Sr^{2+} concentrations in carbonate minerals and rocks: Journal of Sedimentary Petrology, v. 39, p. 486-508.

Kobluk, D. R., and M. A. Lysenko, 1984, Carbonate Rocks and Coral Reefs, Bonaire, Netherlands Antilles: Geological Association of Canada, Mineralogical Association of Canada, Field Trip Guidebook, Field Trip 13, 67 p.

Labaume, P., E. Mutti, and M. Seguret, 1987, Megaturbidites: A depositional model from the Eocene of the SW-Pyrenean foreland basin, Spain: Geo-Marine Letters, v. 7, p. 91-101.

Lees, A., 1975, Possible influence of salinity and temperature on modern shelf carbonate sedimentation: Marine Geology, v. 19, p. 159-198.

Lees, A., and A. T. Buller, 1972, Modern temperate-water and warm-water shelf carbonate sediments contrasted: Marine Geology, v. 13, M67-M73.

Lloyd, R. M., R. D. Perkins, and S. D. Kerr, 1987, Beach and shoreface ooid deposition on shallow interior banks, Turks and Caicos Islands, British West Indies: Journal of Sedimentary Petrology, v. 57, p. 976-982.

Logan, B. W., 1987, The MacLeod Evaporite Basin, Western Australia—Holocene Environments, Sediments and Geological Evolution: AAPG Memoir 44, 140 p.

Loreau, J. P., and B. H. Purser, 1973, Distribution and ultra-structure of Holocene ooids in the Persian Gulf, *in* B. H. Purser, ed., The Persian Gulf: Berlin, Springer-Verlag, p. 279-328.

Loucks, R., and A. Brown, 1991, Sedimentology and geometry of Permian toe-of-slope limestones, member of the Bell Canyon Formation, McKittrick Canyon, Texas, USA. (abstract): Dolomieu Conference on Carbonate Platforms and Dolomitization, Abstracts, Ortisei, Italy, September 16-21, 1991, p. 155.

Loucks, R., and C. R. Handford, 1992, Origin and recognition of fractures, breccias, and sediment fills in paleocave-reservoir networks, *in* M. P. Candelaria and C. L. Reed, eds., Paleokarst, Karst Related Diagenesis and Reservoir Development: Examples from Ordovician–Devonian Age Strata of West Texas and the Mid-Continent: Permian Basin Section-SEPM, Publication No. 92-33, p. 31-44.

Loutit, T. S., J. Hardenbol, P. R. Vail, and G. R. Baum, 1988, Condensed sections: The key to age determination and correlation of continental margin sequences, *in* C. K. Wilgus, B. S. Hastings, C. G. St. C. Kendall, H. W. Posamentier, C. A. Ross, and J. C. Van Wagoner, eds., Sea Level Changes: An Integrated Approach: SEPM Special Publication No. 42, p. 183-213.

Lowenstein, T. K., and L. A. Hardie, 1985, Criteria for the recognition of salt-pan evaporites: Sedimentology, v. 32, p. 627-644.

Lowenstein, T. K., L. A. Hardie, E. Casas, and B. C. Schreiber, 1989, Sub-Mediterranean giant salt: A perennial subaqueous evaporite (abstract): Geological Society of America Abstracts with Programs, v. 21, p. A364.

Lucia, F. J., 1972, Recognition of evaporite–carbonate shoreline sedimentation, *in* J. K. Rigby, and W. K. Hamblin, eds., Recognition of Ancient Sedimentary Environments: SEPM Special Publication No. 16, p. 61-191.

MacIntyre, I. G., 1983, Growth, depositional facies, and diagenesis of a modern bioherm, Galeta Point, Panama, *in* P. M. Harris, ed., Carbonate Buildups—A Core Workshop: SEPM Core Workshop No. 4, p. 578-593.

Magnier, Ph., T. Oki, and L. W. Kartaadiputra, 1975, The Mahakam Delta, Kalimantan, Indonesia: Proceedings, Ninth World Petroleum Congress, v. 2, Geology, p. 239-250.

Maire, R., 1981, Karst and hydrogeology synthesis. Spelunca, Supplement to No. 3, p. 23-30.

Mazzullo, J., A. Malicse, and J. Siegel, 1991, Facies and depositional environments of the Shattuck Sandstone on the Northwest Shelf of the Permian Basin: Journal of Sedimentary Petrology, v. 61, p. 940-958.

McIlreath, I. A., and N. P. James, 1979, Facies models 12. Carbonate slopes, in R. G. Walker, ed., Facies Models: Geoscience Canada Reprint Series 1, p. 133-143.

Meyers, W. J., 1988, Paleokarstic features in Mississippian limestones, New Mexico, in N. P. James and P. W. Choquette, eds., Paleokarst: New York, Springer-Verlag, p. 306-328.

Miall, A. D., 1991, Stratigraphic sequences and their chronostratigraphic correlation: Journal of Sedimentary Petrology, v. 61, p. 497-505.

Milliman, J. D., 1974, Marine Carbonates: New York, Springer-Verlag, 375 p.

Mullins, H. T., 1983, Modern carbonate slopes and basins of the Bahamas, in H. E. Cook, A. C. Hine, and H. T. Mullins, eds., Platform Margin and Deepwater Carbonates: SEPM Short Course 12, p. 4.1-4.138.

Mullins, H. T., and H. E. Cook, 1986, Carbonate apron models: Alternatives to the submarine fan model for paleoenvironmental analysis and hydrocarbon exploration: Sedimentary Geology, v. 48, p. 37-79.

Mullins, H. T., J. Dolan, N. Breen, B. Andersen, M. Gaylord, J. L. Petruccione, R. W. Wellner, A. J. Melillo, A. D. Jurgens, 1991, Carbonate platforms: Response to tectonic processes: Geology, v. 19, p. 1089-1092.

Mullins, H. T., and A. C. Hine, 1989, Scalloped bank margins: Beginning of the end for carbonate platforms? Geology, v. 17, p. 30-33.

Mutti, E., 1985, Turbidite systems and their relations to depositional sequences, in G. G. Zuffa, ed., Provenance of Arenites, NATO Advanced Science Institutes Series C, No. 148: Dordrecht, D. Reidel Publishing Co., p. 65-93.

Nardin, T. R., F. J. Hein, D. S. Gorsline, and B. D. Edwards, 1979, A review of mass movement processes, sediment and acoustic characteristics, and contrasts in slope and base-of-slope systems versus canyon-fan-basin floor systems, in L. J. Doyle and O. H. Pilkey, Jr., eds., Geology of Continental Slopes: SEPM Special Publication No. 27, p. 61-73.

Neumann, A. C., and L. S. Land, 1975, Lime mud deposition and calcareous algae in the Bight of Abaco, Bahamas: A budget: Journal of Sedimentary Petrology, v. 45, p. 763-786.

Newell, N. D., E. G. Purdy, and J. Imbrie, 1960, Bahamian oolitic sand: Journal of Geology, v. 68, p. 481-497.

Palmer, A. N., 1991, Origin and morphology of limestone caves: Geological Society of America Bulletin, v. 103, p. 1-21.

Pinet, P. R., and P. Popenoe, 1985, A scenario of Mesozoic-Cenozoic ocean circulation over the Blake Plateau and its environs: Geological Society of America Bulletin, v. 96, p. 618-626.

Pomar, L., 1991, Reef geometries, erosion surfaces and high-frequency sea level changes, upper Miocene reef complex, Mallorca, Spain: Sedimentology, v. 38, p. 243-269.

Pomar, L., 1993, High-resolution sequence stratigraphy in prograding Miocene carbonates: application to seismic interpretation, this volume.

Posamentier, H. W., D. P. James, and G. P. Allen, 1990, Aspects of sequence stratigraphy: Recent and ancient examples of forced regressions (abstract): AAPG Bulletin, v. 74, p. 742.

Posamentier, H. W., and P. R. Vail, 1988, Eustatic controls on clastic deposition II—sequence and systems tract models, in C. K. Wilgus, B. S. Hastings, C. G. St. C. Kendall, H. W. Posamentier, C. A. Ross, and J. C. Van Wagoner, eds., Sea Level Changes: An Integrated Approach: SEPM Special Publication No. 42, p. 125-154.

Pray, L. C., 1988, Geology of the western escarpment, Guadalupe Mountains, Texas, in J. F. Sarg, C. Rossen, P. J. Lehmann, L. C. Pray, eds., Geologic Guide to The Western Escarpment, Guadalupe Mountains, Texas: Permian Basin Section, SEPM Publication 88-30, p. 1-8.

Purdy, E. G., 1974, Karst-determined facies patterns in British Honduras: Holocene carbonate sedimentation model: AAPG Bulletin, v. 58, p. 825-855.

Read, J. F., 1985, Carbonate platform facies models: AAPG Bulletin, v. 69, p. 1-21.

Roberts, H. H., and S. P. Murray, 1988, Gulf of the northern Red Sea: Depositional settings of abrupt siliciclastic-carbonate transitions, in L. J. Doyle, and H. H. Roberts, eds., Carbonate-Clastic Transitions: Developments in Sedimentology 42, Amsterdam, Elsevier Science Publishers B. V., p. 99-142.

Rossen, C., P. J. Lehmann, and J. F. Sarg, 1988, Trail guide for day 1: Shumard to Bone Canyon traverse, in Sarg J. F., C. Rossen, P. J. Lehmann, L. C. Pray, eds., Geologic Guide to The Western Escarpment, Guadalupe Mountains, Texas: Permian Basin Section, SEPM Publication 88-30, p. 17-60.

Rudolph, K. W., and P. J. Lehmann, 1989, Platform evolution and sequence stratigraphy of the Natuna platform, South China Sea, in P. D. Crevello, J. L. Wilson, J. F. Sarg, and J. F. Read, eds., Controls on Carbonate Platform and Basin Development: SEPM Special Publication No. 44, p. 353-361.

Sarg, J. F., 1988, Carbonate sequence stratigraphy, in C. K. Wilgus, B. S. Hastings, C. G. St. C. Kendall, H. W. Posamentier, C. A. Ross, and J. C. Van Wagoner, eds., Sea Level Changes: An Integrated Approach: SEPM Special Publication No. 42, p. 155-181.

Schlager, W., 1981, The paradox of drowned reefs and carbonate platforms: Geological Society of America Bulletin, v. 92, p. 197-211.

Schlager, W., 1991, Depositional bias and environmental change—important factors in sequence stratigraphy: Sedimentary Geology, v. 70, p. 109-130.

Schlager, W., and O. Camber, 1986, Submarine slope angles, drowning unconformities, and self-erosion of limestone escarpments: Geology, v. 14, p. 762-765.

Schlager, W., and N. P. James, 1978, Low Mg-calcite limestones forming on the deep-sea floor (Tongue of the Ocean, Bahamas): Sedimentology, v. 25, p. 675-702.

Schofield, D. W., R. A. Slater, and C. V. G. Phipps, 1983, Holocene erosion of Elizabeth Reef, Tasman Sea, Australia, in P. M. Harris, ed., Carbonate Buildups—A Core Workshop: SEPM Core Workshop No. 4, p. 558-577.

Schreiber, B. C., G. M. Friedman, A. Decima, and E. Schreiber, 1976, Depositional environments of Upper Miocene (Messinian evaporite deposits of the Sicilian Basin: Sedimentology, v. 23, p. 729-760.

Schreiber, B. C., and K. J. Hsü, 1980, Evaporites, in G. D. Hobson, ed., Developments in Petroleum Geology, v. 2: London, Applied Science Publishers, p. 87-138.

Schwarz, H. U., 1982, Subaqueous slope failures—experiments and modern occurrences, in H. Fuchtbauer, A. P. Lisitzyn, J. D. Milliman, and E. Seibold, eds., Contributions to Sedimentology: Stuttgart, E. Schweizerbart'sche Verlagsbuchhandlung (Nagele u. Obermiller), 116 p.

Scoffin, T. P., 1987, An Introduction to Carbonate Sediments and Rocks: New York, Chapman and Hall, 274 p.

Shaw, A. B., 1964, Time in Stratigraphy: New York, McGraw-Hill, 353 p.

Sheriff, R. E., 1980, Seismic Stratigraphy: Boston, International Human Resources Development Corporation, 227 p.

Tucker, K. E., and R. G. Chalcraft, 1991, Cyclicity in the Permian Queen Formation, U. S. M. Queen Field, Pecos County, Texas, in A. J. Lomando and P. M. Harris, eds., Mixed Carbonate Siliciclastic Sequences, Society for Sedimentary Geology Core Workshop No. 15, p. 385-428.

Tucker, M. E., 1991, Sequence stratigraphy of carbonate-evaporite basins: models and application to the Upper Permian (Zechstein) of northeast England and adjoining North Sea: Journal of the Geological Society, London, v. 148, p. 1019-1036.

Tucker, M. E., and V. P. Wright, 1990, Carbonate Sedimentology: Oxford, Blackwell Scientific Publications, 482 p.

Vail, P. R., 1987, Seismic stratigraphy interpretation using sequence stratigraphy, Part 1; Seismic stratigraphy interpretation procedure, in A. W. Bally, ed., Atlas of Seismic Stratigraphy: AAPG Studies in Geology 27, v. 1, p. 1-10.

Vail, P. R., R. M. Mitchum, Jr., R. G. Todd, J. M. Widmier, S. Thompson, III, J. B. Sangree, J. N. Bubb, W. G. Hatlelid, 1977, Seismic stratigraphy and global changes of sea level, in C. E. Payton, ed., Seismic Stratigraphy—Applications to Hydrocarbon Exploration: AAPG Memoir 26, p. 49-212.

Van Wagoner, J. C., R. M. Mitchum, K. M. Campion, and V. D. Rahmanian, 1990, Siliciclastic Sequence Stratigraphy in Well Logs, Cores, and Outcrops: Concepts for High-Resolution Correlation of Time and Facies: AAPG Methods in Exploration Series, No. 7, 55 p.

Van Wagoner, J. C., H. W. Posamentier, R. M. Mitchum, et al., 1988, An overview of the fundamentals of sequence stratigraphy and key definitions, in C. K. Wilgus, B. S. Hastings, C. G. St. C. Kendall, H. W. Posamentier, C. A. Ross, and J. C. Van Wagoner, eds., Sea Level Changes: An Integrated Approach: SEPM Special Publication No. 42, p. 39-45.

Warren, J. K., 1989, Evaporite Sedimentology, Importance in Hydrocarbon Accumulation: Englewood Cliffs, New Jersey, Prentice Hall, 285 p.

Wendt, J., 1988, Condensed carbonate sedimentation in the late Devonian of the eastern Anti-Atlas (Morocco): Eclogae Geologica Helvetica, v. 81, p. 155-173.

White, W. B., 1988, Geomorphology and Hydrology of Karst Terrains: New York, Oxford University Press, 464 p.

Wilber, R. J., J. D. Milliman, and R. B. Halley, 1990, Accumulation of bank-top sediment on the western slope of Great Bahama Bank: Rapid progradation of a carbonate megabank: Geology, v. 18, p. 970-974.

Wilson, J. L., 1975, Carbonate facies in geologic history: New York, Springer-Verlag, 471 p.

Yose, L. A., 1991, Part 2: Sequence stratigraphy of mixed carbonate/volcaniclastic slope deposits flanking the Sciliar (Schlern)-Catinaccio buildup, Dolomites, Italy, in R. Brandner, E. Flügel, R. Koch, and L. A. Yose, eds., The Northern Margin of the Schlern/Sciliar-Rosengarten/Catinaccio Platform: Guidebook Excursion A, Dolomieu Conference on Carbonate Platforms and Dolomitization, Ortisei, The Dolomites, Italy, p. 17-39.

Chapter 2

◆

Stratigraphic Framework of Productive Carbonate Buildups

Stephen M. Greenlee
Patrick J. Lehmann
Exxon Production Research Company
Houston, Texas, U.S.A.

◆

ABSTRACT

Post-Ordovician carbonate buildups and buildup plays from around the world have been evaluated to determine the distinctive aspects of hydrocarbon-productive buildups. Ninety percent of the approximately 40 billion barrels of recoverable oil equivalent found within carbonate buildup reservoirs exists within strata deposited during just 15% of post-Ordovician geologic time. These time windows correspond to periods of extensive source rock deposition and, with the exception of the late Paleozoic and late Miocene, to the rising sea-level portions of long-term, second-order eustatic highstands. Nearly three-fourths of buildup reserves are found in buildups deposited during the early phases of periods of rapidly increasing rates of relative sea-level rise. These buildups are found in the basal, transgressive portions of thick sedimentary wedges and display progressive areal restriction through time until they are eventually unable to keep up with rapid increases in accommodation (total space available for sedimentation). A younger regressive wedge progrades over the buildups during a later period of more limited increases in accommodation and seals the mounded reservoir in fine-grained basinal sediment. Buildups deposited during this regressive phase carry higher exploration risk due to leaky top and lateral seals, and they generally have smaller trap sizes. These cycles of basin fill are clearly recognizable on seismic and log data and have a distinctive character on geohistory plots. Although the origin of these long-term accommodation changes is difficult to assess, we conclude that basin tectonism is responsible for these changes in most of our examples. Long-term eustasy plays a secondary role, according to our analysis. The model illustrated here can be helpful in both the geologic and risk assessment of new buildup plays and prospects.

INTRODUCTION

Carbonate buildups have been attractive targets for explorationists in both frontier and mature basin settings. Because buildups may be readily apparent on seismic reflection data, they are often among the first prospects to be identified and tested in a new basin. Later, as more geologic control becomes available, new and more subtle carbonate buildup trends become prospective. The predrill prediction of poros-

43

ity, seal, and source is critical to success in buildup plays. This study of carbonate buildups was in response to an effort by the Exxon Corporation to improve buildup exploration success. Figure 1 summarizes the performance of Exxon in buildup plays worldwide during the years 1975 to 1987. Of the 60 exploration wells drilled on seismically identified buildup prospects, about one-half encountered carbonate buildups and slightly fewer than one-third of those encountered were commercial discoveries. The reasons for failure to penetrate buildups include: (1) misinterpretation of other mounded geologic features as carbonate buildups; (2) errors in seismic acquisition, processing, and interpretation; and (3) other factors such as drilling off the seismic line or cases where the origin of the seismic anomaly remains unknown. When the location of a carbonate buildup was successfully predicted, several reasons accounted for failure to establish production. These factors, common to all other plays, include the lack of a mature source rock or migration pathway, porosity, and seal.

Because our failures were due to both misidentification of buildup prospects and poor prediction of play elements, we focused on seismic, sequence stratigraphic, and basin-analysis methods of stratigraphic prediction. We report here on the evaluation of basin tectonism and basin-filling stratal patterns in basins with carbonate buildups as a means of evaluating buildup prospectivity.

In this chapter, we inventory carbonate buildup production worldwide (Table 1, Figure 2), and attempt to recognize trends in the temporal and spatial distribution of productive buildups. We use geohistory analysis techniques tied to basin stratal patterns that were recognized using seismic, well log, core, and outcrop data to isolate factors common to successful buildup plays. Finally, we relate these results to other common stratigraphic traps in carbonate depositional environments. By using information derived from productive buildups, we hope to provide direction and to reduce risk for exploration in buildup-prone basins.

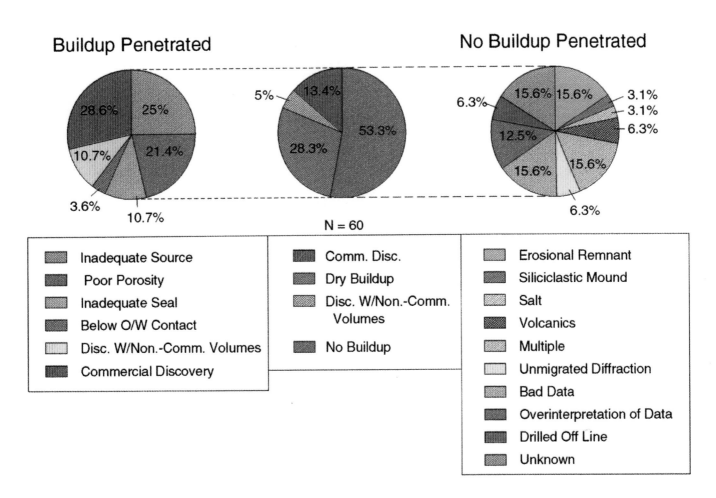

Figure 1. Pie diagrams showing the performance of the Exxon Corporation in discovering productive carbonate buildups. The center diagram illustrates the general outcome of buildup tests. The right diagram shows the reasons for failure to penetrate a buildup and the left diagram shows the results of wells drilled into carbonate buildups.

Table 1. Summary of productive buildup plays.*

Country	Basin/Area	Field/Play	OEB (Billions, EUR)	Age	First Discovery	Depth of Burial (ft)	Seal	Source (Age)	Thickness (ft)	Wedge Position	Reference
Indonesia	North Sumatra	Arun	2.28**	Early Miocene	1971	9500	Basinal Shale	Encasing Bambo and Pentu Shale (Olig. and M. Mio.)	1100	Base	Jordan and Abdullah, 1988
Indonesia	Java	Ardjuna Poleng Bima	1.3**	Miocene						Base	Exxon Files; Woodling et al., 1991
Indonesia	Salawati	Walio Kasim	0.9	Miocene						Base	Exxon Files
Indonesia	Sunda Strait	Krisna Rama	0.6	Miocene						Base	Exxon Files
Malaysia	Central Luconia	F-06 and others	3.9**	Middle–Late Miocene		~4500	Basinal Shale	Coastal Plain Coals and Shale (Olig.)	<1500	Base	Epting, 1980; Carmalt and St. John, 1986
Philippines		Nido	0.02	Early Miocene	1978	6800	Basinal Shale		660	Base	Longman, 1985
Indonesia	Natuna	L-Structure	6.5**	Middle–Late Miocene	1973	8600	Basinal Shale		5150	Base	May and Eyles, 1985; Rudolph and Lehmann, 1989
Indonesia	South Sumatra	Baturaja (Musi, Jene, etc.)	0.17**	Early Miocene	1941	2360–7500	Basinal Shale	Coastal Plain Shale and Coal (Olig.)	<500	Base	Exxon Files
Egypt	Gulf of Suez	Ras Gharib	0.07 (in Reef section)	Miocene	1938	2520	Evaporites	Shale (Pre-Miocene)	754	Evaporites	Khattab and Hadidi, 1961
Burma	Martaban	MOC-13A	0.8	Miocene							Exxon Files
Indonesia	Gulf of Papua	Pasca Pandora Uranu	0.3**	Early Miocene		5000–6000	Basinal		3000	Base	Davies et al., 1989; Exxon Files
Libya	Sirte (Intisar A and B)	Idris	2.8**	Late Paleocene	1967	8620	Basinal Marl		1200	Base	Terry and Williams, 1969; DeBrisay and Daniel, 1971

Table 1. Continued.

Country	Basin/Area	Field/Play	OEB (Billions, EUR)	Age	First Discovery	Depth of Burial (ft)	Seal	Source (Age)	Thickness (ft)	Wedge Position	Reference
Libya	Sirte Basin/Zelten Platform	Defa	2.7	Early Paleocene	1960	4500	Lagoonal/Backreef Limestone		500	Top	Exxon Files; Joiner and Myers, 1972
Libya	Dahra Platform	Hofra, Um Farud, Bu Maras, Bahi, Beda	1.016	Paleocene	1959						Exxon Files
Mexico	Tampico Embayment	Golden Lane	1.9	Early Cretaceous	1908	1910–9295	Basinal Shale	Shale (Jurassic)	5108	?	Viniegra-O and Castillo, 1970; Coogan et al., 1972
USA	South Texas/Gulf Basin	Stuart City	0.025**	Early Cretaceous		13,000–14,000	Basinal Shale	Basinal Shale (Cretaceous)	2200	Top	Bebout et al., 1977
Italy	Pelagean Basin	Vega	0.1	Jurassic		8550				Top	
USA	Midland Basin	Horseshoe atoll	2.2	Late Pennsylvanian Early Permian	1948	6100–9900	Basinal Shale	Basinal Shale (Permian, Penn., Miss.)	2920	Base	Vest, 1970; Burnside, 1959; Stafford, 1959
USA	Permian Basin	Wolfcamp	0.075	Early Permian	1940	10,000	Basinal Shale (Wolfcamp)	Basinal Shale and Limestones (Late Penn.–Early Perm.)	100	Top	Exxon Files; Cys and Mazzullo, 1985
USA	Permian Basin	Strawn	1.4	Late Pennsylvanian	1938	8000–10,000	Basinal Shale	Basinal Shale (Penn.)	100–450	Base	Exxon Files
USA	Paradox Basin	Anneth	0.35	Late Pennsylvanian	1956	5900	Black Shale and Evaporites	Black Shale (Penn.)	160	?	Exxon Files; Peterson and Hite, 1969
USA	Hardeman and Midland Basins	Chappel Mounds	0.2	Early Mississippian	1959	7800–9000	Shale	Shale (Miss.)	80–250	Base	Exxon Files; Ahr and Ross, 1982

Table 1. Continued.

Country	Basin/Area	Field/Play	OEB (Billions, EUR)	Age	First Discovery	Depth of Burial (ft)	Seal	Source (Age)	Thickness (ft)	Wedge Position	Reference
USA	Appalachian Basin	Fort Payne	0.0008	Early Mississippian	1924	1400–1800	Basinal Marl	Off-Reef Shale (Miss.) and Basinal Off-Reef (Dev.–Miss.)	80	Base	Macquown and Perkins, 1982
Canada	Western Canada Basin	Swan Hills Platform	3.2**	Late Devonian	1957	10,000	Slope and Basinal Argillaceous Limestone Slope and Off-Reef	Off-Reef Shaly Limestone (Devonian)	150–250	Base	Hemphill et al., 1970
Canada	Western Canada Basin	Leduc	2.7**	Late Devonian	1947	3200–8440	Basinal Argillaceous Limestone	Shaly Limestone (Devonian)	≤850	Base	ERCB, 1986; Exxon Files
Canada	Western Canada Basin	Nisku	1.0**	Late Devonian	1977	8000–10,000	Basinal Shale and Dolomite	Basinal Shale (Devonian)	35–350	Top	Chevron Staff, 1979; ERCB, 1986; Exxon Files
Canada	Western Canada Basin	Rainbow Zama (Keg River)	1.25	Middle Devonian	1965	4300–6110	Evaporites	Off-Reef Limestone (Devonian)	≤820	Evaporites	Barrs et al., 1970; ERCB, 1986; Exxon Files
Canada	Elk Point Basin	Winnepegosis	0.004	Middle Devonian	1985	7800–8443	Evaporites	Basinal Off-Reef Limestone	≤350	Evaporites	Rosenthal, 1987; Stoakes et al., 1987; ERCB, 1986; Exxon Files
USA	Michigan and Illinois Basins	Niagarian	0.6**	Late Silurian	1889	4000–7000	Evaporites	Off-Reef Limestone (Silurian)	150–800	Evaporites	Bay, 1983; Labo et al., 1981

* Information was gathered from the geologic literature as well as Exxon unpublished reports. References noted are by no means exhaustive but are meant to give an easily accessible overview of the play in terms of the overall geologic setting and description of the trap. Analysis does not include buildups from the former Soviet Union and China.
** Indicates Gas + Oil Conversion = 6000 cubic feet per barrel.

Figure 2. Map illustrating the locations of buildup plays or Exxon buildup prospects analyzed in this study.

Here, buildups are referred to as those carbonate accumulations that formed significant topography on the sea floor, regardless of organic composition, where the topography subsequently formed the trap. This would include isolated buildups as well as shelf-margin buildups where depositional relief exists in a landward direction that results in the updip trap. We focused on these buildups because we were charged with developing guidelines that would improve the success of our exploration efforts associated with seismically defined carbonate buildup prospects. For the purposes of this chapter, we are omitting those deposits that may be low-relief buildups encased in coeval sealing facies whose traps were formed by later structure.

TEMPORAL AND SPATIAL DISTRIBUTION OF CARBONATE BUILDUPS

Although the growth of carbonate buildups occurs throughout the Phanerozoic (James, 1983), those that produce oil and gas are confined to but a few isolated periods of geologic time. In fact, approximately 90% of the nearly 40 billion oil equivalent barrels of recoverable hydrocarbons have been found in reservoirs deposited during just 15% of post-Ordovician time (Figure 3, Table 1). Productive buildups are generally restricted to Middle–Late Silurian, late Middle and Late Devonian, Late Pennsylvanian, and Early Permian, middle Cretaceous, Paleocene, and Miocene time. Forty-three percent of these reserves occur in the Miocene buildups alone. Many of these periods tend to correlate with times of second-order eustatic highstand (Figure 4), with the exception of the Late

Pennsylvanian and late Miocene, which are discussed below. Basins showing productive buildups of these ages were subject to high rates of tectonic subsidence that overwhelmed the effects of falling eustatic sea level. Periods of productive buildup growth also coincide with periods of extensive source deposition. The Silurian, Late Devonian, Pennsylvanian, Lower Permian, middle Cretaceous, and Oligocene–Miocene intervals were all recognized as periods of outstanding global source deposition by Klemme and Ulmishek (1991). Only the Late Jurassic Period, dominated by nonbuildup carbonate production, is recognized as outstanding as a global source contributor but not a major stratigraphic interval of buildup production.

At the present time, modern platform carbonates are being deposited in warm seas between 30° north and south latitudes. By analogy to the modern, we can use plate reconstructions as a first-order check on the viability of and interpretation of a mounded feature as a carbonate buildup. However, several periods of buildup growth occurred significantly above 30° north latitude—most notably in the Permian, Triassic, Paleocene, and Miocene periods. Warmer paleoclimates or ocean circulation patterns are the most likely causes of this.

GEOHISTORY ANALYSIS AND BASIN FILL PATTERNS

Geohistory analysis is a technique used to analyze basin subsidence patterns (Van Hinte, 1978). Geohistory curves from wells in four basins characterized by productive carbonate buildups are shown in Figure 5. Three curves are shown in each plot of

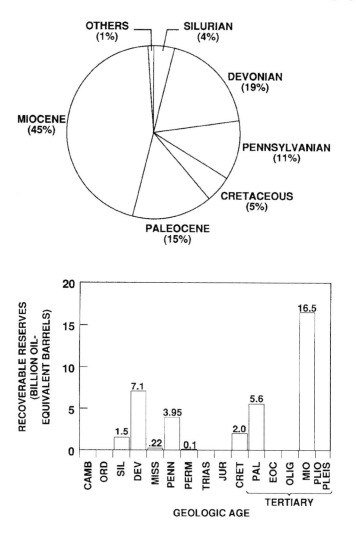

Figure 3. The temporal distribution of known hydrocarbon reserves in recoverable oil equivalent barrels compared to reservoir age. The histogram illustrates total reserves versus geologic age and the pie diagram shows the percentage of the total reserves in buildups with age.

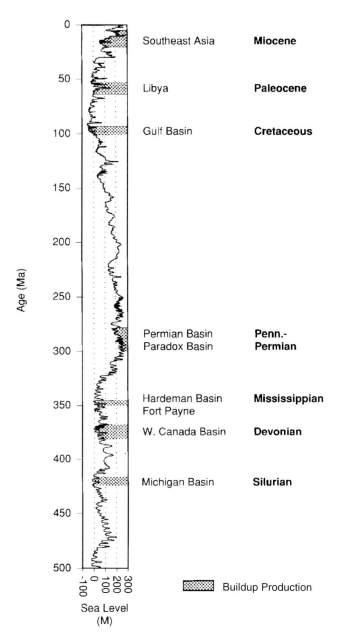

Figure 4. Global eustatic cycle chart from Haq et al. (1987) and S. Schutter (personal communication, 1990) with the times of growth of productive buildups.

depth versus time. "Total" refers to the total subsidence of a point at the base of the sedimentary section corrected for sediment compaction effects. This curve represents the total accommodation due to tectonic processes, including sediment loading, and eustatic sea-level changes. The "tectonic" curve (identical to the R1 curve of Bond et al., 1989) has been corrected for the effects of sediment loading using an Airy-type isostatic model. This curve shows the combined effects of "driving," or nonload-induced subsidence and changes in eustasy. The upper curve, "sed surf," illustrates paleo–water depth of the sediment surface with downward excursions corresponding to increasing paleo–water depth estimates. This curve may be used to evaluate the effect of paleo–water depth estimates on the subsidence curves. No long- or short-

term sea-level estimates have been included in these calculations, thus total accommodation changes reflect the contribution of both subsidence and eustasy. On each geohistory plot, we have shaded the time interval characterized by buildup growth at that location.

The first plot in Figure 5 is from the Virginia Hills field of the Swan Hills Platform area (Late Devonian Period) of the Western Canada Basin. Slow subsidence occurred until the start of Beaverhill Lake (374 Ma) deposition, at which time the rate of subsidence increased dramatically. High subsidence rates contin-

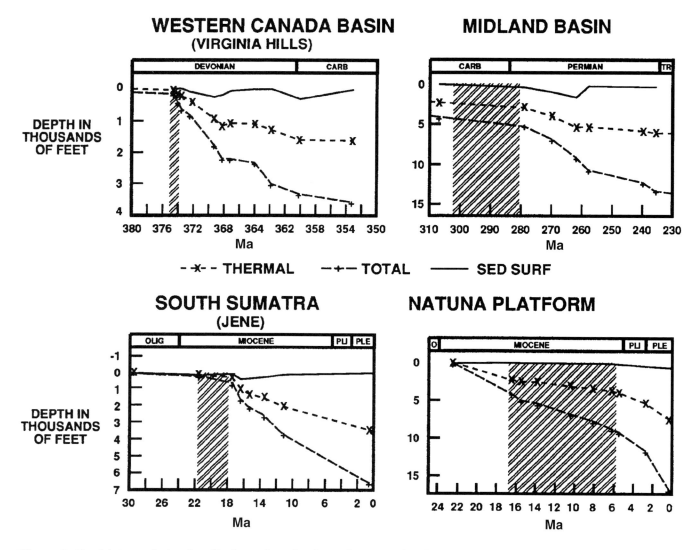

Figure 5. Geohistory plots of wells from four basins where productive buildups occur. In these plots, no eustatic sea-level changes have been incorporated in the calculations, thus, the "tectonic" and "total" curves represent the combined effects of subsidence and eustasy on the net long-term accommodation increase. The "tectonic" curve represents the "total" subsidence corrected for the effects of isostasy and compaction. Paleo–water depth values are illustrated in the "sed surf" curve. The periods of buildup growth are shaded. The time scale on the plot from the Virginia Hills field is from AGAT (Davies, 1985).

ued until the late Frasnian (Nisku, 369 Ma) when rates declined. The effect of this long-term accommodation pattern is reflected in the stratigraphic development of the basin as documented by Imperial Oil (in Davies, 1975) and Wendte and Stoakes (1982) (Figure 6). Beaverhill Lake and Leduc buildup margins display retrogradational geometries, each successive sequence is progressively more areally restricted than the previous one. Platform margins, deposited as the buildups are growing, step progressively landward. Decreasing rates of long-term accommodation after the deposition of Leduc buildups result in a strongly progradational pattern of sedimentation that covers the older buildups with deep-water argillaceous limestones. Here the rapid increase in accom-

modation results in the stratigraphic juxtaposition of shallow-water reservoir carbonates with overlying deeper marine fine-grained sealing facies. The relative contributions of sea-level fluctuation and tectonism to this long-term accommodation increase are difficult to assess; however, a similar stratal configuration documented by Playford (1980) in the Late Devonian Period of the Canning basin in Australia suggests that eustasy may be at least partially responsible. Canning basin carbonates exhibit a lower retrogradational series of sequences deposited during the Frasnian Epoch, which is overlain by prograding Famennian-age reefs. These relationships suggest that this basin fill pattern may be due in part to a eustatic rise in the Frasnian reaching a maximum near the

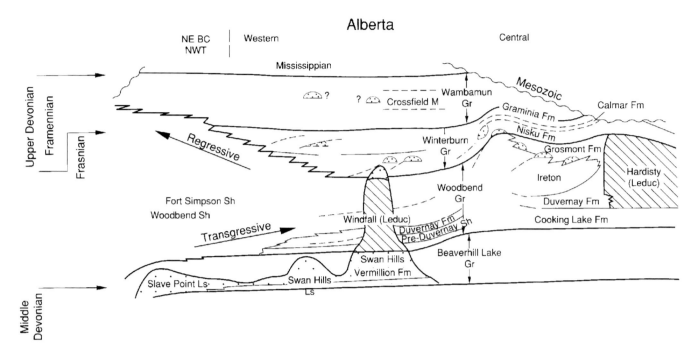

Figure 6. Schematic stratigraphic cross section of the Western Canada Basin from Davies (1975, after Imperial Oil, Ltd.).

Frasnian–Famennian boundary (Johnson et al., 1985; Ross and Ross, 1988).

An additional observation of the geohistory profiles is the high rate of long-term accommodation (close to 100 m per m.y.) associated with the deposition of productive Beaverhill Lake and Leduc buildups in the Western Canada Basin. The effects of this rapid rate of long-term accommodation increase on the effects of third-order depositional sequences will be discussed below.

A very similar accommodation history is found in the late Paleozoic of the Midland basin (Figure 5). This geohistory plot, constructed from a well on the eastern part of the Horseshoe atoll, illustrates rapid and increasing rates of long-term accommodation during and immediately after deposition of the Late Pennsylvanian- to Early Permian-age buildup. The initiation of rapid subsidence in this basin follows a middle Wolfcampian structural event, which is expressed by major erosion of basin margins throughout the basin. The rate of accommodation addition decreases in the latter part of the Permian. The stratal configuration that resulted from this accommodation history is shown dramatically on the seismic line across the eastern edge of the Horseshoe atoll shown in Figure 7. During the buildup deposition phase, buildups become progressively more areally isolated (Vest, 1970), while platform margins backstep (Figure 8). This retrogradational style of deposition is visible on seismic data (Figures 7, 9). The result of this basin fill pattern is a series of stacked Pennsylvanian and lower Wolfcampian buildups deposited in a shallow-water setting sealed by Wolfcampian deep-marine shales.

Two Miocene examples from Southeast Asia are also shown in Figure 5. The first is from the early Miocene (Baturaja Limestone) Jene field of the South Sumatra basin (Figure 5). The geohistory plot shows slow rates of long-term accommodation throughout deposition of the buildup (<30 m per m.y.) followed by markedly increased rates of accommodation. As in the previous examples, the result of this increase in accommodation is a stratigraphic juxtaposition of shallow-water reservoir carbonates overlain by deep-marine sealing shales (Figure 10). The stratal expression of this is a downlap surface over the reservoir, which is imaged by seismic data (Figure 11). An early Miocene sea-level rise may account for marine incursion and carbonate deposition during this time (Fulthorpe and Schlanger, 1989), but cannot explain the large amounts of postreef accommodation necessary to account for the paleo–water depth of the overlying latest early Miocene shales. Basin formation associated with nearby subduction-related tectonism during the middle Miocene is probably responsible for the rapid accommodation increase.

The second Miocene example in Figure 5 is from the L-structure in the Natuna Platform, offshore Indonesia (Rudolph and Lehmann, 1989) (Figure 12). Buildup deposition occurs on a rapidly subsiding horst block and results in a middle and late Miocene gas reservoir up to 6000 ft in thickness. A progressive increase in the rate of accommodation occurs in Pliocene and Pleistocene time in association with

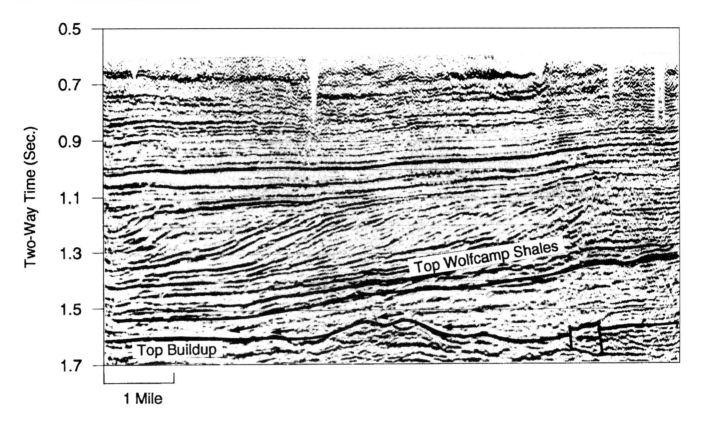

0.5
0.7
0.9
1.1
1.3
1.5
1.7

Two-Way Time (Sec.)

Top Wolfcamp Shales

Top Buildup

1 Mile

Figure 7. Seismic section across the eastern Horseshoe atoll (Scurry and Borden counties) in the Midland basin. Reflections illustrate the dominantly shelly Wolfcampian section and mostly carbonate Leonardian Clear Fork strata prograding across the basin. The Wolfcampian clinoforms downlap over the mounded reservoir interval of Late Pennsylvanian to Early Permian age.

lithospheric flexure in front of the Borneo Trench. This results in rapid subsidence of the buildup below the zone of optimal autochthonous carbonate accumulation. The buildup is then covered and sealed by the distal "toes" of prograding siliciclastic clinoforms. Although the scale and tectonic setting of this buildup are different from the examples noted above, there are several important similarities. The buildup back-steps with each successive sequence through time (Rudolph and Lehmann, 1989) (Figure 13), and subsequently the buildup is encased by deep-marine shales. Although the buildup was deposited during the middle and late Miocene long-term eustatic sea-level fall (Haq et al., 1987), subsidence interpreted to be due to lithospheric flexure (Rudolph and Lehmann, 1989) was apparently great enough to overwhelm the effects of eustasy on the Natuna Platform.

RELATIONSHIP OF BASIN FILL TO TRAP DEVELOPMENT

In each of the above cases, buildups deposited in shallow water are overlain and sealed by prograding clinoform "toes" of younger regressive wedges. Buildups are deposited prior to or during the earliest portions of periods of marked increases in long-term

accommodation rates. The regressive wedges are deposited during subsequent periods of slower long-term accommodation increase. These relationships are illustrated schematically in Figure 14. A schematic geohistory plot showing an increase, then decrease in the rate of accommodation is shown in Figure 14A. Each increment of accommodation is coded to the stratal surface of Figure 14B, which is a cross section illustrating common stratal patterns and buildup plays common to basins with productive carbonate buildups. A similar relationship of carbonate platform development to long-term subsidence patterns was recognized by Davies et al. (1989).

This pattern of relative sea level change, typical of buildup-prone stratigraphic sections, results in the stratigraphic juxtaposition of retrogradational, shallow-water carbonate buildups sealed in deeper marine shales (Figure 14B). During these transgressive–regressive periods of basin fill occurring over time scales of tens of millions of years, buildups are deposited during the basal transgressive, middle aggradational, and uppermost regressive portions of the cycle if conditions for carbonate deposition occur throughout the accommodation cycle. The basal transgressive portions of the cycle are optimal for trap development for two primary reasons. First, they are overlain and sealed by deep-

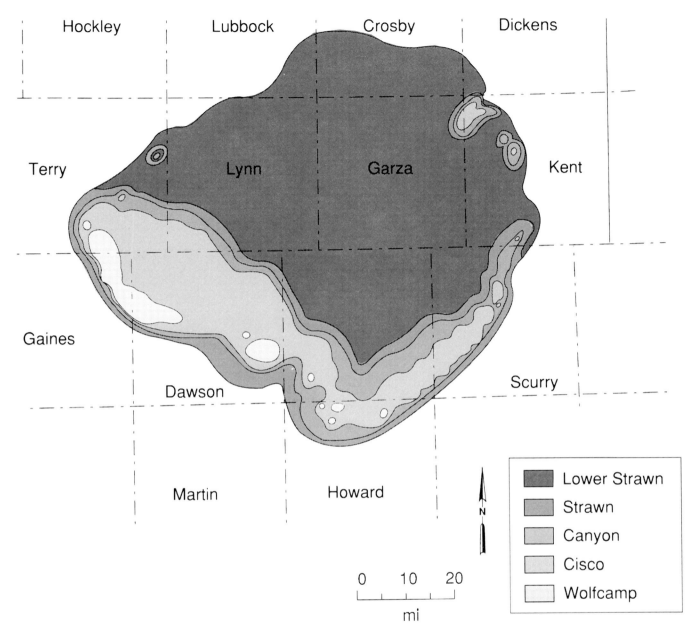

Figure 8. Map from Vest (1970) that shows the progressive areal restriction through time of sequences within the Horseshoe atoll.

marine fine-grained sediments. Second, the escalating rates of accommodation enable the growth of thick, isolated accumulations of carbonate reservoir-prone facies. Buildups deposited during the latter phase of this retrogradational part of the basin-fill cycle are productive except in the most updip areas where they typically have thin, easily breached topseals where they drain directly into the overlying regressive package. We have termed these wedge-base deposits after White (1980), who noted a similar relationship between basin-fill sequences and accumulations in buildups. Wedge-base buildups contain over three-fourths of the reserves noted in Table 1.

The relationship of the buildup reservoirs to the overlying deep-marine sealing rocks and the origin of the surface between these two lithofacies is critical to an understanding of wedge development. Surfaces characterized by platform deposits immediately overlain by deep-marine facies are discussed in several papers by Schlager (Kendall and Schlager, 1981; Schlager, 1981, 1989) who terms these "drowning unconformities." Because of the great disparity of paleo–water depths recognized in many of these wedge-base buildups between the reservoir carbonates (deposited close to sea level) and the encasing deep-marine sediments (deposited in hundreds or

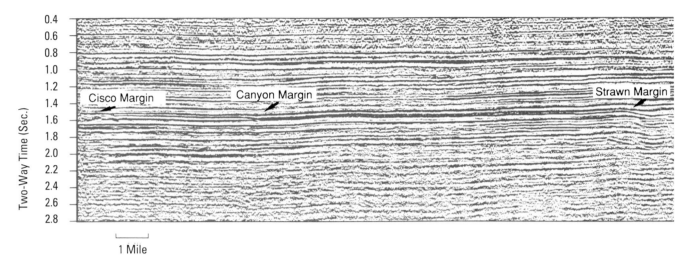

Figure 9. Seismic section from the northwestern margins of the Horseshoe atoll (Terry, Dawson, and Gaines counties) in the Midland basin. This section shows the retrogradational stacking of individual sequences within the atoll.

even thousands of feet of water), we do not interpret the deep-marine deposits as being the cause of the demise of the buildups. If overlying clastics were related to the ultimate cessation of buildup growth, we would expect to see a gradual deepening in the overlying facies, and the overlying stratal contact would not be a downlap surface. Evidence of subaerial exposure of the buildups is commonly present at or slightly below the stratigraphic contact between the

shallow-water and deep-water facies based on core and outcrop control. As buildup growth was terminated, these buildups apparently were stranded in basinal water depths in starved sedimentary conditions, similar to submerged Holocene buildups now ringing many deep shelves, or buildups now found in basinal water depths.

Buildups deposited during the later, regressive phase of sedimentation are often plagued by seal

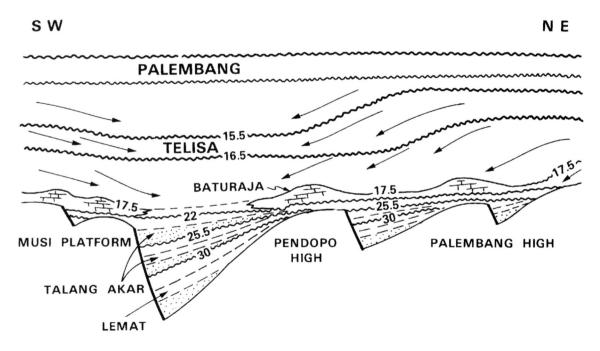

Figure 10. Schematic stratigraphic section across the Musi Platform, South Sumatra Basin, Indonesia. Numbers refer to interpreted sequence boundary ages as correlated to Haq et al.'s (1987) global cycle chart. Fine lines with arrows in Telisa section represent stratal surfaces interpreted from seismic and well data. Stratigraphic section was prepared by geologists from PTSI Stanvac, the Indonesian affiliate of Exxon and Mobil.

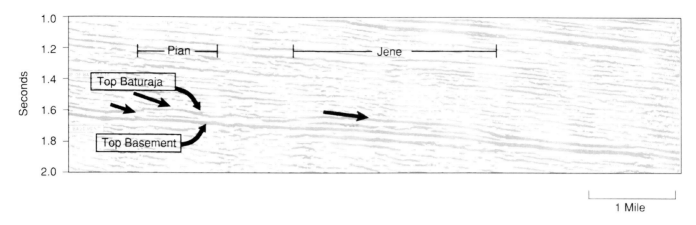

Figure 11. Seismic section across the Jene field, Miocene of the South Sumatra basin. This buildup, typical of Miocene buildups in southeast Asia, was deposited during the early flooding of the platform and then buried in deep marine clastics. This stratal juxtaposition is evident as a downlap surface separating the carbonate section from overlying prograding deltaic deposits.

problems. Buildup plays in these wedge-top stratigraphic intervals consist of shelf margin and slope buildups, as well as low-relief platform interior buildups. In addition, because of the lower rates of accommodation increase, buildups are lower in relief and seldom are deposited for longer than one third-order eustatic cycle. Because of these constraints, only about 15% of buildup reserves are cataloged in a wedge-top position in Table 1.

Basin-filling evaporites make up another common sealing facies for buildups, usually in slope and basinal positions. Rather than being associated with a major accommodation increase, these are associated

with reduced marine-water influx due to eustatic or tectonic isolation of the basin. The best known of these include the Middle Devonian Keg River and Winnepegosis reefs of Western Canada and North Dakota, and the Silurian reefs of the Michigan basin. These slope and basinal pinnacles contain approximately 5% of the recoverable buildup reserves.

CARBONATE BUILDUPS AND STRATIGRAPHIC TRAPS

The relationship of buildup-related traps to other stratigraphic trap types in carbonates is illustrated in

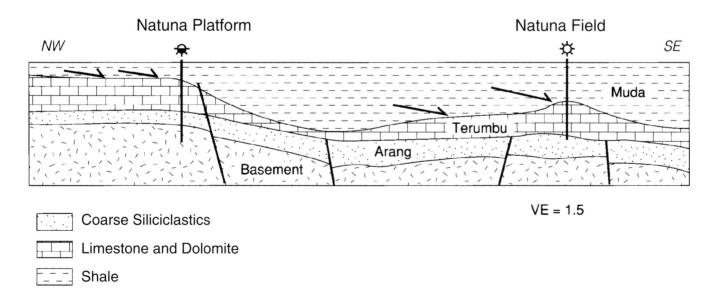

Figure 12. Schematic stratigraphic section across the Natuna Platform (after Rudolph and Lehmann, 1989) showing progressively more areally restricted buildups overlain by the toes of prograding siliciclastics. This thick buildup lies offshore of a wide carbonate platform which continued to grow through several additional third-order cycles of sea-level change until carbonate deposition was terminated in the Pliocene (VE = vertical exaggeration).

Natuna Platform Margin

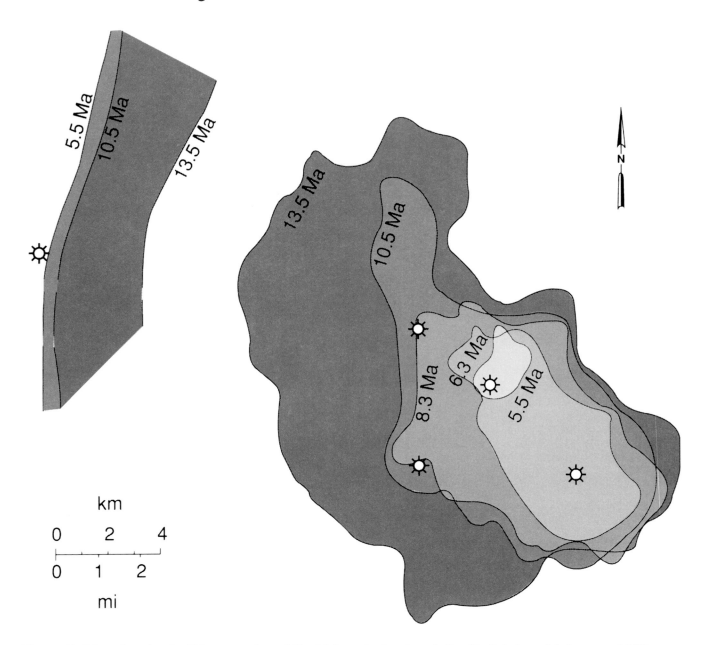

Figure 13. Map showing buildup margins of the Natuna L-structure (after Rudolph and Lehmann, 1989). Shades are keyed to the ages of the overlying sequence boundary. The margins stack retrogradationally through the middle and late Miocene.

Figure 15. Buildup-related stratigraphic traps tend to dominate the lower portions of these sedimentary wedges formed during cycles of relative sea-level change and other types of traps dominate the upper portions of these wedges. Buildups that form traps in the upper portion of these sedimentary wedges include buildup trends focused along the upper slope and shelf margins such as the Nisku trend in Western Canada and platform interior buildups or shoals encased in impermeable lagoonal micrites such as the

Defa field in Libya. Sealing facies are commonly associated with condensed shales deposited during short-term marine flooding events or cemented shallow-marine facies. In addition, mounded accumulations of allochthonous carbonate debris interpreted as lowstand deposits form small reservoirs in the Midland basin (Leary and Feeley, 1991). Unlike the situation in wedge-base deposits, buildups contain a relatively small percentage of the hydrocarbons found in stratigraphic traps within the upper regres-

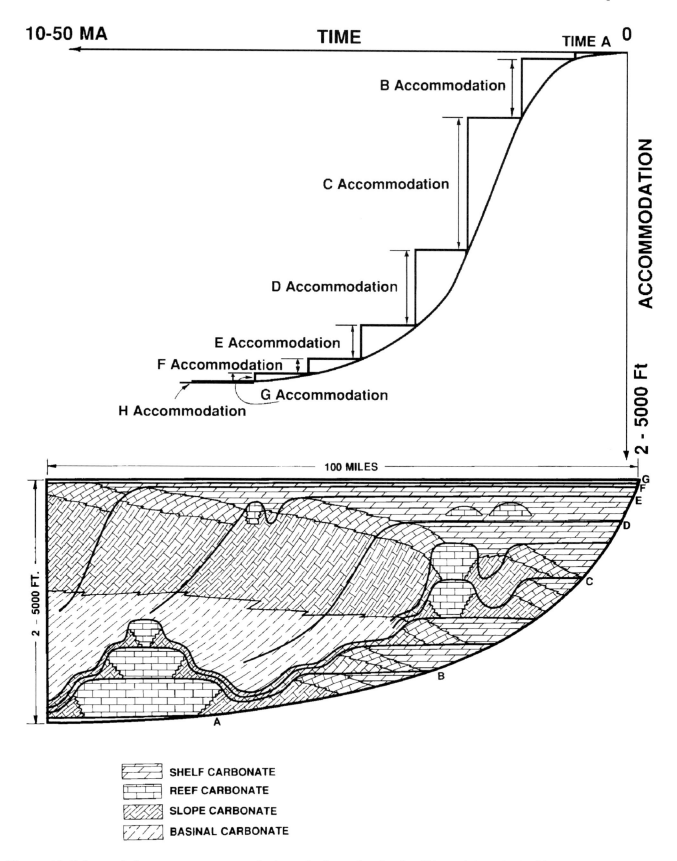

10-50 MA

TIME

TIME A   0

B Accommodation

C Accommodation

D Accommodation

E Accommodation

F Accommodation

G Accommodation

H Accommodation

ACCOMMODATION

2 - 5000 Ft

100 MILES

2 - 5000 FT.

G
F
E

D

C

B

A

SHELF CARBONATE

REEF CARBONATE

SLOPE CARBONATE

BASINAL CARBONATE

Figure 14. Schematic long-term accommodation plot keyed to basin-fill model typical of basins with productive buildups. The curve in the top figure shows a rapidly increasing rate of long-term accommodation increase followed by a decreasing rate of long-term accommodation increase. This is similar to the pattern observed in the geohistory plots in Figure 5.

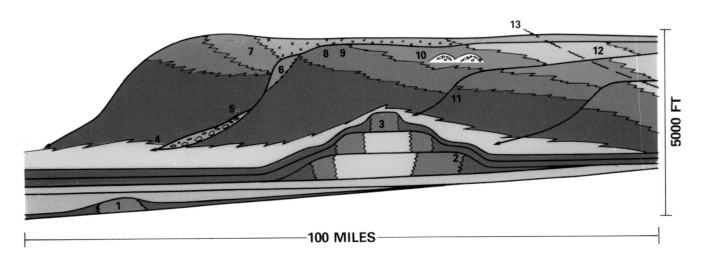

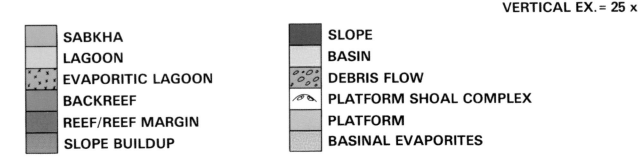

VERTICAL EX. = 25 x

	SABKHA			SLOPE
	LAGOON			BASIN
	EVAPORITIC LAGOON			DEBRIS FLOW
	BACKREEF			PLATFORM SHOAL COMPLEX
	REEF/REEF MARGIN			PLATFORM
	SLOPE BUILDUP			BASINAL EVAPORITES

Figure 15. Common carbonate stratigraphic traps showing the relationship of buildup-related traps to others. Note that buildup traps dominate the base of the sedimentary wedge while updip porosity/permeability barriers are more common in the upper regressive wedges. Buildup traps include: (1) basinal/slope buildups with evaporite seals; (2) retrogradational platform margins with later tilting; (3) buildups with independent closure sealed by downlapping basinal sediments; (6) slope buildups; (8) shelf or platform margin buildups; and (10) platform interior buildups within impermeable backreef or lagoonal facies. Common nonbuildup-related traps include: (4 and 5) allochthonous wedge pinch-out; (7) updip facies change to evaporites; (9) updip facies change to impermeable mud-rick lagoonal rocks; (11 and 12) updip change from porous dolomite to tight limestone; and (13) subcrop beneath an erosional unconformity.

sive deposits of the sedimentary wedges. The bulk of these accumulations are related to updip permeability pinch-outs along the margins of carbonate basins related to facies and diagenetic changes (Figure 15). Volumetrically, these stratigraphic traps contain far more hydrocarbons than buildup-related traps in many basins. These traps tend to rim the basin margins and approximate the updip facies changes, whereas the wedge-base buildup traps tend to lie farther toward the basin center.

SUPERIMPOSITION OF HIGHER ORDER EUSTATIC CYCLICITY

To a large degree, the long-term (tens of millions of years) transgressive to regressive packages produced during these basin-filling cycles resemble the stratal configuration of transgressive and highstand systems

tracts within depositional sequences produced by sinusoidally varying, third- (1 to 5 m.y.) and fourth-order (0.1 to 0.4 m.y.) eustasy (Vail and Todd, 1981; Jervey, 1988; Posamentier and Vail, 1988). This is due to the fact that they both are deposited during similar accommodation histories, however on different time scales. Increasing rates of accommodation during the transgressive systems tract is analogous to that which occurs during the basal transgressive portion of these basin-filling sedimentary wedges. Progressively decreasing rates of accommodation in the highstand systems tract results in a generally regressive pattern of deposition, similar to that recognized in the upper portion of basin-filling wedges. Higher frequency third- and fourth-order cyclicity is superimposed on these long-term basin-filling cycles. The lithologic and diagenetic effects of these cycles depends on the position within the basin-filling wedge, the rate of basin subsidence (or background long-term eustatic change), and rates of short-term eustatic change.

During the early, transgressive phase of deposition, accelerated rates of long-term accommodation increase will tend to enhance transgressive and highstand deposition within buildups. When coupled with rapid rates of basin subsidence and "background" (or nonglacio-eustatic) short-term sea-level fluctuations, type 2 (Vail and Todd, 1981) sequences are deposited. The best example of this is the Late Devonian section of Western Canada. Subaerial exposure at sequence boundaries, the deposition of basin-restricted, autochthonous, lowstand wedge deposits, and ubiquitous progradation of buildup margins are rare. When rapid rates of sea-level change due to glaciation occurs, such as in Pennsylvanian time, subaerial exposure and lowstand deposition is common, even in rapidly subsiding buildups. Such is the case in the Horseshoe atoll example, where prominent exposure surfaces exist within and at the top of the buildup and detrital lowstand carbonates are productive along the buildup margins (Myers et al., 1956; Burnside, 1959). Where buildups are deposited during the initial flooding of more slowly subsiding platforms, type 1 falls are often encountered and early meteoric diagenesis is common, as in the case at Jene field. Explorationists can evaluate these factors from stratal patterns, geohistory analysis, and the global eustatic cycle chart (Haq et al., 1987) early in a frontier exploration program.

In the upper, regressive portion of the basin fill cycle, higher order eustatic fluctuations are superimposed on a lower rate of long-term accommodation increase. This results in more progradation and a higher propensity to subaerial exposure at sequence boundaries. Sequences in this phase of basin fill are similar to those described by Sarg (1988, his figure 1) with a less well-developed transgressive systems tract. Buildups deposited in the wedge-top are more likely to have undergone subaerial exposure.

ORIGIN OF BASIN FILL CYCLES

Our analysis of the stratigraphic framework of productive carbonate buildups has indicated the dominant stratigraphic position of different buildup plays. The eustatic or tectonic origin of these transgressive and regressive cycles is difficult to assess. As noted above, the periods of productive buildup growth cluster near the major second-order highstands recognized on the global eustatic cycle chart (Figure 4); however, the thickness of section involved in the basin fill is, in many cases, an order of magnitude greater than the amount of actual eustatic rise depicted on the Exxon cycle chart and other measurements of ancient sea-level fluctuations. In addition, many of the productive far eastern (Luconia Province from Epting, 1980; Natuna and Sumatra examples; and others), as well as intracratonic (Permian basin, Western Canada Basin, Sirte basin) examples are clearly affect-

ed by contemporaneous basin-forming tectonism. Finally, stratal relationships from the carbonate platforms of northeast Australia show that during the same time period a well-developed wedge base and top can develop with wedge-base productive buildups (in the Pandora Trough) immediately adjacent to shelves characterized by relatively continuous aggradational carbonate deposition (Davies et al., 1989). As in the examples noted above, the Gulf of Papua and the Pandora Trough are adjacent to a buried compressional belt (see Davies et al., 1989, their figure 9A) that is related to increased rates of basin accommodation immediately after buildup deposition. For these reasons, we conclude that the optimal stratigraphic setting for buildup-related traps is primarily a function of basin tectonism, which is then modified by short-term eustatic changes. Nonetheless, the close correspondence between the large-scale stratal configuration of the Late Devonian section of the Western Canada and coeval Canning basin in Australia, as well as the tie between productive buildups and second-order global eustasy, suggest that eustatic changes may be enough to control these stratal geometries in some basins. It is also possible that some of our bias in the interpretation of long-term Paleozoic sea-level fluctuations may be a function of our experience in basins with buildup production that show this tectonic overprint.

CONCLUSIONS

Although carbonate buildups are ubiquitous in the Phanerozoic geologic record, productive carbonate reservoirs are restricted to narrow windows of geologic time. These periods are times of extensive source deposition and, in general, correspond to times of eustatic highstand.

Buildups are deposited throughout long-term cycles of accommodation change; however, the largest hydrocarbon reserves are associated with the basal, transgressive portions of these cycles. These cycles are characterized by an initial phase of retrogradation, which results in the landward translation of platform margins with time and the progressive areal restriction and upward growth of downdip, isolated buildups. Buildups are eventually unable to keep up with relative sea-level rise and are stranded in a basinal position. Eventual burial by downlapping clinoforms of the overlying wedge forms the topseal. Many buildups deposited during the overlying, more accommodation-limited regressive wedge have problems with topseal and smaller trap sizes.

Carbonate buildups are the dominant carbonate stratigraphic trap type in the basal portion of basin-filling sedimentary wedges but are generally subordinate in total reserves to updip porosity pinch-outs in the upper parts of these wedges where carbonate deposition exists throughout.

Higher order eustatic fluctuations affect the parasequence and sequence stacking patterns and early diagenetic histories of buildups differently in (1) basins having different rates of long-term relative sea-level change; (2) different positions of buildups within the basin-filling sedimentary wedge; and (3) periods characterized by high rates of short-term glacio-eustatic sea-level change. These factors are then taken into account with climatic and mineralogic aspects to predict reservoir quality and position.

ACKNOWLEDGMENTS

This work was performed as part of a study conceived and facilitated by John Bubb of Exxon Corporation and John Hickson of Esso Resources Canada Ltd. Exxon Production Research Company and Exxon Corporation–affiliate geologists assisted us throughout the project in sharing their data and experience in buildup plays worldwide. Discussion of buildup plays with K. Rudolph, T. Harding, and J. F. Sarg were especially helpful. Review by E. Oswald, A. Brown, R. Sucheki, and R. Loucks significantly improved the manuscript. We thank Exxon for permission to publish this chapter.

REFERENCES CITED

Ahr, W. M., and Ross, S. L., 1982, Chappel (Mississippian) bioherm al in the Hardeman Basin, Texas: Transactions of the Gulf Coast Association of Geological Sciences, v. 32, p. 185-193.

Barrs, D. L., Copland, A. B., and Ritchie, W. D., 1970, Geology of Middle Devonian reefs, Rainbow Area, Alberta, Canada, in Halbouty, M. T., (ed.), Geology of Giant Petroleum Fields: AAPG Memoir 14, p. 19-49.

Bay, T. A., 1983, The Silurian of the Northern Michigan Basin, in Harris, P. M., (ed.), Carbonate Buildups: SEPM Core Workshop No. 4, p. 53-72.

Bebout, D. G., Schatzing, R. A., and Loucks, R. G., 1977, Porosity distribution in the Stuart City Trend, Lower Cretaceous, South Texas, in Bebout, R. G., and Loucks, R. C., (eds.), Cretaceous Carbonates of Texas and Mexico: Applications to Subsurface Exploration, p. 234-256.

Bond, G. C., Kominz, M. A., Steckler, M. S., and Grotzinger, J. P., 1989, Role of thermal subsidence, flexure, and eustasy in the evolution of early Paleozoic passive-margin carbonate platforms, in Crevello, P. D., Wilson, J. L., Sarg, J. F., and Read, J. F., (eds.), Controls on Carbonate Platform and Basin Development: SEPM Special Publication No. 44, p. 39-61.

Burnside, R. J., 1959, Geology of part of the Horseshoe atoll in Borden and Howard Counties, Texas: U. S. Geological Survey Professional Paper 315-B, 34 p.

Carmalt, S. W., and St. John, B., 1986, Giant oil and gas fields, in Halbouty, M. T., (ed.), Future Petroleum Provinces of the World: AAPG Memoir 40, p. 11-53.

Chevron Standard Limited Exploration Staff, 1979, The geology, geophysics, and significance of the Nisku Reef discoveries, West Pembina Area, Alberta, Canada: Bulletin of Canadian Petroleum Geology, v. 27, p. 326-359.

Coogan, A. H., Bebout, D. G., and Maggio, C., 1972, Depositional environments and geological history of Golden Lane and Poza Rica trends, Mexico, an alternative view: AAPG Bulletin, v. 56, p. 1419-1447.

Cys, J. M., and Mazzullo, S. J., 1985, Depositional and diagenetic history of a Lower Permian (Wolfcamp) phylloid algal reservoir, Hueco Formation, Morton Field, Southeastern New Mexico, in Roehl, P. O., and Choquette, P. W., (eds.), Carbonate Petroleum Reservoirs: Springer-Verlag, p. 277-288.

Davies, G. R., 1975, Introduction to Davies, G. R., (ed.), Devonian Reef Complexes of Canada 1, Rainbow, Swan Hills: Canadian Society of Petroleum Geologists Reprint Series 1, Calgary, Alberta, p. iii-ix.

Davies, G. R., 1985, AGAT Laboratories, Alberta, Chronostratigraphic Chart.

Davies, P. J., Symonds, P. A., Feary, D. A., and Pigram, C. J., 1989, The evolution of the carbonate platforms of northeast Australia, in Crevello, P. D., Wilson, J. L., Sarg, J. F., and Read, J. F., (eds.), Controls on Carbonate Platform and Basin Development: SEPM Special Publication No. 44, p. 233-258.

DeBrisay, C. L., and Daniel, E. L., 1971, Supplemental recovery planning, development and predicted performance—Intisar A and D Reef Fields, Libyan Arab Republic: Society of Petroleum Engineers Paper 3438.

Energy Resources of Canada Bureau, 1986, ERCB Reserve Report Series ERCB-18, Alberta's reserves of crude oil, oil shale, natural gas liquids, and sulphur, December 31, 1986.

Epting, M., 1980, Sedimentology of Miocene carbonate buildups, central Luconia, offshore Sarawak: Geological Society of Malaysia Bulletin, v. 12, p. 17-30.

Fulthorpe, C. S., and Schlanger, S. O., 1989, Paleoceanographic and tectonic settings of early Miocene reefs and associated carbonates of offshore southeast Asia: AAPG Bulletin, v. 73, p. 729-754.

Haq, B. U., Hardenbol, J., and Vail, P. R., 1987, Chronology of fluctuating sea levels since the Triassic: Science, v. 235, p. 1136-1167.

Hemphill, C. R., Smith, R. I., and Szabo, F., 1970, Geology of Beaverhill Lake reefs, Swan Hills Area, Alberta, in Halbouty, M. T., (ed.), Geology of Giant Petroleum Fields: AAPG Memoir 14, p. 50-90.

James, N.P., 1983, Reef environment, in Scholle, P.A., Bebout, D. G., and Moore, C. H., (eds.), Carbonate Depositional Environments: AAPG Memoir 33, p. 346–444.

Jervey, M. T., 1988, Quantitative geological modeling of siliciclastic rock sequences and their seismic expression, in Wilgus, C. K., Hastings, B. S., Kendall, C. G., Posamentier, H. W., Ross, C. A., and Van Wagoner, J. C., (eds.), Sea Level Changes: An Integrated Approach: SEPM Special Publication No. 42, p. 47-69.

Johnson, J. G., Klapper, G., and Sandberg, C. A., 1985, Devonian eustatic fluctuations in Euramerica: Geological Society of American Bulletin, v. 96, p. 567-587.

Joiner, D. S., and Myers, C. E., 1972, Developing a 10 billion barrel reservoir with modern techniques: Society of Petroleum Engineers, paper no. 3739, 14 p.

Jordan, C. F., Jr., and Abdullah, M., 1988, Lithofacies analysis of the Arun reservoir, North Sumatra, Indonesia, in Lomando, A. J. and Harris, P. M., (eds.), Giant Oil and Gas Fields: SEPM Core Workshop No. 12, Houston, Texas, vol. 1, p. 89-118.

Kendall, C. G. and Schlager, W., 1981, Carbonates and relative changes in sea level: Marine Geology, v. 44, p. 181-212.

Khattab, H. A., and Hadidi, T. A., 1961, Abnormal stratigraphic features in Ras Gharib Oilfield: in Proceedings, Third Arab Petroleum Congress, October 16-21, 1961.

Klemme, H. D., and Ulmishek, G. F., 1991, Effective petroleum source rocks of the world: Stratigraphic distribution and controlling depositional factors: AAPG Bulletin, v. 75, p. 1809-1851.

Labo, J., Cousins, J., Werner, W. G., and Pan, P. H., 1981, Exploring for Silurian-Niagaran pinnacle reefs in the southern Michigan Basin: Oil and Gas Journal, v. 77, p. 93-98.

Leary, D. A., and Feeley, M. H., 1991, Seismic expression and sedimentologic characteristics of a Wofcampian (Permian) carbonate lowstand fan, eastern Midland basin, in Weimer, P., and Link, M. A., (eds.), Seismic facies and sedimentary processes of submarine fans and turbidite systems: Springer-Verlag, New York.

Longman, M. W., 1985, Fracture porosity in reef talus of a Miocene pinnacle-reef reservoir, Nido B Field, the Philippines, in Roehl, P. O., and Choquette, P. W., (eds.), Carbonate Petroleum Reservoirs: Springer-Verlag, New York, p. 547-560.

Macquown, W. C., and Perkins, J. H., 1982, Stratigraphy and petrology of petroleum-producing Waulsortian-type carbonate mounds in Fort Payne Formation (Lower Mississippian) of north central Tennessee: AAPG Bulletin, v. 66, p. 1055-1075.

May, J. A., and Eyles, D. R., 1985, Well log and seismic character of Tertiary Terumbu carbonate, South China Sea, Indonesia: AAPG Bulletin, v. 69, p. 1139-1358.

Myers, D. A., Stafford, P. T., and Burnside, R. J., 1956, Geology of the Late Paleozoic Horseshoe atoll in west Texas: Texas Bureau of Economic Geology Publication No. 5607.

Peterson, J. A., and Hite, R. J., 1969, Pennsylvanian evaporite-carbonate cycles and their relation to petroleum occurrence, southern Rocky Mountains: AAPG Bulletin, v. 53, p. 884-908.

Playford, T. E., 1980, Devonian "Great Barrier Reef" of the Canning Basin, Western Australia: AAPG Bulletin, v. 64, p. 814-840.

Posamentier, H. W., and Vail, P. R., 1988, Eustatic controls on clastic deposition II—Sequence and systems tract models, in Wilgus, C. K., Hastings, B. S., Kendall, C. G., Posamentier, H. W., Ross, C. A., and Van Wagoner, J. C., (eds.), Sea Level Changes: An Integrated Approach: SEPM Special Publication No. 42, p. 125-154.

Rosenthal, L. R., 1987, the Winnipegosis Formation of the northeast margin of the Williston Basin, 1987, in Carlson, C. G., and Christopher, J. E., (eds.), Fifth International Williston Basin Symposium, Grand Forks, North Dakota, June 14-17, 1987, p. 37-46.

Ross, C. A., and Ross, J. R., 1988, Late Paleozoic transgressive-regressive deposition, in Wilgus, C. K., Hastings, B. S., Kendall, C. G., Posamentier, H. W., Ross, C. A., and Van Wagoner, J. C., (eds.), Sea Level Changes: An Integrated Approach: SEPM Special Publication No. 42, p. 227-247.

Rudolph, K. W., and Lehmann, P. J., 1989, Sequence stratigraphy of the Natuna D-Alpha Block, offshore Indonesia, in Crevello, P. D., Wilson, J. L., Sarg, J. F., and Read, J. F., (eds.), Controls on Carbonate Platform Development: SEPM Special Publication No. 44, p. 353-361.

Sarg, J. F., 1988, Carbonate sequence stratigraphy, in Wilgus, C. K., Hastings, B. S., Kendall, C. G., Posamentier, H. W., Ross, C. A., and Van Wagoner, J. C., (eds.), Sea Level Changes: An Integrated Approach: SEPM Special Publication No. 42, p. 155-181.

Schlager, W., 1981, The paradox of drowned reefs and carbonate platforms: Geological Society of America Bulletin, v. 92, p. 197-211.

Schlager, W., 1989, Drowning unconformities on carbonate platforms, in Crevello, P. D., Wilson, J. L., Sarg, J. F., and Read, J. F., (eds.), Controls on Carbonate Platform Development: SEPM Special Publication No. 44, p. 15-25.

Stafford, P. T., 1959, Geology of part of the Horseshoe atoll in Scurry and Kent Counties, Texas: U. S. Geological Survey Professional Paper 315-A, 20 p.

Stoakes, F., Campbell, C., Hassler, G., Dixon, R., and Forbes, D., 1987, Sedimentology and hydrocarbon source potential of the Middle Devonian Winnipegosis Formation of southwestern Saskatchewan: Stoakes-Campbell Geoconsulting Ltd., Calgary, Alberta, 100 p.

Terry, C. E., and Williams, J. J., 1969, The Idris "A" bioherm and oilfield, Sirte Basin, Libya—Its commercial development, regional Paleocene geologic setting and stratigraphy, in Exploration for petroleum in Europe and North Africa: Joint meeting of Institute of Petroleum (London) and American Association of Petroleum Geologists, Brighton, England (June 29-July 2, 1969), p. 31-48.

Vail, P. R., and Todd, R. G., 1981, North Sea Jurassic unconformities, chronostratigraphy, and sea level changes from seismic stratigraphy, *in* Proceedings, Petroleum Geology of the Continental Shelf of Northwest Europe, p. 216-235.

Van Hinte, J. E., 1978, Geohistory analysis: Application of micropaleontology in exploration geology: AAPG Bulletin, v. 62, p. 201-222.

Vest, E. L., 1970, Oil fields of Pennsylvanian-Permian Horseshoe atoll, West Texas, *in* Halbouty, M. T. (ed.), Geology of giant petroleum fields: AAPG Memoir 14, p. 185-203.

Viniegra-O., F., and Castillo-Tejero, C., 1970, Golden Lane Fields, Veracruz, Mexico: AAPG Memoir 14, p. 309-325.

Wendte, J. C., and Stoakes, F. A., 1982, Evolution and corresponding porosity of the Judy Creek Complex, Upper Devonian, Central Alberta, *in*

Cutler, W. G. (ed.), Canada's Giant Hydrocarbon Reservoirs: Canadian Society of Petroleum Geologists 1982 Core Conference, p. 63-81.

White, D. A., 1980, Assessing oil and gas plays in sedimentary wedges: AAPG Bulletin, v. 64, p. 1158-1178.

Woodling, G. S., Kaldi, J. G., Roe, R. C., and Oentarsih, K. I., 1991, Multidisciplinary reservoir study of the Bima Field, Offshore N. W. Java, Indonesia, *in* The Integration of Geophysics, Petrophysics, and Petroleum Engineering in Reservoir Delineation, Description, and Management, Proceedings of the First Archie Conference, October 22-25, 1990, Houston, Texas: American Association of Petroleum Geologists, p. 1-36.

Chapter 3

The Drowning Succession in Jurassic Carbonates of the Venetian Alps, Italy: A Record of Supercontinent Breakup, Gradual Eustatic Rise, and Eutrophication of Shallow-Water Environments

William G. Zempolich[1]

Department of Earth and Planetary Sciences
The Johns Hopkins University
Baltimore, Maryland, U.S.A.

ABSTRACT

The Ammonitico Rosso of the western Venetian Alps is a 10- to 25-m-thick, red nodular limestone that overlies thick Late Triassic to Middle Jurassic shallow-water carbonates that form the South Trento Platform. Deposition of the Ammonitico Rosso is thought to represent a Middle–Late Jurassic drowning event whereby the South Trento Platform became a deeply submerged plateau. The Ammonitico Rosso is problematic in that it: (1) overlies a platform-wide unconformity that contains complex brecciated fabrics filled by pelagic-rich *"Posidonia alpina"* sediment and cement; (2) is rich in ammonites and other pelagic fauna; and (3) contains stromatolites, oncolites, and wave-rippled coquinas.

Based on new data from the eastern margin of the South Trento Platform, the drowning succession is interpreted to have a shallow origin. These data include: (1) the discovery of two sponge-coral-stromatoporoid patch reefs within oolitic and peloidal grainstone below the unconformity; (2) cavities and fill associated with the unconformity; and (3) diagenetic fabrics and transition from platform interior to platform margin facies in both the Lower and Intermediate members of the Ammonitico Rosso. Faunal and lithologic similarity of sponge-coral-stromatoporoid reefs with other Lower Jurassic reef complexes suggest that these reefs are Late Pliensbachian in age. Cavities and neptunian dikes within back-reef, reef, and fore-reef sediments are filled by *P. alpina* sediment, rounded lithoclasts, fibrous cement, and crystal silt. Pendant cement and crystal silt found within reef cavities and neptunian dikes overlap deposition of internal sediment.

In a west-to-east transect above the unconformity, stromatolitic and oncolitic mudstone/wackestone in the Lower Ammonitico Rosso grades first into

[1]Present address: Mobil Oil Company, P.O. Box 650232, Dallas, Texas, U.S.A.

thrombolites and stromatolites, then into nodular burrowed wackestone and packstone/grainstone. Packstone and grainstone contain well-preserved ammonites, pelagic bivalves, peloids, belemnites, gastropods, solitary corals, and fibrous cement. In the Intermediate Member, thin-bedded chert-rich limestone grades into event strata (i.e., tempestites) composed of limestone gravel and well-sorted sand, and pelagic-dominated mudstone/wackestone. Gravels are poorly sorted, sometimes imbricated and contain lithoclasts derived from underlying sediments. These lithologies overlie truncation surfaces that include deep irregular excavations, rounded gutters, and gently scoured surfaces and grade upward into sands that possess hummocky, low-angle, and planar cross-stratification. Sands are composed of coarse- to fine-size lithoclasts, peloids, and skeletal grains. Peloidal mudstone and wackestone contain protoglobigerinids, radiolarians, ammonites, pelagic bivalves, belemnites, crinoids, and solitary corals. Solitary corals are found in growth position on ammonite and belemnite substrates.

The drowning succession of the South Trento Platform correlates with long-term eustatic rises and falls of sea level and includes: (1) Upper Pliensbachian deposition and tectonism; (2) a transgressive systems tract and high-stand systems tract (Toarcian); (3) a small-scale type 1 sequence boundary (late Toarcian–Lower Bajocian?); (4) a drowning sequence (Aalenian–Upper Bajocian); and (5) a composite condensed section (Upper Bajocian–Tithonian).

The appearance of pelagic organisms on the South Trento Platform and a biotic succession from (1) sponge-coral-stromatoporoid reefs to (2) bioeroded sponge and hermatypic coral reefs to (3) grainstone composed of ahermatypic suspension/detrital feeders and planktic organisms to (4) "stunted" pelagic and benthic faunas to (5) microbial mat (stromatolite) structures indicates progressive paleoecologic deterioration of shallow-water environments. Analogy of these transitions with those observed on modern "drowned" platforms suggests that the demise of carbonate producing benthos was caused by increasing amounts of nutrients and organic matter (i.e., *trophic resources*) and establishment of oxygen-deficient environments. Faunal transition is coincident with the breakup of Pangea during the Lower–Middle Jurassic, the deposition of organic-rich shale and manganese-rich limestone in periplatform and basinal settings, and eustatic sea-level rise. This suggests that influx of trophic resources was associated with changes in regional circulation patterns and upwelling. Drowning is interpreted to have occurred gradually over time through a combination of eustatic sea-level rise and environmental change.

INTRODUCTION

The presence of stratigraphic gaps and transition to condensed pelagic sequences above karst-like surfaces in platform carbonates presents numerous problems for sedimentologists (Schlager, 1981, 1989; Erlich et al., 1990). Processes or events must have suppressed carbonate production to the point where relative sea-level rise outpaced the accumulation of shallow carbonate sediment. In attempts to explain these drowning events several questions immediately arise:

1. What role, if any, did subaerial exposure play in arresting carbonate production?
2. Are fauna present that would indicate a sudden or gradual change in the environment of deposition?
3. Was change induced by local tectonics or a eustatic rise in sea level?

The answers to these questions are critical when interpreting the paleogeography, sea-level history, and sequence stratigraphy of carbonate platforms. In addition, this information has economic relevance because early formed diagenetic porosity and source rock deposition are intimately linked to eustasy and environmental change.

To investigate drowning unconformities in carbonate stratigraphic sequences, this study examines Jurassic carbonates of the South Trento Platform, Southern Alps, northeast Italy (Figure 1). In this example, thick shallow-water carbonates (Calcari Grigi Formation/San Vigilio Group) are abruptly overlain by condensed pelagic lithologies (Ammonitico Rosso Veronese), which are thought to have been deposited in deep water (Figure 2). The Ammonitico Rosso is composed of "Lower," "Intermediate," and "Upper" members and contains an abundance of pelagic organisms (e.g., ammonites, belemnites, protoglobigerinids, radiolarians). A major stratigraphic problem arises in that, in addition to the pelagic fauna, stromatolite and oncolite assemblages of the Ammonitico Rosso are found in association with tempestites (Massari, 1981, 1983). Interpretation of the transition from shallow platform to pelagic lithologies is further complicated by the presence of an unconformity at the top of the platform, deposition of pelagic-rich crinoidal/brachiopod grainstone (i.e., the "Posidonia Alpina Beds"; Sturani, 1964, 1971), and significant age gaps between the platform and the overlying Ammonitico Rosso.

The transition from platform carbonates to the Ammonitico Rosso has importance not only to the interpretation of the sequence stratigraphy of the South Trento Platform (S.T.P.), but to regional paleogeographic study as well. Transitions from shallow platform to condensed pelagic lithologies occur in many Tethyan Mesozoic sequences (e.g., Garrison and Fischer, 1969; Jenkyns, 1971a, 1980; Bernoulli and Jenkyns, 1974; Bosellini and Winterer, 1975; Farinacci et al., 1981; Fazzuoli et al., 1981) and are key elements in paleogeographic reconstructions and paleoceanographic models. On a more local scale, drowning of the Trento Platform has been cited as the mechanism by which the nearby Friuli Platform became a windward margin during the Middle Jurassic and thus a "super-producing oolite factory" (Bosellini et al., 1981).

Evidence and discussion presented in this chapter suggest that the S.T.P. was still at shallow depth during the transition from neritic to pelagic lithofacies. The presence of shallow-water sedimentary structures, facies transitions, and petrographic fabrics within the Posidonia Alpina Beds, the Lower Ammonitico Rosso, and the Intermediate Member are extremely difficult to explain in deep pelagic depositional models (e.g., Bosellini and Winterer, 1975; Bosellini et al., 1981; Ogg, 1981; Winterer and Bosellini, 1981). These models rely on a Middle Jurassic subsidence "event" which rapidly drowns the platform. In lieu of rapid subsidence, a "shallow-water" model is proposed which contends that platform drowning was brought about by external events (e.g., Cretaceous Adriatic/Dinaric platform; Jenkyns, 1991). The model herein proposes that drowning occurred through a combination of a long-term sea-level rise and faunal response to regional changes in circulation and introduction of trophic resources (e.g., Upper Jurassic platforms of the proto-Atlantic seaway; Poag, 1991). This alternative model is put forth as an initial attempt to: (1) synthesize newly gathered and previously reported field evidence (Zempolich, 1991); (2) place these data into a sequence stratigraphic framework that is consistent with postulated global sea-level change; and (3) relate these findings to study of modern drowned platforms. It is not offered as a unique solution to the many problems presented by this and other condensed sequences.

GEOLOGIC SETTING

During Late Triassic time, a widespread carbonate shelf existed across most of Europe and Africa (Dolomia Principale/Hauptdolomit). Separation of northern Africa and Europe during Early Jurassic time created an extensive transform zone (Weissert and Bernoulli, 1985) and established a horst-and-graben tectonic setting along the southern Tethyan margin (Figure 3). The Trento Platform was the most landward horst block of the Southern Alps and was bounded to the west by the Lombardy Basin and to the east by the Belluno Basin, which separated the Trento Platform from the stable foreland (Friuli Platform).

Paleontologic evidence and lithofacies patterns of Lower Jurassic sediments suggest that the Trento Platform can be divided into at least two major blocks (North and South) across the Valsugana paleolineament (Figure 1A). Study by Masetti and Bottoni (1978) indicates that the North Trento Platform (N.T.P.) began subsiding during Pliensbachian time. Della Bruna and Martire (1985), in the Alpi Feltrine area, identified a hardground of Lower to Middle Toarcian-age, and east to west onlap of overlying Middle Jurassic sediments. These data, when compared to paleontologic and lithologic data from both the South and North Trento Platforms (Sturani, 1971; Masetti and Bottoni, 1978; Clari et al., 1984; Martire, 1988, 1989), suggest that this smaller block behaved independently from the two major north and south platforms (Della Bruna and Martire, 1985).

In addition to the structural definition of local platforms and basins, the Early Jurassic breakup of

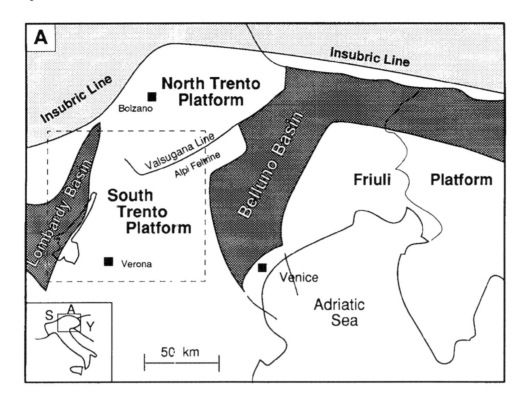

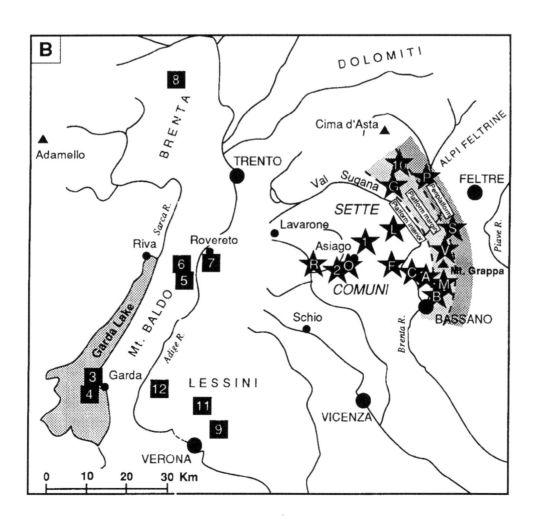

Pangea led to several other important regional changes including long-term sea-level rise, establishment of ephemeral epicontinental seaways, and ocean anoxia. These changes are recorded in the biostratigraphic and lithostratigraphic record (Figure 4) as: (1) an intermigration of molluskan faunas (Hallam, 1977, 1983; Smith, 1983; Smith and Tipper, 1986); (2) the deposition of organic-rich shale and manganese-rich limestone along epicontinental seaways and margins (early Toarcian OAE: Jenkyns et al., 1985; Jenkyns and Clayton, 1986; Jenkyns, 1988; Jenkyns et al., 1991); and (3) regional faunal extinctions (early Toarcian) caused by the spread of anoxic bottom waters and deterioration of shallow-water environments (Hallam, 1987). These studies indicate that environmental change was associated with changing circulation patterns and was widespread during early Toarcian time.

STRATIGRAPHY AND PREVIOUS WORK

Comprehensive reviews of the Jurassic stratigraphy of the Southern Alps is provided by Gaetani (1975) and Winterer and Bosellini (1981). A summary of major lithofacies found on the S.T.P. is given here with emphasis on data which bear on previous interpretations of the unconformity and drowning succession.

Subunconformity Platform Lithofacies

During the Lower Jurassic, a thick (250–600 m) sequence of peritidal carbonates (Calcari Grigi Formation) accumulated on the S.T.P. (Figure 3B). These sediments include cyclic peritidal deposits, oolitic and skeletal grainstone, and lagoonal deposits that contain oysterlike *Lithiotis* banks (Bosellini and Broglio Loriga, 1971). Further lithostratigraphic definition of lithologies located above the Calcari Grigi Formation has been made by Beccarelli Bauck (1988) and Trevisani (1991). A disconformity is present at the top of the Calcari Grigi where *Lithiotis* valves are commonly leached and infilled by Pliensbachian and Toarcian (?) sediment and cement (Sturani, 1971; Clari and Marelli, 1983; Trevisani, 1991; Winterer et

al., 1991). During the Toarcian to Aalenian, the San Vigilio Group (0–250 m), composed of oolite, oncolite, sponge-coral patch reefs, and hemipelagic mudstone, was deposited along the southern and western margins of the S.T.P. (Barbujani et al., 1986; Clari and Marelli, 1983; Sturani, 1964, 1971). "Tepee" structures composed of reworked Pliensbachian sediment within the base of the San Vigilio Group suggest transgression of an exposed surface (Clari and Marelli, 1983). In the platform interior, the San Vigilio Group is missing. Along the eastern margin of the S.T.P., pervasive dolomitization and incomplete stratigraphic sections make correlation of Calcari Grigi and San Vigilio (?) equivalents a difficult task (Trevisani, 1991). Along the platform margins, San Vigilio lithofacies (Aalenian) are transitional with, or are unconformably overlain by, the Posidonia Alpina Beds (Sturani, 1971).

The Unconformity Surface and the Posidonia Alpina Beds

In the platform interior, the karstlike regional unconformity/drowning surface separates the platform succession (San Vigilio Group and the Calcari Grigi Formation) from pelagic lithofacies (Figure 3B). The stratigraphic gap between platform lithologies and pelagic lithofacies is represented, in part, by Middle Jurassic deposition of thin patches of cross-bedded grainstone and fill within breccia, neptunian dikes, and "tepee-like" structures (Figure 5A–D) and thick accumulations (up to 80 m) of cross-stratified crinoidal/brachiopod grainstone along the platform margins (Figure 5E–H). These sediments are rich in ammonites and the pseudoplanktonic bivalve *Bositra buchi* (*Bositra buchi* = *Posidonia alpina*). Sturani (1964, 1971) identified a diverse epifaunal community within the Posidonia Alpina Beds that included crinoids, brachiopods, ammonites, gastropods, bivalves, echinoids, and solitary corals. Ooids are also found. Fossil assemblages found within individual cavities in the platform interior are commonly distinguished by unique ammonite subzones (Figure 5D). Associated gastropods belong to genera (e.g., *Littorina*) that are believed to feed on algae and are commonly found at

◄────────────────────────────────

Figure 1. (A) Early Jurassic paleogeography of the Venetian Alps (modified from Bosellini et al., 1981). The Trento Platform is divided into North and South blocks across the Valsugana paleolineament. Paleontologic and facies evidence suggests that the North block began subsiding during the Lower Jurassic (Masetti and Bottoni, 1978). Paleontologic and facies evidence from the Alpi Feltrine suggests that this smaller block, located along the Valsugana paleolineament, was onlapped east to west during the Middle Jurassic (Della Bruna and Martire, 1985). (B) Outcrop localities of the South Trento Platform (dashed inset in A). Letters and/or stars refer to localities measured and sampled during the course of this study. Numbered localities refer to those studied by Sturani (1964, 1971). Transitions from platform interior to platform margin facies occur in both the Lower and Intermediate members of the Ammonitico Rosso toward the northeast. A = Mt. Asolone (3 sections); B = Baita Camol; C = Col Moschin; F = Foza; G = Grigno; L = Mt. Lisser; M = Mt. Grappa (4 sections); O = Volta Scura Quarry; P = Ponte Serra; R = Rotzo; S = Fontana Secca; V = Valpore di Cima; 1 = Mt. Longara; 2 = Cima Tre Pezzi; 10 = Mt. Agaro.

A

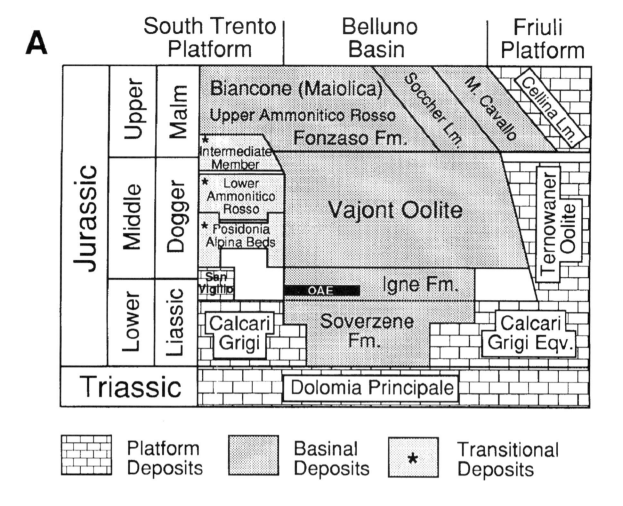

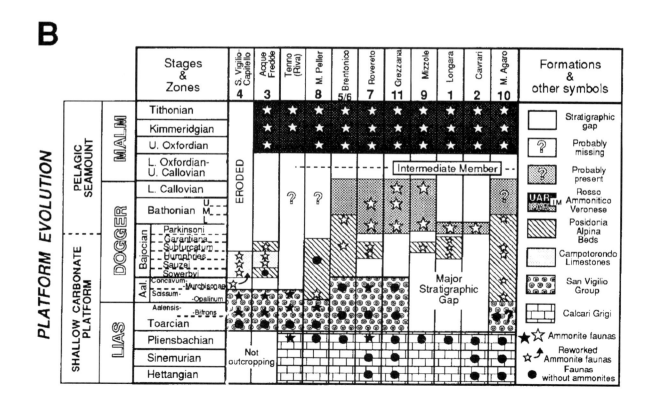

shallow depth. One of the peculiar aspects of these fossil assemblages is that both benthic (e.g., gastropods, brachiopods) and pelagic (e.g., ammonites) organisms are represented by juvenile and small "stunted" adult specimens (Sturani, 1971; Ferrari and Manara, 1972). Full-size adult specimens are rare. Along the southwestern margin of the platform and in the Alpi Feltrine, the Posidonia Alpina Beds grade into, and are partly synchronous with, cream- to red-colored biomicrites (a.k.a. Campotorundo Limestones) that contain normal-size adult ammonite phragmacones and rare juvenile specimens.

Castellarin (1966, 1972), Martire (1990, 1993), Massari (1981), Ogg (1981), Sturani (1971), Trevisani (1991), and Winterer et al. (1991) discuss the origin of cavities, the timing of cementation, the nature of sediment fill, and the accumulation of grainstone for localities found in the platform interior and along the platform margins. Based on petrographic study of underlying platform lithologies and biostratigraphic study of the Posidonia Alpina Beds, Sturani (1971) proposed that: (1) the unconformity was karstic in origin; and (2) epifauna derived from peripheral "rocky shoals" and "algal meadows" and pelagic fauna derived from the open seas were washed into a slightly elevated platform interior during intense storms. These interpretations are supported by the occurrence of breccia and mud cracks that contain Pliensbachian-age microfossils (Calcari Grigi Formation) and epifaunal grainstone along platform margins, which display hummocky, large-scale bimodal planar, trough, and herringbone cross-stratification (Posidonia Alpina Beds). The latter data indicate that some deposition occurred above wave base and within the influence of tides (Massari, 1981; Trevisani, 1991). Other hypotheses that have been proposed for breccia, dike, and cavity formation include syntectonic opening and/or erosion in a marine environment (Castellarin, 1966, 1972; Ogg, 1981) and sliding and slumping of host lithologies in a marine environment (Winterer et al., 1991). Martire (1990, 1993) has documented the presence of crystal silt interlayered with radiaxial-fibrous calcite cement in breccia of the platform interior. These data indicate that some pore fluids were undersaturated with respect to calcite and may suggest a brief period of platform exposure during the Bajocian.

Pelagic Lithofacies

The Ammonitico Rosso Veronese contains a predominantly pelagic fauna and is composed of two formal members (Figure 2), the Lower Ammonitico Rosso (L.A.R.) and the Upper Ammonitico Rosso (U.A.R.). These members are separated by either an iron- and manganese-rich hardground or a chert-rich limestone, informally referred to as the "Intermediate Member." Ammonitico Rosso lithologies and biostratigraphy have been studied in detail by Sturani (1971), Massari (1981, 1983), Ogg (1981), Clari et al. (1984, 1987), Pavia et al. (1987), and Martire (1988, 1989). Together these units represent an enormous time range (Upper Bajocian to Tithonian) but only attain a thickness of up to 25 m across the platform. Sturani (1971) suggested that deposition of the Ammonitico Rosso recorded a drowning event whereby the Trento Platform became a submerged current-swept plateau at a depth below wave base but not deeper than the base of the photic zone (<150 m depth). Bosellini and Winterer (1975), Winterer and Bosellini (1981), and Ogg (1981) concluded that the unconformity and abrupt transition to the Ammonitico Rosso represents tectonic foundering and drowning of the Trento Platform and intermittent sedimentation on a deeply submerged, current-swept plateau that passed through the aragonite lysocline (ALy) and aragonite compensation depth (ACD).

The basal L.A.R. represents the first platform-wide sedimentation event (Upper Bajocian–Bathonian) following the regional unconformity. L.A.R. sediments are similar to and coeval (in part) with those belonging to the Posidonia Alpina Beds. Toward the western platform margin, a transitional contact between the Posidonia Alpina Beds and the L.A.R. is exhibited by alternations of hummocky cross-stratified *P. alpina* grainstone and stromatolite beds (Massari, 1981). In the platform interior, a planar truncation surface and/or hardground separates *P. alpina*-filled breccia and platform lithologies from overlying stromatolite/oncolite assemblages (i.e., microbial fabrics) of the L.A.R. (Figures 2, 3B, 6). The L.A.R. is generally massive and poorly fossiliferous (Clari et al., 1984; Martire, 1989; Massari, 1981, 1983). The L.A.R. contains peloids, ammonites, belemnites, aptychi, pelagic bivalves, crinoids, benthic foraminifers, protoglobigerinids, radiolaria, microcoprolites, and fish teeth.

◄

Figure 2. (A) Stratigraphic-age relations of the South Trento Platform and neighboring domains (modified from Bosellini et al., 1981). Drowning of the South Trento Platform is recognized as a transition from shallow platform to pelagic lithologies during the Middle to early Upper Jurassic. Early Toarcian OAE occurs within the Igne Formation. (B) Biostratigraphy of the South Trento Platform (modified from Sturani, 1964, 1971; for location of sections, see Figure 1). Biostratigraphic analysis of Jurassic sequences identify several significant stratigraphic gaps between platform and pelagic lithologies. In the platform interior, a major gap is represented by an unconformity which spans Toarcian to Bajocian time. Posidonia Alpina Beds of the platform interior occur as isolated patches of coquina and fill within cavities and breccia.

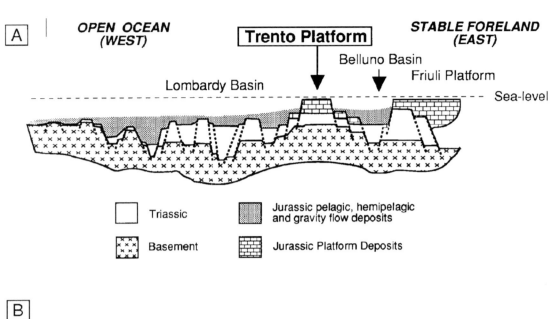

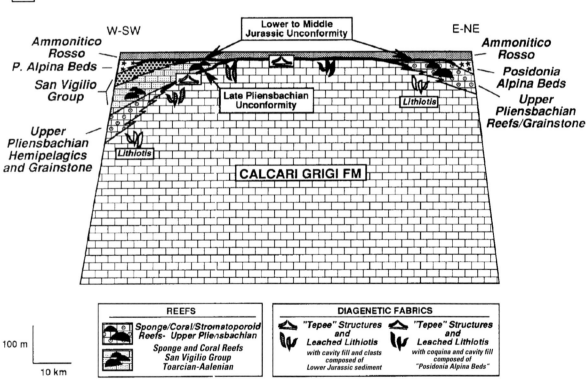

Figure 3. (A) Palinspastic cross section across the Southern Alps at the beginning of early Middle Jurassic time (modified from Bosellini et al., 1981). The Trento Platform is the most landward horst block adjacent to the stable foreland (Friuli Platform). (B) Simplified schematic of the stratigraphy of the South Trento Platform. A thick Lower Jurassic shallow-water platform sequence (Calcari Grigi Formation) is overlain by Lower/Middle Jurassic oolite, reefs, and hemipelagic ramp sediments. Platform lithologies are separated from overlying pelagic lithologies by a Lower to Middle Jurassic regional unconformity. The Posidonia Alpina Beds (Middle Jurassic) occur in the platform interior as patches of cross-bedded grainstone, pelagic-rich coquina, and as sediment fill within breccia and cavities. Along the platform margins, the Posidonia Alpina Beds occur as pelagic-rich, cross-bedded crinoidal-brachiopod grainstone (up to ~80 m). Pelagic-rich, red nodular, and chert-rich limestone (i.e., the Ammonitico Rosso) obtains a maximum thickness of approximately 25 m. New data gathered by this study from outcrops along the eastern platform margin include an Upper Pliensbachian patch reef complex and diagenetic fabrics below the unconformity, and transitions from platform interior to platform margin facies in the "Lower" and "Intermediate" members of the Ammonitico Rosso.

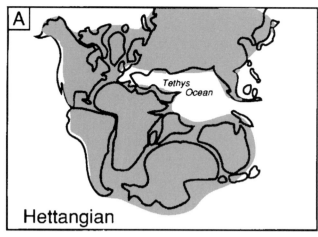

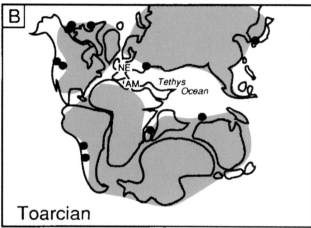

Figure 4. (A) Hettangian and (B) Toarcian paleo-geography based on molluskan biogeography (modified from Hallam, 1983, with distribution of early Toarcian organic-rich shales from Jenkyns, 1988). Continental areas are in stipple pattern. Shale occurrences in northwest Europe and the Alpine-Mediterranean regions are marked NE and AM, respectively. Other shale localities are noted by dots. With separation of Europe and north Africa, the western Tethys and proto Atlantic Ocean were connected via several ephemeral sea-ways. Intermigration of molluskan fauna suggests that these seaways were open by Toarcian time. Distribution of early Toarcian organic-rich shale and regional extinction events suggests that environmental deterioration was associated with the breakup of Pangea and sea-level rise.

Benthic gastropods, gastropod larvae, echinoderms, brachiopods, and rare solitary corals and Porifera are also found. Iron-manganese is present as grain replacements, diagenetic crusts, nodules, and pisolitic ironstone.

Stromatolitic structures of the L.A.R. (Figure 6) include low-relief linked hemispheres (LLH), colum-nar morphologies, stratiform sheets, oncolites, and coatings on a variety of reworked grains (Massari,

1981, 1983). Stromatolites and grains have been bored by (?) Thallophyta and Polychaeta and encrusted by brachiopods and rare serpulid worms. Burrowing, cementation, and compaction have also played major roles in modification and preservation of these fabrics (Clari et al., 1984; Martire, 1989; Massari, 1981, 1983; Ogg, 1981). Recent work on stromatolites collected from Callovian-age horizons of the Ammonitico Rosso yield amino acid distributions that indicate an "algal" and/or "cyanobacterial" origin (Massari et al., 1989). Massari (1981, 1983) described cyclic stromato-lite/oncolite assemblages that are tens of centimeters in thickness (Figure 6F). Erosional surfaces commonly truncate stromatolite assemblages. Discontinuity sur-faces are overlain by lithoclasts and coquina, which display subtle grading and/or plane-lamination and wave-ripple cross-lamination. Wave-ripples are stabi-lized and preserved by stromatolite growth. Massari (1981, 1983) interpreted these structures as tem-pestites. If correct, these data indicate that the L.A.R. was deposited, in part, above storm wave base. The U.A.R. exhibits lateral and vertical transitions from stromatolitic facies to red, thin-bedded, flaser-nodu-lar, and sparsely nodular limestone, with shaly part-ings (Massari, 1981, 1983; Martire, 1988, 1989). Ammonites, belemnites, and aptychi are common. During the Tithonian, the U.A.R. was replaced by a thick, white, geographically widespread cherty lime-stone that contains an abundance of nannofossils (Biancone Formation).

NEW DATA

Correlation and Description of Facies

Sedimentologic and petrographic data have been gathered from below and above the Lower to Middle Jurassic unconformity along the eastern margin of the S.T.P. (Figures 1B, 3B). These data provide new infor-mation which aids in the interpretation of the plat-form drowning succession. New data are: (1) the discovery of two sponge-coral-stromatoporoid patch reefs below the unconformity; (2) cavities and fill associated with the unconformity; and (3) diagenetic fabrics and facies transitions in both the L.A.R. and the Intermediate Member from the platform interior to the eastern platform margin. Outcrops were corre-lated with those found in the platform interior by lithostratigraphic correlation of the unconformity, the Posidonia Alpina Beds, the L.A.R., the chert-rich Intermediate Member, and the U.A.R. Key sections and lithostratigraphic correlation are presented in Figures 7 and 8. Lithostratigraphic correlation with the type section of the platform interior (Martire, 1988, 1989) provides biostratigraphic constraint.

Subunconformity Platform Lithofacies

Two patch reefs were discovered within oolite and peloidal grainstone in the Mt. Grappa area (Figures 1B, 7). These reefs are several meters in thickness and

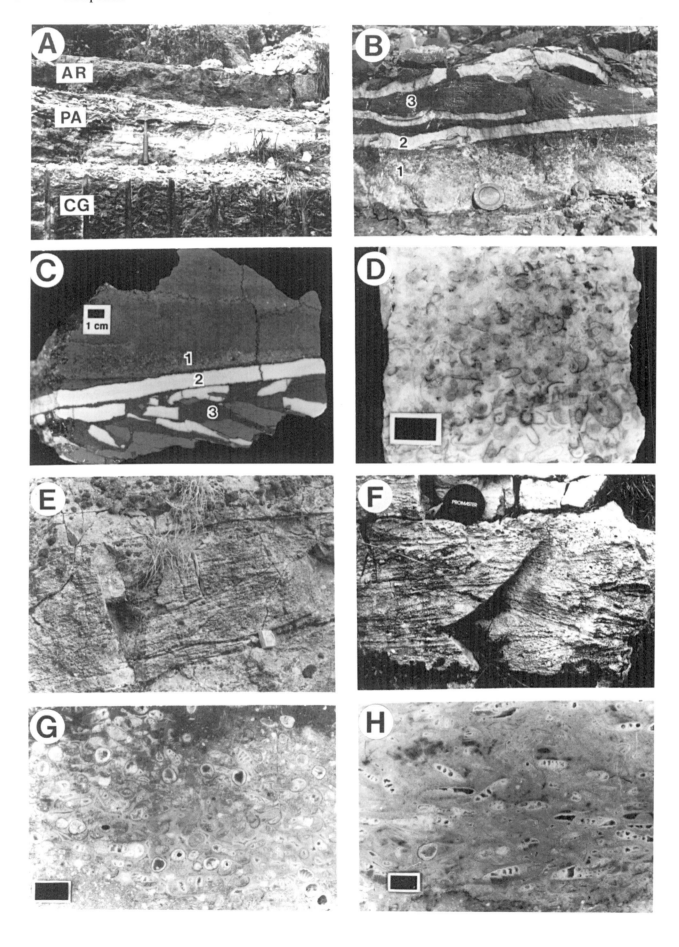

up to tens of meters in width. Patch reefs and grain-stone are located just below the L.A.R. They consist of well-cemented yellow/tan boundstone and flanking fore-reef talus beds (Figure 9A–D). Calcareous sponges [Figure 9A: *Stylothalamia* cf. *budaensis* (Wells, 1934)], solitary and colonial corals (*Oppelismilia?*, *Lepidophyllia?*, *Astrocoenia?*), large skirted stromato-poroids, calcareous algae (*Solenopora*), gastropods, and large bivalve mollusks comprise the core of each reef. Sponges and stromatoporoids can be found in growth position. Boundstone fabrics, including both shelter and intraskeletal pore space, are filled by yel-low/tan peloidal packstone/grainstone and lined by early fibrous cement. Fore-reef talus is composed of sponge and coral debris within a crinoidal–peloidal packstone matrix. These beds gently dip south-south-east in a direction toward open water (i.e., the Belluno Basin). Toward the platform interior, reef limestones pass into fossiliferous packstones, wacke-stones, and mudstones. Several large fist-sized cavi-ties are found within reef and back-reef lithologies (Figure 9E–H).

Each reef is capped by peloidal and oolitic grain-stone, which suggests that reef-grainstone sequences represent meter-scale shallowing-upward cycles. Trough and planar cross-stratified, skeletal-rich, oolitic grainstone, oncolitic packstone, and oncolitic micrite are found 1 km to the east of the reef–grain-stone complex. A series of large neptunian dikes crosscut these grain-rich lithologies, which are situat-ed near the east platform/basin boundary fault (Figure 1A).

Interpretation

Winnowed boundstone fabrics, a diverse shallow-water fauna, and associated cross-bedded grainstone collectively suggest that the reefs at the Mt. Grappa locality were deposited at shallow depth (≤10–20 m). While a precise age assignment for the patch reef complex is beyond the scope of this chapter, the fol-

lowing observations suggest that these lithologies are Upper Pliensbachian in age: (1) Fauna are similar to that described from sponge-coral-stromatoporoid patch reefs found in Pliensbachian sediments of Morocco, Peru, and Central Italy (Le Maître, 1935, 1937; Wells, 1953; du Dresnay, 1971, 1975; Beauvais, 1977, 1984; Pallini and Schiavinotto, 1980). (2) Calcareous sponges (e.g., *Stylothalamia*), solitary corals, and grainstone are similar to fauna and lithologies found above the Calcari Grigi Formation along the western platform margin (Beccarelli Bauck, 1986, 1988). (3) Oolite that encases these reefs is pre-sent elsewhere beneath the unconformity along the eastern platform margin and overlies the Calcari Grigi Formation (Trevisani, 1991). (4) Following the End-Triassic extinction, Lower Liassic reef develop-ment was suppressed over much of the world until the reappearance of small patch reefs during the Pliensbachian (Beauvais, 1984; Stanley, 1988). (5) The basinal Igne Formation (Upper Pliensbachian–Toarcian) overlies down-faulted blocks of the eastern Trento Platform margin (Masetti and Bianchin, 1987; and personal observation). Based on these lithostrati-graphic and biostratigraphic data, it is probable that the reef-grainstone interval found beneath the region-al unconformity at Mt. Grappa is Upper Pliensbachian in age. An exact age assignment awaits detailed paleontologic study.

**The Unconformity Surface
and the Posidonia Alpina Beds**

In the northeast platform interior, *Lithiotis* wacke-stone banks and oolitic-peloidal sediments (Calcari Grigi Formation) are truncated by the regional uncon-formity (Figure 1B: Mt. Lisser). *Lithiotis* valves are commonly leached and geopetally filled by red *P. alpina* sediment (Figure 10A). At the eastern limit of the Sette Comuni, the unconformity truncates peloidal packstone and wackestone (Figure 1B: Foza).

Figure 5. (A) The stratigraphic gap between Lower Jurassic platform lithologies and Middle to Upper Jurassic pelagic lithologies in the platform interior (Cima Tre Pezzi; hammer for scale). Lithiotis wackestone (Calcari Grigi Formation: CG) overlain by breccia composed of the "Posidonia Alpina Beds" (PA) and the Ammonitico Rosso (AR). (B) "Tepee-like" breccia is composed of ammonite grainstone (1), thick linings of radiaxial-fibrous calcite cement (2), and P. alpina sediment that contains radiaxial cement/micrite clasts (3). Coin for scale. (C) Cavity roof (1) is composed of cross-bedded crinoidal grainstone and downward-oriented radiaxial cement (2). Cement/micrite clasts (3) have been brecciated and reworked into the cavity along with P. alpina sediment (polished slab). (D) Graded coquina composed of juvenile ammonites (polished slab; scale bar = 1 cm). Ammonites belong to the Bajocian Subfurcatum zone, Banksi subzone (Sturani, 1971). Brachiopods, gas-tropods, and P. alpina valves are common. (E) Bimodal planar cross-stratified grainstone, Posidonia Alpina Beds (Bajocian; Mt. Agaro. Tape measure for scale ~6 cm). Crinoidal grainstone contains lenses of bra-chiopods, P. alpina valves, and ammonites. (F) Trough cross-stratified brachiopod grainstone, lower Posidonia Alpina Beds (Aalenian; Valpore di Cima. Lens cap for scale). (G) Brachiopod grainstone in (F) is predominantly composed of terebratulid and rhynchonellid brachiopods and cemented by fibrous cement (polished slab). P. alpina valves and ammonites are common. (H) Ludwigia murchisonae grainstone (polished slab), lower Posidonia Alpina Beds (Aalenian; Valpore di Cima). Ammonite phragmacones are found in grain-to-grain contact and aligned on foresets of cross-stratified crinoidal grainstone.

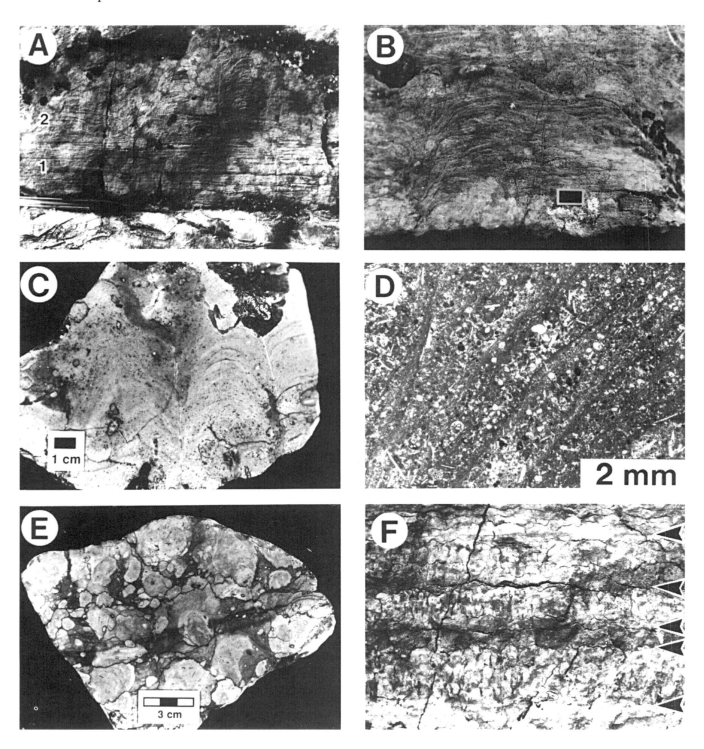

Figure 6. Stromatolite structures found in the Lower Ammonitico Rosso in the platform interior. (A) Truncated breccia (Posidonia Alpina Beds) overlain by stratiform (1) and hemispheric stromatolites (2) (Mt. Longara; pen for scale). (B) Flat-topped laterally linked hemispheric stromatolites (Mt. Longara). Medium to coarse skeletal material fills interdomal space. (C) Columnar stromatolites (polished slab; Mt. Longara). Individual columns are up to 15 cm in height. (D) Columnar stromatolite laminae composed of peloids, fine skeletal debris, protoglobigerinids, radiolaria, and carbonate mud (thin-section photograph). Laminae exceed the angle of repose indicating that sediment was trapped and bound. (E) Nodular stromatolites, oncolites, and coated grains (polished slab; Rotzo). (F) Decimetric cyclic stromatolite assemblages (Volta Scura Quarry). Truncation surfaces (denoted by arrows) commonly cross-cut stromatolite-oncolite assemblages. Renewed stromatolite growth stabilizes pavement composed of lithoclasts, oncolites, and skeletal grains. Coin for scale at bottom of photograph.

Farther eastward, the unconformity truncates oolitic and peloidal sediments and thin beds of skeletal-rich wackestone (Figures 1B, 7: Col Moschin and Mt. Asolone). Oolitic, peloidal, and skeletal-rich lithologies overlie peritidal cycles of the Calcari Grigi Formation at these localities. In the Mt. Asolone area and at other localities along the eastern margin, oolitic-peloidal grainstone is well sorted and rich in crinoids.

Oolite located directly below the unconformity along the eastern platform margin is broken into shallow curvilinear fractures and is abruptly truncated (Figure 10B). These fractures sole out within the oolite and are filled with red *P. alpina* sediment. Offset of oolite across these fractures is minimal; basal décollements were not observed. Back-reef lithologies found directly below the unconformity are brecciated, rounded, and lined by reddish silt (Figure 10C,D). Reef, fore-reef, and back-reef limestone is cut by small neptunian dikes which are lined by multiple generations of radiaxial cement, fine-grained *P. alpina* sediment, and crystal silt. (Figures 9D, 10C). Dikes are aligned vertical to subhorizontal with bedding and exhibit minor offset when crosscutting large stromatoporoids and corals. In several cases these dikes can be traced up to the unconformity where they are abruptly truncated and overlain by the L.A.R. (Figure 10C). Large fist-sized cavities are found within reef and back-reef sediments 5 to 15 m below the unconformity (Figure 9E–H). These large cavities have rounded walls, are lined by radiaxial cement, and are geopetally filled with red peloidal sediment and *P. alpina* biomicrite. In one example, pendant calcite cement overlaps deposition of red peloidal internal sediment, which is found perched on cavity walls.

Large neptunian dike systems, meters to tens of meters in height and tens of centimeters in width, penetrate meter-scale cross-bedded oolitic grainstone and peloidal/oncolitic packstone (Figure 10E–H). Dike walls display rounded, irregular, and sharp fracture geometry. Rounded clasts and sediment are perched on irregularities along dike walls. These features are overlain and filled by radiaxial cement, *P. alpina*, juvenile ammonites, crinoids, and well-rounded oolitic and *P. alpina* grainstone clasts.

Interpretation

Diagenetic features found at the regional unconformity and within the structurally competent Upper Pliensbachian reef-grainstone complex indicate that some of the cavities associated with the unconformity were not created and filled by episodic processes (e.g., slumping). Instead, generation and filling of some of these cavities took place during or following platform exposure. These fabrics and stratal relationships have major implications for sequence stratigraphic interpretation of the S.T.P. because of the following: (1) exposure-related fabrics that separate the Lower Jurassic platform and the Middle Jurassic Posidonia Alpina Beds in both the platform interior and the eastern platform margin imply the presence

of a (small-scale) type 1 sequence boundary (Sarg, 1988); (2) accumulation and aggradation of shallow cross-bedded grainstone seaward of the exposed Upper Pliensbachian reef complex suggest that the lower Posidonia Alpina Beds (Aalenian–Lower Bajocian?) were initially deposited as a lowstand wedge; and (3) subsequent onlap of the upper Posidonia Alpina Beds (Lower–Upper Bajocian) over the exposure surface and into the platform interior signify a thin, transgressive systems tract.

Pendant cement and crystal silt provide constraints on the emplacement of both cement and internal sediment within the reef complex. Downward elongated cements indicate that cementation occurred in a two-phase (air–water) fluid. Pendant cement overlies deposition of red internal sediment, which is also present as matrix within back-reef breccia found at the unconformity. Crystal silt is interlayered with radiaxial cement and *P. alpina* sediment in small neptunian dikes. These data imply that the reef complex was subaerially exposed and that some cementation and sedimentation of cavities occurred in a vadose environment. The presence of large rounded cavities which truncate reef and back-reef fabrics indicates that exposure-related dissolution enhanced boundstone cavities and fractures prior to cementation and infilling of *P. alpina* sediment. Rounded walls, thick linings of radiaxial cement, and *P. alpina* fill in large neptunian dikes indicate that these cavities remained open for some time after their formation during late Pliensbachian (?) tectonism. Well-rounded oolite and *P. alpina* clasts in neptunian dikes, and rounded back-reef clasts and truncation of small neptunian dikes and oolite at the unconformity imply that erosion, cementation, sediment infilling, and truncation of neptunian dikes took place before deposition of the L.A.R. These data constrain the filling of reef cavities and neptunian dikes from Aalenian to Upper Bajocian time. Furthermore, the absence of Aalenian-age brachiopod–crinoidal sediment in cavities of the reef complex (e.g., Figure 10B) suggests that these cavities, like others found in the platform interior, were filled during the Bajocian.

The significance of these diagenetic fabrics and the patch reef complex along the eastern platform margin is threefold. First, the presence of exposure-related fabrics in these reefs and sponge-coral patch reefs within the San Vigilio Group (Clari and Marelli, 1983) suggests that both east and west margins of the platform were exposed at some point during the diagenetic history. Exposure is also implied by leached *Lithiotis* valves which are found in the platform interior and filled by *P. alpina* sediment. These data indicate that the major unconformity in the platform interior was formed by an extensive period of exposure (late Toarcian–Lower Bajocian?) and nondeposition. Second, exposure-related fabrics and significant accumulation of cross-bedded, pelagic-rich brachiopod/crinoidal grainstone seaward of the Upper Pliensbachian reef complex (Figures 1B, 3B, 5F–H: Valpore di Cima/Mt. Agaro) suggest that genesis of the lower Posidonia Alpina Beds (Aalenian–Lower

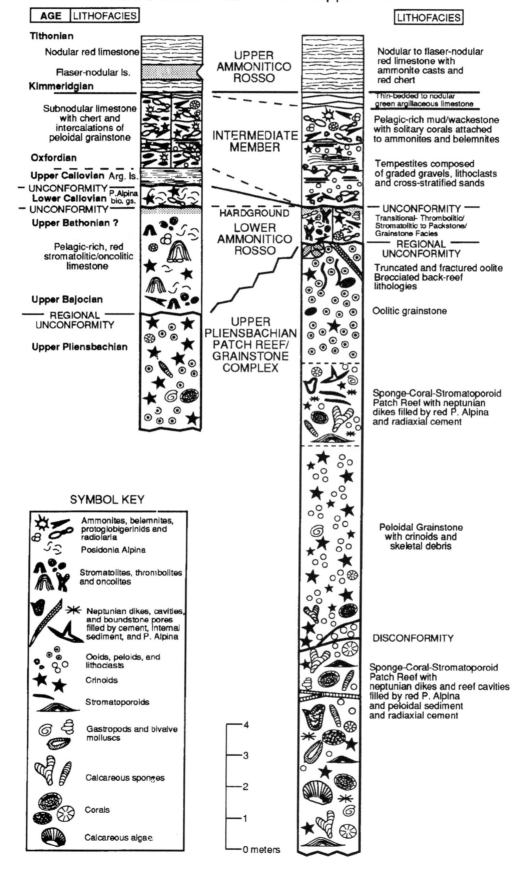

Composite Section Mt. Asolone Area   Composite Section Mt. Grappa Area

| AGE | LITHOFACIES | | LITHOFACIES |

Tithonian

Nodular red limestone

Flaser-nodular ls.

Kimmeridgian

Subnodular limestone with chert and intercalations of peloidal grainstone

Oxfordian

Upper Callovian Arg. ls.

— UNCONFORMITY — P.Alpina
Lower Callovian bio. gs.

— UNCONFORMITY —

Upper Bathonian ?

Pelagic-rich, red stromatolitic/oncolitic limestone

Upper Bajocian

— REGIONAL — UNCONFORMITY

Upper Pliensbachian

UPPER AMMONITICO ROSSO

INTERMEDIATE MEMBER

HARDGROUND

LOWER AMMONITICO ROSSO

UPPER PLIENSBACHIAN PATCH REEF/ GRAINSTONE COMPLEX

Nodular to flaser-nodular red limestone with ammonite casts and red chert

Thin-bedded to nodular green argillaceous limestone

Pelagic-rich mud/wackestone with solitary corals attached to ammonites and belemnites

Tempestites composed of graded gravels, lithoclasts and cross-stratified sands

— UNCONFORMITY —
Transitional- Thrombolitic/ Stromatolitic to Packstone/ Grainstone Facies

— REGIONAL — UNCONFORMITY

Truncated and fractured oolite Brecciated back-reef lithologies

Oolitic grainstone

Sponge-Coral-Stromatoporoid Patch Reef with neptunian dikes filled by red P. Alpina and radiaxial cement

Peloidal Grainstone with crinoids and skeletal debris

DISCONFORMITY

Sponge-Coral-Stromatoporoid Patch Reef with neptunian dikes and reef cavities filled by red P. Alpina and peloidal sediment and radiaxial cement

SYMBOL KEY

- Ammonites, belemnites, protoglobigerinids and radiolaria
- Posidonia Alpina
- Stromatolites, thrombolites and oncolites
- Neptunian dikes, cavities, and boundstone pores filled by cement, internal sediment, and P. Alpina
- Ooids, peloids, and lithoclasts
- Crinoids
- Stromatoporoids
- Gastropods and bivalve molluscs
- Calcareous sponges
- Corals
- Calcareous algae

4
3
2
1
0 meters

Bajocian?) took place along the periphery of a slightly elevated subaerially exposed platform (as suggested by Sturani, 1971). Moreover, the presence of terebratulid and rhynchonellid brachiopod grainstone (de Gregorio, 1885; Botto Micca, 1893) suggests a nearshore coastal setting without reefs (Ager, 1965). These data are consistent with the presence of a small-scale type 1 unconformity and corresponding lowstand wedge. Lowstand grainstones, which display large-scale bimodal planar, trough, and herringbone cross-stratification are strikingly similar to bank margin sand bodies described from the Northern Bahamas and Cat Island platform (Hine, 1983; Dominguez et al., 1988) that were produced by tidal and storm-generated currents in waters several meters to tens of meters deep during sea-level transgression. Third, onlap of the platform is indicated by lowstand deposition along the margins and transgression of the platform interior (Lower–Upper Bajocian). Deposition may have occurred as storm overwash into preexisting cavities in the platform interior (Sturani, 1971), or following sea-level transgression at which time new cavities, platform breccia, and coquina may have been formed and/or reworked in a shallow wave-swept environment. Hummocky cross-stratification (Massari, 1981) and complicated diagenetic fabrics (e.g., Figure 5A–D) that include multiple episodes of brecciation, cementation, and current-influenced sedimentation are consistent with this latter interpretation. Accordingly, sediments found in the platform interior are interpreted as a thin, poorly preserved transgressive system tract. Eventual stabilization of these transgressive deposits is evidenced by transition from the Posidonia Alpina Beds to stromatolitic facies of the L.A.R. (e.g., Colme di Vignola section; Massari, 1981). Stabilization of similar transgressive sand bodies by seagrass and algal mats has been documented by Hine (1983) and Dominguez et al. (1988) in the Bahamas and attributed to a decrease in current and wave energy with increasing water depth (Hine, 1983).

In summary, sedimentary structures and geometries in the Posidonia Alpina Beds indicate that drowning commenced at a relatively shallow depth (tens of meters) along the platform margins. With sea level rise, large-scale bedforms aggraded along and eventually transgressed the platform margin/interior where some sediment accumulated as cross-bedded grainstone, coquina, and sediment/cement fill in cavities and breccias. These bedforms were reworked by waves and currents until they were stabilized by the development of a benthic "floral" cover (i.e., stromatolites) and/or hardgrounds.

Pelagic Lithofacies

The Lower Ammonitico Rosso

In the platform interior, stromatolite/oncolite assemblages overlie a planar truncation surface/hardground which separates *P. alpina*-filled breccia and the Calcari Grigi Formation from the L.A.R. (e.g., Figures 6, 11: Mt. Longara). Stromatolite accumulation through detrital sedimentation and binding is clearly evident in that individual stromatolite laminae exceed the angle of repose and contain fine debris composed of broken pelagic bivalve shells, protoglobigerinids, and peloids (Figure 6C,D). Transitions in stromatolite morphologies from finely laminated stratiform to low-amplitude LLH and flat-topped LLH to columnar morphologies with interbedded grainstone are observed at Mt. Longara (Figures 6A–D, 11). These transitions occur both horizontally and vertically. Sediment between stromatolite heads is composed of coarse material including crinoid debris and pelagic bivalves. Small erosional gullies in stratiform fabrics are covered by draping stromatolitic laminae.

Toward the eastern platform margin, nodular/domal stromatolites and oncolites are found in the L.A.R. at the Col Moschin and Mt. Asolone localities (Figures 1B, 7). At the eastern platform margin, thrombolite morphologies appear as complex digitate and columnar forms with internal lamination and massive texture (Figure 12). Laminae display pinch and swell structures and exceed the angle of repose. These structures are several centimeters in width and up to tens of centimeters in height and commonly bifurcate (Figure 12C,D). Channels between thrombolites are composed of pelagic bivalve and crinoid-rich packstone. Thrombolites exhibit transition upwards into finely laminated stromatolite caps that are several centimeters in thickness (Figure 12E,F). Stromatolite laminae initially drape over the tops of thrombolites and eventually amalgamate into low-amplitude domal stromatolites. Stromatolite caps are infrequently burrowed.

Transition from stromatolite/thrombolite facies to packstone/grainstone facies is observed above the

---

Figure 7. Schematic cross section of key outcrops which are used to define the transition from platform interior to the platform margin facies (see Figure 1B). Lithostratigraphic correlation with the type section of the platform interior (Martire, 1988, 1989) provides biostratigraphic constraint. Lithostratigraphic boundaries include: (1) Upper Pliensbachian reef/grainstone facies, the regional unconformity and scattered patches of *P. alpina* sediment; (2) the basal L.A.R. stromatolite horizon (Upper Bajocian); (3) red to brown mineralized hardgrounds in the middle and top of the L.A.R.; (4) thin beds of *P. alpina*–rich bioclastic grainstone (Lower Callovian) above the upper L.A.R. hardground; (5) dark-brown to dark-red argillaceous limestone (Upper Callovian) at the base of the Intermediate Member; and (6) chert-rich argillaceous limestone (Upper Oxfordian) and flaser-nodular facies of the U.A.R. (Kimmeridgian).

Figure 8. Key stratigraphic relationships used in the correlation of outcrops from the platform interior to the eastern platform margin and facies transitions within the Intermediate Member (see Figure 7). (A) Thin-bedded chert, chert-rich limestone (lower right arrow), and limestone composed of graded peloidal grainstone/mudstone (upper left arrow) in the platform interior (Intermediate Member; Mt. Lisser. Hammer for scale). (B) The Lower Ammonitico Rosso (LAR) overlain by hummocky Upper Callovian argillaceous limestone (AL) and slightly nodular limestone of the Intermediate Member (IM) at Mt. Asolone (hammer for scale). Intermediate Member beds (IM) contain graded peloidal grainstone/wackestone and are transitional with flaser-nodular limestone (Upper Ammonitico Rosso). (C) Crinoidal oolite (UP) overlain by the Lower Ammonitico Rosso (LAR) and the Intermediate Member (IM) at the eastern platform margin (near Mt. Grappa; hammer for scale). A hardground (left arrow) is locally present at the top of the L.A.R. Yellow- to cream-colored limestone (IM) contains thin beds of graded gravels, peloidal sand, and pelagic-rich peloidal mudstone/wackestone.

Figure 9. Fauna and petrographic features found in Upper Pliensbachian sponge-coral-stromatoporoid patch reefs at the eastern platform margin (see Figure 7). (A) Calcareous sponge *Stylothalamia cf. budaensis* (Wells, 1934). (B) Reef boundstone composed of sponges, peloidal packstone, and fibrous calcite cement (polished slab). (C) Fore-reef talus beds composed of corals, sponges, and crinoidal/peloidal packstone (polished slab). (D) Fore-reef packstone cross-cut by a small neptunian dike, which is infilled by multiple generations of fibrous cement, *P. alpina* sediment, and crystal silt. (E–F) Cavities within back-reef wackestone (outcrop and polished slab; pen for scale in E). Wackestone (1) is filled by fibrous calcite cement (2) and red *P. alpina* biomicrite (3). Cavity walls were smoothed (solution-enlarged?), lined by cement, and then filled by *P. alpina* sediment. (G) Fibrous cement (1) and *P. alpina* biomicrite (2) in cavity shown in (E) (thin-section photograph). (H) Large cavity within reef boundstone. Cavity is first floored by two layers of red-colored geopetal silt (1). The second layer is composed of fine peloidal silt and is perched along cavity walls (arrow). Thick, downward-elongated pendant cement has precipitated from irregularities along the cavity ceiling (2) and overlaps introduction of red peloidal sediment (arrow). Pendant cement morphology indicates precipitation in a vadose environment.

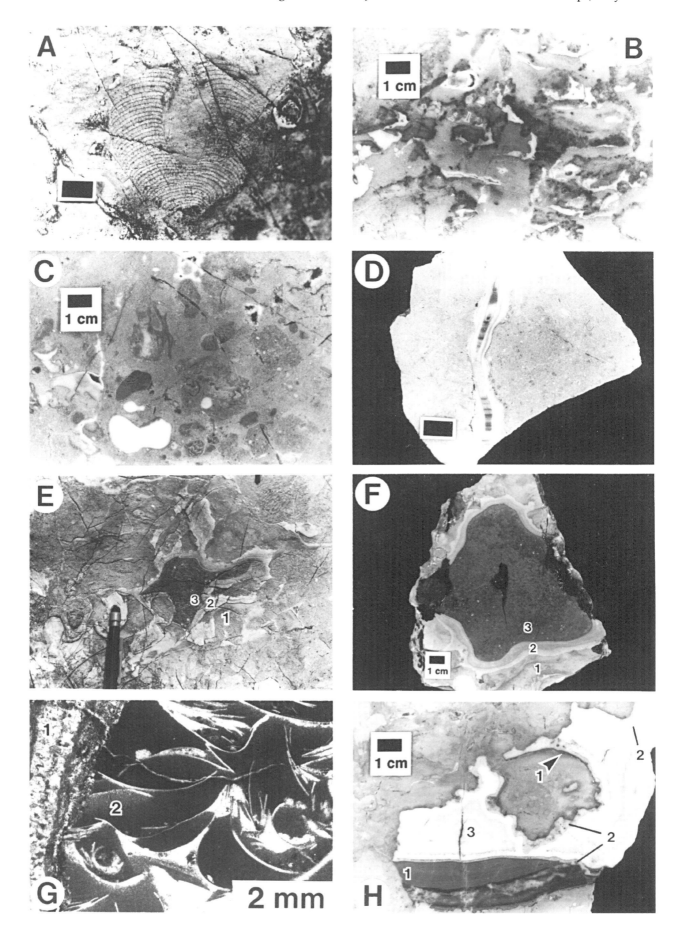

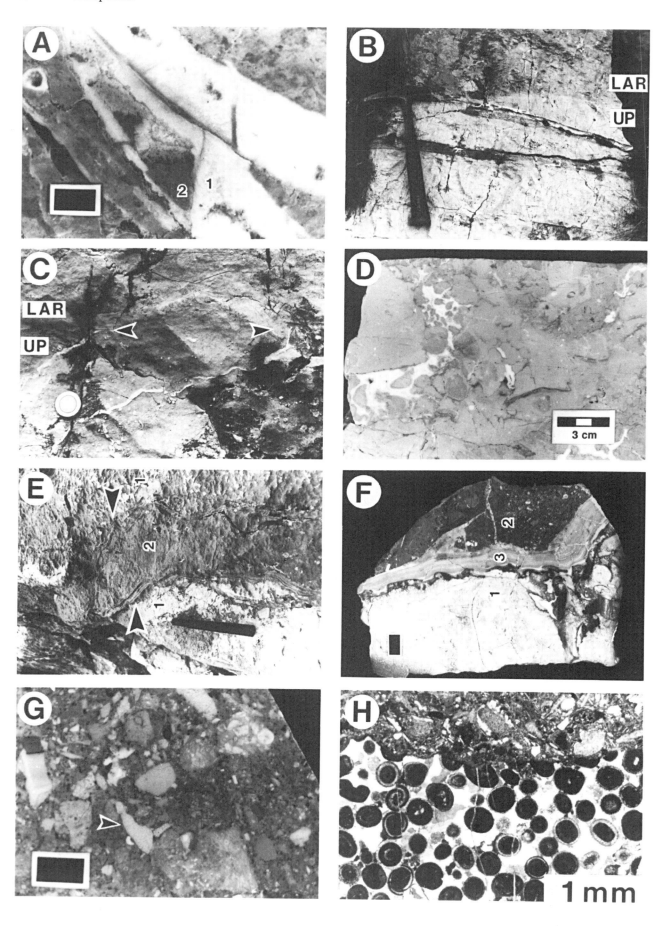

regional unconformity over a lateral distance of several hundred meters (Figure 12A,B: Mt. Grappa). Stromatolites and thrombolites grade laterally first into nodular burrowed wackestone and oncolitic wackestone then into packstone and grainstone. Packstone and grainstone are located above the Upper Pliensbachian patch reef complex and contain ammonites, ammonite fragments, crinoids, pelagic bivalves, belemnites, solitary corals, and rare gastropods (Figure 13). Ammonites are ubiquitous within packstone and grainstone and range in size from 1 to 15 cm in diameter. Intraskeletal pore space within ammonite phragmacones is filled by geopetal sediment, thin isopachous fibrous cement, isopachous radiaxial-fibrous calcite cement, and equant calcite spar (Figure 13B). Ammonite phragmacones have been replaced by neomorphic, micritic, and void-filling calcite, which preserves original shell morphology. Erosional surfaces that truncate macrofossils are common in packstone and grainstone. Successive slab cuts from large samples show that planar erosional surfaces cut across large undeformed ammonite phragmacones (Figure 13C,D). Gastropods are up to several centimeters in length and include an archeogastropod belonging to the Superfamily *Amberleyacea* and possibly the Family *Turbo* (?) (Figure 13E,F). Stromatolitic and grain-rich sediments are often capped by a dark-red hardground that contains small iron-manganese nodules and/or are overlain by thin beds (tens of centimeters) of bivalve grainstone. These beds are rich in *P. alpina*, crinoids, ammonites, and ammonite lithoclasts and correlate with similar bioclastic facies in the platform interior (Lower Callovian: Martire, 1988, 1989).

Platform margin facies of the L.A.R. pass basinward into gray mudstones with burrows filled by pinkish sediment containing protoglobigerinids, pelagic bivalves, fine-size crinoids, and carbonate mud (Ponte Serra section). Slabbed samples collected from the Ponte Serra locality do not display the stratiform, columnar, or thrombolite morphologies which are present in the interior and eastern margin of the platform.

Interpretation

The L.A.R. is interpreted to represent a transition from deposition on a shallow wave-swept bank (Posidonia Alpina Beds) to deposition on a submerged platform located below fair-weather wave base but within the influence of waves and currents (similar to that suggested by Massari, 1981, i.e., above deep-storm wave base). Lateral transition from mud-rich stromatolite/oncolite lithologies to grain-rich lithologies above the Upper Pliensbachian reef complex near Mt. Grappa indicate that this locality was a platform margin through time (Figures 1B, 7). Grain-rich lithologies contain concentrations of ammonites, ammonite fragments, isopachous cement, and planar erosional surfaces. These data indicate that winnowing, abrasion, cementation, and erosion took place early in the diagenetic history in a high-energy environment.

Microbial fabrics in the L.A.R. are problematic. Depositional models of the accumulation of stromatolites and the more general problem that microbial fabrics present (e.g., biotic binding versus precipitation, bacterial or cyanobacterial origin, depth significance, etc.) have recently been reviewed by Ginsburg (1991). In view of the evidence presented for a shallow origin of the regional unconformity and fill, several of these models and other lines of evidence (e.g., pelagic fauna, preservation of ammonites, microbial fabrics) presented in support of a deep pelagic origin for the L.A.R. are reconsidered below.

The reliability of ammonites, belemnites, and plankton as indicators of "deep" pelagic sedimentation—Examples of pelagic organisms in ancient shallow-water facies and Holocene shallow-water environments are common. Several examples are introduced to point out that organisms normally associated with "pelagic" deposits (e.g., ammonites, belemnites, and plankton) do not necessarily imply deep depositional environments and may represent the temporary or prolonged incursion of pelagic organisms into shallow realms (e.g., Jenkyns, 1991)

Pelagic organisms documented in ancient shallow-water facies include ammonite phragmacones in

Figure 10. Exposure-related fabrics observed at the regional unconformity between Lower Jurassic platform sediments and the Lower Ammonitico Rosso. (A) *P. alpina* sediment (2) occurring as geopetal fill within leached Lithiotis valves (1) (Mt. Lisser). (B) Truncated and fractured oolite (UP) overlain by the Lower Ammonitico Rosso (LAR) (near Mt. Grappa; hammer for scale). Subhorizontal fractures are filled by *P. alpina* sediment and sole out at shallow depth within the oolite. (C) Crinoidal wackestone (UP) overlain by the Lower Ammonitico Rosso (LAR) (near Mt. Grappa; coin for scale). Small neptunian dikes within wackestone are filled by *P. alpina* sediment and are truncated at the unconformity (arrows). (D) Brecciated back-reef lithologies are rounded and set in a red silt matrix at the unconformity (polished slab) (near Mt. Grappa). (E) Large neptunian dike (2, between arrows) within oolite (1) at the eastern platform margin (marker for scale) (up is to the left). Dikes are several to tens of meters in height and up to tens of centimeters in width. Dikes are filled by radiaxial calcite cement and *P. alpina* sediment (2). (F) Neptunian dike wall (1) and fill (2) composed of *P. alpina* sediment and radiaxial cement (3) (polished slab) (up is to the left). Clasts and silt are perched on irregularities along rounded dike wall. (G) Neptunian dike fill composed of rounded oolitic clasts (arrow), *P. alpina* sediment, and rounded clasts of *P. alpina* sediment (polished slab). (H) Rounded oolitic clast (thin-section photograph). Ooids and healed fractures are truncated along the periphery of the clast.

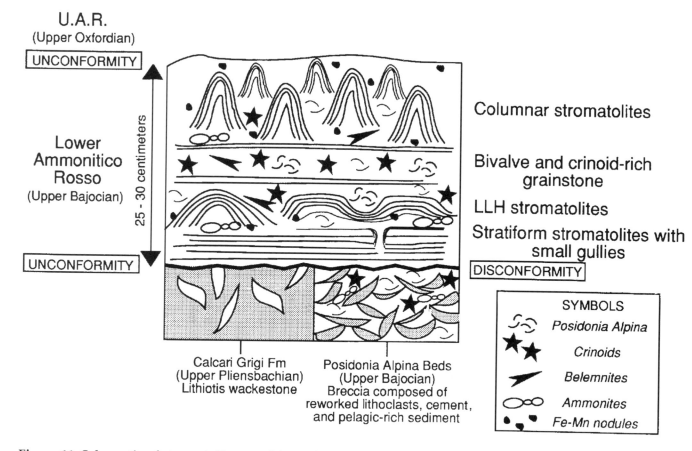

U.A.R.
(Upper Oxfordian)

UNCONFORMITY

Lower
Ammonitico
Rosso
(Upper Bajocian)

25 - 30 centimeters

UNCONFORMITY

Columnar stromatolites

Bivalve and crinoid-rich
grainstone

LLH stromatolites

Stratiform stromatolites with
small gullies

DISCONFORMITY

SYMBOLS

Posidonia Alpina

Crinoids

Belemnites

Ammonites

Fe-Mn nodules

Calcari Grigi Fm
(Upper Pliensbachian)
Lithiotis wackestone

Posidonia Alpina Beds
(Upper Bajocian)
Breccia composed of
reworked lithoclasts, cement,
and pelagic-rich sediment

Figure 11. Schematic of stromatolite transitions observed in the platform interior near Mt. Longara (see Figure 6). Structures found directly above the unconformity show transition from stratiform to laterally linked hemispheres (LLH), and from LLH morphologies to columnar morphologies. LLH morphologies display lateral variation between flat-topped domal and domal morphologies. Erosional gullies within stratiform fabrics are "healed" by stromatolite laminae, which drape over eroded edges.

tepee structures of the Triassic Latemar Platform (Gaetani et al., 1981; Dunn, 1991; and personal observation), concentrations of belemnites intermixed with oolite and large mollusks in shallow channels of the Calcari Grigi Formation (M. Claps, personal communication, 1991, and personal observation), and ammonites, nannoplankton, and vertebrate faunas within the well-known Solnhofen Limestone.

Recent sediment deposited in tens of meters of water within the southern Belize Lagoon contain coccolith ooze, mollusk and foraminiferal marl, pteropod marl, pyrite, and glauconite (?) (Scholle and Kling, 1972; Purdy et al., 1975). Coccolith ooze generally accounts for 10 to 15% of total carbonate mud volume and resembles modern *Globigerina* ooze and some ancient chalks (Scholle and Kling, 1972). Other Holocene examples include coastal upwelling zones off of northwestern South America and northwest Africa which occur over shelves 20 km in width and 150 m depth, and 50 km in width and 100 m depth, respectively. Skeletal plankton and nekton collected from near-surface waters include diatoms, planktonic and benthic foraminifers, radiolarians, larvae of benthic gastropods and bivalves, pelagic gastropods, and

fish bones (Thiede, 1981). Benthic organisms include mollusks, echinoderms, benthic foraminifers, ostracods, siliceous sponges, red algae, ascidian spicules, octocorals, and boring sponges (Fütterer, 1981). Preservation of pelagic and benthic fauna within the underlying sediment is sometimes variable (e.g., Fütterer, 1981). Nevertheless, this faunal assemblage is very similar to that observed in the L.A.R. and suggests that "pelagic" lithofacies may be deposited at relatively shallow depth as a result of upwelling. As a consequence, interpretation of deep pelagic environments (>200 m) based solely on the presence of pelagic fauna may be flawed (as pointed out by Scholle and Kling, 1972).

Early diagenesis of ammonite phragmacones—The cementation, submarine solution, and calcitization of ammonite phragmacones in Mesozoic sequences has been intensely studied and is the subject of much debate (for discussion see Jenkyns, 1971a and Schlager, 1974). Preservation of aragonitic ammonite phragmacones in pelagic lithofacies (including the L.A.R.) has been used as a proxy for depth of oceanic saturation states (e.g., Bosellini and Winterer, 1975). Ammonite phragmacones within packstone/grain-

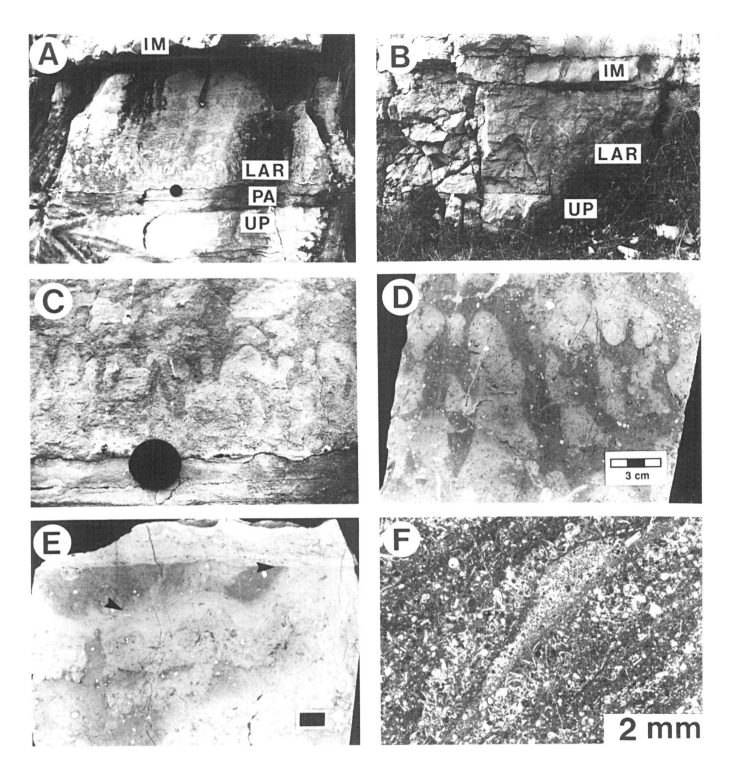

Figure 12. Thrombolites and stromatolites found in the Lower Ammonitico Rosso along the eastern platform margin. Transition from thrombolites/stromatolites into burrowed oncolitic wackestone/packstone and packstone/grainstone occurs over a distance of several hundred meters. (A) Thrombolites (LAR) overlying thin patches of *P. alpina* sediment (PA) and the regional unconformity (UP; oolite). The L.A.R. is separated from the overlying Intermediate Member (IM) by an erosional unconformity (lens cap for scale). (B) Burrowed oncolitic wackestone/packstone and packstone/grainstone (hammer for scale). Note relative absence of microbial structures. Bounding relationships are the same as in (A). (C–D) Thrombolites commonly bifurcate, display internal lamination and massive texture, and have been modified by burrowing organisms (lens cap for scale in C). Channels between thrombolites are filled by crinoid and *P. alpina*-rich sediment. Individual thrombolites may range up to 10 cm in height (outcrop and polished slab). (E) Thrombolites exhibit transition to laminated stratiform morphologies and are truncated by a planar disconformity. (F) Stromatolite caps and thrombolites possess laminae composed of fine skeletal debris, protoglobigerinids, peloids, and carbonate mud. Laminae exceed the angle of repose and display pinch and swell structure.

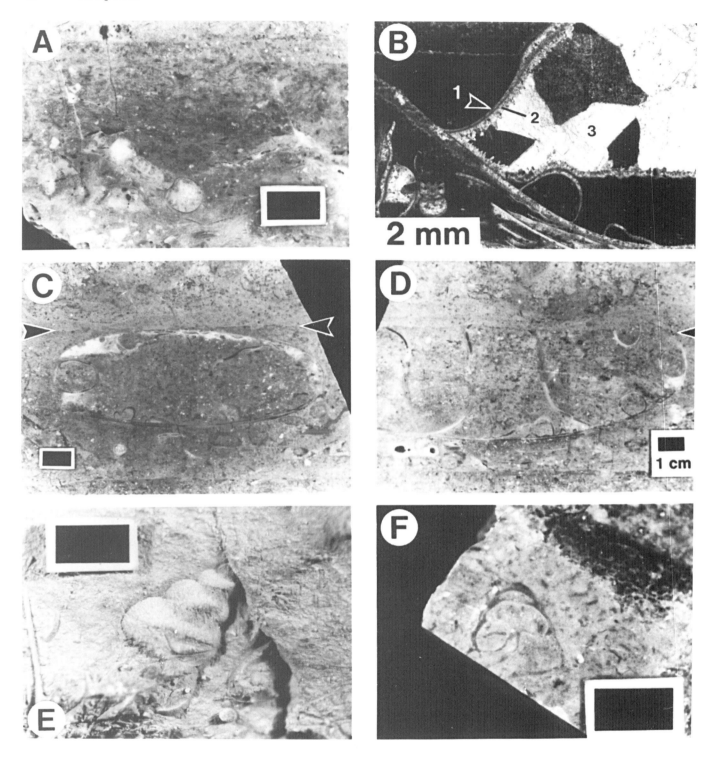

Figure 13. Packstone and grainstone found in the Lower Ammonitico Rosso along the eastern platform margin. (A) Ammonite-rich packstone (polished slab). (B) Thin-section photomicrograph of ammonite phragmacone in (A). Phragmacone (1) is partially filled by internal sediment and lined by thin isopachous fibrous cement (2). Later equant calcite cement fills remaining pore space (3). Ammonite phragmacone is composed of neomorphic, micritic, and void-filling calcite. (C–D) Planar discontinuity surfaces (arrows) are present within packstone/grainstone and truncate ammonite macro conchs. Successive slab cuts (C) and (D) reveal that macro conchs were deposited and cemented within packstone/grainstone lithologies prior to truncation.
(E) Archeogastropod belonging to the Superfamily *Amberleyacea*. Identification made by N. Sohl, U.S.G.S.
(F) Gastropod possibly belonging to the Family *Turbo*(?).

stone of the eastern platform margin are composed of neomorphic, micritic, and void-filling calcite. Phragmacones are undeformed, are cemented by isopachous fibrous cement, and are truncated by planar erosional surfaces. These data suggest that cementation, erosion, and stabilization of aragonite occurred early in the diagenetic history. Ammonite phragmacones within mud-rich sediment of the platform interior show evidence of early selective dissolution and are generally poorly preserved (Clari et al., 1984; Massari, 1981; Martire, 1989). These data suggest that the relative "preservation" of ammonite phragmacones is influenced by the presence or lack of early cement, and is not necessarily dependent on depth below the ALy or ACD (as suggested by Jenkyns, 1971a and Massari, 1981).

Accumulation of stromatolites—Observation of Holocene stromatolites and laboratory study suggest three general depositional models for stromatolite accumulation in ancient sequences: peritidal-restrictive (e.g., Garrett, 1970; Hagan and Logan, 1974; Logan et al., 1974); subtidal high-energy (e.g., Dravis, 1983; Dill et al., 1986); and deep water (e.g., Monty, 1971, 1973, 1977; Williams and Reimers, 1983). The absence of synsedimentary exposure fabrics (e.g., mud-cracked laminites) and other supratidal facies within the L.A.R. rule out a direct analogy with peritidal-restrictive models. Subtidal high-energy models predict that stromatolites should be found in packstone/grainstone facies. This model is contradicted by distribution of L.A.R. stromatolites within mud-dominated lithologies of the platform interior. Models for accumulation of microbial mats at depth indicate that cyanobacteria can operate at deep photic depths (≤150 m), while bacteria may produce stromatolite-like structures in dysaerobic coastal upwelling regimes (<200 m to 650 m) and in aphotic depths as great as 1000 m or more. The latter model is generally accepted as an explanation for stromatolites found in the Ammonitico Rosso and is consistent with the presence of pelagic fauna and iron-manganese crusts, nodules, and grains (Ogg, 1981). However, Drittenbass (1979) has compared iron-manganese deposits of the Ammonitico Rosso to similar occurrences found in present-day oceans and concluded that these features are not diagnostic of depth.

Argument for either a deep- or shallow-water origin for the L.A.R., based on the presence of microbial fabrics, pelagic fauna, or iron-manganese, is equivocal. However, microbial mat morphologies and sedimentary structures are similar to that observed in Holocene shallow-marine environments and shallow coastal upwelling zones. For example, thin sections of microbial mat structures reveal that some accumulation of L.A.R. "stromatolites" took place by detrital sedimentation and microbial binding (Figures 6D, 12F). These structures are remarkably similar to stromatolite structures described from Andros Island and offshore Peru (<200 m depth) that have been attributed to detrital sedimentation and binding by cyanobacteria and bacteria, respectively (see figure 35 of Hardie and Ginsburg, 1977, and figure 2a of

Williams and Reimers, 1983). In addition, the stabilization and preservation of wave ripples by microbial mats (Massari, 1981) implies that binding and agglutination of sediment was relatively rapid. Otherwise, scouring, bioerosion, and bioturbation would likely have destroyed these features (e.g., Dalrymple et al., 1992). Event stratification (i.e., tempestites) described by Massari (1981, 1983) includes disconformity surfaces, lithoclastic lags, graded and plane-laminated beds, and wave ripples. These structures indicate that some deposition of the L.A.R. occurred above storm wave base. Wave-generated sedimentary structures composed of sand- and gravel-size sediment occur on Holocene submerged banks and shelves and provide some estimate of the depth of "storm wave base" (e.g., McMaster and Conover, 1966; Leckie, 1988; Dalrymple et al., 1992). For example, wave-rippled sand and wave-rippled sandy gravels are produced by long-period (12 to 16 s) surface waves at depths ≤160 m and ≤110 m, respectively (Leckie, 1988; Dalrymple et al., 1992). Theoretical calculations suggest that long-period surface waves of 16 s will generate oscillatory-wave motion and bottom effects to water depths of ~200 m (O.M. Phillips, personal communication). This suggests that wave-rippled sand may form in water depths up to a maximum of ~200 m. These data suggest that tempestites in the L.A.R. that contain gravel-size lithoclasts were probably produced by waves at depths ≤110 m while wave-rippled coquinas alone may have formed at depths ≤200 m. Based on these depth constraints, it is probable that microbial mats and coated grains accumulated above deep-storm wave base (≤110–200 m) and were composed of cyanobacteria and/or bacteria operating in the deep photic zone. The presence of small clusters of gastropods within stromatolites, gastropods in packstone/grainstone along the eastern platform margin, and amino acid data indicating cyanobacteria forms (Massari et al., 1989) are consistent with this interpretation.

Lateral and vertical transitions in mud-dominated stromatolite morphologies of the platform interior, and abrupt transition to platform margin facies containing thrombolites and winnowed packstone/grainstone indicate that depositional environments varied across the S.T.P. Similar stromatolite morphologies and transitions occur in Hamelin Pool, Shark Bay, Australia, and are dependent upon various combinations of water depth, wave energy, sedimentation rates, bioerosion, length of exposure, type of cyanobacteria, etc. (Hagan and Logan, 1974; Logan et al., 1974; Hofmann, 1976; Playford and Cockbain, 1976). Of these factors, water depth, wave energy, sedimentation rate, and bioerosion exert the strongest influence on stromatolite morphology. Stromatolite morphologies and sedimentary structures which are similar to those found in the L.A.R. include: (1) subtidal stratiform morphologies, which accumulate in protected embayments by mechanical sedimentation and binding by cyanobacteria films; (2) subtidal low-amplitude LLH morphologies; (3) "calyx" structures (similar to thrombolites of this study), which develop by mechanical and biologic erosion of columnar morphologies; (4) erosional scars and primary sedimentary structures, such as wave ripples, which are preserved by stromatolite growth; (5) truncated stromatolite fabrics attributed to high-energy storm events; and (6) a seaward facies transition to burrowed and nodular skeletal wackestone/grainstone.

Based on observations of pelagic organisms in shallow environments and processes that influence stromatolite morphology in Holocene environments, it is concluded that microbial fabrics, sedimentary structures, and facies variations in the L.A.R. were produced by a combination of currents, storms, detrital sedimentation, cyanobacterial/bacterial binding, early cementation, and bioerosion. Deposition of mud-dominated lithologies in the platform interior probably occurred in an episodically active subtidal environment between fair-weather wave base and deep-storm wave base (≤110–200 m). Deposition of winnowed grain-rich lithologies at the eastern platform margin occurred at similar depths and was influenced by currents and waves.

The Intermediate Member

Intermediate Member lithologies exhibit transition from thin-bedded chert-rich limestone in the platform interior to bedded grain-rich limestone at the eastern platform margin (Figures 1B, 7, 8: Mt. Grappa and Grigno). Bedded limestone contains graded gravels and cross-bedded peloidal grainstone, and burrowed peloidal wackestone and mudstone. Mud-poor and mud-rich lithologies are thought to represent sedimentation "events" and pelagic deposition, respectively.

Event strata—Gravel-rich sands found in the Mt. Grappa area overlie deeply scoured surfaces; contain rip-up clasts and skeletal material derived from underlying beds; contain imbricated belemnite clasts arranged in fanlike geometry; exhibit graded bedding; possess hummocky, low-angle, and planar cross-stratification; and are commonly burrowed (Figures 14; 15A). Erosional surfaces exhibit a wide range in geometry from deep irregular excavations to rounded gutters to shallow scours (Figure 14B,D,H). Ripples occur as gutter casts and exhibit aggradational wave-ripple and lenticular morphology (Figure 14D). Cross-bedded grainstone is either erosionally truncated or gra-

dational with pelagic mudstone/wackestone (Figure 14H). Skeletal grains and fragments include benthic forams, bryozoans, solitary corals, crinoids, and mollusks. Reworked micritized ooids are also present. Grainstones are cemented by isopachous micritic and equant calcite cements. Toward the platform interior, grain-rich graded beds that possess low-angle and hummocky cross-stratification (Figure 1B: Mt. Grappa and Grigno) eventually pass into mud-rich graded beds (Figures 1B, 8A: Baita Camol and Mt. Lisser).

Pelagic sediments—Pelagic and mud-rich lithologies consist of yellow-tan colored peloidal mudstones/wackestones (Figure 15) that contain an abundance of ammonites, belemnites, solitary corals, protoglobigerinids, poorly preserved radiolarians, pelagic bivalve fragments, small benthic forams, recrystallized skeletal grains, and rare *Saccacoma* (pelagic crinoid) fragments. These lithologies are found above and below event strata and have been homogenized by extensive burrowing. Nucleation and growth of small solitary corals occurs exclusively on hard substrates, i.e., ammonite phragmacones and belemnite rostrums (Figure 15C–E). Some corals are found oriented in growth position. Ammonites found in pelagic mudstone/wackestone toward the top of the Intermediate Member are poorly preserved as stylolitic "steinkerns." Mudstone/wackestone lithologies grade into nodular, thin-bedded argillaceous and chert-rich limestone (Figure 7) that contain calcitic crinoids and aptychi.

Platform margin facies of the Intermediate Member pass basinward into graded grainstones, hemipelagic mudstones, and red thin-bedded argillaceous chert (Ponte Serra section). These lithologies contain peloids, radiolarians, belemnites, belemnite fragments, lithoclasts, and scattered skeletal debris (Figure 15F,G). Lithoclasts are composed of yellow- to tan-colored mudstone that contains radiolarians. Lithostratigraphic correlation suggests that some of these lithologies may be basinal equivalents of belemnite-rich gravels and peloidal sands found in the Intermediate Member along the eastern platform margin. Thin-graded beds are composed of fine peloidal grainstone that display ripple cross-lamination and hemipelagic mudstone. Flute marks are found along bedding planes and are oriented toward

Figure 14. Grain-rich sediment found in the Intermediate Member at the eastern platform margin. (A) Graded gravel contains belemnites (right arrow), lithoclastic fragments, and ammonite lithoclasts (left arrow) (polished slab). Gravel overlies scoured pelagic-rich wackestone and grades into peloidal-lithoclastic sand. (B) "Prise-up" scour structure observed at the base of graded lithoclastic gravel (polished slab). (C) Belemnite clasts displaying imbricated fanlike arrangement at the base of graded gravel. (D) "Scour and fill" structure. Scoured depression has been filled by sand-mud couplets that display aggradational wave-ripple and lenticular morphology. (E) Thin gravel lag grades into peloidal sand displaying low-angle hummocky cross-stratification (lens cap for scale). Hummocky cross-stratification is comprised of meter-scale wavelength undulatory laminations that display low-angle discordance. (F) Lithoclastic-peloidal sands contain forams, crinoids, and mollusk fragments. (G) Graded bed disrupted by postdepositional burrowing (polished slab) (up is to the left). (H) Rounded scour surface overlain by graded lithoclastic gravel and planar cross-stratified peloidal sand. Sand is truncated (arrow) and overlain by pelagic-rich wackestone. Reworked grainstone clasts are found in overlying wackestone.

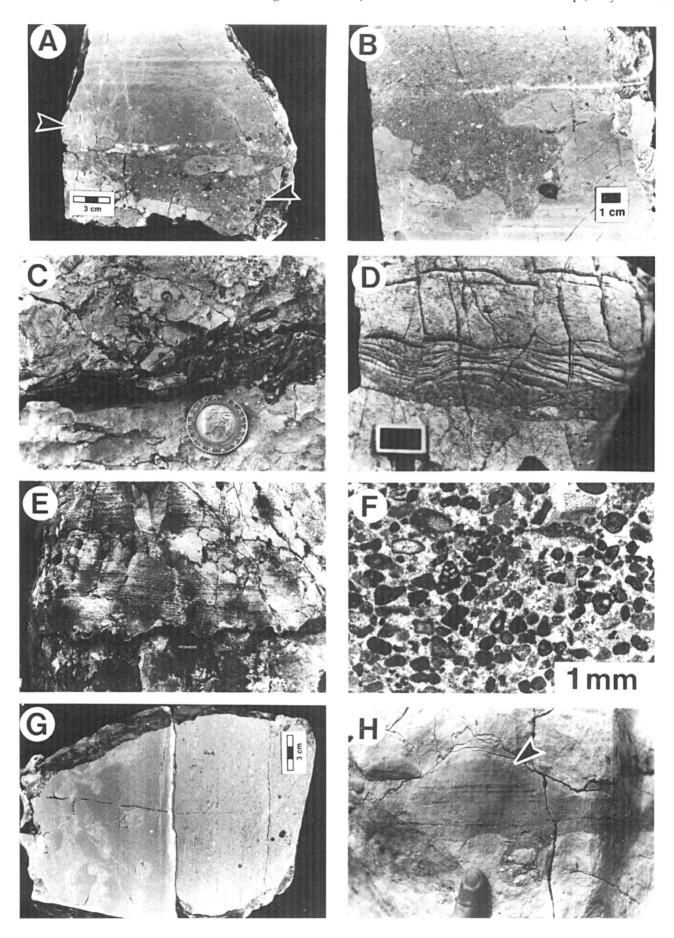

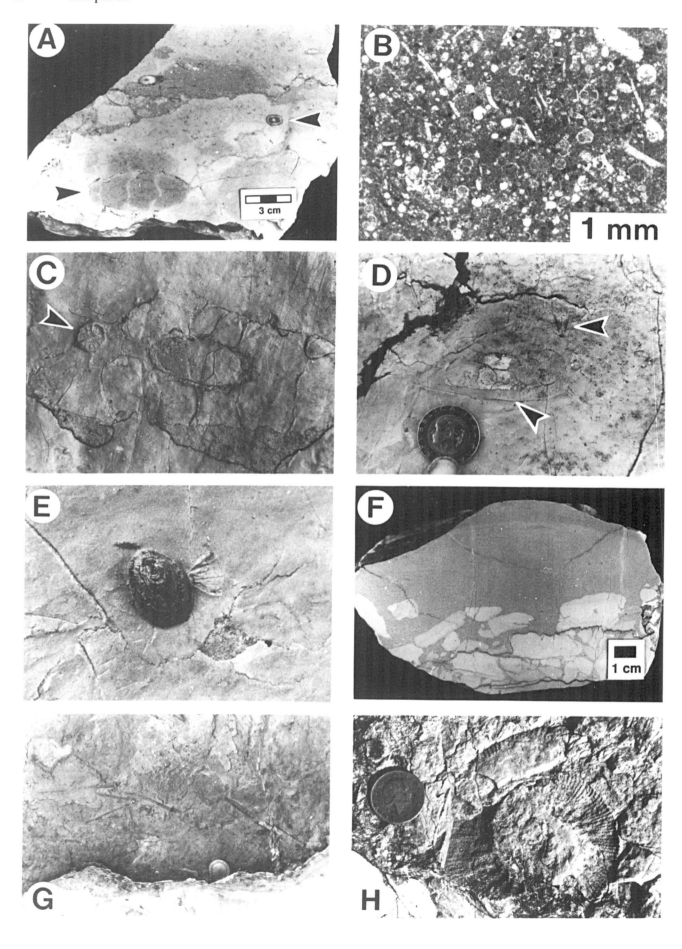

the southeast. Statistical analysis of the orientations of belemnite rostrums and fragments found at the base of two thin turbidites indicates a random distribution.

Interpretation

The Intermediate Member is interpreted here to have been deposited in a pelagic environment near deep-storm wave base. This interpretation is based on platform to margin transition into high-energy winnowed grainstones, tempestite deposition across the platform interior, intense bioturbation of mudstone/wackestone, and evidence for aragonite stability at depth. Lithostratigraphic correlation (Figures 7, 8) and the presence of *Saccacoma* suggests that these sediments are Oxfordian in age.

Lateral transition from thin-bedded chert-rich limestone in the platform interior to thick-bedded grain-rich limestone that contains event strata, scour surfaces, and scour fills suggests that deposition at the eastern platform margin was influenced by bottom currents and storms. Event stratification along the eastern platform suggests two possibilities; graded deposits reflect gravity-flow deposition and allochthonous material derived solely from higher elevations on the Friuli Platform, or tempestite deposition and autochthonous material derived from the platform margin and possibly some parautochthonous material reworked from the periplatform. If a turbidite interpretation is correct then these deposits represent coarse-graded (A) and upper plane bed (B) divisions of a Bouma sequence. Other diagnostic sedimentary structures which would normally be associated with turbidite deposition, i.e., flute marks and current ripples, are absent. If a tempestite interpretation is correct, then sedimentary structures should reflect the waxing and waning of passing storms. Ager (1974), Aigner et al. (1978), Aigner (1982a,b, 1985), and Goldring and Aigner (1982) have described many sedimentary structures that are useful in identification of storm deposits and shelf sedimentation. These include deep "prise-up" structures, "pot and gutter casts," "scour and fill" structures, graded beds, imbricated clasts, hummocky and planar cross-stratification, and wave ripples. Based on similarity with

sedimentary structures described in these studies, and evidence which indicates that some sediment was derived from underlying beds, graded deposits found in the Intermediate Member are interpreted as tempestites. In a study of Intermediate Member lithologies located in the western platform interior, Massari (1981) concluded that reworked nodules of burrowed limestone and packstone/grainstone that exhibit grading, plane-lamination, and wave ripples are indicative of storm erosion and deposition. The presence of similar sedimentary structures in the western platform interior and eastern platform margin suggest that some deposition took place near or at deep-storm wave base ($\leq$110–200 m).

Low-energy, muddy depositional environments are indicated by the presence of pelagic-rich wackestone and mudstone, extensive burrowing, an absence of grain-supported textures and cement, and solitary corals on skeletal substrates. Pervasive recrystallization of skeletal grains and fragments in muddy lithologies suggests that early micritic cementation helped to preserve metastable skeletal fragments within grainstone. The presence of ahermatypic solitary corals on ammonite shells (aragonite) and belemnite rostrums (calcite) suggests that nucleation and growth took place in an environment that, over time, was experiencing little to low sedimentation. Furthermore, if the platform was below the ACD during this time, ammonite shells should have undergone dissolution at the sediment/water interface and, theoretically, would not have provided a sufficient substrate for the encrustation and growth of solitary corals. These data indicate that the sea floor was located above the ACD at some point during the Oxfordian. Vertical transition to nodular thin-bedded argillaceous and chert-rich limestone, and preferential preservation of calcitic skeletal material suggests that aragonitic components were diagenetically removed from these lithologies through compaction and dissolution.

Although belemnites and belemnite fragments in periplatform turbidites display statistically random orientation, some orientations are parallel with flute marks that indicate a northwest provenance. These

Figure 15. Pelagic-rich mudstone and wackestone found in the Intermediate Member at the eastern platform margin. (A) Belemnites (right arrow) and ammonites (left arrow) in burrowed pelagic-rich wackestone matrix (polished slab). Ammonite is filled by matrix. Scour surface separates wackestone from overlying graded gravel. (B) Protoglobigerinids, radiolarians, peloids, forams, crinoids, and recrystallized skeletal clasts are dispersed in wackestone (thin-section photograph). (C–E) Small solitary corals encrust ammonites and belemnites in pelagic-rich wackestone. In (D), bottom arrow points to ammonite body chamber. Top right arrow points to solitary coral which is in growth position. (F) Periplatform sediments found at Ponte Serra. Fine peloidal grainstone and pelagic-rich mudstone clasts (polished slab) are shown. (G) Belemnites, belemnite fragments, and belemnite impressions found at the base of a thin turbidite (Ponte Serra). (H) Stephanoceratid ammonite collected within the Vajont Oolite in the central Belluno Basin. This specimen was collected from a bedding plane 150 m from the base of the formation and is indicative of the Lower Bajocian. It is identified as (?) *Docidoceras* (sp. ind.) belonging to the Discites–Sauzei zones (identification made by G. Pavia and L. Martire, University of Torino).

orientations are opposite to the Vajont/Fonzaso fan system, which was shed southeast to northwest from the Friuli Platform (Bosellini et al., 1981). If belemnite- and gravel-rich tempestites found at the eastern platform margin are correlative, some of these turbidites may have a similar storm event origin and represent material that was reworked on the S.T.P. and transported to a narrow fringing slope system by gravity flows. Sediment production/erosion on "drowned" platforms has been documented by Glaser and Droxler (1991) and Dalrymple et al. (1992). These studies indicate that submerged bank tops can be strongly influenced by wave motion (storm events) and by tidal/bottom currents. Winnowing and scouring results in a minor accumulation of sediment in the platform interior and the transport of sediment to periplatform/slope settings.

DISCUSSION: DROWNING OF THE SOUTH TRENTO PLATFORM

Observation and argument presented by Schlager (1981) points out that under ideal conditions neritic carbonate production and accumulation will exceed the formation of platform accommodation space. Accumulation results from the large difference between growth potential (~1000 m/m.y.) and long-term subsidence rates (10–100 m/m.y.) and sea-level change (~10 m/m.y.). This suggests that drowning of the S.T.P. must have involved environmental deterioration of a productive shallow-marine environment, rapid increase in platform subsidence, and/or eustatic sea-level rise in which normal shallow-marine environments were not established. Stratigraphic, sedimentologic, and petrographic data provide some constraint on the time interval and range of depth over which one or several of these processes must have operated.

Comparison of the Drowning Succession with Postulated Global Sea-Level Change

Subsidence Constraints

Crude estimates of Lower to early Middle Jurassic platform subsidence, using thicknesses of the sub-unconformity platform sequences and estimates of eustatic sea-level rise (approximately +135 m/33 m.y. estimated from Haq et al., 1988), suggest an average subsidence rate of approximately 10.3 m/m.y. (~340 m/33 m.y.). Minimum and maximum subsidence rates are 3.5 m/m.y. and 21.7 m/m.y., respectively (~115 m/33 m.y. and ~715 m/33 m.y.). Exposure-related diagenetic fabrics within the Upper Pliensbachian reef complex and neptunian dikes beneath the unconformity, and seaward accumulation of cross-bedded grainstone which contain a mixed shallow-water and pelagic fauna suggest that the S.T.P. was at or near sea level during initial deposition of the Posidonia Alpina Beds (Aalenian–Lower Bajocian?). Tempestites in the L.A.R. and the Intermediate Member suggest that the platform was above or near deep-storm wave base (≤110–200 m) from Upper Bajocian to Oxfordian time (~15 million yr). These data imply a slow rate of platform subsidence (~5–10 m/m.y.) during genesis of the drowning succession.

Sequence Stratigraphy

S.T.P. stratigraphy and theoretical subsidence were compared with estimates of global sea-level change (Haq et al., 1988) using biostratigraphic information of Sturani (1971) and previous studies by Barbujani et al. (1986) and Martire (1988, 1989). Stratigraphic gaps, depositional geometries, and petrographic fabrics found below and above the Lower to Middle Jurassic unconformity correlate well with long-term eustatic rises and falls of sea level (Figures 16, 17). The drowning succession of the S.T.P. includes Upper Pliensbachian deposition and tectonism, a transgressive systems tract and highstand systems tract (Toarcian), a small-scale type 1 sequence boundary (late Toarcian–Lower Bajocian?), a drowning sequence (Aalenian–Upper Bajocian), and a composite condensed section (Upper Bajocian–Tithonian).

Upper Pliensbachian Deposition and Tectonism— Comparisons of Upper Pliensbachian strata are made with caution as detailed paleontologic data for these intervals are lacking. Evidence presented by Sturani (1971) and Clari and Marelli (1983) suggests that exposure and dissolution of the Calcari Grigi Formation (i.e., leached *Lithiotis* valves) occurred prior to deposition of some late Pliensbachian sediment in the platform interior. These features are indicative of a type 2 unconformity and are probably related to late Pliensbachian sea level fall. Subsequent deposition of reef, grain-rich, and hemipelagic lithologies indicates onlap of both west and east margins of the platform during the late Upper Pliensbachian (Beccarelli Bauck, 1986, 1988; this study). Furthermore, the presence of two shallowing-upwards reef/grainstone couplets on the eastern platform margin indicates deposition of several parasequences prior to late Pliensbachian tectonism.

Small-Scale Type 1 Sequence Boundary—Barbujani et al. (1986) have identified a transgressive systems tract and a highstand systems tract within the lower San Vigilio Group, which precede the regional unconformity. These data are consistent with a late Lower Jurassic tectonic event in which the S.T.P. became a west-facing structural ramp (Barbujani et al., 1986) and early Toarcian eustatic sea-level rise. Onlap of a west-tilted surface is consistent with data that suggest that the eastern platform interior and margin were exposed (this study). The highstand systems tract is composed of reef/grain-rich couplets which probably reflect deposition of several parasequences during the Upper Absaroka B 4.3–4.6 cycles. Importantly, these data indicate that maximum onlap and deposition of the TST/HST was coeval, in part, with deposition of early Toarcian organic-rich shale/manganese-rich limestone in periplatform and basinal settings (Jenkyns et al., 1985; Jenkyns and Clayton, 1986; Jenkyns, 1988).

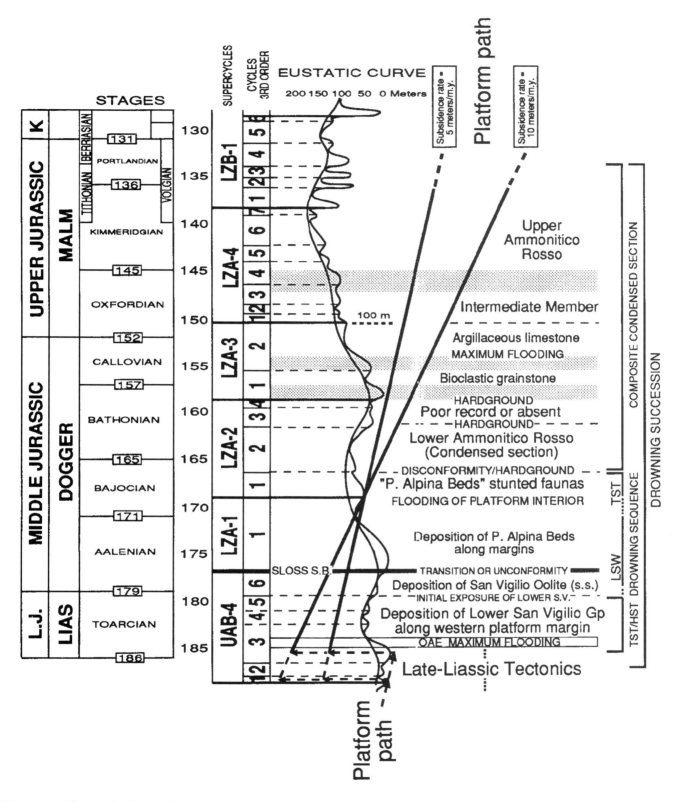

Figure 16. Theoretical subsidence (5–10 m/m.y.), sedimentation history, and sequence stratigraphy of the South Trento Platform plotted against the eustatic curve of Haq et al. (1988). Correlation of the Ammonitico Rosso modified from Martire (1988). Stratigraphic gaps are in shade pattern. The drowning succession of the South Trento Platform correlates well with long-term eustatic sea-level variation (see text for discussion). This model incorporates a late Liassic tectonic event which tilted the platform toward the west-southwest (Barbujani et al., 1986). Note: The chart of Haq et al. (1988) is based on study of sequences located outside of the present study area. Fortunately, this eliminates a circular argument concerning the significance of correlation between stratigraphy of the S.T.P. and the global curve.

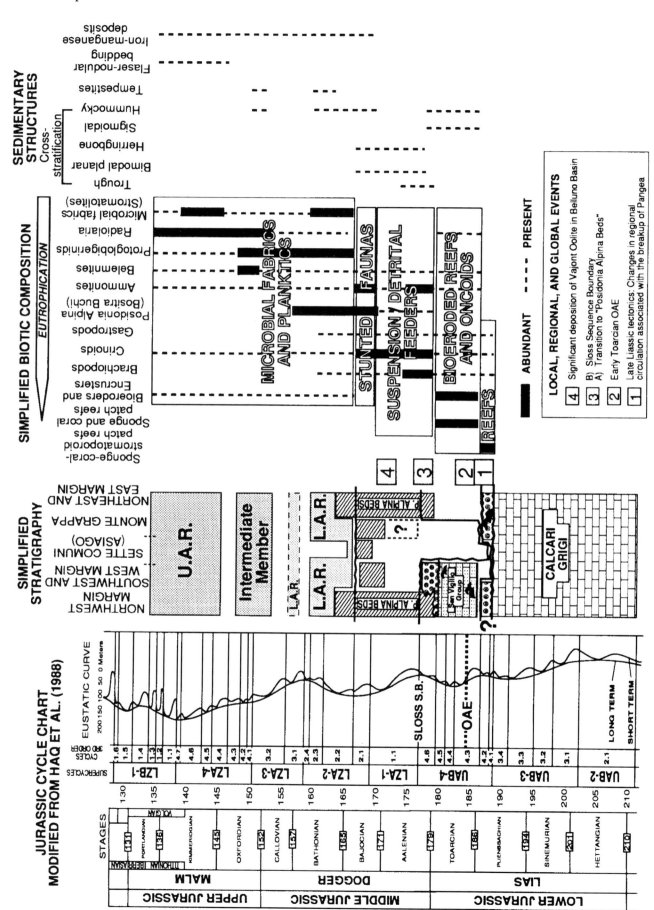

Exposure of San Vigilio reefs (Clari and Marelli, 1983) and lowstand deposition of the San Vigilio Oolite (Aalenian *s.s.*, Barbujani et al., 1986) along the western platform margin are consistent with similar observations on the eastern platform margin (i.e., exposure of Upper Pliensbachian reefs/lowstand deposition of Aalenian grainstone). These data indicate that the regional unconformity, which is present across the S.T.P. initially developed as a result of late Toarcian to Aalenian exposure and is a small-scale type 1 unconformity.

This small-scale type 1 unconformity correlates with a globally recognized second-order sea-level lowstand (Haq et al., 1988). In North America, this lowstand separates the Absaroka and Zuni Supersequences of Sloss (1963). On the S.T.P., it falls at the transition or unconformity between the San Vigilio Group and the lower Posidonia Alpina Beds along the western platform margin, at the unconformity that separates Lower Jurassic platform strata and the upper Posidonia Alpina Beds in the platform interior, and within the lower Posidonia Alpina Beds that are found seaward of the Upper Pliensbachian reef complex along the eastern platform margin.

Drowning Sequence—The drowning sequence (Erlich et al., 1990) of the S.T.P. consists of a lowstand wedge, a transgressive systems tract, and a disconformity/hardground. The drowning sequence defined here consists of those systems tracts which represent the transition from shallow platform lithofacies to condensed pelagic/hemipelagic lithofacies. It is bounded below by a type 1 unconformity and is bounded above by a disconformity/hardground which separates it from the overlying condensed section. The drowning sequence lacks a highstand systems tract which reflects "drowning" of the platform.

Biostratigraphy and cross-stratification suggests that the lower Posidonia Alpina Beds (Aalenian–Lower Bajocian?) accumulated along the platform margins as lowstand deposits during the UAB 4.6 and Lower Zuni A 1.1 cycles. In the platform interior, the Posidonia Alpina Beds contain ammonite assemblages belonging to the Humphries, Subfurcatum, and Garantiana zones (Sturani, 1971) and correlate with those which are used to define the LZA 2.1 third-order cycle. These data suggest that the platform interior began to receive sediment as the result of Bajocian sea-level rise. Initial deposition of the lower Posidonia Alpina Beds along platform margins and subsequent onlap of the upper Posidonia Alpina Beds in the platform interior is consistent with middle

Aalenian to late Bajocian transgression. Following initial platform flooding (LZA 2.1), a highstand systems tract failed to develop. As a consequence, Posidonia Alpina Beds of the platform interior are truncated by a disconformity/hardground (e.g., Figures 6A, 11), which correlates with a lowstand separating the LZA 2.1 and LZA 2.2 cycles (Figures 16, 17). Sedimentologic and petrographic evidence suggests that this disconformity/hardground was the result of winnowing and scouring on a shallow wave- and current-swept platform.

Composite Condensed Section—Despite variable preservation, the Lower, Intermediate, and Upper members of the Ammonitico Rosso comprise a "composite condensed section" (Loutit et al., 1988) that was deposited at increasing depth and with decreasing influence of waves and currents (Figure 17). In the platform interior, Lower Jurassic platform strata and the Posidonia Alpina Beds are unconformably overlain by pelagic- and stromatolite-rich L.A.R. facies (Upper Bajocian–Bathonian). Ammonite assemblages within the base of the L.A.R. (Parkinsoni zone and rare Garantiana specimens: Sturani, 1971; Clari et al., 1984; Martire, 1988, 1989) correlate with the beginning of the LZA 2.2 cycle. These data indicate that initial deposition of the L.A.R. on the S.T.P. was relatively synchronous over a wide geographic area and was associated with eustatic sea-level rise. As a result, the L.A.R. qualifies as a "condensed section" (Loutit et al., 1988). In correlation of the L.A.R., the Intermediate Member, and the U.A.R. with global sea-level change, Martire (1988, 1989) recognized a correlation between the L.A.R. with eustatic rises and falls during the Upper Bajocian–Bathonian (LZA 2.2-2.4 cycles), bioclastic grainstone with transgression during the Lower Callovian (LZA 3.1 cycle), and the base of the Intermediate Member (red argillaceous limestone beds/limy shale) with maximum flooding during the Upper Callovian (LZA 3.2 cycle). Detailed correlation of Oxfordian, Kimmeridgian, and Tithonian sediments with sea-level change becomes increasingly difficult as biostratigraphic data and sections tend to be incomplete. However, vertical transition in the Intermediate Member from grain-rich beds to pelagic wackestone/mudstone to nodular, thin-bedded argillaceous and chert-rich limestone suggests progressive deepening during Oxfordian time. Similarly, lateral and vertical transition from stromatolitic to flaser-nodular and sparsely nodular facies during deposition of the U.A.R. reflects progressive deepening during Kimmeridgian and Tithonian time

Figure 17. Stratigraphy (simplified), biotic composition (simplified), and sedimentary structures of the South Trento Platform versus postulated global eustatic sea-level change of Haq et al. (1988). Paleoecologic deterioration is initiated with introduction of trophic resources during the early Toarcian (see text for discussion). Local and regional events noted in key. Data sources: Barbujani et al., 1986; Beccarelli Bauck, 1986, 1988; Bosellini and Broglio Loriga, 1971; Clari and Marelli, 1983; Clari et al., 1984, 1987; Martire, 1988, 1989; Massari, 1981, 1983; Sturani, 1964, 1971; Trevisani, 1991.

(Sturani, 1971; Massari, 1983; Martire, 1988, 1989). Condensed sedimentation on the S.T.P. ended during the Tithonian with deposition of the thick Biancone Formation. As a result, the Ammonitico Rosso represents a composite condensed section (up to 25 m thick) that encompasses ~25–30 m.y. of sedimentation (Upper Bajocian to Tithonian).

Studies of "drowned" or "incipiently drowned" platforms (e.g., Hine and Steinmetz, 1984; Dominguez et al., 1988; Glaser and Droxler, 1991; Dalrymple et al., 1992) suggest that the effective winnowing and scouring of submerged platforms begins at fairly shallow depth (~20–110 m). In these modern examples, platform interiors are either bare or covered by thin veneers of mud and/or coarse sediment. This suggests that the genesis of "condensed sections" may begin during early platform submergence. In ancient pelagic sequences, sedimentation is generally recorded as thin condensed intervals during relative sea-level rise, while disconformities, poorly preserved sediment records, and/or hardgrounds are produced during relative sea-level fall (Seyfried, 1981; Martire, 1988). Martire (1988) has suggested that bottom currents, possibly responding to eustatic sea-level fluctuation, influenced the formation of hardgrounds and preservation of sediment within the Ammonitico Rosso. Based on study of modern platforms and evidence which suggests that the L.A.R. and Intermediate Member were deposited (in part) above storm wave base, it is concluded here that deposition of the composite condensed section was initiated at relatively shallow depth under the influence of waves and currents. With continued subsidence and eustatic rise, condensed sedimentation continued under the influence of bottom currents. Transition from massive stromatolitic limestone (L.A.R.) containing tempestites to mud- and chert-rich limestone (Intermediate Member) containing tempestites to stromatolitic and thin-bedded flaser-nodular limestone (U.A.R.) is consistent with this interpretation.

In summary, correlation of stratigraphic gaps, depositional geometries, and petrographic fabrics with postulated sea-level change/platform subsidence indicates that relatively shallow depositional depths (near sea level to ≤110–200 m) persisted from late Lower Jurassic to early Upper Jurassic time. Holocene sediment accumulation rates (Schlager, 1981) greatly exceed these subsidence/depth estimates. Comparison of the drowning succession of the S.T.P. with eustatic sea-level change indicates that the critical transition from shallow-water to pelagic lithofacies is recorded by a small-scale type 1 sequence boundary, a drowning sequence (LSW, TST, and a disconformity hardground), and a composite condensed section. Wave-resistant biogenic "rims" and/or highstand systems tracts failed to develop following transgression of the platform. This implies that other "events" must have suppressed carbonate producing benthos to the extent that carbonate accumulation could not keep up with the formation of accommodation space during long-term subsidence and eustatic sea-level rise. Other factors are now explored which may have contributed to suppression of the carbonate factory and genesis of the S.T.P. drowning succession.

Integrated Sea-Level–Trophic Model for the South Trento Platform Drowning Succession

Biofacies Transition and Introduction of Trophic Resources

Carbonate platforms and reefs must be in a healthy productive state if they are to keep up with or outpace relative changes in sea level (Schlager, 1981; Erlich et al., 1990). It is essential that the carbonate factory be equipped with the right organisms and conditions (i.e., light, temperature, nutrient levels, etc.) so that sediment accumulation meets or exceeds the formation of accommodation space. In the case of the S.T.P., the "health" of the platform, prior to deposition of the Ammonitico Rosso, is questionable. This is because major faunal shifts occur from "healthy" reef benthos to "stressed" communities including bioeroded reefs and grainstone composed of ahermatypic suspension/detrital feeders. Initial paleoecologic deterioration on the S.T.P. is coincident with regional extinctions described by Hallam (1987) and the early Toarcian OAE (Jenkyns, 1988). Progressive deterioration occurs in the absence of a major extinction event (Sepkoski, 1989). In light of recent study, it is postulated that paleoecologic deterioration was caused by increasing influx of trophic resources (i.e., inorganic nutrients and organic matter), eustatic sea-level rise, and establishment of oxygen-deficient environments (Figure 18).

As discussed by Hallock and Schlager (1986) and Hallock et al. (1988), shallow carbonate production and bioerosion rates are extremely sensitive to trophic resources. These studies argue that small increases in nutrient levels and plankton densities may drastically affect shallow-water hermatypic coral communities by favoring the growth of ahermatypic suspension feeders and fleshy algae. Many of these competitors are bioeroders which may shift the existing carbonate system from net accumulation to one of net erosion. As a result, platforms may drown at relatively shallow depth (<50 m). These conclusions are supported by an east–west succession of coral, coral-algal, and sponge-algal benthos (including rhodoliths), and increasing amounts of trophic resources across platforms of the Nicaragua Rise. Sponge-algal benthos described by Hine et al. (1988) include the appearance of algal (Halimeda) bioherms and a "planktivory and detrivory" faunal assemblage (sponges, octocorals, and echinoderms) on the flanks of a relatively shallow open seaway (Miskito Channel: 220 m deep, 5–10 km wide, 125 km long). Trophic resources are delivered to these shallow platforms and basins by indirect terrestrial runoff, oceanic upwelling, and local topographic upwelling of the Caribbean current. Similarly, Roberts et al. (1987) postulate that upwelling of cold nutrient-rich water at shallow depth (≥20 m) promoted the growth of *Halimeda* bioherms and suppressed the develoment of reefs along the western and southern margins of Kalukalukuang Bank (eastern Java Sea).

Biofacies of the S.T.P. display a transition from "healthy" sponge-coral-stromatoporoid reefs to bio-eroded sponge and hermatypic coral reefs, to grain-stone composed of ahermatypic suspension/detrital feeders and planktic organisms, to "stunted" pelagic and benthic faunas, and finally to mud-dominated lithologies containing microbial (cyanobacteria?) mat structures (Figures 17, 18).

Significant bioerosion of sponge and hermatypic coral reefs and algal growth is evident in the San Vigilio Group. Sponges, hermatypic corals, and over-sized oncoids are extensively bored by bivalves and encrusted by serpulid worms and bryozoa (see figure 9 and tables 39–41 of Clari and Marelli, 1983). Deposition and bioerosion of these fabrics was con-current with deposition of hemipelagic lithologies in platform margin settings and early Toarcian organic-rich shale/manganese-rich limestone (Jenkyns et al., 1985; Jenkyns and Clayton, 1986; Jenkyns, 1988) in the neighboring Lombardy and Belluno basins. The asso-ciation of organic-rich sediment and oceanic upwelling in both Holocene and ancient settings is well documented (see Thiede and Suess, 1981). Furthermore, Jenkyns and Clayton (1986) postulated that upwelling in this region may have commenced at the Pliensbachian–Toarcian boundary. This implies that nutrients and organic matter were abundant com-ponents of oceanic water masses during deposition of the lower San Vigilio Group. Extensive bioerosion of San Vigilio lithofacies indicates that trophic waters were present at shallow depth (Figures 17, 18B).

Bioeroded reefs and oolite of the San Vigilio Group are overlain by or are transitional with bedded lime-stone and cross-bedded grainstone that contain tere-bratulid and rhynchonellid brachiopods, crinoids, ammonites, and other suspension/detrital feeders (i.e., Aalenian "Bilobate Beds" and lower Posidonia Alpina Beds; Sturani, 1964, 1971) (Figures 17, 18C). This transition is concomitant with the appearance of the planktic bivalve *P. alpina* and the first appearance of stunted benthic faunas along the platform margins (Sturani, 1971; Ferrari and Manara, 1972). With sea level rise, deposition in the platform interior is noted by accumulation of *P. alpina* and both juvenile and stunted ammonites and benthic organisms (upper Posidonia Alpina Beds; Figures 17, 18D). Normal-size adult specimens are found in periplatform areas and the Alpi Feltrine (i.e., Campotorundo Limestones, Figures 1, 2B). *P. alpina* increases in abundance through the Bajocian and ultimately forms thick coquina along the platform margins (e.g., Mt. Agaro: Upper Bajocian–Bathonian, Sturani, 1971).

The association of epifaunal grainstone (lower Posidonia Alpina Beds) and juvenile and stunted organisms (upper Posidonia Alpina Beds) with unique ammonite assemblages (e.g., Figure 5D,H) suggests that mass mortality events occurred on the

margins and in the interior of the S.T.P. The cause of such events may include: (1) upwelling of eutrophic and hypertrophic waters into shallow water (Brongersma-Sanders, 1957); (2) "episodic" coloniza-tion and demise of epifauna in dysaerobic environ-ments (Sageman et al., 1991); (3) the development of inimical bank waters during initial platform flooding (Adey et al., 1977); (4) periods of high environmental stress (e.g., prolonged cold fronts—Roberts et al., 1982; rapid warming—Glynn, 1984); and (5) strand-ing, vertical currents, and/or high-energy storm events (Brongersma-Sanders, 1957; Sturani, 1971). Of these mechanisms, upwelling is an especially appeal-ing analog for the lower Posidonia Alpina Beds (Figures 17, 18C) because cold, oxygen-deficient nutrient-rich waters are known to produce plankton blooms in coastal upwelling regions and associated mass mortality faunas in littoral settings that are com-posed of cephalopods and other invertebrates (Brongersma-Sanders, 1957). Furthermore, persistent upwelling and circulation of cold eutrophic waters over the S.T.P. may have established a dysaerobic environment at relatively shallow depth (Figures 17, 18D). If so, accelerated maturation, small size, and the production of high numbers of young are several strategies that organisms may have adopted in attempts to overcome temporary or prolonged envi-ronmental stress (e.g., Hartman and Barnard, 1958; Berger, 1969; Leppakoski, 1969; Rosenberg, 1977; Dunbar, 1981). Episodic colonization of dysaerobic environments followed by brief periods of anoxia typically result in mass mortality and the deposition of juvenile and stunted faunas (Sageman et al., 1991). These "episodic" faunal accumulations are analogous to the unique assemblages of stunted and juvenile epifaunas found within the upper Posidonia Alpina Beds. These data suggest that "stunted" and "juvenile" pelagic/benthic faunas in the upper Posidonia Alpina Beds reflect an adaptive faunal strategy in response to the sustained upwelling and circulation of cold, oxy-gen-deficient eutrophic waters over the S.T.P.

Overlying the Posidonia Alpina Beds are mud-rich lithologies (L.A.R.) which contain abundant normal-size pelagic fauna (e.g., ammonites, belemnites, pro-toglobigerinids), some benthic fauna (e.g., small gastropods), a variety of microbial fabrics, iron-man-ganese deposits, and tempestites (Figures 17, 18E). As noted earlier, pelagic fauna and microbial fabrics of the L.A.R. are consistent with that observed in shal-low coastal upwelling zones (e.g., Thiede, 1981; Williams and Reimers, 1983). Moreover, sustained upwelling of cold nutrient-rich water and the estab-lishment of low-energy dysaerobic to anaerobic ben-thic environments at photic depths in the platform interior is consistent with the relative absence of shelly benthic macrofauna, the growth and preserva-tion of stratiform and columnar cyanobacteria/bacte-ria mats, the oxidation/deposition of iron-manganese within the sediment/water interface, the accumula-tion of normal-size pelagic organisms, and the pres-ence of thrombolites and a limited benthic fauna (e.g.,

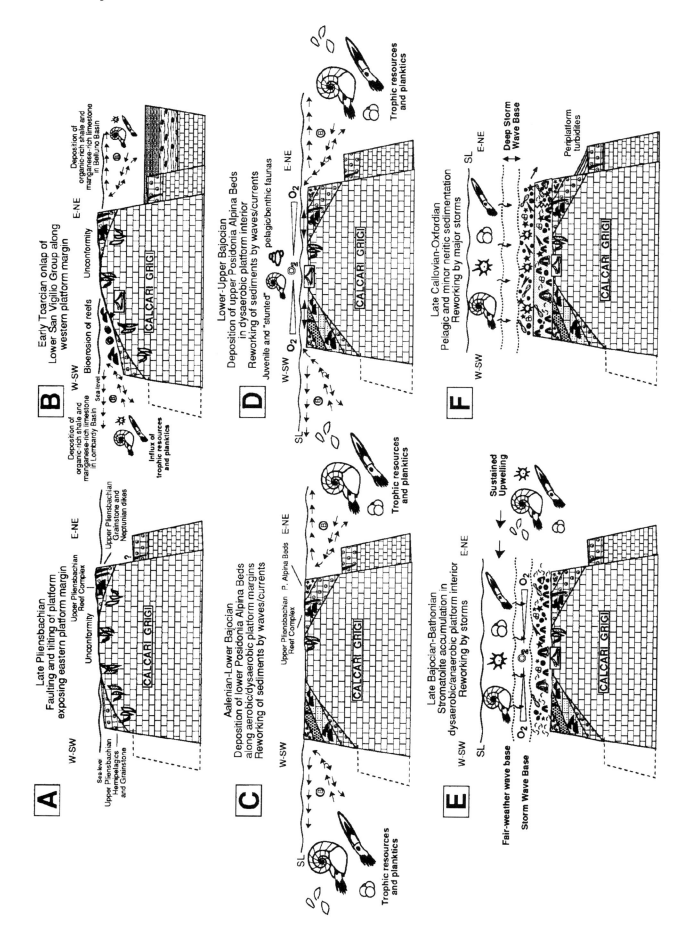

small gastropods) in grain-rich sediments along the platform margins. Flared valves and distribution of thick, time-equivalent *P. alpina* coquina along presumed upwelling regions (i.e., platform margins) may indicate that this organism was equipped to take advantage of the abundant trophic food supply (i.e., swimming capability, Sturani, 1971). With continued subsidence, sea-level rise, and variable upwelling, a predominantly low-energy pelagic environment was established under the decreasing influence of waves and currents. Here mud- and chert-rich lithologies of the Intermediate Member were deposited (Figures 17, 18F) followed by stromatolite-rich and flaser-nodular limestone of the Upper Ammonitico Rosso.

In summary, biofacies data of the S.T.P. indicate that the shift from neritic to pelagic fauna took place gradually from late Pliensbachian to late Bajocian time. Furthermore, the similarity between faunal succession of the S.T.P. and that predicted from study of modern environments, and coincidence of these transitions with known upwelling phenomena (i.e., deposition of organic-rich shale, mass mortality events, deposition of pelagic fauna on shallow shelves) collectively suggest that faunal succession was caused by increasing amounts of trophic resources and environmental stress. As a result, neritic sedimentation was gradually suppressed, wave-resistant margins/highstand systems tracts failed to develop, pelagic sedimentation occurred by default, and sediment accumulation was outpaced by relative sea-level rise causing the South Trento Platform to "drown" at shallow depth. Introduction of trophic resources and environmental deterioration was associated with the breakup of Pangea and changing circulation patterns.

Potential Upwelling Mechanisms

As discussed above, paleoecologic deterioration of the S.T.P. is thought to have been initiated by changing regional circulation patterns during the late Pliensbachian–early Toarcian. Local and regional upwelling patterns may have been established by the interaction of oceanic currents, local margin physiography, and wind patterns (e.g., Fiuza, 1981; Jones et al., 1981; Smith, 1981). In the case of the S.T.P., paleoclimate studies suggest that wind directions during the Pliensbachian were from the north-northeast along the Southern Tethyan margin (Parrish and Curtis, 1982; Chandler et al., 1992). These winds would have been roughly parallel to the margins of the Trento Platform. If winds during the early Toarcian were similar, upwelling may have been induced at the platform margins by surface-generated flow over the submerged N.T.P. and along the partially-exposed S.T.P. (Figure 18). North to south migration of crinoidal sand waves over the N.T.P. (Masetti and Bottoni, 1978) and distribution of organic-rich shale on the eastern edge of the Alpi Feltrine, in the western half of the Belluno Basin, and in the eastern Lombardy Basin (Jenkyns et al., 1985; Jenkyns and Clayton, 1986; Jenkyns, 1988) is consistent with this hypothesis.

Upwelling during the Aalenian-Bajocian may have been induced by surface-generated flow along the platform margins, or possibly enhanced by the flow of oceanic currents through the relatively shallow platform and basin topography. As part of the related study, an ammonite belonging to the Stephanoceratid Family of Lower Bajocian age has been collected within the Vajont Oolite in the central Belluno Basin (Figure 15H). Progressive filling of the Belluno Basin during the Bajocian would have had a major impact on basin topography and may have affected local upwelling patterns through the physical restriction of southerly currents and/or the development of surface-generated flow (Figure 19). A modern analog would be the flow of the Caribbean current through the Nicaraguan Rise (Hallock et al., 1988).

Figure 18. Integrated sea-level–trophic model for the South Trento Platform drowning succession. The transition from shallow platform to pelagic lithofacies occurred gradually from Lower to Middle Jurassic time. Introduction of trophic resources followed tectonism and tilting of the S.T.P. during the late Pliensbachian (A). Ramp geometry and early Toarcian sea-level rise allowed for the establishment of a shallow carbonate depositional system (lower San Vigilio Group) that was under environmental stress (B). Paleogeographic distribution of organic-rich shale/manganese-rich limestone suggests that upwelling along the platform margins was the mechanism by which trophic resources were brought to shallow depth. This sequence was followed by sea-level fall and lowstand deposition (San Vigilio Oolite), and then by gradual sea-level rise which kept the platform at shallow depth (C) but in a eutrophic environment that favored ahermatypic suspension/detrital feeders (lower Posidonia Alpina Beds). Flooding of the platform interior and sustained/seasonal upwelling resulted in establishment of a dysaerobic environment and the deposition of "stunted" and juvenile pelagic/benthic faunas in a wave- and current-swept environment (D). Wave-resistant margins and/or highstand systems tracts failed to develop and the platform "drowned." Subsequent sea-level rise and upwelling allowed for establishment of dysaerobic to anaerobic environments over the platform interior and deposition of stromatolite- and pelagic-rich lithogacies at relatively shallow depth (E; L.A.R.). A predominantly low-energy pelagic environment was eventually established with continued subsidence and sea-level rise (F; Intermediate Member). See text for further discussion.

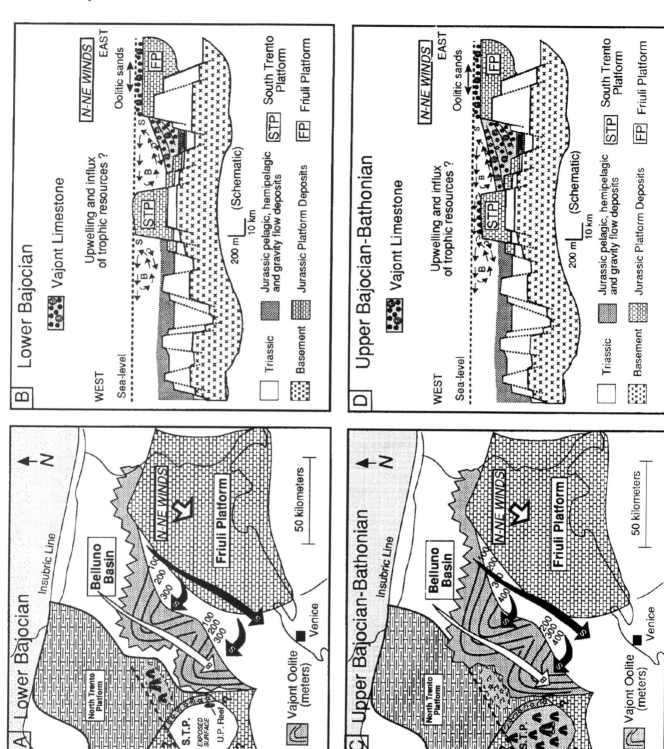

POSSIBLE IMPLICATIONS

Important differences exist between the "gradual" drowning succession observed on the S.T.P. and the "geologically instantaneous" drowning records observed on other platforms/atolls in association with anoxia and sea-level rise (e.g., Matthews et al., 1974; Jenkyns, 1991; see discussions by Hallock and Schlager, 1986 and Erlich et al., 1990). In the present study, it is proposed that rapid sea level rise and introduction of trophic resources followed tectonism and tilting of the S.T.P. Ramp geometry allowed for the establishment of a shallow carbonate depositional system (San Vigilio Group), which was under environmental stress. This sequence was followed first by a major sea-level fall, and then by a gradual sea-level rise, which kept the platform at depths favorable for some carbonate production (Posidonia Alpina Beds), but in a dysaerobic/eutrophic environment that suppressed the return of a "healthy" neritic fauna. Continued sea-level rise resulted in deposition of pelagic-rich lithofacies at shallow depth (L.A.R.). Thus the transition from shallow platform to pelagic lithofacies is observed as a gradual transition lasting from Lower to Middle Jurassic time. Importantly, the late Pliensbachian/ early Toarcian "tectono-oceanographic" event was followed by Aalenian sea-level fall, and the effects of this new circulation did not become fully apparent until sea level rose and flooded the S.T.P. during the Bajocian. If these interpretations are correct, other Lower to Middle Jurassic drowning successions may represent a similar response to trophic resources, eustatic sea-level rise, and changing circulation patterns along the proto-Atlantic–Tethyan seaway.

CONCLUSIONS

Study of Jurassic outcrops located along the eastern margin of the South Trento Platform reveals important information as to genesis of a Lower to Middle Jurassic unconformity and Lower to Upper Jurassic drowning succession.

1. Two sponge-coral-stromatoporoid patch reefs and neptunian dikes are found below a Lower to Middle Jurassic unconformity within peloidal and oolitic grainstone. Faunal and lithologic similarity with other Lower

Jurassic reef complexes suggests that these reefs are Upper Pliensbachian in age. Platform sediments located directly beneath the unconformity are brecciated, truncated, rounded, and/or fractured. Reef cavities and neptunian dikes are filled by red peloidal sediment, *P. alpina* sediment, radiaxial-fibrous calcite cement, pendant cement, crystal silt, and rounded lithoclasts. These data indicate that formation and filling of some reef cavities and neptunian dikes occurred in vadose and shallow-marine environments. Moreover, exposure-related fabrics in the platform interior and deposition of Aalenian grainstone seaward of the Upper Pliensbachian reef complex imply a small-scale type 1 sequence boundary.

2. In an east–west transect above the unconformity, transition is exhibited from platform interior facies to platform margin facies in both the "Lower" and "Intermediate" members of the Ammonitico Rosso. Stromatolite- and oncolite-rich lithologies in the Lower Ammonitico Rosso grade into thrombolitic- and grain-rich lithologies. Thin-bedded chert-rich limestone in the Intermediate Member passes into graded gravels, peloidal-rich grainstone, and pelagic-rich mudstone/wackestone. Gravels contain skeletal clasts and lithoclasts derived from underlying beds and grade into grainstone, which exhibits hummocky, low-angle, and planar cross-stratification. These beds overlie truncation surfaces, which include deep irregular excavations, rounded gutters, and gentle scours, and are interpreted as tempestites. Ammonite phragmacones and belemnites in pelagic mudstone/wackestone are encrusted by solitary corals. These data indicate that deposition occurred, in part, at depths above storm wave base (≤110–200 m) and the aragonite compensation depth during the Oxfordian.

3. Comparison of sedimentary structures, depositional geometries, and petrographic data with postulated global sea-level change shows correlation between hiatuses and major rises and falls of sea level. The drowning succes-

Figure 19. Upwelling along the South Trento Platform may have been induced or enhanced by filling of the Belluno Basin during Bajocian–Bathonian time (general paleogeography and schematic cross section modified from Bosellini et al., 1981). (A–B) Lower Bajocian—Accumulation of the Vajont Oolite in the Belluno Basin was concomitant with deposition of cross-bedded crinoidal/brachiopod grainstone (lower Posidonia Alpina Beds) along the eastern margin of the S.T.P. and microbial fabrics (L.A.R. facies) on the Alpi Feltrine (Della Bruna and Martire, 1985). Significant filling of the Belluno Basin may have induced sustained/ephemeral upwelling along the South Trento Platform through topographic restriction of boundary currents ("B") and/or through surface-generated flow ("S"). (C–D) Upper Bajocian–Bathonian—With sea-level rise and continued filling of the Belluno Basin, persistent upwelling may have been established over the South Trento Platform during deposition of the upper Posidonia Alpina Beds (not shown) and the Lower Ammonitico Rosso.

sion of the South Trento Platform consists of:
(1) a late Pliensbachian tectonic/platform tilting event and exposure of the eastern platform interior and margin; (2) deposition of an
environmentally stressed TST/HST on the
western margin (lower San Vigilio Group,
early Toarcian); (3) a small-scale type 1
sequence boundary (late Toarcian–Lower
Bajocian?); (4) deposition of a drowning
sequence that includes a lowstand wedge
(San Vigilio Oolite *s.s.* and lower Posidonia
Alpina Beds, Aalenian–Lower Bajocian?), a
transgressive systems tract (upper Posidonia
Alpina Beds, Lower–Upper Bajocian), and an
upper sequence boundary represented by an
disconformity/hardground; and (5) deposition of pelagic lithofacies as a composite condensed section (Ammonitico Rosso, Upper
Bajocian-Tithonian).

4. Drowning of the South Trento Platform
occurred gradually over time through a combination of eustatic sea-level rise and environmental change. Beginning in the late
Lower Jurassic, major faunal changes are
denoted by transition from sponge-coral-
stromatoporoid reefs to bioeroded sponge
and hermatypic coral reefs to grainstone
composed of suspension/detrital feeders and
pelagics to "stunted" and "juvenile" pelag-
ic/benthic faunas to microbial mat (stromatolite) structures. Similarity of this transition
with that predicted from study of modern
drowned carbonate platforms and oxygen-
deficient environments suggests that
increased trophic resources, oxygen depletion, and eustatic sea-level rise contributed
to the demise of the S.T.P. Initial paleoecologic deterioration was concomitant with deposition of organic-rich shale/manganese-rich
limestone in basinal settings and opening of
the proto-Atlantic–Tethyan seaway. This suggests that changes in regional circulation and
upwelling were the mechanisms by which
trophic resources were introduced to the
platform at shallow depth.

ACKNOWLEDGMENTS

This chapter is dedicated to Beth Zempolich for her
support and encouragement through the course of
this study. I would like to thank A. Bosellini, M.
Claps, C. Doglioni, B. Lehner, D. Masetti, F. Massari,
and E. Trevisani who provided logistical assistance
and information which enabled the completion of this
study. I am indebted to G. Pavia and L. Martire of the
University of Torino for identification of the
ammonite collected from the Vajont Limestone. Dr.
N. Sohl provided tentative identification of gastropods found in the Lower Ammonitico Rosso. R.N.
Ginsburg is gratefully acknowledged for pointing out
the presence of "pelagic" lithofacies in the Belize
Lagoon. M. Claps, L.A. Hardie, L. Hinnov, O.M.
Phillips, S.M. Stanley, and L. Yose provided helpful
discussion concerning the interpretation and presentation of data. M. Claps, L.A. Hardie, L. Hinnov, and
A.K. Zempolich reviewed early drafts of this manuscript. A.C. Hine, J.R. Markello, and J.F. Sarg provided critical review of the final manuscript. This study
was made possible by grants from the American
Association of Petroleum Geologists, the Geological
Society of America, Sigma Xi, Mobil Oil Company,
the Johns Hopkins University Balk Fund, and the
National Science Foundation (Grant #EAR 910510).

REFERENCES CITED

Adey, W.H., Macintyre, I.G., Stuckenrath, R., and
Dill, R.F., 1977, Relict barrier reef system off St.
Croix: Its implications with respect to late
Cenozoic coral reef development in the western
Atlantic: Third Coral Reef Symposium;
Proceedings, University of Miami, v. 2, p. 15-21.

Ager, D.K., 1965, The adaptation of Mesozoic brachiopods to different environments: Palaeogeography, Palaeoclimatology, Palaeoecology, v. 1,
p. 143-172.

Ager, D.K., 1974, Storm deposits in the Jurassic of the
Moroccan High Atlas: Palaeogeography,
Palaeoclimatology, Palaeoecology, v. 15, p. 83-93.

Aigner, T., 1982a, Calcareous tempestites: storm-dominated stratification in Upper Muschelkalk limestones (Middle Trias, SW-Germany), in Einsele, G.,
and Seilacher, A., eds., Cyclic and Event
Stratification: Springer Verlag, New York, 1982, p.
180-198.

Aigner, T., 1982b, Event stratification in Nummulite
accumulations and in shell beds from the Eocene of
Egypt, in Einsele, G., and Seilacher, A., eds., Cyclic
and Event Stratification: Springer Verlag, New
York, p. 248-262.

Aigner, T., 1985, Storm depositional systems:
Dynamic stratigraphy in modern and ancient shal-
low-marine sequences, in Friedman, G.M.,
Neugebauer, H.J., and A. Seilacher, eds., Lecture
Notes in Earth Sciences: Springer-Verlag, New
York, v. 3, 174 p.

Aigner, T., Hagdorn, H., and Mundlos, R., 1978,
Biohermal, biostromal and storm-generated
coquinas in the Upper Muschelkalk: Neues Jahrb:
Geologie u. Palaontologie Abh., v. 157, p. 42-52.

Barbujani, C., Bosellini, A., and Sarti, M., 1986,
L'oolite di San Vigilio nel Monte Baldo (Giurassico,
Prealpi Venete): Annali dell'Universita di Ferrara,
Sezione IX, Scienze Geologiche e Paleontologiche,
9(2), p 1-47.

Beauvais, L., 1977, Main characteristics of the Liassic
coral fauna from Morocco, in Third International
Coral Reef Symposium, Proceedings: University of
Miami, v. 2, p. 375-378.

Beauvais, L., 1984, Evolution and diversification of
Jurassic scleractinia: Palaeontographica Americana,
no. 54, p. 219-224.

Beccarelli Bauck, L., 1986, Stylothalamien aus dem unterjurassischen Misone-kalk der Südalpen, Italien: Palaeontographica Abt. A., v. 192, p. 1-13.

Beccarelli Bauck, L., 1988, Unter- bis mitteljurassische Karbonatformationen am Westrand der Trento-Plattform (Südalpen, Norditalien): München Geowiss. Abh., v. 13, p. 1-86.

Berger, W.H., 1969, Ecological patterns of living planktonic foraminifera: Deep-Sea Research, v. 16, p. 1-24.

Bernoulli, D., and Jenkyns, H.C., 1974, Alpine, Mediterranean, and Central Atlantic Mesozoic facies in relation to the early evolution of the Tethys, in Dott, R.H., Jr., and Shaver, R.H., eds., Modern and Ancient Geosynclinal Sedimentation: SEPM Special Publication No. 19, p. 29-160.

Bosellini, A., and Broglio Loriga, C., 1971, I Calcari Grigi di Rotzo (Giurassico inferiore, Altopiano di Asiago) e loro inquadramento nello paleogeografia e nella evoluzione tettono-sedimentaria delle Prealpi Venete: Annali dell'Universita di Ferrara, 5(1), p. 1-61.

Bosellini, A., Masetti, D., and Sarti, M., 1981, A Jurassic "Tongue of the Ocean" infilled with oolitic sands: The Belluno Trough, Venetian Alps, Italy: Marine Geology, v. 44, p. 59-95.

Bosellini, A., and Winterer, E.L., 1975, Pelagic limestone and radiolarite of the Tethyan Mesozoic: A genetic model: Geology, v. 3, p. 279-282.

Botto Micca, L., 1893, Fossili degli "Strati a *Lioceras opalinum* Rein. e *Ludwigia murchisonae* Sow," della Croce di Valpore (M. Grappa) Prov. di Treviso: Bollettino della Societa Paleontologica Italiana, v. 12, p. 143-194.

Brongersma-Sanders, M., 1957, Mass mortality in the sea, in Hedgpeth, J.W., ed., Treatise on Marine Ecology and Paleoecology: Geological Society of America Memoir No. 67, p. 941-1010.

Castellarin, A., 1966, Filoni sedimentari nel Giurese di Loppio (Trentino meridionale): Giornale di Geologia, v. 33, p. 528-554.

Castellarin, A., 1972, Evoluzione paleotettonica sinsedimentaria del limite tra "piattaforma veneta" e "bacino lombardo" a nord di Riva del Garda: Giornale di Geologia, v. 38, p. 11-212.

Chandler, M.A., Rind, D., and Ruedy, R., 1992, Pangaean climate during the Early Jurassic: GCM simulations and the sedimentary record of paleo-climate: Geological Society of America Bulletin, v. 104, p. 543-559.

Clari, P., and Marelli, C., 1983, I Calcari oolitici di S. Vigilio nei Lessini settentrionali (Prov. di Verona): Riv. Italiana Paleontologia e Stratigrafia, v. 88, p. 443-476.

Clari, P.A., Marini, P., Pastorini, M., and Pavia, G., 1984, Il Rosso Ammonitico Inferiore (Baiociano-Calloviano) nei Monti Lessini Settentrionali (Verona): Riv. Italiana Paleontologia e Stratigrafia, v. 90, p. 15-85.

Clari, P.A., Martire, L., and Pavia, G., 1987, L'Unita selcifera del Rosso Ammonitico Veronese (Alpi Meridionali): Atti II Conv. Int. F.E.A., Pergola, p. 151-162.

Dalrymple, R.W., LeGresley, E.M., Fader, G.B.J., and Petrie, B.D., 1992, The western Grand Banks of Newfoundland: Transgressive Holocene sedimentation under the combined influence of waves and currents: Marine Geology, v. 105, p. 95-118.

de Gregorio, A., 1885, Fossili del Giura-Lias (Alpiniano De Greg.) di Segan e di Valpore di Cima (Cima d'Asta e M. Grappa): Mem. R. Acc. Sc. Torino, v. 37, no. 2, p. 1-32.

Della Bruna, G., and Martire, L., 1985, La successione giurassica (Pliensbachiano-Kimmeridgiano) delle Alpi Feltrine (Belluno): Riv. Italiana Paleontologia e Stratigrafia, v. 91, p. 15-62.

Dill, R.F., Shinn, E.A., Jones, A.T., Kelly, K., and Steinen, R.P., 1986, Giant subtidal stromatolites forming in normal salinity waters: Nature, v. 324, p. 55-58.

Dominguez, L.L., Mullins, M.T., and Hine A.C., 1988, Cat Island Platform, Bahamas: An incipiently drowned Holocene carbonate shelf: Sedimentology, v. 35, p. 805-819.

Dravis, J.J., 1983, Hardened subtidal stromatolites, Bahamas: Science, v. 219, p. 385-386.

Drittenbass, V.W., 1979, Sedimentologie und Geochemie von Eisen-Mangan führenden Knollen und Krusten im Jura der Trento-Zone (ostliche Südalpen, Norditalien): Eclogae Geol. Helvetiae, v. 72, p. 313-345.

du Dresnay, R., 1971, Extension et développement des phénomenes récifaux jurassiques dans le domaine atlasique marocain, particulierement au Lias moyen: B.S.G.F., no. 1-2, p. 46-56.

du Dresnay, R., 1975, Le milieu récifal fossile du Jurassique inférieur (Lias) dans le domaine des chaînes atlasiques du Maroc: 2nd Sympos. Intern. sur les Coraux et Récifs Coralliens Fossiles. Mém. B.R.G.M. no. 89, p. 296-312.

Dunbar, R.B., 1981, Stable isotope record of upwelling and climate from Santa Barbara Basin, California, in Thiede, J., and Suess, E., eds., Coastal Upwelling: Its Sediment Record—Part B: Sedimentary Records of Ancient Coastal Upwelling: NATO Conference Series IV: Marine Sciences, v. 10B, p. 217-242.

Dunn, P.A., 1991, Cyclic stratigraphy and early diagenesis: an example from the Latemar Platform, northern Italy: Unpublished Ph.D. dissertation, The Johns Hopkins University, Baltimore, 836 p.

Erlich, R.N., Barrett, S.F., and Guo, B.J., 1990, Seismic and geologic characteristics of drowning events on carbonate platforms: AAPG Bulletin, v. 74, p. 1523-1537.

Farinacci, A., Mariotti, N., Nicosia, U., Pallini, G., and Schiavinotto, F., 1981, Jurassic sediments in the Umbro-Marchean Apennines: An alternative model, in Faranacci, A., and Elmi, S., eds., Rosso Ammonitico Symposium Proceedings: Edizioni Tecnoscienza, Rome, p. 335-398.

Fazzuoli, M., Marcucci Passerini, M., and Sguazzoni, G., 1981, Occurrence of "Rosso Ammonitico" and paleokarst sinkholes on the top of the "Marmi Formation," (Lower Liassic), Apuane Alps, Northern Apennines, in Farinacci, A., and Elmi, S., eds., Rosso Ammonitico Symposium Proceedings: Edizioni Tecnoscienza, Rome, p. 400-407.

Ferrari, A., and Manara, C., 1972, Brachiopodi del Dogger Inferiore di Monte Peller- Trentino: Giornale di Geologia, v. 38, p. 253-348.

Fiuza, A.F.G., 1981, Upwelling patterns off Portugal, in Suess, E., and Thiede, J., eds., Coastal Upwelling: Its Sediment Record—Part A: Responses of the Sedimentary Regime to Present Coastal Upwelling: NATO Conference Series IV: Marine Sciences, v. 10A, p. 85-98.

Fütterer, D.K., 1981, The modern upwelling record off northwest Africa, in Thiede, J., and Suess, E., eds., Coastal Upwelling: Its Sediment Record—Part B: Sedimentary Records of Ancient Coastal Upwelling: NATO Conference Series IV: Marine Sciences, v. 10B, p. 105-121.

Gaetani, M., 1975, Jurassic stratigraphy of the Southern Alps: A review, in Souyres, C., ed., Geology of Italy: Earth Sci. Soc. Libyan Arab Rep., Tripoli, p. 377-401.

Gaetani, M., Fois, E., Jadoul, F., and Nicora, A., 1981, Nature and evolution of Middle Triassic carbonate buildups in the Dolomites (Italy): Marine Geology, v. 44, p. 25-57.

Garrett, P., 1970, Phanerozoic stromatolites: Noncompetitive ecologic restriction by grazing and burrowing animals: Science, v. 169, 171-173.

Garrison, R.E., and Fischer, A.G., 1969, Deep-water limestones and radiolarites of the Alpine Jurassic, in Friedman, G.M., ed., Depositional Environments in Carbonate Rocks, A Symposium: SEPM Special Publication No. 14, p. 20–56.

Ginsburg, R.N., 1991, Controversies about stromatolites: Vices and virtues: Controversies in Modern Geology: Academic Press Ltd., New York, p. 25-36.

Glaser, K.S., and Droxler, A.W., 1991, High production and highstand shedding from deeply submerged carbonate banks, northern Nicaragua Rise: Journal of Sedimentary Petrology, v. 61, p. 128-142.

Glynn, P.W., 1984, Widespread coral mortality and the 1982-83 El Niño warming event: Environmental Conservation, v. 11, no. 2, p. 133-146.

Goldring, R., and Aigner, T., 1982, Scour and fill: The significance of event separation, in Einsele, G., and Seilacher, A., eds., Cyclic and Event Stratification: Springer-Verlag, New York, p. 354-362.

Hagan, G.M., and Logan, B.W., 1974, Development of carbonate banks and hypersaline basins, in Evolution and Diagenesis of Quaternary Carbonate Sequences, Shark Bay, Western Australia: AAPG Memoir No. 22, p. 61-139.

Hallam, A., 1977, Biogeographic evidence bearing on the creation of Atlantic seaways in the Jurassic, in Strangeway, D.W., ed., Paleontology and Plate Tectonics with Special Reference to the History of the Atlantic Ocean: Milwaukee Publ. Mus. Spec. Publ. Biol. Geol., v. 2, p. 23-34.

Hallam, A., 1983, Early and Mid-Jurassic molluscan biogeography and the establishment of the central Atlantic seaway: Palaeogeography, Palaeoclimatology, Palaeoecology, v. 43, p. 181-193.

Hallam, A., 1987, Radiations and extinctions in relation to environmental change in the marine Lower Jurassic of northwest-Europe: Paleobiology, v. 13, 152-168.

Hallock, P., Hine, A.C., Vargo, G.A., Elrod, J.A., and Jaap, W.C., 1988, Platforms of the Nicaraguan Rise: Examples of the sensitivity of carbonate sedimentation to excess trophic sources: Geology, v. 16, no. 12, p. 1104-1107.

Hallock, P., and Schlager, W., 1986, Nutrient excess and the demise of coral reefs and carbonate platforms: Palaios, v. 1, p. 389-398.

Haq, B.U., Hardenbol, J., and Vail, P.R., 1988, Mesozoic and Cenozoic chronostratigraphy and cycles of sea level change, in Wilgus, C.K., Hastings, B.S., Ross, C.A., Posamentier, H., Van Wagoner, J., and Kendall, C.G.St.C., eds., Sea Level Changes: An Integrated Approach: SEPM Special Publication No. 42, p. 71-108.

Hardie, L.A., and Ginsburg, R.N., 1977, Layering: The origin and environmental significance of lamination and thin bedding, in Hardie, L.A., ed., Sedimentation on the Modern Carbonate Tidal Flats of Northwest Andros Island, Bahamas: The Johns Hopkins University Studies in Geology, No. 22, p. 50-123.

Hartman, O., and Barnard, J.L., 1958, The benthic fauna of the deep basins off southern California, Parts I and II: Allen Hancock Pacific Expedition, v. 22, 297 p.

Hine, A.C., 1983, Relict sand bodies and bedforms of the Northern Bahamas: Evidence of extensive early Holocene sand transport, in Peryt, T.M., ed., Coated Grains: Springer-Verlag, Berlin, p. 116-131.

Hine, A.C., Hallock, P., Harris, M.W., Mullins, M.T., Belknap, D.F., and Jaap, W.C., 1988, Halimeda bioherms along an open seaway: Miskito Channel, Nicaraguan Rise, SW Caribbean Sea: Coral Reefs, v. 6, p. 173-178.

Hine, A.C., and Steinmetz, J.C., 1984, Cay Sal Bank, Bahamas—a partially drowned carbonate platform: Marine Geology, v. 59, p. 135-164.

Hofmann, P., 1976, Stromatolite morphogenesis in Shark Bay, western Australia, in Walter, M.R., ed., Stromatolites: Developments in Sedimentology 20: Elsevier, New York, p. 261-271.

Jenkyns, H.C., 1971a, The genesis of condensed sequences in the Tethyan Jurassic: Lethaia, v. 4, p. 327-352.

Jenkyns, H.C., 1971b, Speculations on the genesis of crinoidal limestones in the Tethyan Jurassic: Geol. Rdsch., v. 60, p. 471-488.

Jenkyns, H.C., 1980, Tethys: past and present: Proceedings of the Geologists Association, v. 91, p. 107–118.

Jenkyns, H.C., 1988, The Early Toarcian (Jurassic) anoxic event: stratigraphic, sedimentary, and geochemical evidence: American Journal of Science, v. 288, p. 101-151.

Jenkyns, H.C., 1991, Impact of Cretaceous sea-level rise and anoxic events on the Mesozoic carbonate platform of Yugoslavia: AAPG Bulletin, v. 75, p. 1007–1017.

Jenkyns, H.C., and Clayton, C.J., 1986, Black shales and carbon isotopes in pelagic sediments from the Tethyan Lower Jurassic: Sedimentology, v. 33, p. 87-106.

Jenkyns, H.C., Géczy, B., and Marshall, J.D., 1991, Jurassic manganese carbonates of central Europe and the Early Toarcian anoxic event: Journal of Geology, v. 99, p. 137-149.

Jenkyns, H.C., Sarti, M., Masetti, D., and Howarth, M.K., 1985, Ammonites and stratigraphy of Lower Jurassic black shales and pelagic limestones from the Belluno Trough, Southern Alps, Italy: Eclogae Geol. Helvetiae, v. 78, p. 299-311.

Jones, B.H., Brink, K.H., Dugdale, R.C., Stuart, D.W., Van Leer, J.C., Blasco, D., and Kelley, J.C., 1981, Observations of a persistent upwelling center off Point Conception, California, in Suess, E., and Thiede, J., eds., Coastal Upwelling: Its Sediment Record—Part A: Responses of the Sedimentary Regime to Present Coastal Upwelling: NATO Conference Series IV: Marine Sciences, v. 10A, p. 37-60.

Leckie, D., 1988, Wave-formed, coarse-grained ripples and their relationship to hummocky cross-stratification: Journal of Sedimentary Petrology, v. 58, no. 4, p. 607–622.

LeMaître, D., 1935, Description des spongiomorphides et les algues: Notes Mém. Serv. Mines Carte géol. Maroc., 34, p. 17-61.

LeMaître, D., 1937, Nouvelles recherches sur les spongiomorphides et les algues du Lias et de l'Oolithe inférieure: Notes Mém. Serv. Mines Carte géol. Maroc., 43, p. 1-27.

Leppakoski, E., 1969, Transitory return of the benthic fauna of Bornholm Basin after extermination by oxygen insufficiency: Cali. Biol. Mar., v. 10, p. 163–172.

Logan, B.W., Hofmann, P., and Gebelein, C.D., 1974, Algal mats, cryptalgal fabrics, and structures, in Evolution and Diagenesis of Quaternary Carbonate Sequences, Shark Bay, Western Australia: AAPG Memoir No. 22, p. 140-194.

Loutit, T.S., Hardenbol, J., Vail, P.R., and Baum, G.R., 1988, Condensed sections: The key to age determinations and correlation of continental margin sequences, in Wilgus, C.K., Hastings, B.S., Ross, C.A., Posamentier, H., Van Wagoner, J., and Kendall, C.G.St.C., eds., Sea Level Changes: An Integrated Approach: SEPM Special Publication No. 42, p. 183-213.

Martire, L., 1988, Eta', dinamica deposizionale e possibile organizzazione sequenziale del Rosso Ammonitico dell'Altopiano di Asiago (VI): Rend. Soc. Geol. It., v. 11, p. 231-236.

Martire, L., 1989, Analisi biostratigrafica e sedimentologica del rosso ammonitico veronese dell'Altopiano di Asiago (VI): Doctorate Thesis, Department of Earth Science, University of Torino, p. 166.

Martire, L., 1990, Nuovi dati sull'ambiente deposizionale e diagenetico della "Lumachella a Posidonia alpina": Riassunti Posters, 75º Congresso Nazionale Societa Geologica Italiana, p. 96-97.

Martire, L., 1993, New data on the depositional and diagenetic environment of the "Lumachella a Posidonia alpina" (Middle Jurassic, Trento Plateau, NE Italy): Memoire Societa Geologica Italiana, v. 45, in press.

Masetti, D., and Bianchin, G., 1987, Geologia del Gruppo della Schiara (Dolomiti Bellunesi). Suo inquadramento nella evoluzione giurassica del margine orientale della Piattaforma di Trento: Mem. Ist. Geol. Min. Univ. Padova, v. 39, 187-212.

Masetti, D., and Bottoni, A., 1978, L'Encrinite di Fanes e suo inquadramento nella paleogeografia giurassica dell'area Dolomitica: Riv. Italiana Paleontologia e Stratigrafia, v. 84, p. 169-186.

Massari, F., 1981, Cryptalgal fabrics in the Rosso Ammonitico sequences in the Venetian Alps, in Faranacci, A., and Elmi, S., eds., Rosso Ammonitico Symposium Proceedings: Edizioni Tecnoscienza, Rome, p. 435-469.

Massari, F., 1983, Oncoids and stromatolites in the Rosso Ammonitico sequences (Middle-Upper Jurassic) of the Venetian Alps, Italy, in Peryt, T.M., ed., Coated Grains: Springer-Verlag, Berlin, p. 358-366.

Massari, F., Baccelle, L.S., Ballarini, L., and Nardi, S., 1989, Aminoacidi nel Rosso Ammonitico Veronese di S. Ambrogio di Valpolicella (Verona): Bollettino della Societa Paleontologica Italiana, v. 108, p. 545-552.

Matthews, J.L., Heezen, B.C., Catalano, R., Coogen, A., Tharp, M., Natland, J., and Rawson, M., 1974, Cretaceous drowning of reefs on mid-Pacific and Japanese guyots: Science, v. 184, p. 462-464.

McMaster, R.W., and Conover, J.T., 1966, Recent algal stromatolites from the Canary Islands: Journal of Geology, v. 74, p. 647-652.

Monty, C.L.V., 1971, An autoecological approach of intertidal and deep water stromatolites: Ann. Soc. Géol. Belg. Bull., v. 94, p. 265-276.

Monty, C.L.V., 1973, Les nodules de manganese sont des stromatolithes océaniques: C.R. Acad. Sci. Paris, Sér. D, v. 276, p. 3285-3288.

Monty, C.L.V., 1977, Evolving concepts on the nature and the ecological significance of stromatolites, in Flügel, E., ed., Fossil Algae: Springer-Verlag, Berlin, p. 15-35.

Ogg, J., 1981, Middle and Upper Jurassic sedimentation history of the Trento Plateau (Northern Italy), *in* Faranacci, A., and Elmi, S., eds., Rosso Ammonitico Symposium Proceedings: Edizioni Tecnoscienza, Rome, p. 435-469.

Pallini, G., and Schiavinotto, F., 1980, Upper Carixian-Lower Domerian Sphinctozoa and ammonites from some sequences in Central Apennines, *in* Faranacci, A., and Elmi, S., eds., Rosso Ammonitico Symposium Proceedings: Edizioni Tecnoscienza, Rome, p. 521-539.

Parrish, J.T., and Curtis, R.L., 1982, Atmospheric circulation, upwelling and organic-rich rocks in the Mesozoic and Cenozoic Eras: Palaeogeography, Palaeoclimatology, Palaeoecology, v. 40, p. 31-66.

Pavia, G., Benetti, A., and Minetti, C., 1987, Il Rosso Ammonitico dei Monti Lessini Veronesi (Italia NE). Faune ad Ammoniti e discontinuita stratigrafiche nel Kimmeridgiano inferiore: Bollettino della Societa Paleontologica Italiana, v. 26, no. 1-2, p. 63-92.

Playford, P.E., and Cockbain, A.E., 1976, Modern algal stromatolites at Hamelin Pool, a hypersaline barred basin in Shark Bay, western Australia, *in* Walter, M.R., ed., Stromatolites: Developments in Sedimentology 20: Elsevier, New York, p. 389-411.

Poag, C.W., 1991, Rise and demise of the Bahama-Grand Banks gigaplatform, northern margin of the Jurassic proto-Atlantic seaway: Marine Geology, v. 102, p. 63-130.

Purdy, E.G., Pusey, W.C., III, and Wantland, K.F., 1975, Continental shelf of Belize—Regional shelf attributes, *in* Wantland, K.F., and Pusey, W.C., III, eds., Belize Shelf—Carbonate Sediments, Clastic Sediments, and Ecology: AAPG Studies in Geology No. 2, p. 1-52.

Roberts, H.H., Phipps, C.V., and Effendi, L., 1987, Halimeda bioherms of the eastern Java Sea, Indonesia: Geology, v. 15, p. 371–374.

Roberts, H.H., Rouse, L.J., Walker, N.D., and Hudson, J.H., 1982, Cold-water stress in Florida Bay and northern Bahamas: A product of winter cold-air outbreaks: Journal of Sedimentary Petrology, v. 52, p. 145-155.

Rosenberg, R., 1977, Benthic macrofaunal dynamics, production, and dispersion in an oxygen-deficient estuary of west Sweden: Journal of Experimental Marine Biology and Ecology, v. 26, p. 107–133.

Sageman, B.B., Wignall, P.B., and Kauffman, E.G., 1991, Biofacies models for oxygen-deficient facies in epicontinental seas: Tool for paleoenvironmental analysis, *in* Einsele, G., Ricken, W., and Seilacher, A., eds., Cycles and Events in Stratigraphy: Springer-Verlag, New York, p. 542–564.

Sarg, J.F., 1988, Carbonate sequence stratigraphy, *in* Wilgus, C.K., Hastings, B.S., Ross, C.A., Posamentier, H., Van Wagoner, J., and Kendall, C.G.St.C., eds., Sea Level Changes: An Integrated Approach: SEPM Special Publication No. 42, p. 155-181.

Schlager, W., 1974, Preservation of cephalopod skeletons and carbonate dissolution on ancient Tethyan sea floors, *in* Hsü, K.J., and Jenkyns, H.C., eds.,

Pelagic Sediments on Land and Under the Sea: International Association of Sedimentologists Special Publication 1, p. 49-70.

Schlager, W., 1981, The paradox of drowned reefs and carbonate platforms: Geological Society of America Bulletin, v. 92, p. 197-211.

Schlager, W., 1989, Drowning unconformities on carbonate platforms, *in* Crevello, P.D., Wilson, J.L., Sarg, J.F., and Read, J.F., eds., Controls on Carbonate Platform and Basin Development: SEPM Special Publication No. 44, p. 15-25.

Scholle, P.A., and Kling, S.A., 1972, Southern British Honduras: Lagoonal coccolith ooze: Journal of Sedimentary Petrology, v. 42, no. 1, p. 195-204.

Sepkoski, J.J., 1989, Periodicity in extinction and the problem of catastrophism in the history of life: Journal of the Geological Society, v. 146, p. 7-19.

Seyfried, H., 1981, Genesis of "regressive" and "transgressive" pelagic sequences in the Tethyan Jurassic, *in* Faranacci, A., and Elmi, S., eds., Rosso Ammonitico Symposium Proceedings: Edizioni Tecnoscienza, Rome, p. 547-579.

Sloss, L.L., 1963, Sequences in the cratonic interior of North America: Geological Society of America Bulletin, v. 64, p. 93-113.

Smith, P.L., 1983, The Pliensbachian ammonite Dayiceras dayiceroides and early Jurassic paleogeography: Canadian Journal of Earth Science, v. 20, p. 86-91.

Smith, P.L., and Tipper, H.W., 1986, Plate tectonics and paleobiogeography: Early Jurassic (Pliensbachian) endemism and diversity: Palaios, v. 1, p. 399-412.

Smith, R.L., 1981, Circulation patterns in upwelling regimes, *in* Suess, E., and Thiede, J., eds., Coastal Upwelling: Its Sediment Record—Part A: Responses of the Sedimentary Regime to Present Coastal Upwelling: NATO Conference Series IV: Marine Sciences, v. 10A, p. 13-35.

Stanley, G.D., 1988, The history of early Mesozoic reef communities: A three-step process: Palaios, v. 3, p. 170-183.

Sturani, C., 1964, La successione delle faune ad Ammoniti nelle formazioni medio-giurassiche delle Prealpi Venete occidentali (Regione tra il lago di Garda e la valle del Brenta): Mem. Ist. Geol. Min. Univ. Padova, v. 24, p. 1-63.

Sturani, C., 1971, Ammonites and stratigraphy of the "Posidonia alpina" beds of the Venetian Alps (Middle Jurassic): Mem. Ist. Geol. Min. Univ. Padova, v. 28, p. 1-190.

Thiede, J., 1981, Skeletal plankton and nekton in upwelling water masses off northwestern South America and northwest Africa, *in* Suess, E., and Thiede, J., eds., Coastal Upwelling: Its Sediment Record—Part A: Responses of the Sedimentary Regime to Present Coastal Upwelling: NATO Conference Series IV: Marine Sciences, v. 10A, p. 183-207.

Thiede, J., and Suess, E., 1981, eds., Coastal Upwelling: Its Sediment Record—Part B:

Sedimentary Records of Ancient Coastal Upwelling: NATO Conference Series IV: Marine Sciences, v. 10B, p. 610.

Trevisani, E., 1991, Il Toarciano-Aaleniano nei settori centro-orientali della Piattaforma di Trento (Prealpi Venete): Riv. Italiana Paleontologia e Stratigrafia, v. 97, p. 99-124.

Weissert, H.J., and Bernoulli, D., 1985, A transform margin in the Mesozoic Tethys: Evidence from the Swiss Alps: Geol. Rdsch., v. 73, p. 665-679.

Wells, J.W., 1934, A new species of calcisponge from the Buda Limestone of Central Texas: Journal of Paleontology, v. 8, 167-168.

Wells, J.W., 1953, Mesozoic invertebrate faunas of Peru, part 3, Lower Jurassic corals from the Arequipa region: American Museum Novitates, no. 1631, 14 p.

Williams, L. A., and Reimers, C., 1983, Role of bacterial mats in oxygan-deficient basins and coastal upwelling regimes: preliminary report: Geology, v. 11, p. 267–269.

Winterer, E., and Bosellini, A., 1981, Subsidence and sedimentation on Jurassic passive continental margin, Southern Alps, Italy: AAPG Bulletin, v. 65, p. 394-421.

Winterer, E., Metzler, C.V., and Sarti, M., 1991, Neptunian dykes and associated breccias (Southern Alps, Italy and Switzerland): Role of gravity sliding in open and closed systems: Sedimentology, v. 38, p. 381-404.

Zempolich, W.G., 1991, The dilemma of gaps in carbonate stratigraphic sequences: a case history from the Jurassic of the Venetian Alps, Italy: AAPG Bulletin, v. 75, p. 700.

Chapter 4

Timing of Deposition, Diagenesis, and Failure of Steep Carbonate Slopes in Response to a High-Amplitude/ High-Frequency Fluctuation in Sea Level, Tongue of the Ocean, Bahamas

G. Michael Grammer
Robert N. Ginsburg
Comparative Sedimentology Laboratory
University of Miami/RSMAS
Miami, Florida, U.S.A.

Paul M. Harris
Chevron Petroleum Technology Company
La Habra, California, U.S.A.

ABSTRACT

Sequence stratigraphic interpretations of carbonate platform margins are based to a large degree on concepts of variable timing and nature of deposition relative to fluctuations in sea level. Quaternary platform margins, such as those found in the Bahamas, provide a unique opportunity to calibrate the sedimentary record because of the well-constrained nature of sea-level history during this period. Detailed observations and sampling from a research submersible combined with high-resolution radiocarbon dating in the Tongue of the Ocean, Bahamas, have enabled us to document variations in deposition along the upper parts of the marginal slope during the most recent rise in sea level.

We have found that the steep marginal slopes around the Tongue of the Ocean record deposition during the early rise of sea level following the last lowstand some 18–21 Ka. Coarse-grained skeletal sands, gravel, and boulders derived from reefs growing along the overlying escarpment were deposited on slopes of 35–45° and cemented in place within a few hundred years. Deposition by rockfall and grainflow resulted in a series of elongate lenses oriented parallel to the slope. These lenses are generally less than 0.5 m thick and pinch out downslope within tens of meters. Repeated deposition and cementation produced slope deposits that are both laterally discontinuous and internally heterogeneous. Radiocarbon dating of skeletal components and cements indicate that active deposition on the slopes ceased approximately 10,000 years ago as sea level rose above the escarpment and began to flood the top of the Great Bahama Bank. Fine-grained, nonskeletal sands and muds derived from the platform are presently bypassing these slopes and are deposited downslope as a wedge of sediment with slope declivities of 25–28º.

Cracks and slide scars are a common feature of the steep-cemented slopes. The cracks are a few centimeters wide and may extend for tens of meters across the slope with an arcuate, convex-up expression. The slide scars are generally a few meters wide by several meters long and cut back into the slope a few meters to less than 1 m, although one large example is 30 m wide, extends downslope for 75 m, and has exposed 10 m of the interior of the slope. Transects downslope from slide scars show that large blocks of the slope, some in excess of 10 m across, have been transported for tens or hundreds of meters downslope. The release and transport of such blocks may be one mechanism by which turbidity currents are initiated in deeper slope environments.

INTRODUCTION

The upper marginal slopes or foreslopes of carbonate platforms are an important transitional zone between shallow-water platform carbonates and deeper water basinal and distal slope deposits and may contain significant reservoirs of hydrocarbons (Enos, 1977; Enos and Moore, 1983; Krueger and North, 1983; Minero, 1991) or metallic ores (Cruickshank and Rowland, 1983). Understanding of the depositional and early diagenetic processes operating along foreslopes is an integral part of evaluating the evolution of carbonate platforms and may be a key to the interpretation of inclined deposits (clinoforms/clinothems) often observed in outcrop and seismic profiles (e.g., Sarg, 1988).

Despite numerous studies involving steep (30–45º) carbonate slopes in both the ancient (e.g., Yurewicz, 1977; Playford, 1980; Playford et al., 1989; Bosellini, 1984; Haddad et al., 1984; Devaney et al., 1986; Enos, 1986; Ward et al., 1986; Harris, 1988; Garber et al., 1989; Kenter, 1990; and Kenter and Campbell, 1991) and the modern (e.g., Moore et al., 1976; Land and Moore, 1977; James and Ginsburg, 1979; and Mullins, 1983), there are still fundamental questions regarding the timing and processes of formation. One of the most elementary questions is whether 30–45º slopes can truly represent primary depositional slopes or whether the high declivities are due to some postdepositional modification, such as compaction, pressure dissolution, or simple dewatering of the sediment (e.g., Devaney et al., 1986). Additional questions, such as what depositional and/or diagenetic processes are responsible for the formation of the steep slopes and the timing of formation relative to fluctuations in sea level have yet to be fully explained.

Recent work in the southern portion of the Tongue of the Ocean, Bahamas, has provided valuable insight into how and when steep carbonate slopes are formed. The Tongue of the Ocean provides an unique opportunity to study modern carbonate foreslopes for several reasons. First, we know the slopes are forming in a tectonically stable (Dietz et al., 1970; Mullins and Lynts, 1977; Burke, 1988), humid subtropical environment where the more recent fluctuations of sea level are relatively well defined (e.g., Fairbanks, 1989). Second, the physical configuration of the Tongue of the Ocean allows for an examination of both windward and leeward margins rimming the same basin. Third, we know the distribution of surface sediments on the surrounding platform (Palmer, 1979) and can therefore trace the origin of foreslope sediments.

Observations and sampling using a research submersible in the southern portion (or cul-de-sac) of the Tongue of the Ocean has shown that the foreslopes on both windward and leeward margins are characterized by a consistent overall morphology (Grammer et al., 1990). The profile is dominated by a near-vertical escarpment extending down from approximately 50 to 140 m, followed by steeply dipping (35–45º) hard-rock surfaces that may extend down to depths greater than 365 m. These steeply dipping slopes are of primary interest because of their similarity to some of the inclined beds or clinoforms often observed in outcrop or seismic profiles of fossil platform margins.

The primary goals of this chapter are twofold: first, to document the general morphologic features characteristic of the upper slopes around the Tongue of the Ocean; and second, to discuss the mechanisms responsible for the development and subsequent failure of the steep slopes and the timing of formation relative to Quaternary fluctuations of sea level.

METHODS

A total of 134 dives at nine locations were made with the two-person research submersible *Delta* around the southern portion or cul-de-sac of the Tongue of the Ocean (Figure 1). The cul-de-sac was chosen because of the opportunity to evaluate fore-slope characteristics from both windward (east-facing) and open leeward (west-facing) margins of the same basin. The southernmost end of the Tongue of the Ocean (dive sites Thunder Channel and Blossom Channel) is tidal dominated and was given only cursory examination. Observations and measurements of morphologic features to depths of 365 m were documented by videotape and still photography. Slope angles were measured using a hand-held inclinometer from inside the submersible.

A total of 171 rock and sediment samples were collected from eight of the dive sites (Figure 1). Samples of hard rock were dislodged from the cemented slope with small charges of explosives and were retrieved with a mechanical arm attached to the outside of the submersible. Rock samples from 3.5 m into the slope were collected from zones where slide failure had exposed the interior of the cemented slope. All rock samples were slabbed to examine the texture and fabric of the rock and to select representative samples for thin-sectioning. Unconsolidated sediment (sand fraction) was sampled using a battery-powered suction sampler mounted on the submersible. The sediment samples were split, impregnated with polyester resin, and cut into thin-sections. A minimum of 400 points were counted per thin-section to ensure statistical relevance of compositional data following

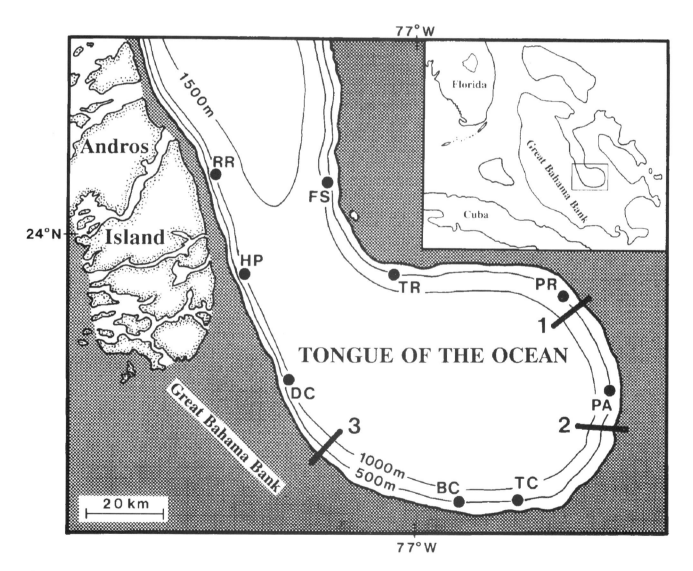

Figure 1. Map showing location of study area, submersible dive sites, and location of seismic profiles (lines 1–3) in the southern portion of the Tongue of the Ocean, Bahamas. Depth in meters. RR—Rock Range; HP—High Point Cay; DC—Dolly Cay; BC—Blossom Channel; TC—Thunder Channel; PA—Palmer Point; PR—Privitt Place; TR—Twin Rocks; FS—False Starter. Rock and/or sediment samples collected from all dive sites except Thunder Channel (see Grammer and Ginsburg, 1992, for exact locations).

guidelines established by Plas and Tobi (1965). The mineralogy of cements was determined through the use of Clayton Yellow stain as described by Choquette and Trusell (1977) and verified by x-ray diffraction.

Samples of skeletal components and cements were extracted by microdrill for radiocarbon age analysis at the NSF Accelerator Facility at the University of Arizona. Radiocarbon age dates were determined by accelerator mass spectrometry (AMS), which differs from conventional radiocarbon dating in that AMS measures the actual number of ^{14}C atoms in a sample rather than the number of disintegrations per minute. Because of this difference, AMS dates can be obtained from samples as small as 5–10 mg, as opposed to the 30+ g required for conventional radiocarbon dating. This not only limits the potential for sampling error but also enables dates to be obtained from small samples that were previously undatable through conventional radiocarbon dating techniques (see, for example, Rucklidge, 1984, and Linick et al., 1986). The original mineralogy of skeletal components was verified by x-ray diffraction prior to dating.

High-resolution seismic profiles were obtained at three sites using boomer and alternate sparker systems (Figure 1). Power levels ranged from 200–400 joules for the boomer and 200–800 joules for the sparker; filter range varied from 200 Hz to 3.5 kHz. The transducer and hydrophone array was towed at a constant speed of 5–6 knots and positioning of the ship was determined by a combination of Loran A, fathometer, and precision depth recorder. Water-depth conversions from two-way travel time sections were made using a seismic velocity of 1500 m/s in water. Unconsolidated sediment thickness was calculated using the seismic velocity of 1650 m/s calculated for south Florida Holocene shelf sediments (Enos and Perkins, 1977).

GENERAL CHARACTERIZATION OF FORESLOPES

Observations of the slopes around the Tongue of the Ocean indicate that the upper slope, or foreslope, is characterized by a distinctive morphology that may be subdivided into four zones (Figure 2).

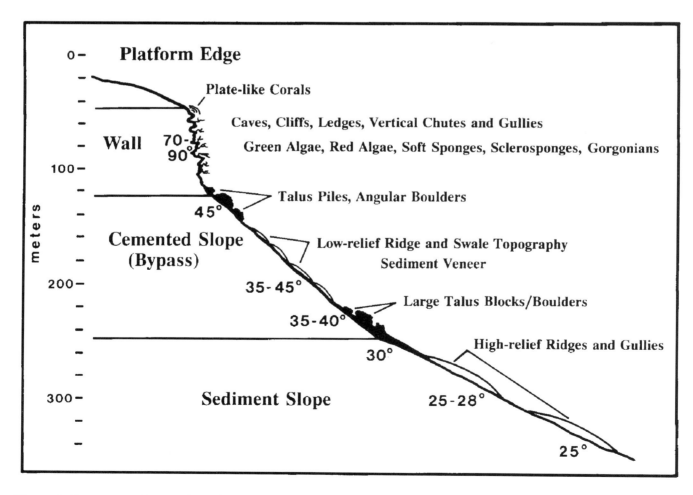

Figure 2. Representative profile of upper slope environments around the cul-de-sac of the Tongue of the Ocean. Profile was constructed from a synthesis of data accumulated during a total of 134 dives during 1987, 1988, and 1990. No vertical exaggeration. Modified from Grammer and Ginsburg (1992).

Platform Edge

The platform edge, which consists of a Holocene sediment wedge, is situated in shallower water bankward of one or more areas of deep (40–60 m) reef growth. These reefs, which are best developed along the windward (east-facing) margin, are dissected by a series of sediment chutes that frequently extend to the marginal escarpment. Seismic reflection profiling reported by Palmer (1979) indicates that these Holocene sediments may be as much as 24 m thick along the open leeward edge of the platform (Figure 3). The sand-sized sediment of the platform edge consists predominantly of ooids, peloids, aggregate grains, and coralgal sands whose distribution varies depending upon the windward or leeward positioning of the margin around the cul-de-sac (Palmer, 1979). The sediment along open leeward margins (e.g., Palmer Point) consists predominantly of nonskeletal (ooids, peloids) sands, whereas along windward margins (e.g., Dolly Cay) they are dominated by skeletal sands (Palmer, 1979).

Wall

The wall is a near-vertical (70–90°) escarpment that begins in water depths of 40–60 m and extends to average depths of 130–140 m. The wall is characterized by numerous pits, caves, ledges, and vertical overhangs that vary in size from a few centimeters to tens of meters that give the surface an overall impression of "karsted" limestone (Figure 4A). Unconsolidated sediment is present on nearly all horizontal and subhorizontal surfaces of the wall (Figure 4B).

The surface of the wall is colonized by a variety of living organisms including deep-water corals (e.g., plate-like *Montastrea annularis* and *Agaricia* spp.), encrusting coralline algae, boring sponges, sclerosponges, and at least three species of *Halimeda* (*H. opuntia*, *H. copiosa*, and *H. cryptica*). The abundance and diversity of organisms is noticeably greater near the top (50–75 m water depth) of the wall at windward dive sites as compared to leeward sites, and decreases markedly with depth. Figure 5 illustrates the vertical distribution of the major living organisms derived from observations around the entire cul-de-sac.

Our observations suggest that the wall is capable of producing copious amounts of silt- to gravel-sized sediment through both biologic production (primarily *Halimeda* spp.) and bioerosion. Some of this sediment is locally trapped in depressions and along horizontal surfaces of the wall, while the remainder (possibly a major proportion) is transported farther downslope.

The base of the wall is defined by an abrupt change in slope that is often marked by the presence of near-horizontal ledges or terraces and flat-floored caves. These ledges and caves, which are laterally continuous for several tens of meters and extend back into the wall for approximately 1–5 m, are especially well developed along the windward margin. The similarity of these caves and ledges to "marine notches" formed along coasts at or near present-day sea level (Pirazzoli, 1986) suggests that they may mark a previous lowstand of sea level. Large talus blocks, some up to 30 m high, are frequently found at the base of the wall. Many of these blocks consist of plate-like and massive corals (observed from the submersible), indicating that they were probably derived through mass-wasting of the wall above.

Cemented Slope

The cemented slope extends from the base of the wall to depths exceeding 365 m and is characterized by a well-lithified surface, steep slope angles, and internal bedding (Figure 4C,D). Only a thin veneer (generally less than 1–2 cm) of unconsolidated sediment is present, primarily in localized topographic lows. The slope declivity of the cemented slope averages about 35–45°, but may be up to 55–60° or more, especially near the base of the wall (Figure 2).

On the lower portions of the cemented slope, below depths of approximately 220–240 m, the surface varies from smooth to irregular and consists of a series of low-amplitude ridges and swales oriented parallel (i.e., dip-parallel) to the slope (Figure 4C). These ridges average 1–3 dm in height, are a few meters wide, and extend up and down the slope for a few meters to several tens of meters.

In localized areas of the lower cemented slope, there are elongate lenses of talus debris perched on the surface (Figures 4E,F). These trains of sediment,

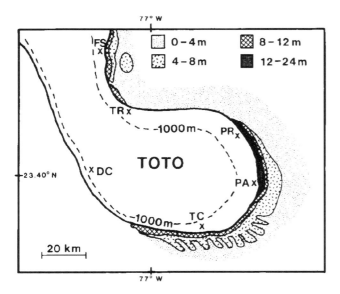

Figure 3. Isopach map of Holocene sediment around leeward (west-facing) and tidal-dominated (southernmost) margins of the southern Tongue of the Ocean. The maximum thickness of Holocene sediment on leeward margins is as much as 24 m. DC—Dolly Cay; TC—Thunder Channel; PA—Palmer Point; PR—Privitt Place; TR—Twin Rocks; FS—False Starter. Modified from Palmer (1979).

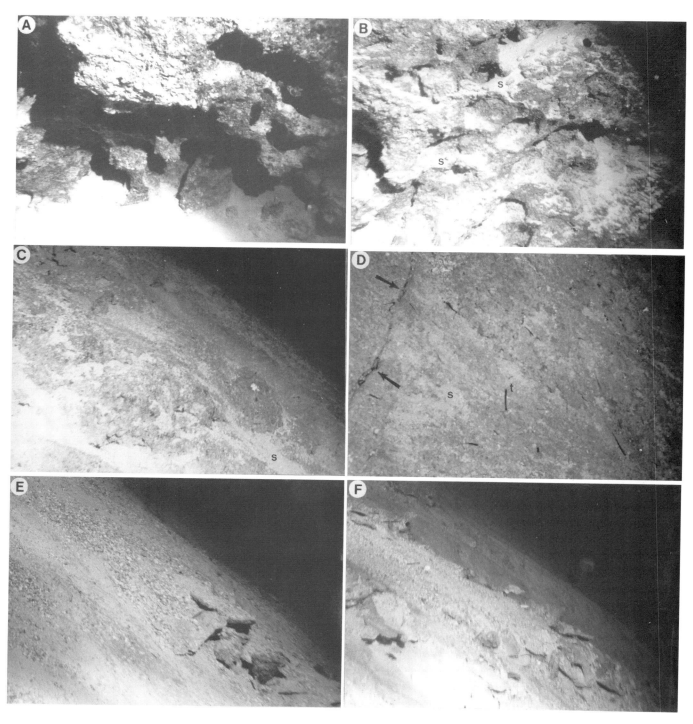

which are typically cemented in place, consist of large blocks or piles of rubble on the downslope end, with coarse-grained sands, gravels, and boulders backed-up behind (i.e., upslope). Most of the sediment is angular and/or elongate in shape (i.e., *Halimeda* flakes, platelike corals, and angular boulders) and tends to fine in an upslope direction. In some of the lenses, platelike slabs are imbricated in an upslope direction. The dimensions of these sediment lenses are very similar to the low-amplitude ridges described above. Upslope, the ridges increase in height to an average of 3–5 dm and the overall surface becomes more irregular with local pits and

depressions (centimeter to decimeter scale), while the ridges appear to coalesce and talus accumulates in topographic lows.

The uppermost areas of the cemented slope, from depths of 120–130 m to approximately 220 m, often exhibit an irregular morphology of rounded to elliptical topographic highs ("moguls") separated by trains of unconsolidated sediment in adjacent lows that give the slope the appearance of a mogul field on an alpine ski slope. Near the base of the wall, the moguls are nearly spherical in shape and range in height from a few decimeters to more than 1 m. Downslope, the moguls become progressively more elliptical or

Figure 4. (A) Photograph taken from submersible showing the lower portion of the wall and the characteristic "karsted" appearance. Horizontal dimension is approximately 2.5 m. High Point Cay, 140 m (460 ft). (B) Photograph taken from the submersible showing small pits and ledges on the surface of the wall. Sediment (s) accumulates on virtually all horizontal and subhorizontal surfaces of the wall giving the appearance of a "dusting of snow." Horizontal dimension is approximately 2 m. High Point Cay, 130 m (426 ft). (C) Photograph of the lower portion of the cemented slope showing irregular rock surface partially obscuring low-amplitude ridge and swale topography. Ridges are oriented parallel to slope (i.e., dip-parallel). A thin veneer of sediment (s) tends to accumulate in topographic lows. Horizontal dimension is approximately 5 m. False Starter, 240 m (787 ft). (D) Photograph showing the typical hard-rock or pavement-like surface of the cemented slope with only a thin veneer of localized sediment (s). Observations suggest that because of the relatively fine grain size and steep declivity of the slopes, most sediment is presently bypassing the cemented slopes. Cracks in the rock surface are relatively common (see arrows at left). These cracks are generally a few centimeters wide and may extend across the slope, parallel to contours, for up to 75 m. The majority of the cracks have an arcuate, convex-up expression on the surface of the slope. Blades of the turtle grass *Thalassia* (t) derived from the platform identified for scale. Horizontal dimension is 2–2.5 m. Thunder Channel, 250 m (820 ft). (E) Photograph taken from the submersible showing a localized lens of sediment on the cemented slope. The sediment lens consists of talus slabs (mostly platelike coral) with coarse sand (primarily *Halimeda* flakes) dammed up behind the blocks. As with most of the other lenses on the cemented slope, there is an upslope decrease in grain size. Horizontal dimension is approximately 5 m. Twin Rocks, 274 m (900 ft). (F) Photograph showing a series of linear sediment/talus ridges oriented parallel to slope (i.e., dip parallel). Ramming of the submersible into these apparently loose sediment lenses confirms that the majority of them are well cemented. Repeated deposition and subsequent cementation of similar sediment lenses leads to an amalgamation of linear deposits that are laterally discontinuous in both strike and dip direction. Horizontal dimension is approximately 5–7 m. Twin Rocks, 305 m (1000 ft).

elongate in shape and gradually merge into the ridge and swale topography described above. The surfaces of the moguls, especially in the depth range of 120 m to approximately 160–170 m, are typically veneered with encrusting coralline algae along with lesser amounts of bryozoans and serpulid worm tubes. In addition to the moguls, large promontories project out from the upper slope like large proboscises. These promontories range from approximately 1–10 m in height, and are generally several meters wide and extend downslope for several meters to tens of meters.

Arcuate, convex-up cracks that run parallel to the contours of the slope are relatively common on the surface of the cemented slope and may represent areas of incipient slope failure. The cracks were observed at five of the dive sites and are most common on the 35–45° portions of the cemented slope. These cracks are a few centimeters wide (Figure 4D) and have been traced laterally across the slope for as much as 75 m.

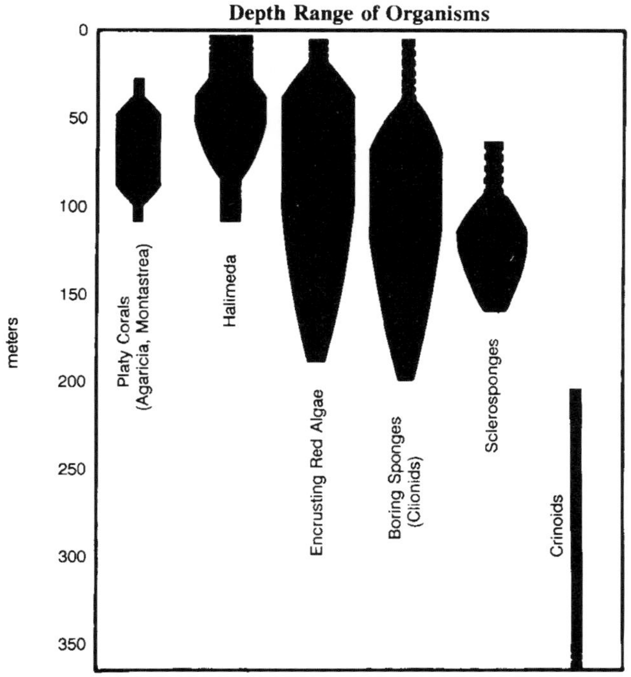

Depth Range of Organisms

Figure 5. General distribution of major living organisms that may contribute to the sediment record through either growth (accretion, encrustation) and/or production of sediment (silt-boulder size). As can be seen from the depth distribution, the majority of the organisms are found living on the wall. Encrusting coralline algae, boring sponges, and sclerosponges extend to the upper portions of the cemented slope and stalked crinoids were observed down to the limit of our dives. The abundance of organisms is based upon observations made during submersible dives coupled with an analysis of videotape documentation and indicates the *relative* abundance of organisms vertically with no attempt to compare actual species numbers.

Evidence for localized slide failure of the steep slopes is also relatively common and was observed at all dive sites with the exception of Dolly Cay. These areas are arcuate in outline (convex upslope) and are generally a few meters wide by several meters long in a downslope direction. The majority of the failure zones have incised only a meter or less of the slope but larger areas have exposed up to several meters of the interior of the cemented slope. The best-developed and largest zones of slope failure were observed at Privitt Place where the largest slide scar is up to 30 m across, extends downslope for 75 m, and has exposed up to 10 m of internal slope deposits (Figure 6A). A slightly smaller scar, exposing 3–3.5 m of the interior of the slope, was found approximately 300 m away from and at the same depth horizon as the larger slide. Observation and mapping of slope deposits from these natural exposures indicates that the cemented slope is internally composed of lenticular beds that average 1–5 dm in thickness and tend to pinch out within tens of meters in dip direction (Figure 6A,B).

Samples from the interior of the slope were obtained from the smaller of the two large slides at Privitt Place and at a smaller failure zone at Rock Range (Figure 1). The samples at Privitt Place were obtained from as much as 3–3.5 m beneath the surface of the slope while the samples from Rock Range are from a maximum of 2–2.5 m within the slope. Samples from the surface of the cemented slope were collected at Privitt Place, Palmer Point, Rock Range, Dolly Cay, and Blossom Channel (Figure 1).

Sediment Slope

The sediment slope consists of a wedge of unconsolidated sediment with slope angles of 25–28° that onlaps the base of the cemented slope. This sediment wedge exhibits varying surface morphologies ranging from a smooth sediment blanket to a series of large sediment ridges and troughs. These ridges and troughs are up to several meters high and wide and may extend downslope (i.e., dip parallel) for many tens of meters. Active bioturbation in this zone is indicated by the presence of numerous tracks, trails, and burrow mounds on the surface of the unconsolidated sediment.

The sediment consists of a mixture of skeletal and nonskeletal sands (dominated by *Halimeda* and hardened peloids, respectively) as well as varying amounts of carbonate mud. Compositional analysis of the sand fraction along the slopes of both windward and leeward margins indicates that it is derived from both the wall and the outermost margin of the platform with only minor contribution from the interior of the platform (Grammer and Ginsburg, 1992). Available sampling methods prevented recovery of the mud fraction but analyses of sediment cores obtained farther basinward in the cul-de-sac (Droxler and Schlager, 1985) indicate that it probably consists of carbonate mud derived from the platform with subordinate amounts of planktonic material.

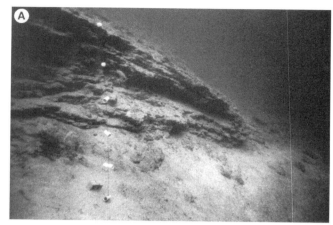

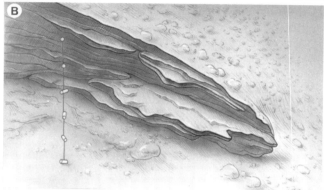

Figure 6. (A) Photograph taken from submersible in ambient light at a depth of 160 m (525 ft) showing the internal bedding of the cemented slope. The interior of the slope has been exposed through localized slide failure. Close examination of the beds from the submersible indicate that they are a few decimeters to approximately 0.5 m thick and that many of them exhibit a lenticular geometry (see also 6B). The dip of beds (38–40°) is consistent with the present surface of the cemented slope and represents the primary depositional slope angle for these deposits. Scale increments = 1 m. Privitt Place, 160 m (525 ft). (B) Line drawing of the exposed beds in the cemented slope shown in the photograph of 6A. Note lenticular nature of beds and how they pinch out within several meters in a dip direction. Drawing is based on submersible observations and bed mapping during six dives combined with black-and-white photos and videotape footage. Illustration by Barbara Cousins.

Variation Between Windward and Leeward Margins

Our observations in the Tongue of the Ocean indicate that there are distinctive patterns of Holocene sediment accumulation related to the windward and leeward positioning of the margins. The most striking feature is the variable thickness in the wedge of unconsolidated, fine-grained (sand-mud) sediment onlapping the cemented slope. Leeward transects in the submersible show a significantly higher upper limit of sediment onlap (depth about -240 m) than the windward margin (depth >360 m) suggesting that the thickness of platform-derived Holocene sediment is considerably greater along the leeward margin. Variation in the thickness of Holocene sediment is confirmed by seismic reflection profiling indicating that the wedge of sediment may be up to 80 m thick along the leeward margin (Figure 7), whereas it is less than 20 m thick on the windward side. These variations in sediment thickness are presumably a result of greater amounts of sediment being swept off the open leeward platform, and are similar to results reported by Wilber et al. (1990) for the leeward margin of Great Bahama Bank. Evidence for fine-grained sediment transport off the leeward margin was observed during a few dives at Privitt Place (Figure 1) where carbonate mud was being actively swept over the escarpment in response to 20 knot winds from the east-southeast that had been sustained for 4–5 hours (Grammer and Ginsburg, 1992). The sand-sized fraction on the leeward side of the bank is probably only mobilized during storms as suggested by Hine et al. (1981).

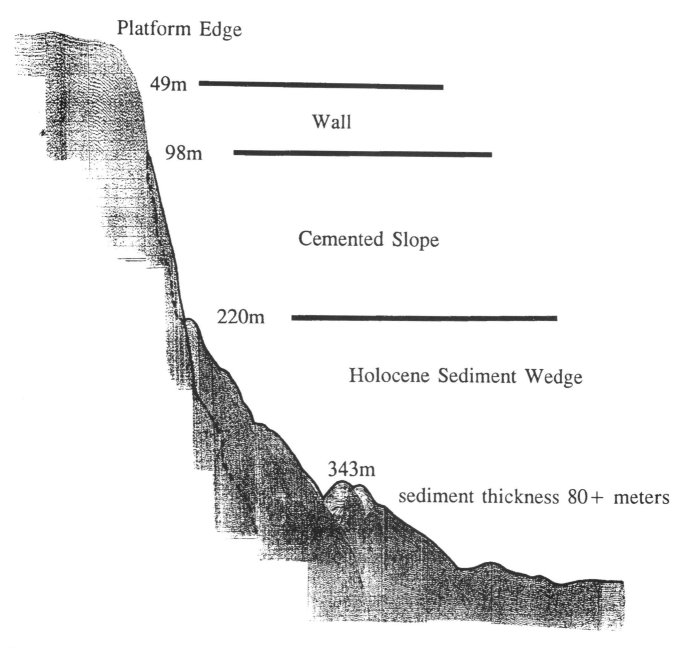

Figure 7. High-resolution seismic profile (line 1 shown in Figure 1) from the leeward margin of the Tongue of the Ocean. Onlap of Holocene sediment wedge is marked by a distinct decrease in slope declivity compared with the updip cemented slope. Thickness of sediment wedge is in excess of 80 m.

PETROGRAPHY AND COMPOSITION OF STEEP SLOPE LIMESTONES

Interior of Cemented Slope

The rocks from the interior of the cemented slope generally consist of well- to very well-indurated, poorly sorted, skeletal packstones–grainstones (Figure 8A–D). The texture is locally variable to a wackestone. Grain size in most of the samples varies from silt to pebble size but clasts of coral up to 30 cm long are present in some hand samples. Many of the silt-sized (20–40 µm) fragments are faceted and are identifiable as chips excavated by boring *Clionid* sponges. Skeletal components make up 94% (by thin-section point-count) of the sand and coarser fraction.

Compositionally, the rocks are dominated by *Halimeda* (48%), with lesser amounts of coralline algae (13%), coral (11%), mollusks (8%), benthic foraminifera (5%), bryozoans (3%), serpulid worm tubes (3%), echinoderms (2%), and planktonic foraminifera (2%). The percentages of coral and mol-

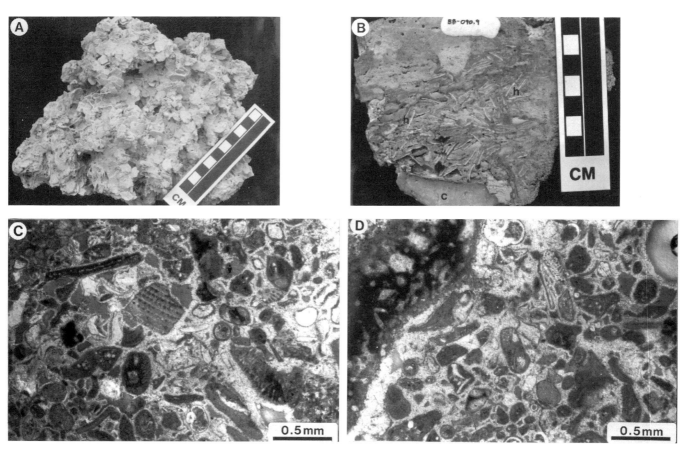

Figure 8. (A) Well-indurated to somewhat friable *Halimeda* packstone from the interior of the cemented slope. The *Halimeda* plates are up to 15–17 mm in diameter and are probably from the deep-water species *H. copiosa*. Note the lack of any preferential alignment of the plates. Visual porosity is estimated at 20–30%. Privitt Place, 168 m (550 ft). (B) Slab of well-indurated to friable (bottom), poorly sorted skeletal packstone from the interior of the cemented slope. Grain size of skeletal components varies from silt to pebble size. Thin-section analysis indicates that 70–80% of the skeletal fragments are angular to subangular or elongate in shape. The thin white lines (h) are *Halimeda* plates (probably from the deep-water species *H. copiosa*). Portion of plate-like coral (*Agaricia* spp.) is at bottom (c). Rock Range, 152 m (500 ft). (C) Thin-section photomicrograph (plane-polarized light) of skeletal grainstone from the interior of the cemented slope. Sample is cemented primarily with abundant isopachous bladed Mg-calcite cement, which is volumetrically the most important type of cement in skeletal packstones and grainstones of the interior of the cemented slope. Skeletal grains are dominated by *Halimeda*, mollusks, and coralline algae. The majority of the rounded grains are fragments of *Halimeda* plates. Rock Range, 152 m (500 ft). (D) Thin-section photomicrograph (plane polarized light) of skeletal grainstone from the interior of the cemented slope. Sample is cemented primarily with isopachous bladed Mg-calcite. The poor sorting and variation in grain size from fine sand to pebble size is typical in samples from the interior of the cemented slope. Note also how the majority of skeletal fragments are angular to subangular or elongate in shape. Rock Range, 152 m (500 ft).

lusk fragments determined by point-counting are probably underestimated for two reasons: first, the silt-sized sediment fraction, which may make up more than 50% (visual estimate) of some samples, is dominated by coral and molluskan fragments that were not counted during point-counting of the sand and coarser fraction; second, large clasts of coral that are cobble to boulder size were collected but not included in the thin-section point-count data because of their large size.

The majority of skeletal fragments (70–80% by visual estimate) are subangular to angular or elongate in shape and appear to have been subjected to intense bioerosion rather than to anything other than minimal mechanical abrasion (Figure 8A–D). Large *Halimeda* plates up to 12–17 mm long and probably from the deep-water species *Halimeda copiosa* are abundant throughout many of the samples. No preferential alignment of the *Halimeda* plates in the samples was observed except on a very localized (<1 cm) scale. Many of the plates are encrusted with coralline algae or bryozoans.

Numerous voids ranging in size from <1 mm to greater than 40 mm are present in all but a few of the samples. Some of the pores are shelter cavities beneath *Halimeda* plates or occasionally coral fragments, others are identifiable as galleries created by boring (Clionid) sponges or boring (e.g., *Lithophaga*) mollusks. Many of the voids are irregular in shape and unidentifiable as to origin and probably represent the coalescing of several borings. In most of the samples, several of the voids are partially filled with internal sediment forming geopetal structures. The geopetals within a particular sample are always aligned in the same orientation. Limitations inherent to the sampling method, which excavated piles of rubble from the exposed sides of the slope, precluded the possibility of comparing geopetals from sample to sample or to the declivity of the slope.

The void-filling internal sediments generally consist of laminated silt- to fine-sand-sized material. The dark laminae consist of silt-sized (average 20–40 μm) peloidal micrite (Mg-calcite) cemented with bladed Mg-calcite and the light laminae consist of silt- to fine-sand-sized skeletal fragments. As mentioned above, the majority of the silt-sized skeletal fragments are faceted and are interpreted as being chips excavated by boring sponges. Individual laminae are generally 50–250 μm thick but may be in excess of 1 mm. The individual laminae exhibit both normal and inverse grading, although the majority are nongraded. Inverse grading appears to be most common in the peloidal micrite laminae. Nonlaminated internal sediments, consisting of a homogeneous mixture of peloidal and clotted micrite and silt- to sand-sized skeletal fragments, are also present but not as abundant as the laminated sediments. As with the laminated internal sediments, the peloidal micrite is always Mg-calcite, whereas the clotted micrite is occasionally aragonite.

The overwhelming percentage and the compositional distribution of skeletal fragments in these sam-

ples suggest that the rocks were derived almost exclusively from a reef or reeflike environment. Both the morphology and type of corals (platy *Agaricia* spp.; platy and massive *Montastrea annularis*; *Madracis mirabilis*, *Madracis decactis*, and *Siderastrea siderea*) as well as the dominant species of *Halimeda* (primarily *Halimeda copiosa*) suggest that this reef was growing in a relatively deep (greater than 20 m) environment (Blair and Norris, 1988; Johns and Moore, 1988; Suchanek, 1989). Abundant evidence of bioerosion and the presence of cavity-filling internal sediments, consisting at least in part of sponge chips, is consistent with this type of environment.

Surface of Cemented Slope

In contrast to samples collected from the interior of the slope, samples collected from the surface and near-surface of the cemented slope consist of well-indurated to somewhat friable, bioturbated, peloidal-skeletal wackestones and packstones. The most striking difference is that these samples have been subjected to several generations of endolithic boring followed by sediment infill that has totally destroyed the original fabric of the rock (Figure 9A,B) The minor amounts of sediment on the surface combined with the extensive bioerosion suggest that the surface of the cemented slope is a submarine hardground. The composition of the rocks is also somewhat different than the samples from the interior of the slope, with a significant nonskeletal fraction.

The average composition of the sand and coarser fraction derived from thin-section point-count is as follows: peloid (31%); coral (14%); serpulid worm tubes (12%); mollusk (12%); *Halimeda* (9%); ooid (5%); coralline algae (5%); benthic foraminifera (4%); echinoderm (2%); bryozoan (1%); and planktonic foraminifera (1%). Peloids range in size from coarse silt (approximately 50 μm) to coarse sand (700–800 μm) although the majority of peloids (60–70% by visual estimate) are very fine- to fine-sand size (100–150 μm). In thin-section, there are a few aggregate grains consisting of peloids and ooids that may be from intraclasts. The extensive boring combined with the degree of lithification, however, made it impossible to determine whether these are actually intraclasts or just pieces of the rock dislodged through bioerosion. No identifiable intraclasts were observed in the hand specimens.

Numerous voids ranging in size from <1 mm to more than 4 cm that are open or partially filled with mud-silt-sized sediment are present in most of the samples (Figure 9B). The void-filling internal sediment varies from laminated to nonlaminated and consists of mud- to fine-sand-sized material. The internal sediment is generally well lithified, although some of the nonlaminated sediment is only partially lithified (Figure 9B). The composition of the void-filling sediment consists primarily of a mixture of mud and very fine- to fine-sand-sized detrital peloids or peloidal micrite with average diameters of 20–50 μm that are cemented with bladed Mg-calcite (Figure 9C). No grading of internal sediment was observed.

The surfaces of most of the samples are extensively encrusted with serpulid worm tubes with lesser amounts of bryozoans and benthic foraminifera. A reddish-brown iron oxide stain is frequently observed on the surface of hand specimens and sometimes extends for several centimeters into the sample. The majority of skeletal grains are extensively micritized. The degree of bioerosion and boring activity decreases from the exposed surface of the rocks downward to a point where in a few samples, a relatively unaltered skeletal packstone to grainstone, similar to that observed from the interior of the cemented slope, is still present.

Multiple generations of boring activity are visible in both hand samples and thin-sections. Individual burrows or borings are still visible in the mottled matrix of some samples. In many cases, these original burrows or borings were infilled with sediment and then rebored after the sediment lithified. Evidence that the majority of bioturbation occurred subsequently to lithification (i.e., endolithic borings rather than burrows) is seen by the sharp boundaries of the boring traces as shown in Figure 9B and by the fact that many of the borings cut across individual grains (Figure 9D). Some of the samples exhibit several generations of sediment infill–lithification boring within an individual cavity (Figure 9B).

CEMENTATION OF STEEP-SLOPE LIMESTONES

General

One of the most striking things about samples from the cemented slope is the abundance of cement, especially in the rocks recovered from the interior of the slope. As previously stated, samples vary from being somewhat friable to being extremely well indurated, so much so that the sample will "ring" when hit with a rock hammer. There is a distinct difference in the types and relative abundance of cement, depending upon whether the samples were collected from the interior of the slope or from the surface of the cemented slope. The thoroughly bioturbated, matrix-rich wackestones and packstones from the surface and near-surface of the cemented slope are generally cemented with matrix micrite and peloidal micrite (approximately 80–90% of cement by visual estimate), with lesser amounts of bladed Mg-calcite (approximately 10%) and fibrous aragonite (<5% by visual estimate).

The average distribution of cements in samples from the interior of the cemented slope is approximately 31% bladed Mg-calcite, 18% matrix micrite (Mg-calcite), 13% fibrous aragonite, 12% botryoidal aragonite, and 12% peloidal micrite (Mg-calcite). Thin-sections of representative samples from the interior of the cemented slope were point-counted for cement abundance but because of the numerous pos-

sibilities of error discussed by Halley (1978), visual estimations of cement abundance were also used. Total abundance of cement varies from approximately 15–50% with an average of 20–30% (by visual estimate) and is less abundant in the packstones than in the grainstones as expected. There was no consistent pattern in the distribution of the cements to unequivocally determine a paragenetic trend. Textural evidence suggests that pore-filling botryoidal aragonite cements were one of the last cements to precipitate. The similarity of radiocarbon ages between some botryoidal cements and coexisting skeletal fragments discussed below (see also Table 1), as well as data presented on the growth rates of botryoidal aragonite cements (Grammer et al., 1993), suggests that the majority of cementation took place very early, probably within a few hundred years. Porosity within the samples varies from near zero to approximately 25% (combined point-count and visual estimates).

Micritic Mg-Calcite Cement

Microcrystalline Mg-calcite cement, or micrite, is widely considered to be the most common type of magnesian calcite in both modern and ancient submarine environments (Ginsburg et al., 1971; Alexandersson, 1972; Schroeder, 1973; Friedman et al., 1974; James et al., 1976; MacIntyre, 1977, 1985; Longman, 1980; Friedman, 1985). Micrite occurs as thin layers around the edges of some skeletal grains and more commonly as micrite envelopes where borings on the surface of the grain by endolithic algae and/or fungi (Bathurst, 1975) have been infilled with microcrystalline Mg-calcite cement. Micrite envelopes are relatively common in samples from the interior of the slope, especially on mollusk fragments. The majority of skeletal grains in the surface samples are completely micritized.

Peloidal micrite, consisting of rounded, "clotted" grains of micrite that average 20–40 μm in diameter, is relatively common in both interior and surface slope samples. The peloids are always composed of Mg-calcite and are typically cemented with finely crystalline (20–75 μm) bladed Mg-calcite. Peloidal micrite is found only as a pore-filling cement, typically infilling borings and other sheltered cavities in a rock (Figure 9C). Peloidal cement can generally be differentiated from detrital peloids using guidelines suggested by MacIntyre (1985), which include a general spherical shape, a size restriction normally in the 20–50 μm range, and isolated sites of deposition/precipitation such as internal cavities and borings.

Matrix micrite is locally present, primarily as a void-fill in samples from the interior of the slope and is abundant in the wackestones that make up the surface of the cemented slope. As discussed by Friedman (1985), it is generally quite difficult to determine unequivocally whether this type of micrite is precipitated as cement or deposited mechanically as mud. In some instances, the coexistence of micrite with silt-sized chips derived from boring *Clionid* sponges

(Rutzler, 1975) is suggestive of a depositional origin, but this evidence is still not definitive. In other cases, staining clearly shows that much of the matrix or cavity-filling micrite is aragonite (rather than Mg-calcite)

suggesting a depositional origin for the micrite as mud. This is especially true for samples from the surface of the slope where as much as 30–40% (visual estimate) of the micrite is aragonite.

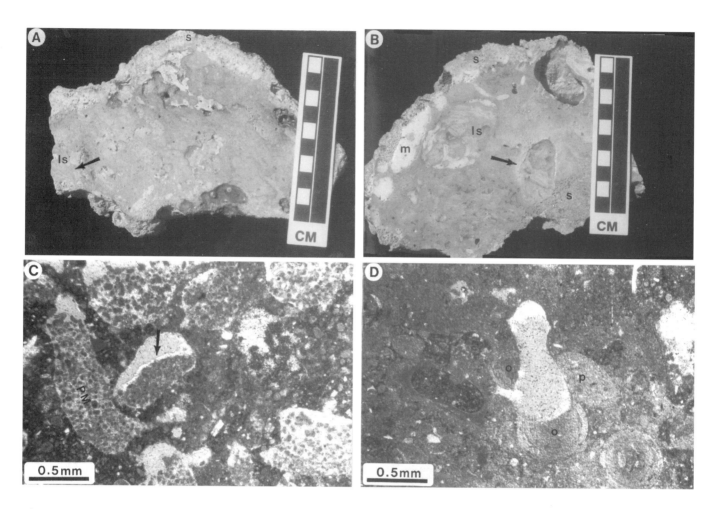

Figure 9. (A) Slab from the surface of the cemented slope exhibiting characteristic features of a submarine hardground surface. Multiple generations of cementation-boring-sediment infill-cementation have destroyed the original fabric of the rock leaving only a mottled appearance. Individual borings can no longer be identified in most instances because they have been cross-cut repeatedly by subsequent borings. The surface of the rock has been extensively bored by *Clionid* sponges (s). Some of the voids have been infilled with laminated internal sediment (ls; arrow) that is now well lithified. Scale is in centimeters. Privitt Place, 168 m (550 ft). (B) Slab from near the surface of the cemented slope. The exposed edges of the slab have been bored by *Clionid* sponges (s) that are still open and larger borings and voids that have been filled but only partially lithified (m). Large void at right is partially filled with geopetal sediment (arrow). Large void at left represents the multiple generations of boring and sediment infill typical of the hardground surface of the cemented slope. Pore has been infilled by at least three generations of laminated sediment (ls), which was subsequently reworked (mottled) before lithification in places and bored after lithification in others. Note also the presence of unlithified to partially lithified mud in white. Privitt Place, 168 m (550 ft). (C) Thin-section photomicrograph (plane-polarized light) from hardground surface of the cemented slope with several borings infilled with peloidal Mg-calcite cement (PM). Genesis as a cement, as opposed to a mechanically deposited peloidal grain, is indicated by the isolated sites of occurrence, the general spherical shape, and the small size (20–30 µm). Partial geopetal void-fill (in center) of Mg-calcite micrite (cement or sediment?) is followed by a later stage of bladed Mg-calcite cement (arrow). Palmer Point, 122 m (400 ft). (D) Thin-section photomicrograph (plane polarized light) of sample from the surface of the cemented slope showing evidence of postlithification boring in hardground surface. Note how boring has cut across two ooids (o) and a large peloid (p). Compare the size of mechanically deposited peloids (average 100–250 µm) with peloidal micrite cement (20–30 µm) in Figure 9C. Palmer Point, 122 m (400 ft).

120   Grammer et al.

Table 1. Radiocarbon dates of skeletal components and pore-filling botryoidal aragonite cements from the surface and the interior of the cemented slope. Data show that samples from the interior of the cemented slope vary in age from approximately 11.3–14.0 Ka, while rock samples collected from the surface of the slope are clustered around 10.5 Ka. Samples marked with (*) are from the wall, not the cemented slope. The Rock Range samples marked with (1) and the Privitt Place samples marked with (2) indicate those samples where the ages of a skeletal fragment and pore-filling botryoidal aragonite cement from the same rock were determined.

I. Loose Surface Material

Location	Depth	Component Dated	Age (ybp)
Privitt Place	168 m	Coral (*A. cervicornus*)	365 +/– 50
Palmer Point	122 m	Conch	894 +/– 95
Dolly Cay*	116 m	Coral (*M. annularis*)	1560 +/– 95
Palmer Point	122 m	Conch	4490 +/– 140
Privitt Place	174 m	Conch	4535 +/– 100
Palmer Point	122 m	Conch	6830 +/– 100

II. Indurated Samples Near Surface (+/– 0.5 m) of Cemented Slope

Location	Depth	Component Dated	Age (ybp)
Blossom Channel	168 m	Coral (*M. annularis*)	10,300 +/– 160
False Starter*	100 m	Sclerosponge	10,400 +/– 105
Privitt Place	168 m	Coral (*M. annularis*)	10,470 +/– 140
Blossom Channel	168 m	Coral (*M. annularis*)	10,560 +/– 230
Blossom Channel	168 m	Coral (*M. annularis*)	10,640 +/– 115
Rock Range	152 m	Cemented Internal Sediment	10,660 +/– 110
Privitt Place	174 m	Sclerosponge	10,710 +/– 120

III. Samples from Interior (0.5–3.5 m) of Cemented Slope

Location	Depth	Component Dated	Age (ybp)
Privitt Place	174 m	Sclerosponge	11,330 +/– 130
Privitt Place	174 m	Botryoidal Aragonite Cement	11,435 +/– 105
Rock Range	152 m	Coral (*M. annularis*)	11,470 +/– 105
Rock Range[1]	152 m	Botryoidal Aragonite Cement	11,680 +/– 90
Rock Range	152 m	Botryoidal Aragonite Cement	12,005 +/– 105
Rock Range	152 m	Botryoidal Aragonite Cement	12,075 +/– 115
Rock Range[1]	152 m	Coral (*Agaricia* spp.)	12,240 +/– 100
Rock Range	152 m	Botryoidal Aragonite Cement	12,290 +/– 100
Rock Range	152 m	Botryoidal Aragonite Cement	12,300 +/– 110
Rock Range	152 m	Botryoidal Aragonite Cement	12,380 +/– 110
Rock Range	152 m	Botryoidal Aragonite Cement	12,475 +/– 115
Rock Range	152 m	Botryoidal Aragonite Cement	12,585 +/– 100
Rock Range	152 m	Botryoidal Aragonite Cement	12,645 +/– 220
Rock Range	152 m	Botryoidal Aragonite Cement	12,660 +/– 115
Privitt Place[2]	174 m	Botryoidal Aragonite Cement	12,705 +/– 110
Privitt Place[2]	174 m	Coral (*Madracis decactis*)	13,405 +/– 90
Rock Range	152 m	Botryoidal Aragonite Cement	13,685 +/– 115
Rock Range	152 m	Botryoidal Aragonite Cement	13,935 +/– 115

Bladed Mg-Calcite Cement

Bladed Mg-calcite is the most common type of cement in the packstones and grainstones that comprise the interior of the cemented slope. The cement occurs as elongated crystals oriented perpendicular to the substrate and typically forms isopachous rims around interparticle and intraparticle voids as shown in Figure 10A and B, but also occurs as interparticle cement between peloidal micrite. The isopachous rims average 100–150 μm in thickness but may be as much as 2.0 mm thick and visible in hand specimen. The crystals generally have a poorly defined scalenohedral morphology and are consistently 4–7 μm wide, regardless of length. A finer crystalline (narrower) form of bladed Mg-calcite, consisting of partial spherulites or splays, grows exclusively as cavity-lining isopachous rims in the skeletal packstones and grainstones from the interior of the cemented slope (Figure 10C). In both types of isopachous rim cements, there are frequently several "dust lines" made up of solid, dark, irregularly shaped inclusions that average 1–10 µm in diameter that probably represent discontinuity surfaces in the growth of the cement. As many as 14 generations or pulses of cement growth were observed within a single isopachous rim.

Fibrous Aragonite Cement

Fibrous aragonite cements grow in an elongate or acicular morphology with high length-to-width ratios similar to those described from shallow marine settings (Shinn, 1969; Ginsburg et al., 1971; Alexandersson, 1972; Schroeder, 1973; Friedman et al., 1974; James et al., 1976; Sandberg, 1985). Fibrous aragonite occurs most commonly as an intraparticle, void-filling cement within skeletal grains, especially corals and *Halimeda* (Figure 10D). It also occurs as an epitaxial fringe, typically discontinuous, that grows on skeletal fragments (primarily molluskan) and less frequently as an interparticle cement. Individual crystals are generally 3–10 µm wide and may be in excess of 200 µm long although the average length is probably around 50–75 µm. Crystal terminations are both pointed and blunt or chisel-shaped. Individual crystals of pore-filling fibrous aragonite grow approximately perpendicular from the substrate but the crystals are only weakly oriented as they extend into the pore.

Botryoidal Aragonite Cement

Botryoidal aragonite is a spectacular pore-filling type of aragonite cement present only in the coarse skeletal packstones and grainstones from the interior of the cemented slope. The term "botryoidal" was originally defined by Ginsburg and James (1976) to describe "individual and coalescing mamelons of compact, fibrous aragonite" found within various cavities in forereef deposits in Belize (Ginsburg and James, 1976). Individual masses range in size from a few hundred microns to more than 15 mm long. They consist of fans of elongate euhedral fibers and exhibit a characteristic sweeping extinction in crossed polarized light, and occur as both individual and compound growths (Figure 10C). The botryoids show no preferential growth direction and grow out into pores from the floors, walls, and ceilings of cavities. Many of the botryoids exhibit what appear to be discontinuity horizons marked by a "dustline" of small (1–3 µm), solid, dark, irregularly shaped inclusions and/or slight irregularities in crystal growth. Some of these discontinuity surfaces are marked by small (5–10 µm) patches or thin layers (<10 µm) of Mg-calcite micrite. These horizons probably represent interruptions in the growth of the cement. As many as 38 discontinuity horizons were observed within a single

fan. Discontinuity horizons from separate fans within a single pore do not match up one for one, suggesting that the fans may initiate growth at different times. In addition, there is no consistent relationship between the number of discontinuity horizons and the size of the fan, even for those growing in the same pore.

Grammer et al. (1993) have reported exceptionally high growth rates for pore-filling botryoidal aragonite cements in marginal slope deposits from the Tongue of the Ocean and Belize. Radiocarbon dating of samples obtained from the base and top of several botryoids indicates that the cements grew at average rates of 8–10 mm/100 yr with maximum rates in excess of 25 mm/100 yr. In addition, microsampling of the botryoids in a series of transects indicated little variation in the $\delta^{13}C$ and $\delta^{18}O$ composition of the cements, supporting the conclusion based on radiocarbon dating that cementation was rapid.

RADIOCARBON DATING OF CEMENTED SLOPE LIMESTONES

Thirty-one samples of various skeletal components (corals, sclerosponges, large gastropods) and pore-filling cements from the surface and interior of the cemented slope were radiocarbon dated to establish the age of the deposits (Table 1). The dates were obtained by accelerator mass spectrometer because of the advantage it has over conventional radiocarbon dating techniques in requiring significantly smaller samples. Three general groupings of radiocarbon ages can be seen in Table 1.

The youngest radiocarbon ages (365–6830 ybp*) are from large gastropod (conch) shells and coral samples collected from the surface of the cemented slope. Most of these samples were only loosely attached to the surface of the slope but had some sediment cemented onto the surface of the sample. The exceptions to this were the two youngest samples, a stick of *Acropora cervicornus* (age: 365 ybp) collected on the floor of the large slide scar at Privitt Place and a conch (age: 894 ybp) collected on the surface of the slope at Palmer Point. Neither of these samples had any lithified sediment either on the surface or within the skeleton.

The second group of samples, with ages ranging from 10,330–10,710 ypb, were collected either at the surface of the slope or within approximately 0.5 m of the surface (average depth excavated by an explosive charge set on the surface of the cemented slope). These tightly grouped samples were all completely enclosed by lithified sediment.

The third group, containing the oldest radiocarbon ages (11,330–13,935 ybp), consist of samples collected from the interior of the cemented slope. These samples are from the two locations (Privitt Place and Rock

Range) where localized slope failure provided the opportunity to sample as much as 3.5 m into the interior of the slope. The use of small charges of explosives to dislodge rock samples from the walls of the exposures has the unfortunate disadvantage of break-

ing out rocks in a radial pattern up to 1 m or more away from the blast site. The result is that it is impossible to reconstruct the stratigraphic position of the samples with respect to one another, and the only definitive conclusion that can be made is that the

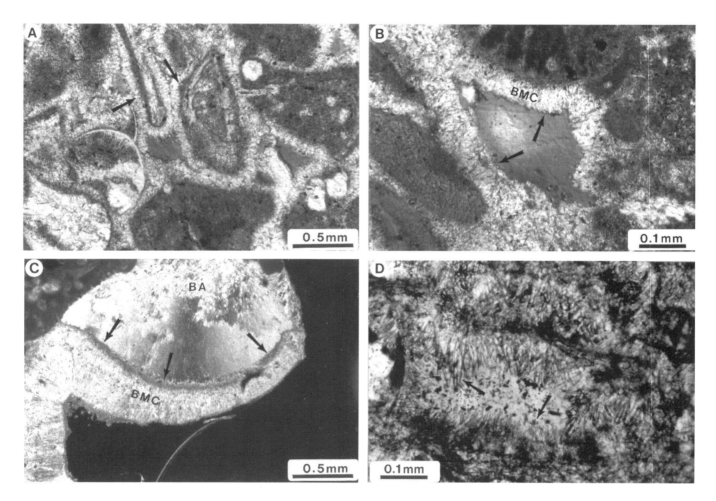

Figure 10. (A) Thin-section photomicrograph (plane-polarized light) showing skeletal grainstone from the interior of the cemented slope cemented with intergranular and void-lining, isopachous bladed Mg-calcite cement. Note early stage micrite envelopes of micritic Mg-calcite around the perimeters of some grains (arrows). Micrite envelopes are most commonly developed on mollusk fragments but are also found on pieces of *Halimeda*, coral, and benthic foraminifera. Rock Range, 152 m (500 ft). (B) High-magnification thin-section photomicrograph (plane-polarized light) of interparticle and void-lining, isopachous bladed Mg-calcite cement (BMC) partially filling void (arrows). Isopachous rim is up to 75 μm thick. Rock Range, 152 m (500 ft). (C) Thin-section photomicrograph (crossed nicols) of sample from the interior of the cemented slope showing pore-filling botryoidal aragonite (BA) cement terminating against isopachous bladed Mg-calcite (BMC) cement growing into same pore. The thin dark line (arrows) is a layer of Mg-calcite micrite between the two pore-filling cements. Note how the botryoidal aragonite initiated from the other side of the pore from the bladed Mg-calcite. The bladed Mg-calcite is a finer crystalline cement than most of the other bladed Mg-calcites (see for example A and B), with crystals that are narrower and that tend to grow in a more spherulitic form. While it is fairly common to have both botryoidal aragonite and bladed Mg-calcite precipitated within the same pore, one cement was never observed to have precipitated on top of the other. Observations suggest that the botryoidal aragonite cements are one of the latest stages of cementation in these rocks. Privitt Place, 168 m (550 ft). (D) Thin-section photomicrograph (plane-polarized light) of sample from the interior of the cemented slope showing fibrous aragonite cement (arrows) partially filling intraskeletal voids in coral skeleton. Fibrous aragonite occurs most commonly as an intragranular pore-filling cement within coral, *Halimeda*, and gastropods. It also occurs less frequently as an epitaxial fringe on some skeletal fragments, especially mollusks. Individual crystals are 40–125 μm long. Privitt Place, 168 m (550 ft).

samples range in age from approximately 11,000–14,000 ybp.

The radiocarbon age data clearly show that deposition of the steep slopes was actively taking place from at least 14 Ka. The presence of similar deposits observed in deep slide scars that extend back as far as 10 m into the slope suggests that deposition may have been taking place even earlier (Figure 6A,B). Radiocarbon ages determined for corals and botryoidal aragonite cements coexisting in the same rock indicate that cementation of the interior slope deposits occurred within a relatively short time (i.e., hundreds of years) after deposition (Table 1).

The tight clustering of radiocarbon ages of samples collected from the surface of the cemented slope suggests that most deposition ceased rather abruptly around 10.5 Ka. The presence of younger skeletal material on the slope (i.e., less than approximately 6800 yr) indicates that some deposition has taken place more recently, but this latest phase of deposition is interpreted to be minor because of the localized nature of the clasts. The presence of only a thin veneer of localized sediment on the surface of the slope suggests that the majority of sediment is currently bypassing the steep slopes. Combined with textural evidence indicating that the surface of the slope is a submarine hardground (e.g., multiple generations of boring), the radiocarbon data support the conclusion that, for the most part, the present surface of the cemented slope is basically one of nondeposition.

DISCUSSION OF SLOPE DEVELOPMENT AND SEA-LEVEL FLUCTUATION

Development of Steep Slopes

The questions of how and when the cemented slope formed are of primary interest because of the apparent similarity to steeply dipping slope deposits documented from the fossil record. Carbonate slopes from the Permian of west Texas and New Mexico (Yurewicz, 1977; Ward et al., 1986; Garber et al., 1989), the Devonian of Western Australia (Playford, 1980; Playford et al., 1989), the Triassic Dolomites of northern Italy (Bosellini, 1984; Harris, 1988), the Cretaceous of east-central Mexico (Enos, 1986), and the Miocene of the Gulf of Suez (Haddad et al., 1984) all exhibit primary depositional slopes of 30–40º. In addition to slope declivity, the geometry and thickness of beds as well as the dominant texture of the slope deposits in the Tongue of the Ocean are also similar to these ancient examples. Steep-slope profiles similar to those observed in outcrop are also frequently observed in seismic profiles from ancient carbonate platforms in the subsurface (e.g., Sarg, 1988).

Process

Researchers working on both modern and ancient carbonate slopes have suggested a myriad of downslope, gravity-induced mechanisms for the deposition of sand-sized and coarser sediments. On modern slopes, Cook and Mullins (1983) and Enos and Moore (1983) indicate that relatively large-scale turbidity currents and debris flows appear to be the dominant mechanisms for the downslope transport of coarse detritus. On ancient carbonate slopes, all types of sediment gravity flows have been proposed, but again the predominant depositional mechanisms are interpreted to be debris flows or turbidity currents (Cook, 1983).

Observations from the Tongue of the Ocean, however, suggest that the steep slopes formed by an alternative mechanism. The lenses of lithified sediment that are locally present on the surface of the lower cemented slope presumably represent some of the last depositional events on the slope (Figure 4E,F). The apparent lack of matrix and poor sorting of the deposits, combined with upslope imbrication of clasts, suggests that deposition took place through a combination of rockfall and grainflow processes (Lowe, 1979; Cook, 1983). The similarity in the geometry (i.e., elongate lenses) and scale of these talus trains to the low-amplitude ridges that are ubiquitous to the lower cemented slope suggests that deposition of the steep slopes may have taken place as a series of individual and localized pulses of sediment rather than by emplacement of large-scale mass-flow deposits. Repeated episodes of deposition and cementation would result in the amalgamation of individual sediment packages that are linear in shape and discontinuous in both lateral and vertical dimensions along the slope. Both the texture of internal slope deposits, which are dominated by poorly sorted skeletal packstones and grainstones made up of angular to elongate silt- to pebble-sized grains (Figure 8A,B), and the lenticular nature of the bedding observed in slide scars (Figure 6A,B) are consistent with this proposed mode of slope development.

The coarse grain size and angularity of the sediments combined with syndepositional cementation resulted in both the formation and preservation of steep (35–45º) primary depositional slopes. Chorley et al. (1984) and Kirkby (1987) have shown that angles of repose for sediments consisting of a mixture of sand- to gravel-sized clasts are typically 35–45º. The high angles of repose are due both to the coarse grain size and angularity of clasts, as well as an increase in clast interlocking associated with the large variation in grain size (Kirkby, 1987). Although the above-referenced studies were for subaerial, dry sediments, angles of repose for submerged granular sediments have been shown to not deviate by more than 1–2° from dry sediments (Chowdury, 1978; Feda, 1982). Another possibility is that the steep depositional slopes are preserved at their original angle of initial yield. Theoretically, granular materials build up to their angle of initial yield, then undergo shear failure that results in the sediment being left at the more stable angle of repose (Allen, 1985). Rapid growth rates of pore-filling cements on the order of 8–10 mm/100 yr, reported by Grammer et al. (1993), indicate that at least some of the slope deposits have undergone extremely rapid cementation and suggest that the deposits may have been "frozen" at their angle of initial yield.

Timing of Development of Steep Slopes

Radiocarbon dating (Table 1) of skeletal components and cements from samples collected from the interior of the slope indicate that deposition of the steep slopes was active from at least 14 Ka. The oldest dates are from samples collected approximately 2.5–3.5 m below the present surface of the cemented slope. The presence of up to 10 m of bedded slope deposits observed in deep slide scars, however, indicates that deposition probably commenced much earlier (Grammer and Ginsburg, 1992).

Comparison of the radiocarbon dates to the high-precision sea level curve for the Caribbean over the last 18 k.y., shown in Figure 11, indicates that deposition of the steep slopes took place during the initial rise of sea level following the last lowstand (Figure 12A). The abundance of reef-building corals, coralline algae, and *Halimeda* in samples collected from the interior of the cemented slope indicate that the sediment was probably derived from a fringing reef growing along the vertical escarpment. Normal erosion due to storm activity, bioerosion, and oversteepening of the reef would have provided coarse skeletal sands, gravels, and boulders to the slopes below (Figure 12B).

The tight clustering of dates from samples collected near the surface of the cemented slope (top 0.5 m) indicates that most deposition ceased abruptly approximately 10.5 Ka, probably due to drowning of the fringing reef along the escarpment (Figure 12C). Fairbanks (1989) has shown that during the period from 10 to 9 Ka, sea level rose more than 20 m in response to a "melt-water pulse" related to the termination of the Younger Dryas event (Kennett, 1990). This rapid rise in sea level was approximately twice as fast as the maximum keep-up rates (8–12 mm/yr) reported for modern coral-algal reefs (Grigg and Epp, 1989) and would have most likely drowned any fringing reefs along the escarpment. Drowning of the reefs during this time is even more likely because of an earlier period of rapid rise that would have put the reefs in a "catch-up" phase. This earlier melt-water pulse was centered at approximately 12 Ka and resulted in a rise of sea level of approximately 24 m in less than 1000 years (Fairbanks, 1989). Textural evidence of multiple generations of endolithic boring-sediment infill-cementation confirms that the surface of the slope is presently an inactive depositional surface or submarine hardground. The resulting diastem (i.e., surface of the cemented slope) would therefore represent a type of drowning unconformity that is similar in many ways to those defined by Schlager and Camber (1986) and Schlager (1989).

Another possibility is that the sudden shut-down of deposition along the steep slopes resulted from a rapid shift in the locus of sediment production and deposition from the escarpment to the edge of the platform. Figure 11 shows that at 10 Ka, sea level was

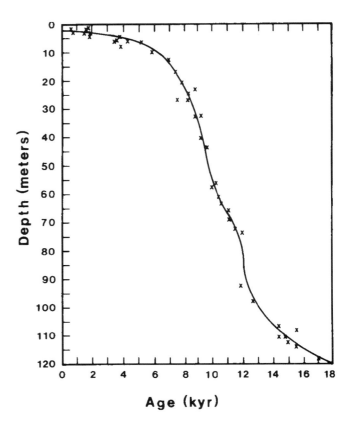

Figure 11. Sea-level curve for the Caribbean region showing rise of sea level from last lowstand at approximately 18 Ka. Curve is based primarily upon radiocarbon-dated *Acropora palmata* from Barbados. Modified from Fairbanks (1989).

approximately 60 m lower than it is today. This depth correlates well with the top of the near-vertical escarpment at all dive sites around the Tongue of the Ocean (Grammer and Ginsburg, 1992) and presumably marks the edge of the exposed platform during the last lowstand. As sea level began to flood the platform, the primary carbonate factory may have shifted away from the escarpment and onto the platform top. Initial production of sediment probably would have been relatively minor while the system adjusted to the rapid rise of sea level (Figure 12D). The combination of minor sediment production and the likelihood that some sediment was trapped along the upper margin would result in minimal sediment transport to the slopes below.

Sedimentation on the foreslope is presently dominated by highstand shedding of platform-derived sand, silt, and mud (Figure 12E). The sediment bypasses the steep cemented slopes because slope angles are higher than angles of repose for these finer grained sediments, and are deposited downdip as the more gently dipping "sediment slope" (Grammer and Ginsburg, 1992). Figure 13 illustrates the present-day, three-dimensional distribution of the various marginal slope environments around the Tongue of the Ocean.

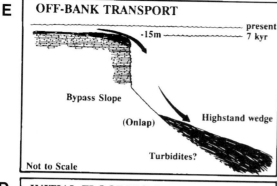

E   OFF-BANK TRANSPORT

present
-15m — 7 kyr

Bypass Slope

(Onlap)   Highstand wedge

Turbidites?

Not to Scale

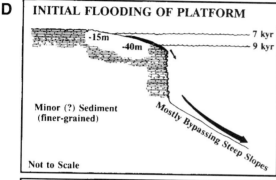

D   INITIAL FLOODING OF PLATFORM

-15m — 7 kyr
-40m — 9 kyr

Minor (?) Sediment
(finer-grained)

Mostly Bypassing Steep Slopes

Not to Scale

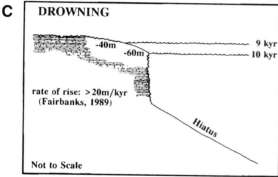

C   DROWNING

-40m — 9 kyr
-60m — 10 kyr

rate of rise: >20m/kyr
(Fairbanks, 1989)

Hiatus

Not to Scale

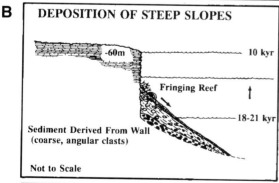

B   DEPOSITION OF STEEP SLOPES

-60m — 10 kyr

Fringing Reef   ↑

— 18-21 kyr

Sediment Derived From Wall
(coarse, angular clasts)

Not to Scale

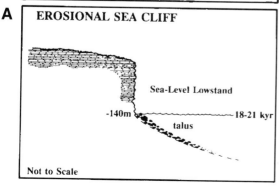

A   EROSIONAL SEA CLIFF

Sea-Level Lowstand

-140m — 18-21 kyr
talus

Not to Scale

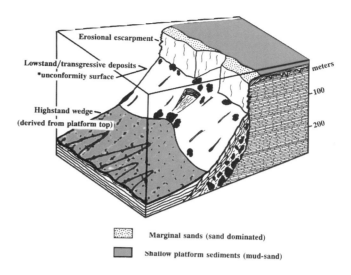

Erosional escarpment

Lowstand/transgressive deposits
*unconformity surface

Highstand wedge
(derived from platform top)

meters
100
200

▦ Marginal sands (sand dominated)

▨ Shallow platform sediments (mud-sand)

Figure 13. Block diagram illustrating the present-day surface characteristics and inferred three-dimensional structure of the Tongue of the Ocean foreslopes. Modified from Grammer and Ginsburg (1992).

Failure of Steep Slopes

The presence of numerous cracks visible on the surface of the cemented slope suggests the slope is undergoing incipient failure (Figure 4D). As previously discussed, these cracks are arcuate (convex-up) and run across the slope, parallel to contours, for up to 75 m. At the surface of the cemented slope, they are typically a few centimeters wide. The cracks have been observed at five of the dive sites on slopes varying from 30–50°, and appear to be most prevalent on slopes of 35–45°.

Similar cracks have been described from numerous fossil slope deposits, including the Late Cambrian and Early Ordovician of Nevada (Cook and Taylor, 1977; Cook, 1979; Cook and Mullins, 1983), the Devonian of the Canning basin in Australia (Playford, 1980, 1984; Kerans et al., 1986; Playford et al., 1989), the Silurian of Indiana (Devaney et al., 1986), and the Cretaceous of Mexico (Enos, 1977, 1986). In many of these ancient examples, the cracks are implicated as being precursors to slope failure (e.g., Playford et al., 1989, and Cook and Mullins, 1983).

A causal relationship between cracks in the slope and failure of the slope is clearly suggested on the steep slopes in the Tongue of the Ocean. Local areas of slide failure occurring along the slope and the up-slope geometry (i.e., arcuate, convex-upslope) of the

Figure 12. Sequence of diagrams illustrating model for the development of the Tongue of the Ocean foreslopes (see text for discussion). Modified from Grammer and Ginsburg (1992).

126   Grammer et al.

failed surface is very similar to the cracks on the surface (Figure 14). In at least three locations, Privitt Place, Palmer Point, and Rock Range, these slides have excavated several meters to several tens of meters of cemented slope deposits. The largest zone of failure, illustrated in Figure 14, is 10 m deep at the upslope end, up to 30 m across, and extends downslope for approximately 75 m. The presence of a "weathering profile" in the exposed beds within these failure zones, characterized by what appears to be resistant beds visible in Figure 6A and B, may provide a clue as to why these slopes fail. The variation in resistance to erosion is likely due to variations in cementation, possibly representing differences in depositional fabric (i.e., variations in the amount of cementation between coarser grained packstones and grainstones relative to finer grained deposits). Zones of more weakly cemented slope deposits would likely provide planes of weakness along which masses of the slope could fail.

Timing of Slope Failure

The timing of slope failure in the Tongue of the Ocean is clearly postlithification, as shown by the

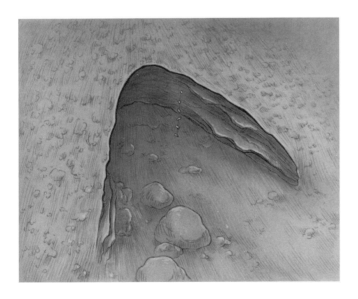

Figure 14. Line drawing of large slide at Privitt Place made from a combination of observational notes and videotape footage. Upslope terminus of failure is at a depth of 146 m (480 ft). Scale is in meter increments. Oblique angle of drawing underestimates depth of slide into slope. Measurements at the upslope end indicated the failure was close to 10 m deep (i.e., back into the cemented slope). Note competent blocks (5–8 m in diameter) derived from slide. A vertical transect directly downslope from the failure found blocks up to 10 m in diameter that had impounded into the downdip sediment slope at a depth of 311 m (1020 ft). Illustration by Barbara Cousins.

rigid walls of the slides and the competence of the blocks of bedded deposits moved downslope. The fact that the uppermost surface of the cemented slope is intact around the perimeter of the slides indicates that failure took place after deposition had ceased and the deposits had become lithified (Figure 14). Based on the radiocarbon ages of both cements and skeletal components from the cemented slope (Table 1), failure must have taken place in the last 10 k.y. or so, which correlates with a relative highstand of sea level (i.e., sea level had begun to flood the top of the platform). The timing is notable because most of the literature on carbonate slope failure suggests that failure takes place most often during relative lowstands of sea level where the slopes may be subjected to erosion by wave activity and/or storm surges (e.g., Cook, 1983; Sarg, 1988; Cook and Taylor, 1991). This is especially true if the area is not susceptible to tectonic activity (i.e., earthquakes) as in the Bahamas (Dietz et al., 1970).

Mechanism of Slope Failure and Affect on Depositional Record

The mechanisms believed to be most likely responsible for failure of carbonate slope deposits and the generation of sediment gravity flows have been summarized by Cook (1983) as follows:
1. earthquake shocks;
2. action of storm waves or tsunamis on a buildup margin (i.e., hydraulic shock);
3. gravity acting on an unstable, overloaded, or oversteepened slope or buildup margin, which could be fractured in place;
4. fractures may propagate during movement of a slide or slump mass, leaving the slope susceptible to further failure;
5. subaerial erosion during a long-term relative sea-level lowstand, creating a karsted and therefore structurally weakened slope or margin;
6. diagenetic factors, in particular differential cementation; and
7. increased pore pressures within deposits due to rapid sedimentation rates.

The most likely mechanisms for failure of the Tongue of the Ocean slopes are probably gravitational force and water loading on the steep to possibly oversteepened slopes combined with differential cementation. Earthquakes would not be expected in the tectonically inactive Bahamas (Dietz et al., 1970), and effects of storm surge is equally unlikely as the slopes are presently in approximately 120–300+ m of water. Even if failure took place 10 Ka, which would be the earliest possible time based on radiocarbon dates and observational evidence, the slopes would still have been in water depths 60–240+ m and beyond the influence of storm activity (Figure 11). In addition, as discussed above, failure of the slopes clearly did not occur during or subsequent to a lowstand of sea level that had exposed the slopes to subaerial or shoreline erosional processes.

The affect of failure along the steep Tongue of the Ocean slopes to the resulting sedimentary record is twofold. First, the slides will result in intraformational truncation surfaces similar to those reported in ancient slopes (e.g., McIlreath, 1977; Playford, 1980). The second effect is due to the downslope transport of material, much of it boulder sized. Observations from failure zones on the Tongue of the Ocean slopes have shown an accumulation of large blocks and boulders extending downslope from the slides. Vertical transects along the slope have shown that blocks up to 10 m in diameter have been moved downslope from the zones of failure for tens or hundreds of meters. Some of these blocks came to rest on the lower portions of the cemented slope while others impounded into the sediment slope farther downdip. Blocks of this scale are remarkably similar to megabreccias described in the literature (e.g., Cook et al., 1972; Hubert et al., 1977; McIlreath, 1977; Cook, 1983; Cook and Mullins, 1983; Saller et al., 1989) and could conceivably initiate turbidity flows when they impound into the sand and mud making up the sediment slope. This would provide one mechanism for the initiation of "highstand" turbidites (e.g., Crevello and Schlager, 1980), at least on a local scale.

IMPLICATIONS TO FOSSIL RECORD

Although the marginal slope deposits around the Tongue of the Ocean exhibit similarities in depositional style (fabric, texture, and geometry) to many of the ancient slope deposits mentioned above, the presence of a high vertical escarpment appears to be an anomaly and brings up an interesting question. Is it possible that some of the fossil examples of steep, bioclastic-rich, coarse-grained slope deposits were derived from an escarpment such as we presently see in the Tongue of the Ocean? This question deserves some thought because we would likely miss the presence of such an escarpment in outcrop. A near-vertical to vertical escarpment, such as that in the Tongue of the Ocean, today would be preserved as an extremely narrow, near-vertical plane only a few tens of meters wide in dip direction. The chance of missing such a surface in outcrop, after the basin has been filled, cannot be overestimated. To expose such a surface so that it could be recognized would take a near-perfect dip-oriented section and exceptional rock preservation (e.g., the Middle Cambrian example described by McIlreath, 1977, or the Devonian [Late Frasnian] margin described by Playford et al., 1989). Such a surface would be virtually undetectable in the subsurface.

The question of whether some steep fossil slopes may have been derived from an escarpment could have important implications when determining depositional histories, especially related to fluctuations in relative sea level. Recognition that slope deposits consisting of coarse bioclastic debris were derived from an escarpment during a rise of sea level rather than from the margin during a relative highstand could

change an interpretation of the sequence stratigraphic framework of a basin.

Figures 12 and 13 illustrate important stratigraphic relationships across a carbonate shelf margin that have implications for both the interpretation of ancient examples and also the application of sequence stratigraphic models in carbonate depositional settings. The rapidity with which facies change across the steep escarpment and also the complexity of the stratigraphic section seaward of the escarpment would make such a margin that much more difficult to interpret in the fossil record. Because the Tongue of the Ocean margin shares several depositional, diagenetic, and evolutionary aspects with other carbonate margins, the stratigraphic relations illustrated here should not be viewed as a unique case but as one type of pattern that is no doubt repeated in other margins, both modern and ancient. The complex pattern documented here, formed as a result of rising sea-level flooding a steep-sided, flat-topped margin, is the important lesson. Considering the application of sequence stratigraphy "models" during the interpretation of fossil examples, one must be skeptical of applying too general and simplistic a model in light of this complexity.

The stratigraphic relationships documented here for the Tongue of the Ocean are the result of the most recent rise in sea level from a lowstand of some 120–130 m approximately 18–20 Ka, to the present time where sea level is flooding the bank top. Considering the well-documented sea-level curve for the Pleistocene, we know not only that the sea-level amplitude was varying greatly, but that the variations were of short duration, i.e., occurring at a high frequency. Furthermore, the well-constrained Pleistocene curve suggests that 100 k.y. eccentricity-driven variations were well developed and the latest sea level rise may be the initial phase of such a 100-k.y. cycle. Considering the complexity of the stratigraphy that can be directly linked to one of these high-amplitude, high-frequency sea-level variations in our example, we suggest that some knowledge of stratigraphic relations on a similar high-frequency scale is essential to unraveling the true stratigraphy across ancient shelf-to-basin carbonate transitions. Indeed, flooding, aggradation, progradation, erosion, and bypassing are high-frequency processes. Considering that these relations may be at or beyond the limit of recognition from seismic data, the importance of well-studied Holocene and ancient examples to serve as better models for subsurface interpretation cannot be overstated.

CONCLUSIONS

The upper slopes around the Tongue of the Ocean provide a modern example of how steeply dipping foreslopes may develop along carbonate platform margins and are remarkably similar to the clinoforms often described from fossil platforms. The awareness that primary depositional slopes of 35–45° were deposited during the early rise of sea level following the last lowstand and that the steep slopes were apparently stabilized by syndepositional cementation provides valuable insight into how and when some fossil slopes may have formed. In addition, the recognition that the Tongue of the Ocean slopes formed by the amalgamation of localized lenses and not large-scale mass-flow deposits may have important implications to the understanding of steep carbonate slopes in the subsurface. Realization of the internal heterogeneity of slope deposits and discontinuous nature of the lenticular beds may be a critical component to the accurate evaluation of possible reservoir facies. Evidence for "highstand" failure of the slopes provides an alternative to the accepted dogma of lowstand failure and may represent one means by which highstand turbidites are initiated.

ACKNOWLEDGMENTS

This project has been funded by NOAA's National Undersea Research Program (SU-2-0887, SU-2-0888, 90[89]-OR-SUB/GINS to RNG) and by research grants from the American Association of Petroleum Geologists, and the Geological Society of America (to GMG). Radiocarbon dating was performed by Tim Jull at the University of Arizona. Nancy Budd helped with the coral identifications. We would like to acknowledge the valuable assistance of the captain and crew of the R/V *J. W. Powell*, and by the *Delta*'s support team, especially pilots Rich Slater and Doug Privitt. We are grateful to D. McNeill for valuable discussion on platform evolution and to W. Schlager for comments on an earlier version of the chapter. We thank our shipboard colleagues during three cruises (A. Bosellini, A. Brown, P. Crevello, J. Dunham, G. Eberli, J. Kenter, I. MacIntyre, D. McNeill, S. Murray, H. Roberts, A. Saller, S. Schumm, P. Swart, and S. Williams). The manuscript was improved by thoughtful reviews from J. Kenter and H. Mullins. Sustaining support for our research comes from the Industrial Associates of the Comparative Sedimentology Laboratory. The contents of this chapter were derived primarily from the senior author's Ph.D. dissertation.

REFERENCES CITED

Alexandersson, T., 1972, Intragranular growth of marine aragonite and Mg-calcite: Evidence of precipitation from supersaturated sea water: Journal of Sedimentary Petrology, v. 42, p. 441-460.

Allen, J. R. L., 1985, Principles of Physical Sedimentology: Allen and Unwin, London, 272 p.

Bathurst, R. G. C., 1975, Carbonate Sediments and Their Diagenesis: Developments in Sedimentology 12, 2nd edition, Elsevier, New York, 658 p.

Blair, S. M., and J. N. Norris, 1988, The deep-water species of *Halimeda Lamouroux* (Halimedaceae, Chlorophyta) from San Salvador Island, Bahamas: Species composition, distribution and depth records: Coral Reefs, v. 6, p. 227-236.

Bosellini, A., 1984, Progradation geometries of carbonate platforms: Examples from the Triassic of the Dolomites, northern Italy: Sedimentology, v. 31, p. 1-24.

Burke, K., 1988, Tectonic evolution of the Caribbean: Annual Review Earth and Planetary Sciences, v. 16, p. 201-230.

Choquette, P. W., and F. C. Trussel, 1977, A procedure for making the Titan-Yellow stain for Mg-calcite permanent: Journal of Sedimentary Petrology, v. 47, p. 639-641.

Chorley, R. J., S. A. Schumm, and D. E. Sugden, 1984, Geomorphology: Methuen and Company, New York, 605 p.

Chowdury, R. N., 1978, Slope Analysis, Developments in Geotechnical Engineering, v. 22: Elsevier Scientific Publishing Company, Amsterdam, 423 p.

Cook, H. E., 1979, Ancient continental slope sequences and their value in understanding modern slope development, in Doyle, L. J., and O. H. Pilkey, (eds.), Geology of Continental Slopes: SEPM Special Publication No. 27, p. 287-305.

Cook, H. E., 1983, Ancient carbonate platform margins, slopes, and basins: Platform margin and deep water carbonates: SEPM Short Course No. 12, p. 5.1-5.189.

Cook, H. E., P. N. McDaniel, E. W. Mountjoy, and L. C. Pray, 1972, Allochthonous carbonate debris flows at Devonian bank ("reef") margins Alberta, Canada: Bulletin of Canadian Petroleum Geology, v. 20, p. 439-497.

Cook, H. E., and H. T. Mullins, 1983, Basin Margin, in Scholle, P. A., D. G. Bebout, and C. H. Moore, (eds.), Carbonate Depositional Environments: AAPG Memoir 33, p. 539-619.

Cook, H. E., and M. E. Taylor, 1977, Comparison of continental slope environments in the Upper Cambrian and lowest Ordovician of Nevada, in Cook, H. E., and P. Enos, (eds.), Deep-water Carbonate Environments: SEPM Special Publication No. 25, p. 51-81.

Cook, H. E., and M. E. Taylor, 1991, Carbonate-slope failures as indicators of sea level lowerings: AAPG Bulletin, v. 75, p. 556.

Crevello, P. D., and W. Schlager, 1980, Carbonate debris sheets and turbidites, Exuma Sound, Bahamas: Journal of Sedimentary Petrology, v. 50, p. 1121-1148.

Cruickshank, M. J., and T. J. Rowland, 1983, Mineral deposits at the shelfbreak, in Stanley, D. J., and G. T. Moore, (eds.), The Shelfbreak: Critical Interface on Continental Margins: SEPM Special Publication No. 33, p. 429-436.

Devaney, K. A., B. H. Wilkinson, and R. Van der Voo, 1986, Deposition and compaction of carbonate clinothems: The Silurian Pipe Creek Junior complex of east-central Indiana: Geological Society of America Bulletin, v. 97, p. 1367-1381.

Dietz, R. S., J. C. Holden, and W. P. Sproll, 1970, Geotectonic evolution and subsidence of Bahama Platform: Geological Society of America Bulletin, v. 81, p. 1915-1928.

Droxler, A. W., and W. Schlager, 1985, Glacial versus interglacial sedimentation rates and turbidite frequency in the Bahamas: Geology, v. 13, p. 799-802.

Enos, P., 1977, Tamabra Limestone of the Poza Rica Trend, Cretaceous, Mexico, in Cook, H. E., and P. Enos, (eds.), Deep-water Carbonate Environments: SEPM Special Publication No. 25, p. 273-314.

Enos, P., 1986, Diagenesis of Mid-Cretaceous rudist reefs, Valles Platform, Mexico, in Schroeder, J. H., and B. H. Purser, (eds.), Reef Diagenesis: Springer-Verlag, Berlin, p. 160-185.

Enos, P., and C. H. Moore, 1983, Fore-reef Slope, in Scholle, P. A., D. G. Bebout, and C. H. Moore, (eds.), Carbonate Depositional Environments: AAPG, Memoir 33, p. 507-539.

Enos, P., and R. D. Perkins, 1977, Quaternary sedimentation in South Florida: Geological Society of America, Memoir 147, 198 p.

Fairbanks, R. G., 1989, A 17,000-year glacio-eustatic sea level record: Influence of glacial melting rates on the Younger Dryas event and deep-ocean circulation: Nature, v. 342, p. 637-642.

Feda, J., 1982, Mechanics of particulate materials, the principles: Elsevier Scientific Publishing Company, Amsterdam, 447 p.

Friedman, G. M., 1985, The problem of submarine cement in classifying reefrock: An experience in frustration, in Schneidermann, N., and P. M. Harris, (eds.), Carbonate Cements: SEPM Special Publication No. 36, p. 117-121.

Friedman, G. M., A. J. Amiel, and N. Schneidermann, 1974, Submarine cementation in reefs—Example from the Red Sea: Journal of Sedimentary Petrology, v. 44, p. 816-825.

Garber, R. A., G. A. Grover, and P. M. Harris, 1989, Geology of the Capitan Shelf Margin—subsurface data from the Northern Delaware Basin, in Harris, P. M., and G. A. Grover, (eds.), Subsurface and outcrop examination of the Capitan Shelf Margin, Northern Delaware Basin: SEPM Core Workshop No. 13, p. 3-268.

Ginsburg, R. N., and N. P. James, 1976, Submarine botryoidal aragonite in Holocene reef limestones, Belize: Geology, v. 4, p. 431-436.

Ginsburg, R. N., D. S. Marszalek, and N. Schneidermann, 1971, Ultrastructure of carbonate cements in a Holocene algal reef of Bermuda: Journal of Sedimentary Petrology, v. 41, p. 472-482.

Grammer, G. M., and R. N. Ginsburg, 1992, Highstand vs. lowstand deposition on carbonate platform margins: Insight from Quaternary foreslopes in the Bahamas: Marine Geology, v. 103, p. 125-136.

Grammer, G. M., R. N. Ginsburg, and D. F. McNeill, 1990, Morphology and development of modern carbonate foreslopes, Tongue of the Ocean, Bahamas, in Larue, D. K., and G. Draper, (eds.), Transactions of the 12th Caribbean Geological Conference, St. Croix, U.S.V.I.: Miami Geological Society, Miami, Florida, p. 27-32.

Grammer, G. M., R. N. Ginsburg, P. K. Swart, D. F. McNeill, A. J. T. Jull, and D. R. Prezbindowski, 1993, Rapid growth rates of syndepositional marine aragonite cements in steep marginal slope deposits, Bahamas and Belize: Journal of Sedimentary Petrology, v. 63, no. 5.

Grigg, R. W., and D. Epp, 1989, Critical depth for the survival of coral islands: Effects on the Hawaiian Archipelago: Science, v. 243, p. 638-641.

Haddad, A., M. D. Aissaoui, and M. A. Soliman, 1984, Mixed carbonate-siliciclastic sedimentation on a Miocene fault-block, Gulf of Suez: Sedimentary Geology, v. 37, p. 182-202.

Halley, R. B., 1978, Estimating pore and cement volumes in thin section: Journal of Sedimentary Petrology, v. 48, p. 642-650.

Harris, M. T., 1988, Margin and foreslope deposits of the Latemar Carbonate Buildup (Middle Triassic), The Dolomites, Northern Italy: Unpublished Ph.D. dissertation, Johns Hopkins University, Baltimore, 473 p.

Hine, A. C., R. J. Wilber, J. M. Bane, A. C. Neumann, and K. R. Lorenson, 1981, Offbank transport of carbonate sands along open, leeward bank margins; northern Bahamas: Marine Geology, v. 42, p. 327-348.

Hubert, J. F., R. K. Suchecki, and R. K. M. Callahan, 1977, The Cow Head Breccia: Sedimentology of the Cambro-Ordovician continental margin, Newfoundland, in Cook, H. E., and P. Enos, (eds.), Deep-Water Carbonate Environments: SEPM Special Publication No. 25, p. 125-154.

James, N. P., and R. N. Ginsburg, 1979, The seaward margin of Belize barrier and atoll reefs: International Association of Sedimentologists, Special Publication no. 3, 191 p.

James, N. P., R. N. Ginsburg, D. S. Marszalek, and P. W. Choquette, 1976, Facies and fabric specificity of early subsea cements in shallow Belize (British Honduras) reefs: Journal of Sedimentary Petrology, v. 46, p. 523-544.

Johns, H. D., and C. H. Moore, 1988, Reef to basin sediment transport using Halimeda as a sediment tracer, Grand Cayman Island, West Indies: Coral Reefs, v. 6, p. 187-193.

Kennett, J. P., 1990, The Younger Dryas cooling event: An Introduction: Paleoceanography, v. 5, p. 891-895.

Kenter, J. A. M., 1990, Carbonate platform flanks: slope angle and sediment fabric: Sedimentology, v. 37, p. 777-794.

Kenter, J. A. M., and A. E. Campbell, 1991, Sedimentation on a Jurassic carbonate platform flank: Geometry, sediment fabric and related depositional structures (Djebel Bou Dahar, High Atlas, Morocco): Sedimentary Geology, v. 72, p. 1-34.

Kerans, C., N. F. Hurley, and P. E. Playford, 1986, Marine diagenesis in Devonian reef complexes of the Canning Basin, Western Australia, in Schroeder, J. H., and B. H. Purser, (eds.), Reef Diagenesis: Springer-Verlag, Berlin, p. 357-380.

Kirkby, M. J., 1987, General models of long-term slope evolution through mass movement, in Anderson, M. G., and K. S. Richards, (eds.), Slope Stability: John Wiley and Sons, New York, p. 359-379.

Krueger, W. C., and F. K. North, 1983, Occurrences of oil and gas in association with the paleo-shelf-break, in Stanley, D. J., and G. T. Moore, (eds.), The Shelfbreak: Critical Interface on Continental Margins: SEPM Special Publication No. 33, p. 409-427.

Land, L. S., and C. H. Moore, 1977, Deep forereef and upper island slope, north Jamaica: AAPG, Studies in Geology No. 4, p. 53-65.

Linick, T. W., A. J. T. Jull, L. J. Toolin, and D. J. Donahue, 1986, Operation of the NSF-Arizona Accelerator Facility for Radioisotope Analysis and results from selected collaborative research projects: Radiocarbon, v. 28, p. 522-533.

Longman, M. W., 1980, Carbonate diagenetic textures from near-surface diagenetic environments: AAPG Bulletin, v. 64, p. 461-487.

Lowe, D. R., 1979, Sediment gravity flows: Their classification and some problems of application to natural flows and deposits, in Doyle, L. J., and O. H. Pilkey Jr., (eds.), Geology of Continental Slopes: SEPM Special Publication No. 27, p. 75-82.

MacIntyre, I. G., 1977, Distribution of submarine cements in a modern Caribbean fringing reef, Galeta Point, Panama: Journal of Sedimentary Petrology, v. 47, p. 503-516.

MacIntyre, I. G., 1985, Submarine cements—the peloidal question, in Schneidermann, N., and P. M. Harris, (eds.), Carbonate Cements: SEPM Special Publication No. 36, p. 109-116.

McIlreath, I. A., 1977, Accumulation of a Middle Cambrian, deep-water limestone debris apron adjacent to a vertical, submarine carbonate escarpment, Southern Rocky Mountains, Canada, in Cook, H. E., and P. Enos, (eds.), Deep-Water Carbonate Environments: SEPM Special Publication No. 25, p. 113-124.

Minero, C. J., 1991, Sedimentation and diagenesis along open and island-protected windward margins of the Cretaceous El Abra Formation, Mexico: Sedimentary Geology, v. 71, p. 261-288.

Moore, C. H., E. A. Graham, and L. S. Land, 1976, Sediment transport and dispersal across the deep fore-reef and island slope (–55m to –305m), Discovery Bay, Jamaica: Journal of Sedimentary Petrology, v. 46, p. 174-187.

Mullins, H. T., 1983, Modern carbonate slopes and basins of the Bahamas: SEPM Short Course No. 12, p. 4.1-4.138.

Mullins, H. T., and G. W. Lynts, 1977, Origin of the northwestern Bahama Platform: Review and reinterpretation: Geological Society of America Bulletin, v. 88, p. 1447-1461.

Palmer, M., 1979, Holocene facies geometry of the leeward bank margin of Tongue of the Ocean: Unpublished M.S. Thesis, University of Miami, FL, 199 p.

Pirazzoli, P. A., 1986, Marine notches, in Van de Plassche, O., (ed.), Sea Level Research: A manual for the collection and evaluation of data: IGCP, p. 361-400.

Plas, L. van der, and A. C. Tobi, 1965, A chart for judging the reliability of point counting results: American Journal of Science, v. 263, p. 87-90.

Playford, P. E., 1980, Devonian "Great Barrier Reef" of Canning Basin, Western Australia: AAPG Bulletin, v. 64, p. 814-840.

Playford, P. E., 1984, Platform-margin and marginal slope relationships in Devonian reef complexes of the Canning Basin: Proceedings of the Geological Society of Australia and Petroleum Exploration Society of Australia Symposium, Perth, Western Australia, p. 189-214.

Playford, P. E., N. F. Hurley, C. Kerans, and M. F. Middleton, 1989, Reefal platform development, Devonian of the Canning Basin, Western Australia, in Crevello, P. D., J. L. Wilson, J. F. Sarg, and J. F. Read, (eds.), Controls on Carbonate Platform and Basin Development: SEPM Special Publication No. 44, p. 187-202.

Rucklidge, J. C., 1984, Radioisotope detection and dating with particle accelerators, in Mahaney, W. C., (ed.), Quaternary Dating Methods, Developments in Paleontology and Stratigraphy 7: Elsevier, New York, p. 17-32.

Rutzler, K., 1975, The role of burrowing sponges in bioerosion: Oecologia, v. 19, p. 203-216.

Saller, A. H., J. W. Barton, and R. E. Barton, 1989, Slope sedimentation associated with a vertically building shelf, Bone Spring Formation, Mescalero Escarpe Field, southeastern New Mexico, in Crevello, P. D., J. L. Wilson, J. F. Sarg, and J. F. Read, (eds.), Controls on Carbonate Platform and Basin Development: SEPM Special Publication No. 44, p. 275-288.

Sandberg, P., 1985, Aragonite cements and their occurrence in ancient limestones, in Schneidermann, N., and P. M. Harris, (eds.), Carbonate Cements: SEPM Special Publication No. 36, p. 33-57.

Sarg, J. F., 1988, Carbonate sequence stratigraphy, in Wilgus, C K., B. S. Hastings, C. G. St. C. Kendall, H. W. Posamentier, C. A. Ross, and J. C. Van Wagoner, (eds.), Sea Level Changes—An Integrated Approach: SEPM Special Publication No. 42, p. 155-181.

Schlager, W., 1989, Drowning unconformities on carbonate platforms, in Crevello, P. D., J. L. Wilson, J. F. Sarg, and J. F. Read, (eds.), Controls on Carbonate Platform and Basin Development: SEPM Special Publication No. 44, p. 15-25.

Schlager, W., and O. Camber, 1986, Submarine slope angles, drowning unconformities, and self-erosion of limestone escarpments: Geology, v. 14, p. 762-765.

Schroeder, J. H., 1973, Submarine and vadose cements in Pleistocene Bermuda reef rock: Sedimentary Geology, v. 10, p. 179-204.

Shinn, E. A., 1969, Submarine lithification of Holocene carbonate sediments in the Persian Gulf: Sedimentology, v. 12, p. 109-144.

Suchanek, T., 1989, A guide to the identification of the common corals of St. Croix, *in* Hubbard, D. K., (ed.), Terrestial and Marine Geology of St. Croix, U. S. Virgin Islands: West Indies Laboratory, Special Publication no. 8, p. 197-213.

Ward, R. F., C. G. St. C. Kendall, and P. M. Harris, 1986, Upper Permian (Guadalupian) facies and their association with hydrocarbons—Permian Basin, West Texas and New Mexico: AAPG Bulletin, v. 70, p. 239-262.

Wilber, R. J., J. D. Milliman, and R. B. Halley, 1990, Accumulation of Holocene banktop sediment on the western margin of Great Bahama Bank: Geology, v. 18, p. 970-974.

Yurewicz, D. A., 1977, Origin of the massive facies of the Lower and Middle Capitan Limestone (Permian), Guadalupe Mountains, New Mexico and West Texas: SEPM (Permian Basin Section) Guidebook No. 77-16, p. 45-92.

Chapter 5

Influence of Sediment Type and Depositional Processes on Stratal Patterns in the Permian Basin-Margin Lamar Limestone, McKittrick Canyon, Texas

A. A. Brown
R. G. Loucks
ARCO Exploration and Production Technology
Plano, Texas, U.S.A.

ABSTRACT

Basin-margin, Late Guadalupian carbonate strata contain units with alternating baselap patterns. Downlap occurs at the base of steeply dipping boulder-conglomerate lower slope strata; onlap occurs at the base of gently dipping wackestone-rich toe-of-slope strata. The downlap and onlap stratal patterns reflect differences in gravity-flow deposition in response to different sediment type supplied to the basin margin. The lower slope was too steep for deposition of peloidal carbonate muds transported by low-density turbidity currents, so gravity flows bypassed the slope and deposited carbonate mud–rich beds at the toe-of-slope, where they onlapped lower slope boulder conglomerates. Matrix-poor boulder conglomerates terminate by downlap in talus cones because the rock falls and low-matrix debris flows responsible for their transport reached a slope too gentle for continued transport. In contrast, steeply dipping, matrix-rich boulder-bearing debris-flow deposits interfinger downslope into finer grained debris-flow and turbidity current deposits concordant with underlying strata.

The sediment type delivered to the basin margin varied systematically during the Late Guadalupian due to sea-level fluctuations. Relative sea-level stand was interpreted from correlation to the relative sea-level record of contemporaneous shelf strata and from analogy to the Bahamian Quaternary carbonate sediment history. Siliciclastic silts were deposited in the basin during lowstands as onlapping strata. Downlapping silt-matrix boulder conglomerates (units 1 and 7) were deposited during the early transgression, when fringing reefs supplied boundstone boulders and siliciclastic silt could be transported across the emergent shelf. Downlapping matrix-poor boulder conglomerates (unit 2) were deposited during the late transgression, when fringing reefs were still growing and the shelf was flooded enough to stop the basinward transport of quartz silt but was not flooded enough to produce significant quantities of carbonate mud. Onlapping toe-of-slope wacke-stones (unit 3) were deposited during early highstand, when the flooded shelf was producing carbonate mud and the reefs either had not caught up with sea level rise or had stepped back onto the shelf. Boulder conglomerates with lime-mud matrixes (units 4, 6, and 7) formed after reefs had caught up with sea-level rise or had prograded to the shelf edge (late highstand).

The stratal patterns in these Permian, basin-margin carbonates are different from those commonly interpreted in generalizations of siliciclastic sequences (e.g., Vail et al., 1984). The Late Guadalupian stratal patterns are caused by the change in carbonate sediment type and quantity with change of relative sea level, rather than by proximity to shoreline with changing relative sea level. This study demonstrates the importance of understanding the relationship between sea level, sediment supply, and depositional mechanisms before using stratal patterns to interpret relative sea levels in carbonate basin margins.

INTRODUCTION

Late Guadalupian (Permian) lower slope and toe-of-slope carbonates exposed at McKittrick Canyon, in Guadalupe Mountains National Park (Figure 1), were deposited predominantly by gravity flows originating at the shelf margin and on the slope. These deposits contain correlatable surfaces of onlap and downlap separating units with different styles of gravity-flow sedimentation. These surfaces can be used to identify sequences in the sense of Mitchum et al. (1977); that is, the surfaces are unconformities or are correlatable to unconformities, and they separate relatively conformable strata. The relative sea-level history of these sequences can be determined by biostratigraphic correlation to the sea level record of equivalent shelf strata and by the types of sediment supplied to the basin margin. The McKittrick Canyon exposures thus provide a basin-margin carbonate system in which stratal patterns (patterns of systematic bed termination, convergence, or concordance) can be related to sediment types, depositional mechanisms, and relative sea-level positions.

The objectives of this study are to interpret the physical stratigraphy and determine the controls of stratal patterns in basin-margin deposits of this Permian carbonate system. To meet these objectives, the stratal patterns and lithology distribution of the basin-margin Lamar Limestone and equivalent strata exposed near McKittrick Canyon were described by measuring sections, walking out key beds and surfaces of discordance, and mapping stratal patterns from photographs. Depositional mechanisms were interpreted from sedimentary structures, rock textures, and bed continuity. The basin-margin section was correlated to the shelf using published biostratigraphic information, and relative sea level of the shelf section was interpreted from published descriptions. These data are available on file at the Guadalupe Mountains National Park.

GEOLOGIC SETTING

The field area is located at the mouth of McKittrick Canyon, on the east side of the Guadalupe Mountains in Guadalupe Mountains National Park, Texas (Figure 1). The east side of this mountain range is a late Tertiary erosional scarp that exposes the Late Guadalupian shelf margin on the northwest side of the Delaware basin. McKittrick Canyon cuts the shelf margin almost parallel to depositional dip. The youngest reef and much of the shelf carbonate section have been eroded; however, the youngest Guadalupian lower slope and toe-of-slope deposits are well exposed at the mouth of McKittrick Canyon and just north of the canyon. The stratigraphic interval studied contains the Lamar Limestone of the Bell Canyon Formation, unnamed siltstone members of the Bell Canyon Formation above and below the Lamar Member, and the shelf-margin (Capitan Formation) and shelf (Tansill Formation and uppermost Yates Formation) stratigraphic equivalents of the upper Bell Canyon Formation (Figure 2). More extensive discussion of the regional setting of

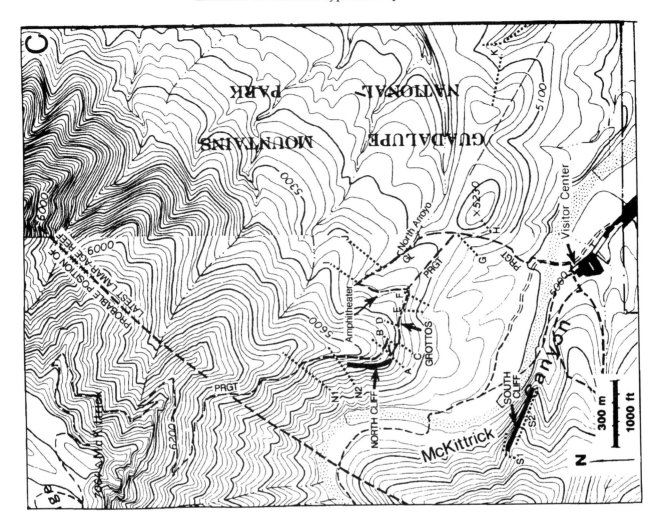

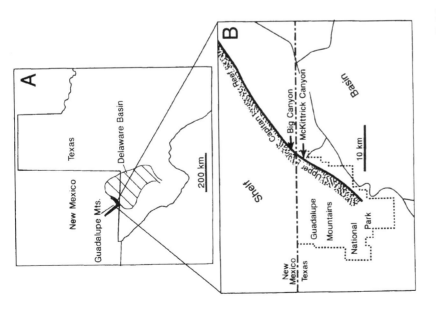

Figure 1. (A) Location of the Guadalupe Mountains field area relative to present geography (shown by state boundary outlines) and to the Delaware basin. (B) Location of the McKittrick Canyon study area in Guadalupe Mountains National Park. (C) Location of measured sections (short dashed lines) and some geographic features at the mouth of McKittrick Canyon. The youngest Guadalupian shelf margin is eroded; its approximate position is marked by the dashed line. Older Capitan shelf margins are northwest of and parallel to the dashed line. McKittrick Canyon cuts a dip section through these shelf margins. Trails are marked by the long dashed lines. PRGT is the Permian Reef Geology Trail and GL is the Geology Loop.

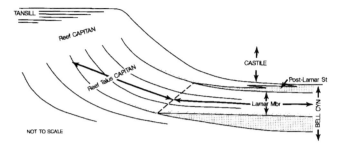

Figure 2. Schematic bedding relationships between time equivalents of the Capitan Formation. Shelf formations (Yates and Tansill formations) are distinguished by horizontal bedding planes. Reef facies of the Capitan Formation is characterized by the absence of bedding, whereas reef-talus facies is characterized by inclined bedding planes. This terminology follows Newell et al. (1953) and does not necessarily have environmental significance. As used in this study, the Bell Canyon Formation extends to the updip siltstone pinch-outs and a line connecting these pinch-outs separates the Lamar Limestone from the Capitan Formation.

Guadalupian deposition is presented in Ward et al. (1986).

The Lamar Limestone in the study area was deposited in deep water just seaward of the Capitan reef. Present-day topographic relief between basin and equivalent shelf strata exceeds 600 m (Figure 3A). True paleobathymetric relief was probably greater because the relief has not been corrected for the substantial compaction in lower slope carbonates. Figure 3 summarizes the paleoenvironmental terminology used in this study. Basin margin refers to the lower slope, toe-of-slope, and proximal basin (Cook and Mullins, 1983), shelf foreslope refers to the steeply dipping upper slope (following Davies, 1977), and shelf margin refers to the reef complex and shelf edge (Wilson, 1975). All strata examined in the study area are basin-margin strata deposited by various mechanisms of gravity-flow deposition. Although some pelagic and planktonic organisms are identified in the basin-margin strata, most of the carbonate sediment was produced on the shelf margin and shelf foreslope.

Sediment type, fossils, and depositional slope change abruptly with paleobathymetric position (Figure 3B). In general, the basinal Lamar Limestone thins and becomes less grainy toward the center of the Delaware basin. Grainstones and cobble- and pebble-intraclastic packstones are present only within about 3 km of the shelf edge. Skeletal packstones extend about 8 km from the shelf (Brown and Loucks, 1987). Carbonate mudstones and wackestones dominate the section beyond 1.5 km from the shelf edge, and constitute the entire section beyond 8 km from

the shelf edge. The wackestones and mudstones become more argillaceous and silty toward the basin center. Beyond about 40 km from the shelf margin, all carbonates contain siliciclastic material (Tyrrell, 1969). This pattern reflects the fact that deep-water carbonate sediment was derived from the shelf and shelf-margin systems. There were no abundant calcareous planktonic organisms during Permian time (L. Babcock, 1977). The deep-water siliciclastics were derived from the shelf siliciclastic system.

LITHOLOGY AND STRATAL PATTERNS OF BASIN-MARGIN STRATA

Stratigraphic sections were measured near the mouth of McKittrick Canyon to determine the distribution of lithofacies in basin-margin strata. The carbonate lithofacies were grouped into packstones (subdivided by their largest clast size), wackestones (subdivided into burrow-mottled wackestone and sparsely to unburrowed wackestone), and grainstones [Table 1, Figure 4 (foldout, following p. 138)]. The more common rock types are figured in Tyrrell (1962, 1969). The depositional characteristics and depositional mechanisms (interpreted using criteria of Lowe, 1976, 1979, 1982, and Nardin et al., 1979) are summarized in Table 1. Correlations between measured sections were determined by walking out key beds, observing bedding relationships with binoculars from across the canyon, tracing beds on photo mosaics, and correlating lithostratigraphy to sections on the south wall of the canyon and to sections north of the north wall. Many bedding terminations were identified from photo mosaics and binocular examination from the opposite side of the canyon and confirmed by walking out strata. Some bed terminations could be determined only by walking out strata.

STRATAL PATTERNS

Basin-margin strata were divided into seven stratigraphic units (Table 2, Figure 4). Units are separated by surfaces which show a systematic pattern of bed termination (such as downlap, onlap, or truncation) or by surfaces near which bedding converges. As in all basin-margin strata, individual beds terminate at almost all stratigraphic horizons. The features which characterize the unit bounding surfaces are the systematic termination or convergence pattern and correlatability of the surfaces.

One of the key concerns of this study was distinguishing bed termination from bed convergence. The concepts of downlap, onlap, and truncation are well discussed in the sequence stratigraphic literature (e.g., Mitchum et al., 1977). The terms imply that the observed lapout is the true lapout and that the nondepositional hiatus lies beyond the lapout. On the scale of observation from well logs or seismic, this can

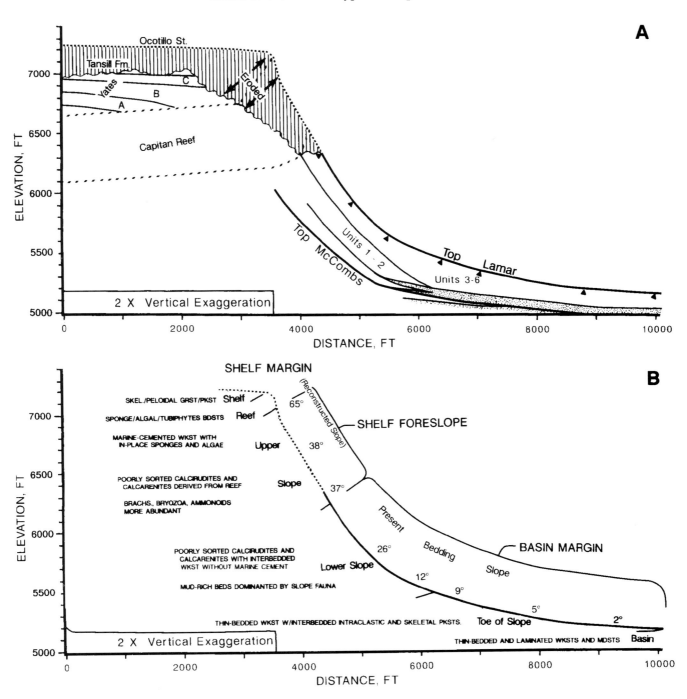

Figure 3. Paleotopography and paleoslopes of the Capitan and Lamar limestones. (A) Dip cross section showing paleotopography reconstructed along the north wall of McKittrick Canyon, based on present-day elevations of control points on the top of the Lamar Limestone projected onto the dip line (triangles). Vertical lines indicate eroded shelf and shelf-margin strata. The thicknesses of these units and position of the youngest Lamar shelf edge were reconstructed from Dark Canyon data (Tyrrell, 1962; Tyrrell et al., 1978). (B) Shelf-margin, shelf-foreslope, and basin-margin physiographic environments and generalized sediment types projected onto the paleoslope of the top of the Lamar Limestone and its shelf equivalent. Basin margin refers to the lower slope, toe-of-slope, and proximal basin (Cook and Mullins, 1983); shelf foreslope refers to the steeply dipping upper slope (following Davies, 1977); and shelf margin refers to the reef complex and shelf edge (Wilson, 1975). True Lamar basin strata are not exposed at McKittrick Canyon. Data for the upper slope and reef lithologies are from J. Babcock (1977) and Mruk (1989), with additional observations by the authors.

Table 1. Interpreted depositional mechanisms.

PACKSTONES AND GRAINSTONES

Characteristics	Depositional Mechanism
Largest Clast > 0.5 m diameter:	
Matrix-poor boulder orthoconglomerate of sponge boundstone clasts; ungraded, angular clasts.	Rock falls or matrix-poor debris flows derived from the upper slope and reef and deposited as proximal slope talus cone deposits.
Wackestone-matrix boulder ortho- and paraconglomerate of sponge boundstone clasts; some with inversely graded bases, boulders protrude from top of beds, some with thin (<5 cm) packstone/grainstone cap. Matrix may also be silty or dolomitic.	Debris flows derived from reef and upper slope.
50 cm > Largest Clast > 5 cm:	
Packstone-matrix ortho- and paraconglomerate with boundstone and slope wackestone clasts. Ungraded, clasts protrude from top of bed, some with 1–3-cm-thick grainstone cap.	Debris flows derived from shelf foreslope modified by incorporation of slope clasts and muddy matrix.
Wackestone- or packstone-matrix ortho- and paraconglomerate with wackestone and packstone clasts. Ungraded, structureless, some clasts protrude from top of bed.	Debris flows derived from the basin margin.
5 cm > Largest Clast > 0.5 cm:	
Thin to medium beds of packstone with large bioclasts and small wackestone clasts. Ungraded, structureless, with no selective orientation of shells. Some beds abruptly overlain by graded packstone with ripple structures.	Debris flows.
Thin to medium beds of normally or inversely graded skeletal packstone. Rare beds with multiple inversely graded intervals. Faint low-angle cross-strata. Matrix content small.	High-density turbidity currents.
0.5 cm > Largest Clast > 0.5 mm:	
Thin beds of skeletal-peloidal grainstone with normal grading.	Turbidity currents.
Medium- to thick-bedded, well-sorted, ungraded, skeletal grainstone with high-angle, basinward-dipping, tabular cross-beds in dip-oriented channels.	High-density turbidity currents.

Table 1. Continued.

Characteristics	Depositional Mechanism
Thin beds of ungraded, skeletal packstone with abrupt upper and lower contacts and no sedimentary structures, usually immediately underlying a graded peloidal wackestone bed.	Slope-derived debris flows.
Thin beds of skeletal, peloidal packstone without sedimentary structures and irregular upper and lower contacts.	Turbidity currents (?)
WACKESTONES	
Laminated wackestone with graded lamina, some with millimeter-thick packstones of very fine-grained skeletal and peloidal grains at the base. Deep-water fauna.	Low-density turbidity currents derived from the lower slope.
Thin-bedded skeletal, peloidal wackestone, most with basal thin (2–5 mm thick) packstone, grading at the top and base, and faint, low-angle cross-lamination and climbing ripples.	Low-density turbidity currents.
Thin-bedded skeletal, peloidal wackestone with irregular upper and lower contacts, scattered feeding traces, and absence of sedimentary structures. Some beds are graded near the top.	Low-density turbidity current deposits (?) modified by burrowing and differential compaction in a dysaerobic setting.
Thin- to medium-bedded, skeletal, peloidal wackestone with numerous *Planolites* burrows and indistinct bed contacts. No grading or primary sedimentary structures.	Low-density turbidity-current deposits (?) modified by extensive burrowing in an aerobic setting.
Thick-bedded, skeletal, peloidal wackestone with mottled fabric interpreted to be caused by bioturbation.	Low-density turbidity-current-deposits (?) modified by extensive burrowing in an aerobic setting.
Thick-bedded, skeletal-peloidal wackestone with soft-sediment deformation fabrics, or faint compacted breccia fabric. Hydrodynamic sedimentary structures and grading are absent.	Turbidity-current (?) deposits homogenized by extensive deformation near glide planes. Some may be mud-matrix, mud-clast debris-flow deposits.

Table 2. Stratal relationships of units.

	Proximal	Medial	Distal
UNIT 1 Basal contact:	Downlap	Concordant	Concordant
Lithology:	Conglomerate with bound-stone boulders and silt matrix	Silty packstones with bound-stone clasts	Skeletal packstone and siliciclastic siltstone
UNIT 2 Basal contact:	Downlap	Absent	Absent
Lithology:	Matrix-poor, boundstone-boulder conglomerate with grainstones		
UNIT 3 Basal contact:	Absent	Shelfward onlap	Concordant
Lithology:		Wackestones, skeletal pack-stones and grainstones with-out boundstone clasts	Wackestone with rotational slumps and glide-plane deformation
UNIT 4 Basal contact:	Apparent downlap	Concordant	Concordant
Lithology:	Wackestone-matrix, bound-stone-boulder conglomerate	Boundstone-boulder and wackestone-cobble conglomerate interbedded with thin-bedded wackestone with interbedded skeletal packstone; shelfward converging beds (apparent onlap) in middle of unit	Thin-bedded and laminated wackestone ith interbedded packstones in lower part of unit
UNIT 5 Basal contact:	Shelfward onlap (or glide-plane truncation)	Concordant	Concordant
Lithology:	Thin-bedded wackestone	Thin-bedded wackestone with wackestone-clast pack-stone and skeletal-peloidal packstone	Laminated and thin-bedded wackestone, skeletal-peloidal packstone
UNIT 6 Basal Contact:	Basinward downlap	Shelfward onlap	Concordant
Lithology:	Wackestone-matrix, bound-stone-boulder conglomerate with interbedded burrowed wackestone	Burrowed wackestone cut by major packstone-filled slope gully	Burrowed wackestone cut by packstone-filled gully
UNIT 7 Basal contact:	Apparent downlap	Onlap	Concordant
Lithology:	Silty wackestone matrix, boundstone-boulder conglomerate	Silty wackestone matrix bound-stone-boulder conglomerate with silty wackestones in lower part of unit	Silty wackestone matrix boundstone-boulder con-glomerate interbedded with siltstone and silty wackestone

rarely be demonstrated for downlap surfaces because the nondepositional hiatus is represented by a condensed zone of finite thickness, which is unresolvable from seismic and internally uncorrelatable on well logs. Operationally, downlap surfaces and onlap surfaces are recognized by seaward convergence of dipping beds onto beds with a lower dip and landward convergence of gently dipping beds onto beds with steeper dips. In this study, beds were walked out to determine if they truly terminate or if they just thin and change dip. True downlap and onlap surfaces with bed termination will be called downlap and onlap surfaces, respectively. Surfaces with basinward convergence without true bed termination will be called apparent downlap surfaces, and surfaces with shelfward convergence without true bed termination will be called apparent onlap surfaces. These surfaces are not true baselap surfaces; however, on the coarser resolution of well-log or seismic data, they may easily be interpreted as baselaps.

Downlap Surfaces

True downlap occurs at the base of units 1, 2, and possibly unit 6 in the section proximal to the slope (Figure 4). Downlap occurs in beds of matrix-poor boulder conglomerate in units 1 and 2 (Figure 5). These deposits are interpreted to be talus cones deposited by rock falls and possibly debris flows at the inflection of slope between the lower slope and the proximal toe-of-slope settings. Because these deposits are localized at this inflection, the downlap was probably caused by the abrupt change in slope. Slopes were sufficiently steep on the lower slope for transport of rock falls and matrix-poor debris flows, but the gentler slopes on the toe-of-slope was insufficient for transport farther basinward. No correlative secondary flows were identified downslope from these deposits. It is speculated that the sparsity of fine-grained material prevented formation of secondary gravity flows which could transport at the gentle slopes characteristic of the toe-of-slope setting.

Downlap at the base of unit 6 occurs where unit 6 buries a bathymetric high on the top of unit 5 (Figures 4, 5). This downlap may be caused by truncation by a low-angle glide-plane cutting parallel to the contact instead of sedimentologic processes, because the contact is poorly exposed and glide planes are common in this part of the basin margin. This is a feature localized by the thickening of unit 5 on the north side of McKittrick Canyon. On the south side of the canyon, unit 5 does not thicken as noticeably and true downlap at the base of unit 6 is not observed (Figure 6).

Apparent Downlap Surfaces

Apparent downlap occurs near the base of units 4, 6, and 7 and near the top of unit 1, all at positions proximal to the slope (Figure 4). These surfaces are characterized by convergence of steeply dipping beds against an underlying, less steeply dipping surface similar to true downlap. However, near the lower surface, the dipping beds thin and become parallel to the underlying surface. This results in a less distinctive surface, which is geometrically similar to the condensed zones interpreted in sequence stratigraphic analysis of well-log data. This transition is characterized by a change from very thick-bedded wackestone-matrix, boulder conglomerates to thick-bedded intraclast packstones and wackestone-matrix cobble conglomerates interbedded with thin-bedded wackestone. In unit 1, siliciclastic silt is present in the matrix instead of carbonate mud. Some individual beds appear to be traceable across the transition from boulder conglomerate to intraclast packstone. All of the coarse-grained lithologies have characteristics of debris-flow deposits—absence of grading and sedimentary structures (except for inverse grading at the base of some deposits), poor sorting, and clasts protruding from the upper surface of the deposit. It is speculated that the high matrix content of the debris flows allowed formation of secondary flows without boulders which could have transported along the gentler slopes on the toe-of-slope, whereas the low matrix content of debris flows and rock falls in unit 2 and parts of unit 1 precluded formation of secondary flows.

Onlap Surfaces

Surfaces of true onlap occur at the base of unit 3 near its updip limit, at the base of unit 5 at its updip limit, at the base of unit 6 in the medial part of the section, and at the base of unit 7 in the medial part of the section (Figure 4). Onlap of units 3 and 5 is caused by termination of gently dipping wackestone and packstone beds against steeply dipping boulder conglomerate beds (Figure 5). The thin-bedded wackestones are interpreted as low-density turbidity current deposits. Wackestone termination against the steeper dipping boulder conglomerates probably results from carbonate mud bypassing the steep slope because the slope exceeded the angle of repose of carbonate mud and steep slopes resulted in turbidity current velocities great enough to keep the mud in suspension.

Onlap of unit 6 occurs on the downslope end of the paleohigh, which developed on the toe-of-slope during deposition of unit 5 (Figures 4, 5). Like the true downlap at the base of unit 6, this baselap is a local feature restricted to areas downdip of thickening of unit 5. The baselap pattern at the base of unit 6 is interpreted to be caused by burial of a paleotopographic feature in unit 5 by burrowed carbonate muds of unit 6 because position of the baselap terminations is controlled by the position of the thick portion of unit 5.

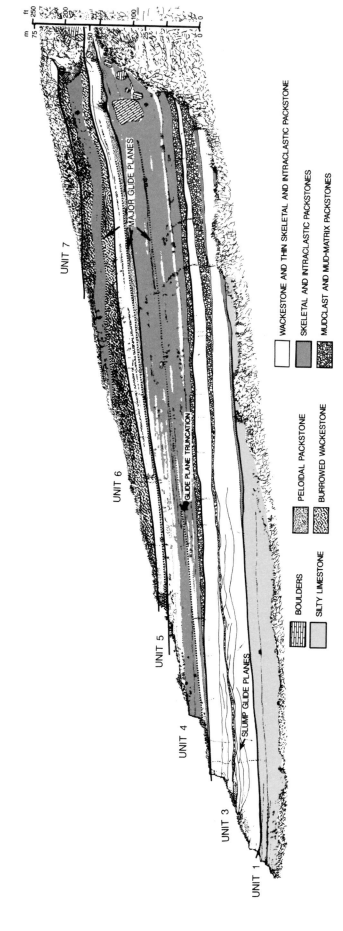

Figure 6. Line drawing of lithology and bedding relationships on the south cliff, on the south side of McKittrick Canyon. Translational glide planes (dashed lines) and slump glide planes truncate bedding (broad arrow). Upper part of unit 4 shows shelfward backstepping of the packstone-to wackestone transition, which represents apparent onlap. Thinning and termination of unit 5 is affected by glide planes around large boundstone boulders, so nature of onlap cannot be determined. Bedding pattern interpreted from photo mosaics. Lithology was determined from measured sections at each end of the cliff and by binocular identification of wackestone lithology by its characteristic thin-bedded weathering pattern.

Onlap of unit 7 onto unit 6 is characterized by onlap of silty carbonate wackestone and intraclast packstone onto intraclast packstone at the top of unit 6. Carbonates in unit 7 have a somewhat gentler dip than underlying carbonates of unit 6. Where onlap occurs, both units contain wackestone and packstone beds, although the wackestones of unit 7 contain siliciclastic silt.

Apparent Onlap

Apparent onlap occurs in unit 4 in the upper part of the toe-of-slope and possibly at the base of unit 7 in the downdip part of the medial toe-of-slope (Figure 4). Apparent onlap in unit 4 occurs where thick units of gently dipping, thin-bedded wackestones thin as they interfinger upslope with somewhat steeper dipping, thick beds of intraclastic packstone (Figure 6). The packstone beds abruptly thin in a down-paleoslope direction. The transition between the wackestone- and packstone-dominated facies shifts shelfward in stratigraphically higher beds. The resulting overall geometry is a basinward-thinning wedge of packstone beds overlain by a shelfward thinning wedge of wackestone (Figure 4, between sections A and E). If the contemporaneity of the beds could not be demonstrated by the interbedded facies change from wackestone to packstone, this pattern might be interpreted as onlap. This geometry may not be coincidental; it could be caused by the different response of the packstone and grainstone depositional mechanisms to small changes in paleoslope. Downslope thickening of the wackestone facies is probably caused by the decreasing paleoslope, which aids suspension fallout from the low-density turbidity currents responsible for wackestone deposition. Individual packstone beds abruptly thin and terminate down paleoslope as paleoslope decreases. Downslope limit to the packstone facies is probably caused by freezing of the debris flows as slope decreases. The shelfward shift in the packstone to wackestone facies boundary is interpreted to reflect decreasing amounts of muddy carbonate sand and increasing amounts of sandy carbonate mud caused by sediment supply.

Some of the onlapping packstones at the base of unit 7 in the medial toe-of-slope setting (downdip from section I) might actually be concordant with the packstone at the top of unit 6 because the terminations of beds were difficult to identify and the character of the packstone at the top of unit 6 is variable. If so, then this onlap surface is characterized by apparent onlap in addition to true onlap observed farther updip (near and updip from section I). However, most of unit 6 is not contemporaneous with unit 7 because lower unit 7 beds are contemporaneous only to the amalgamated upper part of the packstone bed at the top of unit 6 and not to the bioturbated wackestone.

Truncation

Truncation results from channel-base erosion and from structural truncation by low-angle glide planes. Areally extensive surfaces of erosional truncation were not observed. Small-scale erosional truncation was observed at the base of dip- and oblique-oriented channels in all units. The largest area of channel base erosional truncation occurs at the base of a large slope gully developed in units 5 and 6 (Figure 7). The slope gully is about 600 m wide and is oriented almost directly down paleoslope. It was probably a low-relief feature which channelized coarse-grained debris flows, high-density turbidity flows, and other currents, while carbonate muds were deposited and burrowed on surrounding highs. Erosional relief decreases in a downdip direction.

Structural truncation is more widespread. Flat, low-angle basinward-dipping glide planes truncate underlying beds in a basinward direction. Overlying beds terminate at the truncation surface in a pattern identical with onlap (Figure 6). Structural origin of these features can be recognized by close inspection of the surface (Figure 8). Concave-upward rotational-glide planes also truncate strata (Figure 6, unit 3).

RELATIVE SEA-LEVEL HISTORY

Two approaches were used to determine relative sea-level history. First, sediment type on the basin margin supplied from the shelf margin indicates the type of sediment produced on the shelf margin and shelf. Quantity and type of sediment production is assumed to respond to relative sea level in a manner similar to the response of Quaternary carbonates to sea-level changes. Second, shelf sediments directly record sea-level changes by subaerial exposure and landward and seaward shifting of depositional facies. The sea-level record for the shelves was interpreted from published measured sections and the resultant sea-level history was correlated to the basin margin by fusulinid biostratigraphy.

Sea Level Determined from Basin-Margin Sediment Type

There are systematic differences in the total amount of carbonate mud and reef-derived carbonate boulders in Late Guadalupian basin-margin rocks deposited at different times (Figure 4). These differences are interpreted to result from differences in shallow-water production and delivery of sediment to the basin margin in response to changing sea level. Quaternary carbonates provide an analog for this process that can be adapted to evaluate Late Guadalupian sea-level changes.

Quaternary Carbonate Sediment Supply Patterns

Quaternary shelf-derived carbonate mud is deposited in periplatform areas mainly during high-

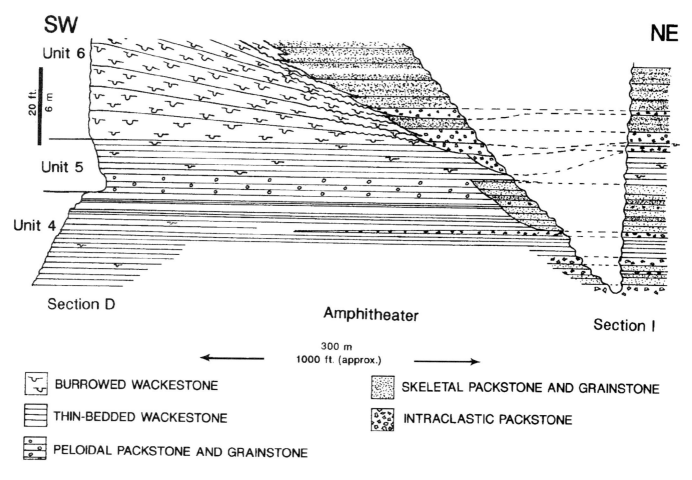

Figure 7. Schematic cross section of south margin of the slope gully in units 5 and 6 showing interbedding of gully packstones and surrounding toe-of-slope wackestones. The section is a strike section extending from section D at the Grotto, through the Amphitheater, to section I at the North Arroyo (see Figure 1). Scale is approximate.

stands (e.g., Droxler and Schlager, 1985; Wilber et al., 1990; Glaser and Droxler, 1991; Grammer and Ginsburg, 1992; Grammer et al., 1993). During lowstands, platforms are exposed, and little carbonate mud is produced. Lowstand, deep-water, fine-grained carbonate is dominantly pelagic in origin (Droxler and Schlager, 1985). Transgressive deposits also lack significant carbonate mud because the tops of carbonate platforms are not flooded until late in the transgression (e.g., Grammer et al., 1993).

Supply of coarse-grained reef debris to the basin margin is controlled by reef growth and proximity of the reef to the basin margin in Quaternary platforms. Unfortunately, Quaternary, late-highstand and low-stand proximal basin-margin sediments have not been investigated, so supply of coarse-grained, reef debris during these sea-level stands remains speculative. Low sea-level stand constricts reef growth to the narrow exposed slope proximal to the basin margin. The growth area is considerably reduced because of the steeper slopes, but the proximity may allow easier transport to the basin margin. In the absence of a Quaternary analog, the interaction of these effects

cannot be evaluated. Early transgressive deposits show that reef growth and sediment supply to the basin margin is prolific (Grammer et al., 1993). Evidently, reef growth can keep up or exceed sea-level rise during early transgression. Late transgression is recorded by a hiatus representing limited reef growth in the Tongue of the Ocean, when reef growth could not keep up with rapid sea-level rise (Grammer et al., 1993). Highstand reef growth is limited by shelf flooding. In Quaternary systems, many shelf-margin reefs are incipiently drowned when the shallow shelf behind the reef is flooded during early highstand (Lighty, 1985). These drowned reefs either continue slow growth as deep relict features, or they are buried by grain-dominated margins (e.g., Hine and Neumann, 1977). Active reef growth may have stepped back onto the platform, so transport of coarse-grained reef debris to the basin margin is difficult even where reef growth is present. This is the case at the present-day Tongue of the Ocean (Grammer et al., 1993). During late highstand, reefs might prograde back to the shelf margin or overcome highstand drowning as the shelves fill and shallow.

Figure 8. Glide plane truncation near the base of unit 5 on the North Cliff, on the northern side of McKittrick Canyon. Channeling by peloidal grainstones in peloidal packstone sediment is evident in the lower part of unit 5. The upper part of unit 5 (top of photo) shows pinch-and-swell bedding representing creep or incipient slumping of thin-bedded wackestone.

The reefs could supply coarse-grained debris to the basin margin during late highstand, although no example of late highstand Quaternary carbonates are available to document this. From this analysis, reef debris is most likely transported to the basin margin during the early stages of transgression (corresponding to the late lowstand and transgressive systems tracts) and during late highstand.

Interpreted Permian Sediment Supply Patterns

The Quaternary analog can be applied to the Permian Delaware sea with only a few modifications (Figure 9). There are four general types of Late Guadalupian basin-margin rocks (Table 3): (1) siliciclastic coarse-grained siltstone and very fine-grained sandstone; (2) carbonate mud; (3) carbonate boulders with little carbonate mud; and (4) carbonate boulders with carbonate mud.

1. Siliciclastic deposition corresponds to relative lowstand below the shelf edge because this is the only time during which carbonate productivity could have been reduced to such a degree that little or no carbonate sediment was supplied to the basin margin (Reeckmann, 1986). Siliciclastic siltstone and sandstone are most easily transported across the shelf by fluvial or eolian processes during times when the shelf is subaerially exposed (Fischer and Sarnthein, 1988).

2. Because few Permian pelagic organisms produce carbonate mud (L. Babcock, 1977), basin-margin muds must have been derived almost entirely from the shelf. This means that Permian-age basin-margin carbonate mud-rich strata must be deposited when the shelf was flooded. The carbonate mud-dominated intervals are therefore highstand deposits. Absence of reef boundstone debris in basin-margin mud-rich deposits indicates that reef growth was suppressed or had stepped back onto the shelf so that debris could not be supplied to the basin margin. Flooding of shelves may have suppressed reef growth (Lighty, 1985) or rapid relative sea-level rise may have drowned the reefs. Reef backstepping may also be related to rapid sea-level rise. All of these causes of reef suppression are related to rise in relative sea level, either during early highstand or during renewed shelf flooding during highstand.

3. Reef boulders were derived from actively growing reefs, not from eroding subaerially exposed reefs. The boundstone boulders show no macroscopic evidence of subaerial or meteoric diagenesis prior to transport and deposition. If boulders were derived primarily from erosion during lowstand, they should underlie or interfinger with the lowstand siltstone below unit 1, yet they are absent from the siltstone and downlap onto it. If this analysis is correct, then boulder-rich Lamar

deposits indicate times of active reef growth on the shelf margin. In contrast, boulders in other members of the Bell Canyon Formation have features indicative of subaerial exposure prior to transport down slope (Reeckmann, 1986). Even so, subaerial exposure prior to downslope transport may not indicate contemporaneous subaerial exposure on the shelf. Although erosion during lowstand could deliver subaerially altered reef debris to the basin margin (James and Ginsburg, 1979), large-scale slope failure of the shelf margin (such as that interpreted by Mullins and Hine, 1989) could deliver subaerially altered boulders during any stand of sea level.

Carbonate boulder-conglomerates with little carbonate mud matrix are interpreted to be transgressive deposits. Reef growth was active enough to indicate that the shelf margin was flooded, yet the shelf was not supplying mud to the shelf margin and was probably still subaerially exposed. The presence of silt matrix in reef boundstone boulders of unit 1 indicates that seaward transport of silt was not yet interrupted by shelf flooding.

4. Units with both carbonate mud and reef boundstone debris were deposited when reef growth was active and the shelves were flooded. This clearly represents highstand deposition. Reef growth contemporaneous with shelf flooding corresponds to a time when either reefs have caught up with sea-level rise, or prograded back to the shelf margin. This is interpreted to occur later in the highstand depositional cycle, but before drop of relative sea level, because even a small drop of relative sea level during a highstand would subaerially expose the shelf and sharply reduce carbonate mud supply to the basin margin.

Late Guadalupian, basin-margin, relative sea-level history (Figure 10) can be reconstructed by using the sediment-supply criteria developed above. The pre-Lamar siltstone corresponds to lowstand because shallow-water carbonate production was minimal, and silt was supplied to the basin. Unit 1 is an early transgressive deposit formed while silt was actively transported over the subaerially exposed shelf and shelf-margin reef growth was active. Unit 2 is a late transgressive deposit, deposited when silt supply was interrupted by shelf flooding. However, incipient shelf flooding was insufficient to decrease reef productivity or to contribute significant amounts of carbonate mud to the basin margin. Unit 3 is an early highstand deposit. Shelf production of mud was high and reefs were drowned or growing far enough from the shelf edge that they were not contributing sediment to the basin margin. Unit 4 may represent late highstand because carbonate mud was produced on

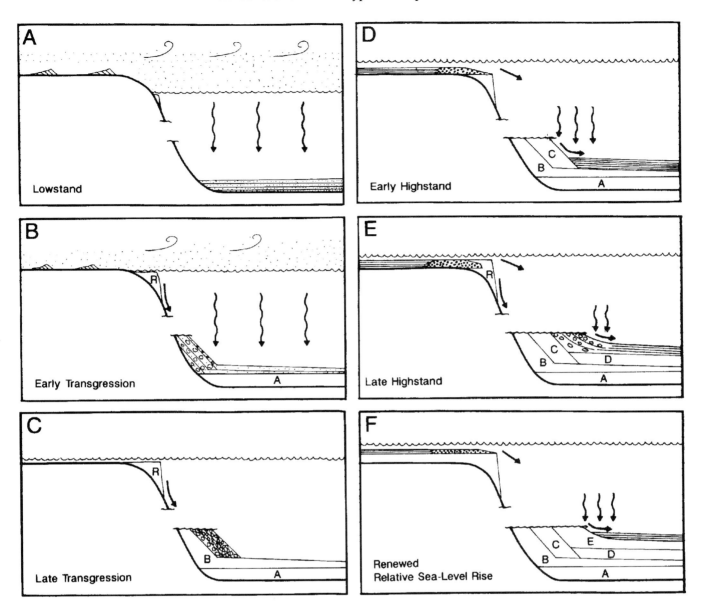

Figure 9. Interpreted control of relative sea level on shelf-margin sedimentation and basin-margin sediment supply. Deep foreslope and middle slope parts of the system are omitted so that the shelf margin and basin margin can be expanded. (A) Relative lowstand exposes shelf and dramatically reduces carbonate production. Siliciclastic sand and silt are transported across the exposed shelf. Arid paleosols form on the shelf. (B) Early transgression results in flooding of the seaward margin of the shelf margin, which increases reef productivity and supplies boundstone boulders to the basin margin. Sand is still transported across the exposed shelf and mixed with carbonate in the basin margin. (C) Late transgression results in incipient flooding of the shelf and termination of silt transport to the basin margin. Reef growth is active because shelf poisoning effect is not significant. Basin margin is mud- and silt-poor boulder conglomerates. (D) Early highstand results in flooding of the shelf for significant carbonate-mud production and drowning of the reefs. Basin margin has exclusively carbonate-mud and sand deposition. (E) Late highstand results from catch-up of sea level by reefs, so that both boundstone boulders and carbonate mud are supplied to the basin margin. (F) Renewed relative sea-level rise results in drowning of reefs and increased mud production, so onlapping, carbonate mud- and sand-rich strata are deposited.

Table 3. Stratal Relationships in the Lamar basin-margin setting.

Sea-Level Stand	Sediment Source	Sediment Type	Stratal Relationship
Lowstand	Eolian sand flat	Siliciclastics	Onlap
Transgressive	Active reef growth, shelf not flooded	Matrix-poor or silty, conglomeratic deposits	True downlap
Early highstand	Flooded shelf, reef drowned	Lime mud–rich, fine-grained deposits	Onlap
Late highstand	Flooded shelf with active reef growth	Lime mud–rich, conglomeratic deposits	Apparent downlap

the flooded shelf and reef boulders were supplied from reefs which had caught up to sea level or prograded to the shelf edge. Increasing carbonate mud toward the top of the section in unit 4 may represent renewed relative sea-level rise. Unit 5 is a highstand deposit because carbonate mud is abundant. The shelf-derived grains in the lower part of the unit is not consistent with a highstand. We originally interpreted the lower part of unit 5 as a minor sea-level fall which aided grain transport to the shelf edge by promoting progradation of carbonate sand bodies to the shelf edge. More likely, the lower part of unit 5 represents a time when reefs were not active, so the shelf margin was characterized by grain-shoal deposition. This may be a culmination of the sea-level rise observed in the upper part of unit 4, or it may be an ecologic effect having no relationship to sea level. Unit 6 represents renewal of reef-boulder supply to the basin, indicating active shelf-margin reefs during relative highstand. Unit 7 is a siliciclastic lowstand deposit overlain by a highstand deposit of mud-matrix boulder conglomerate. Distinct transgressive deposits and early highstand deposits were not recognized in unit 7, possibly because the toe-of-slope part of the unit was too thin.

Relative Sea Level Determined from Shelf Sediment Record

Relative sea-level history interpreted for the basin margin can be tested by correlation to the relative sea level reconstructed from the shelf record. The sea-level history is interpreted from the data available in published measured sections (e.g., Tyrrell, 1962; Pray et al., 1977; Parsley, 1988) and unpublished investigations of the same outcrops by the authors. Subaerial exposure surfaces are interpreted as evidence of lowstands. Shelf siliciclastic siltstone beds are nonmarine deposits formed when the shelf was exposed during the lowstand and the initial stages of transgression. The seaward margins of the siliciclastic units may be reworked in a marine environment during transgression. Shelf-marine or marginal-marine carbonate deposits indicate highstand deposition. Small-scale, relative sea-level changes during highstands are inter-

preted from facies patterns on the shelf. The generalized facies tracts of Upper Guadalupian carbonates landward from the shelf margin are sponge-algal boundstone; skeletal packstone/grainstone; nonskeletal, fenestral wackestone/packstone; and calcisphere wackestone/mudstone (Tyrrell, 1962; Pray et al., 1977; Parsley, 1988). Sea-level changes were interpreted where the facies pattern abruptly shifted seaward (minor regression) or stepped backward (minor transgression).

The Yates upper Triplet siltstone comprises interpreted eolian sand-flat and wadi deposits on the shelf, which were reworked on the outer shelf by marine processes (see Borer and Harris, 1991, for further discussion on shelf-sandstone depositional environments and their interaction with sea level). The subjacent Triplet dolomite is capped by a prominent subaerial exposure surface. This surface is interpreted as a lowstand event, and the siltstone is interpreted as an early transgressive or later lowstand deposit.

The Basal Dolomite member of the Tansill Formation represents flooding of the shelf during early highstand. This unit is characterized by meter-thick, shoaling-upward beds possibly reflecting minor base-level changes. The correlatable, thin siltstone bed near the middle of the unit and associated pisolitic packstones may represent a minor sea-level drop, but a subaerial exposure surface is not present on well-exposed outcrops. Maximum faunal diversity on the shelf occurs near the top of the member.

The Walnut View member is a sparsely fossiliferous mudstone to wackestone several kilometers behind the shelf edge, but time-equivalent strata near the shelf edge are dominantly skeletal and nonskeletal packstone. Its basal contact is the restricted mudstone overstepping fenestral packstones in a seaward direction; this is indicative of progradation during stillstand of relative sea level (a drop in sea level associated with this progradation would result in subaerial exposure). A tongue of skeletal grainstone projecting shelfward in the middle part of the Walnut View member (best evident in section 3 of Tyrrell, 1962) represents facies backstepping onto the shelf indicative of a minor relative sea-level rise during the late highstand. Restricted mudstone in the upper part

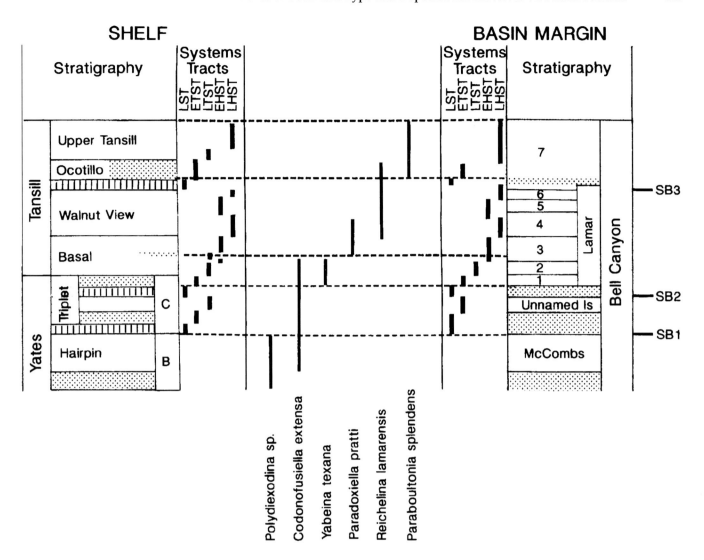

Figure 10. Comparison and correlation of shelf (left) and basin-margin (right) strata and systems tracts. The two sections are correlated using fusulinid biostratigraphy (center). Dashed lines are approximate biostratigraphic datums used to correlate the sections. The sequence and positions of systems tracts on the shelf and basin margin are strikingly similar. The biostratigraphic control is sufficient to demonstrate that the relative sea-level history interpreted for the basin is generally the same as that of the shelf. The vertical lines in the shelf stratigraphic column mark subaerial exposure and missing shelf section. The subaerial unconformity below the Ocotillo Siltstone is speculative. The range of *Y. texana* is interpreted to extend down into the upper Triplet siltstone, but is not observed there due to unfavorable depositional environment. The vertical scale is not linear to time or to stratigraphic thickness. Shelf biostratigraphic ranges are from Tyrrell et al. (1978), correlated to the lower Tansill member names using stratigraphy in Tyrrell (1962) and Parsley (1988). Basin biostratigraphic ranges are from Newell et al. (1953), Wilde (1955), and Skinner and Wilde (1955). The ranges were tied to our basin-margin units by correlating to the locations given in these references. Fusulinid populations were not analyzed in our measured sections.

of the Walnut View member oversteps the grainstone during subsequent seaward progradation.

The Ocotillo Member of the shelf interior has sedimentary structures similar to those in Yates siltstones interpreted to indicate wadi and eolian sand-flat deposition (figured in Garber et al., 1989). On the outer shelf, the Ocotillo siltstones are siltstones interbedded with marine carbonates (Parsley, 1988) similar to outer shelf Yates siltstones. Subaerial expo-

sure has not been recognized in association with the Ocotillo Member, but outcrops of the base of the siltstone are poorly exposed, so the contact may not have been recognized. Despite the lack of documented subaerial exposure, the Ocotillo interval is interpreted to be an early transgressive (or reworked lowstand) deposit over a lowstand subaerial hiatus postulated in analogy with Yates siltstones. The carbonate overlying the Ocotillo Member is a skeletal-rich packstone similar to the Basal Dolomite member (Tyrrell, 1962; Parsley, 1988). This unit is interpreted to be an early highstand deposit. The uppermost exposed Tansill Formation shows an uninterrupted progradational pattern (Parsley, 1988) indicative of stillstand or initial sea-level fall during late highstand.

Based on these interpretations, two transgressive–regressive intervals are identified on the shelf (Figure 10). The lower interval consists of the following: (1) the lowstand below the upper Triplet siltstone is recorded by the paleosol on the Triplet dolomite; (2) the upper Triplet siltstone records late transgression prior to flooding of the shelf; (3) the Tansill Basal Dolomite is an early highstand deposit with a short regression near the middle of the member; and (4) the Walnut View dolomite is a late highstand deposit with a probable small transgression in the middle of the member.

The upper transgressive-regressive interval consists of the following: (1) the Ocotillo Member interval is a transgressive deposit. Underlying subaerial exposure corresponding to the lowstand was not documented, so subaerial exposure may have been brief or absent. (2) The basal skeletal part of the post-Ocotillo Tansill Formation represents early highstand. (3) The upper part of the Tansill Formation shows continued highstand progradation on the shelf.

Correlation of Shelf to Basin

The correlation of the Lamar Limestone to time-equivalent strata on the shelf is based on biostratigraphy because neither the siltstone lithostratigraphic markers nor bedding planes can be traced through the massive upper Capitan Formation. Tyrrell et al. (1978) used fusulinids to correlate shelf units to the basin. They report occurrences of fusulinids in cores from wells at Dark Canyon, which can be correlated to nearby lithostratigraphic sections of Parsley (1988). The ranges of fusulinids in the Lamar Limestone were determined by correlating the occurrences of the species reported by Skinner and Wilde (1955) and Wilde (1955) to our measured sections at McKittrick Canyon.

Figure 10 shows a correlation of basin and shelf section based on available fusulinid data. The fusulinid stratigraphy is accurate enough to substantiate the similarity of the shelf relative sea-level positions to those determined by sediment types in the basin margin (Figure 10). Both shelf and basin record two transgressive/regressive packages. In addition, the small transgression recorded by the upper part of unit 4 and unit 5 appears to correlate with the transgressive backstepping observed in the shelf Walnut View member. The small regression recorded by the thin siltstone in the Tansill Basal Dolomite member on the shelf is not evident in the basin-margin record. The biostratigraphic correlation is consistent with the lithostratigraphic correlation between shelf and basinal siltstones.

SEQUENCE STRATIGRAPHY AND STRATAL PATTERNS

Although the sequences observed in McKittrick Canyon are thin, they are true sequences in the sense that bounding surfaces correlate to unconformities and that the sequence can be divided into systems tracts. The McKittrick Canyon study area is a large enough area to identify consistent, large-scale stratal patterns, yet small enough to allow identification or inference of depositional processes and sediment source for most of the beds. Thus, the relative influence of depositional process and proximity to sediment supply on baselap stratal patterns in a basin-margin carbonate setting can be evaluated. Toplap stratal patterns were not examined in this study because the study was restricted to the basin margin.

Sequence Stratigraphy and Systems Tracts

The sediment supply and correlated shelf sea-level history can be used to interpret sequence boundaries and systems tracts (shown schematically on Figure 11). Below the interval of interest, a sequence boundary is interpreted on the top of the McCombs Member, based on correlation with the subaerial exposure surface on top of the Hairpin dolomite. Bell Canyon siliciclastics above the McCombs Limestone

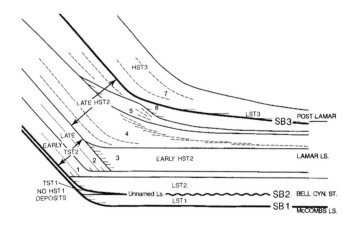

Figure 11. Our interpretation of the sequence stratigraphy of upper Bell Canyon sedimentation using stratal relationships and lithology data. Sequence 1 is composed of the lower Bell Canyon siltstone and the unnamed limestone tongue. Sequence 2 includes the Bell Canyon siltstone above the unnamed limestone tongue and the Lamar Limestone section. Sequence 3 consists of the post-Lamar section. As discussed in the text, complex bedding termination pattern within the late highstand part of the second sequence is related to small-scale sea-level fluctuations which do not expose the shelf.

are lowstand deposits which onlap clinoforms equivalent to the McCombs Member (Figure 4). The unnamed limestone tongue downlaps onto underlying siltstone. It has the characteristics of an early transgressive deposit, due to the short duration of shelf flooding and lack of significant carbonate mud production.

The sequence which includes most of the Lamar Limestone has its lower boundary on top of the unnamed limestone tongue and correlates to the unconformity at the top of the Triplet dolomite. The Bell Canyon siliciclastic unit above the unnamed limestone tongue is the lowstand deposit for this sequence; beds onlap clinoforms equivalent to the unnamed tongue. Downlapping deposits of unit 1 and unit 2 were deposited during initial transgression of the shelf. Unit 3, representing early highstand, onlaps transgressive deposits of unit 2. Unit 4, representing late highstand, has apparent downlap onto unit 3, and apparent onlap near the top of the unit. Unit 5 (renewed transgression during highstand) onlaps unit 4, and unit 6 (late highstand) has apparent downlap, and buries topography of unit 5 by onlap and downlap. The transgression recorded in the upper part of unit 4 and in unit 5 does not represent a distinct sequence because there was no intervening correlatable unconformity necessary to define a sequence boundary.

The next sequence boundary is near the top of the Lamar Limestone at the base of unit 7. This sequence boundary correlates to the base of the Ocotillo Member of the Tansill Formation, which has an interpreted subaerial exposure surface. Lowstand deposits are onlapping siltstones in the lower post-Lamar interval, and highstand deposits are the boulder conglomerates and correlative carbonates on the toe-of-slope. Distinct transgressive deposits are not recognized.

Controls on Stratal Patterns

The stratal patterns in McKittrick Canyon basin-margin carbonate transgressive and highstand strata are controlled by changing depositional mechanisms. Most carbonates on basin margins are deposited by gravity flows. Gravity-flow mechanisms are controlled by slope and by sediment type (Nardin et al., 1979). Matrix-poor boulder conglomerates were deposited in steep, proximal talus cones because the matrix-poor gravity flows or rock falls which deposited them could not maintain flow on gentle slopes. Carbonate wackestones were deposited as turbidites on the gently dipping toe-of-slope because the lower slope was too steep for deposition from low-density turbidity currents, and the modest slopes of the toe-of-slope was gentle enough to deposit sediment. Fine-grained sediment also has a lower angle of repose than coarse-grained sediment, so any fine-grained sediments deposited on steep slopes may be remobilized by slope failure (Kenter, 1990). Together, the processes of selective deposition and selective slope

failure results in a characteristic steep slope for boulder-rich, coarse-grained carbonates and gentle slope for mud-rich, fine-grained carbonates. Where mud-dominated sediments alternate with coarse-grained carbonates, gentle dips will alternate with steep dips. The result is alternating onlap and downlap patterns.

Changing depositional mechanisms result in systematic bedding termination patterns only when sediment supply changes significantly. Increased carbonate mud supply and decreased supply of coarse-grained carbonate (units 3 and 5) changed the dominant deposit type from lower slope, boulder conglomerates to toe-of-slope wackestones. This resulted in onlap or apparent onlap during highstands. Decreased carbonate mud supply with continued coarse-grained carbonate supply caused lower slope deposits to prograde over gently dipping toe-of-slope deposits. This resulted in downlap during transgression (units 1 and 2) and apparent downlap during late highstand (unit 4, unit 6, and the upper part of unit 7).

The McKittrick Canyon basin-margin carbonates were indirectly but effectively controlled by relative sea level. Relative sea-level controls degree of shelf flooding, which controls sediment type transported to the basin margin. Sediment type controls the dominant depositional mechanism on the basin margin, and depositional mechanisms control stratal patterns.

DISCUSSION

Scale of Sequences

The Late Guadalupian sequences identified in McKittrick Canyon have boundaries correlative to unconformities, so they are sequences in the sense of Mitchum et al. (1977). Amplitudes of sea-level changes were adequate to completely expose the shelf for periods of time long enough to develop soil zones in platform strata. The sequences are interpreted as a product of a third- or fourth-order sea-level cycle.

Meter-thick shoaling-up carbonate cycles in the Tansill Basal Dolomite member may be parasequences in the sense of Van Wagoner et al. (1990). High-frequency relative sea-level changes on the shelf, such as the flooding event in the middle of the Walnut View member or the regression in the Basal Dolomite member, may also be parasequences. The pattern of high-frequency parasequence events superimposed on the third-order sea-level changes is similar to the multiple frequency sea-level model proposed by Borer and Harris (1991) for the Yates Formation. The mixed carbonate wackestone and siliciclastic siltstone at the base of unit 7 may be caused by analogous high-frequency sea-level oscillations during deposition of the Ocotillo Member. Although gradual changes in sediment supply and interbedded packstone and wackestone deposition (such as in unit 4) may be related to the transgression recognized in the Walnut View member or high-frequency sea-level oscillations, no parasequences on a similar scale to the

Tansill Basal Dolomite parasequences were recognized in correlative basin-margin strata.

Comparison to Other Sequence Stratigraphic Models

The baselap stratal patterns in basin-margin deposits at McKittrick Canyon (shown schematically in Figure 11) can be compared to the generalized stratal patterns in systems tracts schematically shown by Haq et al. (1988) for siliciclastic strata and by Sarg (1988) for carbonates (Figure 12). This model is widely referenced in the literature, and is in many cases used as a template for sequence systems tract stratal patterns. The Haq et al. (1988) systems tract model will be referred to as the "idealized" siliciclastic sequence model in the following discussion.

Some systems tracts have geometries and stratal patterns similar to those of the idealized siliciclastic model; others do not. The onlap pattern in the Bell Canyon Siltstones is similar to the onlapping pattern of the lowstand wedge in the idealized siliciclastic model. These Permian lowstand strata contain insignificant quantities of carbonate, so if lowstand siliciclastic deposition did not take place, sea-level lowstand would be represented by a hiatus. Transgressive systems tract in the idealized siliciclastic model are onlapping strata located predominantly on the shelf, and basin-margin deposits do not have recognizable transgressive deposits. The McKittrick Canyon basin-margin transgressive systems tracts are thick, downlapping wedges of carbonate and silty carbonate. Highstand deposits in the basin part of the idealized siliciclastic model are fine-grained siliciclastics deposited as a condensed interval at apparent downlap surfaces or as lower slope clinoforms. Identical features are evident in the lower part of unit 4, in unit 6, and the upper part of unit 7, in the McKittrick Canyon basin margin. However, fine-grained deposits of the early highstand systems tract may also onlap boulder conglomerates, as in units 3 and 5. This geometry is not recognized in the highstand deposits of the idealized siliciclastic model. Overall, coarse-grained carbonate is delivered farther into the basin during periods of highstand (such as units 4 and 7) than during transgression or lowstand.

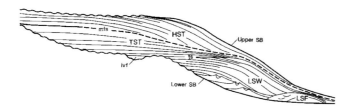

Figure 12. Stratal patterns for siliciclastic systems tracts. Similar stratal patterns are presented in Sarg (1988) and in Jervey (1988). Modified from Haq et al. (1988).

This is opposite of the pattern expected from the idealized siliciclastic model, when coarse-grained material extends farthest basinward during lowstand deposition.

The differences between the McKittrick Canyon basin-margin systems tracts and the idealized siliciclastic model systems tracts indicates that stratal patterns in the two models have different controls. The idealized siliciclastic model is based on the concept of systematic filling of accommodation space as it becomes available from sea-level change and subsidence (Jervey, 1988). Sediment is assumed to fill the accommodation space nearest the shore first because this process requires the least transport distance. The effect of sea-level change is to shift the locus of sedimentation by shifting the shoreline from the inner shelf to the shelf edge without major changes in sediment delivery to the shore. Overall, the major control on locus of sedimentation is proximity to shoreline. This model is a good, first-order approximation of the response of siliciclastic systems to relative sea-level changes. However, carbonates are grown, not transported from subaerial source areas, so they can accumulate substantial deposits away from the shoreline and changes in relative sea level can result in substantial changes in the amount of carbonate produced (Handford and Loucks, 1991, 1993). The amount and type of carbonate sediment are still controlled by relative sea level, but control of sedimentation is through growth area, bathymetry, and physical oceanographic changes and not necessarily proximity to shore line. This means that there is no *a priori* reason for the Permian carbonate strata to respond to sea-level change as predicted by the idealized siliciclastic model.

Proximity of sediment supply does not affect carbonate sedimentation on the steep Late Guadalupian basin margins as much as the depositional mechanism does. One possible explanation for observed stratal patterns different from that predicted from the accommodation-space pattern of the idealized siliciclastic model is the presence of steep paleoslopes on the Capitan shelf margin. Carbonate mud–rich sediments cannot be deposited or remain stable on steep paleoslopes, so shelf-derived muds are constrained to onlap the slope, no matter what the proximity of sediment supply or what the sea level. Mud-poor, grain-flow, and talus deposits are restricted to steeper depositional slope because they cannot be transported on low depositional slopes farther down paleoslope.

Some other carbonate systems appear to follow the predictions of stratal geometry from the idealized siliciclastic model (e.g., Jacquin et al., 1991). Differences in depositional slope may explain the different stratal patterns. On the gentle slopes of carbonate ramps, all beds may be deposited at relatively low dips and relatively minor sea-level change may substantially shift the locus of sedimentation. In this case, proximity to sediment source may outweigh the small differences in angle of repose and depositional slope of gravity flows on the gently dipping ramp. If so, the resulting stratal pattern could be very similar to the stratal pattern predicted by the idealized siliciclastic model, but depositional dips in all units would probably be very gentle.

If similar sea-level histories produce different stratal patterns in siliciclastic and carbonate systems and in carbonate systems of different types, then stratal patterns may have limited utility for interpreting sea-level history by themselves. If the McKittrick Canyon section was imaged by seismic and the systems tracts were identified from the stratal patterns predicted by the idealized siliciclastic model with no additional information, the sea-level history would be misinterpreted. Stratal patterns must be used in combination with other data types in order to interpret the relative sea-level history.

Late Guadalupian Sea-Level History

Late Guadalupian sea-level histories have also been proposed by Sarg et al. (1988) and Reeckmann (1986). Sarg et al. (1988) interpreted the McCombs Limestone as a lowstand deposit, the sub-Lamar siltstone as a transgressive deposit, the lower Lamar Limestone as a highstand deposit, and the upper Lamar Limestone and post-Lamar member as a falling sea-level (late highstand) deposit. These interpretations are shown in their figures but not discussed in their text. Reeckmann interpreted the McCombs Limestone as a small sea-level rise, the sub-Lamar siltstone as a lowstand, the basal Lamar Limestone (probably equal to parts of units 1 and 2) as a transgressive system, and the remainder of the Lamar Limestone as a highstand. The relative sea level interpreted here is consistent with that proposed by Reeckmann (1986) for the Lamar Member. The sequence representing the post-Lamar member was previously unrecognized.

This study found no evidence of early subaerial exposure prior to transport of Lamar-age reef blocks to the basin. In contrast, Reeckmann (1986) interpreted features in boulders within the Rader slide exposed at the highway 62–180 road cut as evidence of pretransport meteoric diagenesis. Also, meteoric phreatic diagenesis was interpreted in some McCombs grainstones and packstones. She generalized these findings to conclude that all sand-matrix boulder conglomerates were derived from subaerially exposed reef. Because these lower intervals were not studied, the interpretations of Reeckmann cannot be verified or refuted. However, the Lamar-equivalent boulders (even those in unit 1 with silty matrix) show no evidence of early subaerial exposure, so generalizations about subaerial exposure of boulders should be avoided.

Scale and Applicability

Although the sequences identified here meet the criteria to be called "sequences," the thickness (approximately 100 m) and time interval (probably less than 1 m.y.) represented by the sequences are smaller than those of most third-order sequences interpreted in the literature (e.g., Vail et al., 1984; Haq et al., 1988). There is a significant question if the observations of depositional processes on this scale are applicable to the thicker sequences resolvable on seismic sections. Also, stratal patterns observed on these outcrops may not be imaged in sufficient detail by standard seismic techniques to be correctly identified.

From a theoretical point of view, there is no reason to believe that depositional mechanisms cannot control stratal patterns in a thicker sequence with longer duration. The thickness and duration of the sequences examined is great enough to demonstrate that the depositional controls are not due to a single fortuitous depositional event such as slope failure of the shelf margin or some exceptionally thick turbidity current deposit. If the general pattern of smaller depositional events can endure for the period examined here, there is no reason to think that they could not form thicker deposits over longer periods of time.

The low-frequency information of seismic data and wide spacing of well-log data precludes the distinction of downlap from apparent downlap and onlap from apparent onlap. In the case of downlap, this has long been recognized; in fact, the concept of the condensed interval is based on the tacit assumption that beds do not terminate at the downlap surface; deposition continues into the basin as thin beds parallel to the basin floor. In this study, true downlap was observed only under special circumstances—where beds downlapped against a topographic high (base of unit 6) or where coarse grain size precluded transport farther into the basin.

However, it is not as widely recognized that apparent onlap can occur in settings where it can sometimes cause confusion of relative age of strata. Where apparent onlap results from thinner slope beds than bottomset beds, as in unit 4, slope and bottomset beds might be interpreted as distinct deposits of different age if the interfingering of the slope and bottomset beds could not be imaged. Where apparent onlap is caused by a thin plastering of strata onto a clinoform during subsequent deposition of bottomset beds, as at the base of unit 7, confusion of apparent and true onlap would not lead to erroneous interpretation of age relationships.

If there were no onlapping basinal siliciclastics, then the stratal patterns would be even more difficult to reconcile with the stratal patterns of the idealized siliciclastic model because lowstands mark the seaward shift of sedimentation characteristic of the lower sequence boundary. In the absence of a distinct onlapping lowstand systems tract, onlapping beds of highstand, mud-rich wedges such as those observed in McKittrick Canyon may be interpreted as lowstand deposits where lithology data are not available. The similarity between onlapping stratal patterns of lowstand deposits and onlapping fine-grained highstand sediments has been discussed by Schlager and Camber (1986) and by Schlager (1989).

CONCLUSIONS

Deposition of Upper Guadalupian, basin-margin strata at McKittrick Canyon was investigated to determine controls on stratal patterns in carbonate basin margins. The Lamar Limestone is not a simple tongue of limestone extending into the basin; rather, it is composed of depositional units with systematic changes in sediment types, depositional mechanisms, and bedding relationships. The distribution of depositional mechanisms is controlled by the depositional slope and the type of sediment supplied to the basin. All depositional facies are dominated by gravity-flow deposits. Proximal, steep-dipping deposits are dominantly low-mud-matrix talus cones and high-mud-matrix, cohesive debris-flow deposits. Most distal, gently dipping deposits are low-density turbidity current deposits of fine-grained carbonate. High-density turbidity current deposits are channelized coarse-grained packstones, which are present in the slope and toe-of-slope setting.

Sea level position during basin-margin sedimentation was determined by the types of sediment supplied to the margin and confirmed by correlation to the shelf relative sea-level curve. Lowstands are characterized by deposition of siliciclastic siltstone in the basin margin, transgressions are characterized by deposition of carbonate boulder-conglomerates with relatively little carbonate mud, and highstands are characterized by deposition of carbonate wackestones with or without carbonate boulders.

Late Guadalupian basin-margin deposits have alternating onlap and downlap patterns similar to patterns documented in siliciclastic sequences. However, equivalent systems tracts have stratal patterns different from those of the commonly referenced idealized siliciclastic model. Lowstand carbonates are absent. Downlapping transgressive deposits not predicted from the standard sequence stratigraphic models are voluminous. Early highstand deposits onlap, whereas late highstand carbonate units downlap. The patterns observed here are caused by changes in depositional process, not by proximity to sediment source. The changes in depositional processes are controlled by the types and amount of sediment supplied to the basin margin, which is controlled by relative sea level.

All carbonate basin margins do not necessarily respond to sea-level changes in a manner similar to that outlined for the Late Guadalupian basin margin at McKittrick Canyon. Sediment distribution and bedding patterns are ultimately controlled by the mechanisms of transportation and deposition. Generalizing the relationships between relative sea-level changes and bedding patterns assumes that depositional mechanisms will be the same in each case examined. This study demonstrates the importance of understanding the depositional process that deposited the sediment before using stratal patterns to interpret relative sea-level changes in carbonates.

ACKNOWLEDGMENTS

The authors wish to acknowledge the cooperation of the United States National Park Service and the rangers at Guadalupe Mountains National Park for allowing access and sampling of rocks. We also thank ARCO Exploration and Production Technology Company for permission to publish. This research is a byproduct of the Permian Reef Geology Trail description project (organized by D. G. Bebout and C. Kerans, Bureau of Economic Geology, the University of Texas at Austin). Discussions with authors of other parts of the trail guide helped clarify general stratigraphic relationships of the Capitan Formation. Special thanks to R. K. Suchecki, C. R. Handford, and S. Ng (ARCO Exploration and Production Technology) and R. Sarg (Mobil Oil Company) for reviews and comments.

REFERENCES CITED

Babcock, J., 1977, Calcareous algae, organic boundstones, and the genesis of the upper Capitan Limestone (Permian, Guadalupian), Guadalupe Mountains, West Texas and New Mexico, *in* M. Hileman, and S. J. Mazzullo, eds., Upper Guadalupian Facies, Permian Reef Complex, Guadalupe Mountains, New Mexico and West Texas: Permian Basin Section—SEPM Publication No. 77-16, p. 3-44.

Babcock, L. C., 1977, Life in the Delaware Basin: the Paleoecology of the Lamar Limestone, *in* M. Hileman, and S. J. Mazzullo, eds., Upper Guadalupian Facies, Permian Reef Complex, Guadalupe Mountains, New Mexico and West Texas: Permian Basin Section—SEPM Publication No. 77-16, p. 357-390.

Borer, J. M., and P. M. Harris, 1991, Lithofacies and cyclicity of the Yates Formation, Permian basin: Implications for reservoir heterogeneity: AAPG Bulletin, v. 75, p. 726-779.

Brown, A. A., and R. G. Loucks, 1987, Comments on Lamar limestone road cut, *in* S. T. Reid, R. O. Bass, and P. Welch, eds., Guadalupe Mountains Revisited: West Texas Geological Society 1988 Field Seminar, West Texas Geologic Society Publication No. 88-84, p. 9-10.

Cook, H. E., and H. T. Mullins, 1983, Basin Margin Environments, *in* P. A. Scholle, D. G. Bebout, and C. H. Moore, eds., Carbonate Depositional Environments: AAPG Memoir 33, p. 539-617.

Davies, G. R., 1977, Turbidites, debris sheets, and truncation structures in Upper Paleozoic deep-water carbonates of the Sverdrup Arctic Archipelago, *in* H. E. Cook, and P. Enos, eds., Deep-Water Carbonate Environments: SEPM Special Publication No. 25, p. 221-249.

Figure 4. Structural dip section of the Lamar
Limestone and overlying strata along the north wall of
McKittrick Canyon showing distribution of generalized lithology and bedding terminations. The distribution of packstone facies between measured sections
is schematic. The wackestone interval contains many
thin (<5 cm) beds of skeletal packstone not shown on
this section. Thin wackestone beds also occur
interbedded with the packstones of units 3, 4, 6, and 7.
Locations of measured section are shown in Figure 1.
Table 1 summarizes characteristics of named lithologies shown on this section.

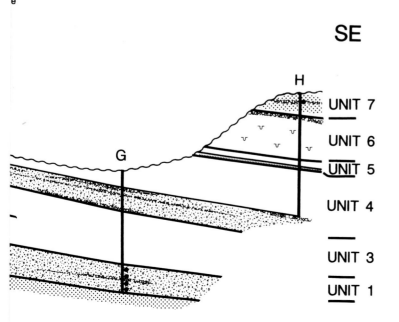

SE

A

B

C

Siltstone

Droxler, A., and W. Schlager, 1985, Glacial versus interglacial sedimentation rates and turbidite frequency in the Bahamas: Geology, v. 13, p. 799-802.

Fischer, A. G., and M. Sarnthein, 1988, Airborne silts and dune-derived sands from the Permian of the Delaware Basin: Journal of Sedimentary Geology, v. 58, p. 637-643.

Garber, R. A., G. A. Grover, and P. M. Harris, 1989, Geology of the Capitan shelf margin—subsurface data from the northern Delaware basin, in P. M. Harris, and G. A. Grover, eds., Subsurface and Outcrop Examination of the Capitan Shelf Margin, Northern Delaware Basin: SEPM Core Workshop 13, San Antonio, April 23, p. 3-269.

Glaser, K. S., and A. W. Droxler, 1991, High production and highstand shedding from deeply submerged carbonate banks, Northern Nicaragua Rise: Journal of Sedimentary Geology, v. 61, p. 128-142.

Grammer, G. M., and R. N. Ginsburg, 1992, Highstand versus lowstand deposition on carbonate platform margins: Insight from Quaternary foreslopes in the Bahamas: Marine Geology, v. 103, p. 125-136.

Grammer, G. M., R. N. Ginsburg, and P. M. Harris, 1993, Timing of deposition, diagenesis, and failure of steep carbonate slopes in response to a high-amplitude/high-frequency fluctuation in sea level, Tongue of the Ocean, Bahamas, this volume.

Handford, C. R., and R. G. Loucks, 1991, Unique signature of carbonate strata and the development of depositional sequence on system tract models for ramps, rimmed shelves, and detached platforms (Abstract): AAPG Bulletin, v. 75, no. 3, p. 588-589.

Handford, C. R., and R. G. Loucks, 1993, Carbonate depositional sequences and systems tracts—responses of carbonate platforms to relative sea-level changes, this volume.

Haq, B. U., J. Hardenbol, and P. R. Vail, 1988, Mesozoic and Cenozoic chronostratigraphy and eustatic cycles, in C. Wilgus, B. Hastings, C. Kendall, H. Posamentier, C. Ross, and J. Van Wagoner, eds., Sea Level Changes: An Integrated Approach: SEPM Special Publication No. 42, p. 71-108.

Hine, A. C., and A. C. Neumann, 1977, Shallow carbonate bank margin growth and structure, Little Bahama Bank, Bahamas: AAPG Bulletin, v. 61, no. 3, p. 376-406.

Jacquin, T., A. Arnaud-Vanneau, H. Arnaud, C. Revenne, and P. R. Vail, 1991, Systems tracts and depositional sequences in a carbonate setting: A study of continuous outcrops from platform to basin at the scale of seismic lines: Marine and Petroleum Geology, v. 8, p. 122-138.

James, N. P., and R. N. Ginsburg, 1979, The Seaward Margin of Belize Barrier and Atoll Reefs: Special Publication No. 3, International Association of Sedimentologists, 191 p.

Jervey, M. T., 1988, Quantitative geological modeling of siliciclastic rock sequences and their seismic expression, in C. Wilgus, B. Hastings, C. Kendall, H. Posamentier, C. Ross, and J. Van Wagoner, eds., Sea Level Changes: An Integrated Approach: SEPM Special Publication No. 42, p. 47-70.

Kenter, J. A., 1990, Carbonate platform flanks: Slope angle and sediment fabric: Sedimentology, v. 37, p. 777-794.

Lighty, R. G., 1985, Preservation of internal reef porosity and diagenetic sealing of submerged early Holocene barrier reef, Southeast Florida shelf, in N. Schneidermann and P. M. Harris, eds., Carbonate Cements: SEPM Special Publication No. 36, p. 123-151.

Lowe, D. R., 1976, Grain flow and grain flow deposits: Journal of Sedimentary Geology, v. 46, p. 188-199.

Lowe, D. R., 1979, Sediment grain flows: Their classification and some problems of application to natural flows and deposits, in L. Doyle, and O. H. Pilkey, eds., Geology of Continental Slopes: SEPM Special Publication No. 27, p. 75-82.

Lowe, D. R., 1982, Sediment gravity flows: II. Depositional models with special reference to the deposits of high-density turbidity currents: Journal of Sedimentary Geology, v. 52, p. 279-297.

Mitchum, R. M., P. R. Vail, and S. Thompson, 1977, Seismic stratigraphy and global changes in sea level, part two: The depositional sequence as a basic unit for stratigraphic analysis, in C. E. Payton, ed., Seismic Stratigraphy—Applications to Hydrocarbon Exploration: AAPG Memoir 26, p. 63-83.

Mruk, D., 1989, Diagenesis of the Capitan Limestone, Upper Permian, McKittrick Canyon, West Texas, in P. M. Harris, and G. A. Grover, eds., Subsurface and Outcrop Examination of the Capitan Shelf Margin, Northern Delaware Basin: SEPM Core Workshop 13, San Antonio, April 23, p. 387-406.

Mullins, H. T., and A. C. Hine, 1989, Scalloped bank margins: Beginning of the end for carbonate platforms?: Geology, v. 17, no. 1, p. 30-33.

Nardin, T. R., F. J. Hein, D. S. Gorsline, and B. D. Edwards, 1979, A review of mass movement processes, sediment and acoustic characteristics, and contrasts in slope and base-of-slope systems versus canyon-fan-basin floor systems, in L. Doyle, and O. H. Pilkey, eds., Geology of Continental Slopes: SEPM Special Publication No. 27, p. 61-73.

Newell, N. D., J. K. Rigby, A. G. Fischer, A. J. Whiteman, J. E. Hickox, and J. S. Bradley, 1953, The Permian Reef Complex of the Guadalupe Mountains Region, Texas and New Mexico; A study in Paleoecology: W. H. Freeman and Co., San Francisco, 236 p.

Parsley, M. J., 1988, Deposition and diagenesis of an upper Guadalupian barrier-island complex from the Middle and Upper Tansill Formation, east Dark Canyon, Guadalupe Mountains, New Mexico: M.A. thesis, The University of Texas at Austin, 246 p.

Pray, L., J. Babcock, M. Esteban, D. Neese, J. F. Sarg, A. Schwartz, and D. Yurewicz, 1977, Road logs and locality guides, *in* M. Hileman, and S. J. Mazzullo, eds., Upper Guadalupian Facies, Permian Reef Complex, Guadalupe Mountains, New Mexico and West Texas: Permian Basin Section—SEPM Publication No. 77-16, v. 2, 194 p.

Reeckmann, S. A., 1986, Geology of the foreslope to basinal transition, Permian Reef Complex, McKittrick Canyon, Guadalupe Mountains: Special Study for Guadalupe Mountains National Park, 21 p.

Sarg, J. F., 1988, Carbonate sequence stratigraphy *in* C. Wilgus, B. Hastings, C. Kendall, H. Posamentier, C. Ross, and J. Van Wagoner, eds., Sea Level Changes: An Integrated Approach: SEPM Special Publication No. 42, p. 155-182.

Sarg, J. F., C. Rossen, P. Lehman, and L. C. Pray, eds., 1988, Geologic Guide to the Western Escarpment, Guadalupe Mountains, Texas: Permian Basin Section—SEPM Publication No. 88-30, 60 p.

Schlager, W., 1989, Drowning unconformities on carbonate platforms, *in* P. Crevello, J. Wilson, J. F. Sarg, and J. F. Read, eds., Controls on Carbonate Platform and Basin Development: SEPM Special Publication No. 44, p. 15-26.

Schlager, W., and O. Camber, 1986, Submarine slope angles, drowning unconformities, and shelf-erosion of limestone escarpments: Geology, v. 14, p. 762-765.

Skinner, J. W., and G. L. Wilde, 1955, New fusulinids from the Permian of West Texas: Journal of Paleontology, v. 29, p. 927-940.

Tyrrell, W. W., 1962, Petrology and stratigraphy of near-reef Tansill–Lamar strata, Guadalupe Mountains, Texas and New Mexico, *in* G. Wilde, S. D. Kerr, T. McClarin, and E. Mears, eds., Permian of the Central Guadalupe Mountains, Eddy County, New Mexico, Field Trip Guidebook and Geological Discussions: West Texas Geological Society Publication No. 62-48, p. 59-75.

Tyrrell, W. W., 1969, Criteria useful in interpreting environments of unlike but time-equivalent carbonate units (Tansill-Capitan-Lamar), Capitan Reef complex, West Texas and New Mexico, *in* G. M. Friedman, ed., Depositional Environments in Carbonate Rocks: SEPM Special Publication No. 14, p. 80-97.

Tyrrell, W. W., D. H. Lokke, G. A. Sanderson, G. J. Verville, 1978, Late Guadalupian correlations, Permian reef complex, West Texas and New Mexico: New Mexico Bureau of Mines and Mineral Resources Circular 159, p. 84-85.

Vail, P. R., J. Hardenbol, and R. G. Todd, 1984, Jurassic unconformities, chronostratigraphy, and sea level changes from seismic stratigraphy and biostratigraphy, *in* J. S. Schlee, ed., Interregional Unconformities and Hydrocarbon Accumulations: AAPG Memoir 36, p. 129-144.

Van Wagoner, J. C., R. M. Mitchum, K. M. Campion, and V. D. Rahmanian, 1990, Siliciclastic sequence stratigraphy in well logs, cores, and outcrops: Concepts for high-resolution correlation of time and facies: AAPG Methods in Exploration Series, No. 7, 55 p.

Ward, R. F., Kendall, C. G. St. C., and P. M. Harris, 1986, Upper Permian (Guadalupian) facies and their association with hydrocarbons—Permian Basin, West Texas and New Mexico: AAPG Bulletin, v. 70, no. 3, p. 239-262.

Wilber, R. J., J. D. Milliman, and R. B. Halley, 1990, Accumulation of bank-top sediment on the western slope of Great Bahama Bank: Rapid progradation of a carbonate megabank: Geology, v. 18, p. 970-974.

Wilde, G. L., 1955, Permian fusulinids of the Guadalupe Mountains: Permian Field Conference to the Guadalupe Mountains, October 1955, Permian Basin Section: SEPM, 71 p.

Wilson, J. L., 1975, Carbonate Facies in Geologic History: Springer-Verlag, New York, 471 p.

Chapter 6

Reciprocal Lowstand Clastic and Highstand Carbonate Sedimentation, Subsurface Devonian Reef Complex, Canning Basin, Western Australia

Peter N. Southgate
John M. Kennard
Michael J. Jackson
Phillip E. O'Brien
Michael J. Sexton
Australian Geological Survey Organisation
Canberra, Australia

ABSTRACT

Integrated sequence analysis of seismic and well data of the Givetian to Tournaisian sedimentary succession on the outer margin of the Lennard Shelf and adjacent Fitzroy Trough has recognized 18 third-order stratigraphic sequences in a major transgressive-regressive facies cycle. A Givetian phase of crustal extension initiated the transgressive-regressive cycle. The cycle terminated in the Tournaisian, following a phase of slow thermal subsidence. The transgressive half-cycle comprises at least four third-order sequences; an initial Frasnian–Givetian sequence followed by three back-stepping Frasnian sequences. The regressive half-cycle comprises 14 basinward-advancing third-order sequences. A period of tectonic uplift and erosion immediately prior to the Frasnian–Famennian boundary resulted in a major basinward shift in coastal onlap in the initial stages of the regressive half-cycle. Systems tracts and facies are partitioned according to their position on the transgressive-regressive cycle; in the transgressive half-cycle, lowstand deposits are subdued and transgressive and highstand deposits accentuated. In the regressive half-cycle, lowstand deposits are accentuated and provide a foundation for the overlying transgressive and highstand deposits. The extent of basinward highstand progradation was limited by the break point of their underlying lowstand deposits.

Sequence and systems tract geometries, their stacking patterns, and component facies define two distinct styles of sedimentation: (1) a Givetian to early Famennian, reef-rimmed platform complex; and (2) a late Famennian to Tournaisian mixed carbonate and siliciclastic ramp complex. The reef complex demonstrates marked reciprocal sedimentation. During lowstands, terrigenous sediments by-passed the exposed platform to be deposited in the basin as basin floor fans, slope fans, and prograding complexes. During transgressions and highstands, carbonate sediments were deposited on the platform, allodapic carbonate particles were shed into proximal marginal-slope settings, and clastics were trapped on the inner platform. Lowstand carbonate production occurred locally in areas starved of terrigenous clastic influx. The recognition of relative sea-level cycles in these strata has led to a subsurface model of reef development, which provides new insights into the third-order cyclicity within the larger Pillara and Nullara cycles recognized from previous outcrop studies. The outcrop model of backstepping and advancing reef complexes emphasizes transgressive and highstand depositional systems, and fails to recognize phases of lowstand deposition.

INTRODUCTION

The well-exposed Late Devonian reef complexes in the Canning basin in Western Australia represent one of the world's best-preserved Paleozoic reef successions. The outcropping and subsurface reef complexes provide an opportunity to combine surface and subsurface sequence stratigraphy in an area where steep depositional slopes result in sequence geometries being especially well preserved. The Canning basin also provides excellent examples of reciprocal sedimentation controlled by base-level changes, and of carbonate deposition in the presence of an abundant supply of coarse terrigenous sediment. The pronounced depositional relief of the platform margins and the tendency for carbonate and siliciclastic facies to interfinger, together result in lithic units being markedly diachronous.

In this study, we have utilized sequence stratigraphic concepts and models (Vail, 1987; Posamentier and Vail, 1988; Posamentier et al., 1988; Sarg, 1988; Van Wagoner et al., 1990; Wornardt, 1991) to interpret seismic and well data and to generate a new model for the development of the subsurface reef complex and its evolution into a mixed carbonate and siliciclastic ramp. This sequence analysis provides a new understanding of carbonate and siliciclastic facies distribution in the reef and ramp, and a broader perspective for previous and future outcrop studies.

PREVIOUS STUDIES

The exhumed Devonian reef complex is located on the inner portion of the Lennard Shelf adjacent to the Proterozoic Kimberley block (Figure 1). The complex has been extensively studied in outcrop where it reaches a maximum thickness of 2000 m (Playford, 1980, 1982, 1984; Playford et al., 1989). Previous studies have led to a widely accepted model for genesis of the complex: a Givetian–Frasnian Pillara cycle characterized by low-relief carbonate banks, stromato-

poroid-coral-cyanobacterial reefs, vertical platform growth and widespread platform drowning and backstepping; and a Famennian Nullara cycle characterized by cyanobacterial reefs and strongly advancing platforms. Reef growth ceased in the late Famennian when mixed carbonate and siliciclastic rocks of the late Famennian–Tournaisian Fairfield Group were deposited. Most sedimentologic work has been lithostratigraphic based, focusing on carbonate depositional facies (Playford and Lowry, 1966; Playford, 1980, 1984; Hall, 1984) and their diagenetic history (Kerans et al., 1986; Hurley and Lohmann, 1989; Wallace et al., 1991). The lithostratigraphic units are markedly diachronous and contrast with our new chronostratigraphic sequences. Playford et al. (1989) relate reef growth to an almost continuous rise in relative sea level interrupted by a brief regression, minor subaerial exposure of platform facies, and mild karst at the Frasnian–Famennian boundary.

In areas adjacent to the Proterozoic Kimberley Block, thick masses of basement-derived conglomerate and sandstone interfinger with the reef complex (Playford, 1980; Botten, 1984). Playford et al. (1989) interpret these terrigenous facies as accumulating in proximal alluvial fans, fan deltas, and submarine fans. Where the terrigenous sediments occur in marginal-slope facies, they form debris-flow deposits that contain large blocks of reef and platform facies (Playford, 1984). In contrast to the well-exposed platform and marginal-slope carbonate facies, basinal facies crop out poorly and comparatively little information is available regarding their composition and origin. Mixed carbonate and siliciclastic rocks of the Fairfield Group overlie sediments of the reef complex. Lithofacies subdivision of these strata suggests deposition in shallow marine and peritidal environments (Druce and Radke, 1979). Poor outcrop has severely hampered a regional understanding of these sediments.

Prior to the late 1970s, ten petroleum exploration wells had been drilled in the study area (Mayne,

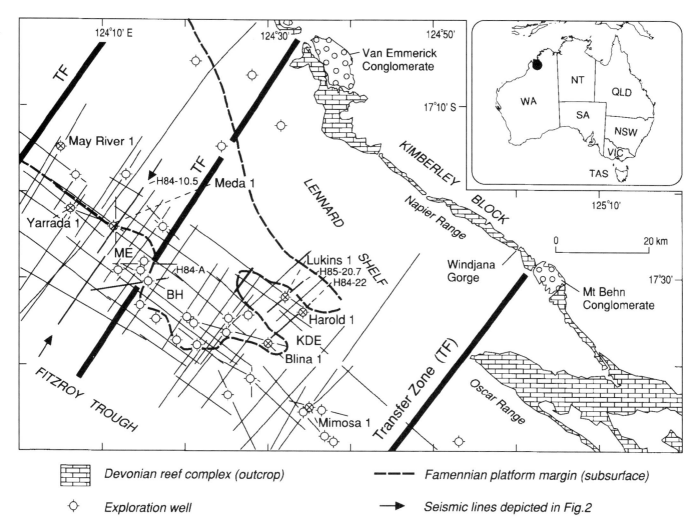

Figure 1. Location map showing outcropping reef complex, seismic lines, principal structural elements, and wells. ME—Meda Embayment; BH—Blackstone High; KDE—Kimberley Downs Embayment.

1975). In the late 1970s and 1980s, petroleum companies acquired 4000 km of seismic data and drilled 29 wells, many of which targeted structural closures in platform carbonates of Frasnian and Famennian age. The Blina oil field was discovered in 1981. It has proven reserves of approximately 2 MMbbl and occurs in carbonate rocks of Famennian age. Following the Blina discovery, several smaller fields were found in the overlying Permian–Carboniferous clastic strata.

Because most of this exploration was based on the outcrop-derived model for reef growth and analogy to petroleum exploration in the Devonian reef complexes of Alberta, Western Canada (Playford, 1982), comparatively little exploration activity has been focused on basinal deposits outboard of the reef-rimmed platform. In order to assess the petroleum potential of these sediments and further understand processes of platform evolution, the Australian Geological Survey Organisation (formerly the

Australian Bureau of Mineral Resources, Geology and Geophysics) undertook a subsurface study of the outer Lennard Shelf and adjacent Fitzroy Trough (Figure 1). Previous seismic interpretations on the Lennard Shelf are given in Middleton (1987) and Kemp and Wilson (1990). Kennard et al. (1992) discuss some highlights of the present study and their implications to some key outcrop localities, especially with reference to the Frasnian–Famennian boundary; new petroleum play concepts are described in Jackson et al. (1992).

REGIONAL SETTING

The Canning basin is a vast sedimentary basin in northwestern Australia that contains an unmetamorphosed Paleozoic and Mesozoic section in excess of 12,000 m thick (Brown et al., 1984; Purcell, 1984; Drummond et al., 1991). Covering an onshore area of 430,000 km², it is one of Australia's largest onshore basins with a promising, but as yet largely unful-filled, petroleum potential. Shallow-water carbonates, evaporites, and aeolian clastics of Ordovician age were deposited in an epeiric setting and represent the oldest rocks in the basin. Rocks of Middle Devonian to Early Carboniferous age unconformably overlie the Ordovician strata. These sediments accumulated in an extensional basin with a northwest-trending depocenter (the Fitzroy Trough, Figure 1). The trough developed as a number of interconnected asymmetri-cal half-grabens bounded on their southwestern mar-gin by a northeast-dipping listric normal fault (the Fenton Fault) that represents a major basement detachment surface some 10–12 km below the study area (Drummond et al., 1988, 1991). The Lennard Shelf separates the Fitzroy Trough from the Proterozoic Kimberley block. The shelf forms a 50–60-km-wide platform along the hinged northeast margin of the Canning basin (Figure 1) and is delineated along its southern margin by the Pinnacle and Harvey faults, a system of antithetic faults linked, in part, to the Fenton Fault. The Middle Devonian to Early Carboniferous reef and ramp successions (the subject of this study) accumulated on the Lennard Shelf and in the adjacent Fitzroy Trough. Transfer faults subdi-vide the Fitzroy Trough and Lennard Shelf into a series of compartments (Drummond et al., 1988) which are marked by offsets in the reef trend (Figure 1) and abrupt changes in structural and stratigraphic geometries and facies (Figure 4). Structural subdivi-sions in the area (e.g., Meda Embayment, Blackstone High, and Kimberley Downs Embayment) partly result from differential movement along transfer faults (Figure 1). Transpression in the mid-Late Carboniferous and again during the Early Jurassic reactivated some of these faults to generate large anti-clines and flower-type structures. Many of these fault systems have undoubtedly influenced the migration and trapping of hydrocarbons on the Lennard Shelf and in the Fitzroy Trough.

SEISMIC AND WELL LOG DATA

Variations in the quality of seismic data necessitat-ed two levels of sequence interpretation. Vibroseis™ data acquired in 1982 (source frequencies of 15–65 Hz) are not sufficient to resolve systems tract and some sequence boundaries. Thus, in areas covered by this data, only major sequence boundaries have been mapped. However, in areas covered by the 1984 and more recent Vibroseis data (Meda Embayment and Kimberley Downs Embayment, Figure 1), source fre-quencies of 12–90 Hz permit the resolution of sequence and systems tract boundaries. In basinal set-tings, systems tract boundaries are tied on all dip and strike lines; however, in platform settings, structural complexities and attenuation of seismic data beneath platform carbonates prevent the interpretation of sys-tems tracts on most strike lines.

Because most wells drilled in the study area target-ed platformal reef plays, comparatively little data are

available on the composition of the voluminous basi-nal deposits. Wells drilled in the Meda Embayment intersected predominantly platform successions; in the Kimberley Downs Embayment, a few wells inter-sected basinal deposits. Synthetic seismograms and time depth plots have been used to tie wells to seis-mic. Age determinations are primarily based on con-odont and microfloral data from cores and cuttings in the Meda 1, Langoora 1, May River 1, Blackstone 1, and Mimosa 1 wells (Kennard et al., 1992; Jones and Young, 1993). Most of the wells drilled in the 1980s have microfloral determinations from sidewall cores. Meda 1, Yarrada 1, May River 1, Lukins 1, and Blina 1 (Figure 1) provide examples of depositional systems on the Lennard Shelf. Yarrada 1 is located on the outer margin of the Lennard Shelf; May River 1 is the most inboard of all the wells studied; and Lukins 1, Blina 1, and Mimosa 1 intersect basinal deposits.

Eighteen sequences, interpreted as third-order rel-ative sea-level cycles (Vail et al., 1977), have been defined and mapped throughout the study area. Each sequence is identified by a number preceded by an appropriate stage name. Sequence and systems tract geometries and their stacking patterns define two dis-tinct styles of sedimentation (Figure 2): (1) a Givetian to early Famennian reef-rimmed platform complex (Frasnian–Givetian sequence, Frasnian sequences 1–4, the Frasnian–Famennian sequence, and Famennian sequences 1–2B); and (2) a late Famennian to Tournaisian mixed carbonate and siliciclastic ramp complex (Famennian sequences 3A–4 and Tournaisian sequences 1–6). The Frasnian–Givetian sequence and Frasnian sequences 1–4 correlate with the Pillara cycle, the Frasnian–Famennian sequence and Famennian sequences 1, 2A, and 2B correlate with the Nullara cycle of Playford (1980).

Reef-Rimmed Carbonate Platform

Sequences in the reef-rimmed carbonate platform complex are grouped into two sets: the lower set dis-plays successively backstepping platform margins (the Frasnian–Givetian sequence and Frasnian sequences 1–3); the upper set displays basinward advancing platform margins (Frasnian sequence 4, the Frasnian–Famennian sequence, and Famennian sequences 1–2B (Figures 2–4). A marked basinward shift in coastal onlap occurs at the base of the Frasnian–Famennian sequence. This shift reflects the onset of tectonic uplift and erosion in the Meda–Blackstone–Blina area (Figure 5). A relatively minor basinward shift of facies separates Famennian sequences 2A and 2B. This shift in facies arrested development of the Famennian 2A highstand deposits so that its platform margin is slightly back-stepped from the underlying Famennian 1 platform margin. Famennian sequence 2B is characterized by a shelf margin systems tract, lacks lowstand deposits, and is interpreted as overlying a type 2 sequence boundary. Thick lowstand deposits occur in the upper set of sequences where they display advancing

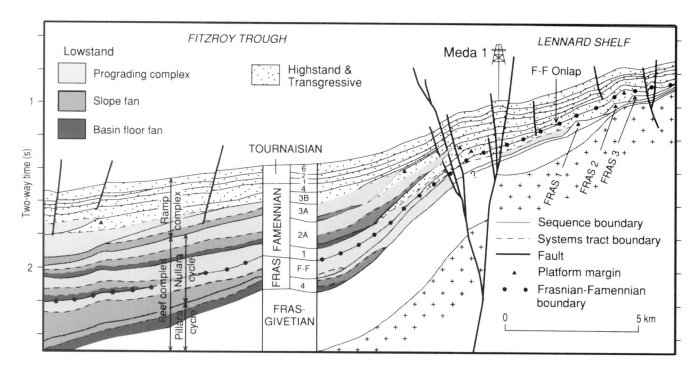

Figure 2. Line diagram depicting sequence and systems tract development in the Meda Embayment (modified from Kennard et al., 1992). Note coastal onlap of the Frasnian–Famennian sequence (arrow).

platform margins (Frasnian sequence 4, the Frasnian–Famennian sequence, and Famennian sequences 1 and 2A). In the lower set of sequences, lowstand deposits are subdued (Figures 2, 3, 7). Well data (Figures 7–12) indicate that platform carbonates dominate transgressive and highstand deposits and basinal siliciclastic facies dominate lowstand deposits. This temporal and spatial separation of lithologies indicates reciprocal lowstand-highstand sedimentation (Meissner, 1972).

Siliciclastic Lowstand Deposits

In Frasnian sequence 4, the Frasnian–Famennian sequence and Famennian sequences 1 and 2A lowstand deposits form a wedge-shaped body of sediment that lies basinward of the previous platform margin (Figures 13B, 14B) and locally fills valleys eroded into the platform. These deposits reach a maximum thickness of 120–200 ms two-way time within 5–8 km of the previous platform margin, after which they gradually thin in a basinward direction. Lowstand deposits are subdivided into three seismically defined depositional systems: (1) basin floor fan; (2) slope fan; and (3) prograding complex (Figures 2–6). A fourth depositional system, shingled turbidites, is present at the toes of some prograding complexes.

Basin floor fans form extensive sheetlike deposits or aprons tens of kilometers across and between 10–60 ms two-way time thick (Figures 4–6). They onlap the sequence boundary basinward of the platform margin and generally have upper boundaries

defined by high-amplitude reflections. The amplitude of this reflection commonly diminishes near the point of onlap (Figures 4–6). Most basin floor fans lack internal reflections, but where present, they are discontinuous and of low amplitude. Basin floor fans have been intersected in the Frasnian–Givetian and Frasnian–Famennian sequences in Mimosa 1 (Figure 12). In cores and cuttings, they are composed of poorly sorted very fine to medium sandstone, with scattered coarse grains and a calcareous and clay-silt matrix. On spontaneous potential and gamma logs, the sandstone has a uniform blocky signature (Figure 12). Seismic mapping shows that both of these fans were intersected in a distal position, close to their basinal pinchout. The high-amplitude reflection at the upper surface of a basin floor fan is interpreted to represent the contact between a sand-prone facies in the basin floor fan and a mud-silt prone facies in the overlying slope fan.

Slope fan deposits form thick wedge-shaped units above basin floor fans, and pinchout and onlap the sequence boundary at the slope of the previous platform margin. In contrast to the clearly defined lower boundary described above, their upper surfaces vary from indistinct transitional intervals to high-amplitude reflections. Discontinuous and hummocky, low-amplitude reflections that locally display "gull-wing" patterns characterize the internal parts of these deposits (Figures 4–6). In Lukins 1 (Figure 10) and Mimosa 1 (Figure 12), slope fans are composed of siltstone, very fine to fine-grained sandstone, and occasionally medium-coarse-grained sandstone and

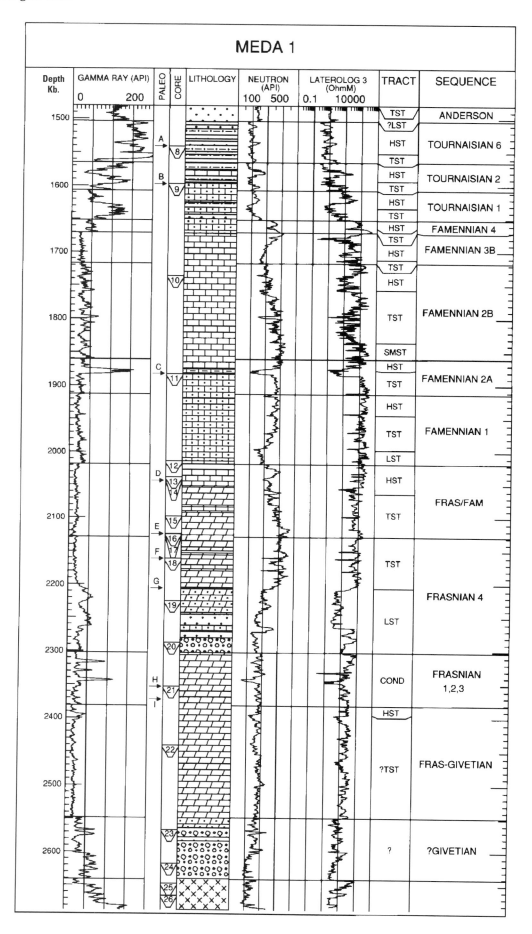

MEDA 1

minor shale. The spontaneous potential and gamma logs have a spiky motif arranged in crescentic patterns with a 20–50 m wavelength. These patterns are interpreted to represent deposition in channel-levee complexes.

Prograding complexes usually comprise the thickest depositional system within the reef-rimmed complex. They form thick progradational and aggradational lenses with locally well-defined sigmoidal clinoforms that onlap the sequence boundary at or near the previous platform margin, and downlap basinward onto the slope fans (Figures 3–6). Slightly mounded to hummocky low-amplitude reflections locally disrupt the sigmoidal clinoforms (Figures 4, 6). These zones may represent slumps on the inclined surface of the prograding complex. The lower boundaries of the prograding complex are locally indistinct downlap surfaces that "climb" basinward. Such surfaces are marked by the transition from high-amplitude inclined reflections at the toes of the sigmoidal clinoforms to discontinuous horizontal reflections of variable amplitude that extend basinward from these clinoforms (Figures 5, 6). These relationships are interpreted to represent shingled turbidites. Shingled turbidites intersected in Lukins 1 (Figure 6) are composed of siltstone and display crescent-shaped log motifs similar to the underlying slope fans. In distal basinal settings, the base of the prograding complex may be locally defined by a high-amplitude reflection (Figures 4, 5, Famennian sequences 1 and 2A). This relationship occurs as the prograding complex thins and may mark an interval of sediment starvation and/or increased diagenetic alteration.

In Lukins 1 (Figures 6, 10) and Mimosa 1 (Figure 12), the prograding complex comprises siltstone, shale, and fine-grained sandstone. In Mimosa 1, the gamma and spontaneous potential logs are relatively uniform and lack distinctive trends. However, in Lukins 1, the gamma, sonic, and resistivity logs display several upward-fining trends. In the gamma log, each upward-fining motif is arrested by a short negative trend that coincides with a sharp increase in velocity on the sonic log. Well cuttings indicate that these intervals are composed of bioclastic wackestone, mudstone, and calcareous shale; these calcareous intervals are interpreted as flooding surfaces that mark periods of relative terrigenous sediment starvation and increased carbonate production.

In Meda 1 and Yarrada 1 (Figures 7, 8), conglomerate composed of pebbles and boulders of quartzite, granite, chert, mica schist, and rare carbonate lithoclasts overlies the Frasnian 4 sequence boundary. The conglomerate has a poorly sorted calcareous and dolomitic sandstone matrix and grades upward into pebbly and sandy limestone and dolostone. In Figure 3, the conglomerates are included at the base of the prograding complex. However, they may represent proximal feeder channels for the basin floor and slope fans recognized in a basinward position from Meda 1. The sandy and pebbly limestone and dolostone overlying these conglomerate deposits represent proximal facies of the prograding complex (see following section). A 4-m-thick conglomeratic lowstand deposit also overlies the Famennian 1 sequence boundary in Meda 1 (core 12), but is too thin to be shown in Figure 7.

Calcareous Lowstand Deposits

In the Kimberley Downs Embayment, significant lowstand carbonate production occurred around topographic highs in areas of reduced clastic influx (Figures 5, 6). The Famennian 1 sequence boundary is overlain by an interval of discontinuous low-amplitude reflections that define sigmoidal sediment wedges. These rocks are intersected in Blina 1 and Harold 1 where they consist of pale red to brown limestone, sandy limestone, silty limestone, and calcareous siltstone. In Famennian sequence 2A (Figure 5), lowstand carbonates are interpreted to occur on the flanks of the platform within a seismically transparent zone 10–20 ms two-way time thick that mantles the sequence boundary. High-amplitude clinoforms extend basinward from the transparent zone and are interpreted as tongues of carbonate-rich sediment. In Lukins 1, silty and argillaceous bioclastic mudstone and wackestone containing crinoid ossicles, tubular spines, and pelecypod fragments intersected at the base of the Famennian 2A sequence (Figures 6, 10) probably represent the distal toe of a lowstand carbonate unit present in one of these seismically transparent zones. Interfingering facies relationships between the lowstand siliciclastic and calcareous facies in Famennian sequences 1 and 2A produce an unusual set of onlap relationships (Figure 5). Onlap of lowstand carbonate units and the basin floor fan takes place on the sequence boundary. However, the remaining third-order silici-

Figure 7. Well logs and systems tract subdivision of Meda 1 showing the locations of core samples and biostratigraphic control points. A—Tournaisian spores, *G. spiculifera* Zone; B—Tournaisian conodonts, *isosticha-Upper crenulata* Zone and spores, *G. spiculifera* Zone; C—middle to late Famennian conodonts, *marginifera* to Early *praesulcata* Zone; D—early Famennian conodonts, Middle to Late *triangularis* Zone; E—early Famennian conodonts, *triangularis* Zone; F—late Frasnian ostracods and brachiopods; G—late Frasnian conodonts; H—lower Frasnian conodonts; I—Frasnian to Givetian conodonts. A legend for the lithology column is included in Figure 9.

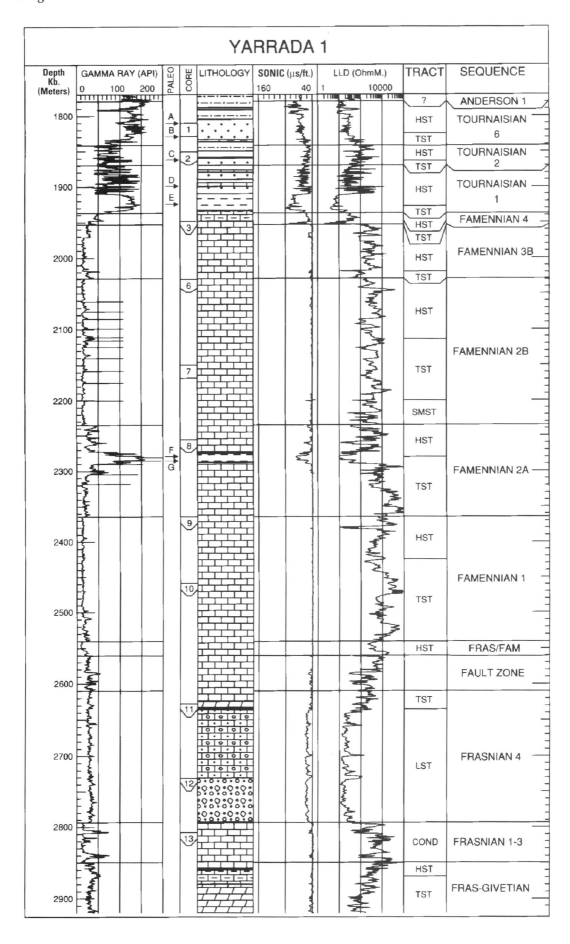

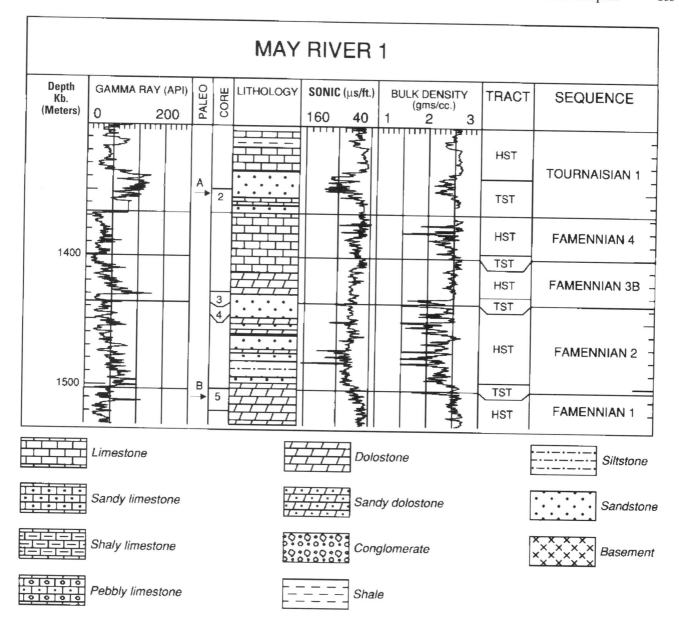

Figure 9. Well logs and systems tract subdivision of May River 1, drilled on the inner platform, showing the location of cores and biostratigraphic control points. A—Tournaisian spores, *G. spiculifera* Zone; B—Famennian conodonts.

clastic lowstand units onlap and bury these lowstand carbonates. For this relationship to occur, carbonate production would have preceded siliciclastic deposition and the interfingering relationship probably represents reciprocal sedimentation at a fourth-order scale of cyclicity. We postulate that encroaching lowstand siliciclastic deposits led to the demise of lowstand carbonate production.

Transgressive Deposits

Although transgressive and highstand deposits are the dominant systems tracts preserved in outcrop,

Figure 8. Well logs and systems tract subdivision of Yarrada 1 (TD 3295 m) showing the location of cores and biostratigraphic control points. A to E—Tournaisian spores, *G. spiculifera* Zone; F and G—Famennian spores, *R. lepidophyta* Zone. This well was drilled on the Famennian 2B platform margin. Parasequence cycles for sequences 2A and 2B are shown in the gamma log column. A legend for the lithology column is included in Figure 9.

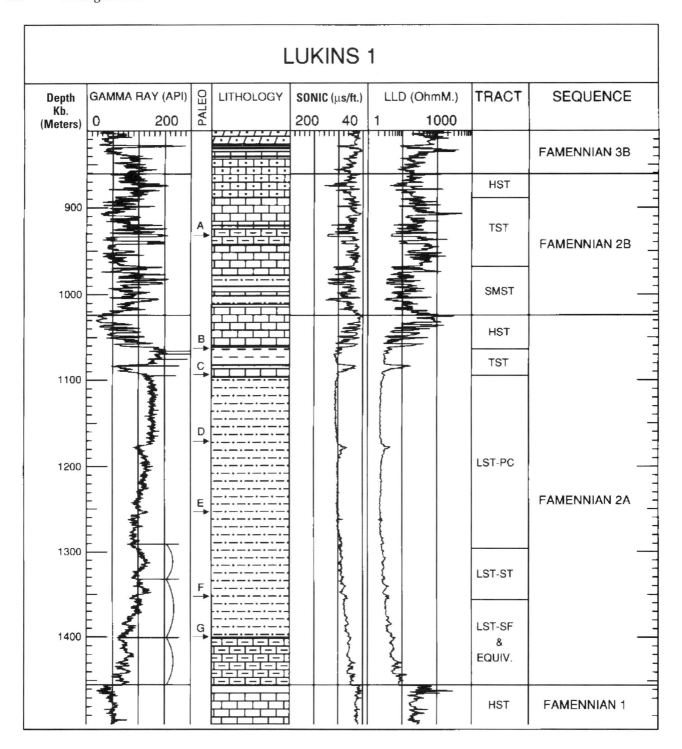

Figure 10. Well logs and systems tract subdivision of Lukins 1 showing the location of biostratigraphic control points. A to G—Famennian spores, *R. lepidophyta* Zone. Parasequence cycles (transgressive deposits) and crescent-shaped patterns (slope fan and shingled turbidites) are shown in the gamma log column. A legend for the lithology column is included in Figure 9.

Figure 11. Well logs and systems tract subdivision of Blina 1 showing the location of cores and biostratigraphic control points. A—Tournaisian spores, *G. spiculifera* Zone; B to E—Famennian spores, *R. lepidophyta* Zone; F—Frasnian spores. Parasequence cycles for the transgressive deposits in sequence 2A are shown in the gamma log column. A legend for the lithology column is included in Figure 9.

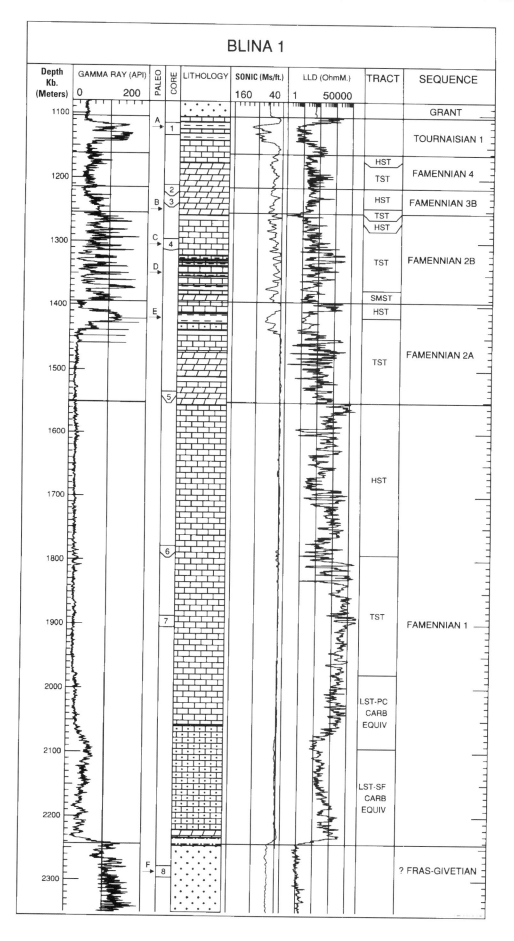

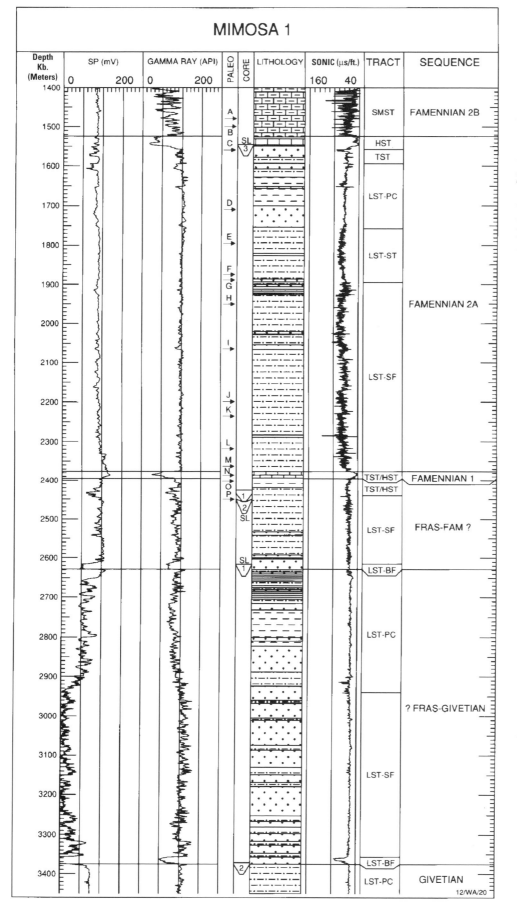

Figure 12. Well logs and systems tract subdivision of Mimosa 1 showing the location of cores and biostratigraphic control points. A to M—Famennian spores, *R. lepidophyta* Zone; N—mid-Famennian ostracods, *eocostata* Zone, correlates with Uppermost *marginifera* to Lower *postera* conodont zones; O—? late Frasnian conodonts; P—Frasnian spores. A legend for the lithology column is included in Figure 9.

seismic data show that in the late Frasnian and Famennian they are volumetrically minor compared to their related lowstand deposits. On seismic sections, transgressive and highstand deposits often are difficult to distinguish from each other. However, well log motifs permit the identification of two types of transgressive deposits: (1) backstepping; and (2) keep up. Backstepping transgressive deposits are best developed in Famennian sequence 2A. In Yarrada 1 (Figure 8), a subtle but sharp positive shift in the gamma log, coincident with a change in its character, marks the Famennian 2A sequence boundary (2364 m). Between 2280–2364 m, five upward-thinning progradational parasequences define a distinctive backstepping motif on the gamma log (Figure 8). Each progradational parasequence is sharply truncated by an abrupt positive shift in the gamma log interpreted to mark a flooding surface and shaly limestone deposition. The maximum flooding surface occurs within the interval of shale (2270–2285 m) interpreted to represent the condensed section. Similar backstepping gamma log motifs are present in Blina 1 and Lukins 1 (Figures 10, 11). In Blina 1, five backstepping parasequences define the transgressive depositional system. In Lukins 1, the transgressive surface lies at the base of a prominent upward-cleaning, progradational gamma log motif (1080–1100 m, Figure 10). Cuttings from this interval are composed of bioclastic mudstone and wackestone. Shale and shaly limestone overlie the progradational unit. These shaly sediments (1063–1080 m) have a backstepped log motif interpreted to indicate rapid drowning, and minor carbonate sediment production, features that are consistent with a condensed section at the time of maximum flooding. The absence of a well-developed backstepping parasequence set in Lukins 1 compared with Blina 1 and Yarrada 1, reflects the basinal position of this well.

On seismic sections, it is not possible to resolve backstepping parasequences. However, an impedance contrast associated with the low-velocity shale and shaly limestone deposits of the condensed section and the encasing higher velocity limestones results in a prominent high-amplitude reflection that onlaps the sequence boundary in a platform position. This high-amplitude reflection enables the maximum flooding surface to be mapped. In basinal positions, the transgressive deposits pinch out and the maximum flooding surface merges with the top of the prograding complex (Figure 6).

In contrast to the distinctive log patterns described above, backstepping log motifs and condensed section deposits are subdued or lacking in keep-up transgressive deposits. In consequence, criteria for delineating the boundary between transgressive and highstand deposits in keep-up depositional systems

are equivocal. In Blina 1 (Figure 11), keep-up transgressive deposits are interpreted to occur in the Famennian 1 sequence. Lowstand carbonates with a cleaning-upward progradational gamma log motif occur between 1980–2090 m. At 1980 m, a small but sharp negative shift in the gamma log may mark the base of the transgressive deposits. Between 1550–1980 m, the gamma log has a uniformly low signature. Core and cuttings data in this interval indicate that the rocks are composed of recrystallized limestone with relict peloids and bioclast molds. Minor positive shifts in the gamma log, slight velocity changes in the sonic log, and major changes in the trend of the deep resistivity log occur between 1770–1830 m. In the absence of a well-defined backstepping log motif, the boundary between the transgressive and highstand deposits is tentatively placed in this interval. Seismic data (Figure 5) show that this interval coincides with a horizon defined by low-amplitude discontinuous reflections.

A similar gamma log motif occurs in Yarrada 1 in Famennian sequence 2B. Limestones between 2030–2200 m have a subdued, uniformly low gamma log motif interrupted by peaks that reflect minor positive shifts in the log. Core 8, although in the underlying highstand deposits, indicates that these peaks coincide with shaly mudstone deposits interpreted as marking parasequence flooding surfaces (Figure 8). Parasequence stacking patterns in the interval 2030–2200 m show uniform thicknesses between 2120–2200 m and thinning in the interval 2100–2120 m. Between 2030–2100 m, parasequences are not resolved. Throughout the interval 2030–2200 m, the dipmeter log has a random "bag of nails" motif; core 7 comprises microbial boundstone and skeletal wackestone with *Renalcis* sp and fragments of ostracods, bivalves, and solitary corals, and core 6 comprises fenestral peloid wackestone (microbial boundstone) and peloid packstone. Collectively, these data suggest that the interval 2030–2200 m represents submergent to semi-emergent reef facies. Parasequence cycles of uniform thickness (2120–2200 m) beneath a subdued condensed section (2110–2120 m) are interpreted to represent reef growth in a keep-up transgressive depositional system. Parasequences in the overlying highstand deposits are thinner than those in the transgressive depositional system. Other sequences displaying keep-up transgressive deposits include the Frasnian–Givetian, Frasnian 4, Frasnian–Famennian and Famennian 1 sequences in Meda 1 (Figure 7), the Frasnian–Givetian and Fammenian 1 sequence in Yarrada 1 (Figure 8), and the Famennian 2B sequence in Meda 1 and Blina 1 (Figures 7, 11).

On seismic sections, keep-up transgressive deposits are 20–50 ms two-way time thick (Figures

3–5). These deposits display similar seismic reflection characteristics to the overlying highstand deposits and only where a downlap surface is recognizable at the base of the highstand is it possible to bracket the underlying transgressive deposits between this downlap surface and the underlying sequence boundary or transgressive surface.

Highstand Deposits

Highstand sediments form elongate, lenticular bodies that gradually thicken toward the platform margin, where they are 20–50 ms two-way time thick, and then thin rapidly basinward to marginal-slope deposits, thus resulting in steep platform margins (Figures 13A, 14A). Across the platform, highstand deposits display discontinuous parallel reflections, and at the platform margin, sigmoid-oblique clinoforms (Figures 3, 4). Where highstand deposits overlie backstepping transgressive deposits, they downlap onto the high-amplitude reflection marking the maximum flooding surface. Where keep-up transgressive deposits underlie highstand deposits, it is often difficult to separate highstand clinoforms from transgressive clinoforms (e.g., Frasnian sequence 4, Figure 3). In outer platform settings, highstand deposits have well-defined toplap surfaces; in inner and middle platform settings, the upper boundary of the highstand is a discontinuous and irregular, high-amplitude reflection that may indicate karst. Highstand deposits comprise inner platform, back-reef, reef, and fore-reef carbonate facies.

May River 1 provides an example of highstand facies development on the inner platform (Figure 9). The most distinctive feature of inner platform depositional systems is a ratty and at times spiky gamma log motif that defines a coarsening-upward trend in each highstand deposit (Figure 9). Core and cuttings indicate that the principal lithologies are coarse to very fine sandstone, sandy dolostone, and dolostone. Minor shale and siltstone occur at the base of each progradational trend.

The Famennian 2A sequence in Yarrada 1 provides an example of marginal-slope facies above shale and shaly limestone deposits of a condensed section (Figure 8). Six upward-cleaning parasequences (2233–2270 m) define a progradational log pattern (Figure 8). Each progradational trend is terminated by a sharp positive shift in the gamma log, interpreted as shaly limestone at a flooding surface. In core 8, these sediments consist of nodular mudstone and wackestone with scattered crinoid, bryozoan, and brachiopod debris (progradational carbonates), and shale (flooding surface). Dipmeter signatures within the interval 2233–2270 m are uniformly to the southwest at 10–30°, consistent with the interpretation of marginal-slope facies. Highstand carbonates deposited in outer platform settings occur in the Frasnian–Famennian sequence in Meda 1, the Famennian 1 sequence in Yarrada 1 and Blina 1, the Famennian 2A sequence in Meda 1, and the Famennian 2B sequence in Yarrada 1. These deposits have a uniformly low gamma signature and are interpreted as reef and back-reef facies.

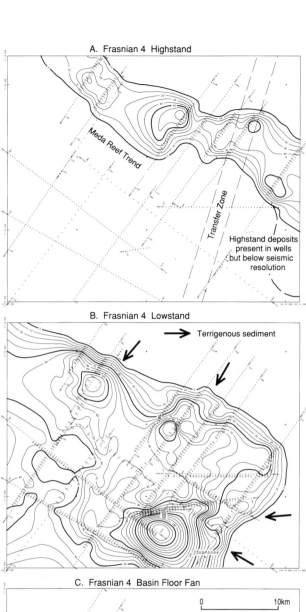

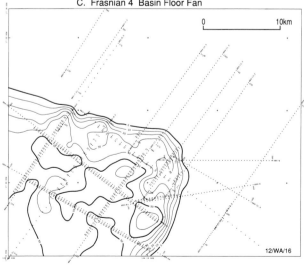

Figure 13. Isopach maps of systems tracts in Frasnian sequence 4. (A) highstand, (B) lowstand, (C) basin floor fan. Lowstand and highstand isopach maps demonstrate reciprocal sedimentation and indicate the steep gradients developed at the platform margin.

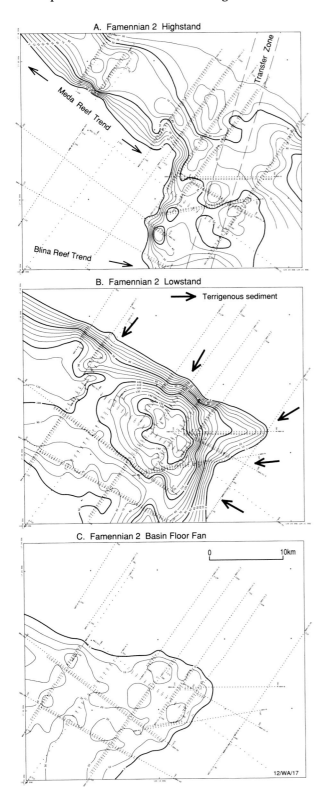

Figure 14. Isopach maps of systems tracts in Famennian sequences 2A and 2B. (A) highstand of 2A and all of 2B, (B) lowstand of 2A, (C) basin floor fan. Comparison of the highstand isopach map of Frasnian sequence 4 (13A) with the highstand isopach map for Famennian sequence 2A and 2B (14A) shows the extent of reef and platform advancement during the Famennian.

Shelf Margin and Associated Transgressive Deposits

A shelf margin depositional system occurs at the base of Famennian sequence 2B. The type 2 sequence boundary is distinguished in wells by a sharp change in the gamma log motif and on seismic by a distinctive onlap surface (Figure 5). In seismic, the overlying shelf margin systems tract is difficult to resolve. However, in Blina 1, Lukins 1, and Mimosa 1 (Figures 9–12), a sharp change in the baseline of the gamma log from low values (20–30 API) to higher values (50–130 API) coincides with a change in the logs character from uniform to spiky and irregular. This change in log character is interpreted to represent the rapid transition from reef and back-reef facies of the outer platform to middle and inner platform carbonate facies. In Yarrada 1, the same event is marked by an abrupt change from marginal-slope facies of the Famennian 2A highstand to reef facies of Famennian sequence 2B. The presence of a basinward shift in facies and absence of an associated lowstand wedge is interpreted to represent a significant lowering of relative sea level and a type 2 sequence boundary.

The overlying transgressive deposits initially display stacked aggradational patterns (e.g., Yarrada 1, Figure 8; Blina 1, Figure 11) characteristic of a keep-up transgressive depositional system followed by a subdued backstepping pattern beneath the maximum flooding surface. In Yarrada 1 (Figure 8), the transgressive deposits are represented by reef facies; in Meda 1, Lukins 1, and Blina 1, they are middle platform facies (Figures 7, 10, 11).

Frasnian–Givetian Sequence and Frasnian Sequences 1–3

Seismic data did not permit the regional mapping of depositional systems in the Frasnian–Givetian sequence and Frasnian sequences 1–3. Carbonates of the Frasnian–Givetian sequence have only been intersected in Meda 1 and Yarrada 1 (Figures 7, 8). Frasnian sequences 1–3 are recognized in both of these wells and on seven dip lines parallel to the line on which Meda was sited (Figures 2, 3). Interpretations in this part of the section, particularly with respect to the Frasnian–Givetian sequence, are guided by the occurrence of backstepping platforms in outcrop (Playford et al., 1989).

The Frasnian–Givetian sequence in Meda 1 and Yarrada 1 (Figures 7, 8) is interpreted as a keep-up transgressive depositional system. We are unsure where to place the sequence boundary at the base of this sequence and for this reason conglomerate intersected below 2550 m in Meda 1 lacks a systems tract interpretation. Similar conglomerates also occur in Yarrada 1, but are below the depth shown in Figure 8. Three backstepping sequences (Frasnian sequences 1–3) occur slightly basinward of the prominent Frasnian 4 highstand buildup (Figures 2, 3). Each backstepping sequence is defined by a set of low-amplitude discontinuous clinoforms and a prominent onlap surface. In Meda 1 and Yarrada 1 (Figures 7, 8),

Frasnian sequences 1–3 are intersected in a basinward position some 3–4 km outboard of the Frasnian 1 platform margin. In both wells, Frasnian sequences 1–3 comprise a 60–80 m thick succession of predominantly dolostone and dolomitic limestone, with some sandy dolostone and dolomitic silty sandstone in Yarrada 1 (core 13, Figure 8). This succession overlies an interval of dolostone interpreted to comprise a keep-up transgressive depositional system capped by a prominent maximum flooding surface at 2395 m in Meda 1 and 2870 m in Yarrada 1, and underlies conglomerate at the base of Frasnian sequence 4 (Figures 7, 8). In both wells this Frasnian succession contains three prominent peaks on the gamma log. Cuttings descriptions in Meda 1 emphasize that these peaks do not correspond with shale, but instead represent a radioactive mineral (Pudovskis, 1962). Based on the Meda 1 and Yarrada 1 log correlations and their location outboard from the platform margins of Frasnian sequences 1–3, this succession is interpreted as a condensed, slope to basinal section, equivalent to the three backstepping Frasnian sequences.

Support for this interpretation may be provided by the backstepping Frasnian succession in the Teichert Hills (approximately 180 km southeast of Windjana Gorge). Here, finely laminated phosphatic crusts (formerly recognized as a bed of micritic limestone) were recently identified by the authors. The crusts mantle an erosion surface developed on stromatoporoid-bearing limestone, and separate these shallow-water reef deposits from recessive deeper water, basinal, and marginal-slope facies (Playford, 1981). Playford (1981) interprets the bed of micritic limestone (phosphatic crust) as accumulating in relatively deep water immediately after platform drowning. The three gamma log peaks present in the Meda 1 and Yarrada 1 well logs may represent phosphatic crusts similar to those present in the Teichert Hills. Uranium commonly occurs in the carbonate-fluorapatite (phosphate) lattice where it substitutes for calcium ions (Kolodny and Kaplan, 1970). The gamma peaks therefore may represent radioactivity associated with enhanced phosphate concentrations in these sediments.

Ramp Complex

The late Famennian–Tournaisian ramp complex comprises nine basinward advancing sequences (Famennian sequence 3A to Tournaisian sequence 6, Figure 2). On seismic sections, each of these sequences comprises a lowstand clastic wedge with sigmoidal internal reflections and a thin and more laterally extensive, mixed carbonate and siliciclastic highstand deposit lacking internal reflections. Slope fans are restricted to Famennian sequences 3A, 3B, and 4 (Figure 2) and represent the gradual transition from a reef-rimmed platform to a distally steepened ramp (cf. Read, 1985). Lowstand prograding complexes overlie the fans and are the sole recognized lowstand component of Tournaisian sequences 1–6. Each highstand deposit extends several kilometers across its underlying lowstand and thins basinward once it

reaches the break point of the underlying prograding complex.

Wells intersecting the ramp complex are sited in middle to inner ramp positions. Consequently, lowstand deposits have not been intersected and their lithologies and component facies remain unknown. Transgressive and highstand deposits are intersected in many wells; however, only rocks belonging to Famennian sequences 3B and 4 and Tournaisian sequences 1, 2, and 6 are identified. The remaining sequences are interpreted to have either already pinched out or to be so thin that they cannot be resolved on well logs.

In Famennian sequences 3B and 4, well logs show transgressive deposits to be thin and composed of two or three backstepping progradational parasequences. In cuttings and cores, these rocks consist of sandy, silty, and muddy dolostone deposited in peritidal settings. In middle ramp settings (e.g., Yarrada 1 and Meda 1), highstand deposits are composed of bioclastic ooid grainstone and microbial boundstone deposited in shallow-water environments. A ratty and at times spiky gamma log motif that defines a forestepping progradational trend characterizes highstand deposits in inner ramp positions (Figure 9). Parasequences are thin and composed of interbedded coarse to very fine sandstone, sandy dolostone, and dolostone deposited in nearshore peritidal settings.

In Tournaisian sequences 1 and 6, marked landward shifts in coastal onlap are associated with relatively thick transgressive deposits. On the middle-inner ramp, the sequence boundary and associated maximum flooding surface at the base of Tournaisian sequences 1 and 6 are imaged as "booming" reflections that form key marker horizons (Figures 3–5). In well logs these major transgressions are represented by prominent backstepping gamma log patterns culminating in the deposition of 10–20 m thick shale units that mark the condensed sections (Figures 7–9, 11). Oils in the Sundown, West Terrace, and Lloyd fields are probably sourced from the shale at the base of Tournaisian sequence 1 (Goldstein, 1989). Highstand deposits in Tournaisian sequences 1 and 2 have a spiky gamma log motif and consist of bioclastic grainstone, calcareous sandstone, siltstone, and shale. In inner ramp positions of the Meda Embayment, continued progradation of Tournaisian sequences 3–5 (Figure 4) results in either their absence or their thinning below seismic and well resolution.

Isopach maps showing the regional distribution of lowstand and highstand depositional systems in Tournaisian sequence 3 are shown in Figure 15. The highstand deposits form a comparatively thin, sheet-like unit that thins in its updip position below seismic resolution. The basinward extent of progradation apparently was largely controlled by the break point for the underlying lowstand deposits.

Conceptual Development of a Reef Sequence

This section generalizes the previous descriptive data to outline possible processes involved in the for-

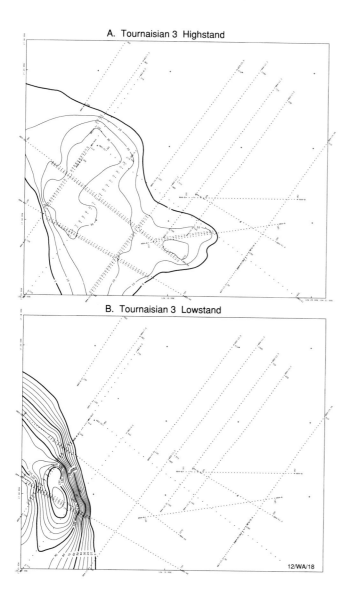

A. Tournaisian 3 Highstand

B. Tournaisian 3 Lowstand

12/WA/18

Figure 15. Isopach maps of systems tracts in Tournaisian sequence 3. (A) highstand, (B) lowstand.

mation of one third-order, basinward-advancing sequence in the reef-rimmed platform complex. During lowstands, streams transported terrigenous sediments across the platform to the former platform margin. Initial stream rejuvenation resulted in conglomeratic and sandy sediments being supplied to the exposed platform edge. However, as gradients declined and the supply of coarse clastic sediment previously trapped in the coastal and terrestrial belts diminished, finer grained sediments were the dominant component transported to the platform edge. At the same time, the subaerially exposed carbonates were subject to meteoric diagenesis and the former platform margin began to collapse, shedding allochthonous blocks of reef and platform facies into deeper water environments. In deep-water settings,

lowstand sediment accumulation commenced as mass flows, grain flows, and turbidity currents transported the sand-prone terrigenous sediments down the slope and onto the basin floor to form a slightly mounded sandy apron in front of the previous platform margin (Figures 13C, 14C). This resulted in the accumulation of sand-prone basin floor fans above the sequence boundary (Figures 3, 4, 12). Following this initial phase of lowstand deposition, debris flows, grain flows, and probable slumps transported terrigenous sediments onto the slope, where they were mixed with carbonate detritus shed from the eroding platform to form calcareous sandy conglomerate and sandstone deposits (e.g., Frasnian sequence 4, Yarrada 1, and Meda 1, Figures 7, 8). Turbidity currents transported the finer grained sand, silt, and mud down the slope into proximal basinal settings to form slope fans (Figure 12). In platform and slope environments isolated from significant terrigenous detritus (e.g., Lukins 1, Figures 6, 10; and Blina area, Figures 5, 11), lowstand carbonate production resulted in aggradational and progradational sediment wedges composed of red-brown, sandy, silty, and muddy bioclastic limestone (Figures 5, 11).

During the final phases of lowstand sedimentation, fine-grained sand, silt, and shale deposits accumulated as a wedge-shaped body of sediment to form a prograding complex (Figures 3–6, 10). This prograding and aggrading unit downlapped onto the earlier deposited basin floor and slope fans, or in areas of reduced terrigenous flux, argillaceous carbonates. Clear-water carbonate production continued on platforms isolated from terrigenous influx (e.g., Blina area, Figures 5, 11). On the platform terrigenous clastics were deposited in fluvial valleys to form pebbly sandstone and polymict conglomerate indicative of local basement provenance.

During the ensuing transgression, terrigenous clastics were trapped in coastal facies of the inner platform, facilitating carbonate production on the outer and middle platform. In backstepping transgressive depositional systems, progradational parasequences composed of bioclastic and glauconitic peloidal grainstone and wackestone accumulated in outer platform settings (Famennian sequence 2A, Figures 7, 8). These backstepping transgressive carbonate units were eventually drowned and buried by shales of the condensed section. In keep-up transgressive depositional systems, reef and back-reef facies accumulated in outer platform settings (e.g., Famennian sequence 1, Figures 5, 11). In inner platform settings, shallow-water, transgressive peritidal carbonate deposits were mixed with coastal, terrigenous sediments. Shales of the condensed section do not extend to the inner platform.

During highstands, clear-water carbonate production took place on the outer platform and led to the progradation of reef and back-reef facies over their associated marginal slope deposits (see Playford, 1980, 1984 for discussion of these facies in outcrop). On the middle parts of the platform, carbonate production dominated, but sporadic influxes of terrigenous clay and silt resulted in the accumulation of silty and muddy carbonates. On the inner platform, dolomitic peritidal carbonates accumulated. In nearshore settings, periodic influxes of terrigenous sand and silt resulted in sandy dolostone and sandstone deposits, and during the late highstand, environments were dominated by calcareous sandstone and siltstone. We postulate that a significant volume of coarse clastic sediment accumulated in the hinterland as alluvial fans and fan deltas during highstand periods.

DISCUSSION

Sequence analysis has provided new insights into the third-order cyclicity within the Pillara and Nullara reef cycles recognized from outcrop studies (Figure 12 in Playford et al., 1989). Whereas the outcrop model depicts phases of transgressive and highstand sedimentation, our subsurface model indicates that highstand reef growth was repeatedly punctuated by periods of lowstand clastic deposition (Figure 16). In our subsurface model, an initial phase of

aggradational and backstepping sequences (Frasnian–Givetian sequence and Frasnian sequences 1–3) is followed by a phase of basinward-advancing sequences (Frasnian sequence 4, the Frasnian–Famennian sequence, and Famennian sequences 1–2B). A major and abrupt basinward shift in coastal onlap occurs at the base of the Frasnian–Famennian sequence. This downward shift records the initial phases of uplift in the Meda–Blackstone–Blina area. Continued uplift heralded a phase of erosion that removed most of the Frasnian–Givetian section from the Blackstone High. On the platform, significant erosion in the Meda area and the western flank of the Blackstone High are shown by an incised surface at the base of the Famennian 1 sequence (Figure 3, landward of the Frasnian 4 platform margin; and Figure 4, above the undifferentiated section in the transfer zone). In the Blina area, tectonic uplift and erosion removed most of the Frasnian section (truncation beneath the Famennian 1 sequence boundary, Figure 5). In the study area, the lower Famennian *rhomboidea* and *crepida* conodont zones are not recorded in any wells and their absence is interpreted as a tectonic

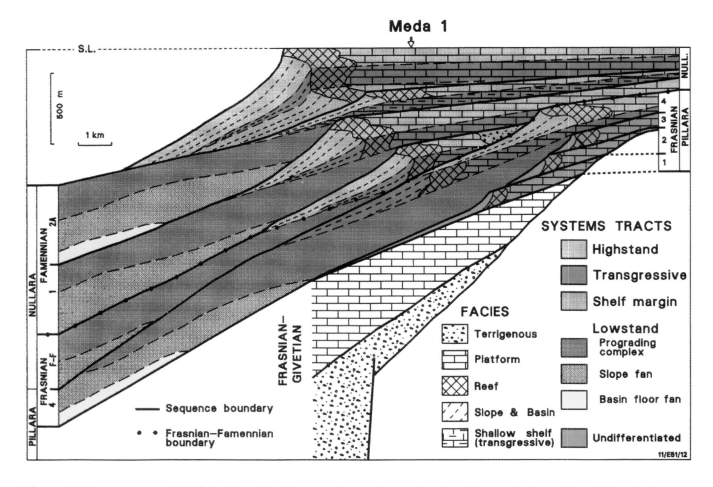

Figure 16. Sequence model of the Frasnian–Famennian reef complex based on seismic and well data in the Meda area (location of Meda 1 is an approximation). Facies are shown for transgressive and highstand deposits (modified from Kennard et al., 1992).

erosional event (Figure 17). Because both of these conodont zones are present in the outcropping reef complex (Nicoll and Playford, 1993), the outcropping inner reef complex preserves a more complete sedimentary record of transgressive and highstand deposits than the subsurface study area.

Playford (1980, 1982) defined the Pillara and Nullara cycles of platform development on the basis of two criteria: (1) a minor regression and subaerial exposure of the platform at or near the Frasnian–Famennian boundary; and (2) changes in the reef-building community from a biota dominated by stromatoporoids, cyanobacteria, and corals to one dominated by cyanobacteria. The Givetian–Frasnian Pillara cycle is characterized by vertical platform growth followed by widespread drowning and backstepping; the Famennian Nullara cycle is characterized by strongly advancing platforms (Playford et al., 1989). Note in this subsurface study that the phase of uplift associated with the tectonically enhanced Frasnian–Famennian sequence boundary probably occurs within or near the base of the *linguiformis* conodont zone (Figure 17) and thus we place the Pillara–Nullara boundary in the latest Frasnian (Figures 16, 17) rather than at the Frasnian–Famennian boundary.

Progressive changes in sequence and systems tract geometries in the reef and ramp complexes define a major tectonic transgressive-regressive facies cycle (Figure 17). This transgressive-regressive cycle was initiated in the Givetian by a phase of crustal extension (Drummond et al., 1991) and terminated in the Tournaisian following a phase of slow thermal subsidence. The transgressive half-cycle comprises at least four third-order sequences—the Frasnian–Givetian sequence and Frasnian sequences 1–3; the regressive half-cycle comprises the remaining 14 third-order sequences. A major basinward shift in coastal onlap associated with tectonic uplift and erosion immediately prior to the Frasnian–Famennian boundary accentuates the progressive basinward migration of facies of the regressive half-cycle.

In the initial parts of the transgressive half-cycle, carbonate production was able to keep up with the rate of accommodation change (the Frasnian–Givetian sequence) and a thick aggradational carbonate platform composed of keep-up transgressive deposits was able to develop. However, as the rate of accommodation change continued to increase, third-order rises in relative sea level were accentuated, the reefs and associated carbonate platforms were no longer able to keep up with the rise of relative sea level and they were successively drowned and backstepped. During third-order falls in relative sea level, the magnitude and rate of base level shift was subdued; siliciclastic lowstand deposits are poorly developed in this part of the transgressive half-cycle. As the rate of accommodation change started to decrease, regressive sedimentation commenced. In the regressive half-cycle, third-order falls in relative sea level resulted in significant base level shifts and

the accumulation of major lowstand clastic deposits, which provided a foundation for the overlying transgressive and highstand carbonates (Figure 16). During third-order increases in relative sea level, transgressive deposits were able to keep up and highstand deposits prograded to successively more basinward positions (Figure 16). The extent of basinward progradation was limited by the break point of the underlying prograding complex. Throughout the regressive half-cycle, decreases in the rate of accommodation change are related to declining rates of thermal subsidence. Relatively high subsidence rates in the early parts of the regressive half-cycle are associated with the high-relief, reef-rimmed platform complex; slow subsidence rates in the latter parts of the regressive half-cycle are associated with the ramp complex.

Sequences in the regressive half-cycle are grouped into three shorter term accommodation cycles (Figure 17). Each of these cycles commences with an episode of backstepping transgressive deposits leading to widespread marine shale deposition and the accumulation of a relatively thick condensed section (Famennian sequence 2A, Tournaisian sequences 1 and 6). During these times, the rate of accommodation change outstripped sediment accumulation rates. Subsequent sequences lack well-developed backstepping deposits and are dominated by keep-up transgressive deposits and progradational highstands. In these sequences, sediment accumulation rates kept pace with, and finally exceeded, the rate of accommodation increase. In Famennian sequence 2A, an acceleration in the rate of accommodation change led to minor backstepping of the highstand platform margin and a subdued third-order relative sea-level fall, which resulted in a type 2 sequence boundary at the base of Famennian sequence 2B. The subsequent development of stacked shelf margin, keep-up transgressive deposits, and highstand reef deposits resulted in enhanced topographic relief of the Famennian 2B platform margin (Figure 16). The overlying Famennian 3A sequence boundary is characterized by the most prominent onlap surface in the study area.

Preliminary field studies (Holmes, 1991; Kennard et al., 1992) have identified several types of siliciclastic-rich deposits in the outcropping reef complex: (1) massive pebble and boulder conglomerate incised into the platform (e.g., Van Emmerick and Mt. Behn conglomerates, Figure 1); (2) siliciclastic-rich mass-flow deposits containing abundant allochthonous blocks of reef and platform facies (e.g., Dingo Gap, 50 km southeast of Windjana Gorge); and (3) interbedded conglomerate and back-reef facies (e.g., Stoney Creek, 100 km southeast of Windjana Gorge). Playford et al. (1989) attribute the deposits in each of these categories to normal "highstand" processes of platform growth. Earthquakes are considered to have been the principal mechanism for triggering the debris flows and promoting collapse of the platform margins to produce allochthonous blocks of reef and platform facies origin. Our subsurface sequence

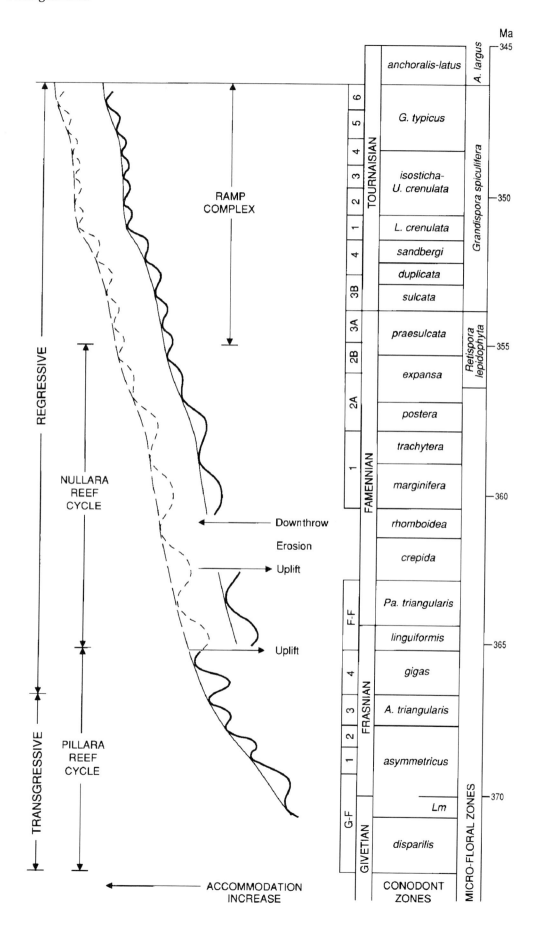

analysis has shown that substantial phases of uplift and renewed subsidence affected the upper Devonian section during development of the reef complex, and earthquakes would have been associated with these events. However, superimposed on the Pillara–Nullara cycles and the transgressive-regressive facies cycle are third-order sequences that display significant changes in relative sea level. These third-order relative sea-level oscillations clearly controlled the relative abundance of siliciclastics in the reef complex. We suggest that the conglomerates at Mt. Behn and Van Emmerick represent lowstand incised valley-fill deposits, and that the siliciclastic-rich mass flow deposits with abundant allochthonous blocks of reef and platform facies origin are also lowstand deposits and need to be differentiated from marginal-slope facies dominated by in situ carbonate buildups and allodapic carbonate components (transgressive and highstand deposits). In contrast, the interbedded conglomerate and back-reef facies at Stoney Creek probably represent back-reef highstand carbonates interspersed with terrigenous fan deltas in a fringing reef setting.

The platform margin unconformity at the classic face in Windjana Gorge is of late Frasnian age (Playford, 1981) and, based on conodont data (Nicoll, 1991), is correlated with the major sequence boundary at the base of the Van Emmerick Conglomerate (Figure 4A in Kennard et al., 1992) and Frasnian sequence 4. Boulder conglomerate exposed at the eastern end of Windjana Gorge (Playford and Lowry, 1966) is interpreted here as a lowstand deposit overlying the platform margin unconformity (sequence boundary). These deposits thus correlate with conglomerates at the base of Frasnian sequence 4 in Meda 1, Yarrada 1, and the Van Emmerick Conglomerate.

CONCLUSIONS

Sequence stratigraphy has enhanced our understanding of the development and origin of the Late Devonian to Early Carboniferous succession in the Canning basin, Western Australia, and has provided new insights into third-order relative sea-level cyclicity within the larger Pillara and Nullara reef cycles recognized from outcrop studies. Nine third-order relative sea-level cycles occur in the Givetian to early Famennian reef complex and nine in the mid-Famennian to Tournaisian ramp complex. Each reef sequence is characterized by patterns of marked reciprocal sedimentation. We interpret highstand carbonate deposition to be repeatedly punctuated by episodes of lowstand clastic-dominated sedimentation when the reef-rimmed carbonate platform was subaerially exposed and incised by streams. These streams transported terrigenous sediments across the platform to basin floor and slope fans outboard of the platform margin. As relative sea level began to rise, terrigenous sediments accumulated as a thick prograding and aggrading wedge (prograding complex) immediately outboard of the platform margin. The rate of relative sea-level rise is interpreted to have then increased, and this wedge and the exposed platform were transgressed leading to widespread carbonate deposition in keep-up settings and back-stepping carbonate deposits and shales in drowned settings. As the rate of relative sea-level rise abated, highstand carbonate production continued and the platform prograded across the drowned lowstand deposits until carbonate sedimentation was once again terminated by the next relative fall in sea level. The subsurface study has also identified areas of local lowstand carbonate production. Where exposed carbonate platforms were isolated from the influx of terrigenous sediments, lowstand carbonate sediments were deposited on the flanks of the platform.

Sequence and systems tract geometries, their stacking patterns, and component facies are partitioned according to their position on a major transgressive-regressive facies cycle. In the initial parts of the transgressive half-cycle, carbonate productivity was able to keep up with the rate of accommodation change. However, as the rate of accommodation change increased, third-order rises in relative sea level were accentuated, the reefs and associated carbonate platforms were no longer able to keep up with the rise of relative sea level and they were successively drowned and backstepped. Lowstand deposits are subdued in the transgressive half-cycle. In the regressive half cycle, third-order falls in relative sea level were accentuated and resulted in the accumulation of major lowstand clastic deposits. During third-order increases in relative sea level, transgressive deposits

◄

Figure 17. Relative tectonic-eustatic curve for the Givetian–Tournaisian succession on the outer Lennard Shelf and adjacent Fitzroy Trough in the Meda area showing the major transgressive-regressive facies cycle (smooth curve) and the shorter wavelength third-order relative sea-level cycles. In the regressive half cycle, the curves are offset by fault-movements and erosion in the latest Frasnian and the early Famennian. The dashed curves are corrected for these movements. Note the three second-order accommodation cycles in the regressive half cycle with maximum flooding surfaces in Famennian sequence 2A and Tournaisian sequences 1 and 6. Recent biostratigraphic data indicates an early Tournaisian age for Famennian sequences 3B and 4; in order to keep sequence nomenclature consistent with Kennard et al. (1992) and Jackson et al. (1992), the original Famennian sequence nomenclature is retained.

were able to keep up and highstand deposits prograded to successively more basinward positions.
Throughout the regressive half-cycle, decreases in the
rate of accommodation change are related to declining rates of thermal subsidence. Relatively high subsidence rates in the early parts of the regressive
half-cycle are associated with the high-relief, reef-
rimmed platform complex; slow subsidence rates in
the latter parts of the regressive half-cycle are associated with the ramp complex.

Differences between our subsurface sequence
model for development of the reef complex and previous outcrop-based interpretations of reef evolution
indicate the need for future sequence stratigraphic
interpretations of the outcropping reef complex.
Whereas the previously established outcrop model
depicts two phases of platform growth (Pillara and
Nullara cycles) each controlled by a gradual rise in
relative sea level and separated by a brief regression
and minor subaerial exposure at the Frasnian–
Famennian boundary, our subsurface sequence
model depicts a more dynamic depositional system
subject to at least nine third-order relative sea-level
cycles. The subsurface model provides a predictive
guide for future petroleum exploration in the basin.
The identification of coarse-grained sediments in
basinal environments suggests that many untested
petroleum plays occur outboard of the reef-rimmed
highstand carbonate deposits.

ACKNOWLEDGMENTS

Many people have contributed toward our understanding of sequence stratigraphy and the Canning
basin; in particular we thank Rick Sarg, Nicholas
Christie-Blick, and Ann E. Holmes for sharing their
ideas and experiences in sequence stratigraphy, both
on the outcrop and in the office. Some of the ideas
presented here were generated during a field trip to
the Canning basin conducted by Phillip E. Playford in
1990. We are indebted to Phil for sharing his knowledge of the reef complex outcrops with us. Robert S.
Nicoll, Peter J. Jones, and Gilbert Klapper are thanked
for discussions of biostratigraphic data and reexami-
nation of microfaunal samples. Comments by AAPG
reviewers Charlie Kerans and Gregor Eberli significantly improved an earlier draft of the chapter.
Discussions with colleagues in the BMR, particularly
Peter J. Davies, David A. Feary, John F. Lindsay,
Chris J. Pigram, and Philip A. Symonds are also
gratefully acknowledged. Particular thanks are due to
Bruce Cotton, Natasha Kozin, Martyn Moffat, and
Trevor Brown for the final drafting of the figures. We
thank Petroleum Securities for access to the data necessary for this study.

REFERENCES CITED

Botten, P., 1984, Uranium exploration in the Canning
   basin: A case study, in Purcell, P.G., ed., The
   Canning basin, Western Australia: Proceedings of
   the Geological Society of Australia and Petroleum
Exploration Society of Australia Symposium, p.
   485-501.
Brown, S.A., I.M. Boserio, K.S. Jackson, and K.W.
   Spence, 1984, The geological evolution of the
   Canning basin—Implications for petroleum explo-
   ration, in Purcell, P.G., ed., The Canning basin,
   Western Australia: Proceedings of the Geological
   Society of Australia and Petroleum Exploration
   Society of Australia Symposium, p. 85-96.
Druce, E.C., and B.M. Radke, 1979, The geology of the
   Fairfield Group, Canning basin, Western Australia:
   Bureau Mineral Resources Geology and
   Geophysics, Australia, Bulletin 200.
Drummond, B.J., M.A. Etheridge, P.J. Davies, and
   M.F. Middleton, 1988, Half-graben model for the
   structural evolution of the Fitzroy Trough,
   Canning basin, and implications for resource
   exploration: Australian Petroleum Exploration
   Association Journal, 28, (1), p. 76-86.
Drummond, B.J., M.J. Sexton, T.J. Barton, and R.D.
   Shaw, 1991, The nature of faulting along the mar-
   gins of the Fitzroy Trough, and implications for the
   tectonic development of the trough: Exploration
   Geophysics, v. 22, p. 111-116.
Goldstein, B.A., 1989, Waxings and wanings in
   stratigraphy, play concepts and prospectivity in
   the Canning basin: Australian Petroleum
   Exploration Association Journal 29, (1), p. 466-
   508.
Hall, W.D.M., 1984, The stratigraphic and structural
   development of the Givetian–Frasnian reef com-
   plex, Limestone Billy Hills, western Pillara Range,
   W.A., in Purcell, P.G., ed., The Canning basin,
   Western Australia: Proceedings of the Geological
   Society of Australia and Petroleum Exploration
   Society of Australia Symposium, p. 215-222.
Holmes, A.E., 1991, Late Devonian carbonate-con-
   glomerate association in the Canning basin,
   Western Australia (abstract): AAPG Bulletin, v. 75,
   p. 596.
Hurley, N.F., and K.C. Lohmann, 1989, Diagenesis of
   Devonian reefal carbonates in the Oscar range,
   Canning basin, Western Australia: Journal of
   Sedimentary Petrology, v. 59, p. 127-146.
Jackson, M.J., L.J. Diekman, J.M. Kennard, P.N.
   Southgate, P.E. O'Brien, and M.J. Sexton, 1992,
   Sequence stratigraphy of the Devonian–
   Carboniferous of the Canning basin and its use in
   the identification of basin floor fans and other
   petroleum plays: Australian Petroleum
   Exploration Association Journal, v. 32 p. 214-230.
Jones, P.J., and G.C. Young, 1993, Updated biostrati-
   graphic summaries of selected wells on the
   Lennard Shelf, in Jones, P.J., and G.C. Young, eds.,
   Summary of Phanerozoic Biostratigraphy and
   Paleontology of the Canning basin (Lennard Shelf),
   Appendix: Australian Geological Survey
   Organisation, Record, 62 p.
Kemp, G.J., and B.L. Wilson, 1990, The seismic
   expression of Middle to Upper Devonian reef com-
   plexes, Canning basin: Australian Petroleum
   Exploration Association Journal, v. 30, p. 280-289.

Kennard, J.M., P.N. Southgate, M.J. Jackson, P.E. O'Brien, N. Christie-Blick, A.E. Holmes, J.F. Sarg, 1992, A new sequence perspective on the Devonian reef complex and the Frasnian–Famennian boundary, Canning basin, Australia: Geology, v. 20, p. 1135–1138.

Kerans, C., N.F. Hurley, P.E. Playford, 1986, Marine diagenesis in Devonian reef complexes of the Canning basin, Western Australia, *in* Schroeder, J.H., and B.H. Purser, eds., Reef Diagenesis: Springer-Verlag, Berlin, p. 357-380.

Kolodony, Y., and I.R. Kaplan, 1970, Uranium isotopes in sea floor phosphorites: Geochimica et Cosmochimica Acta, v. 34, p. 3-24.

Mayne, S.J., 1975, History of petroleum search in the Canning basin, Western Australia to 31 December 1974: Australian Bureau of Mineral Resources, Record 1975/109, unpublished.

Meissner, F.F., 1972, Cyclic sedimentation in the Middle Permian strata of the Permian basin, West Texas and New Mexico, *in* Elam, J.C., and S. Chuber, eds., Cyclic Sedimentation in the Permian Basin, second edition: West Texas Geological Society, p. 203-232.

Middleton, M.F., 1987, Seismic stratigraphy of Devonian reef complexes, northern Canning basin, Western Australia: AAPG Bulletin, v. 71, p. 1488-1498.

Nicoll R.S., 1991, Conodonts from BMR samples in the Napier Range area, Canning basin, Western Australia: Bureau of Mineral Resources, Geology and Geophysics, Australia, Professional Opinion 1991/006, 3 p., unpublished.

Nicoll, R.S., and P.E. Playford, 1993, Upper Devonian iridium anomalies, conodont zonation and the Frasnian–Famennian boundary in the Canning basin, Western Australia: Paleogeography, Paleoclimatology, Paleoecology, in press.

Playford, P.E., 1980, Devonian "Great Barrier Reef" of the Canning basin, Western Australia: AAPG Bulletin, v. 64, p. 814-840.

Playford, P.E., 1981, Devonian reef complexes of the Canning basin Western Australia: Field Excursion Guidebook, Fifth Australian Geological Convention, Geological Society of Australia, 64 p.

Playford, P.E., 1982, Devonian reef prospects in the Canning basin: Implications of the Blina oil discovery: Australia Petroleum Exploration Association Journal, v. 22, p. 258-271.

Playford, P.E., 1984, Platform-margin and marginal-slope relationships in Devonian reef complexes of the Canning basin, *in* Purcell, P.G., ed., The Canning basin, Western Australia: Proceedings of the Geological Society of Australia and Petroleum Exploration Society of Australia Symposium, p. 190-214.

Playford, P.E., N.F. Hurley, C. Kerans, and M.F. Middleton, 1989, Reefal platform development, Devonian of the Canning basin, Western Australia, *in* Crevello, P.D., J. Wilson, J.F. Sarg, and J.F. Read, eds., Controls on Carbonate Platform and Basin development: SEPM Special Publication No. 44, p. 187-202.

Playford, P.E., and D.C. Lowry, 1966, Devonian reef complexes of the Canning basin, Western Australia: Geological Survey of Western Australia Bulletin 118, 150 p.

Posamentier, H.W., M.T. Jervey, and P.R. Vail, 1988, Eustatic controls on clastic deposition I—Conceptual framework, *in* Wilgus, C.K., B. Hastings, C. Kendall, H. Posamentier, C. Ross, and J. Van Wagoner, eds., Sea-level Changes: An Integrated Approach: SEPM Special Publication No. 42, p. 109-124.

Posamentier, H.W., and P.R. Vail, 1988, Eustatic controls on clastic deposition II—sequence and systems tract models, *in* Wilgus, C.K., B. Hastings, C. Kendall, H. Posamentier, C. Ross, and J. Van Wagoner, eds., Sea-level Changes: An Integrated Approach: SEPM Special Publication No. 42, p. 125-154.

Pudovskis, V., 1962, Meda no. 1 well, Western Australia: Petroleum Search Subsidy Acts Publication No. 7, Bureau of Mineral Resources Geology and Geophysics, 51 p.

Purcell, P.G., 1984, The Canning basin, Western Australia: Proceedings of the Geological Society of Australia and Petroleum Exploration Society of Australia Symposium.

Read, J.F., 1985, Carbonate platform facies models: AAPG Bulletin, v. 69, p. 1-21.

Sarg, J.F., 1988, Carbonate sequence stratigraphy, *in* Wilgus, C.K., B. Hastings, C. Kendall, H. Posamentier, C. Ross, and J. Van Wagoner, eds., Sea-level Changes: An Integrated Approach: SEPM Special Publication No. 42, p. 155-181.

Vail, P.R., 1987, Seismic stratigraphy interpretation using sequence stratigraphy, part 1: Seismic stratigraphy interpretation procedure, *in* Bally, A.W., ed., Atlas of Seismic Stratigraphy: AAPG Studies in Geology, v. 27, pt. 1, p. 1-10.

Vail P.R., R.M. Mitchum, R.G. Todd, J.M. Widmier, S. Thompson III, J.B. Sangree, J.N. Budd, and W.G. Hatfield, 1977, Seismic stratigraphy and global changes of sea level, *in* Payton, C.E., ed., Seismic Stratigraphy—Applications to Hydrocarbon Exploration: AAPG Memoir 26, p. 49-212.

Van Wagoner, J.C., R.M. Mitchum, K.M. Campion, and V.D. Rahmanian, 1990, Siliciclastic sequence stratigraphy in well logs, cores, and outcrops: Concepts for high-resolution correlation of time and facies: AAPG Methods in Exploration Series, No. 7, 55 p.

Wallace, M.W., C. Kerans, P.E. Playford, and A. McManus, 1991, Burial diagenesis in the Upper Devonian reef complexes of the Geikie Gorge region, Canning basin, Western Australia: AAPG Bulletin, v. 75, p. 1018-1038.

Wornhardt, W.W., compiler, 1991, Sequence stratigraphy concepts and applications (wall chart), Version 1.0: Micro-Strat Inc., Houston.

Chapter 7

Origin of Sedimentary Cycles in Mixed Carbonate–Siliciclastic Systems: An Example from the Canning Basin, Western Australia

Ann E. Holmes
Nicholas Christie-Blick
Department of Geological Sciences and
Lamont-Doherty Earth Observatory of Columbia University
Palisades, New York, U.S.A.

ABSTRACT

The separation of eustatic, tectonic, and other controls on the development of sedimentary cyclicity is difficult. In mixed carbonate–siliciclastic successions, the conventional interpretation of unconformity-bounded depositional sequences is that they are due to reciprocal sedimentation in response to relative changes of sea level. According to this view, transgressive and highstand systems tracts are composed primarily of carbonate rocks, and lowstands of siliciclastic rocks. The application of this model to the interpretation of cyclic carbonate and siliciclastic rocks in the Upper Devonian of the Canning basin, Western Australia, presents a paradox because expected evidence for subaerial exposure of the platform is not well developed.

Sequence stratigraphic studies in outcrop at two localities along the northern margin of the Canning basin confirm a complex relation between carbonate and siliciclastic conglomerate. At Stony Creek, the carbonate rocks are interpreted to represent an assemblage of reef, foreslope floatstones, and backreef carbonate-conglomerate cycles that accumulated in a shallow marine environment along the margin of a fan-delta. Conglomerates and sandstones inferred to onlap the reefal foreslope are interpreted tentatively as fluvial, and the contact is interpreted as a sequence boundary. In the Van Emmerick Range, foreslope floatstones are onlapped along the margin of an incised valley with at least 10 m of relief by conglomerate and sandstone of probable marine origin, and overlain by a transgressive fossiliferous limestone. The age of the sequence boundary at Stony Creek is not well established, but probably early Frasnian. The age of the sequence boundary in the Van Emmerick Range is better constrained as late Frasnian, and this surface appears to correlate with the base of the Frasnian 4 sequence identified by Southgate et al. (1993) on the basis of subsurface data. Both surfaces are thought to have involved base-level lowering, but the evidence is equivocal.

Evaluation of available data indicates that subaerial exposure is required for only one of eight potential sequence boundaries in the Frasnian–Famennian interval, a surface that corresponds locally to minor karstification, and which is dated as latest Frasnian. Several explanations are proposed as working hypotheses. The development of thick lowstand deposits coeval with flooding events on the platform is consistent with continued extension and tilting of a fault block. Alternatively, exposure may have been restricted to topographic highs remaining after extension had ceased. If subaerial exposure was widespread, diagenetic effects may have been limited or not preserved. The development of onlap surfaces within the basin may be related in part to variations in sediment flux, and the distribution of siliciclastic sediments influenced by the geologic structure, especially the configuration of accommodation or transfer zones. Further work is needed to resolve uncertainties in the existing sequence stratigraphic interpretation, to improve the calibration of individual boundaries, and to evaluate these ideas. Ultimately, comparisons with coeval successions on other continents will be needed to evaluate the possible role of eustasy in the development of the observed sequences.

INTRODUCTION

It is commonly assumed in the interpretation of both siliciclastic- and carbonate-dominated successions that the development of sedimentary cyclicity and the formation of sequence boundaries is due primarily to changes in "relative" sea level (Vail, 1987; Van Wagoner et al., 1987; Posamentier et al., 1988; Sarg, 1988; Christie-Blick, 1991). The term relative sea level is normally used to imply some loosely defined combination of the effects of both eustasy and subsidence. These controls are partitioned with difficulty (Pitman, 1978; Officer and Drake, 1985; Schlanger, 1986; Burton et al., 1987; Cloetingh, 1988; Kendall and Lerche, 1988; Cathles and Hallam, 1991; Christie-Blick, 1991), and the tendency to interpret high-frequency cyclicity as largely of eustatic origin is in part due to the absence of appropriate mechanisms for short-term change in the rate of subsidence. In addition, some facies cyclicity may be related not to sea-level change but to changes in sediment flux or production, as a result, for example, of climatic variations or tectonic events distant from the site of sedimentation (Basu, 1976; Mack and Suttner, 1977; Chafetz, 1980; Suttner et al., 1981; Manspeizer, 1985; Schlager, 1989, 1991; Galloway, 1989).

The Canning basin, a large sedimentary basin of Paleozoic and Mesozoic age in Western Australia (Brown et al., 1984; Purcell, 1984), provides an opportunity to address these issues. Seismic and well data are locally abundant, and outcrops along the northern margin of the basin, although of limited extent, have been thoroughly studied, particularly in the

Devonian part of the section (Playford et al., 1989). A recent subsurface study by geologists at the Australian Geological Survey Organisation (ASGO) has resulted in the tentative identification of at least seven unconformity-bounded depositional sequences in the Frasnian and Famennian section of the Upper Devonian (Southgate et al., 1991, 1993; Jackson et al., 1992; Kennard et al., 1992). These sequences are expressed by the alternation of "transgressive" and "highstand" platformal carbonate rocks with off-platform terrigenous rocks ("lowstands"), and they are interpreted by these authors as a response to relative sea-level change. This is an attractive hypothesis because from a geometric standpoint it seems to be a classic case of reciprocal sedimentation controlled by changes in depositional base level. However, the conclusions are based largely on a detailed seismic stratigraphic interpretation of a limited part of the basin. Few cores are available to check the interpretation of facies, and in the absence of adequate dating, correlation with available outcrop is uncertain in detail. If the observed cyclicity is due to base-level change, evidence for subaerial exposure ought to be abundant. Yet in spite of more than a quarter century of sedimentologic research in the carbonate rocks (Playford and Lowry, 1966; Playford, 1980, 1984; Playford et al., 1989), documented evidence for such exposure is surprisingly limited. As part of a related study in cooperation with the AGSO, we examined outcrop relations between carbonate and siliciclastic rocks of Late Devonian age at two localities at the northern edge of the basin. The purpose of this chapter is to describe and interpret these rocks in a sequence stratigraphic context, and to suggest ways of resolving the apparent paradox.

SEQUENCE STRATIGRAPHIC MODELS

Sequence stratigraphy is the study of rock relationships in a chronostratigraphic framework in which successions are divided into intervals of genetically related strata bounded by unconformities or their correlative conformities (Vail et al., 1977; Vail, 1987; Sarg, 1988; Van Wagoner et al., 1988, 1990). The sequence is the fundamental unit of sequence stratigraphy and is composed of systems tracts, subdivisions defined on the basis of their position within the sequence and stacking patterns of facies. The transgressive systems tract is characterized in both siliciclastic and carbonate settings by an overall deepening-upward of both lithofacies and biofacies, although it is usually composed at a smaller scale of parasequences, which shoal upward and are bounded above and below by marine flooding surfaces. Flooding surfaces are typically marked by an accumulation of fossils or a pebble lag, and by a decrease in the rate of sediment accumulation. The top of the transgressive systems tract is the maximum flooding surface. In this part of a sequence, sediment accumulates preferentially in nearshore areas. Coeval basinal areas tend to be sediment-starved (condensed), and characterized by the development of hardgrounds, the presence of abundant and diverse fossils, the accumulation of organic matter and authigenic minerals such as glauconite and phosphates, and intense burrowing (Loutit et al., 1988). The maximum flooding surface and condensed section may also correspond to a downlap surface associated with the progradation of overlying highstand sediments. The highstand systems tract and the lowstand systems tract both shoal upward overall as well as at a parasequence scale, and they are differentiated on the basis of an intervening sequence boundary. Lowstand sediments accumulate preferentially in deeper parts of the basin where they may include relatively coarse-grained sediments such as turbidites, debris-flow deposits, and platform-derived allodapic carbonate breccias. The sequence boundary is characterized at least locally by, and identified on the basis of, a downward shift in onlap, fluvially incised valleys, and a facies discontinuity involving abrupt upward-shoaling. Critical facies evidence tends to be obscured where sequence boundaries merge with flooding surfaces associated with an overlying transgressive systems tract. In the absence of such evidence, it may be difficult to establish that depositional base level was actually lowered, and therefore that a sequence boundary is present. This is especially the case if only basinal sediments are preserved (stacked "lowstands") or if inferences are made primarily on the basis of seismic geometry, and without adequate ties to the rock record.

Many of the characteristics of sequences and their included systems tracts have been idealized for specific paleogeographic settings and sediment types

(e.g., Vail, 1987; Posamentier et al., 1988; Sarg, 1988; Van Wagoner et al., 1990; Jacquin et al., 1991). In reality, sedimentary successions commonly reflect a combination of siliciclastic and carbonate deposition (El Haddad et al., 1984; Mount, 1984; Dorobek and Read, 1986; Mack and James, 1986; Purser et al., 1987; Dolan, 1989; Harris and Sarg, 1989; Yose and Heller, 1989; García-Mondéjar and Fernández-Mendiola, 1993). A model for mixed carbonate–siliciclastic sedimentation developed by Vail (1987) addresses some of the interactions between two markedly different depositional systems, one involving the in situ biogenic production of carbonate, and the other, the deposition of silica- and clay-rich sediment derived from continental erosion and, in some cases, far-traveled (Figure 1). In this and most other existing models, the geometry of depositional sequences and the arrangement of facies within them are attributed for the most part to a balance between subsidence and eustasy, with only a subsidiary role ascribed to factors such as sediment supply and climate. Carbonates are inferred to accumulate primarily during transgressive and highstand parts of cycles, and siliciclastic sediments preferentially during lowstands, when the carbonate factory is effectively shut off. At those times, terrigenous sediment can be transported across carbonate platforms by a combination of fluvial and eolian processes, and deposited in adjacent basins. It is this model that provides the conceptual context for the chapter by Southgate et al. (1993) on the seismic stratigraphy of the Devonian reef complex in the Canning basin.

TECTONIC AND STRATIGRAPHIC SETTING OF THE CANNING BASIN

The Canning basin is an intracratonic basin of extensional origin located in northwestern Australia between Precambrian crystalline and metasedimentary rocks of the Kimberley block to the north and the Pilbara block to the south (Guppy et al., 1958; Veevers and Wells, 1961; Forman and Wales, 1981; Towner and Gibson, 1983; Brown et al., 1984; Purcell, 1984; Begg, 1987). The basin has several prominent structural elements (Figure 2). From north to south these are the Lennard Shelf and related structurally uplifted areas (e.g., Oscar Range), the Fitzroy graben and Gregory subbasin (the deepest parts of the basin), the Broome horst (a composite horst in the central part of the basin), and the Willara and Kidson subbasins in the south. Most of the normal faulting that created sedimentary accommodation in the Canning basin occurred during the Late Silurian(?) to Late Devonian and is documented by evidence in seismic reflection profiles for growth of stratigraphic units across normal faults and footwall uplift (Begg, 1987; Drummond et al., 1988, 1991). Along the Lennard shelf, much of the faulting appears to have ended during Middle Devonian time, although some evidence exists locally for continued activity on faults during early Late Devonian time in the Pillara Range

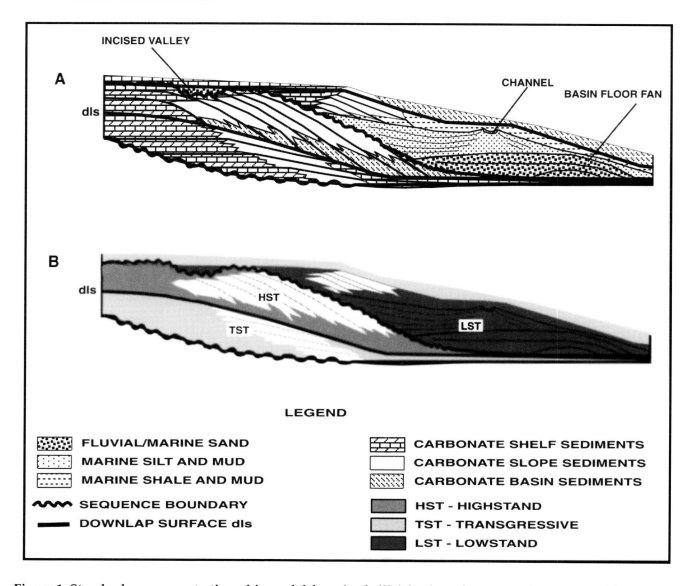

Figure 1. Standard sequence stratigraphic model for mixed siliciclastic-carbonate environments with a well-developed shelf-slope break (modified from Vail, 1987); (A) illustrates lithology and (B) illustrates the systems tracts and bounding surfaces.

(Hall, 1984). There was renewed extension during the Permian, and an episode of compressional strike–slip deformation during the Mesozoic that resulted in the development of en echelon folds and right-lateral wrenching of preexisting structures (Rattigan, 1967; Yeates et al., 1984; Drummond et al., 1991).

The sedimentary section is predominantly of Paleozoic to early Mesozoic age, and in the Fitzroy graben is as much as 12 km thick (Drummond et al., 1991). The lower part of the succession (Ordovician to Lower Carboniferous) is composed largely of mixed siliciclastic and carbonate rocks, with evaporitic deposits of Late Silurian(?) to Early Devonian age preserved especially along the Broome horst and within the southern subbasins. Carbonate platforms developed during the Middle Devonian (Givetian)

and continued through the Late Devonian (Frasnian and Famennian) on faulted Precambrian blocks and, in some areas, on Ordovician rocks (Playford and Lowry, 1966; Druce and Radke, 1979; Playford, 1980, 1981, 1984, 1991; Lehmann, 1984; Hurley, 1986; Playford et al., 1989). Upper Carboniferous to Cretaceous strata are predominantly siliciclastic, including more than 2 km of Permian–Carboniferous glacial sedimentary rocks (Towner and Gibson, 1983; Yeates et al., 1984).

PREVIOUS STUDIES OF DEVONIAN ROCKS IN THE CANNING BASIN

Our sequence stratigraphic studies in the Devonian of the Canning basin build upon an excel-

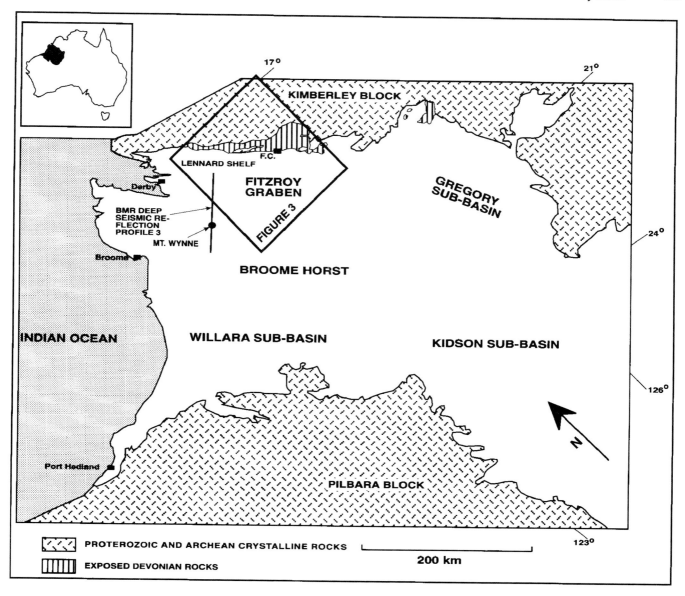

Figure 2. Location map of the Canning basin in Australia (inset) and key structural elements within the basin. Details of the exposed Devonian rocks in the Lennard shelf area are shown in Figure 3. One of the three deep seismic reflection lines, BMR88.03, acquired in 1988 by the Australian Geological Survey Organisation (formerly the Bureau of Mineral Resources) is noted on the northern part of the map as BMR profile 3. F.C.—Fitzroy Crossing.

lent geologic framework established in the course of several decades of effort by others. Early outcrop studies by Wade (1924) and Teichert (1949) and much subsequent research focused on reef morphology and related lithostratigraphy, petrology, paleontology, biostratigraphy, paleomagnetics, and diagenesis (Playford and Lowry, 1966; Hill and Jell, 1970; Read, 1973a,b; Druce, 1976; Logan and Semeniuk, 1976; Cockbain, 1984; various papers in Purcell, 1984; Kerans, 1985; Hurley, 1986; Kerans et al., 1986; Hurley and Van der Voo, 1987, 1990; Wallace, 1987). Much of the seminal work on the Upper Devonian reefal platforms along the northern

margin of the Canning basin has been done by P.E. Playford and colleagues (Playford and Lowry, 1966; Playford, 1980, 1982, 1984, 1991; Kerans and Playford, 1984; Playford et al., 1989). With respect to the platform development, they recognized two depositional cycles: (1) a Givetian–Frasnian cycle (termed the Pillara), which is predominantly transgressive with backstepping and drowning of platforms; and (2) a latest Frasnian–Famennian cycle (the Nullara), which is predominantly progradational. The boundary between these cycles is a physical erosion surface on the platform with a few meters relief, and interpreted to be due to an abrupt, short-lived sea-level fall (Playford, 1984; Playford et al., 1989; Playford, 1991).

Coarse-grained siliciclastic rocks that in many places intertongue with the Devonian carbonate rocks have not been documented in equivalent detail. Apart from the work of Botten (1984), significant accumulations of conglomerates and sandstones have been discussed only in general terms (Playford and Lowry, 1966; Playford, 1984; Yeates et al., 1984; Playford et al., 1989). Botten (1984) recognized conglomerate of Frasnian to Tournaisian age at several localities along the northern margin of the Canning basin and he attributed the rocks to alluvial fan, fluvial–deltaic, and shallow marine sedimentation. Locally, however, the conglomerates extend across the platforms and appear to have accumulated in deeper basinal settings (Playford et al., 1989). Small-scale carbonate-terrigenous sandstone cycles in the Pillara Limestone (Givetian–Frasnian) were studied by Read (1973a, b). He divided the Pillara into three members, which vary in composition from cyclically bedded siliciclastic rocks (closest to the shoreline) to cyclically interbedded carbonate and siliciclastic rocks, and finally to carbonate-dominated rocks (farthest offshore paleogeographically). Read interpreted the cycles in terms of gradual upward-shoaling with biostrome growth under conditions of relatively stable sea level, punctuated by rapid submergence.

NEW OBSERVATIONS IN THE CANNING BASIN

In contrast to previous studies, our recent work in the Canning basin and that of the AGSO has been in a sequence stratigraphic rather than lithostratigraphic context. This has led to some fundamental changes in perception, and to the recognition of important stratigraphic discontinuities, whatever their origin (Holmes, 1991; Southgate et al., 1991, 1993; Jackson et al., 1992; Kennard et al., 1992). Our interest in the relation between the carbonate rocks and conglomerates was spurred not only by the lack of information on the latter but by the idea that the contact relations between these very different facies might provide clues about the location of sequence boundaries within the platforms, and hence a way of evaluating emerging subsurface interpretations. A three-month field season undertaken in 1990, in cooperation with the AGSO, was focused on two localities in particular, Stony Creek in the Margaret embayment and the Van Emmerick Range (Figure 3; Holmes, 1991). These sites were selected for the quality of exposure and because reconnaissance indicated contrasts in the relations between carbonate and conglomerate facies—interfingering at Stony Creek, and unconformable in the Van Emmerick Range. Subsequently, sequence boundaries of probable Frasnian age were discovered at both localities, although they appear to be different boundaries. Several other localities that we visited briefly, and which have influenced our thinking, are indicated in Figure 3, but are not discussed in detail in this chapter. These include the Napier Range (Windjana Gorge, Mt. Behn, and Dingo Gap), Oscar Range and Pillara Range, McWhae Ridge, Sparke Range, Mt. Elma, and the Barramundi Range.

STONY CREEK

Stony Creek, a tributary to the Fitzroy River, is located at the erosional edge of the Canning basin in the Margaret embayment. Here, sedimentary rocks are preserved along a fault contact with the Proterozoic–Archean granites and granitic gneisses of the Kimberley block, and offset to the north from the trend of the Napier Range. At Stony Creek, the rocks include cyclically stratified carbonate and siliciclastic conglomerate (backreef and fan-delta), reef-margin carbonate and associated foreslope floatstones, and conglomerate and sandstone (fluvial?) that are thought to overlie the reef at a sequence boundary (Figure 4). Thin beds of limestone that in another part of the outcrop onlap the same sequence boundary have yielded conodonts of early Frasnian age (Nicoll, 1992). The rocks at Stony Creek are characterized by open folds, and structural dips are for the most part less than 15°. Locally, structural dips are as large as 70° and minor bedding-parallel faulting complicates the stratigraphic interpretation of backreef and foreslope beds.

Cyclic Carbonate and Siliciclastic Conglomerate

Seventeen carbonate-conglomerate cycles are identified over a 230-m-thick section at the Stony Creek locality, on the basis of two sections that were measured through the cyclic interval and lower part of the overlying floatstone interval (Figure 5). A typical cycle consists of fossiliferous carbonate overlain by clast-supported conglomerate and sandstone with a microcrystalline carbonate cement (Figure 6). Cycles vary in thickness from less than 1 m to about 10 m, thicker examples being characterized by the presence of biostromal or biohermal buildups in the basal carbonate interval. Siliciclastic intervals are typically not well exposed because dissolution of cement frees the clasts from the rock to produce an unconsolidated veneer of pebbles and cobbles in a sea of spinifex grass. Contacts can nevertheless be observed at least locally. The bases of cycles tend to be sharp and irregular in places with lags of gastropods and other fossil fragments preserved in depressions. Contacts between carbonate and conglomerate within cycles are also locally undulous but evidence for significant erosion or subaerial exposure has not been observed. The cycles are therefore relatively symmetrical and the selection of cycle boundaries is somewhat arbitrary.

The carbonate half-cycle is composed primarily of micritic floatstones and rudstones containing hemispheric stromatoporoid coenostea; delicate branching, laminar and irregular digitate stromatoporoids; gastropods, stromatolites, *Receptaculites* (sponge?) bioherms, and rare stromatoporoid bioherms (Figure 7).

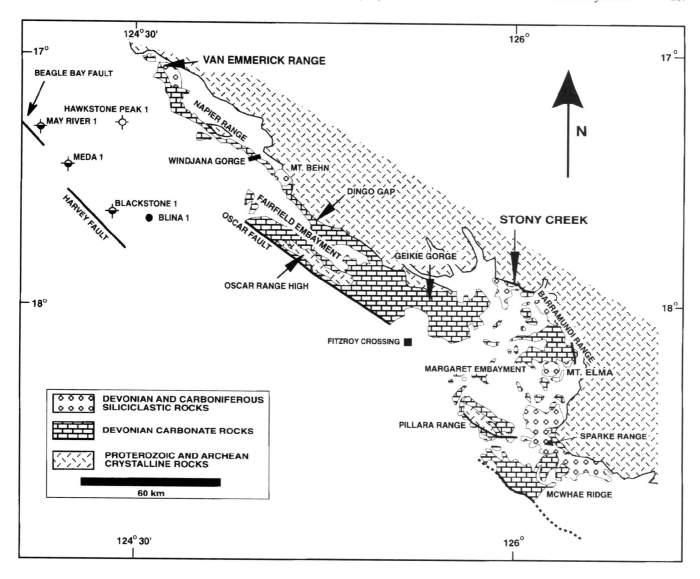

Figure 3. Map of the exposed Devonian rocks in the Lennard Shelf area showing important structural elements, together with localities and wells, cited in this chapter.

The siliciclastic half-cycle consists mainly of poorly stratified conglomerate composed of rounded pebbles, cobbles, and occasional boulders as much as 3 m in diameter of gneiss, granite, vein quartz, muscovite schist, and quartzite in a poorly sorted coarse-grained sandy matrix (see Figure 4C for the results of a representative clast count). Clasts are commonly encrusted by micritic carbonate rinds, and in at least one case, by a stromatoporoid (Figure 8). The matrix of the conglomerate and interstratified sandstone consists of an immature assemblage of angular to subangular particles of quartz, feldspar, mica, and rock fragments. Both high-spired and low-spired gastropods are also common in sandstone beds. Where cement is intact, conglomerate is characterized by median-axis imbrication (Figure 9). Azimuths of imbrication exhibit broad dispersion but are consistent with a preferred sediment transport direction to the southwest (rose diagram in Figure 5). A comparison of measurements

on a bed by bed basis reveals no systematic variations within the stratigraphic section.

Interpretation

The cyclic carbonate and conglomerate at Stony Creek are inferred to represent the progradation of a fan-delta into a shallow, quiet, backreef environment with normal marine salinity. The fan-delta interpretation is supported by the coarse grain size, textural and compositional immaturity, and poorly developed stratification of the siliciclastic rocks, by the broad dispersion of imbrication azimuths, and especially by the abrupt lateral transition between conglomerate and carbonate rocks (less than 1 km; Figures 4, 5; Ethridge and Wescott, 1984; Nemec and Steel, 1988). The lack of evidence for subaerial exposure, the presence of bioclastic sediment in the conglomerate, and the abundance of micritic carbonate rinds on clasts indicate that the conglomerate accumulated in a sub-

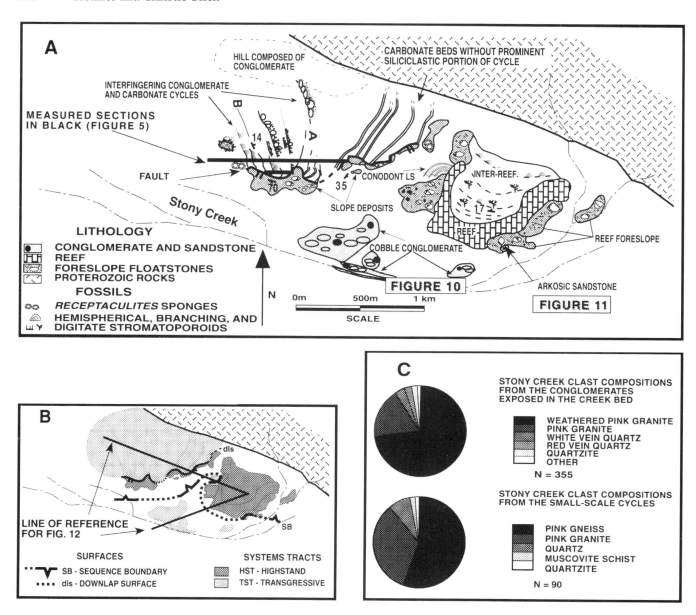

Figure 4. Maps of Stony Creek locality that illustrate (A) lithology, facies, and fossil distribution; (B) a sequence stratigraphic interpretation of facies; and (C) two pie diagrams of clast compositions from creek outcrops and from the small-scale cycles of interbedded siliciclastic conglomerate and carbonate rocks. Most of the clasts from both outcrops were derived from the Kimberley block to the north. In (A), two of the seventeen mixed carbonate–conglomerate cycles are labeled A and B for correlation purposes and are similarly identified in the measured sections shown in Figure 5. Figure 7 is located in cycle A; Figure 8 between cycle A and cycle B; and Figure 9 below cycle B. Locations for Figures 10 and 11 are shown in (A).

aqueous setting. The gravel could have been introduced by either streamflow (during floods) or sediment gravity flow processes, but in either case, sedimentation is likely to have been rapid. Similar coarse-grained sediments have accumulated along the margins of the Red Sea during the Quaternary, and are attributed by El Haddad et al. (1984) and Purser et al. (1987) also to flash flooding. We infer on paleoecologic grounds that in the case of the Stony Creek deposits, the influx of fresh water was sufficiently localized and short-lived that normal marine conditions were not significantly disrupted.

There is no evidence at the scale of these cycles for significant lowering of depositional base level. Indeed, faunal changes in the carbonate half-cycles indicate a gradually deepening marine environment from low-energy, shallow marine conditions populated by hemispheric and delicate branching stromatoporoids and gastropods to somewhat deeper marine conditions that favored stromatolites, *Receptaculites* bioherms, and digitate stromatoporoids, which grew vertically from thin, irregular laminar stromatoporoids (Pratt and Weissenberger, 1989).

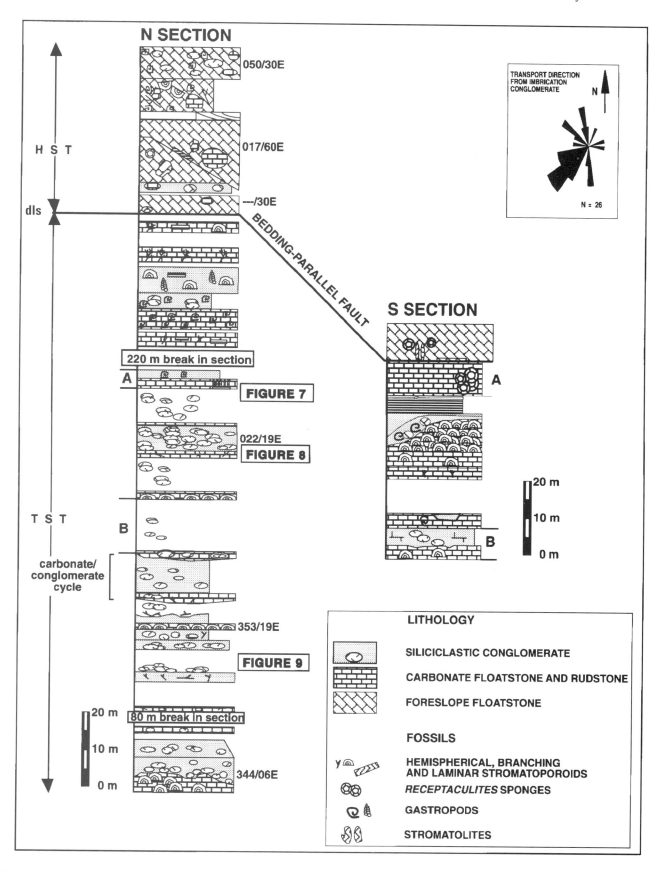

Figure 5. Two stratigraphic sections through the transgressive systems tract (TST) and part of the highstand systems tract (HST) at Stony Creek. Two cycles, labeled A and B, are identified for correlation purposes. The downlap surface (dls) appears diachronous because of displacement along a bedding-parallel fault. Figures 7, 8, and 9 are located on the northern section. The locations of these sections are shown in Figure 4A.

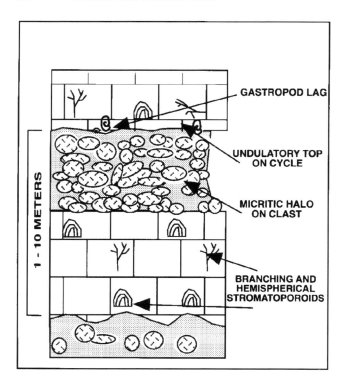

Figure 6. Generalized carbonate–conglomerate cycle from the transgressive systems tract at the Stony Creek locality. The cycles typically are 1–2 m thick, but where bioherms are present the thickness approaches 10 m. The carbonate was deposited in a shallow, quiet marine environment, punctuated by intervals of high siliciclastic influx. Many clasts have micritic haloes. The bases of carbonate units are typically irregular, with lags in topographic lows.

Reef and Associated Foreslope Floatstones

A horseshoe-shaped reef, present in the eastern part of the Stony Creek area (Figure 4), is composed of stromatoporoid and *Renalcis* boundstone. Examination of water-polished walls within a modern cave system in the reef indicates the presence of open spaces in the boundstone framework that became lined with ostracods and filled with equant marine cements (P.J. Jones, ASGO, personal communication, 1991). The boundstone is also cut by neptunian dikes and sills that are oriented north to south and filled with quartz and feldspar sand. Within the area of the horseshoe are outcrops of fossiliferous floatstone and packstone composed primarily of delicate branching stromatoporoid fragments, together with biostromes of more robust hemispheric stromatoporoids and beds of fossiliferous limestone with fenestral fabric. These inter-reef beds are today moderately dipping near the reef margin (up to 20°) and decrease in dip from the reef in all directions toward the center. Geopetal indicators (gastropods) and the lack of a preferred orientation in fragments of branching stromatoporoids (*Amphipora?*) indicate that very little of this dip is depositional, but

instead is due to the differential compaction adjacent to the early cemented reef margin (Playford, 1984; Kerans et al., 1986). However, the presence of truncation surfaces indicates that in detail the history of these inter-reef rocks may be quite complex.

On the outer part of the horseshoe adjacent to the reef margin, and at several other places within the area studied, are carbonate floatstones locally containing terrigenous sediment as large as pebbles and cobbles. Interstratified with these beds are occasional laminar stromatoporoids and *Receptaculites* bioherms preserved in place. The lithology and relation of these rocks to the reef indicates that they are coeval foreslope deposits, and in easternmost outcrops they appear to rest depositionally on the cyclic carbonate and conglomerate described above.

Unfortunately, this relationship is complicated by a bedding-parallel fault that cuts out several tens of meters of section due west of the horseshoe (Figures 4, 5). Initially, this contact was interpreted as a downlap surface, consistent with the recognition of downward termination of strata against the surface at one outcrop. Reevaluation of the stratigraphic data indicates that the apparent termination also of strata beneath the contact (Figure 4) cannot be explained simply in terms of a marked difference in dip across a fold-axial surface, but instead requires the existence of a structural discontinuity along the limb of the fold (Figure 5).

The western part of the reef margin is onlapped by several thin beds (less than 10 cm) of fine-grained limestone and shale. Some of the limestone beds contain biostromes of large hemispheric stromatoporoids; others yielded conodonts of early Frasnian age (Nicoll, 1992).

Isolated Conglomerates and Sandstones

Clast-supported conglomerate and sandstone is exposed in isolated outcrops in Stony Creek itself (Figures 4, 10), where they dip to the south. Clasts are composed mainly of granite and granitic gneiss, consistent with a Kimberley provenance (Figure 4C). Rare red carbonate clasts containing *Renalcis* are of uncertain provenance. The matrix is coarse-grained, poorly sorted, and arkosic. Vague trough cross-stratification is present in some beds, and median-axis imbrication indicates a southerly paleocurrent direction. Isolated outcrops of sandstone are preserved also along the southeastern side of the reef (Figure 11). Here, coarse-grained, poorly sorted, arkosic sandstones containing rounded clasts of granite and gneiss, as well as rare clasts of the same red carbonate, overlie foreslope floatstones.

The conglomerate in Stony Creek is tentatively interpreted as fluvial on the basis of vague cross-bedding and median-axis clast imbrication, and the absence of evidence, either lithic or biologic, that might support a marine interpretation. However, we cannot completely eliminate the possibility that the sediment accumulated in a shallow marine setting. The isolated sandstone in contact with the foreslope is of uncertain origin.

Figure 7. Photograph of digitate stromatoporoids growing vertically from irregular laminar stromatoporoids. Relatively deep-water conditions and the influx of siliciclastic sediment resulted in environmental stresses thought to be responsible for the unusual growth forms. Laterally equivalent beds contain *Receptaculites* sponge(?) mounds and stromatolites.

Sequence Stratigraphic Interpretation

The large-scale facies relations at Stony Creek are used to develop a sequence stratigraphic interpretation (Figure 12). The cyclic carbonate and conglomerate is interpreted as a transgressive systems tract on the basis of the overall deepening-upward trend. The reef and foreslope floatstones are interpreted as a highstand systems tract. The contact between them, although complicated by faulting, is inferred at least locally to be a downlap surface (dls in Figure 4B), and this is supported by the local presence at the contact of stromatolites thought to represent a condensed section. A sequence boundary is inferred on the basis of the onlap of conodont-bearing limestones and shales and the projected onlap geometry of the fluvial(?) conglomerates that outcrop in Stony Creek.

Figure 8. Photograph of a rolled clast with an encrusting stromatoporoid (toppled after encrustation and growth). Light-colored material is carbonate; dark-colored material is siliciclastic. Note micrite rinds on most cobbles (visible in the lower left corner and middle right edge of photograph). Micritic rinds are thought to be due to microbial encrustation. Scale is 20 cm long.

In offering this interpretation, we recognize several uncertainties. First, the significance of the conglomerates can be questioned (1) because they are not actually in contact with the reef margin or the dated limestones; (2) because their paleoenvironmental interpretation is equivocal; and (3) because, although presumed to be Devonian (Derrick and Gellatly, 1971; Derrick et al., 1971), they have not been dated directly. It is possible, for example, that they are as young as Carboniferous. A younger age (e.g., Tertiary) seems unlikely because the dip of the beds (about 20°) is similar to that of much of the Devonian in the area, and because the rocks lack the distinctive iron-rich cement that characterizes terrigenous rocks of Tertiary age in the Canning basin. Second, if the interpretation of the conglomerate is disregarded, it can be argued that onlap of marine limestone against the inferred sequence boundary is not by itself sufficient to demonstrate that a sequence boundary exists because although this is a reasonable interpretation, lowering of depositional base level is not necessarily required.

VAN EMMERICK RANGE

The Van Emmerick Range, located at the northwestern end of the Napier Range (Figure 3), consists of a series of low hills underlain by Upper Devonian to Carboniferous conglomerate. As at Stony Creek, the rocks crop out at the erosional edge of the Canning basin and are in fault contact with the Proterozoic–Archean rocks of the Kimberley block. In the Van Emmerick Range, conglomerate and sandstone partially fill a valley incised at least 10 m through foreslope floatstones into a poorly exposed limestone unit that has not been studied in detail (Figure 13). The siliciclastic rocks are tentatively interpreted to have accumulated by sediment gravity flow in a shallow marine environment. The foreslope deposits have yielded conodonts of late Frasnian age (Nicoll, 1991). Fossiliferous limestone that overlies the sequence boundary, but which is not in direct contact with the conglomerates, is inferred to be no younger than Frasnian on the basis of stromatoporoids and

Figure 9. Photograph of bed of carbonate-cemented siliciclastic conglomerate and sandstone at Stony Creek. An example of median-axis imbrication can be seen in three clasts immediately above the white ruler (length of ruler is 20 cm). The rocks are cemented with microcrystalline carbonate. The quality of exposure correlates with the degree of cementation.

tabulate corals. The area studied appears to be relatively free from structural complications.

Foreslope Floatstones

Two units were mapped in the floatstones (Figures 13, 14). Both are dolomite matrix-supported with both rounded and subangular carbonate blocks and fossil fragments that show no preferred orientation or layering. Within each of the units, *Receptaculites* bioherms and patch reefs of recrystallized colonial organisms (either tabulate corals or stromatoporoids) are preserved in situ. Each of the units has a silty micaceous cap tens of millimeters thick (the upper cap is shown in Figure 13), and the upper 1 m of the lower unit contains abundant fine-grained quartz sand. Large, predominantly angular clasts of limestone are present in the lowest 2 m of the lower floatstone unit, together with a few pebbles of quartzite and granite. The floatstones downlap against poorly exposed limestone at the base of the section.

Conodonts obtained from these rocks indicate a late Frasnian age (Nicoll, 1991).

On the basis of these observations, the floatstones are interpreted as prograding foreslope deposits. Abundant in situ bioherms of *Receptaculites* indicate a relatively deep-water environment in which the sediment supply was intermittent. The mapped units with silty caps can be interpreted as discrete debris flows.

Conglomerates and Sandstones

Conglomerate and sandstone is preserved locally as a veneer against the vertical contact with the floatstones but is otherwise poorly exposed (Figure 14). The conglomerate consists of pebbles and cobbles of quartzite, granite, vein quartz, and mica schist in a poorly sorted feldspathic matrix. Long-axis imbrication is present in several beds and indicates sediment transport to the south. Stratification ranges from well developed to indistinct, and in places there is vague

Figure 10. Photograph of conglomerate and sandstone outcropping in Stony Creek. Note the trough cross-stratification in the clast-supported conglomerate. These rocks are tentatively interpreted as fluvial and are assigned to a transgressive systems tract.

cross-stratification. Both normal and reverse grading is observed. Poorly stratified sandstone also with vague cross-stratification locally contains vertical lined burrows (Figure 15).

These rocks are tentatively ascribed to sediment gravity flow in a marine environment. This interpretation is based on the presence of long-axis imbrication and reverse grading in the conglomerate, and the absence of features such as median-axis imbrication and well-developed cross-stratification that might support a fluvial origin. The presence of burrows is consistent with a marine origin for at least some of the rocks. Nevertheless, given the limited exposure, we cannot entirely eliminate a fluvial interpretation.

Fossiliferous Limestone

The most prominent pinnacle of modern karst at the Van Emmerick Range locality (visible in Figure 17 below) is composed of fossiliferous limestone with a composite thickness of more than 30 m. The lowest 2 m is poorly exposed limestone with abundant quartz, feldspar, hornblende, and prismatic tourmaline, together with mostly disarticulate megalodont bivalve shells (Figures 14, 16, 18), which tend to be aligned parallel to bedding. Rare articulate megalodonts are in growth position. Disarticulate megalodonts are also present as basal lags in three overlying beds and associated with varying amounts of quartz and feldspar sand. Similar facies are preserved at the base of a second karst pinnacle (visible on the left side of Figure 17). There, megalodonts are present in a dolomitic matrix in the lower part of the section, along with colonial tabulate coral and laminar stromatoporoid biostromes and small *Receptaculites* bioherms. The limestone becomes less fossiliferous upward and is characterized by in situ "barrel" sponges as much as 30 cm across and straight nautiloids.

As they did at Stony Creek, fossils proved to be useful indicators of depositional environment and water depth for the limestone outcropping at the Van Emmerick Range. Megalodonts indicate the shallowest water depths, with stromatoporoids, corals,

Figure 11. Photograph of sandstone and conglomerate with occasional reef-derived cobbles that onlap the eastern reef margin and underlying sequence boundary (white line). Scale is 20 cm long.

Receptaculites, "barrel" sponges, and nautiloids representing progressively deeper water. Megalodont bivalves are taxonomic precursors to Cretaceous reef-building rudist bivalves, and their presence indicates high-energy environments of deposition. Stromatoporoid growth forms can be used to infer relative water depth. Hemispheric coenostea are considered to have populated shallower water than laminar forms, whereas digitate or branching forms are thought to have been restricted to quiet and/or deep-water environments. It is not known if these organisms were symbiotic with algae, thereby restricting them to the photic zone, but sedimentologic evidence does not support truly deep-water conditions for these organisms. *Receptaculites* bioherms appear to have flourished in deeper water below wave base in the carbonate slope environment, and the presence of large in situ "barrel" sponges and nautiloid tests, coupled with the absence of stromatoporoids and tabulate corals, indicates still deeper open marine conditions. Thus, faunal variation in the limestone records environmental changes from a shallow, high-energy setting (megalodont assem-

blages) to relatively deeper water (stromatoporoid and tabulate coral assemblages), and successively more open marine conditions ("barrel" sponge communities and straight nautiloids).

Sequence Stratigraphic Interpretation

These observations lead to the following sequence stratigraphic interpretation (Figure 19). Poorly exposed limestone at the base of the section is interpreted as a transgressive systems tract solely on the basis of its position beneath a downlap surface. Foreslope floatstones are assigned to the highstand systems tract, although no upward-shoaling is observed in the available outcrop. The erosion surface that separates these rocks from the fossiliferous limestone has at least 10 m relief, as determined from the elevation to which the conglomerate veneer is present, and is interpreted as a sequence boundary. The conglomerate is tentatively assigned to the lowstand systems tract, although if marine, it could conceivably represent the lowest part of the transgressive systems tract. The marked facies discontinuity and erosional

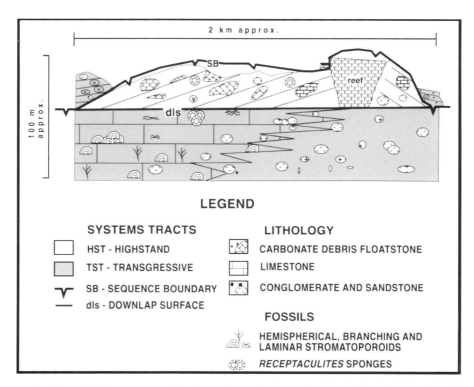

Figure 12. Sequence stratigraphic interpretation based on observations made at Stony Creek. The older transgressive systems tract contains the carbonate-siliciclastic cycles, and is overlain by a highstand systems tract that contains downlapping carbonate floatstone derived primarily from reef debris, but with locally high siliciclastic content. Superimposed is a younger transgressive systems tract composed of onlapping fluvial siliciclastic rocks from Stony Creek itself, and thin, conodont-bearing marine rocks (both onlapping against the reef above the sequence boundary). A line of reference for this figure is shown in Figure 4B. The bedding-parallel fault shown in Figures 4A and 5 is not included in this sequence stratigraphic model.

relief at this contact are consistent with the lowering of depositional base level and with subaerial exposure of the platform. However, we recognize that this interpretation is equivocal if the conglomerate is marine rather than fluvial. It is also possible, if not especially likely, that the conglomerate is considerably younger than the valley, in which case its paleoenvironmental interpretation would not be relevant to the origin of the erosion surface. The well-defined, deepening-upward trend in the fossiliferous limestone indicates that this unit belongs to the transgressive systems tract.

DATING AND CORRELATION OF SEQUENCE BOUNDARIES

The ages of sequence boundaries at Stony Creek and in the Van Emmerick Range are compared in Figure 20 with ages of boundaries inferred by Kennard et al. (1992) and Southgate et al. (1993) on the basis of their subsurface studies. The goniatite zonation is from detailed outcrop studies by Becker et al. (1991, 1993), and can be tied to the subsurface only indirectly through inferences about the physical stratigraphy. Two conodont zonation schemes are shown: the "old" scheme of Ziegler (1962, 1971) and the "new" scheme of Ziegler and Sandberg (1990), the version that is currently in use in the Canning basin. Tentative correlations between goniatites and conodonts are from Becker et al. (1993). The potential for very detailed zonation has made the Frasnian and Famennian especially attractive for paleontologic research. Ongoing studies in the Canning basin and elsewhere will undoubtedly lead to revisions of the schemes shown, and especially in the correlations between them (G. Klapper, personal communication, 1991). Significant uncertainties also remain in the sequence stratigraphic interpretation of the Canning basin. We hope that some of the uncertainties in both physical stratigraphy and calibration will be resolved in the course of continuing subsurface and outcrop studies.

The age of the sequence boundary at Stony Creek is not well established. Conodont-bearing limestones that onlap the boundary provide only a broad younger limit that spans the *Palmatolepis transitans* and *P. punctata* zones (Figure 20; Nicoll, 1992). This surface could therefore be compared with any of several lower Frasnian sequence boundaries interpreted by Kennard et al. (1992) and Southgate et al. (1993), and none of which is well dated. In their work on goniatite biostratigraphy in the vicinity of the Margaret embayment (Figure 3), Becker et al. (1993) describe a number of transgressive–regressive cycles in the lower Frasnian that can be reinterpreted in terms of unconformity-bounded depositional se-

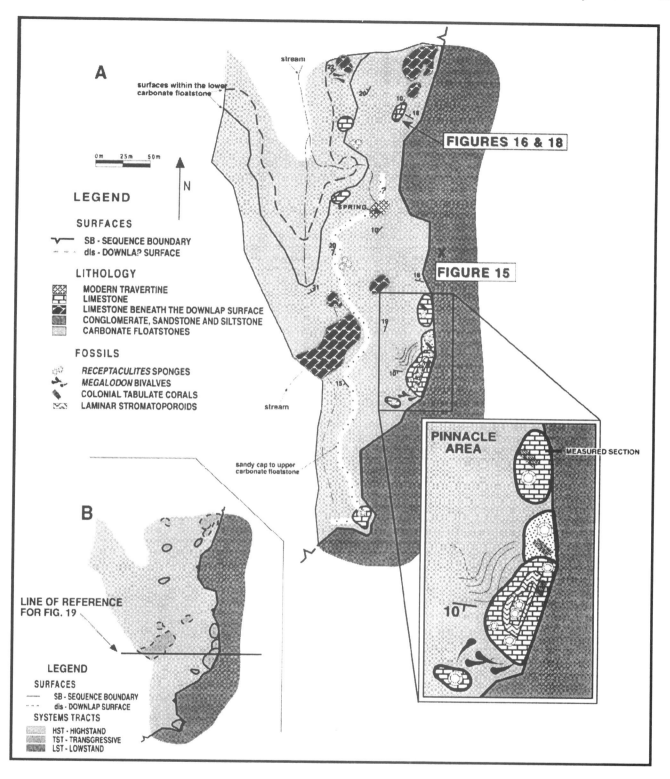

Figure 13. Maps of the Van Emmerick Range that illustrate (A) lithology, facies, and fossil distribution and (B) a sequence stratigraphic interpretation of facies. A sequence boundary coincides with the western edge of an incised valley filled with siliciclastic conglomerates and sandstones. Note location of measured section (shown in Figure 14) along the margin of the incised valley. The carbonate rocks of the older transgressive systems tract are in the dark diagonal limestone pattern and separated by a downlap surface (dls) from debris float-stones of the overlying highstand systems tract. Two distinct carbonate debris flows are mapped in this systems tract with lines tracing surfaces within them; one of the micaceous caps is mapped as a stippled white line. The younger transgressive systems tract limestones are in the light limestone pattern. Remnants of this limestone form modern karst towers (visible in the photograph of Figure 17). The locations of Figures 15, 16, and 18 are noted on (A). The line of reference for the sequence stratigraphic model of Figure 19 is shown in (B).

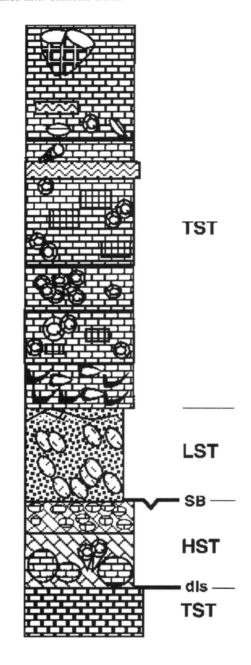

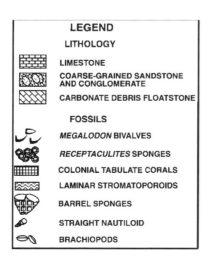

quences (Christie-Blick and Holmes, 1992). Four sequence boundaries are tentatively inferred in the *P. transitans-punctata* interval, with the most prominent dated as within the *Prochorites* (F) goniatite zone (upper *P. punctata* zone). Again, there is not yet sufficient information to make more detailed comparisons.

The age of the sequence boundary in the Van Emmerick Range is somewhat better constrained. Foreslope floatstones from beneath the sequence boundary have yielded conodonts with ranges spanning the *Palmatolepis rhenana* and *P. linguiformis* zones (Nicoll, 1991). Stromatoporoids and tabulate corals from fossiliferous limestone above the sequence boundary are also of Frasnian age (that is, not younger than the *P. linguiformis* zone). The boundary may therefore correlate with the base of either the Frasnian–Famennian sequence of Kennard et al. (1992) and Southgate et al. (1993) or the Frasnian 4 sequence (Figure 20). In the Meda-1 well (Figure 3), the base of the Frasnian–Famennian sequence is dated by conodonts as older than the lowermost *P. triangularis* zone, and on the basis of ostracods and brachiopods as no older than late Frasnian (Jones and Young, 1990). Kennard et al. (1992) tentatively correlate the sequence boundary with a surface within the *P. linguiformis* zone at McWhae Ridge (Figures 3, 20; Becker et al., 1991). The base of the Frasnian 4 sequence is poorly dated as mid-Frasnian (Jones and Young, 1990). Southgate et al. (1993) favor the correlation of this surface with the sequence boundary in the Van Emmerick Range, in part because of the presence of a prominent section of conglomerate at this horizon in the Meda-1 well, conglomerate that is reminiscent of siliciclastic rocks cropping out along the Napier Range and in the Margaret embayment (Figure 3). Support for the interpretation is provided again by the work of Becker et al. (1993). Reinterpretation of their stratigraphy at McWhae Ridge (Figure 3) suggests the presence of a prominent sequence boundary in the lowermost *Archoceras* (K) goniatite zone (lower *P. rhenana* zone). A more cogent argument for the correlation of the Frasnian 4 sequence boundary with the one in the Van Emmerick Range is that according to the seismic stratigraphic interpretation presented, the Fras-

Figure 14. A 25-m section measured along the edge of the incised valley (see Figure 13 for location). Included are the carbonate platform (interpreted here as a transgressive systems tract—TST) overlain by the downlapping carbonate debris of the highstand systems tract (HST), and onlapped by the incised valley fill (lowstand systems tract—LST) and by a younger transgressive systems tract (TST) limestone. The downlap surface (dls) is the surface between the lower TST and the HST; the sequence boundary (SB) is the surface between the HST and the LST. Only one of the debris flows of the HST is exposed in the measured section; the second is covered by the LST veneer (see Figure 13).

Figure 15. Photograph of vertical burrow in lowstand systems tract sandstone at the Van Emmerick Range. The presence of such burrows suggests marine/brackish water conditions.

nian–Famennian sequence pinches out between the Meda-1 well and the adjacent outcrop and merges with the Famennian 1 sequence boundary (Kennard et al., 1992; Southgate et al., 1993). The simplest interpretation of available data is therefore that the sequence boundary in the Van Emmerick Range is located within the lower *P. rhenana* zone.

Of the remaining depositional sequences interpreted by Kennard et al. (1992) and Southgate et al. (1993), none is well dated (Figure 20). The bases of the Famennian 1 and 2 sequences are both constrained in the Meda-1 well only as older than a horizon dated as between the upper *P. marginifera* zone and the lower *P. expansa* zone and younger than the *P. triangularis* zone. The base of the Famennian 3 sequence is constrained as younger than the horizon in the *P. marginifera-expansa* interval and pre-Tournaisian (Jones and Young, 1990). The Famennian 2 and 3 sequences are further subdivided by Southgate et al. (1993) into 2A, 2B, 3A, and 3B. These additional divisions are not included in Figure 20. Also, sequence Famennian 3B and sequence Famennian 4 of Kennard et al. (1992) are now assigned entirely to the Tournaisian (Southgate et al., 1993).

ORIGIN OF SEQUENCE BOUNDARIES IN THE CANNING BASIN

A combination of subsurface and outcrop studies indicates the existence of at least seven unconformity-bounded sequences or potential sequences and parts of two others in the Frasnian–Famennian interval (Figure 20), a span of approximately 15 m.y. according to the time scale of Harland et al. (1990). Outcrop evidence for two sequence boundaries (Stony Creek and the Van Emmerick Range) has been described in this paper. Both surfaces are thought to have involved the lowering of depositional base level, but the evidence is equivocal. Another sequence boundary that is prominent locally in outcrop in the Oscar and Napier ranges (Figure 3) corresponds to the base of the Frasnian–Famennian sequence (Playford et al., 1989; Playford, 1991; Kennard et al., 1992; Southgate et al., 1993). This boundary displays evidence for subaerial erosion and minor karstification, and where such evidence exists, Famennian rocks rest directly on Frasnian ones. A sequence boundary inferred from the work of Becker et al. (1993) in the Margaret embayment to be present in the *Prochorites* (F) goni-

Figure 16. Photograph of predominantly single valve *Megalodon* accumulations in the basal unit of the younger transgressive systems tract at the Van Emmerick Range. These robust bivalves indicate a high-energy environment. Most are disarticulate, concave up, and filled with immature coarse sandstone. There are several articulate bivalves in growth position within these beds.

atite zone (upper *P. punctata* zone) corresponds at McWhae Ridge (Figure 3) to a prominent onlap surface between the Sadler Limestone (and laterally correlative Gogo Formation) and Virgin Hills Formation. At nearby McIntyre Knolls, the same surface is associated with debris flows and huge allochthonous blocks, as much as 200 m across, of reef-margin and reef-flat facies (Playford, 1981; M.R. House, personal communication, 1991). The blocks were attributed by Playford to the collapse of the reef margin (no sequence boundary inferred), and while such an event might have been triggered by subaerial exposure of the reef, no definitive evidence for such exposure has yet been documented. We conclude that of the eight potential sequence boundaries in the Frasnian–Famennian section, the lowering of depositional base level is required for only one, but suggested for at least three others.

In considering the origin of the inferred sequence boundaries, we are aware of several biases in the available data. (1) Outcrop studies have been done almost entirely in a lithostratigraphic, not sequence stratigraphic, context. Sequence boundaries that are likely to be subtle at the scale of an outcrop may easily have gone unrecognized. (2) As a result of subsequent erosion, Famennian platforms in the Canning basin are preserved preferentially in foreslope deposits, which would likely not have been exposed subaerially during sequence boundary development. (3) Seismic evidence indicates that individual sequence boundaries tend to amalgamate up dip and in the direction of the outcrop. It might therefore be anticipated that fewer major sequence boundaries would be identifiable in outcrop than in the subsurface. On the other hand, much greater resolution is possible in the outcrop, and many more surfaces may be present than can be mapped in seismic data, especially in the Frasnian Stage, for which seismic resolution is generally poor. (4) Sequence boundaries tend to be best developed where sedimentation is strongly progradational and sequences are arranged in a forestepping motif. In contrast, sequence boundaries

Figure 17. Photograph of view northward toward the Kimberley block of the western edge of the Van Emmerick Range incised valley. The white line traces the contact between the carbonate rocks to the left (west) and the siliciclastic rocks to the right (east). The siliciclastic conglomerates and sandstones are interpreted as lowstand systems tract. The pinnacle in the back left of the photograph is the uppermost transgressive limestone unit of the measured section (shown in Figure 14).

tend to be poorly developed where sequences are dominated by transgression and backstepping. For this reason, sequence boundaries would be expected to be best developed in the Nullara reef cycle (latest Frasnian–Famennian) than in the Pillara cycle (Givetian–Frasnian), an expectation that is borne out by the seismic reflection data.

Although not acquired in a sequence stratigraphic context, interpretations of the diagenetic history of the Upper Devonian carbonate platforms are relevant to the origin of the inferred sequence boundaries. In a detailed study of the Oscar Range (Figure 3), Hurley and Lohmann (1989) concluded that evidence for syndepositional subaerial exposure of the platform may

Figure 18. Photograph of highstand systems tract carbonate debris floatstone facies (limestone clasts in a sandy dolomite matrix) overlain by transgressive systems tract fossiliferous (*Megalodon*-bearing) sandy lime-stones that onlap the sequence boundary (white line). Scale is 20 cm long.

be present in samples from both the uppermost Frasnian and lowest Famennian but not in other parts of the section. In those samples, early marine cements are overgrown by blocky nonferroan, banded, bright- and dull-luminescing calcites inferred to be of either meteoric-phreatic or marine-burial origin. However, in a comprehensive regional report on the petrography and diagenesis of Devonian and Carboniferous carbonate rocks in the Canning basin, Kerans (1985) concluded that these mixed cements are due to shal-low burial diagenesis. We note that all of the critical samples in the Hurley and Lohmann study appear to have been obtained from close to the prominent sequence boundary at the Frasnian–Famennian contact, which is inferred from outcrop evidence to be an exposure surface. So the available evidence does not bear on cryptic unconformities that might be expect-ed at other horizons. Kerans (1985) found no evidence for pre-Carboniferous exposure in any of the numer-ous samples that he studied.

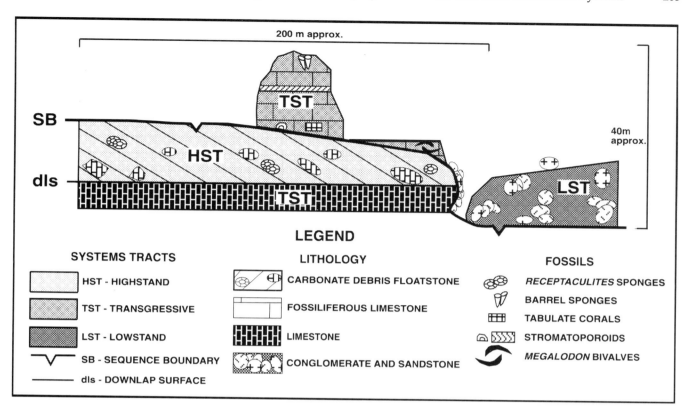

Figure 19. Sequence stratigraphic interpretation based on observations made at the Van Emmerick Range. The older transgressive systems tract is poorly exposed and was not studied in detail. It is overlain by a highstand systems tract that contains downlapping carbonate debris floatstones and in situ *Receptaculites* sponge (?) bioherms. Superimposed is a younger transgressive systems tract composed of fossiliferous limestones. Basal units of these limestones are megalodont- and silica-rich and onlap the highstand systems tract. Higher in the limestone succession are tabulate corals, laminar stromatoporoids, *Receptaculites* sponge bioherms, and barrel sponges in growth position. The veneer of siliciclastic rocks against the valley wall is the site of most of the detailed observations on the lowstand deposits. Line of reference for this figure is shown in Figure 13B.

These data present a paradox. The conventional sequence stratigraphic interpretation of numerous subsurface discontinuities in terms of reciprocal sedimentation controlled by changes in relative sea level appears to require intermittent subaerial exposure of the platform (Kennard et al., 1992; Southgate et al., 1993). Yet with the exception of a single surface, available sedimentologic evidence is equivocal and expected diagenetic evidence lacking, in spite of two comprehensive studies. In the remainder of this section, we discuss a number of ways in which the paradox might be resolved. These are presented as hypotheses to be evaluated rather than firmly established interpretations.

Sequence Boundary Development Related to Extension

Much of the normal faulting along the Lennard shelf appears to have ceased prior to deposition of the Upper Devonian carbonate platforms, and until recently, we favored the idea that the platforms developed in the earliest part of the postrift phase. Published interpretations, based largely on subsur-face data, implying the continuation of extension through the Late Devonian (e.g., Drummond et al., 1991), were disregarded because of evident uncertainty in the correlation and calibration of seismic reflections. In the light of the stratigraphic paradox, we reconsider the timing of extension.

A preliminary interpretation of a deep seismic reflection profile acquired by the AGSO (BMR88.03, located in Figure 2; figure 2 of Drummond et al., 1991) suggests that in the vicinity of a transect through the Meda-1 well and the Napier Range (Figure 3), Late Devonian subsidence may have been controlled by down-to-the-north displacement on a normal fault located beneath the Mt. Wynne anticline (Figure 2). Strata of Devonian through Permian age tend to thicken towards this structure, implying that from at least Late Devonian time onwards, the Pinnacle fault on the southwestern edge of the Lennard Shelf was little more than a subsidiary antithetic feature. Indeed, documented offsets in younger strata may be due largely to differential compaction. Southwestward stratigraphic thickening of the Frasnian and most of the Famennian is shown in greater detail in figure 2 of Kennard et al. (1992) and

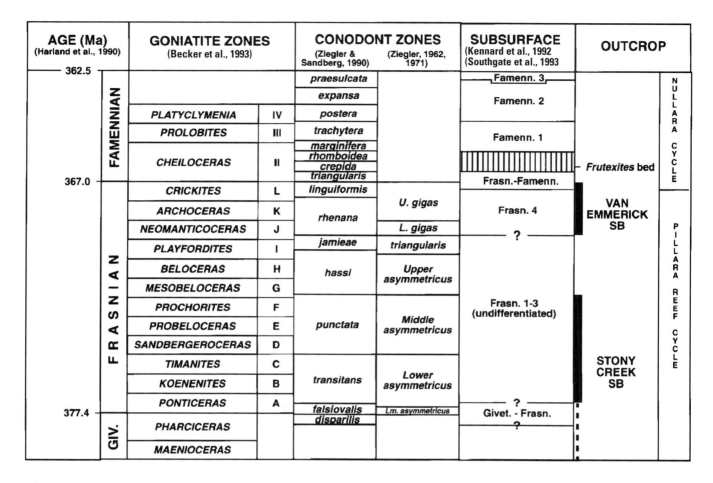

AGE (Ma) (Harland et al., 1990)		GONIATITE ZONES (Becker et al., 1993)		CONODONT ZONES		SUBSURFACE (Kennard et al., 1992 (Southgate et al., 1993	OUTCROP	
				(Ziegler & Sandberg, 1990)	(Ziegler, 1962, 1971)			
— 362.5	**FAMENNIAN**			*praesulcata*		Famenn. 3		N U L L A R A C Y C L E
				expansa		Famenn. 2	*Frutexites bed*	
		PLATYCLYMENIA	IV	*postera*				
		PROLOBITES	III	*trachytera*		Famenn. 1		
		CHEILOCERAS	II	*marginifera*				
				rhomboidea				
				crepida				
— 367.0				*triangularis*		Frasn.-Famenn.		
	FRASNIAN	CRICKITES	L	*linguiformis*	U. gigas	Frasn. 4	VAN EMMERICK SB	P I L L A R A R E E F C Y C L E
		ARCHOCERAS	K	*rhenana*				
		NEOMANTICOCERAS	J		L. gigas	?		
		PLAYFORDITES	I	*jamieae*	*triangularis*			
		BELOCERAS	H	*hassi*	Upper asymmetricus			
		MESOBELOCERAS	G					
		PROCHORITES	F	*punctata*	Middle asymmetricus	Frasn. 1-3 (undifferentiated)		
		PROBELOCERAS	E					
		SANDBERGEROCERAS	D					
		TIMANITES	C	*transitans*	Lower asymmetricus		STONY CREEK SB	
		KOENENITES	B					
		PONTICERAS	A			?		
— 377.4	**GIV.**	PHARCICERAS		*falsiovalis*	Lm. asymmetricus	Givet. - Frasn.		
		MAENIOCERAS		*disparilis*		?		

Figure 20. Biostratigraphic correlation chart illustrates relations among seven subsurface sequences (Kennard et al., 1992; Southgate et al., 1993), goniatite biozonation developed from exposures in the McWhae Ridge area (Becker et al., 1992), conodont biozonation (Ziegler 1962, 1971; Ziegler and Sandberg, 1990), and our interpretation of the ages of sequence boundaries at the Van Emmerick Range and Stony Creek. Ages (in million years) are based on the time scale of Harland et al. (1990). The *Frutexites* bed is an iridium-rich stromatolitic layer some 4 m above the Frasnian–Famennian boundary in a basinal succession at the McWhae Ridge locality. It is thought to be part of the condensed interval for the Frasnian–Famennian sequence of Kennard et al. (1992).

figure 2 of Southgate et al. (1993). Previously, we would have interpreted this section in terms of differential thermal subsidence, accentuated by the filling of bathymetry remaining after extension had ended, and by a combination of differential loading and compaction of syn-rift sediments. The main difficulty with this interpretation is that whereas the Upper Devonian rocks are strongly wedge-shaped, latest Fammenian and Tournaisian strata thicken only slightly across the Pinnacle fault (the ramp complex phase of Southgate et al., 1993). Assuming a thermal time constant of about 60 m.y., this would imply that a significant portion of the observed Late Devonian subsidence is due to differential loading, which, given the thickness of the sediments (in excess of 1 km), seems to require unreasonably deep water at the end of Middle Devonian time.

In the light of these observations, an alternative interpretation is that extension in the Fitzroy graben continued episodically until latest Famennian time,

and that some of the observed sequence boundaries may represent rift-onset unconformities for individual phases of extension and tilting. This interpretation is consistent with very limited evidence for continued faulting along the margin of the Lennard Shelf, as well as for stratigraphic anomalies ascribed by Southgate et al. (1993) to tectonic events. Localized subaerial exposure can easily be accommodated, but in spite of the impressive accumulation of "lowstand" sediments, a prediction of the model is that sequence boundaries would be expressed in many cases by onlap against marine flooding surfaces. Such a scenario is well documented in the early Cretaceous of the Jeanne d'Arc basin of eastern offshore Canada (Driscoll, 1992), and we thank our colleague for drawing attention to the potential similarities between the two basins. The surfaces are nevertheless interpreted as sequence boundaries because they would pass laterally into subaerial exposure surfaces generated by concomitant footwall uplift.

The model is quite different from the standard interpretation of sequence boundaries in terms of relative sea-level fall (e.g., Southgate et al., 1993). In the case of a basin subsiding differentially at a uniform rate, sequence boundaries would conventionally be ascribed to an increase in the rate of relative sea-level fall, a comparable change characterizing all points on the profile. In the case of a rift-onset unconformity, part of a profile experiences a decrease in the rate of subsidence (an increase in the rate of relative sea-level fall), but much of it experiences an *increase* in the subsidence rate.

One of the reasons for postulating exposure of the platform is to account for the accumulation of "lowstand" sediments. Such arguments depend of course on the assumption that terrigenous sediment is transported to the basin parallel to a particular profile or cross section in which lowstands are being interpreted. This is not necessarily the case. Sediment is undoubtedly eroded from newly emergent blocks (e.g., the footwall of an active normal fault), as well as from any residual topography (which in this case might be the Kimberley block), but it might very well be transported considerable distances parallel to the basin.

Explanations Not Involving Continued Extension

The stratigraphic paradox presented by insufficient evidence for subaerial exposure at postulated sequence boundaries can potentially be resolved in several other ways not necessarily involving renewed extension during Late Devonian time. For example, if extension had already ceased by the end of the Middle Devonian, residual topography related to tilted fault blocks would still exist during the Late Devonian. This means that during times of sea-level fall (of whatever origin), the up-tilted corners of fault blocks such as the Oscar Range block would preferentially be exposed and eroded, whereas deeper parts of half-grabens might remain submerged below sea level. An alternative possibility is that significant areas of the platform were indeed exposed intermittently, but that the diagenetic effects were minimal or not preserved. Diagenetic alteration during subaerial exposure depends on a number of factors, including the degree and duration of exposure, climate (Chow and James, 1992), the composition of the carbonate rock (Jones and Desrochers, 1992), and on whether appreciable erosion took place during subsequent transgression. Meteoric diagenesis is inhibited under arid conditions. Plate reconstructions for Late Devonian time place the Canning basin in an arid subequatorial belt at about 15° S (Habicht, 1979; Scotese, 1984; Hurley and Van der Voo, 1987; Boucot, 1988), consistent with the presence of fossil reefs, red beds, and evaporites.

Another factor that is difficult to evaluate in the Canning basin is the possible role of variations in the flux of siliciclastic sediment in the development of marine onlap surfaces (Schlager, 1989, 1991). Outcrop studies indicate that in many cases the carbonate platforms have steep margins, a circumstance that favors the development of prominent stratigraphic discontinuities between carbonate and siliciclastic sediments. Changes in sediment supply can be induced at a range of time scales by climatic variations or by tectonic events that may be distant from the site of sedimentation. In some cases, such events play an important role in the development of sedimentary cyclicity (e.g., Galloway, 1989). Although such considerations are usually assigned a secondary role, they need to be taken seriously in the Canning basin precisely because evidence for repeated subaerial exposure of the platforms is lacking.

Available evidence also indicates that the distribution of siliciclastic sediment was influenced by the geologic structure. The thickness of lowstand units is therefore not necessarily related to the degree of platform exposure, and on the platforms, some topographic lows filled with conglomerate may be of depositional origin and not incised valleys. The northern margin of the Canning basin is characterized by a series of tilted normal fault blocks separated by accommodation zones or transfer faults (Begg, 1987; Drummond et al., 1988, 1991; J. Braun and H. McQueen, 1991, unpublished data). The precise geometry of these features is not yet well established, but their existence is strongly implied by offsets in gravity anomalies, mapped faults (in outcrop and the subsurface), and the trends of reefs and other facies belts (Anfiloff, 1984; Craig et al., 1984; Purcell and Poll, 1984; Tucker et al., 1984; Wilson et al., 1984; Kennard et al., 1992; Southgate et al., 1993). Figure 21 illustrates two structural models that are mainly conceptual, although representative of the gross segmentation of the basin. In Figure 21A, accommodation zones are assumed to be similar to features described from areas of relatively small crustal extension, such as the East African rift (Rosendahl, 1987; Morley et al., 1990; Peacock and Sanderson, 1991; Nelson et al., 1992; Jackson and Leeder, 1993). In this model, the amount of local basin subsidence and corresponding footwall uplift decreases toward the tips of faults, producing topographic lows in the vicinity of so-called intrabasin highs (Anders, 1991). Drainage patterns are controlled by these topographic lows. In Figure 21B, segmentation of normal fault blocks is accomplished by somewhat better defined transfer faults, as described by Gibbs (1984) and Etheridge et al. (1987), and applied to the Canning basin by Begg (1987) and Drummond et al. (1988). In either case, the accommodation zones or transfer faults are assumed to align approximately with areas in the Canning basin at which conglomerate appears to cut across the Upper Devonian platforms (e.g., Van Emmerick Range, Mt. Behn, and Stony Creek, and other sites in the Margaret embayment; Figure 3).

CONCLUSIONS

Sequence stratigraphic studies in rocks of Late Devonian age at Stony Creek and in the Van

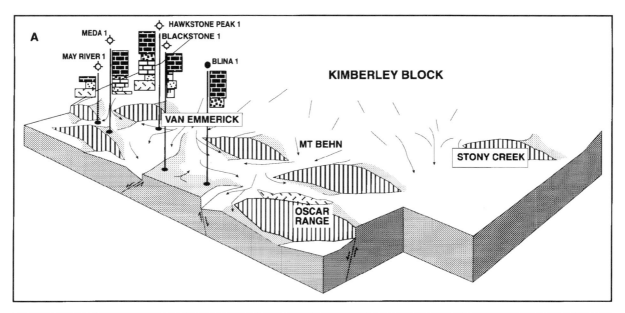

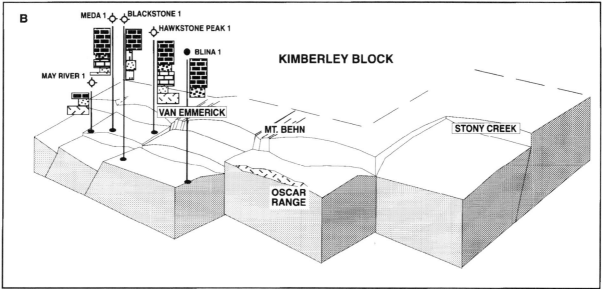

LITHOLOGY SYMBOLS FOR BOREHOLES	
	FAMENNIAN CARBONATES
	FRASNIAN CLASTICS
	FRASNIAN CARBONATES
	M. DEVONIAN EVAPORITES
	M. DEVONIAN CLASTICS
	M. DEVONIAN SILTSTONE
	PRECAMBRIAN BASEMENT

Figure 21. A schematic drawing illustrating general structural variations along the northern margin in the Canning basin. These variations provided localized subaerial and submarine topographic relief in the form of intrabasin highs and footwall uplift. Lithologies penetrated by the wells are shown as unscaled columns adjacent to wellbore location. Thickness of Devonian section and total depths (TD) reached by each well are as follows: Blina 1—1200 m Devonian, TD 2498 m; Blackstone 1—730 m Devonian, TD 3050 m; Hawkstone Peak 1—750 m Devonian, TD 1188 m; May River 1—160 m Devonian, TD 1678 m; and Meda 1—1100 m Devonian, TD 2685 m. Our data suggest that at least some of the depocenters for siliciclastic sedimentation were controlled by paleotopographic features associated with accommodation zones and footwall uplift. The positive paleotopography also acts to focus the exposure surfaces. It may help to explain why we do not see numerous extensive unconformities developed in the platform in the carbonate platforms along the northern margin of the Canning basin.

Emmerick Range along the northern margin of the Canning basin confirm a complex relation between carbonate and coarse-grained siliciclastic sedimentary rocks, one that varies from interfingering to unconformable. At Stony Creek, a sequence boundary is interpreted on the basis of an inferred onlap relation between fluvial(?) sandstone and conglomerate and a reefal foreslope, and dated as early Frasnian(?), with a younger age limit that spans the *Palmatolepis transitans* and *P. punctata* conodont zones. In the Van Emmerick Range, conglomerate of probable marine origin onlaps the margin of an incised valley that cuts through at least 10 m of highstand foreslope and underlying transgressive(?) carbonate rocks. The valley corresponds to a sequence boundary dated as late Frasnian on the basis of conodonts corresponding to the *P. rhenana* and *P. linguiformis* zones from beneath the boundary and the presence of stromatoporoids and tabulate corals of Frasnian affinity above the boundary. The sequence boundary at Stony Creek can be correlated with any of several lower Frasnian boundaries interpreted on the basis of subsurface data by Kennard et al. (1992) and Southgate et al. (1993), and none of which is well dated. The sequence boundary in the Van Emmerick Range appears to correlate with the base of the Frasnian 4 sequence of Kennard and Southgate.

A combination of subsurface and outcrop data indicates the existence of at least eight potential sequence boundaries in the Frasnian–Famennian interval of the Canning basin. Of these, subaerial exposure is required for only one (at the base of the Frasnian–Famennian sequence). In the case of three others, including the two documented in this chapter, evidence for the lowering of depositional base level is permissive but equivocal. These data present a paradox because the conventional interpretation of the sequence boundaries in terms of reciprocal sedimentation controlled by changes in relative sea level appears to require intermittent subaerial exposure of the platform. We suggest a number of ways of resolving the paradox, presented here as working hypotheses to be evaluated in the course of continuing research. (1) Normal faulting in the Canning basin may have persisted through much of Late Devonian time. In this case, some of the observed sequence boundaries may represent rift-onset unconformities for individual phases of extension and tilting, an interpretation that is consistent with the development of thick lowstand deposits coeval with flooding of a tilted fault block. (2) During times of sea-level fall (of whatever origin) subaerial exposure may have been restricted to topographic highs remaining after extension had ceased. (3) Subaerial exposure was indeed widespread, but diagenetic effects were minimal or not preserved. (4) The development of onlap surfaces imaged in subsurface data was due in part to variations in the flux of siliciclastic sediment. (5) The distribution of siliciclastic sediment was influenced by the geologic structure.

ACKNOWLEDGMENTS

This work was done in conjunction with the ASGO, which generously provided logistical support in the Canning basin. We thank many people from that organization for sharing their finely honed bush skills and geological expertise, especially field technicians Dick Brown, Christian Thun, Inge Zeilinger, and Bruno Zimmerman, and geologists Jim Jackson, John Kennard, Russell Shaw, and Peter Southgate. We are grateful also to Phil Playford of the Western Australia Geological Survey (WAGS) for providing color aerial photographs of Stony Creek and for sharing his considerable experience in the geology of the Devonian carbonate rocks of the Lennard shelf; Tim Griffin and Ian Tyler (WAGS) for advice in the field; Peter Muhling (Broken Hill Petroleum, Perth) for loaning some indispensable aerial photographs; Annette and John Henwood (Fossil Downs station) and Bob McCallie (Kimberley Downs station) for permission to conduct field work at Stony Creek and in the Van Emmerick Range, respectively. Reviews of various drafts by Nancye Dawers, Neal Driscoll, Peter Flemings, Neal Hurley, Charlie Kerans, Marge Levy, Phil O'Brien, Paul Olsen, Peter Southgate, and Marc Spiegelman greatly improved the manuscript. We thank Rick Sarg for his input in the field and for his editorial patience. Funding for this work was provided by National Science Foundation grant (EAR-8917400 to N. Christie-Blick and G.D. Karner), by the Donors of the Petroleum Research Fund administered by the American Chemical Society (PRF 25986-AC8 to NC-B), and by AAPG, Sigma Xi, and GSA Grants-in-Aid-of-Research (to A.E. Holmes). Additional support was provided by the Arthur D. Storke Memorial Fund of the Department of Geological Sciences, Columbia University, Lamont-Doherty Contribution No. 5120.

REFERENCES CITED

Anders, M.H., 1991, Are normal fault rupture segments persistent?: Geological Society of America, Abstracts with Programs, v. 23, p. A432.

Anfiloff, V., 1984, The gravity field of the Canning basin, *in* Purcell, P.G., ed., The Canning Basin, W.A.: Proceedings of the Geological Society of Australia and Petroleum Exploration Society of Australia Symposium, p. 287-298.

Basu, A., 1976, Petrology of Holocene fluvial sand derived from plutonic source rocks: Implications to paleoclimatic interpretation: Journal of Sedimentary Petrology, v. 46, p. 694-709.

Becker, R.T., M.R. House, and W.T. Kirchgasser, 1993, Devonian goniatite biostratigraphy and timing of facies movements in the Frasnian of the Canning Basin, Western Australia, *in* Hailwood, E.A., and R.B. Kidd, eds., High Resolution Stratigraphy: Geological Society of London Special Publication No. 70, p. 293–321.

Becker, R.T., M.R. House, W.T. Kirchgasser, and P.E. Playford, 1991, Sedimentary and faunal changes across the Frasnian/Famennian boundary in the Canning Basin of Western Australia: Historical Biology, v. 5, p. 183-196.

Begg, J., 1987, Structuring and controls on Devonian reef development on the north-west Barbwire and adjacent terraces, Canning basin: Australian Petroleum Exploration Association Journal, v. 27, p. 137-151.

Botten, P., 1984, Uranium exploration in the Canning Basin: a case study, in Purcell, P.G., ed., The Canning Basin, W.A.: Proceedings of the Geological Society of Australia and Petroleum Exploration Society of Australia Symposium, p. 485-501.

Boucot, A.J., 1988, Devonian biogeography: an update, in McMillan, N.J., A.F. Embry, and D.J. Glass, eds., Devonian of the World: Canadian Society of Petroleum Geologists Memoir 14, v. 3, p. 211-228.

Brown, S.A., I.M. Boserio, K.S. Jackson, and K.W. Spence, 1984, The geological evolution of the Canning Basin—implications for petroleum exploration, in Purcell, P.G., ed., The Canning Basin, W.A.: Proceedings of the Geological Society of Australia and Petroleum Exploration Society of Australia Symposium, p. 85-96.

Burton, R., C.G.St.C. Kendall, and I. Lerche, 1987, Out of our depth: On the impossibility of fathoming eustasy from the stratigraphic record: Earth-Science Reviews, v. 24, p. 237-277.

Cathles, L.M., and A. Hallam, 1991, Stress-induced changes in plate density, Vail sequences, epeirogeny, and short-lived global sea level fluctuations: Tectonics, v. 10, p. 659-671.

Chafetz, H.S., 1980, Evidence for an arid to semi-arid climate during deposition of the Cambrian System in central Texas, U.S.A.: Palaeogeography, Palaeoclimatology, Palaeoecology, v. 30, p. 83-95.

Chow, N., and N.P. James, 1992, Synsedimentary diagenesis of Cambrian peritidal carbonates: Evidence from hardgrounds and surface paleokarst in the Port au Port Group, western Newfoundland: Canadian Petroleum Geology Bulletin, v. 40, p. 115-127.

Christie-Blick, N., 1991, Onlap, offlap, and the origin of unconformity-bounded depositional sequences: Marine Geology, v. 97, p. 35-56.

Christie-Blick, N., and A.E. Holmes, 1992, Upper Devonian sequence stratigraphy of the Canning basin, Australia: A comparison with coeval carbonates in southwestern Alberta: Abstract, AAPG, Calgary, Alberta Meeting Program, p. 19–20.

Cloetingh, S., 1988, Intraplate stresses: A tectonic cause for third-order cycles in apparent sea level?, in Wilgus, C.K., B.S. Hastings, C.G.St.C. Kendall, H.W. Posamentier, C.A. Ross, and J.C. Van Wagoner, eds., Sea-Level Changes: An Integrated Approach: SEPM Special Publication No. 42, p. 19-29.

Cockbain, A.E., 1984, Stromatoporoids from the Devonian reef complexes, Canning Basin, Western Australia: Geological Survey of Western Australia Bulletin 129, 108 p.

Craig, J., J.W. Downey, A.D. Gibbs, and J.R. Russell, 1984, The application of Landsat imagery in structural interpretation of the Canning Basin, W.A., in Purcell, P.G., ed., The Canning Basin, W.A.: Proceedings of the Geological Society of Australia and Petroleum Exploration Society of Australia Symposium, p. 57-72.

Derrick, G.M., and D.C. Gellatly, 1971, Lennard River 1:250 000 geological sheet SE 51-8, 2nd edition: Bureau of Mineral Resources, Geology and Geophysics, Canberra.

Derrick, G.M., D.C. Gellatly, and A.S. Mikolajczak, 1971, Lansdowne 1:250 000 geological sheet SE 52-5, 2nd edition: Bureau of Mineral Resources, Geology and Geophysics, Canberra.

Dolan, J.F., 1989, Eustatic and tectonic controls on deposition of hybrid siliciclastic/carbonate basinal cycles: Discussion with examples: AAPG Bulletin, v. 73, p. 1233-1246.

Dorobek, S.L., and J.F. Read, 1986, Sedimentology and basin evolution of the Siluro-Devonian Helderberg Group, central Appalachians: Journal Sedimentary Petrology, v. 56, p. 601-613.

Driscoll, N.W., 1992, Tectonic and depositional processes inferred from stratal relationships: Ph.D. dissertation, Columbia University, New York, 464 p.

Druce, E.C., 1976, Conodont biostratigraphy of the Upper Devonian reef complexes of the Canning Basin, Western Australia: Bureau of Mineral Resources, Geology and Geophysics, Australia, Bulletin 158, 303 p.

Druce, E.C., and B.M. Radke, 1979, The geology of the Fairfield Group, Canning Basin, Western Australia: Bureau of Mineral Resources, Geology and Geophysics, Australia, Bulletin 200, 62 p.

Drummond, B.J., M.A. Etheridge, P.J. Davies, and M.F. Middleton, 1988, Half-graben model for the structural evolution of the Fitzroy trough, Canning Basin, and implications for resource exploration: Australian Petroleum Exploration Association Journal, v. 28, p. 76-86.

Drummond, B.J., M.J. Sexton, T.J. Barton, and R.D. Shaw, 1991, The nature of faulting along the margins of the Fitzroy Trough, Canning Basin, and implications for the tectonic development of the trough: Exploration Geophysics, v. 22, p. 111-116.

El Haddad, A., D.M. Aissaoui, and M.A. Soliman, 1984, Mixed carbonate-siliciclastic sedimentation on a Miocene fault-block, Gulf of Suez, Egypt: Sedimentary Geology, v. 37, p. 185-202.

Etheridge, M.A., J.C. Branson, and P.G. Stuart-Smith, 1987, The Bass, Gippsland and Otway basins, southeast Australia: A branched rift system formed by continental extension, in Beaumont, C., and A.J. Tankard, eds., Sedimentary Basins and Basin-forming Mechanisms: Canadian Society of Petroleum Geologists Memoir 12, p. 147-162.

Ethridge, F.G., and W.A. Wescott, 1984, Tectonic setting, recognition and hydrocarbon potential of fan-delta deposits, *in* Koster, E.H., and R.J. Steel, eds., Sedimentology of Gravels and Conglomerates: Canadian Society of Petroleum Geologists Memoir 10, p. 217-235.

Forman, D.J., and D.W. Wales, compilers, 1981, Geological evolution of the Canning basin, Western Australia: Bureau of Mineral Resources, Geology and Geophysics, Australia, Bulletin 210, 91 p.

Galloway, W.E., 1989, Genetic stratigraphic sequences in basin analysis I: Architecture and genesis of flooding-surface bounded depositional units: AAPG Bulletin, v. 73, p. 125-142.

García-Mondéjar, J., and P.A. Fernández-Mendiola, 1993, Sequence stratigraphy and systems tracts of a mixed carbonate and siliciclastic platform-basin setting: The Albian of Lunada and Soba, northern Spain: AAPG Bulletin, v. 77, p. 245–275.

Gibbs, A.D., 1984, Structural evolution of extensional basin margins: Geological Society of London Journal, v. 141, p. 609-620.

Guppy, D.J., A.W. Lindner, J.H. Rattigan, and J.N. Casey, 1958, The Geology of the Fitzroy Basin, Western Australia: Bureau of Mineral Resources, Geology and Geophysics, Australia, Bulletin 36, 116 p.

Habicht, J.K.A., 1979, Paleoclimate, paleomagnetism, and continental drift: AAPG Studies in Geology No. 9, 31 p.

Hall, W.D.M., 1984, The stratigraphic and structural development of the Givetian-Frasnian reef complex, Limestone Billy Hills, western Pillara Range, W.A., *in* Purcell, P.G., ed., The Canning Basin, W.A.: Proceedings of the Geological Society of Australia and Petroleum Exploration Society of Australia Symposium, p. 215-222.

Harland, W.B., R.L. Armstrong, A.V. Cox, L.E. Craig, A.G. Smith, and D.G. Smith, 1990, A Geologic Time Scale 1989: Cambridge, Cambridge University Press, 263 p.

Harris, D.G., and J.F. Sarg, 1989, Seismic modeling of a mixed carbonate-siliciclastic prograded system, middle Permian, Permian Basin, *in* Flis, J.E., R.C Price, and J.F. Sarg, eds., Search for the subtle trap: Hydrocarbon exploration in mature basins: West Texas Geological Society Symposium, v. 89-85, p. 119-138.

Hill, D., and J.S. Jell, 1970, Devonian corals from the Canning Basin, Western Australia: Geological Survey of Western Australia Bulletin 121, 153 p.

Holmes, A.E., 1991, Late Devonian carbonate-conglomerate association in the Canning Basin, Western Australia: AAPG Bulletin, v. 75, p. 596.

Hurley, N.F., 1986, Geology of the Oscar Range Devonian reef complex, Canning Basin, Western Australia: Ph.D. dissertation, University of Michigan, Ann Arbor, 269 p.

Hurley, N.F., and K.C. Lohmann, 1989, Diagenesis of Devonian reefal carbonates in the Oscar Range, Canning Basin, Western Australia: Journal of Sedimentary Petrology, v. 59, p. 127-146.

Hurley, N.F., and R. Van der Voo, 1987, Paleomagnetism of Upper Devonian reefal limestones, Canning basin, Western Australia: Geological Society of America Bulletin, v. 98, p. 138-146.

Hurley, N.F., and R. Van der Voo, 1990, Magnetostratigraphy, Late Devonian iridium anomaly, and impact hypotheses: Geology, v. 18, p. 291-294.

Jackson, J., and M. Leeder, 1993, Drainage systems and the evolution of normal faults: An example from Pleasant Valley, Nevada, submitted for publication.

Jackson, M.J., L.J. Diekman, J.M. Kennard, P.N. Southgate, P.E. O'Brien, and M.J. Sexton, 1992, Sequence stratigraphy, basin-floor fans and petroleum plays in the Devonian-Carboniferous of the northern Canning basin: Australian Petroleum Exploration Association Journal, v. 32, p. 214-230.

Jacquin, T., A. Arnaud-Vanneau, H. Arnaud, C. Ravenne, and P.R. Vail, 1991, Systems tracts and depositional sequences in a carbonate setting: A study of continuous outcrops from platform to basin at the scale of seismic lines: Marine and Petroleum Geology, v. 8, p. 122-139.

Jones, B., and A. Desrochers, 1992, Shallow platform carbonates, *in* Walker, R.G., and N.P. James, eds., Facies Models—Response to Sea-Level Change: Geological Association of Canada, p. 277-301.

Jones, P.J., and G.C. Young, 1990, Biostratigraphic summary of Meda No. 1 Well, Canning basin, Western Australia: Bureau of Mineral Resources, Geology and Geophysics, Professional Opinion 1990/002, 14 p. (unpublished).

Kendall, C.G.St.C., and I. Lerche, 1988, The rise and fall of eustasy, *in* Wilgus, C.K., B.S. Hastings, C.G.St.C. Kendall, H.W. Posamentier, C.A. Ross, and J.C. Van Wagoner, eds., Sea-Level Changes: An Integrated Approach: SEPM Special Publication No. 42, p. 3-17.

Kennard, J.M., P.N. Southgate, M.J. Jackson, P.E. O'Brien, N. Christie-Blick, A.E. Holmes, and J.F. Sarg, 1992, New sequence perspective on the Devonian reef complex and the Frasnian-Famennian boundary, Canning basin, Australia: Geology, v. 20, p. 1135–1138.

Kerans, C., 1985, Petrology of Devonian and Carboniferous carbonates of the Canning and Bonaparte basins, Western Australia: Western Australian Mining and Petroleum Research Institute, Report 12, 203 p.

Kerans, C., N.F. Hurley, and P.E. Playford, 1986, Marine diagenesis in Devonian reef complexes of the Canning basin, Western Australia, *in* Schroeder, J.H., and B.H. Purser, eds., Reef Diagenesis: Berlin, Springer-Verlag, p. 357-380.

Kerans, C., and P.E. Playford, 1984, Scheck breccias from Devonian reef complexes of Canning basin, Western Australia: AAPG Bulletin 68, p. 495.

Lehmann, P.R., 1984, The stratigraphy, palaeogeography and petroleum potential of the Lower to lower Upper Devonian sequence in the Canning Basin, *in* Purcell, P.G., ed., The Canning Basin, W.A.: Proceedings of the Geological Society of Australia and Petroleum Exploration Society of Australia Symposium, p. 253-275.

Logan, B.W., and V. Semeniuk, 1976, Dynamic metamorphism: Processes and products in Devonian carbonate rocks, Canning Basin, Western Australia: Geological Society of Australia, Special Publication No. 6, 138 p.

Loutit, T.S., J. Hardenbol, P.R. Vail, and G.R. Baum, 1988, Condensed sections: The key to age determination and correlation of continental margin sequences, in Wilgus, C.K., B.S. Hastings, C.G.St.C. Kendall, H.W. Posamentier, C.A. Ross, and J.C. Van Wagoner, eds., Sea-Level Changes: An Integrated Approach: SEPM Special Publication No. 42, p. 183-213.

Mack, G.H., and W.C. James, 1986, Cyclic sedimentation in the mixed siliciclastic-carbonate Abo-Hueco transitional zone (Lower Permian), southwestern New Mexico: Journal of Sedimentary Petrology, v. 56, p. 635-647.

Mack, G.H., and L.J. Suttner, 1977, Paleoclimate interpretation from a petrographic comparison of Holocene sands and the Fountain Formation (Pennsylvanian) in the Colorado Front Range: Journal of Sedimentary Petrology, v. 47, p. 89-100.

Manspeizer, W., 1985, The Dead Sea Rift: Impact of climate and tectonism on Pleistocene and Holocene sedimentation, in Biddle, K.T., and N. Christie-Blick, eds., Strike-Slip Deformation, Basin Formation, and Sedimentation: SEPM Special Publication No. 37, p. 143-158.

Morley, C.K., R.A. Nelson, T.L. Patton, and S.G. Munn, 1990, Transfer zones in the East African rift system and their relevance to hydrocarbon exploration in rifts: AAPG Bulletin, v. 74, p. 1234-1253.

Mount, J.F., 1984, Mixing of siliciclastic and carbonate sediments in shallow shelf environments: Geology, v. 12, p. 432-435.

Nelson, R.A., T.L. Patton, and C.K. Morley, 1992, Rift-segment interaction and its relation to hydrocarbon exploration in continental rift systems: AAPG Bulletin, v. 76, p. 1153-1169.

Nemec, W., and R.J. Steel, eds., 1988, Fan Deltas: Sedimentology and Tectonic Settings: Glasgow, Blackie, 444 p.

Nicoll, R.S., 1991, Conodonts from BMR samples in the Napier Range area, Canning Basin, Western Australia: Bureau of Mineral Resources, Geology and Geophysics, Professional Opinion 1991/006, 3 p. (unpublished).

Nicoll, R.S., 1992, Examination of conodonts from Canning Basin samples submitted by Annie Holmes: Bureau of Mineral Resources, Geology and Geophysics, Australia, Professional Opinion 1992/010, 3 p. (unpublished).

Officer, C.B., and C.L. Drake, 1985, Epeirogeny on a short geological time scale: Tectonics, v. 4, p. 603-610.

Peacock, D.C.P., and D.J. Sanderson, 1991, Displacements, segment linkage and relay ramps in normal fault zones: Journal Structural Geology, v. 13, p. 721-733.

Pitman, W.C., III, 1978, Relationship between eustasy and stratigraphic sequences of passive margins: Geological Society of America Bulletin, v. 89, p. 1389-1403.

Playford, P.E., 1980, Devonian "Great Barrier Reef" of Canning basin, Western Australia: AAPG Bulletin, v. 64, p. 814-840.

Playford, P.E., 1981, Devonian reef complexes of the Canning Basin, Western Australia: Geological Society of Australia, Fifth Australian Geological Convention, Field Excursion Guidebook, 64 p.

Playford, P.E., 1982, Devonian reef prospects in the Canning Basin: Implications of the Blina oil discovery: Australian Petroleum Exploration Association Journal, v. 22, p. 258-272.

Playford, P.E., 1984, Platform-margin and marginal-slope relationships in Devonian reef complexes of the Canning Basin, in Purcell, P.G., ed., The Canning Basin, W.A.: Proceedings of the Geological Society of Australia and Petroleum Exploration Society of Australia Symposium, p. 189-214.

Playford, P.E., 1991, Reef development and extinction in response to sealevel change in the Devonian of the Canning basin: AAPG Bulletin, v. 75, p. 654.

Playford, P.E., N.F. Hurley, C. Kerans, and M.F. Middleton, 1989, Reefal platform development, Devonian of the Canning Basin, Western Australia, in Crevello, P.D., J.L. Wilson, J.F. Sarg, and J.F. Read, eds., Controls on Carbonate Platform and Basin Development: SEPM Special Publication No. 44, p. 187-202.

Playford, P.E., and D.C. Lowry, 1966, Devonian reef complexes of the Canning Basin, Western Australia: Geological Survey of Western Australia, Bulletin 118, 150 p.

Posamentier, H.W., M.T. Jervey, and P.R. Vail, 1988, Eustatic controls on clastic deposition I—conceptual framework, in Wilgus, C.K., B.S. Hastings, C.G.St.C. Kendall, H.W. Posamentier, C.A. Ross, and J.C. Van Wagoner, eds., Sea-Level Changes: An Integrated Approach: SEPM Special Publication No. 42, p. 109-124.

Pratt, B.R., and J. Weissenberger, 1989, Fore-slope receptaculitid mounds from the Frasnian of the Rocky Mountains, Alberta, in Geldsetzer, H.H.J., N.P. James, and G.E. Tebbutt, eds., Reefs, Canada and Adjacent Areas: Canadian Society of Petroleum Geologists Memoir 13, p. 510-513.

Purcell, P.G., ed., 1984, The Canning Basin, W.A.: Proceedings of the Geological Society of Australia and Petroleum Exploration Society of Australia Symposium, 582 p.

Purcell, P.G., and J. Poll, 1984, The seismic definition of the main structural elements of the Canning basin, in Purcell, P.G., ed., The Canning Basin, W.A.: Proceedings of the Geological Society of Australia and Petroleum Exploration Society of Australia Symposium, p. 73-84.

Purser, B.H., M. Soliman, and A. M'Rabet, 1987, Carbonate, evaporite, siliciclastic transitions in Quaternary rift sediments of the northwestern Red Sea: Sedimentary Geology, v. 53, p. 247-267.

Rattigan, J.H., 1967, Fold and fracture patterns resulting from basement wrenching in the Fitzroy depression, Western Australia: Proceedings Australia Institute Mining and Metallurgy, v. 223, p. 17-22.

Read, J.F., 1973a, Carbonate cycles, Pillara Formation (Devonian), Canning basin, Western Australia: Bulletin of Canadian Petroleum Geology, v. 21, p. 38-51.

Read, J.F., 1973b, Paleo-environments and paleography, Pillara Formation (Devonian), Western Australia: Bulletin of Canadian Petroleum Geology, v. 21, p. 344-394.

Rosendahl, B.R., 1987, Architecture of continental rifts with special reference to East Africa: Annual Review of Earth and Planetary Sciences, v. 15, p. 445-503.

Sarg, J.F., 1988, Carbonate sequence stratigraphy, in Wilgus, C.K., B.S. Hastings, C.G.St.C. Kendall, H.W. Posamentier, C.A. Ross, and J.C. Van Wagoner, eds., Sea-Level Changes: An Integrated Approach: SEPM Special Publication No. 42, p. 155-181.

Schlager, W., 1989, Drowning unconformities on carbonate platforms, in Crevello, P.D., J.L. Wilson, J.F. Sarg, and J.F. Read, eds., Controls on Carbonate Platform and Basin Development: SEPM Special Publication No. 44, p. 15-25.

Schlager, W., 1991, Depositional bias and environmental change—important factors in sequence stratigraphy: Sedimentary Geology, v. 70, p. 109-130.

Schlanger, S.O., 1986, High frequency sea-level fluctuations in Cretaceous time: An emerging geophysical problem, in Hsü, K.J., ed., History of Mesozoic and Cenozoic Oceans: American Geophysical Union Geodynamic Series 15, p. 61-74.

Scotese, C.R., 1984, An introduction to this volume: Paleozoic paleomagnetism and the assembly of Pangea, in Van der Voo, R., C.R. Scotese, and N. Bonhommet, eds., Plate Reconstruction from Paleozoic Paleomagnetism: American Geophysical Union Geodynamics Series 12, p. 1-10.

Southgate, P.N., J. Jackson, J.M. Kennard, P.E. O'Brien, V.L. Passmore, J.F. Lindsay, A.E. Holmes, and N. Christie-Blick, 1991, Subsurface sequence stratigraphy of Devonian carbonates, Canning basin, Western Australia: AAPG Bulletin, v. 75, p. 674-675.

Southgate, P.N., J.M. Kennard, M.J. Jackson, P.E. O'Brien, and M.J. Sexton, 1993, Reciprocal lowstand clastic and highstand carbonate sedimentation, subsurface Devonian reef complex, Canning basin, Western Australia, this volume.

Suttner, L.J., A. Basu, and G.H. Mack, 1981, Climate and the origin of quartz arenites: Journal of Sedimentary Petrology, v. 51, p. 1235-1246.

Teichert, C., 1949, Stratigraphy and palaeontology of Devonian portion of the Kimberley division, Western Australia: Bureau of Mineral Resources, Geology and Geophysics, Australia, Report 2, 55 p.

Towner, R.R., and D.L. Gibson, 1983, Geology of the onshore Canning Basin, Western Australia: Bureau of Mineral Resources, Geology and Geophysics, Australia, Bulletin 215, 51 p.

Tucker, D.H., M. Bacchin, and R. Almond, 1984, Significance of airborne magnetic and gamma spectrometric anomalies over the eastern margin of the Canning basin, in Purcell, P.G., ed., The Canning Basin, W.A.: Proceedings of the Geological Society of Australia and Petroleum Exploration Society of Australia Symposium, p. 299-318.

Vail, P.R., 1987, Seismic stratigraphy interpretation using sequence stratigraphy. Part I: Seismic stratigraphy interpretation procedure, in Bally, A.W., ed., Atlas of Seismic Stratigraphy: AAPG Studies in Geology No. 27, v. 1, p. 1-10.

Vail, P.R., R.M. Mitchum, Jr., R.G. Todd, J.M. Widmier, S. Thompson, III, J.B. Sangree, J.N. Bubb, and W.G. Hatlelid, 1977, Seismic stratigraphy and global changes of sea level, in Payton, C.E., ed., Seismic Stratigraphy—Applications to Hydrocarbon Exploration: AAPG Memoir 26, p. 49-212.

Van Wagoner, J.C., R.M. Mitchum, K.M. Campion, and V.D. Rahmanian, 1990, Siliciclastic Sequence Stratigraphy in Well Logs, Cores, and Outcrops: Concepts for High-Resolution Correlation of Time and Facies: AAPG Methods in Exploration Series, No. 7, 55 p.

Van Wagoner, J.C., R.M. Mitchum, Jr., H.W. Posamentier, and P.R. Vail, 1987, Seismic stratigraphy interpretation using sequence stratigraphy. Part II: Key definitions of sequence stratigraphy, in Bally, A.W., ed., Atlas of Seismic Stratigraphy: AAPG Studies in Geology No. 27, v. 1, p. 11-14.

Van Wagoner, J.C., H.W. Posamentier, R.M. Mitchum, P.R. Vail, J.F. Sarg, T.S. Loutit, and J. Hardenbol, 1988, An overview of the fundamentals of sequence stratigraphy and key definitions, in Wilgus, C.K., B.S. Hastings, C.G.St.C. Kendall, H.W. Posamentier, C.A. Ross, and J.C. Van Wagoner, eds., Sea-Level Changes: An Integrated Approach: SEPM Special Publication No. 42, p. 39-45.

Veevers, J.J., and A.T. Wells, 1961, The Geology of the Canning Basin, Western Australia: Bureau of Mineral Resources, Geology and Geophysics, Australia, Bulletin 60, 323 p.

Wade, A., 1924, Petroleum Prospects, Kimberley district of Western Australia and Northern Territory: Melbourne, Commonwealth of Australia, 63 p.

Wallace, M.W., 1987, Origin of dolomitization on the Barbwire terrace, Canning basin, Western Australia: Sedimentology, v. 37, p. 105-122.

Wilson, P., I.J. Tapley, and F.R. Honey, 1984, Structural observations of the Canning basin by NOAA-AVHRR satellite imagery, in Purcell, P.G., ed., The Canning Basin, W.A.: Proceedings of the Geological Society of Australia and Petroleum Exploration Society of Australia Symposium, p. 535-544.

Yeates, A.N., D.L. Gibson, R.R. Towner, and R.W.A. Crowe, 1984, Regional geology of the onshore Canning basin, W.A., *in* Purcell, P.G., ed., The Canning Basin, W.A.: Proceedings of the Geological Society of Australia and Petroleum Exploration Society of Australia Symposium, p. 23-55.

Yose, L.A., and P.L. Heller, 1989, Sea-level control of mixed-carbonate-siliciclastic, gravity-flow deposition: Lower part of the Keeler Canyon Formation (Pennsylvanian), southeastern California: Geological Society of America Bulletin, v. 101, p. 427-439.

Ziegler, W., 1962, Taxionomie und Phylogenie Oberdevonischer Conodonten und ihre stratigraphische Bedeutung: Abhandlungen des Hessischen Landesamtes für Bodenforschung, v. 38, 166 p.

Ziegler, W., 1971, Conodont stratigraphy of the European Devonian, *in* Sweet, W.C., and S.M. Bergstrom, eds., Symposium on Conodont Biostratigraphy: Geological Society of America Memoir 127, p. 227-284.

Ziegler, W., and C.A. Sandberg, 1990, The Late Devonian standard conodont zonation: Courier Forschungsinstitut Senckenberg, 121, 115 p.

Chapter 8

Upper Pennsylvanian Seismic Sequences and Facies of the Eastern and Southern Horseshoe Atoll, Midland Basin, West Texas

Lowell E. Waite
Mobil Research and Development Corporation
Mobil Exploration and Producing Technical Center
Dallas, Texas, U.S.A.

ABSTRACT

Upper Pennsylvanian carbonate platform, bank, and reef-mound complexes of the Horseshoe atoll constitute major oil reservoirs within the northern Midland basin of west Texas. Analyses of over 200 mi of seismic data, constrained by fusulinid biostratigraphy, allow seismic sequences and facies to be identified for the eastern and southern portions of the atoll. The reef complex in these regions is composed of four third-order (1–10 m.y.) seismic sequences, including, from oldest to youngest: (1) *Strawn* (Desmoinesian) sequence; (2) *Canyon A* (early-early Missourian) sequence; (3) *Canyon B* (middle-early to early-middle Missourian) sequence; and (4) *Canyon C/Cisco* (late-middle Missourian–early Virgilian) sequence. The seismic sequences are composed of one to five parasequence sets, and display a retrogradational geometry in cross section and map view. Additional third-order sequences may be present in the Desmoinesian and Virgilian intervals, but are unresolved seismically in the study area.

The Strawn seismic sequence is characterized by the occurrence of discontinuous, mounded reflectors interpreted to represent amalgamated phylloid-algal mound complexes. The lower Strawn sequence boundary represents the eroded surface of the Absaroka I cratonic subsequence, a type 1 sequence boundary. The upper Strawn sequence boundary appears conformable with the overlying Canyon A sequence, although the exact nature of the upper Strawn sequence boundary is equivocal. The Canyon A sequence is characterized by internal sigmoid geometries interpreted as prograding clinoforms, indicative of oolitic and skeletal grainstone-bearing units. The Canyon B and Canyon C/Cisco sequences, which are generally restricted to the topographically highest portions of the atoll, are characterized by coherent mound facies that are interpreted to represent heterogeneous reef complexes. Adjacent to many of the larger reef masses, the presence of reef-debris facies of late Canyon age is indicated by reflector packages that show offlaping geometries and basinward downlap. The upper Canyon B sequence boundary shows evidence of significant erosion associated with mass wasting of the atoll bank margin, perhaps related to an atoll-wide exposure event. The top of the atoll appears to be a maximum-flooding surface that is unconformably overlain by onlapping Upper Pennsylvanian and Lower Permian shales.

The documentation of third-order seismic sequences and facies of the Horseshoe atoll is important, not only because it provides an internally consistent stratigraphic framework for stratigraphic analysis, but also because it serves as an ancient example of a detached, retrogradational carbonate system. Seismic analysis of the atoll illustrates that the geometries of these types of systems differ substantially from that of attached, prograding systems. The arrangement of systems tracts within retrogradational systems may also differ. The stratigraphic architecture of third-order seismic sequences and facies displayed by the Horseshoe atoll may represent a recurring depositional pattern that arose during Late Pennsylvanian time, a consequence of the geographic, climatic, oceanographic, and tectonic setting of the developing Pangea supercontinent. If so, such a pattern may be anticipated in other time-equivalent carbonate regions.

INTRODUCTION

The Horseshoe atoll (Figure 1) is a massive, isolated, carbonate complex of Late Paleozoic age covering an area of approximately 6300 mi² (16,300 km²) within the subsurface of west Texas. The complex consists of a heterogeneous assemblage of shallow-marine carbonate bank and reefal[1] buildups that developed on the southwestern margin of the northern Gondwana (Laurentian) shelf during Late Pennsylvanian and Early Permian time (Figure 2). Numerous fields within the atoll, including the giant Kelly–Snyder complex in Scurry County, have together produced over 2000 million barrels of oil (Galloway et al., 1983). Development of these hydrocarbon reserves has provided an abundance of well log, core, and seismic data to characterize the large, multicomponent carbonate body. The physical size of the atoll, combined with its depositional complexities and economic importance, makes it a unique subsurface laboratory for the investigation of late Paleozoic carbonate reservoir systems.

This chapter focuses on the large-scale, seismic-stratigraphic aspects of the Horseshoe atoll, based on interpretations of over 200 mi of modern, multichannel, reflection-seismic data. It provides documentation of recognizable, biostratigraphically constrained, third-order (i.e., 1–10 m.y.) seismic sequences and associated seismic facies of the eastern and southern atoll. Included is an interpretation of the large-scale

seismic stratigraphic architecture of the atoll, together with an assessment of the regional distribution of Upper Pennsylvanian carbonates on the atoll and adjacent shelf areas.

SIGNIFICANCE OF SEISMIC APPROACH

Recent applications of seismic and sequence stratigraphy in a variety of carbonate settings have demonstrated that this technology can be used to characterize depositional architecture and lithofacies (Sarg, 1988). Although the Horseshoe atoll has been recognized as a stratigraphic entity and studied for over four decades, a detailed seismic sequence approach of the complex has not been documented. Previously published reports have provided regional geologic descriptions of the complex (e.g., Myers et al., 1956; Vest, 1970; Galloway et al., 1983), or concentrate on specific geologic and/or engineering aspects of individual oil fields (e.g., Heck et al., 1952; Anderson, 1953; Rothrock et al., 1953; Burnside, 1959; Stafford, 1959; Crawford et al., 1984; Schatzinger, 1988; Walker et al., 1990). Sequence stratigraphic analyses of time-equivalent units from nearby regions have been presented (e.g., Brown et al., 1990; Boardman and Heckel, 1989), but specific studies of the Horseshoe atoll have not been considered or remain unpublished.

As mentioned, a seismic stratigraphic approach focuses on the identification of third-order (i.e., 1–10 m.y.) depositional events. Previous stratigraphic studies of Carboniferous units normally concerns smaller scale, fourth- to fifth-order, glacio-eustatic–induced depositional cycles. An analysis of third-

[1]Unlike most modern coral reefs, organic buildups of the Horseshoe atoll are dominated by nonframework carbonate facies. Terms such as "reef-mound," "reef-knoll," "mound," and "reef" have been used to describe carbonate buildups of the atoll. In the present study, these terms are used interchangeably to identify carbonate buildups of the atoll.

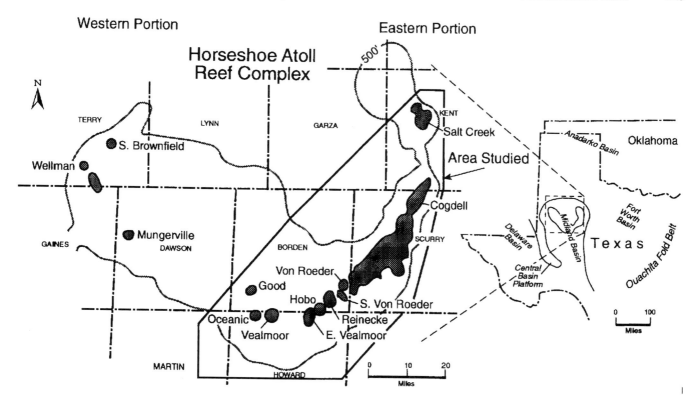

Figure 1. Location of Horseshoe atoll, west Texas. Main area of study includes the eastern and southern portions of the atoll (modified from Walker et al., 1990, with permission to modify original illustration and republish courtesy of The West Texas Geological Society).

order sequences is important because the larger scale sequences are more likely to be the product of global, rather than local, changes in sea level (Vail and Mitchum, 1977; Ross and Ross, 1987). The definition of third-order depositional sequences should, therefore, be useful in refining global correlations and assessing the timing and extent of eustatic sea-level changes. Locally, the establishment of third-order sequences provides an internally consistent stratigraphic framework to constrain wireline-log correlations and refine reservoir/trap relationships within the atoll complex.

DATA AND METHODS

A 200-mi grid of modern (post-1984), multichannel, reflection-seismic data was selected and interpreted for a large portion of the eastern and southern Horseshoe atoll (Figure 3). Initial analyses of seismic data involved the identification of seismic horizons representing the top of the atoll complex (top reef) and the base of the Pennsylvanian section (base Strawn = top Mississippian). For stratigraphic control, other regional pre-Pennsylvanian and Lower Permian horizons were also correlated (Figure 4). All seismic horizons were correlated on positive polarity sections at a scale 1 in. = 7.5 s (positive polarity is defined here as shale-to-limestone transition = trough). The large contiguous grid of data in the

southwestern region of the atoll (Figure 3) was looptied to ensure correlation away from synthetic seismogram control. Correlation between noncontiguous lines was made using a jump-correlation method and verified using synthetic seismograms.

Sequence boundaries and seismic facies were defined using the criteria and descriptions of Mitchum et al. (1977), Sarg (1988), and Van Wagoner et al. (1990). In the study area, Pennsylvanian sequence boundaries are defined by low-angle downlap surfaces, low- to high-angle onlap surfaces, or high-angle lateral truncation events (Figure 5). Log tops and rock/paleontologic data were correlated into the seismic sections through the use of synthetic seismograms. Stratigraphic ages of sequences were determined by the application of fusulinid biostratigraphy. Geologic data from Rothrock et al. (1953), Myers et al. (1956), Schatzinger (1988), Reid and Reid (1991), and company files also were used to constrain the age of seismic sequences and to interpret the depositional nature of seismic facies.

APPLICATION OF FUSULINID BIOSTRATIGRAPHY

Previous geologic studies (e.g., Rothrock et al., 1953; Hollingsworth, 1955; Myers et al., 1956; Stafford, 1959) have shown that fusulinid foraminiferal biostratigraphy can be used to define and subdivide the

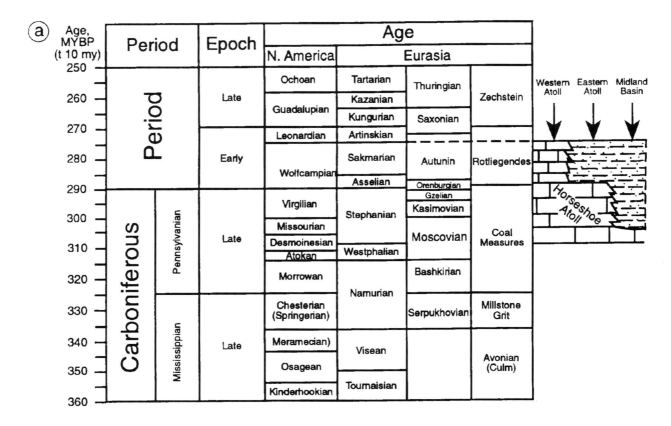

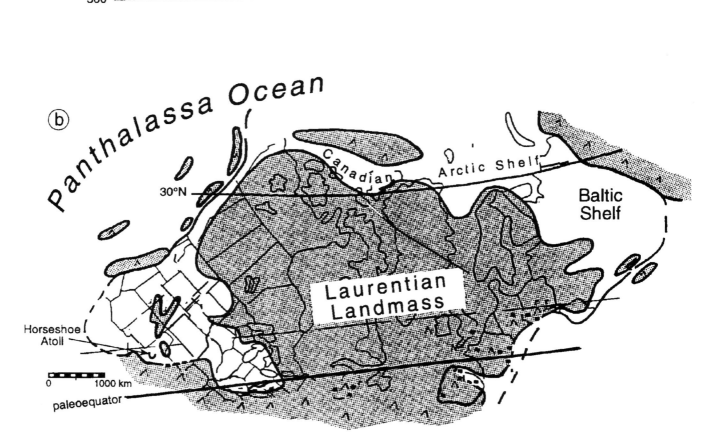

Figure 2. (a) Generalized stratigraphic column showing age of the Horseshoe atoll. (b) Paleogeographic location of Horseshoe atoll during Late Pennsylvanian time (modified from Witzke, 1990, with permission to modify original illustration and republish courtesy of The Geological Society, London, and the author).

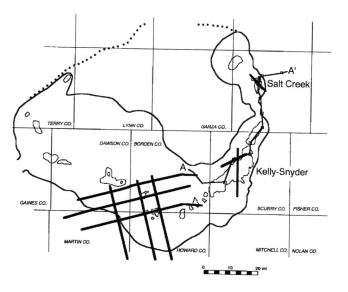

Figure 3. Seismic data coverage (lines #1–9) for present study. Geologic cross section A–A' shown in Figure 8.

major lithostratigraphic units (i.e., Strawn, Canyon, Cisco, Wolfcamp units) of the Horseshoe atoll. Application of fusulinid biostratigraphy is necessary to constrain log correlations because the massive atoll carbonates are depositionally heterogeneous and lack regionally extensive, internal log markers (Rothrock et al., 1953). Recent refinements in Midland basin fusulinid zonation now allow a much more detailed subdivision of the major lithostratigraphic units (Reid et al., 1988; Jensen and Walker, 1990; Walker et al., 1990; Wilde, 1990). Currently recognized fusulinid zones within the Late Pennsylvanian and Early Permian periods are on the order of 1 m.y. or less in duration, depending on the absolute time scale used for calibration. Therefore, fusulinid zones constrain the relative ages of third-order genetic time-rock units within the atoll and provide more precise global correlation to other late Paleozoic stratigraphic sequences (Jensen and Walker, 1990; Walker et al., 1990).

The fusulinid zonation of Reid et al. (1988) is utilized in this study (Figure 6). In this zonation, both the Strawn and Canyon stages are subdivided into seven zones, each representing approximately 1 m.y.

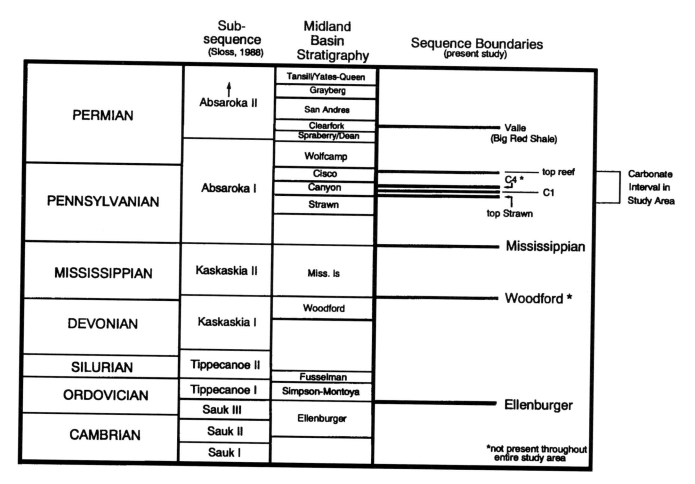

Figure 4. Stratigraphic position of seismic sequence boundaries identified for the eastern and southern Horseshoe atoll. Pennsylvanian sequences are interpreted to represent third-order (1–10 m.y.) sequences within the Absaroka I cratonic subsequence.

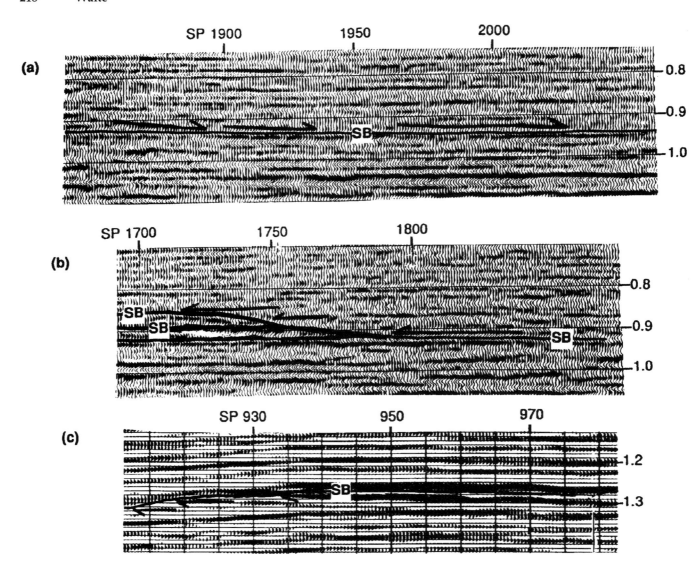

Figure 5. Examples of sequence boundaries of the Horseshoe atoll: (a) low-angle downlap, which occurs at the top of the reef and is similar in geometry to a maximum flooding surface; (b) low-angle onlap; (c) high-angle lateral truncation. SB—sequence boundary; seismic two-way time in seconds.

in duration. Cisco and Wolfcamp zonations were not used due to the paucity of post-Canyon fusulinid data within the study area. Biostratigraphic data from core and cutting samples were provided by A.M. Reid and S.T. Reid.

SEISMIC SEQUENCES OF THE EASTERN AND SOUTHERN HORSESHOE ATOLL

Four third-order Upper Pennsylvanian seismic sequences can be recognized within the study area. These seismic sequences include, from oldest to youngest: (1) Strawn; (2) Canyon A; (3) Canyon B; and (4) Canyon C/Cisco sequences (Figure 6). Additional sequences may be present within the Strawn but could not be delineated in the study area. The recognized sequences include one to seven fusulinid zones and therefore range in duration from

approximately 1 to 7 m.y. The sequences vary considerably in thickness and lateral extent as they pass from the main portions of atoll buildup into the Midland basin (Figures 7, 8). The thickness, duration, and lateral extent of these Upper Pennsylvanian seismic sequences indicate that they represent both third-order stratigraphic cycles as defined by Miall (1990), and depositional sequences as defined by Haq et al. (1988). Individual seismic sequences therefore presumably represent deposition that occurred during one complete cycle of global sea-level change (Van Wagoner et al., 1990). The third-order seismic sequences are composed of assemblages of smaller, higher frequency (fourth- and fifth-order) genetic units that include parasequences (i.e., shoaling-upward cycles, transgressive–regressive cycles) and parasequence sets (Vail and Mitchum, 1977; Ross and Ross, 1987; Haq et al., 1988). Detailed correlation of

NE→

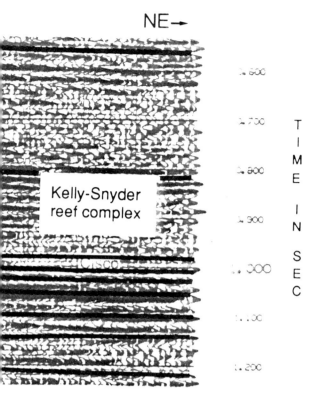

T
I
M
E

I
N

S
E
C

1.600

1.700

1.800

1.900

1.000

1.100

1.200

Kelly-Snyder
reef complex

E→

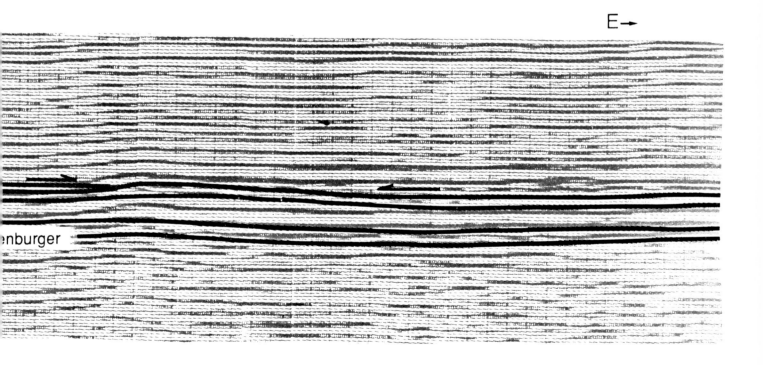

enburger

A'
Northeast

m: **- 4000 subsea**

5,000   10,000 15,000 ft.

Vertical exaggeration = 33x

Figure 8. Cross section A–A' along crest of Horseshoe atoll; datum is –4000 ft subsea. Location of cross section shown in Figure 17.

SEISMIC
SEQUENCE

Stage		Hollingsworth (circa 1959)		Reid et al. (1988)	In-house usage	
CANYON	LATE	Home Creek		late Late Canyon	MC7	—SB—
		Ranger		early Late Canyon	MC6	Canyon C-Cisco
	MIDDLE	Adams Branch		late Middle Canyon	MC5	
		Palo Pinto		early Middle Canyon	MC4	—SB—
	EARLY	Lowermost Canyon "Fusulina-Triticites"		late Early Canyon	MC3	Canyon B
				middle Early Canyon	MC2	
				early Early Canyon	MC1	— SB — Canyon A — SB —
STRAWN	LATE	Upper	Marmaton	late Late Strawn	DS7	Strawn
				early Late Strawn	DS6	
	MIDDLE			late Middle Strawn	DS5	
		— ? —		early Middle Strawn	DS4	
	EARLY	Lower	Upper Cherokee	late Early Strawn	DS3	
				middle Early Strawn	DS2	Not present in study area
			Lower Cherokee	early Early Strawn	DS1	

Figure 6. Fusulinid zonation for Strawn and Canyon units showing position of Pennsylvanian seismic sequences of the eastern and southern atoll. SB—sequence boundary.

higher frequency depositional cycles of the atoll to the third-order seismic sequence framework defined here awaits further study.

Strawn Sequence

The Strawn sequence includes all sediments from the top of the Mississippian unconformity to the top of the Strawn section (Figures 7, 8). The sequence is defined by Desmoinesian fusulinid zones DS-3 to DS-7 (Figure 6). In the study area, fusulinid zones DS-1 and DS-2 are missing due to erosion and/or nondeposition. The lower portion of the Strawn sequence forms a regionally extensive, low-relief carbonate platform upon which the younger, isolated bank and reef-mound sediments of the Horseshoe atoll developed (Vest, 1970). In the northern Midland basin, the Strawn ramp margin can be characterized as a distally steepened ramp (Mazzullo and Reid, 1988), which is consistent with the seismic signature observed within the study area. Seismic and geologic correlations indicate the Strawn sequence is 400 to 600 ft thick.

The Strawn sequence is composed of 1 to 3 reflector packages that constitute a single third-order seismic sequence. However, where the Strawn

stratigraphic interval thickens beneath the crest of the atoll, it is possible that more than one seismic sequence is present. The definition of additional Strawn seismic sequences is complicated, however, by the presence of discontinuous, mounded reflectors (discussed below), and by the lack of Strawn well control in many of the reef fields. For these reasons only one Strawn seismic sequence is defined in the present report. The study of additional core, logs, and synthetic seismograms may ultimately resolve additional third-order sequences within the Strawn interval. Analysis of wireline-log and sample data from Scurry County shows that the Strawn seismic sequence is composed of a minimum of five parasequence sets, one of which is capped by a possible subaerial exposure surface (Figure 9).

The sequence boundary at the base of the Strawn represents the interregional unconformity at the base of the Absaroka cratonic sequence, which is characterized by widespread karsting and subaerial exposure (Sloss, 1988). In the study area, this surface is defined by seismic onlap of the lower Strawn reflectors (fusulinid zones DS-2 and DS-3) onto the underlying top Mississippian erosional surface (Figure 7). The sequence boundary at the top of the Strawn is

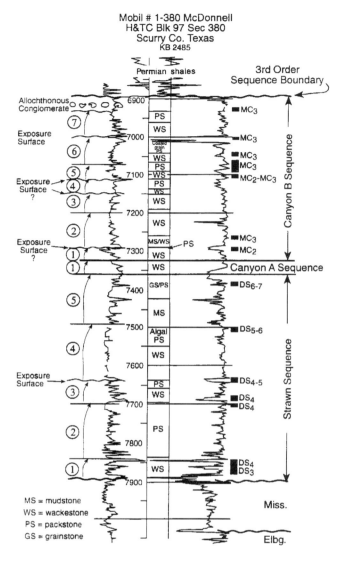

Figure 9. Depositional cycles (circled numbers) within Strawn and Canyon sequences, Scurry County. Sequences represent third-order events composed of smaller scale upward-shoaling cycles representing fourth- and fifth-order events (parasequence sets and parasequences). Fusulinid biostratigraphic control is indicated by vertical black bars using notation of Figure 6. Despite the thinness of the Canyon A sequence in this location, it is a mappable seismic package here and throughout the study area.

usually characterized by a strong, coherent, seismic amplitude. In basinal regions where the top of the Strawn sequence represents the top of the carbonate interval, the upper Strawn sequence boundary is downlapped by the overlying Upper Pennsylvanian and Lower Permian shales.

Whereas the lower Strawn sequence boundary is a type 1 sequence boundary, the exact nature of the upper Strawn sequence boundary in the study area is questionable. The Strawn seismic sequence appears seismically conformable with the overlying Canyon A

sequence, implying either the upper Strawn sequence boundary was not subaerially exposed, or else the exposure signature is beyond the resolution of the seismic. Available wireline-log and sample data from the northwest margin of Kelly–Snyder field (Figure 9) show the presence of argillaceous mudstones and wackestones, interpreted as deeper water (>10 m) subtidal facies, and imply depositional continuity across the Strawn-Canyon sequence boundary. In this interpretation, if an unconformity is present, it is more likely to be a subaqueous type (e.g., marine hardground), more indicative of a type 2 sequence boundary. Many workers, however, report an alternative interpretation of the upper Strawn sequence boundary on the atoll and adjacent regions. For example, Myers et al. (1956), Burnside (1959), and Stafford (1959) all report that the Strawn–Canyon boundary on the atoll is characterized by a significant erosional unconformity. Reid and Reid (1991) note that the top of the upper Strawn interval at Cogdell field in Kent and Scurry counties shows evidence of a major unconformity marked by subaerial exposure and erosion, associated with glacio-eustatic sea-level changes. Boardman and Barrick (1989) report a type 1 unconformity at the Strawn–Canyon boundary in the U.S. mid-continent and north-central Texas. Given these reports and the evidence cited above, it can be concluded that the nature of the upper Strawn sequence boundary on the atoll is equivocal, and that further study of the Strawn–Canyon contact is warranted before the exact nature of the upper Strawn sequence boundary can be specified.

Canyon A Sequence

The Canyon A sequence includes units of early-early Canyon age, defined by the Missourian MC-1 fusulinid zone (Figure 6). Compared to other seismic sequences of the eastern and southern atoll, the Canyon A sequence is shortest in duration (i.e., 1 m.y. or less), encompassing a single fusulinid zone. Seismic analysis indicates the sequence is 50 to 100 ft thick and is composed of one parasequence set (Figures 7–9). At Salt Creek and Cogdell fields in Kent and Scurry counties, however, at least two parasequence sets of equivalent age are present (Reid and Reid, 1991; Walker et al., 1991; J.M. Jensen, personal communication). Nevertheless, the total number of parasequences within the Canyon A sequence are contained within a single, well-developed, seismic wavelet package (Figure 7).

The Canyon A seismic sequence is thickest directly beneath the main atoll buildups, whereas it is extremely thin, or not present, between buildups (e.g., between Salt Creek and Cogdell fields in Kent county, Figure 8). The limits of the Canyon A sequence are indicated by a rapid increase in basinward dip of the sequence at the bank edge, and/or a downlap of the Canyon A sequence boundary onto the underlying Strawn sequence. In basinal positions off the southern edge of the atoll, the Canyon A sequence can occur in aerially limited, structurally low positions, and consists of autochthonous sea-level lowstand deposits (Reid et al., 1990a,b).

Seismic and paleontologic data suggest that the Canyon A sequence is conformable with the overlying Canyon B sequence within the study area. Recent study of Cogdell field in Kent and Scurry counties, however, indicates that a lowering of sea level occurred during the end of early-early Canyon time, exposing and karsting the top of the sequence (Reid and Reid, 1991). Either this exposure surface is local in extent and is not covered by the current data set, or its signature is beyond seismic resolution.

Canyon B Sequence

The Canyon B sequence includes units of middle-early Canyon, late-early Canyon, and early-middle Canyon age, encompassing Missourian fusulinid zones MC-2 to MC-4 (Figure 6). Stratigraphic correlations indicate the sequence ranges in thickness from 0 to 400 ft in the study area and is composed of 1 to 2 seismic wavelet packages. Wireline-log and sample data from Scurry County suggest the sequence is composed of a minimum of seven parasequence sets, containing several internal unconformities (Figure 9). Many of these unconformities may represent subaerial exposure surfaces concomitant with lowering of relative sea level during middle- to late-early Canyon and early-middle Canyon time. For example, Burnside (1959) postulated a widespread sea-level lowering and associated exposure in the southern atoll region during middle Canyon time. Boardman and Barrick (1989) noted a type 1 unconformity separates the lower and middle Canyon units of the U.S. mid-continent and north-central Texas. Reid et al. (1990a) noted the presence of five separate depositional cycles within the late-early Canyon in Howard County. Each cycle begins with a relative lowering of sea level, exposing the main atoll buildups and shifting shallow-water carbonate deposition to a position immediately basinward of the atoll. Continued sea-level lowering ultimately exposes the early lowstand carbonates. During exposure, karsting and leaching of the carbonates occurs. The ensuing rapid sea-level rise associated with the transgressive systems tract ultimately submerges these deposits, followed by a new cycle of highstand deposition. Evidence for these cycles within the Canyon B sequence are beyond seismic resolution, but their presence is consistent with the interpretation of multiple exposure events observed in the Scurry County well (Figure 9).

The upper boundary of the Canyon B sequence (top of lower-middle Canyon = MC-4 zone) shows clear seismic evidence of being a regional truncation surface (Figure 7). In many locations the erosive surface cuts downward into the early-early Canyon section. In some locations it is estimated that as much as 200 to 300 ft of sediment may have been eroded from the late-early and early-middle Canyon bank margins. These eroded margins are associated with the occurrence of large, allochthonous, upper Canyon debris wedges off the margin flanks. The nature of

these debris wedges and the significance of the truncation surface are described in more detail below.

Canyon C/Cisco Sequence

The Canyon C/Cisco seismic sequence includes all carbonates of late-middle Canyon to Cisco age within the study area, including Missourian fusulinid zones MC-5 to MC-7 (Figure 6). Geologic and seismic correlations indicate that it ranges from 0 to 250 ft in thickness over the study area. When present, the Canyon C/Cisco seismic sequence is composed of one or more seismic wavelet packages that delineate relatively small, mound-shaped complexes.

The top of the Canyon C/Cisco sequence corresponds to the highest structural portions of the atoll complex. The upper surface appears to be a maximum flooding surface that is onlapped by black shales and other fine terrigenous clastics of latest Pennsylvanian and Early Permian age. Lack of fusulinid data at the top of the sequence precludes an exact age-date for the youngest atoll carbonates, but Reid and Reid (1991) indicated an early Cisco age for the highest portions of Cogdell field in Kent and Scurry counties (Figure 8). By correlation, the uppermost portion of the Canyon C/Cisco sequence in the study area is considered to be early Cisco in age. It is important to note, however, that fusulinids of Wolfcampian age have been reported by earlier workers from structurally high carbonates in the eastern and southern portions of the atoll (Heck et al., 1952; Myers et al., 1956; Burnside, 1959; Stafford, 1959). It is therefore probable that one or more upper Cisco–Wolfcamp sequences are present in portions of the atoll not covered by the current seismic data base.

SEISMIC FACIES OF THE EASTERN AND SOUTHERN HORSESHOE ATOLL

Each of the Late Pennsylvanian seismic sequences of the eastern and southern Horseshoe atoll are characterized by one or two distinct seismic facies. Late Pennsylvanian seismic facies on the atoll can be defined on the basis of internal geometry of seismic reflectors, nature of reflector termination, and external geometry relative to adjoining reflector packages. Recognized seismic facies includes a mounded discontinuous facies, characteristic of the Strawn sequence; a shingled clinoform facies, common in the Canyon A sequence and occasionally observed in the upper part of the Strawn sequence; and a mounded coherent facies, restricted to the Canyon B and Canyon C/Cisco sequences. In addition, a lensoid/chaotic facies of post–early Canyon age is also recognized.

Mounded Discontinuous Facies

Mounded discontinuous facies (Figure 10) occurs within the Strawn sequence. The mounded discontinuous facies is identified by a series of divergent, discontinuous reflectors that outline a series of larger, convex-upward or mound shapes. Typically, the

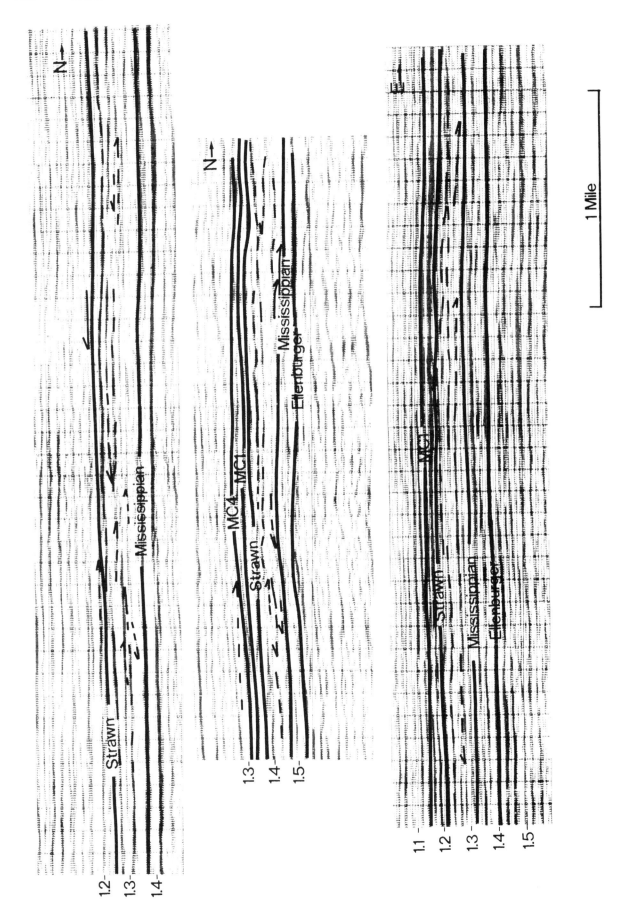

Figure 10. Examples of mounded discontinuous seismic facies that is characteristic of the Strawn sequence. Seismic two-way time in seconds.

flanks of larger mound structures are developed in a stratigraphically lower wavelet package while the crest is contained in a higher package. Available seismic control suggests the individual mound packages are elongate features, 1 to 2 mi wide and up to several miles in length. Within the mounded discontinuous facies, individual internal reflector packages often show numerous onlapping and downlapping geometries within a small lateral distance (i.e., hundreds of feet). Mound packages are separated from one another by a series of continuous to discontinuous, horizontal, parallel reflectors.

The mounded discontinuous facies is best developed in the southern portion of the atoll complex, where the Strawn sequence is 400 ft or greater in thickness. Here, the facies is fairly contiguous with the overlying Canyon bank and reef-mound complexes. Basinward, the facies does not extend far beyond the Canyon bank edge, where it is replaced by horizontal, parallel reflectors. The mounded discontinuous facies was not observed on the eastern side of the atoll, where the Strawn sequence is generally reduced in thickness to one wavelet package. If the facies is present on the eastern side of the atoll, it is beyond seismic resolution.

Based on seismic character, distribution, and age (middle-late Desmoinesian), the mounded discontinuous facies is interpreted to represent amalgamated phylloid algal and sponge/bryozoan mound complexes. Based on reported occurrences, algal mounds were well developed and widespread in North America during Desmoinesian time (West, 1988). Desmoinesian phylloid algal mounds are known from the surface and subsurface of Texas, New Mexico, Colorado, Utah, and the mid-continent (Wilson, 1975). In the subsurface of west Texas, Desmoinesian phylloid algal mound complexes associated with sponges and bryozoa are reported from the Nena Lucia field in Nolan County and from the Goen Limestone in Runnels and Concho counties (Toomey and Winland, 1973; Marquis and Laury, 1989). On the Horseshoe atoll, phylloid algal mounds have been identified from upper Desmoinesian units at Diamond M field in Scurry County (Schatzinger, 1988). These occurrences suggest that phylloid algal mounds are likely a common component of the Strawn throughout the atoll. The seismic signature of the mounded discontinuous facies indicates the facies consists of a number of smaller, individual algal mounds that coalesce to form larger mound complexes. The horizontal, parallel seismic reflectors of intermound areas are interpreted to represent deeper water carbonate mudstones and skeletal wackestones.

Clinoform Facies

The clinoform facies occurs primarily in the Canyon A sequence, deposited during the early portion of the early Canyon time. It is occasionally observed in the uppermost portions of the Strawn sequence. The facies is identified by a series of distinct, subparallel, shingled clinoforms (Figure 11). Stratigraphic orientation of the clinoforms indicates a progradation of the facies to the south and west, from the interior of the atoll complex toward the Midland basin. The limits of the clinoform facies correspond approximately to the limits of the early-early Canyon bank margins. The facies is particularly well developed along the southwest margin of the atoll in Borden and Howard counties and at Salt Creek field in Kent County.

The clinoform facies is interpreted to represent a series of relatively shallow-water, prograding carbonates. Lack of upper Strawn and lower Canyon core control where the seismic facies is observed prevents an assessment of the exact lithologic nature of the facies, but it is likely to contain both high-energy skeletal and oolitic grainstones and an assemblage of skeletal packstones and mudstones. Previous studies indicate that oolitic and skeletal grainstones within the Canyon interval are abundant within certain areas of the atoll. For example, oolitic grainstones characterize the lower Canyon units at Wellman field in Terry County (Andersen, 1953), and Schatzinger (1988) noted that Canyon grainstones are more abundant at Cogdell and North Snyder fields in Kent and Scurry counties, while mud-bearing units are more common to the south at Kelly and Diamond M fields. Walker et al. (1990) reported that many Canyon depositional cycles at Salt Creek field in Kent County are capped by ooid grainstones, their distribution on the eastern side of the atoll controlled by paleogeographic and paleoclimatic factors. Areas where the clinoforms are best developed on seismic (discussed below) may indicate regions with higher grainstone content, although the seismic facies may contain a variety of carbonate rock types.

Mounded Coherent Facies

The mounded coherent facies occurs within the Canyon B and Canyon C/Cisco sequences. It consists of large, individual, coherent mound structures showing internal, high-angle, discordant truncations (Figure 12). The top of the facies is denoted by low-angle seismic onlap of younger units and by seismic downlap onto the underlying Canyon A sequence. The mounded coherent facies differs from the previously described mounded discontinuous facies in several ways. The structures of the mounded coherent facies are younger, generally more laterally extensive, and show a higher degree of internal and external reflector continuity than the discontinuous mounded facies (compare Figures 11, 13). Mounded coherent facies are restricted to the crestal portions of the atoll, coincident with the limits of the Canyon B and Canyon C/Cisco sequences. Most, if not all, of the current reservoirs of the atoll produce from this seismic facies.

Based on age, aerial extent, lithologic character, and internal seismic character, the mounded coherent

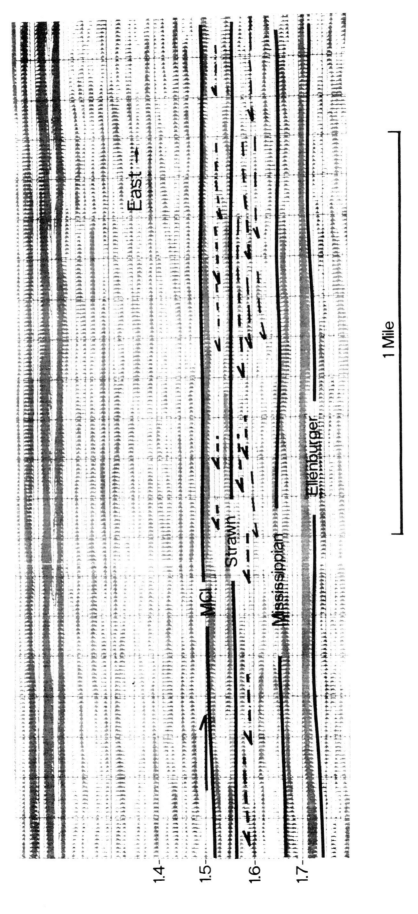

Figure 11. Examples of clinoform seismic facies that is characteristic of the Canyon A sequence. The facies is also observed within the Strawn sequence. Seismic two-way time in seconds.

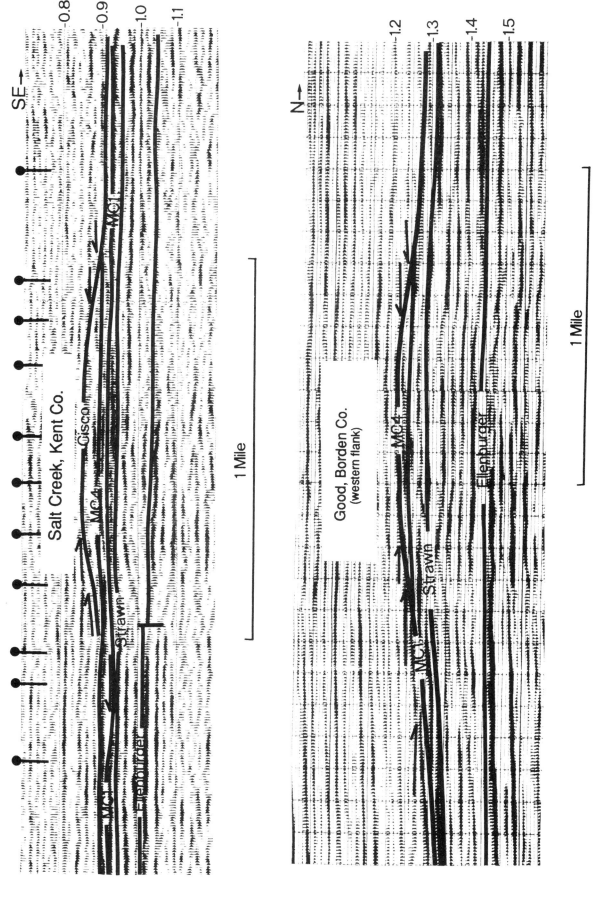

Figure 12. Examples of mounded coherent seismic facies that is characteristic of the Canyon B and Canyon C/Cisco sequences. Seismic two-way time in seconds.

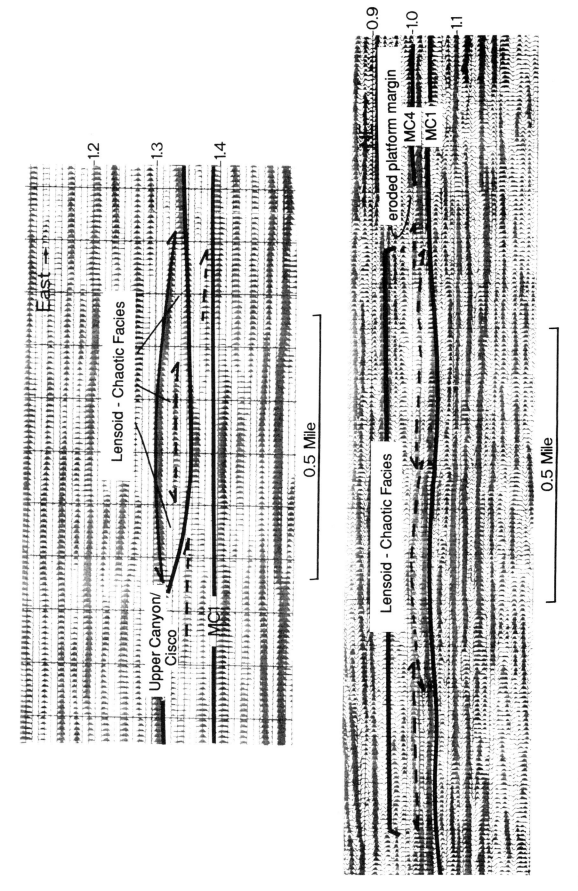

Figure 13. Examples of lensoid-chaotic seismic facies. This facies occurs adjacent to large reef-mound complexes and adjacent to eroded platform margins. Seismic two-way time in seconds.

facies is interpreted to represent large, isolated reef-mound complexes composed of a variety of carbonate facies. In the present study area, sample control tied to seismic data on the northwest side of Kelly–Snyder field indicates the middle- to late-early Canyon interval consists of cyclical, subtidal mudstone-grainstone packages containing crinoids, fusulinids, chaetetid sponges, and coated grains. The occurrence of a variety of facies within this time interval has also been noted by previous workers. For example, at Kelly–Snyder reef in Scurry County, Schatzinger (1988) documented a number of lower and upper Canyon facies including carbonate mudstones, phylloid algal-bearing packstones, sponge-bryozoan boundstones, and oolitic and skeletal grainstones. He interpreted a number of subtidal and peritidal environments of deposition from these occurrences. Walker et al. (1990) observed similar depositional facies for depositional cycles within the upper Canyon interval at Salt Creek reef. Crawford et al. (1984) noted that the upper Canyon and Cisco facies at Reinecke reef are dominated by phylloid algal packstones and grainstones. This information suggests that the mounded-coherent facies consists of a variety of laterally adjacent, heterogeneous, subtidal-to-peritidal carbonate facies. Dominant rock-forming constituents in the mounded-coherent facies include phylloid algae, crinoids, bryozoa, sponges, and foraminifera. The mounds are most likely dominated by muddy subtidal facies, but skeletal grainstones, boundstones, and muddy peritidal sediments may be locally abundant.

Lensoid/Chaotic Facies

The lensoid/chaotic facies occurs as a separate, distinct seismic facies adjacent to the mounded coherent facies. It is characterized by either discrete, doubly convex to convex-downward–shaped reflector packages or by a series of reflectors showing chaotic arrangement (Figure 13). The discrete, lensoid seismic packages onlap the reef-mound complexes and downlap onto the underlying carbonate banks, often showing internal stratification. The facies tends to occur over small, fan-shaped areas adjacent to the highest portions of the main atoll complex. When present over a larger area, individual seismic reflectors appear less coherent and discontinuous. In some cases, the lensoid/chaotic faces occurs adjacent to obviously eroded carbonate bank margins (Figure 14). Wireline-log and sample control on the northwest margin of Kelly–Snyder field indicates the lensoid/chaotic facies consists of transported, conglomeratic, and brecciated mound material in a dark shale matrix (Figure 15).

Based on lateral extent, seismic geometry, and general lithologic and sedimentologic features, the lensoid/chaotic seismic facies is interpreted to be eroded, brecciated, allochthonous deposits. Sediments of this nature are noted in previous studies ("calirudite" of Myers et al., 1956; "breccia facies" of Schatzinger, 1988; "foreshelf conglomerate" of Reid et

al., 1990a). Schatzinger (1988) and Reid et al. (1990a) noted the presence of at least two different types of debris facies based on lithoclast content and proximal/distal position. The debris aprons of the atoll are generally low in porosity and permeability, and in some instances serve as reservoir seals (Myers et al., 1956; Reid et al., 1990a). Many workers believe the debris units are generated during periods of erosion associated with sea-level lowstand (Myers et al., 1956; Stafford, 1959; Reid et al., 1990a).

The exact age of the lensoid/chaotic facies is equivocal. It is probable that allochthonous debris facies was generated several times throughout the history of the atoll, particularly during more intense periods of bank-margin erosion (e.g., early-middle Canyon). If so, the lensoid/chaotic seismic facies may vary from early Canyon to Wolfcamp in age, depending on the shifting location of previous and contemporaneous sites of carbonate buildup. Myers et al. (1956) noted that debris material was most likely generated during the erosion of older reef carbonates during lowstands of sea level, stating that several such events occurred during Strawn–Wolfcamp time. Schatzinger (1988), citing Myers et al. (1956), stated that most of the debris material on the eastern atoll was generated during early Canyon, late Canyon, Cisco, and Wolfcamp time. Reid et al. (1990a) indicated that conglomeratic material off the southern edge of the atoll contains fusulinids of early-middle, late-early, and middle-early Canyon age. Reid and Reid (1991) noted a major erosive event at the end of late Canyon time at Cogdell field in Kent and Scurry counties. Seismic and fusulinid data from the current study indicate an early-middle to late Canyon age for the debris facies.

SEISMIC-STRATIGRAPHIC ARCHITECTURE AND DEPOSITIONAL HISTORY OF THE EASTERN AND SOUTHERN HORSESHOE ATOLL

The combination of Late Pennsylvanian seismic sequences and seismic facies allows the large-scale stratigraphic architecture of the eastern and southern Horseshoe atoll to be defined. In these regions, the atoll can be described as a southwestward-dipping, southwestward-thickening, retrogradational, homoclinal, carbonate mass that developed over a pre-Desmoinesian hinge line (Figure 16). The trend of the hinge line bisects the atoll, coinciding approximately with an increase in structural dip of the Early Ordovician (Ellenburger) unconformity surface, expansion of the pre-Mississippian stratigraphic interval, and increased thickening of the atoll carbonates. To the west of this structural hinge, carbonate deposition on the atoll extends into the Lower Permian (Wolfcamp) section (Vest, 1970). The hinge line also marks the approximate position of the middle-upper Permian shelf edge. Regional seismic data demonstrate that the Horseshoe atoll is essentially a

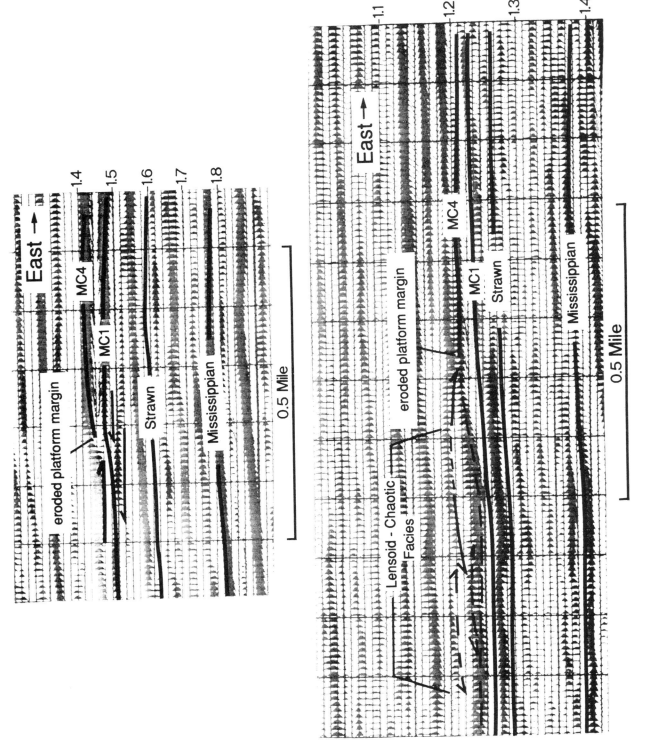

Figure 14. Seismic expression of eroded platform margins. Seismic two-way time in seconds.

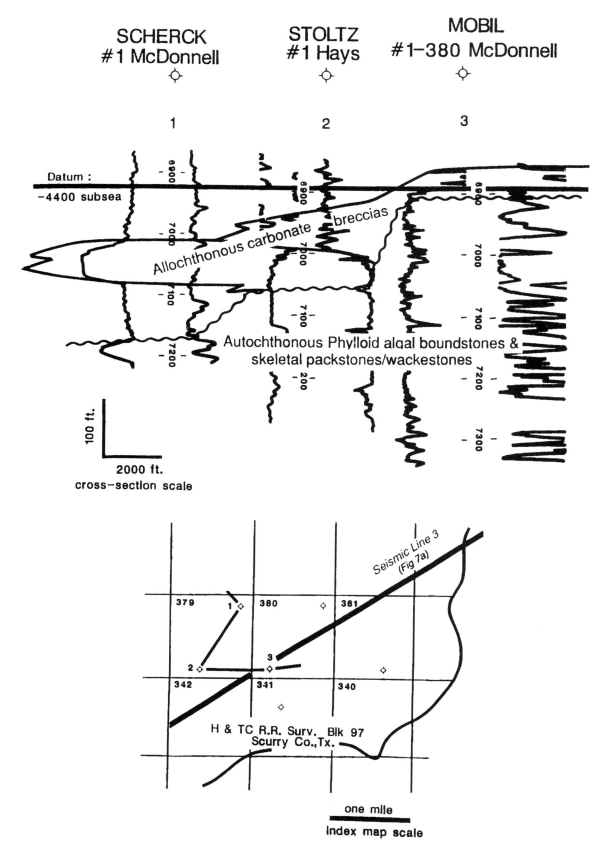

Figure 15. Detailed geologic cross section along northwest flank of Kelly–Snyder field, Scurry County. In this region, the flank of the atoll is characterized by allochthonous, conglomeratic, carbonate breccias in a dark-mudstone matrix. These deposits constitute the lensoid-chaotic seismic facies.

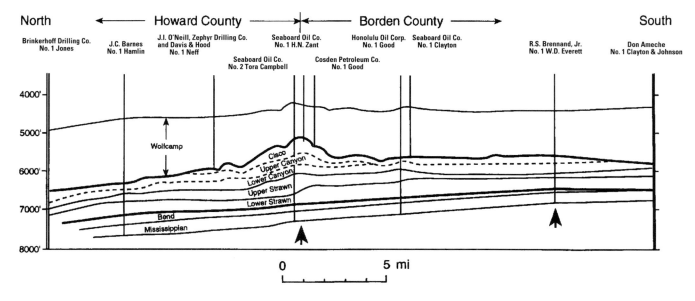

Figure 16. Regional north–south cross section across western Borden and Howard counties illustrating south-western thickening of pre-Strawn and Strawn units under main reefal buildups. Arrows indicate region where sub-Strawn units expand stratigraphically into Midland basin, interpreted as a deep-seated structural hinge zone. Location of cross section shown on Figure 17 (modified from Burnside, 1959).

large combination structural/stratigraphic trap that formed near the edge of a tectonically active cratonic margin.

The distribution of Late Pennsylvanian seismic sequences on the eastern and southern Horseshoe atoll illustrates the results of long-term (10–20 m.y.) retrogradational or transgressive conditions on an offshore carbonate platform. Seismic interpretations suggest the Horseshoe atoll developed as a series of isolated, backstepping, Late Pennsylvanian carbonate bank and reef-mound complexes, consisting of at least six time-stratigraphic depositional facies. All depositional facies except the debris facies are constructional, consisting of relatively shallow-water, coarsening-upward, aggradational to progradational, autochthonous depositional cycles. The debris facies are destructional and allochthonous, resulting in basinward transportation and redistribution of eroded sections of bank-margin and reef-mound carbonates.

The large-scale retrogradational geometry of the Horseshoe atoll is evident by the fact that successive seismic sequences occupy less area than preceding sequences. (Figures 17, 18). Possible reasons for this large-scale geometry are discussed below. The Strawn seismic sequence represents a regional, low-relief platform (ramp to distally steepened ramp) that extends beyond the atoll complex. Compared with the Strawn sequence, each successive Upper Pennsylvanian sequence on the atoll decreases in aerial extent. The Canyon A sequence is limited in occurrence to a minimum of two large carbonate banks that define the general U-shape of the atoll, including an isolated bank in Kent County, and a large, semicircular bank comprising the remaining eastern and south-

ern margins of the atoll. The Canyon A sequence is also present in structurally low mounds or allochthonous blocks off the southern edge of the atoll. The occurrence of the Canyon B sequence is limited to a minimum of four, less-extensive carbonate masses, including a small bank delineating the Salt Creek field, Kent County; a large bank encompassing the Kelly–Snyder/Diamond M field complex and the Von Roeder, Vealmore, East Vealmore, and Hobo fields, Scurry, Borden, and Howard counties; and a large bank including the Oceanic and Good fields in Howard, Borden, and Dawson counties. In addition, seismic data indicate the presence of a small, isolated late-early Canyon bank immediately west of the Kelly–Snyder bank. The Canyon C/Cisco sequence is limited to areas near the crestal portions of the atoll and incorporates a minimum of six bank or reef-mound complexes in the eastern and southern regions. These include a single, small mound in Kent County that constitutes the southwest portion of Salt Creek field; a large Canyon C/Cisco bank in Scurry, Borden, and Howard counties; a series of isolated, small- to medium-size mounds at Oceanic and Good fields in Borden and Howard counties; and a larger, nonproductive mound in southeast Dawson County.

Seismic and geologic data indicate that formation of the Horseshoe atoll began during late-early Strawn time with the deposition of a regionally extensive carbonate bank (ramp to distally steepened ramp) (Figure 19a). On the atoll, this bank constitutes all but the lowermost portions of the Strawn sequence and is composed primarily of phylloid algal and sponge-bryozoan mound complexes and intermound skeletal wackestones/mudstones. In north-central Howard

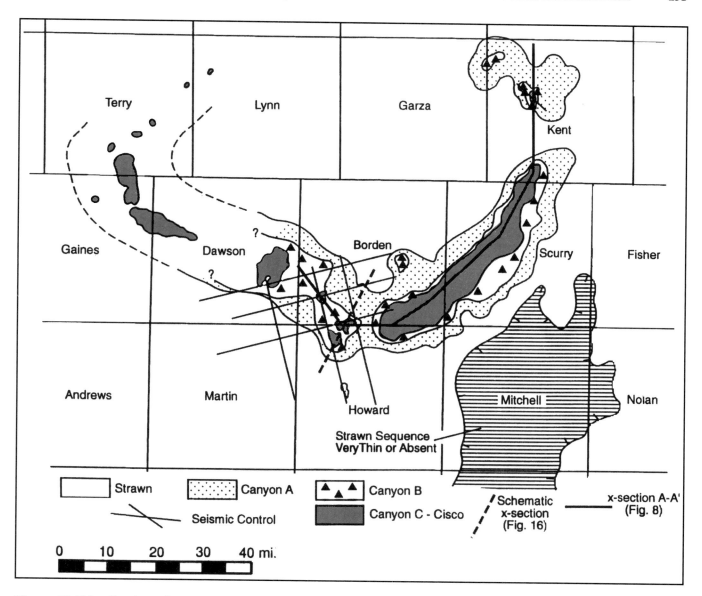

Figure 17. Distribution of Late Pennsylvanian seismic sequences, eastern and southern Horseshoe atoll.

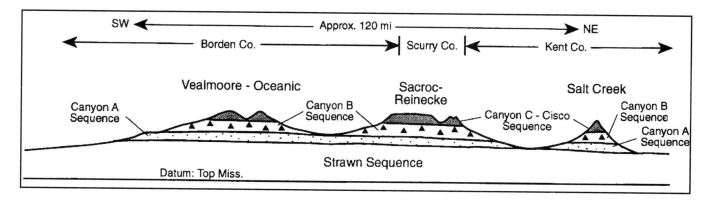

Figure 18. Diagrammatic southwest–northeast stratigraphic cross section across the Horseshoe atoll showing retrogradational arrangement of seismic facies (based on cross section A–A', Figure 8, and Figure 17).

County, and in the southern portion of Salt Creek field in Kent County, the presence of the seismic clinoform facies in the Strawn sequence may indicate a local occurrence of grainstone-rich intervals. Late Strawn grainstones and phylloid algal mounds are present in Cogdell field, Kent and Scurry counties (Reid and Reid, 1991). Facies distinction within the Strawn sequence at Cogdell field and the adjacent Kelly–Snyder complex is beyond the resolution of seismic due to the reduced thickness of the Strawn sequence (Figure 19a).

Seismic data indicate that stratigraphic and morphologic differentiation of the atoll complex began in latest Desmoinesian and early-early Canyon time. At this time, the main portion of the southern and eastern atoll became isolated from the Salt Creek complex and from other carbonate buildups on adjacent shelves (Figure 19b). At least two relatively large carbonate banks developed on the atoll during the early part of early Canyon time, on top of the submerged upper Strawn ramp system. The widespread occurrence of the clinoform facies during the early-early Canyon interval suggests that the banks are primarily composed of southwestward-prograding, oolitic and skeletal packstone/grainstone–bearing units. As mentioned, however, a number of muddy rock types may also occur within this facies. The occurrence of both oolites and phylloid algal mounds has been reported from early-early Canyon deposits at Codgell field (Reid and Reid, 1991) and at Kelly–Snyder field (Schatzinger, 1988).

Morphologic differentiation of the atoll complex continued during middle-early Canyon to early-middle Canyon time, as indicated by the distribution of the Canyon B sequence (Figure 19c). Continued subsidence of the eastern and southern atoll resulted in the deposition of at least three separate banks growing on the older, submerged ramp and bank sequences. The development of these banks represents the onset of Canyon reef-mound growth, composed of at least five shallowing-upward parasequence sets, some of which were terminated with brief periods of subaerial exposure. A variety of shallow- to moderately deep-water carbonate facies containing abundant crinoids and fusulinids comprise the majority of bank sediments. In Howard County, a series of small, in-place, late-early Canyon reefs developed to the south and immediately basinward of the developing bank system (Figure 19c). These lowstand features are located several hundreds of feet below the main bank margin (Reid et al., 1990a,b).

Evolution of the eastern and southern Horseshoe atoll climaxed during late Canyon and early Cisco time with the deposition of the Canyon C/Cisco sequence. During this time, reef-mound complexes were deposited in the southern portion of Salt Creek field, Kent County, and in separate complexes in Scurry, Borden, Howard, and Dawson counties (Figure 19d). Morphologic differentiation within the northeastern portions of large Scurry County bank occurred during this time with the formation of small,

steep-sided, pinnacle-like buildups (Schatzinger, 1988; Reid and Reid, 1991). This overall stage of atoll development is characterized by deposition of a wide variety of carbonate facies, including phylloid algal mounds, sponge-bryozoan reefs, oolitic crinoid-rich skeletal packstones, and carbonate mudstones. In addition, deposition of intraclastic, conglomeratic units associated with severe erosion and mass wasting of the reef-mounds and bank complex occurred repeatedly during late Canyon and early Cisco time. The allochthonous units occupy relatively small, point-sourced regions adjacent to the crests of many reef-mound complexes (Figure 19d).

By late Cisco time, carbonate production had ceased over most of the eastern and southern portions of the atoll (Vest, 1970). Carbonate deposition continued, however, in limited areas on the western atoll throughout the remainder of Pennsylvanian time and into Early Permian time (Figure 20; Myers et al., 1956; Burnside, 1959; Stafford, 1959; Vest, 1970). On the eastern and southern atoll, reef carbonates are unconformably overlain by terrigenous shales of Wolfcamp age. Based on the seismic signature and the known burial history of the east side of the atoll, the top of the carbonate (top "reef") represents a major drowning surface. It is not well known, however, if the top of atoll was subaerially exposed prior to drowning. At Cogdell field in Kent and Scurry counties, the upper reef surface contains fractures and solution-enhanced porosity with cave fill, indicating that the upper reef surface was subaerially exposed at least once before final burial (Reid and Reid, 1991).

DISCUSSION

The documentation of third-order seismic sequences for the Horseshoe atoll is useful not only because it provides an internally consistent framework for stratigraphic analyses, but because it represents an excellent case study of the geometry of a retrogradational sequence on an isolated, offshore carbonate platform. As discussed above, the third-order, Upper Pennsylvanian carbonate sequences of the Horseshoe atoll display a backstepping, or retrogradational, geometry on a regional scale (Figures 17, 18). With time, each successive third-order carbonate sequence occupies less aerial extent than the preceding sequence. A possible explanation for this geometry is suggested by examining the geologic processes that were active within the second- to third-order time frame—global sea-level change and basin subsidence. Many workers now agree that Late Pennsylvanian time was characterized by a second-order (i.e., 10–100 m.y.) relative rise in global sea level beginning in Desmoinesian time (Figure 21a). In addition, reconstructions of simple burial history curves for the Permian basin show the northern Midland basin underwent a period of relatively rapid, mostly uninterrupted subsidence during this time (Figure 21b). The resultant effect of both long-term conditions would result in an overall increase in sediment accommodation space with time. Although local carbonate production most likely kept up with short-term increases in accommodation space, the retrogradational pattern of seismic sequences shows the atoll was successively drowned (even though it was subject to multiple, shorter duration periods of subaerial exposure during the development of individual stratigraphic cycles). A long-term pattern of platform subsidence coupled with a second-order rise in sea level can account for the overall transgressive pattern of third-order seismic sequences of the atoll. In comparison, the interplay of subsidence on a more local scale, coupled with the effects of frequent glacio-eustatic sea level changes, evolving biota, amount of terrigenous clastic influx, and paleogeography, was probably more important in controlling the distribution of progradational, aggradational, and retrogradational carbonate facies within the larger third-order sequences.

Based on the Horseshoe atoll example, it is obvious that the geometry of these types of systems can differ significantly from that of a "standard" or model sequence developed for attached, progradational shelf systems (Figure 22). For example, deposition within the most aerially restricted sequences of isolated, retrogradational platforms (i.e., the youngest reef-mound complexes) may be largely represented by highstand and transgressive system tracts, while off-mound sequences are dominated by lowstand system tracts. In detached, relatively pure carbonate settings like the Horseshoe atoll, lowstand fan deposits may be relatively void of terrigenous clastic material, and be composed of limestone breccias and conglomerates. Also, recent work on Holocene detached platforms (Bahamas) has indicated that a significant amount of carbonate material can be derived from the platform margin and deposited in slope and basinal environments during sea-level highstands (Wilber et al., 1990). In addition, the construction of onlap curves, an important step in assessing the relative sea-level history of a region, may prove difficult during relative highstands in an isolated, detached setting, due to the lack of "coastline" (Eberli and Ginsburg, 1989). These examples point to potentially significant differences in the sequence development of detached, retrogradational carbonate platforms vs. attached, progradational platforms. Such differences should be kept in mind before the application of model concepts in sequence analysis.

It bears emphasizing that a detailed example of the link between internal geometries of parasequences, parasequence sets, and systems tracts to larger sequence packages for retrogradational, isolated carbonate platforms such as the Horseshoe atoll remains to be documented. Sarg (1988) has presented some general observations on a Miocene buildup from Indonesia that are applicable to this geologic setting, and more specific conceptual models for detached carbonate platforms are now emerging (e.g., Handford and Loucks, 1993). The late Paleozoic Horseshoe atoll provides one example where enough

data are available to carry out a more detailed sequence stratigraphic analysis, assessing the linkage between third-, fourth-, and fifth-order carbonate sequences. A complete sequence stratigraphic analysis integrating more rock data into the seismic interpretation is needed to better define the exact nature of the Upper Pennsylvanian sequence boundaries and to appraise the current interpretation of seismic facies. This type of information will not only refine our current knowledge of reservoir/trap relationships to assess the remaining hydrocarbon potential of the atoll, but will yield a better understanding of the number, magnitude, and correlation of Late Pennsylvanian sea-level changes, and the nature of large, retrogradational carbonate systems in general.

Irrespective of more detailed analyses, the knowledge of third-order seismic sequence geometries of the Horseshoe atoll can be incorporated into a depositional model that can assist regional exploration efforts. It is apparent that the pattern of Late Pennsylvanian seismic sequences and facies documented for the Horseshoe atoll extends to adjacent regions. For example, the eastern shelf of the Midland basin, which is directly east of the atoll, contains a number of Upper Pennsylvanian reef mounds that are nearly identical to those of the atoll (Figure 23). Although slightly smaller in physical scale, the Upper Pennsylvanian buildups of the eastern shelf share many geologic and producing characteristics with those of the atoll. Analyses of several buildups on the eastern shelf show they are composed of a similar arrangement of seismic sequences and facies, emphasizing the basic pattern seen on the atoll (Figure 23). Reef mounds, including Ocho Juan, Claytonville, Rowan & Hope, Nena Lucia, and Jameson/Millican/IAB complex, all occur as isolated mounds developed near the edge of the lower Strawn ramp margin. They are composed of Strawn, lower Canyon, and middle-upper Canyon seismic sequences that display retrogradational geometries similar to those of the atoll. They also contain a heterogeneous assemblage of depositional facies similar to those described for the atoll, including crinoid-fusulinid wackestones, phylloid algal packstones, skeletal and oolitic grainstones, and sponge-bryozoan boundstones. The timing and development of these reef mounds are coeval with those of the atoll. Development of most of the downdip, isolated buildups on the eastern shelf was terminated by early Cisco time, although a series of younger, isolated carbonate bank systems continued to develop in updip regions on the adjacent Concho platform (Figure 23).

Based on these similarities, it is interesting to speculate that the pattern of sequence deposition documented for the Horseshoe atoll and eastern shelf may apply to other sequences of Late Pennsylvanian–Early Permian age (upper portion of the Absaroka I subsequence). This pattern, if seen on other Late Pennsylvanian and Early Permian carbonate shelves, could be attributed to the unique and specific set of tectonic, climatic, oceanographic, and biologic condi-

234   Waite

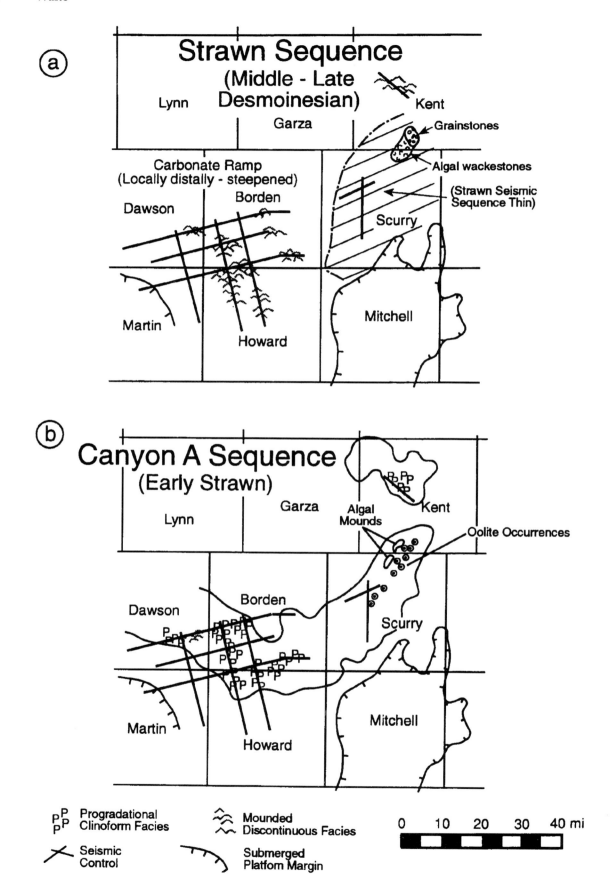

Figure 19. Distribution of seismic sequences: (a) Strawn sequence; (b) Canyon A sequence (oolite occurrences from Schatzinger, 1988).

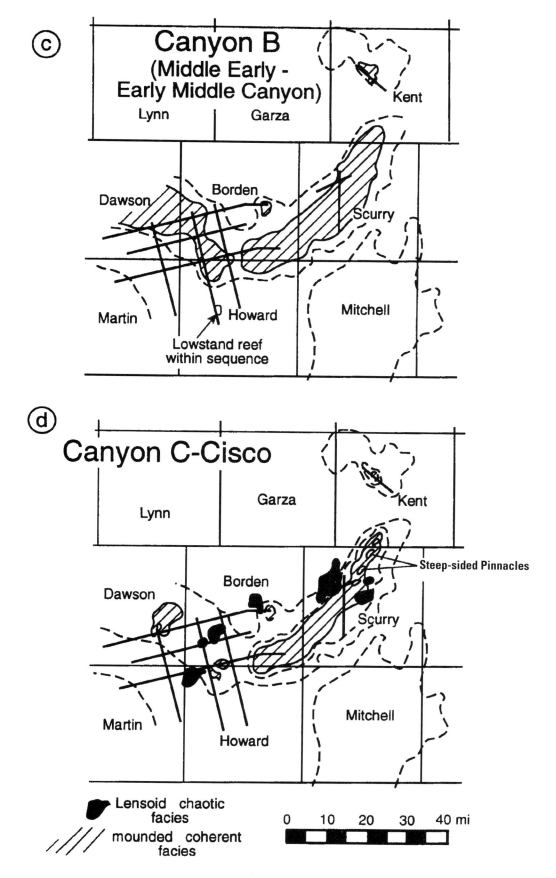

Figure 19 (continued). Distribution of seismic sequences: (c) Canyon B sequence; (d) Canyon C/Cisco sequences.

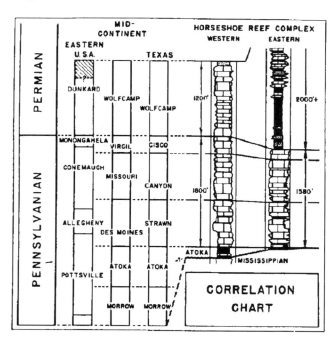

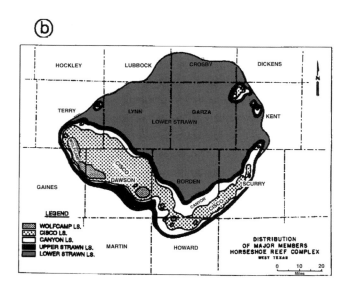

Figure 20. (a) General stratigraphy, eastern vs. western atoll (from Vest, 1970); (b) schematic distribution of major time-rock units of the Horseshoe atoll (modified from Vest, 1970).

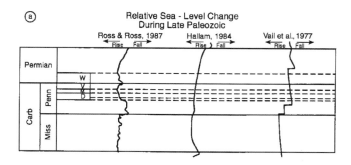

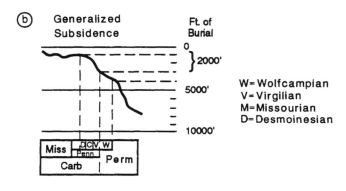

Figure 21. (a) Global, long-term (third-order) relative sea-level changes for the late Paleozoic; (b) generalized subsidence curve for Late Pennsylvanian units in Scurry County (not corrected for compaction).

sylvanian and Early Permian hydrocarbon reserves in west Texas, but will provide important analogs that can be applied in the search for other late Paleozoic carbonate reservoirs.

CONCLUSIONS

1. Late Pennsylvanian carbonates of the eastern and southern Horseshoe atoll are organized into four, biostratigraphically constrained, third-order (1–10 m.y.) seismic sequences, including Strawn; Canyon A (early-early Canyon); Canyon B (middle-early Canyon–early-middle Canyon); and Canyon C/Cisco (late Canyon–early Cisco) sequences. Additional third-order sequences may be present within the Strawn and Cisco sections, but cannot be resolved seismically in the study area.

2. The exact nature of the Late Pennsylvanian sequence boundaries is equivocal and requires further integration of core and sample data. On seismic, the Strawn and Canyon A sequences appear conformable with overlying sequences, and their respective sequence boundaries are consistent with type 2 (subaqueous) surfaces. Evidence for relatively short-lived, glacially induced sea-level lowerings resulting in subaerial exposure and erosion may be beyond the resolution of current seismic. The top Canyon B sequence boundary shows widespread evidence of erosion on seismic, associated with mass wasting of

tions that characterize this particular portion of geologic time. Certainly local geologic conditions would add variations to the general theme, but the basic geometries of third-order stratigraphic sequences documented for the atoll may be developed in most low-latitude shelf regions of Pangea. If so, the seismic geometries documented for the Horseshoe atoll will not only help to exploit remaining Late Penn-

the platform margins. The Canyon B sequence boundary may have been generated during long-term subaerial exposure and formation of a type 1 (subaerial) surface. Where present, the Canyon C/Cisco sequence boundary represents the final stages of reef development and may also represent a type 1 boundary, unconformably overlain by onlapping terrigenous clastics of Early Permian age.

3. Each seismic sequence is characterized by one or more unique, nonparallel seismic facies. The Strawn sequence is marked by a mounded, discontinuous facies, interpreted to represent amalgamated phylloid algal and sponge-bryozoan mud mounds. The Canyon A sequence is characterized by the presence of distinct, subparallel, shingled clinoforms, interpreted to represent basinward-prograding sedimentary packages containing skeletal and oolitic grainstones. The Canyon C/Cisco sequence is denoted by the occurrence of aerially restricted, mounded coherent reflectors interpreted to represent heterogeneous reef-mound buildups. In addition, a lensoid chaotic facies of post–early Canyon age occurs adjacent to many of the large reef-mound complexes or eroded

platform margins. These facies are interpreted to represent allochthonous debris sediments associated with mass-wasting of the platform margin.

4. The combination of third-order Late Pennsylvanian seismic sequences and facies illustrates the large-scale stratigraphic architecture of the eastern and southern Horseshoe atoll. The seismic stratigraphic sequence of the atoll can be described as a southwest-thickening mass of carbonate platform, bank, and reef-mound complexes. The Strawn sequence, representing a broad carbonate platform of low relief, has the largest aerial distribution. Successive sequences display an evolution from platform to bank to reef mound, decreasing in aerial extent through time.

5. The three-dimensional arrangement of seismic sequences and facies of the Horseshoe atoll is consistent with deposition on an isolated, offshore carbonate platform experiencing long-term retrogradation. The physical distribution of third-order seismic sequences was most likely controlled by differential rates of tectonic subsidence along the southern margin of the North American craton, coupled with the

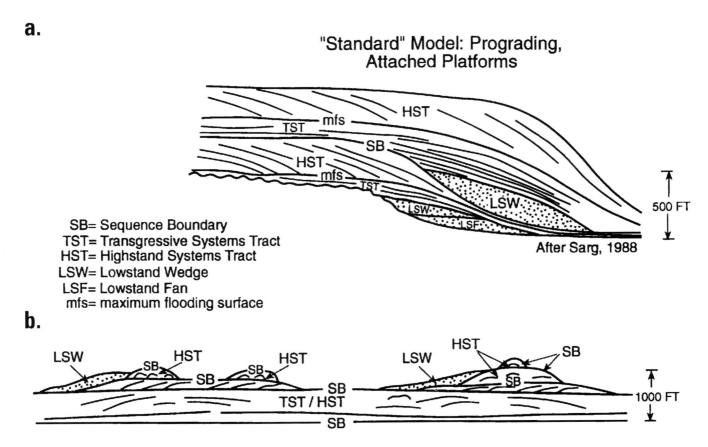

Figure 22. Comparison of sequence-stratigraphic models: (a) "standard" model, based on study of prograding, attached carbonate platforms (after Sarg, 1988); (b) model for Horseshoe atoll, a retrogradational, detached carbonate platform.

long-term relative rise in Late Pennsylvanian eustatic sea level. In contrast, the distribution of smaller scale parasequence sets and parasequences that comprise the third-order seismic sequences may have been more controlled by factors relating to frequent glacio-eustatic sea-level changes, local subsidence, Late Pennsylvanian biotic evolution, and paleogeography.

6. The large-scale sequence-stratigraphic geometries of detached, retrogradational carbonate platforms like the Horseshoe atoll, lacking significant terrigenous clastic input, suggest that the nature and arrangement of systems tracts will differ from the "standard" sequence model based on attached, progradational systems. Further study of the linkage between large-scale, third-order sequences and constituent parasequence sets and parasequences will help elucidate these differences.

7. The pattern of third-order, seismic-sequence geometries and facies documented for the Horseshoe atoll extends to at least one adjacent late Paleozoic shelf region. The documentation of third-order seis-

mic sequences for the Horseshoe atoll not only provides an internally consistent framework for more detailed analyses of smaller scale depositional cycles and associated reservoir/trap/source relationships on the atoll, but may represent a primary depositional pattern inherent to low-latitude carbonate systems of Late Carboniferous Pangea. If so, the Horseshoe atoll is a useful exploration analog that can be applied in geographically separated, yet time-equivalent regions.

ACKNOWLEDGMENTS

The author wishes to thank R.B. Koepnick, J.R. Markello, and L.J. Weber, who read early versions of the chapter and greatly improved the final text, and to R. Loucks, who reviewed the manuscript. The author benefited from discussions with the following people: D.A. Walker, J.M. Jensen, A.W. Small, R.B. Koepnick, J.R. Markello, L.J. Weber, J.F. Sarg, and A.M. Reid. A.L. Hadik provided log and sample data from Scurry County. M.K. Lindsey drafted the fig-

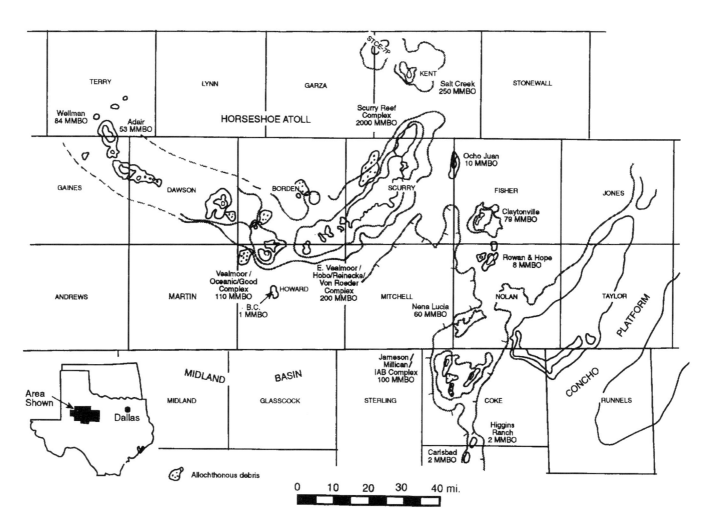

Figure 23. Distribution of Late Pennsylvanian carbonate sequences, Horseshoe atoll, and Eastern Shelf, west Texas. Production figures represent cumulative oil produced. (MMBO = million barrels of oil.) (Compiled from numerous sources.)

ures. The author is grateful to Mobil Oil Corporation for permission to publish.

REFERENCES CITED

Anderson, K. C., 1953, Wellman field, Terry County, Texas: AAPG Bulletin, v. 37, p. 509-521.

Boardman, D. R., II, and J. E. Barrick, 1989, Glacial-eustatic control of faunal distribution in Late Pennsylvanian strata of the midcontinent—implications for biostratigraphy and chronostratigraphy, *in* Franseen, E. K., and W. L. Watney, eds., Sedimentary Modeling: Computer Simulation of Depositional Systems: Kansas Geological Survey, Subsurface Geology Series 12, p. 79-81.

Boardman, D. R., II, and P. H. Heckel, 1989, Glacial-eustatic sea level curve for early Late Pennsylvanian sequence in north-central Texas and biostratigraphic correlation with curve for midcontinent North America: Geology, v. 17, p. 802-805.

Brown, L. F., Jr., R. F. Solis-Iriarte, and D. A. Johns, 1990, Regional depositional systems tracts, paleogeography, and sequence stratigraphy, Upper Pennsylvanian and Lower Permian strata, north- and west-central Texas: University of Texas at Austin, Bureau of Economic Geology, Report of Investigations 197, 116 p.

Burnside, R. J., 1959, Geology of part of the Horseshoe atoll in Borden and Howard counties, Texas: U.S. Geological Survey, Professional Paper 315-B, 34 p.

Crawford, G. A., G. E. Moore, and W. Simpson, 1984, Depositional and diagenetic controls on reservoir development in a Pennsylvanian phylloid algal complex: Reinecke field, Horseshoe atoll, west Texas: AAPG, Southwest Section Transactions, p. 81-90.

Eberli, G. P., and R. N. Ginsberg, 1989, Cenozoic progradation of northwestern Great Bahama Bank, a record of lateral platform growth and sea level fluctuations, *in* Wilgus, C. K., B. S. Hastings, C. G. St. C. Kendall, H. W. Posamentier, C. A. Ross, and J. C. Van Wagoner, eds., Sea Level Changes: An Integrated Approach: SEPM Special Publication No. 44, p. 339-351.

Galloway, W. E., T. E. Ewing, C. M. Garret, N. Tyler, and D. G. Bebout, 1983, Atlas of major Texas oil reservoirs: University of Texas at Austin, Bureau of Economic Geology, 139 p.

Hallam, 1984, Pre-quanternary sea-level changes: Annual Review Earth and Planetary Sciences, v. 12, p. 205–243.

Handford, C. R., and R. G. Loucks, 1993, Carbonate depositional sequences and systems tracts—responses of carbonate platforms to relative sea-level changes, this volume.

Haq, B. U., J. Hardenbol, and P. K. Vail, 1988, Mesozoic and Cenozoic chronostratigraphy and cycles of sea level change, *in* Wilgus, C. K., B. S. Hastings, C. G. St. C. Kendall, H. W. Posamentier, C. A. Ross, and J. C. Van Wagoner, eds., Sea Level Changes: An Integrated Approach: SEPM Special Publication No. 44, p. 71-108.

Heck, W. A., K. A. Yenne, and L. G. Henbest, 1952, Boundary of the Pennsylvanian and Permian(?) in the subsurface Scurry reef, Scurry County, Texas: University of Texas at Austin, Bureau of Economic Geology, Report of Investigations No. 13, 17 p.

Hollingsworth, R. V., 1955, Evolution of the Fusulinidae: unpublished report, Paleontological Laboratory, Midland, Texas, 19 p.

Jensen, J. M., and D. A. Walker, 1990, Transgressive-regressive sequences in the Canyon Formation (Missourian) at the Salt Creek field, Kent County, Texas, correlate to worldwide depositional events (abstract): AAPG Bulletin, v. 74, p. 685.

Marquis, S. A., Jr., and R. L. Laury, 1989, Glacio-eustasy, depositional environments, diagenesis, and reservoir character of Goen Limestone cyclothem (Desmoinesian), Concho Platform, central Texas: AAPG Bulletin, v. 73, p. 166-181.

Mazzullo, S. J., and A. M. Reid, 1988, Stratigraphic architecture of Pennsylvanian and Lower Permian facies, northern Midland basin, Texas, *in* Cunningham, B. K., ed., Permian and Pennsylvanian Stratigraphy, Midland Basin, West Texas: Studies to Aid Hydrocarbon Exploration: SEPM, Permian Basin Section, Publication No. 88-28, p. 1-6.

Miall, A. D., 1990, Principles of Sedimentary Basin Analysis (second edition): Springer-Verlag, New York, 668 p.

Mitchum, R. M., Jr., P. R. Vail, and S. Thompson, III, 1977, Seismic stratigraphy and global changes of sea level, part 2: The depositional sequence as a basic unit for stratigraphic analysis, *in* Payton, C. E., ed., Seismic Stratigraphy—Applications to Hydrocarbon Exploration: AAPG Memoir 26, p. 53-62.

Myers, D. A., P. T. Stafford, and R. J. Burnside, 1956, Geology of the late Paleozoic Horseshoe atoll in west Texas: University of Texas at Austin, Bureau of Economic Geology Publication No. 5607, 113 p.

Reid, A. M., D. C. Mozynski, and W. C. Robinson, 1990a, B.C. Canyon field: A low sea level stand, Early Canyon buildup, *in* Flis, J. E., and R. C. Price eds., Permian Basin Oil and Gas Fields: Innovative Ideas in Exploration and Development: West Texas Geological Society, Publication No. 90-87, p. 119-128.

Reid, A. M., S. T. Reid, and S. J. Mazzullo, 1990b, Lowstand carbonate reservoirs: Upper Pennsylvanian sea level changes and reservoir development adjoining the Horseshoe atoll (abstract): AAPG Bulletin, v. 74, p. 221.

Reid, A. M., S. T. Reid, S. J. Mazzullo, and S. T. Robbins, 1988, Revised fusulinid biostratigraphic zonation and depositional sequence correlation, subsurface Permian basin (abstract): AAPG Bulletin, v. 72, p. 102.

Reid, A. M., and S. A. Tomlinson Reid, 1991, The Cogdell field study, Kent and Scurry counties, Texas: A post-mortem, *in* Candelaria, M. P., ed.,

Permian Basin Plays—Tomorrow's Technology Today: SEPM, Permian Basin Section, Publication No. 91-89, p. 39-66.

Ross, C. A., and J. R. P. Ross, 1987, Late Paleozoic sea levels and depositional sequences, *in* Ross, C. A. and D. Haman, eds., Timing and Depositional History of Eustatic Sequences: Constraints on Seismic Stratigraphy: Cushman Foundation for Foraminiferal Research, Special Publication No. 24, p. 137-149.

Rothrock, H. E., R. E. Bergenback, D. A. Myers, P. A. Stafford, and R. T. Terriere, 1953, Preliminary report on the geology of the Scurry reef in Scurry County, Texas: U.S. Geological Survey, Oil and Gas Investigations Map OM-143.

Sarg, J. F., 1988, Carbonate sequence stratigraphy, *in* Wilgus, C. K., B. S. Hastings, C. G. St. C. Kendall, H. W. Posamentier, C. A. Ross, and J. C. Van Wagoner, eds., Sea Level Changes: An Integrated Approach: SEPM Special Publication No. 44, p. 155-181.

Schatzinger, R. A., 1988, Changes in facies and depositional environments along and across the trend of the Horseshoe atoll, Scurry and Kent counties, Texas, *in* Cunningham, B. K., ed., Permian and Pennsylvanian Stratigraphy, Midland Basin, West Texas: Studies to Aid Hydrocarbon Exploration: SEPM, Permian Basin Section, Publication No. 88-28, p. 79-107.

Sloss, L. L., 1988, Tectonic evolution of the craton in Phanerozoic time, *in* Sloss, L. L., ed., Sedimentary cover—North American craton, U.S.: Geologic Society of America, D-2, p. 25-51.

Stafford, P. T., 1959, Geology of part of the Horseshoe atoll in Scurry and Kent counties, Texas: U.S. Geological Survey, Professional Paper 315-A, 18 p.

Toomey, D. F., and D. Winland, 1973, Rock and biotic facies associated with Middle Pennsylvanian (Desmoinesian) algal buildup, Nena Lucia field, Nolan County, Texas: AAPG Bulletin, v. 57, p. 1053-1074.

Vail, P. R., and R. M. Mitchum, Jr., 1977, Seismic stratigraphy and global changes of sea level; part 1: Overview, *in* Payton, C. E., ed., Seismic Stratigraphy—Applications to Hydrocarbon Exploration: AAPG Memoir 26, p. 51-52.

Vail, P. R., R. M. Mitchum, Jr., and S. Thompson, III, 1977, Seismic stratigraphy and global changes of sea level, part 4: Global cycles of relative changes in sea level, *in* Payton, C. E., ed., Seismic Stratigraphy—Applications to Hydrocarbon Exploration: AAPG Memoir 26, p. 83–97.

Van Wagoner, J. C., R. M. Mitchum, K. M. Campion, and V. D. Rahmanian, 1990, Siliciclastic sequence stratigraphy in well logs, cores, and outcrops: Concepts for high-resolution correlation of time and facies: AAPG Methods in Exploration Series, No. 7, 55 p.

Vest, E. L., Jr., 1970, Oil fields of Pennsylvanian-Permian Horseshoe atoll, west Texas, *in* Halbouty, M. T., ed., Geology of Giant Petroleum Fields: AAPG Memoir 14, p. 185-203.

Walker, D. A., J. Golonka, A. M. Reid, and S. A Tomlinson Reid, 1991, The effects of late Paleozoic paleogeography on carbonate sedimentation in the Midland basin, west Texas, *in* Candelaria, M. P., ed., Permian Basin Plays—Tomorrow's Technology Today: SEPM, Permian Basin Section, Publication No. 91-89, p. 141-162.

Walker, D. A., J. M. Jensen, S. P. Zody, and S. T. Reid, 1990, Pennsylvanian cycle stratigraphy and carbonate facies control of reservoir development in the Salt Creek field, Kent County, Texas, *in* Flis, J. E. and R. C. Price, eds., Permian Basin Oil and Gas Fields: Innovative Ideas in Exploration and Development: West Texas Geological Society, Publication No. 90-87, p. 107-112.

West, R. R., 1988, Temporal changes in Carboniferous reef mound communities: Palaios, v. 3, p. 152-169.

Wilber, R. J., J. D. Milliman, and R. B. Halley, 1990, Accumulation of bank-top sediment on the western slope of Great Bahama Bank: Rapid progradation of a carbonate megabank: Geology, v. 18, p. 970-974.

Wilde, G. L., 1990, Practical fusulinid zonation: The species concept; with Permian basin emphasis: West Texas Geological Society Bulletin, v. 29, no. 7, p. 5-34.

Wilson, J. L., 1975, Carbonate Facies in Geologic History: Springer-Verlag, New York, 471 p.

Witzke, B. J., 1990, Palaeoclimatic constraints for Palaeozoic palaeolatitudes of Laurentia and Euramerica, *in* McKerrow, W. S., and C. R. Scotese, eds., Palaeozoic Palaeogeography and Biogeography: Geological Society of London, Memoir 12, p. 57-73.

Chapter 9

Response of Carbonate Platform Margins to Drowning: Evidence of Environmental Collapse

R. N. Erlich
A. P. Longo, Jr.
Amoco Production Company
Houston, Texas, U.S.A.

S. Hyare
GeoCenter, Incorporated
Houston, Texas, U.S.A.

ABSTRACT

The morphology of platform margins as depicted on high-resolution seismic data and the development of outer shelf, eutrophic ecologic assemblages can be used as keys in understanding the evolution of carbonate platforms. The Baltimore Canyon platform, offshore United States East Coast, and the Liuhua platform, Pearl River Mouth Basin, People's Republic of China, exhibited the effects of environmental collapse prior to their extinction by drowning.

Reprocessed seismic data, lithologic data, and biostratigraphic data from the Baltimore Canyon area show that environmental deterioration of the platform immediately preceded or was coincident with deltaic progradation. This implies that slope-front fill seaward of the platform is probably coeval with platform deposits, and that a previously identified carbonate sequence boundary may actually be an older drowning sequence. A seismic sequence boundary should be placed at the top of the youngest drowning sequence.

Late-growth reefs appear to be discontinuous along both the Baltimore Canyon and Liuhua platform margins. The proximity of prograding deltas appears to be the main control on the location of late-growth reefs on both platforms, though tectonic subsidence (local faulting) may govern their distribution along the southern part of the Liuhua platform margin.

The horizontal–planar onlap of basinal shales onto the Liuhua platform margin could be misinterpreted as always representative of an unconformable contact with the platform sequence. In reality, local differences in highstand off-bank transport of platform and platform margin sediments may have produced a progradational fore-slope (a wedge of fore-slope debris) in some areas, and only horizontal–planar onlap in others.

INTRODUCTION

Since the modern application of stratigraphic principles to seismic data was first widely popularized (see Vail et al., 1977), geoscientists have increasingly focused their investigations of carbonate platforms on the response of platforms to sea-level fluctuations. Much emphasis has been placed on the importance of platform geometry and seismic reflector character as keys to understanding platform evolution (for example Schlager and Camber, 1986; Sarg, 1988; Schlager, 1989; Erlich et al., 1990), though a consensus interpretation of these variables has not been reached.

The controversy over which interpretation is correct has produced the positive side effect of more detailed and integrated studies. One of the most important concepts to emerge as a result of this work is the concept of drowning of carbonate platforms (Schlager, 1981). The realization that reefs and carbonate platforms could drown without undergoing subaerial exposure caused many workers to reexamine previously studied examples, and more carefully examine others (see Hine and Steinmetz, 1984; Dominguez et al., 1988; Grigg, 1988; Whalen, 1988; Meyer, 1989; Rudolph and Lehmann, 1989; Glaser and Droxler, 1991; Ludwig et al., 1991).

Recent studies have also gone beyond the general description of seismic reflectors and lithofacies of drowned platforms and focused on the environmental factors associated with platform drowning (Hallock and Schlager, 1986; Hallock, 1988; Hallock et al., 1988; Erlich et al., 1991). The recognition that the environmental consequences of platform drowning can very often yield a predictable lithologic–ecologic sequence (termed "drowning sequence" by Erlich et al., 1990) on platforms of different ages is an important clue in interpreting the growth history of platforms. It is also important to note that drowning sequences are defined based on lithologic and biologic parameters, as well as their seismic response, and are therefore different from seismic sequences as defined by Van Wagoner et al. (1987).

In this study, a link is forged between the lithologic and seismic evidence of platform drowning and the environmental factors that produced it. Previously unpublished seismic data from the Late Jurassic to Early Cretaceous Baltimore Canyon platform, United States East Coast, and the early Miocene Liuhua platform of the Pearl River Mouth Basin, offshore People's Republic of China, are combined with detailed lithologic data to illustrate the effects of changing environmental conditions on platform growth and margin morphology.

Seismic Data Processing

Seismic data used in this study were acquired in the Baltimore Canyon area by Grant Norpac for Amoco in 1983. Seismic data in the Pearl River Mouth Basin were acquired by Geco for Amoco in

1986–1987. Generally, seismic data from the Pearl River Mouth Basin were converted to zero-phase wavelet data and displayed in a relative amplitude format. These data were reprocessed by Amoco. Synthetic seismograms were used to tie lithologic and biostratigraphic data from both areas to seismic.

Baltimore Canyon seismic data were recently reprocessed by GeoCenter, Inc., for Amoco, with the objective of providing higher frequency content to obtain better stratigraphic information. Velocity analyses were done in shelf-margin areas on a one mile spacing so a more accurate model of the platform–slope–basin (carbonate to shale) transition could be obtained. F-K multiple attenuation was followed by a normal move out correction with trace muting. A 60-fold CDP stack was combined with time variant filtering and scaling to yield unmigrated sections. F-K migration was then applied and the data were plotted according to Amoco internal parameters. The resulting frequency content was improved from 65 Hz to 100 Hz in shallow shelf areas (to 1000 ms) and from 40 Hz to 60 Hz in deep water areas (to 4500 ms).

A detailed discussion of the original processing parameters of the Baltimore Canyon and Pearl River Mouth datasets can be found in Erlich et al. (1988, 1990).

BALTIMORE CANYON PLATFORM

The Baltimore Canyon platform is part of an extensive platform system that parallels the East Coast of North America from Florida to the Scotian Shelf of Canada, ranging in age from Early Jurassic to Early Cretaceous (Figure 1). Three wells drilled by Shell, Amoco, and Sun during 1983–1984 penetrated the platform without finding significant signs of hydrocarbons. Although commercially unsuccessful, lithologic and seismic data from these wells have yielded important new information regarding the evolution of carbonate platforms on subsiding continental margins.

The first of the wells is located in Block 587 and is less than 2 km (1.2 mi) from the platform margin. The Block 587 well drilled a conformable sequence of Tithonian (Late Jurassic) to Berriasian (Early Cretaceous) stromatoporoid–sponge packstones and grainstones, interpreted as back-reef and platform interior sediments by Meyer (1989), Erlich et al. (1990), and Prather (1991). These sediments are conformably overlain by a drowning sequence consisting of fine-grained, tubiphytes–sponge wackestones and outer shelf shales of Berriasian age (Figure 2A).

The second well, drilled in Block 586, penetrated a mixed carbonate–siliciclastic section in a platform interior setting, and was not part of this study (see Prather, 1991 for more information). The third well, in Block 372, drilled a late-growth shelf-margin reef, which was extensively cored (Figures 1, 2B). The Block 372 well drilled a sequence of lower Aptian

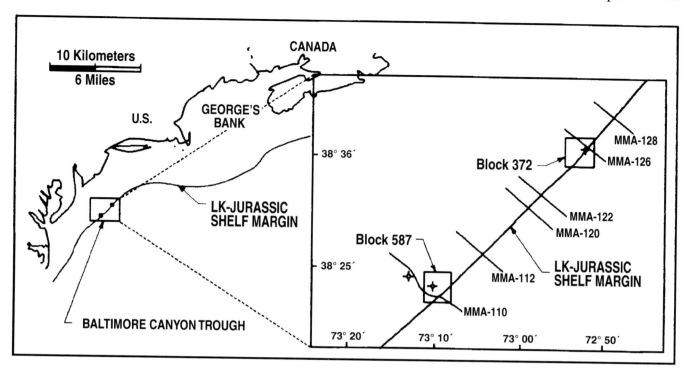

Figure 1. Location map for seismic and wells, Baltimore Canyon area.

(Early Cretaceous) high-energy, stromatoporoid–sponge–coral boundstones and grainstones, which are conformably overlain by lower to middle Aptian low-energy, open shelf, glauconitic–algal wacke-stones and siltstones. A detailed description of the lithofacies of the Block 587 and 372 wells can be found in Meyer (1989), Erlich et al. (1990, 1991), and Prather (1991).

The late-growth reef drilled by the Block 372 well was interpreted by Erlich et al. (1988, 1990, 1991) as a constructional feature and a common product of the drowning sequence of the platform. Lithologic data show that the shelf-margin boundstones and grain-stones underwent extensive submarine cementation, and show no evidence of subaerial exposure prior to drowning of the platform (Figure 3). The high-energy deposits show only a gradual deepening and fining-upwards sequence, composed of algal intraclasts and rhodoliths, glauconite pellets, planktonic fora-minifera, and quartz silts (Figure 4).

Seismic data from the platform-margin area show that late-growth reefs occur in a discontinuous line along a small part of the margin. For example, seismic line MMA-110 (Figure 5A) crosses the margin at the Block 587 well location, where no late-growth reef is present but the drowning sequence is still visible. The drowning sequence at this location, as in other shelf-margin areas, is noted as a pair of high-amplitude reflectors at the top of the carbonate sequence, which change gradationally into a zone of acoustically transparent, low-amplitude reflectors in the platform inte-

rior. This characteristic change was interpreted by Erlich et al. (1990) to represent a facies change from the dense, fine-grained carbonates of the drowning sequence into outer shelf marine shales of the plat-form interior.

Less than 10 km (6 mi) to the northeast, a late-growth reef can be seen at the margin edge on seismic line MMA-112 (Figure 5B). Other lines in the dataset show that the late-growth reef trend appears to be better developed to the north (seismic lines MMA-120, 122, 126, 128; Figures 6, 7). A geologic interpreta-tion of lines MMA-120 and MMA-126 (Figure 8) suggests that the platform margin may have under-gone at least one prior drowning event during its growth history.

High-amplitude slope to basin reflectors common-ly interpreted as the carbonate sequence boundary (Meyer, 1989; Prather, 1991) are here suggested to represent an earlier, intra-platform drowning event (Figure 8). This set of reflectors is very consistent on each line in the dataset, and can be easily traced into the upper part of the margin section.

The terminal drowning event on the Baltimore Canyon platform occurs at the top of the late-growth reef system, and can be recognized by downlapping of the drowning sequence onto a planar sequence boundary seaward of the margin (Figure 8). The seis-mic data show that the upper drowning sequence actually post-dates much of the slope-front fill, previ-ously interpreted to be younger than the upper drowning sequence (Meyer, 1989; Prather, 1991).

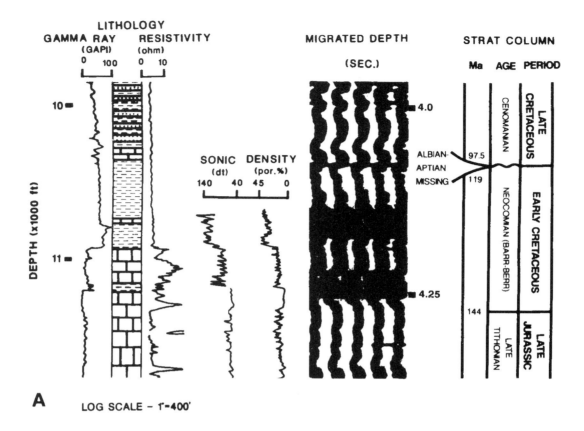

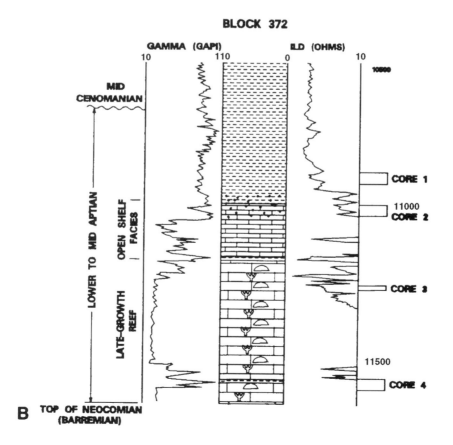

Figure 2. Ages, lithology, wireline logs, and synthetic seismogram from the Block 587 well and the Block 372 well, Baltimore Canyon Trough.

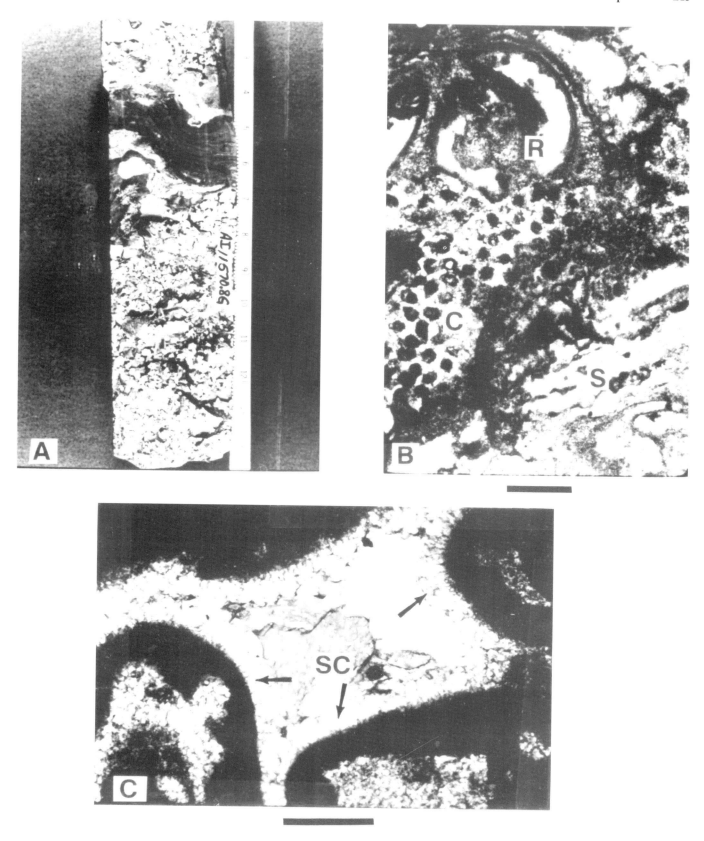

Figure 3. (A) Core 4, Block 372, 11,570.8 ft. Sponge-stromatoporoid boundstone/grainstone. Scale = 12 cm.
(B) Thin-section photo of Figure 3A. C = coral; R = rudist; S = stromatoporoid. Scale bar = 250 microns.
(C) 11,564.5 ft. Thin-section photo of isopachous submarine cement (SC). Scale bar = 125 microns.

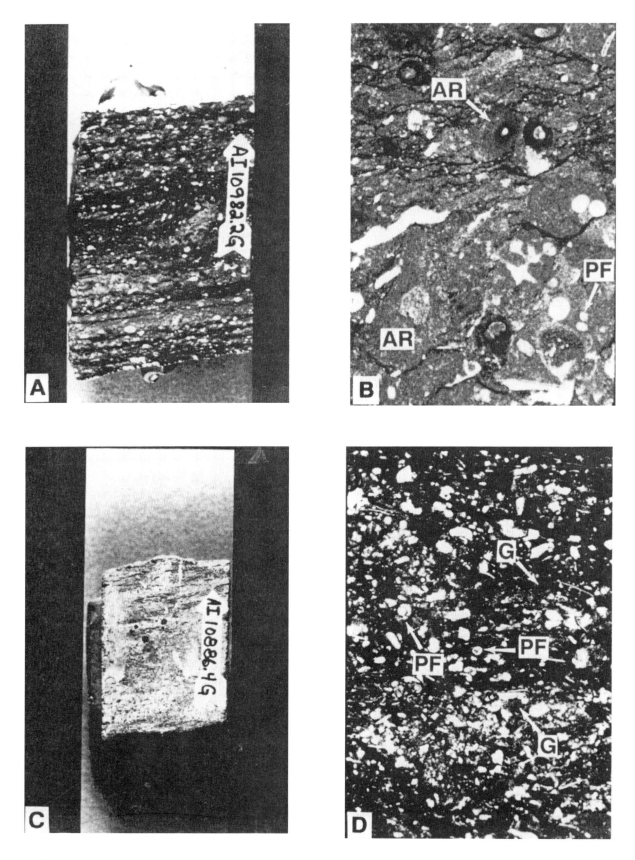

Figure 4. (A) Core 2, Block 372, 10,982.2 ft. Algal-intraclast wackestone. Section shown = 6 cm. (B) 10,981 ft. Thin-section photo of algal-intraclast wackestone. AR = algal rhodolith; PF = planktonic foram. Scale bar = 250 microns. (C) 10,886.4 ft. Glauconitic marly siltstone. Section shown = 6 cm. (D) Thin-section photo of Figure 4C. G = glauconite. Scale bar = 250 microns.

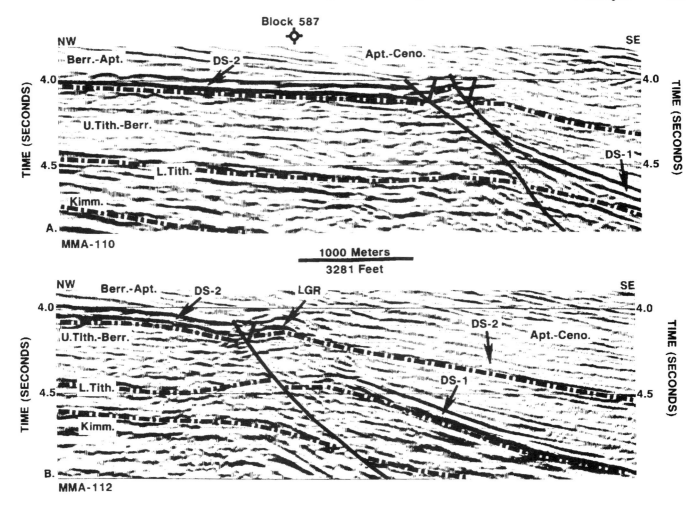

Figure 5. (A) Seismic line MMA-110. (B) Seismic line MMA-112. Vertical scale is in two-way travel time for all seismic sections. Symbol key applies to all Baltimore Canyon seismic lines: DS = drowning sequence; Kimm. = Kimmeridgian; L. Tith. = Lower Tithonian; U. Tith.-Berr. = Upper Tithonian-Berriasian; Berr.-Apt. = Berriasian-Aptian; Apt.-Ceno. = Aptian-Cenomanian; LGR = late-growth reef.

Paleogeography

Paleogeographic maps of the Baltimore Canyon area were constructed using proprietary well and seismic data, and similar published data for platform interior and slope to basinal areas (Libby-French, 1981; Schlee and Hinz, 1987; Prather, 1991). Depositional environments in the Block 587 and 372 well areas at the end of Late Jurassic time were dominated by moderate to high-energy bioclastic carbonate sediments (Figure 9A). This facies graded laterally into platform interior areas dominated by mixed siliciclastic and carbonate sediments. Deltaic sedimentation in the area appears to have occurred 25 km (16 mi) or more to the west of the Block 587 well, and may have been as much as 50 km (31 mi) to the west of the Block 372 well.

Late Berriasian depositional environments reflect the onset of major deltaic progradation, despite a relative rise in sea level (Figure 9B). The maximum landward extent of the drowning sequence appears to have been only about 11 km (7 mi) to the west of the shelf margin, reflecting more than a 50% decrease in the area of carbonate sedimentation. In addition, regional seismic stratigraphic work suggests that drowning sequence carbonates were not widespread north or south of the central Baltimore Canyon area. Fine-grained siliciclastics unconformably overlie Tithonian and Berriasian high-energy carbonates in many areas, or bypassed the then extinct shelf margin and living late-growth reefs through a series of breaks or channels in the margin, to be deposited in slope and basinal areas (Schlee and Hinz, 1987).

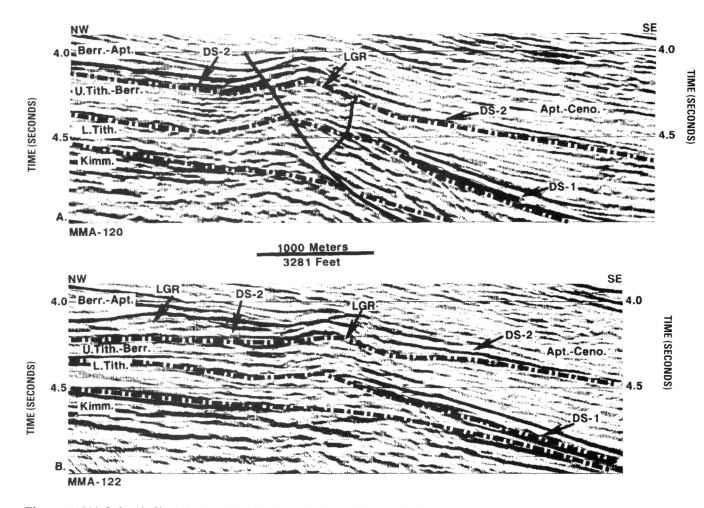

Figure 6. (A) Seismic line MMA-120. (B) Seismic line MMA-122. Note the second late-growth reef near the northwestern end of the line. This feature was found only on this line, and may be a small shelfal buildup that managed to develop in a similar manner as the more common shelf margin buildups.

During late middle Aptian time, delta front and prodelta siliciclastics may have prograded directly over the old Tithonian to Berriasian margin north of the Block 372 well (Figure 9C). Late-growth reef deposition had nearly ceased as prograding shelf sands and muds reached the Block 587 well area. Siliciclastic sediments continued to bypass the margin and were deposited in upper and lower slope settings (Schlee and Hinz, 1987). By early Cenomanian time (early Late Cretaceous), prodelta muds had completely buried the late-growth reefs and any vestige of the exposed Tithonian to Berriasian shelf margin (Figure 9D).

PEARL RIVER MOUTH BASIN

The development of Miocene reefs and carbonate platforms in the South China Sea has been summarized by Hao (1988) and Fulthorpe and Schlanger (1989). Fulthorpe and Schlanger (1989) developed facies and depositional models to explain the distribution of these carbonates in light of their structural

settings. They also postulated the existence of small carbonate buildups and platforms on basement highs within the Pearl River Mouth Basin (Figure 10), though they had little direct evidence.

The existence of a major system of early Miocene carbonate platforms developed on preexisting basement highs was confirmed in 1987 when Amoco and Nanhai East Oil Corporation discovered the giant in-place oil accumulation at Liuhua 11-1 field (Tyrrell and Christian, 1992). The discovery well drilled a sequence of shallow-water, moderate to high energy, lower Miocene limestones, which overlie shallow-marine to nonmarine upper Oligocene sandstones. Conformably overlying the limestones are lower to middle Miocene outer shelf shales. The entire sequence sits unconformably on Paleogene to Cretaceous volcanic basement.

A thin (7 m) drowning sequence conformably overlies the carbonate platform, sealing the oil accumulation at Liuhua 11-1 field, and grades conformably into the overlying shale. Erlich et al. (1990) interpreted the drowning sequence in the Liuhua

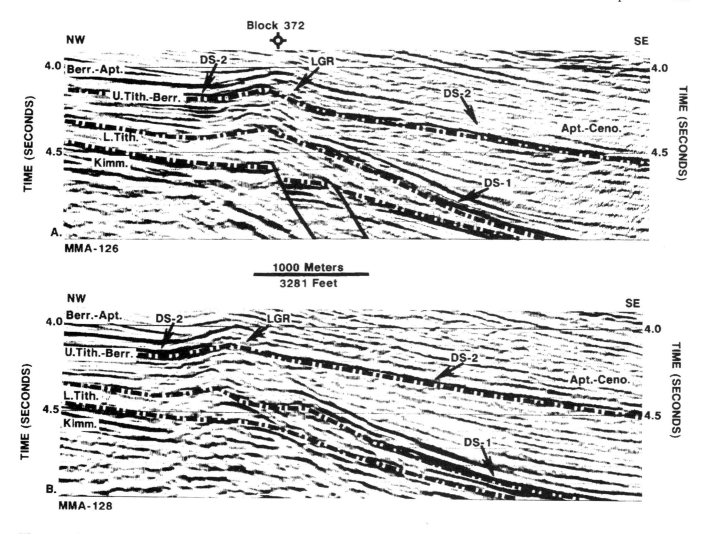

Figure 7. (A) Seismic line MMA-126. (B) Seismic line MMA-128.

platform to represent a condensed sequence, spanning up to 3.3 m.y. within the late early Miocene (Figure 11).

The first well to core the drowning sequence was the LH 11-1-3, drilled in a platform interior setting (Figure 10). The LH 4-1-1 well was the first well to penetrate and core a late-growth reef at the shelf margin (Figure 12). A detailed discussion of the lithofacies of these wells can be found in Erlich et al. (1990).

In general, the lithology of sediments from the LH 4-1-1 well indicates that the late-growth reef is composed predominantly of red algal and hermatypic coral boundstones (Figure 13). Although the top of the drowning sequence was not cored in this well, wireline logs and petrographic observations suggest that the section is much finer-grained toward the top (Figure 12A).

The drowning sequence was completely cored in the LH 11-1-3 well, and the lithologic data clearly illustrate the rapid vertical facies changes associated with platform drowning in the Pearl River Mouth Basin (Figure 12B). Lower parts of the well, within the platform sequence, show that the section is dominat-

ed by red algal packstones and wackestones, with common algal rhodoliths (Figure 14). At the base of the drowning sequence, the section becomes finer-grained and contains more abundant planktonic foraminifera and algal rhodoliths.

Near the top of the drowning sequence, the rhodolith-bearing wackestones grade upwards into glauconitic, shaly limestones–siltstones (Figure 15). A submarine omission surface may be present at the contact between these two units, however, biostratigraphic data are insufficient to resolve any time-rock gap.

Seismic data from the Liuhua platform (Figure 10) show a series of rapidly changing platform margin profiles over a distance of about 30 km (19 mi). The narrow northwestern edge of the platform suggests that the dominant wind and wave energy were from the west–southwest, and that platform limestones grade northeastward into lagoonal–back-reef shales (Figure 16). The platform is generally oriented northwest–southeast.

This pattern remains constant along the entire length of the platform, though some variation in loca-

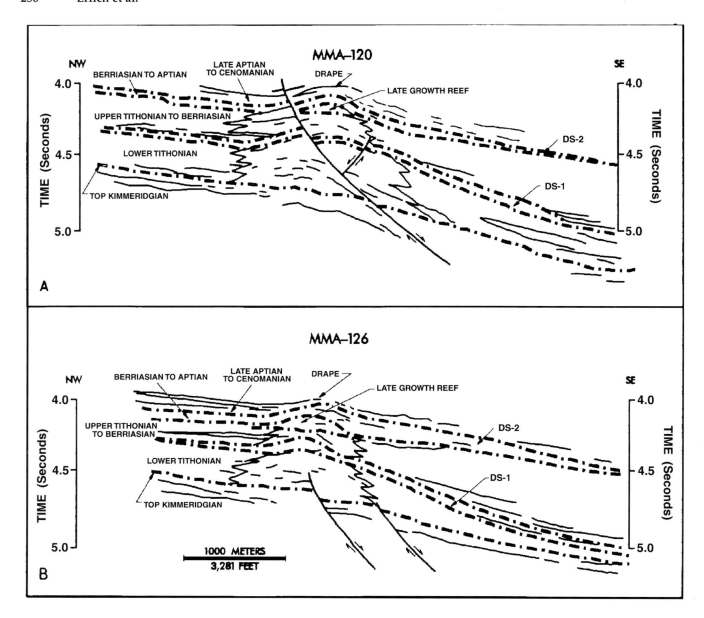

Figure 8. (A) Line tracing and geologic interpretation of MMA-120. (B) Line tracing and geologic interpretation of MMA-126.

tion and morphology of late-growth reefs is apparent (Figures 17–21). Late-growth reef development appears to have been restricted or poorly developed in the northern part of the margin, reaching maximum development at the LH 4-1-1 well (seismic line T14-19; Figure 18A), before disappearing to the south (line T14-31; Figure 20A).

The discontinuous nature of late-growth reefs on the Liuhua platform margin is well illustrated by their presence and absence from seismic lines only 1000 m (3281 ft) apart (lines T14-16 and T14-17, Figure 17; or lines T14-24 and T14-26, Figure 19). South of line T14-26 (Figure 19B), however, the platform margin begins to take on a very different appearance. In the area of line T14-33 (Figure 20B), the margin, fore-

reef slope, and basinal areas show indications of submarine scouring or erosion. The margin also changes orientation abruptly, jutting out to the southwest approximately 2.5 km (1.6 mi) before turning back to a northwest–southeast trend (Figure 10).

Seismic lines T14-41 and T14-43 (Figure 21) show that the southeastern part of the platform margin and the basin were heavily faulted and scoured during and after platform development. No late-growth reefs developed in these southern areas.

Paleogeography

Paleogeographic reconstructions of Liuhua platform lithofacies at the time of maximum develop-

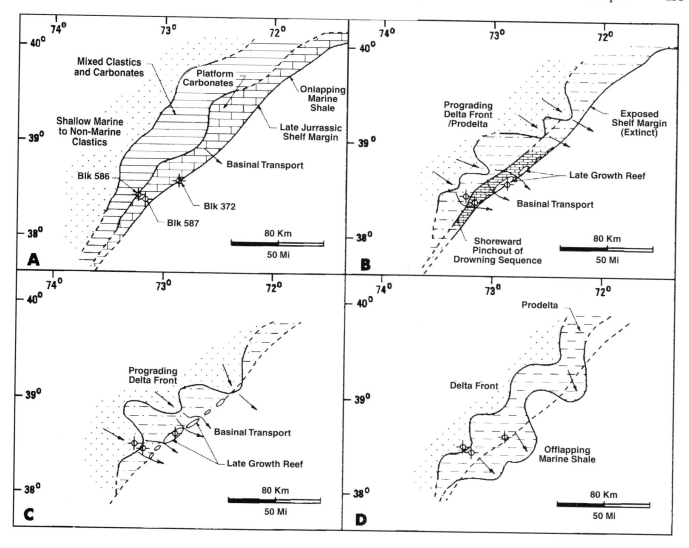

Figure 9. (A) Late Tithonian paleogeography. Note the width of the carbonate platform. (B) Late Berriasian paleogeography, time of maximum development of the drowning sequence above the older Jurassic platform. (C) Late middle Aptian paleogeography. Late-growth reefs were near death as nearby deltaic clastics began to prograde out to the shelf margin. (D) Early Cenomanian paleogeography. Prograding prodelta and delta front deposits had buried the remaining late-growth reefs.

ment of the drowning sequence show a major decrease in platform area (Figure 22) when compared to pre-drowning reconstructions (see Erlich et al., 1990). High-energy carbonate sedimentation probably continued on nearby platforms (Figure 22A) during drowning of the Liuhua platform. However, carbonate deposition in the area was dominated by the fine-grained glauconitic, algal rhodolith wacke-stones of the Liuhua platform. Late-growth reefs dominated only the northwestern part of the platform margin.

Deltaic clastics had prograded to within 20 km (12.5 mi) of the Liuhua platform during maximum development of the late-growth reef system (Figure 22A). Biostratigraphic data suggest that erosional reworking of deltaic deposits was coincident with

extinction of the late-growth reef system, though carbonate sedimentation may have continued until the end of early Miocene time on other nearby platforms (Figure 22B).

DISCUSSION

Why Platforms Drown: Evidence of Environmental Collapse

Seismic data from the Baltimore Canyon and Liuhua platforms show a common characteristic of many drowned carbonate platforms from tectonically passive margins, that of a wedge-shaped cross-sectional profile (Figure 23). Wedge-shaped cross-sectional profiles are a result of two factors: (1)

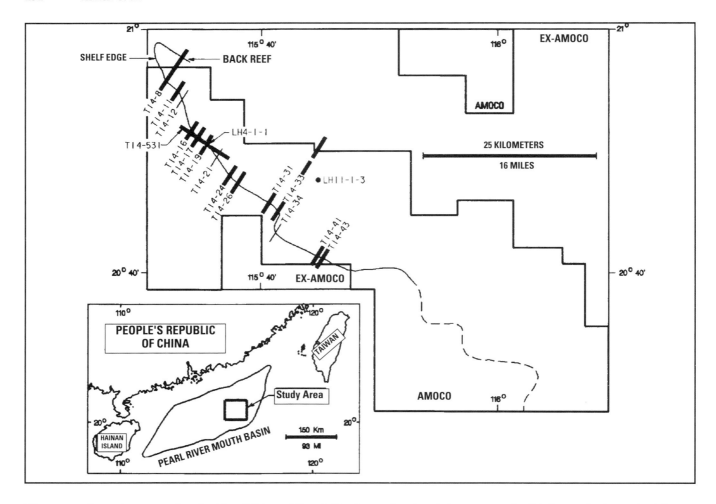

Figure 10. Location map for the Pearl River Mouth Basin, and seismic base map for Liuhua platform. Bold lines are shown in this report.

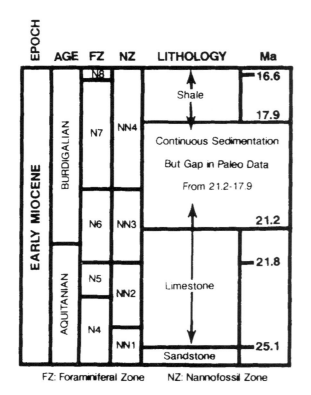

unidirectional (seaward) subsidence; and (2) inherently higher carbonate growth potential of platform margins (Hallock and Schlager, 1986; Erlich et al., 1991).

The development of this type of profile suggests that these platforms were able to keep pace with the combination of subsidence and relative sea-level rise. In order for this to occur, the environmental conditions necessary to allow healthy platforms and platform margins to develop apparently were in place and stable until the onset of drowning. The first evidence for the onset of drowning in both platforms is indicated by the shift from reef building colonial

Figure 11. Biostratigraphy and lithology of the Liuhua platform area. Paleogene–Cretaceous volcanic basement underlies the sandstone.

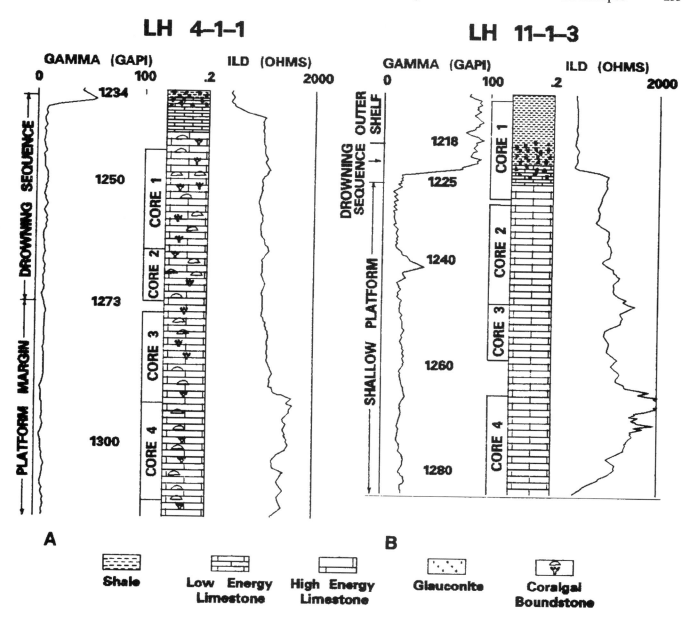

Figure 12. (A) Lithology and wireline logs from the LH 4-1-1 well. (B) Lithology and wireline logs from the LH 11-1-3 well.

organisms (stromatoporoids, sponges, corals) to fine-grained rocks dominated by algae and foraminifera (Figures 3, 4, 13–15).

This lithologic–ecologic change is not coincidental. Hallock (1985, 1988) showed that the change from upward accreting, coral-dominated sediments to sediments composed mostly of large benthic foraminifera and calcareous algae marked a change from mildly oligotrophic (mildly stressed) to mildly eutrophic (highly stressed) environmental conditions. The environmental conditions necessary for reef growth (nutrient level and light availability) shifted from near optimum to highly negative.

The abundance of algal rhodoliths within the drowning sequences of both platforms may also indi-

cate adverse environmental conditions. Hallock et al. (1988) and Peebles et al. (1990) have suggested that algal rhodoliths on isolated carbonate platforms of the Nicaraguan Rise indicate that energy and nutrient levels on those platforms exceed the growth tolerance limits for hermatypic corals. They believe that these platforms are presently being drowned, though Glaser and Droxler (1991) disagree. Hine and Steinmetz (1984), however, found a similar facies relationship on the Cay Sal Bank, Bahamas, and concluded that platform was partially drowned.

The rapid shift from reef-forming ecologic assemblages in platform margin facies of the Block 372 and LH 4-1-1 wells to environmentally stressed drowning sequence assemblages therefore suggests that nutri-

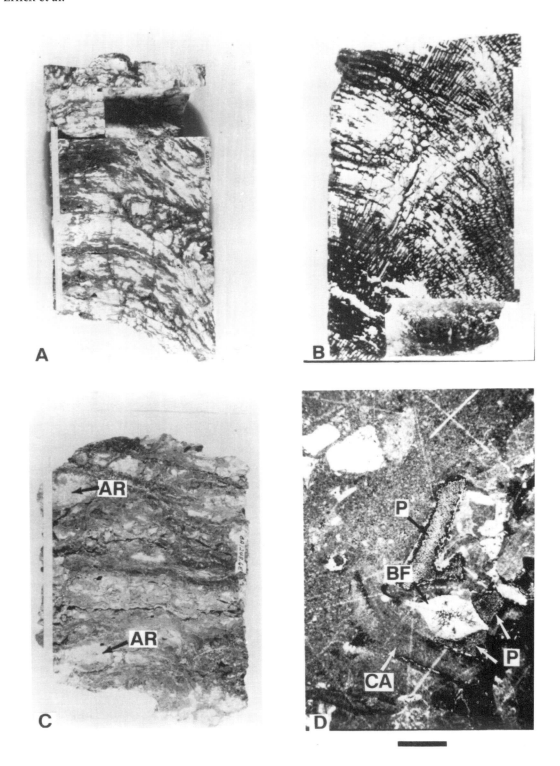

Figure 13. (A) Core 3, LH 4-1-1, 1277.2 m. Red algal boundstone. Scale for all core photos = 15 cm. (B) 1313.1 m. Hermatypic coral boundstone. (C) Core 1, 1248.6 m. Red algal packstone/boundstone. AR = algal rhodolith. (D) 1246.4 m. Thin-section photo of algal-foram packstone. P = porosity; BF = benthic foram; CA = calcareous red algae. Scale bar = 250 microns.

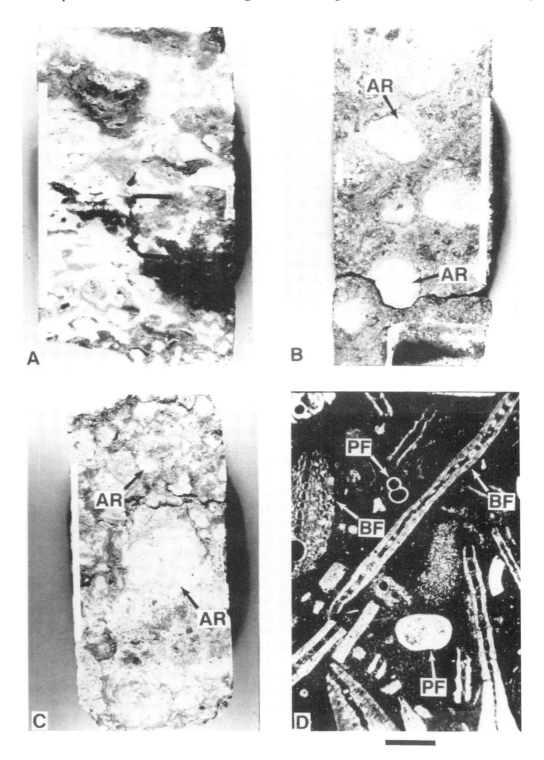

Figure 14. (A) Core 2, LH 11-1-3, 1244.8 m. Red algal packstone/wackestone. Scale in all core photos = 15 cm. (B) Core 1, 1231.7 m. Algal rhodolith packstone. (C) 1223.6 m. Algal rhodolith wackestone/packstone. (D) 1223.5 m. Thin-section photo of foraminiferal wackestone/packstone. Scale bar = 250 microns.

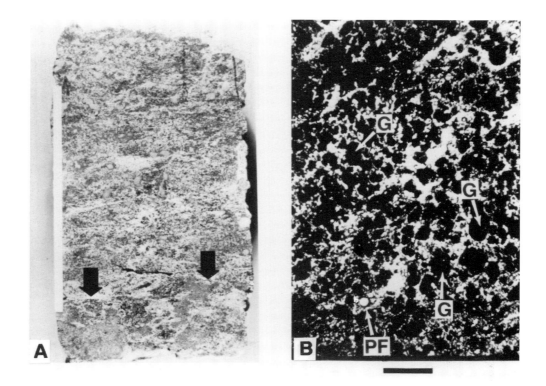

Figure 15. (A) Core 1, LH 11-1-3, 1223.1 m. Glauconitic shaly limestone/siltstone. Arrows mark the contact with the underlying algal rhodolith wackestone/packstone. A submarine omission surface may be present at this contact, but cannot be defined with the biostratigraphic data available. Scale = 15 cm. (B) Thin section of Figure 17A. G = glauconite; PF = planktonic foram. Scale bar = 250 microns.

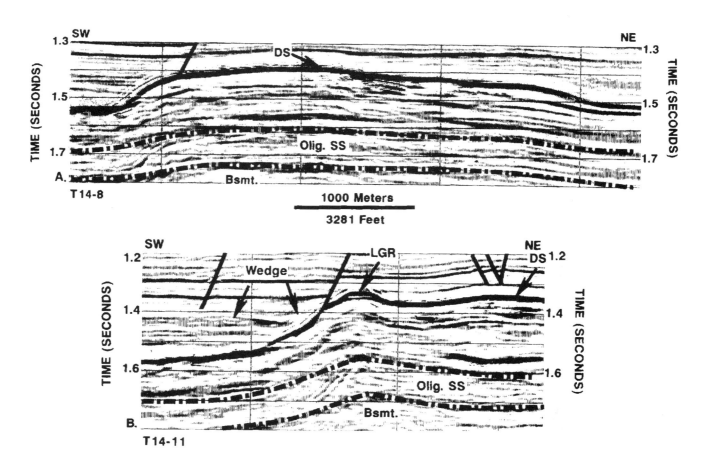

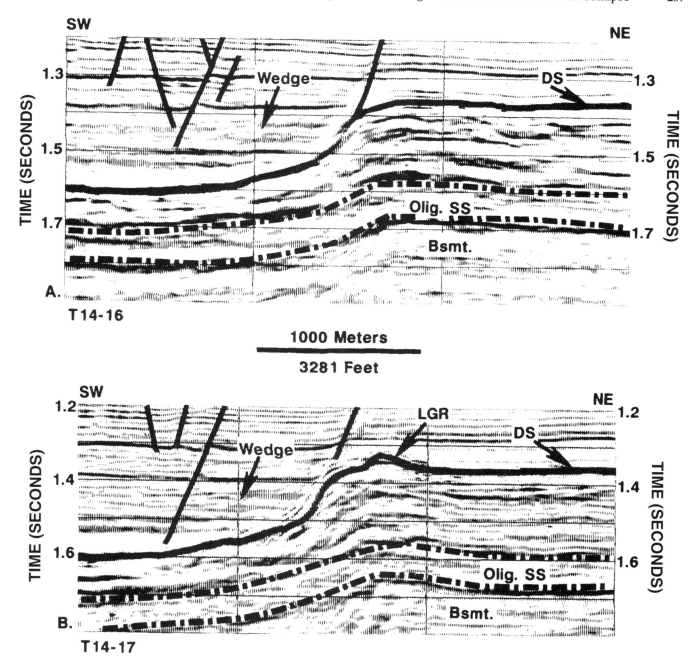

Figure 17. (A) Seismic line T14-16. (B) Seismic line T14-17.

Figure 16. (A) Seismic line T14-8. Note the comparatively narrow width of the platform. Symbol key applies to all Liuhua seismic lines: DS = drowning sequence; LGR = late-growth reef; Olig. SS = Oligocene sandstone; Bsmt. = volcanic basement; Wedge = fore-slope sediment wedge. (B) Seismic line T14-11. The northern limit of the late-growth reef system is shown on this line.

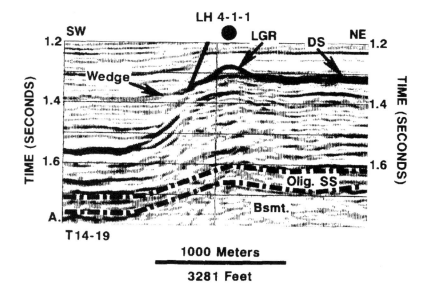

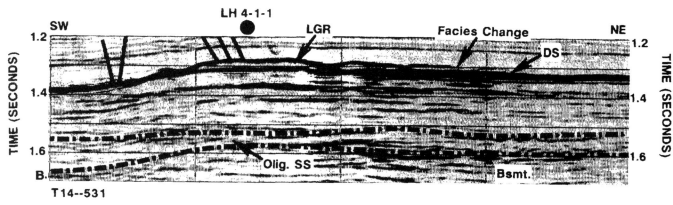

Figure 18. (A) Seismic line T14-19. (B) Seismic line T14-531. This line is oriented oblique to the shelf margin, and ties line T14-19 at the LH 4-1-1 well location. Note the facies change of the late-growth reef system to shale to the southeast.

ent levels and light availability had been radically altered on those platforms. Paleogeographic maps of both areas show that the platforms had been increasingly influenced by prograding deltaic clastics (Figures 9, 22). Prior to the development of the drowning sequences and the coincident deltaic progradation, both platforms had apparently been thriving and capable of maintaining themselves within the photic zone, despite rapid subsidence and relative sea-level rise. After development of the drowning sequences, late-growth reefs on both platform margins had been buried by prodelta and/or marine muds.

Considering the geologically instantaneous increase in nutrient supply from fluvial runoff and the probable decrease in light availability (especially below normal wave base), it is surprising that any reef growth was maintained at the platform margins.

Although the margins continued to be the most favorable ecologic sites for reef development on the platforms, the availability of these locations was apparently not uniform.

Interpreting the Drowning Signal

Seismic data from both areas show that late-growth reefs appear to be better developed almost arbitrarily in central platform locations along the margins. In the Baltimore Canyon area, as well as on the Liuhua platform, deltaic progradation from the north seems to have been an important barrier to northward colonization of the platform margins. This may also explain the lack of late-growth reefs along the southern part of the Baltimore Canyon platform (Figure 9).

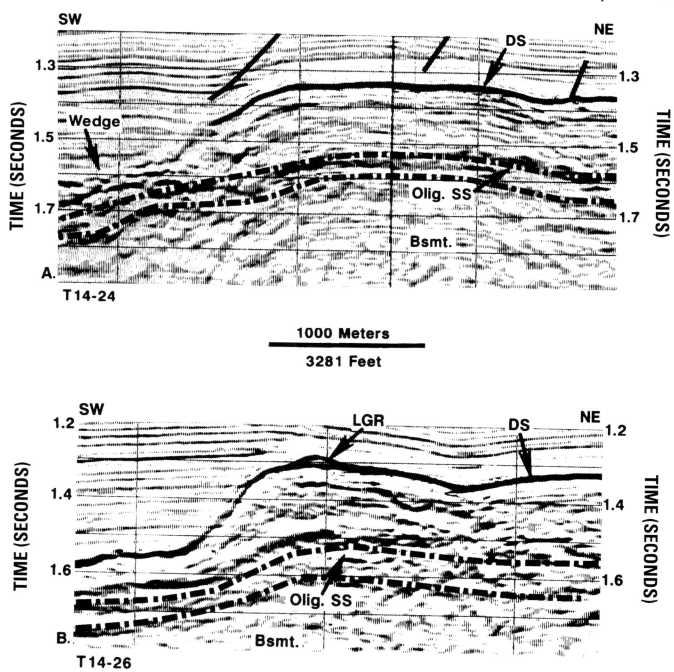

Figure 19. (A) Seismic line T14-24. This is the first line south of T14-21 to show no late-growth reef development (confirmed by the southeastward facies change to shale seen on line T14-531). (B) Seismic line T14-26. This line marks the southern limit of the late-growth reef system.

A similar cause, however, cannot be used to explain the disappearance of late-growth reefs along the southern part of the Liuhua platform margin (Figure 22). Why then were late-growth reefs not developed along the entire Liuhua platform margin? One answer can be found in the tectonic history of the Pearl River Mouth Basin. Tectonic subsidence in the basin appears to have increased during the end of early Miocene time, causing the southern and southwestern parts of the Liuhua platform to fragment and subside more rapidly than areas to the north and northwest (Taylor and Hayes, 1983; Ru and Piggott, 1986; Fulthorpe and Schlanger, 1989). Some support for this idea can be seen from the seismic data (lines

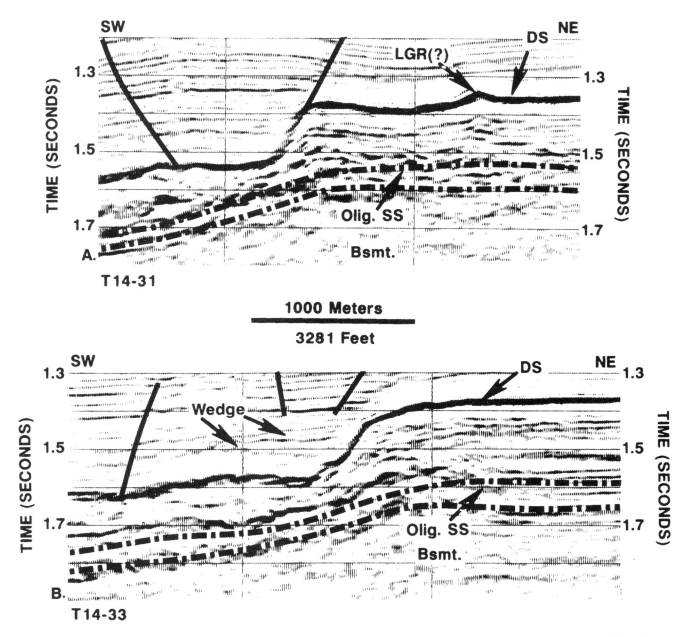

Figure 20. (A) Seismic line T14-31. Shelfal feature at the northeast part of the line may either be diffractions from out of the plane, or an outlier of the late-growth reef system. (B) Seismic line T14-33.

T14-41 and T14-43, Figure 21), which show that the area was tectonically active during Miocene time. Fulthorpe and Schlanger (1989) also suggest that eustatic sea level was rising in the western Pacific during the middle part of early Miocene time (21–18 m.y.), roughly the same time during which the Liuhua platform drowning sequence developed (Figure 11).

The combination of nutrient poisoning, rapid tectonic subsidence, and rising eustatic sea level may have acted to prevent continued colonization of the southern Liuhua platform margin once drowning began. After large parts of the margin had subsided below the photic zone, the dominant process that occurred in those areas was one of erosion, as

inferred by the scour features interpreted from seismic lines T14-41 and T14-43 (Figure 21).

Seismic Sequence Stratigraphy as Applied to Carbonate Platform Margins

Drowning of the Baltimore Canyon platform can now be directly correlated to the progradation of deltaic clastics during a relative sea-level rise, and the environmental deterioration that preceded and accompanied it. Meyer (1989) suggested that prograding clastics were the reason for the demise of the platform, though this study suggests that nutrient

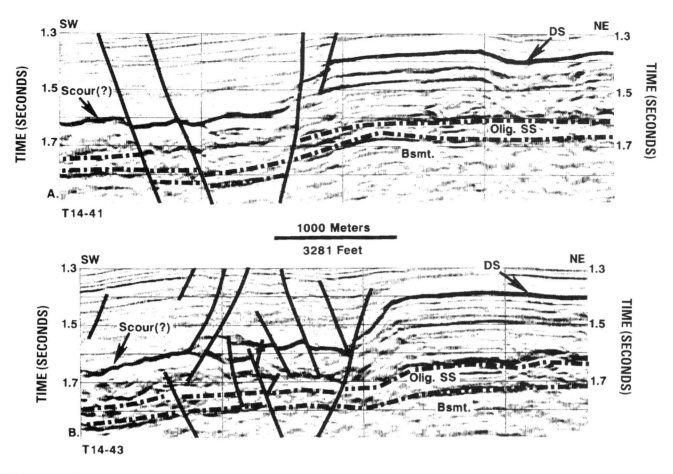

Figure 21. (A) Seismic line T14-41. The platform is highly faulted and segmented from this point south. (B) Seismic line T14-43.

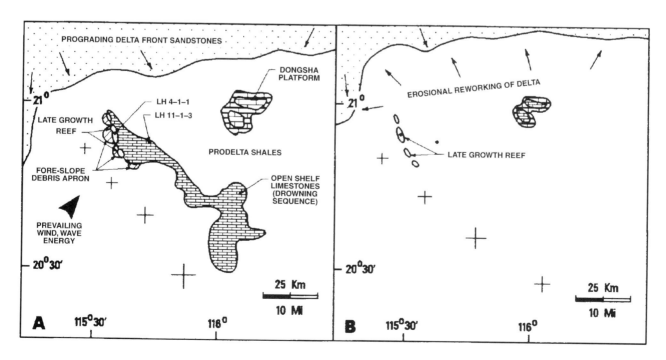

Figure 22. (A) Paleogeography of the Liuhua platform area at the time of maximum development of the drowning sequence. (B) Paleogeography of the area at the end of late-growth reef development.

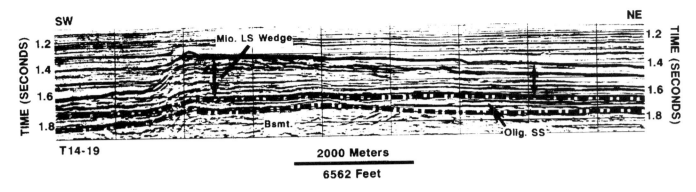

Figure 23. Wedge-shaped cross-sectional profile of the Liuhua platform as seen on seismic line T14-19. Tectonic subsidence was greater to the southwest where the platform aggraded vertically. Mio. LS Wedge = wedge-shaped profile of the Miocene carbonate platform.

poisoning was probably the main factor in the termination of carbonate sedimentation on the platform. The last vestiges of carbonate growth on the platforms, the late-growth reefs, apparently were long dead prior to their burial by siliciclastics.

If the interpretations of the data presented are correct, some conflicts with earlier work are unavoidable. In addition, some problems exist with standard seismic sequence stratigraphy as applied to carbonate platforms.

Previous studies of the Baltimore Canyon area have suggested that the fore-reef slope was essentially devoid of sediments, or experienced very slow rates of sedimentation during drowning of the platform (Meyer, 1989; Schlager, 1989; Prather, 1991). Such a conclusion is inconsistent with the data presented in this study, and is inconsistent with the geologic history of the area. One inconsistency involves the age of slope-front fill adjacent to the Baltimore Canyon platform. Using the Early Cretaceous carbonate platforms of the northern Gulf of Mexico as an analogy (see Winker and Buffler, 1988), combined with biostratigraphic ages established for these sediments in more landward locations (see age–lithology data from Erlich et al., 1988 and Prather, 1991), it is not unreasonable to suggest that at least some of the slope-front fill of the Baltimore Canyon platform is coeval with platform sediments.

Prograding prodelta silts and muds eventually buried the Tithonian to Berriasian carbonate platform, and appear to thin and pinch out just landward of the margin edge (see seismic line MMA-120; Figure 6A). Erlich et al. (1990) showed that a number of submarine erosional unconformities are recorded in the Lower Cretaceous siliciclastic section from the Block 587 well (just behind the margin), indicating that some unknown amount of sediment had been removed prior to the death of the late-growth reefs. This eroded material (probably prodelta or marine muds) was transported somewhere during this process, and the most likely location is the fore-reef slope. It should also be noted that the presence of the

drowning sequence, directly linked to the prograding delta system, suggests that some contemporaneous prodelta mud could easily have bypassed the shelf and been deposited in fore-reef and basinal areas.

Based on the regional geology alone, it is possible to conclude that the slope-front fill is equivalent to the platform section. Seismic data (Figures 5–7) strongly support this interpretation. As a result, previous estimates of 3000 m (over 9800 ft) of shelf to basin relief appear unrealistic, and do not conform to the geologic history of the area (Meyer, 1989). The estimate of 580 m (about 1900 ft) of maximum relief on the margin made by Erlich et al. (1990) therefore may be quite reasonable.

Assuming a smaller shelf to basin profile and a causal link between drowning of the platform and deltaic progradation implies that the high-amplitude base-of-slope reflector found seaward of the Tithonian shelf margin may not be a true seismic sequence boundary (Figure 8). This reflector package, now interpreted as upper lower Tithonian age, is part of an earlier drowning sequence, and therefore is not the upper carbonate sequence boundary (Meyer, 1989; Prather, 1991).

Analogies from modern drowned or highstand carbonate platforms support this interpretation. Data from the Cay Sal Bank, Bahamas (Hine and Steinmetz, 1984), numerous modern platforms (Schlager and Camber, 1986), both sides of the Great Bahama Bank (Wilber et al., 1990; Ginsburg et al., 1991), and the Nicaragua Rise (Glaser and Droxler, 1991) show that highstand flooding of a carbonate platform causes much of the platform interior and platform margin sediment to be transported to slope settings, where it is rapidly submarine cemented. This lithified pavement becomes part of the slope, and is one process by which carbonate platforms aggrade and prograde.

During drowning, carbonate sediment production drops sharply (Schlager, 1981) and transport of platform interior sediments to the fore-slope is minimal. Slope cementation processes, however, continue.

The modern examples of drowned or highstand platforms are presently unaffected by siliciclastics, as was the lower Tithonian part of the Baltimore Canyon platform at the time of the first drowning event. It is very likely that the high-amplitude fore-slope reflectors are submarine cemented hardgrounds, similar to those reported from modern platforms.

The cause of the first drowning event is difficult to determine without supporting lithologic and biostratigraphic data. In general, it appears that drowning of the lower Tithonian platform occurred rapidly, possibly as a result of increased tectonic subsidence and/or relative sea-level rise. Maximum relief between the shelf and the basin probably occurred at this time.

This type of intra-platform drowning event is not unusual, and has been inferred from stratigraphic data for the Early Cretaceous platforms of the northern Gulf of Mexico (Winker and Buffler, 1988). In the case of the Baltimore Canyon platform, a specific environmental cause cannot be determined from the available data.

After drowning, carbonate sedimentation was quickly reestablished, and continued through the end of Jurassic time and into Early Cretaceous time (Figure 8). Late Jurassic and Early Cretaceous offbank transport of platform debris combined with onlapping prodelta and marine shales to reduce the platform to basin relief, prior to the second drowning event. The sequence boundary for the entire platform sequence (which includes the drowning sequence) should therefore be regarded as the top of the second drowning sequence (Figure 8).

Demonstrating an equivalency between slope/basin and platform deposits in the Liuhua platform is not as clear cut. In the Liuhua platform, one of the criteria used to determine this equivalency can be seen on several of the seismic lines from areas adjacent to the late-growth reefs (lines T14-11, Figure 16B; T14-16 and T14-17, Figure 17; T14-19, Figure 18A; T14-24, Figure 19A; and T14-33, Figure 20B). Wedge-shaped reflectors that thicken towards the platform margin and downlap/thin onto basinal shales are interpreted as fore-slope debris, derived from the platform and platform margin. No siliciclastic source for this sediment is possible, since the platform was isolated in an open ocean setting and not connected to a siliciclastic source area.

This pattern, however, is not present on all of the seismic lines, and is notably absent north and south of the main late-growth reef trend (lines T14-8, Figure 16A; T14-26, Figure 19B; T14-31, Figure 20A; and T14-41 and T14-43, Figure 21). In these areas, most basinal reflectors abut the platform margin in a planar or horizontal attitude, suggesting that no sediment was shed off the platforms, or that the basinal fill postdates the development of the platform (i.e., is unconformable with the platform).

If reduced volumes of sediment are shed from the platform to the fore-slope area during drowning (as suggested by analogy with modern platforms), then it is relatively easy to understand why some parts of the fore-slope may demonstrate equivalency with platform deposits, and other parts may not. In the first instance, platform and platform margin debris may have been episodically shed over the margin to the fore-slope and basinal areas as the platform aggraded (Figure 24A). Northern parts of the Liuhua platform margin may have been more prolific transport and production sites even before late-growth reefs developed, possibly accounting for the multiple fore-slope sediment wedges found in those areas (see line T14-17, Figure 17B).

After drowning, the late-growth reefs would have been the only remaining significant sediment production sites on the platform, and therefore could be expected to show more evidence of fore-slope debris fans.

Those areas of the platform that were environmentally less favorable were either more subject to erosion (as in lines T14-41 and T14-43, Figure 21), or had much reduced platform to basin sediment transport (Figure 24B). The minimal initial transport from these sites would have nearly ceased after drowning. Horizontal/planar onlapping basinal muds would have been deposited immediately adjacent to the platform in those areas, suggesting an unconformable relationship between the platform and basin. In reality, this type of pattern should be expected in areas with little or no offbank sediment transport, and would be contemporaneous with the nearby fore-slope wedges and platform deposits (Figure 24).

Further support for equivalency between the basinal fill and platform deposits can often be found in back-reef/lagoonal areas. The youngest platform limestones can be clearly seen changing facies to shale (see figure 10 of Erlich et al., 1990). In some areas, a similar facies change can also be seen within the older platform deposits, demonstrating equivalency between the platform and at least the lower parts of the basinal fill (Figure 25A).

Additional Complications

While the examples presented in this study can be regarded as "ideal" or "textbook" examples of drowned carbonate platforms, even these are not without complications. For example, the Liuhua platform apparently underwent at least one prior episode of subaerial exposure before platform carbonate sedimentation was reestablished (Turner and Zhong, 1991). This exposure event is not readily visible from the 60 Hz seismic data, though additional earlier drowning events (back-stepping of the platform) can be found on some lines (Figure 25A as noted above). Drowning sequences much thicker than the one drilled by the LH 11-1-3 well may also be present on tectonically more active parts of the platform, especially towards the south (Figure 25B). These sequences resemble the infilling of antecedent topography (channels?), and may have accumulated at much higher rates than previously estimated (Erlich et al., 1990).

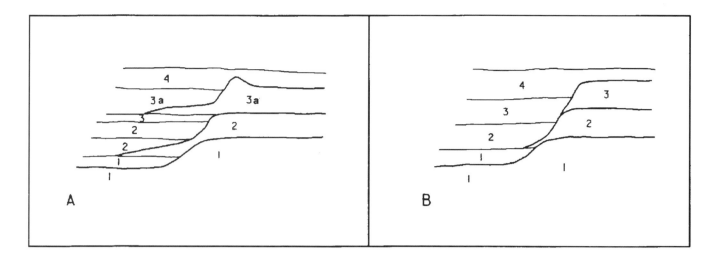

Figure 24. Development of contemporaneous but different margin profiles along the same shelf margin. (A) Platform growth is matched by the deposition of less rapidly accumulating, but time equivalent basin fill. Numbers suggest an equivalency between the platform and basin. (B) Mainly horizontal–planar basinal onlap as a result of little or no offbank sediment transport. Basinal shales are still accumulating as the platform develops, but appear as an unconformable contact.

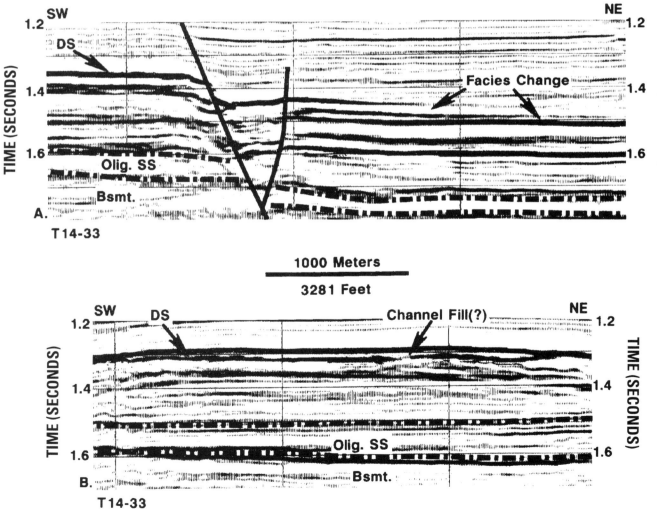

CONCLUSIONS

- At least two drowning sequences can be seen within the Baltimore Canyon platform. Slope-front fill originally interpreted to be part of the post-carbonate sequence is reinterpreted to be composed of contemporaneous platform debris and prodelta/marine shale.
- Late-growth reefs of the Baltimore Canyon and Liuhua platforms are discontinuous and not uniformly well developed. This may be a function of their proximity to prograding siliciclastics, combined with rapid tectonic subsidence and rapid relative sea-level rise.
- Excess nutrient levels, resulting from deltaic progradation, initiated the environmental collapse that eventually led to the termination of carbonate sedimentation and the drowning of both platforms.
- The lithologic–ecologic record of nutrient-poisoned platforms is expressed as a shift from reef-forming ecologic assemblages (stromatoporoids–sponges–coral boundstones, and algal–hermatypic coral boundstones) to sediments dominated by algal rhodoliths and large benthic and planktonic foraminifera. Drowning sequences deepen upwards biostratigraphically, and fine upwards sedimentologically.
- Horizontal/planar basinal onlap may not always postdate carbonate platform development. The deposition of fore-slope debris fans may not necessarily be uniform along the depositional strike of the platform margin, causing some areas to have wedge-shaped fore-slope debris fans, and others to have horizontal/planar basinal onlap that appears, but is not unconformable.

ACKNOWLEDGMENTS

The authors wish to thank E. R. Shaw, E. K. Chau, D. G. Benson, Jr., D. S. Van Nieuwenhuis, and R. A. Salomon for contributing their expertise and time to this project over the years. N. L. Turner provided core and thin section photographs from the Liuhua wells. W. S. Hale-Erlich reviewed an earlier version of the manuscript, and H. V. Nguyen drafted most of the figures. We are also indebted to Amoco Production Company, Nanhai East Oil Corporation, and GeoCenter, Incorporated, for permission to publish this study. The Offshore Technology Conference kindly provided permission to reprint several of the figures. This contribution is dedicated to the memory of R. A. Belmont, who will always be a source of inspiration to those who knew him.

REFERENCES CITED

Dominguez L. L., H. T. Mullins, and A. C. Hine, 1988, Cat Island platform, Bahamas: An incipiently drowned Holocene carbonate shelf: Sedimentology, v. 35, p. 805-819.

Erlich, R. N., S. F. Barrett, and B. J. Guo, 1990, Seismic and geologic characteristics of drowning events on carbonate platforms: AAPG Bulletin, v. 74, p. 1523-1537.

Erlich, R. N., S. F. Barrett, and B. J. Guo, 1991, Drowning events on carbonate platforms: A key to hydrocarbon entrapment?: Proceedings of the 23rd Annual Offshore Technology Conference, Houston, May 6-9, p. 101-112.

Erlich, R. N., K. P. Maher, G. A. Hummel, D. G. Benson, Jr., G. J. Kastritis, H. D. Linder, R. S. Hoar, and D. H. Neeley, 1988, Baltimore Canyon Trough, Mid-Atlantic OCS: Stratigraphy of Shell/Amoco/Sun wells, in Bally, A. W., ed., Atlas of Seismic Stratigraphy, v. 2: AAPG Studies in Geology 27, p. 51-66.

Fulthorpe, C. S., and S. O. Schlanger, 1989, Paleo-oceanographic and tectonic settings of early Miocene reefs and associated carbonates of offshore Southeast Asia: AAPG Bulletin, v. 73, p. 729-756.

Ginsburg, R. N., P. M. Harris, G. P. Eberli, and P. K. Swart, 1991, The growth potential of a bypass margin, Great Bahama Bank: Journal of Sedimentary Petrology, v. 61, p. 976-987.

Glaser, K. S., and A. W. Droxler, 1991, High production and highstand shedding from deeply submerged carbonate banks, northern Nicaragua Rise: Journal of Sedimentary Petrology, v. 61, p. 128-142.

Grigg, R. W., 1988, Paleoceanography of coral reefs in the Hawaiian-Emperor chain: Science, v. 240, p. 1737-1743.

Hallock, P., 1985, Why are larger foraminifera large?: Paleobiology, v. 11, p. 195-208.

Hallock, P., 1988, The role of nutrient availability in bioerosion: Consequences to carbonate buildups: Paleoceanography, Paleoclimatology, Paleoecology, v. 63, p. 275-291.

Hallock, P., A. C. Hine, G. A. Vargo, J. A. Elrod, and W. C. Jaap, 1988, Platforms of the Nicaraguan Rise: Examples of the sensitivity of carbonate sedimentation to excess trophic resources: Geology, v. 16, p. 1104-1107.

Figure 25. (A) Part of seismic line T14-33 that crosses into the back-reef area of the platform. Note the platform limestones changing facies to shale in two different areas of the platform; this is an example of the back-stepping that occurs during drowning of isolated open ocean atolls. (B) A section of seismic line T14-33 that lies between Figure 25A and Figure 20B, showing a thicker part of the drowning sequence, possibly infilling a channel.

Hallock, P., and W. Schlager, 1986, Nutrient excess and the demise of coral reefs and carbonate platforms: Palaios, v. 1, p. 389-398.

Hao, F., 1988, Cenozoic reefs—new targets for oil fields in the northern part of the South China Sea, *in* Wagner, H. C., L. C. Wagner, F. F. H. Wang, and F. L. Wong, eds., Petroleum Resources of China and Related Subjects: Circum-Pacific Council for Energy and Mineral Resources, Earth Science Series, v. 10, p. 199-215.

Hine, A. C., and J. C. Steinmetz, 1984, Cay Sal Bank, Bahamas—a partially drowned carbonate platform: Marine Geology, v. 59, p. 135-164.

Libby-French, J., 1981, Lithostratigraphy of Shell 272-1 and 273-1 wells: Implications as to depositional history of Baltimore Canyon Trough, Mid-Atlantic OCS: AAPG Bulletin, v. 65, p. 1476-1484.

Ludwig, K. R., B. J. Szabo, J. G. Moore, and K. R. Simmons, 1991, Crustal subsidence rate off Hawaii determined from $^{234}U/^{238}U$ ages of drowned coral reefs: Geology, v. 19, p. 171-174.

Meyer, F. O., 1989, Siliciclastic influence on Mesozoic platform development: Baltimore Canyon Trough, western Atlantic, *in* Crevello, P. D., J. L. Wilson, J. F. Sarg, and J. F. Read, eds., Controls on Carbonate Platform and Basin Development: SEPM Special Publication No. 44, p. 211-232.

Peebles, M. W., P. Hallock, and A. C. Hine, 1990, Rhodolite and encrusted-grain sedimentation, Thunder and Lightning Knolls, southwest Caribbean Sea: AAPG Bulletin, v. 74, p. 737-738.

Prather, B. E., 1991, Petroleum geology of the Upper Jurassic and Lower Cretaceous, Baltimore Canyon Trough, western North Atlantic Ocean: AAPG Bulletin, v. 75, p. 258-277.

Ru, K., and J. D. Piggott, 1986, Episodic rifting and subsidence in the South China Sea: AAPG Bulletin, v. 70, p. 1136-1155.

Rudolph, K. W., and P. J. Lehmann, 1989, Platform evolution and sequence stratigraphy of the Natuna platform, South China Sea, *in* Crevello, P. D., J. L. Wilson, J. F. Sarg, and J. F. Read, eds., Controls on Carbonate Platform and Basin Development: SEPM Special Publication No. 44, p. 353-361.

Sarg, J. F., 1988, Middle-Late Permian depositional sequences, Permian Basin, West Texas and New Mexico, *in* Bally, A. W., ed., Atlas of Seismic Stratigraphy, v. 3: AAPG Studies in Geology 27, p. 140-154.

Schlager, W., 1981, The paradox of drowned reefs and carbonate platforms: GSA Bulletin, v. 92, p. 197-211.

Schlager, W., 1989, Drowning unconformities on carbonate platforms, *in* Crevello, P. D., J. L. Wilson, J. F. Sarg, and J. F. Read, eds., Controls on Carbonate Platform and Basin Development: SEPM Special Publication No. 44, p. 15-25.

Schlager, W., and O. Camber, 1986, Submarine slope angles, drowning unconformities, and self erosion of limestone escarpments: Geology, v. 14, p. 762-765.

Schlee, J. S., and K. Hinz, 1987, Seismic stratigraphy and facies of continental slope and rise seaward of Baltimore Canyon Trough: AAPG Bulletin, v. 71, p. 1046-1067.

Taylor, B., and D. E. Hayes, 1983, Origin and history of the South China Sea basin, *in* Hayes, D. E., ed., The tectonic and geologic evolution of Southeast Asian seas and islands, part 2: American Geophysical Union Monograph 27, p. 23-56.

Turner, N. L., and Zhong, H. P., 1991, Lower Miocene Liuhua carbonate reservoir, Pearl River Mouth Basin, offshore People's Republic of China: Proceedings of the 23rd Annual Offshore Technology Conference, Houston, May 6-9, p. 113-123.

Tyrrell, W. W., Jr., and H. E. Christian, Jr., 1992, Exploration history of Liuhua 11-1 field, Pearl River Mouth Basin, China: AAPG Bulletin, v. 76, p. 1209-1223.

Vail, P. R., R. G. Todd, and J. B. Sangree, 1977, Seismic stratigraphy and global changes of sea level, part 5: Chronostratigraphic significance of seismic reflections, *in* Payton, C. E., ed., Seismic Stratigraphy—Applications to Hydrocarbon Exploration: AAPG Memoir 26, p. 99-116.

Van Wagoner, J. C., R. M. Mitchum, Jr., H. W. Posamentier, and P. R. Vail, 1987, Seismic stratigraphy interpretation using sequence stratigraphy, part 1: Key definitions of sequence stratigraphy, *in* Bally, A. W., ed., Atlas of Seismic Stratigraphy: AAPG Studies in Geology 27, v. 1, p. 11-14.

Whalen, M. T., 1988, Depositional history of an Upper Triassic drowned carbonate platform sequence: Wallowa Terrane, Oregon and Idaho: GSA Bulletin, v. 100, p. 1097-1110.

Wilber, R. J., J. D. Milliman, and R. B. Halley, 1990, Accumulation of banktop sediment on the western slope of Great Bahama Bank: Rapid progradation of a carbonate megabank: Geology, v. 18, p. 970-974.

Winker, C. D., and R. T. Buffler, 1988, Paleoceanographic evolution of early deep-water Gulf of Mexico and margins, Jurassic to Middle Cretaceous (Comanchean): AAPG Bulletin, v. 72, p. 318-346.

Chapter 10

Sequence Stratigraphy of Aggrading and Backstepping Carbonate Shelves, Oligocene, Central Kalimantan, Indonesia

Arthur Saller
Richard Armin
Unocal Science and Technology Division
Unocal Corporation
Brea, California, U.S.A.

La Ode Ichram
Unocal Indonesia
Jakarta, Indonesia

Charlotte Glenn-Sullivan[1]
University of Houston
Houston, Texas, U.S.A.

ABSTRACT

Four major Oligocene carbonate sequences were studied in the Teweh area of Central Kalimantan, Indonesia, to better understand how they might serve as reservoirs for hydrocarbons in the area. Each sequence (200–500 m thick) was delineated in outcrops and/or on seismic lines: (1) early Oligocene (34.0–36.5 Ma); (2) middle Oligocene (29.7–32.0 Ma); (3) early late Oligocene (28.0–29.7 Ma); and (4) middle to late late Oligocene (N3; >24–28.0 Ma). In landward areas to the south, sequence 1 consists mainly of sandstones and shales with thin limestone beds. Isolated carbonate buildups and shales occur in basinal areas to the north in sequence 1. An erosional unconformity separates sequences 1 and 2. During deposition of sequences 2–4, carbonate shelves developed in the southern part of the Teweh area, while shales were deposited in basinal environments to the north. The carbonate shelf margin of sequence 2 was established along a structural hinge line. Boundaries between sequences 2–4 do not show onlap or erosional truncation in this area. On seismic lines, boundaries between carbonate sequences 2–4 are defined by surfaces of renewed carbonate growth (mounding and/or downlap) on the shelf immediately above the sequence boundary. Subaerial unconformities were not found in or between sequences 2–4 on outcrop, so boundaries between sequences 2, 3, and 4 were

[1]Present address: Penzoil Corporation, Houston, Texas, U.S.A.

placed where strata first indicated a substantial deepening of depositional environments. Rapid rises in relative sea level (subsidence + eustatic sea level) resulted in drowning and "backstepping" of carbonate shelf margins in some locations, and stacking of shelf margins in other locations.

Internally, the carbonate shelves of sequences 2 and 3 are characterized by vertically building shelf margins with landward-dipping (south-dipping), shingled clinoforms indicating progradation of shallow carbonate environments from the shelf margin into the lagoon. Sequences 2 and 3 have well-developed transgressive systems tracts overlain by highstand systems tracts. In outcrop, the transgressive systems tracts contain interbedded large-foram wackestones/packstones and coral wackestones/packstones with poorly defined facies belts. The highstand systems tracts are characterized by well-developed facies belts which include from the basin shelfward: (1) shale and carbonate debris flows deposited on the lower slope; (2) argillaceous large-foram wackestones on the upper slope; (3) discontinuous coral wackestones and boundstones in bioclastic packstones on the shelf edge; (4) coralline-algae large-foram packstones and grainstones on back-reef flats and shelf-margin shoals; and (5) thin-branching coral and foraminiferal wackestones and packstones in the lagoon. Seismic lines show the carbonate shelf of sequence 4 as a massive buildup which thins substantially into the basin.

Our interpretation suggests that in some circumstances, the definition of sequences requires more flexibility than that given in Van Wagoner et al. (1988). In carbonate systems during times of rapid subsidence and low-amplitude sea-level fluctuations, sea level may not drop below the shelf, and subaerial unconformities will not be present on the shelf to separate different sequences of deposition. The Haq et al. (1987) sea-level curve may also require modification, at least with regard to magnitudes and rates of eustatic sea-level rise. Deposition of carbonate shelves in Central Kalimantan spans the large mid-Oligocene (29.5–30.0 Ma) eustatic sea-level drop of Haq et al. (1987). Shallowing and subaerial exposure of these deposits might be expected during that large eustatic sea-level drop, however none was observed. Instead, deepening and local drowning of the carbonate shelf were observed at 29.5–30.0 Ma.

INTRODUCTION

Subaerial exposure and associated fresh-water leaching (including karstification) were important to the modification and preservation of porosity in many carbonate reservoirs throughout the world. Therefore, prediction of sea-level lowstands from seismic data and sea-level curves may aid in forecasting porosity in subsurface carbonates. However, interpretations of Oligocene limestones from Indonesia suggest that caution should be used in predicting sea-level lowstands and associated fresh-water diagenesis from seismic reflection patterns or from the eustatic sea-level curve of Haq et al. (1987).

Oligocene limestones were studied in outcrop and subsurface in Central Kalimantan, Indonesia, to establish regional models for carbonate deposition which might aid in the prediction of porosity in the subsurface. Carbonates were put in a sequence-stratigraphic framework to develop stratigraphic models. The area of study (approximately 150 km by 150 km) is in and around the Teweh Block concession as it was defined in 1987 (Figure 1). The concession area has been modified substantially since 1987, but the 1987

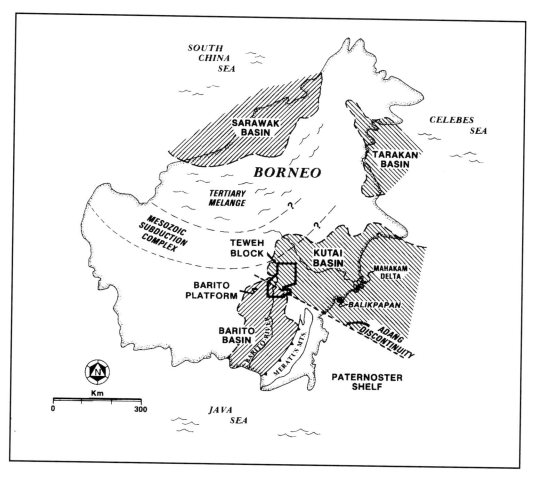

Figure 1. Map of Borneo showing Barito platform, Adang Discontinuity, Kutai Basin, and location of Teweh Block as it was in 1987 (modified from Hamilton, 1979).

boundary of Teweh Block is used as a convenient reference (Figure 2).

METHODS

This project contains two main sources of data—subsurface and outcrop. The subsurface analysis was based on seismic data, well logs, and sidewall cores. Wells were correlated to seismic lines using checkshot data and lithologic velocity data. Seismic data were analyzed by establishing onlap, downlap, and erosional truncation surfaces, which were then correlated through the seismic grid. Reflector patterns within each sequence were examined to help interpret depositional systems.

Outcrops were examined during several field seasons. Oligocene carbonates (Berai Limestone) were studied in detail in three main areas—the Jaan River, Lemo River, and Gunung Anga (Figure 2). Thirty measured sections along the northern part of the Jaan River (Figure 2) extend across a carbonate shelf to the basin along a 15 km south–north transect. Exposed intervals range from a few meters to almost 300 m

thick and were assembled in five composite sections. The transect along the Lemo River included 73 sections and four composite sections extending from west-southwest to east-northeast (shelf to basin) obliquely crossing the carbonate shelf margin (Figure 2). Outcrops in the Gunung Anga area (Figure 2) were studied on gunungs (low mountains), and along trails and small rivers. In many Gunung Anga platform interior sections, the top and bottom of the Berai Limestone were not observed. Additional outcrops were studied at many other locations in the Teweh area.

Biostratigraphic and strontium isotope analyses were used to determine the ages of sedimentary rocks. Planktonic foraminifera, calcareous nannofossils, and larger benthic foraminifera were used for biostratigraphy. Because Oligocene representatives of all three groups tend to be long ranging and facies-controlled, no one group provides sufficient chronostratigraphic control. An integrated time scale for late Eocene to early Miocene was constructed based on chronostratigraphic correlations of Berggren et al. (1985) for planktonic foraminiferal and calcareous

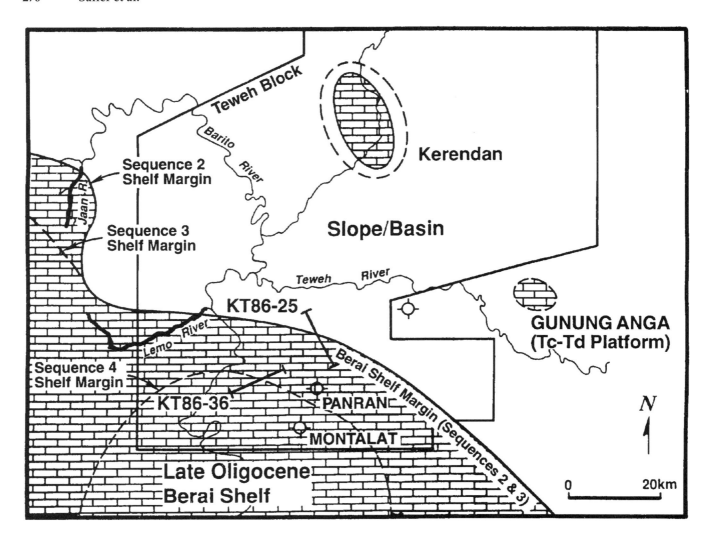

Figure 2. Paleogeography of the Teweh area during the Oligocene. The boundary of the Teweh Block is shown as it was in 1987. It has been substantially changed since then. A series of carbonate shelves developed in the southern part of the study area during the Oligocene. Locations of carbonate shelf margins for sequences 2-4 are shown. Shelf margins of sequences 2 and 3 either stack on each other, or the sequence 3 shelf margin "steps back" landward (to the south) from the sequence 2 shelf margin. The sequence 4 shelf margin "steps back" substantially landward from the location of the sequence 3 shelf margin. The Kerendan platform accumulated carbonate sediment from early Oligocene to early late Oligocene time (28.7 Ma). The Gunung Anga platform accumulated carbonates in early Oligocene time (34.0–36.5 Ma). Outcrops were studied along the parts of the Jaan River and Lemo River that are highlighted on the map. Parts of seismic line KT86-25 are shown in Figures 4 and 16, and part of seismic line KT86-36 is shown in Figure 6. Locations of the Panran and Montalat wells are also shown.

nannoplankton zones (Figure 3). The correlation of the larger foraminiferal zones (East Indian Letter "Stages") to Berggren et al. (1985) is based on Adams' (1984) correlations with the time scale of Hardenbol and Berggren (1978) (Figure 3). The ranges used are from Blow (1979), Berggren et al. (1985), Bolli and Saunders (1985), and Berggren and Miller (1988).

Selected large forams were drilled out of slabs and analyzed for strontium isotopes. Strontium analyses were performed at Washington University in St. Louis, Missouri, under the supervision of Frank

Podosek using methods outlined in Popp et al. (1986). The Oligocene is a good interval of time to date marine sedimentary rocks using strontium isotopes because the $^{87}Sr/^{86}Sr$ ratio in seawater increased steadily throughout the epoch (Burke et al., 1982). Conversion of $^{87}Sr/^{86}Sr$ ratios to numerical ages is based on a linear regression in Miller et al. (1988) from Oligocene data at DSDP site 522. The linear regression has a slope of 0.000035/m.y. in the Oligocene. The internal stratigraphic consistency of the data suggests that the standard deviation of

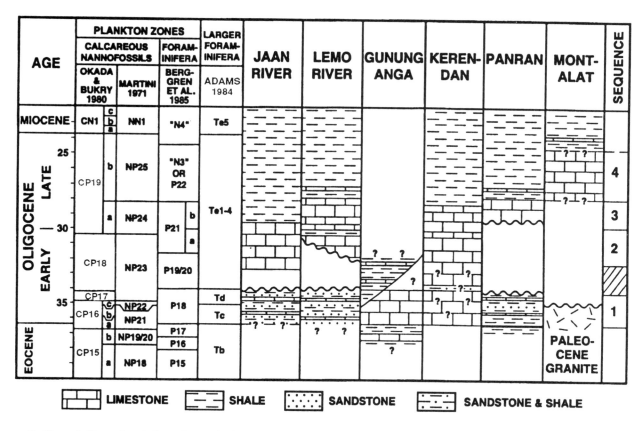

Figure 3. Correlation chart showing relative ages of biostratigraphic zones, ages of stratigraphic sections in various areas, and ages of sequences discussed in this study. Eocene clastics are part of the Tanjung Formation. Oligocene carbonates are part of the Berai Formation.

the analytical precision (1σ; 0.000010) is a good estimate of the precision in this study; therefore, we estimate precision at +/− 0.3 m.y. (0.000010/(0.000035/m.y.)).

The strontium ages are very consistent with biostratigraphic ages especially where biostratigraphic ages were well constrained. For example, five strontium isotope determinations from Tc–Td strata gave ages between 34.1 and 35.1 Ma (Table 1). The Tc–Td biostratigraphic interval has a range of 34.0–36.5 Ma (Figure 3). Another example is outcrops in the Lemo River area, where a shale has a biostratigraphic age of 30–30.5 Ma based on overlapping ranges. A limestone approximately 30 m above that shale has a strontium isotope age of 29.9 Ma.

GENERAL SETTING

The Teweh area straddles the Barito platform and Kutai Basin, which are separated by a basement flexure accompanied locally by high-angle faults which may have some strike-slip displacement (Figure 1). This structural hingeline exerted a strong influence on the types and thicknesses of Tertiary sedimentary rocks. Basement rocks in the Teweh area consist of paraschists with radiometric ages ranging from

Carboniferous through the Mesozoic, slightly metamorphosed Upper Jurassic–Lower Cretaceous sedimentary rocks, and undeformed Late Cretaceous and Paleocene granitic rocks (Van de Weerd et al., 1987).

The earliest deposits in the Tertiary Kutai basin are coarse alluvium (basal Tanjung Formation) eroded from margins of the basin and local highs within the basin. The undated alluvium is overlain by the rest of the Tanjung Formation (Eocene) which is composed of fine-grained sandstones, siltstones, shale, volcaniclastics, thin coal beds, and thin limestone beds that were deposited in marginal-marine, deltaic, estuarine, and shallow-marine environments (Van de Weerd et al., 1987). Rapid late Eocene and early Oligocene subsidence resulted in deposition of deep-marine shales of the lower Bongan Formation in deeper parts of the Kutai Basin. At the same time, fine-grained sandstones, siltstones, shales, and thin interbedded limestones were deposited in the southern part of the Teweh area. Carbonate shelves developed in middle Oligocene time in areas to the south, and deposition of basinal shales continued to the north (Figures 2, 4). Limestones of Oligocene age in the Teweh area are referred to as the Berai Limestone. Minor tectonism during the early to middle (?) Oligocene tilted and erosionally beveled parts of the

Table 1. Strontium Isotope Analyses of Teweh Samples.

Sample	87Sr/86Sr	Age (Ma)	Location
Jaan River			
JN1-7	0.707945	29.9	CS 1, 35 m below top of lms
JN4-6	0.707916	30.8	CS 1, 250 m below top of lms
JN5-17	0.707885	31.7	CS 1, 50 m above top sequence 1
JN8-4	0.707950	29.8	CS 2, 5 m below top of lms
JN8-14	0.707943	30.0	CS 2, 80 m below top of lms
JN8-38	0.707878	31.9	CS 2, 3 m above top sequence 1
JN13-8	0.707903	31.2	CS 3, 60 m above top sequence 1
JN17-9	0.707960	29.5	CS 3, 60 m below top of lms
JN19-1	0.707863	32.3	Onlapping strata CS 4
JN26-1	0.707847	32.7	CS 5, 15 m above top of sequence 1
JN28-1	0.707876	31.9	Slope adjacent to upper part CS 4
Lemo River			
LM2-2	0.70793	30.4	CS 1, highest sample
LM3-2	0.707933	30.3	CS 1, middle sample
LM4-1	0.707931	30.4	CS 1, middle sample
LM5-2	0.707927	30.5	CS 1, lowest sample
LM30-2	0.707983	28.9	CS 2, highest sample
LM23-9	0.707948	29.9	CS 2, middle sample
LM20-1	0.707928	30.4	CS 2, lowest sample
LM60-3	0.708012	28.0	CS 3, highest sample
LM61-1	0.70799	28.7	CS 3, middle sample
LM63-1	0.707956	29.6	CS 3, lowest sample
LM65-3	0.708005	28.2	CS 4, highest sample
LM72-1	0.707976	29.1	CS 4, middle sample
LM66-1	0.707947	29.9	CS 4, middle sample
LM68-1	0.707849	32.7	CS 4, lowest sample
LM70-12	0.707958	29.6	CS 5, high sample
LM69-1	0.707951	29.8	CS 5, low sample
Gunung Anga			
AS16-12	0.707769	35.0	Gunung Tukau, 70 m above base
AS25-1	0.707775	34.8	Upper part of CS A
AS27-1	0.707771	35.0	Upper part of CS A
AS34-1	0.707800	34.1	Gunung Wageng, 30 m above base
AS53-6	0.707766	35.1	Gunung Saheng, 35 m above base
Panran #1 Well			
3916 ft	0.70802	27.8	5 m below top of sequence 3 lms
4128 ft	0.70802	27.8	65 m below top of sequence 3 lms

CS = Composite section
All analyses are of samples of large forams drilled from slabs.
Analytic precision of 87Sr/86Sr ratios are +/- 0.000010 at 1σ. Age determinations were made using curve of Miller et al. (1988). The above analytical precision (1σ) is approximately +/- 0.3 Ma.

Barito platform edge, and gently folded rocks at the Kutai basin margin (Van de Weerd et al., 1987).

Deltaic deposits began filling the Kutai basin in late Oligocene time. Deltaic lobes prograded southward and eastward across the Teweh area from source areas in central Borneo. The lowest deltaic deposits are largely basin-filling prodelta mudstones. A thick succession prodelta to delta plain sediments filled the eastern part of the Kutai Basin and prograded over most of the Oligocene shelf in the early Miocene (Figures 2, 4; Van de Weerd et al., 1987). Compressional tectonics inverted some depocenters

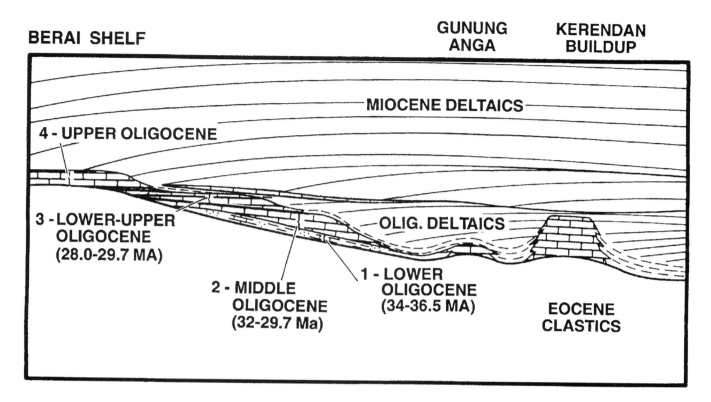

SOUTH

NORTH

BERAI SHELF

GUNUNG ANGA

KERENDAN BUILDUP

MIOCENE DELTAICS

4 - UPPER OLIGOCENE

3 - LOWER-UPPER OLIGOCENE (28.0-29.7 MA)

OLIG. DELTAICS

2 - MIDDLE OLIGOCENE (32-29.7 Ma)

1 - LOWER OLIGOCENE (34-36.5 MA)

EOCENE CLASTICS

Figure 4. Schematic cross section across the Teweh area showing Oligocene sequences 1–4 and the distribution of carbonate shelves and isolated platforms in those sequences.

of the Kutai Basin and uplifted parts of the Barito platform during the middle to late Miocene. Erosion following this deformation and regional uplift produced present-day rock exposures.

Four sequences containing Oligocene carbonates were studied in detail in the southern part of the Teweh area (Figure 4). During deposition of sequences 2–4, carbonate shelves developed in the southern part of the Teweh area, while deeper water shales were deposited in more basinal environments to the north (Figures 2, 4). The boundaries between sequences 2–4 are not classic sequence boundaries described by Van Wagoner et al. (1988) because no erosional truncation, subaerial unconformities, or onlap were observed. Rather, boundaries between sequences 2–4 are the base of strata indicating deepening and/or renewed rapid aggradation of carbonates on the shelf with associated downlap (Figures 5, 6).

SEQUENCE 1

Sequence 1 is a mixture of carbonates and clastics deposited in the early Oligocene (Tc–Td; 34.0–36.5 Ma). In landward (shoreward) areas to the south, sequence 1 consists mainly of sandstones and shales

with thin limestone beds. Isolated carbonate buildups and shales occur in basinal areas to the north in sequence 1. Outcrops of sequence 1 were studied in the Jaan River, Lemo River, and Gunung Anga areas.

Subsurface Data

The upper and lower boundaries of sequence 1 are characterized on seismic lines by erosional truncation of strata below the sequence boundary and onlap of strata immediately above the sequence boundary (Figure 5). A distinctive carbonate shelf margin is not present in sequence 1 in seismic data in the southern part of the Teweh area (Figure 5). Internally, sequence 1 is characterized by parallel, continuous to discontinuous reflectors with no distinct shelf margin. Sequence 1 thins onto a high to the south (Figures 5, 6). Erosional truncation at the top of sequence 1 is quite distinct in areas of structural flexure (around shot point [SP] 951 on Figure 5). Sequence 1 was penetrated in the Panran well (Figure 2) where it consists of interbedded sandstones, shales, and limestones dated biostratigraphically as Tc–Td (34.0–36.5 Ma) (Figure 7).

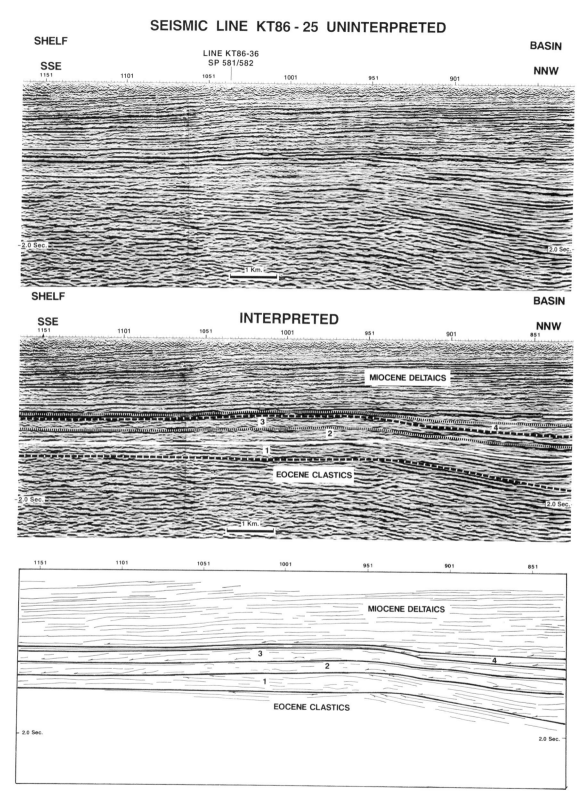

SEISMIC LINE KT86 - 25 UNINTERPRETED

Figure 5. Central part of seismic line KT86-25, uninterpreted, interpreted, and interpreted line drawing. Eocene strata below sequence 1, and upper strata in sequence 1 are truncated at the shelf margin flexure (SP 951). The shelf to basin transition is shown for sequences 2 and 3. Above the sequence 3 shelf, sequence 4 is thin (condensed). Lagoonward-dipping clinoforms are well developed in the southern part of sequence 3 as shown on this section. Lagoonward-dipping clinoforms are present, but more poorly developed in the southern parts of sequence 2. Oligocene deltaics in sequence 4 baselap against the sequence 3 shelf margin and slope. Location of this seismic line is shown on Figure 2.

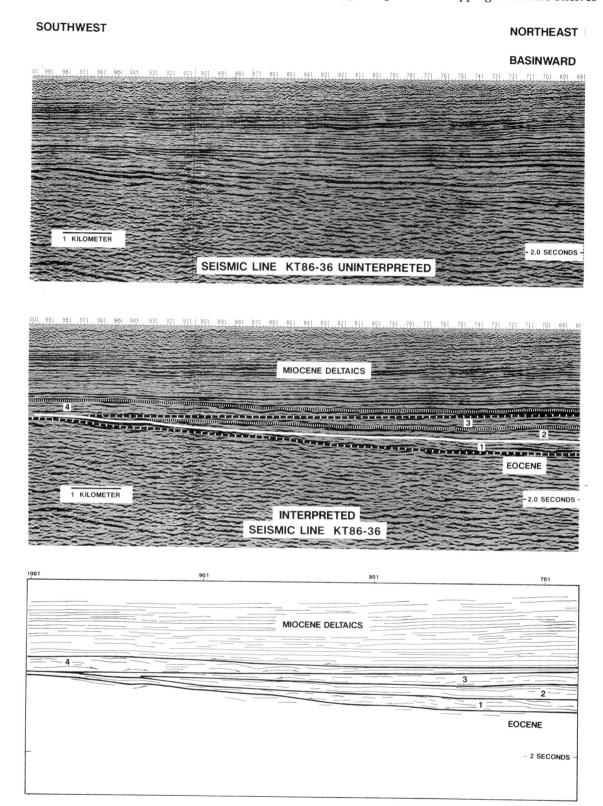

Figure 6. Seismic line KT86-36 (uninterpreted, interpreted, and interpreted line drawing) in the southern part of the study area. Sequences 1–3 thin as they onlap a paleohigh to the southwest. The carbonate shelf of sequence 4 "stepped back" from its location in sequence 3 and built up on the paleohigh (southwest half of the line). Strata in sequence 4 thin off structure (to the northeast) into a condensed sequence in more basinward locations. Miocene deltaics baselap against the sequence 4 shelf margin. Location of line is shown on Figure 2.

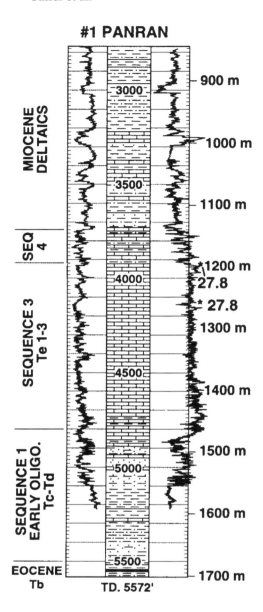

Figure 7. Gamma ray log (left), induction log (right), lithologies (center), and sequences for the lower part of the #1 Panran well. See Figure 2 for location of well. Asterisks indicate location of strontium isotope analyses. Numbers associated with asterisks indicate age in Ma (+/- 0.3 Ma).

Sequence 1 at Jaan River

Sequence 1 consists of a shale-dominated unit below the main Berai Limestone in the Jaan River area (Figure 8). The shale unit contains the large benthic foraminifer *Nummulites fichteli* and rare small specimens of the planktonic foraminifer *Globorotalia increbescens* indicating Tc–lowermost Td age, correlative to the lower three-fourths of the P18 zone of Berggren et al. (1985) (Figure 3). Sequence 1 also contains thin sandstone and limestone beds in the Jaan River area. The shales are generally gray and calcareous, while the sandstones are generally green, poorly

sorted, fine- to coarse-grained, and thin (0.5–5 m thick). Limestones are argillaceous, large-foram wackestone/packstones which occur in beds 0.5–5 m thick. Rare coals are interbedded with the shales.

Sequence 1 at Lemo River

Sequence 1 in the Lemo River area is below the main Berai Limestone (Figure 9) and contains common sandstone and shale with minor thin limestones and rare beds of coal and volcanic tuff. Contacts between sandstones, limestones, and shales are commonly gradational, resulting in many argillaceous limestones, calcareous sandstones, and calcareous shales. Thin limestones contain *Nummulites fichteli* and *Nummulites vascus/striatus* indicating a Tc–Td age (34.0–36.5 Ma) (Figure 3). Sandstones are generally medium- to coarse-grained with ripple-scale cross-beds, flaser bedding, and burrowing. Some sandstones are unfossiliferous, but many contain fossil fragments including large forams. Most limestone beds are fossiliferous wackestones or packstones 0.5–3 m thick that contain large rotaline forams, coral, coralline algae, miliolid forams, and/or echinoderm fragments. Some limestone beds have wackestone/packstones grading upward to bioclastic grainstones.

Sequence 1 in the Gunung Anga Area

Most limestones which crop out in the Gunung Anga area are early Oligocene (Tc-Td; 34.0–36.5 Ma) and hence part of sequence 1. Eocene limestones were observed in the stratigraphically lowest parts of the section at the eastern and western ends of the Gunung Anga outcrop belt. Those areas contain *Biplanispira*, *Discocyclina*, *Nummulites*, and *Asterocyclina*, which indicate a late Eocene age (upper part of the Tb; correlative to mid P15-mid P17 planktonic foraminiferal zones). The rest of the limestone sections at Gunung Anga contain *Nummulites fichteli/vascus/striatus*, indicating an early Oligocene age (Tc–Td) (Figure 3). The occurrence of *Nummulites* with *Sorites sp.* at one location indicates the presence of the uppermost Td. Strontium isotope analyses of Gunung Anga platform carbonates give ages of 34.0–35.1 Ma, which are consistent with a Tc–Td age. The upper parts of westernmost outcrops contain interbedded shales, calcareous turbidites, and debris flows that are probably Te in age.

A summary of depositional facies is shown in Figure 10. Sections within and adjacent to the platform can be separated into four depositional assemblages: (1) platform interior; (2) platform-rim bioclastic grainstone belt; (3) coral-rich platform-edge carbonates; and (4) slope shales and carbonates.

Platform-Interior Strata

The platform-interior sections contain a variety of interbedded lithologies including thin-branching coral wackestones and bafflestones and coralline-algae, large-foram wackestones, packstones, and

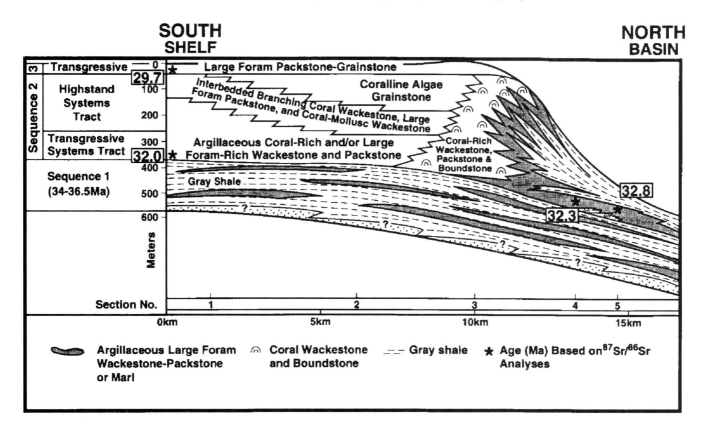

Figure 8. Facies cross section from shelf to basin along the Jaan River. Strontium isotope ages are shown by asterisks and numbers in rectangles (ages in Ma). The age of the top of the limestone is constrained by three similar ages from samples in composite sections 1, 2, and 3 (Table 1). Five more strontium isotope ages between 30 and 32.0 Ma were obtained from limestones within the carbonate shelf strata (sequence 2; Table 1). Very thin, platy large forams and planktonic forams increase in abundance upward in the large-foram packstones/grainstones (lowest sequence 3), suggesting a deepening of depositional environments near the end of carbonate sedimentation on the shelf. Uppermost shelfal limestones (large-foram packstones/grainstones) and slope strata are overlain by shales with abundant planktonic forams suggesting a drowning of the carbonate platform at approximately 29.7 Ma. The lower limits of sequence 1 were not found in measured sections in this area. Composite sections 1 and 2 extend from the upper part of sequence 1 to shales overlying the large foram packstone/grainstone. Composite section 3 extends from the upper part of sequence 1 to the large foram packstone/grainstone, but does not include the overlying shale. Composite section 4 extends from sequence 1 to slightly above the asterisk. Composite section 5 includes 100 m in sequence 1 and the entire thickness of shale and thin limestones shown in sequence 2.

grainstones (Figure 11). Coral wackestones and bafflestones, and coralline-algae, large-foram wackestones/packstones were probably deposited in low-energy, subtidal waters, perhaps 5 to 30 m deep. Bioclastic grainstones were probably deposited in shallow, high-energy environments, 0–10 m deep. A progression from coral wackestone or boundstone (locally argillaceous) up to coralline-algae packstone and grainstone was observed in 8–20-m-thick, shallowing-upward cycles in areas near the platform-rim grainstone belt (Figure 11).

Platform-Rim Bioclastic Grainstone Belt

Thick sections of coralline-algae, large-foram grainstones and packstones (up to 130 m thick) with rare wackestone interbeds were observed along the

rim of the platform (Figures 10, 12). A lack of micrite, the presence of fragmented fossils, and the occurrence of robust (equant), large forams suggest deposition of the coralline-algae, large-foram grainstones in shallow (less than 10 m), high-energy environments, probably carbonate sand shoals and grainstone flats along the platform rim. Few distinct shallowing-upward cycles were recognized in the thick grainstone packages (Figure 12).

Coral-Rich, Platform-Edge Carbonates

Coral wackestones and bafflestones and large-foram wackestones and packstones are present on the eastern and western edges of the Gunung Anga platform (Figure 10). Corals commonly include massive and thick-branching forms. Interbedded coralline-

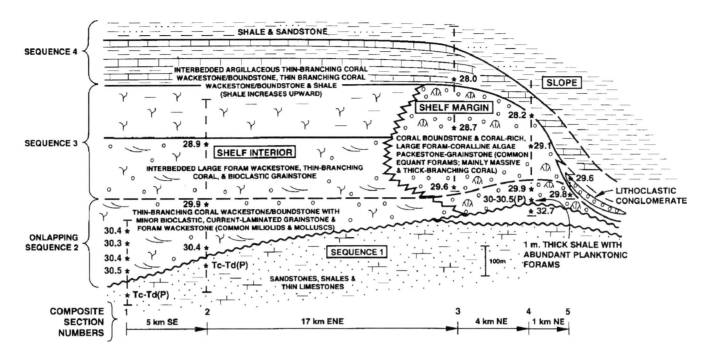

Figure 9. Facies cross section of carbonate strata outcropping along the Lemo River. Numbers beside asterisks are strontium isotope ages (Ma) except where marked "(P)", which indicates a biostratigraphic age. Short horizontal lines connected by dashed lines indicate the stratigraphic location of composite sections.

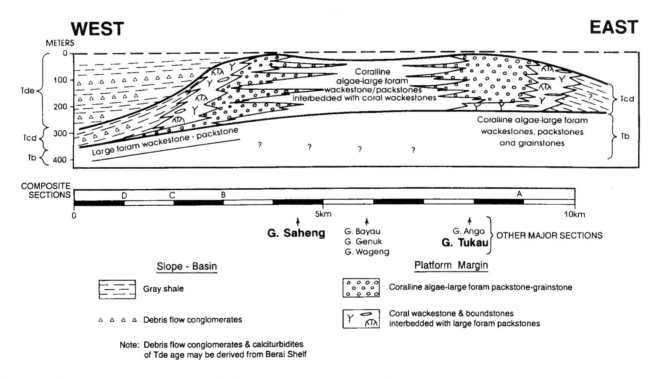

Figure 10. Facies cross section of the isolated carbonate platform in the Gunung Anga area. The platform is mainly Tc–Td in age (part of sequence 1) and is onlapped and apparently overlain by slope strata. Locations of composite sections A–D are shown. Composite section A extends from the Eocene to the top of the limestone shown. Composite sections B–D extend from the Eocene to the overlying Te shales and limestones. Location of other platform sections are also shown. Those sections are entirely within the Tc–Td carbonate platform. Sections from G. Saheng and G. Tukau are shown in Figures 12 and 13, respectively.

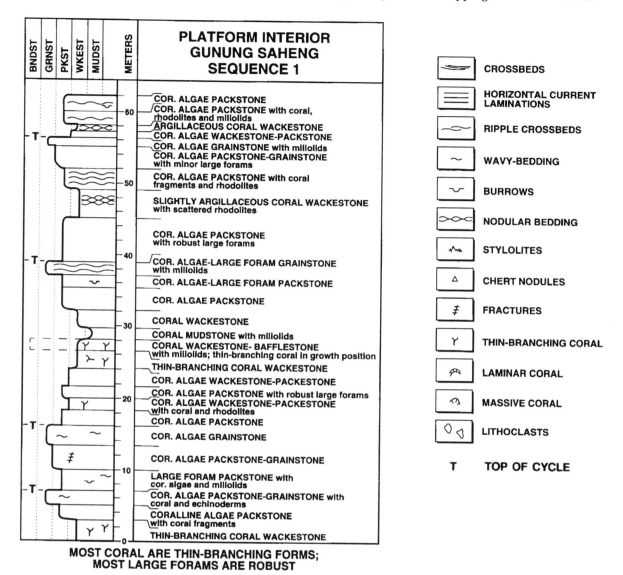

**MOST CORAL ARE THIN-BRANCHING FORMS;
MOST LARGE FORAMS ARE ROBUST**

Figure 11. Detailed stratigraphic section of Gunung Saheng near the margin of the Gunung Anga platform interior. See Figure 10 for relative location. This section is Tc–Td in age (sequence 1). Some distinct shallowing-upward cycles capped by coralline-algae large-foram grainstones are present. "T" indicates top of cycle. Coralline-algae packstones and wackestones are common in the lower parts of cycles. Thin-branching coral wackestones to boundstones are common in the middle of cycles.

algae (rhodolite) grainstones are also present in lower parts of platform-margin sections. Most of these platform-edge carbonates accumulated at moderate to shallow water depths (10–60 m). Foraminifera in platform-edge strata are mainly thin to moderately robust *Nummulites, Heterostegina,* and *Amphistegina* characteristic of slightly deeper water (Hallock and Glenn, 1986). Coralline-algae (rhodolite) grainstones, which were deposited in shallow, high-energy environments, commonly cap cycles that become more grain-rich upward and apparently shallow upward. However, on a larger scale, bioclastic grainstones and grainstones-capped cycles decrease upward, suggesting a larger-scale deepening trend (backstepping

parasequences). In the westernmost sections, shales interbedded with carbonate debris flows overlie well-bedded platform-margin strata, indicating a deepening and backstepping of the platform.

Slope Shales and Carbonates

The lateral transition from carbonate platform to slope deposits can be observed in the western part of the Gunung Anga area (Figure 10). Up section and to the west, platform carbonates grade into argillaceous, platform-margin carbonates and then into gray, calcareous shales with interbedded carbonate conglomerates and graded packstones. Lithoclasts of coral boundstone, fragments of corals, and large forams are

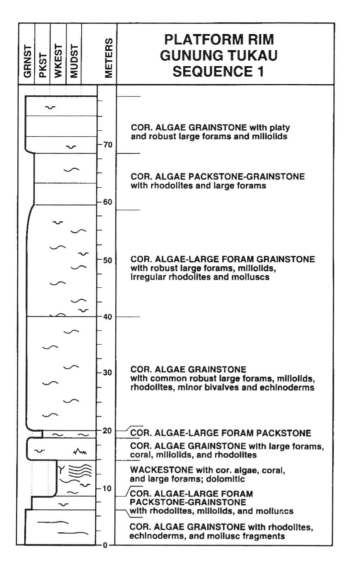

**PLATFORM RIM
GUNUNG TUKAU
SEQUENCE 1**

COR. ALGAE GRAINSTONE with platy
and robust large forams and miliolids

COR. ALGAE PACKSTONE-GRAINSTONE
with rhodolites and large forams

COR. ALGAE-LARGE FORAM GRAINSTONE
with robust large forams, miliolids,
irregular rhodolites and molluscs

COR. ALGAE GRAINSTONE
with common robust large forams, miliolids,
rhodolites, minor bivalves and echinoderms

COR. ALGAE-LARGE FORAM PACKSTONE

COR. ALGAE GRAINSTONE with large forams,
coral, miliolids, and rhodolites

WACKESTONE with cor. algae, coral,
and large forams; dolomitic

COR. ALGAE-LARGE FORAM
PACKSTONE-GRAINSTONE
with rhodolites, miliolids, and molluscs

COR. ALGAE GRAINSTONE with rhodolites,
echinoderms, and mollusc fragments

Figure 12. Detailed stratigraphic section of Gunung
Tukau on the Gunung Anga platform rim. It is
Tc–Td in age (sequence 1). The section is dominantly packstone to grainstone, making cycles difficult
to define in most of the section. See Figure 10 for
relative location and Figure 11 for legend.

common in the carbonate conglomerates. The shales
and interbedded lithoclastic carbonates were apparently deposited in a slope environment. Conglomerates and finer packstones shed off the
carbonate shelf and platform margins were carried
downslope by slumps, debris flows, and turbidity
currents. Some of these slope deposits appear to overlie or onlap shallow-marine carbonates in the Gunung
Anga area, indicating a drowning of the platform
(Figures 4, 10). Shallow-water carbonate clasts in
debris flows of Te age near Gunung Anga were probably derived from the Berai shelf to the south after
the Gunung Anga platform drowned (during
sequences 2 and 3).

Depositional Summary of Sequence 1

Sequence 1 accumulated during regional subsidence in early Oligocene (Tc–Td) time (34.0–36.5 Ma).
In more basinal areas to the north, isolated carbonate
platforms developed on basement highs, while shale
was deposited in surrounding deep water. Sequence
1 thins as it onlaps a paleohigh to the south. In the
southern part of the study area, interbedded sandstones, shales and limestones of sequence 1 were
deposited in several different environments. Rare
coal beds suggest deposition in deltaic or coastal-
plain environments, while sandstones and shales
were deposited in a variety of marine, marginal-
marine, and deltaic environments. *Nummulites* and
small planktonic foraminifera in large-foram wacke-
stones/packstones (limestones) and calcareous shales
indicate deposition in slope and carbonate bank environments in areas of low terrigenous influx (water
depths of 20–100 m). Minor uplift, folding, and erosion occurred between deposition of sequences 1 and
2, causing local erosional truncation of the upper part
of sequence 1 and a depositional hiatus on the shelf.

SEQUENCE 2

Sequence 2 contains a well-developed carbonate
shelf in the southern part of the Teweh area and basinal shales to the north (Figures 2, 4). Shelfal carbonates of sequence 2 are 32.0–29.7 Ma in age. Carbonates
onlapping the slope at the bottom of sequence 2 have
ages of 32.8 and 32.3 Ma. Outcrops of sequence 2 carbonates were studied along the Jaan and Lemo rivers
(Figure 2).

Subsurface Data

The lower boundary of sequence 2 is recognized
on seismic lines by erosional truncation of underlying
strata and onlap of overlying reflectors (Figure 5).
Reflectors onlap the slope and pass upward into a
zone of discontinuous and mounded reflectors at the
shelf margin, which was established on or north of
the hinge line of a structural flexure (SP 951 on Figure
5). Landward of the shelf margin, sequence 2 contains
vague downlapping reflectors dipping landward
away from the shelf margin (SP 1060-1151 on Figure
5). The top of sequence 2 is a relatively flat, high-
amplitude reflector in more lagoonward areas. A carbonate shelf sequence exposed at Jaan River is
correlated with seismic sequence 2 and has a strontium isotope age of 32.0–29.7 Ma. The Panran or
Montalat wells are in more landward (southern) locations (Figure 2), and sequence 2 was not deposited at
those two locations.

Sequence 2 in the Jaan River Area

Strontium isotope and biostratigraphic data indicate that thick shelf carbonates (Berai Limestone) in
the Jaan River area are late early Oligocene to earliest

late Oligocene in age (32.0–29.7 +/- 0.3 Ma) (Figure 8). The main Berai Limestone at Jaan River contains a diverse larger-foraminiferal fauna. *Heterostegina cf. H. borneensis* is present and is restricted to the Oligocene (Te1–4). Larger rotaline forams near the base of the main Berai Limestone on the shelf yield strontium isotope ages of 31.0–32.0 Ma (+/- 0.3 Ma) (Table 1). Larger rotaline forams near the top of the main Berai Limestone yield strontium isotope ages of 29.5–29.9 Ma (+/- 0.3 Ma). The main Berai Limestone in the Jaan River area unconformably overlies sequence 1, and correlates mainly with seismic sequence 2 (Figures 3, 8). The upper 20–30 m of the main Berai Limestone at Jaan River is a large-foram packstone/ grainstone which is the lowest (transgressive) part of sequence 3 (Figure 8).

Biostratigraphic and strontium isotope ages indicate that interbedded shales and thin argillaceous limestones (large-foram wackestones/packstones) north of the main shelfal carbonates are stratigraphically equivalent to the main shelfal limestones (Figure 8). The top of the dominantly shale section contains the planktonic foraminifer *Chiloguembelina cubensis* and the calcareous nannofossil *Sphenolithus ciperoensis*. The presence of these two microfossils restricts the age of the uppermost shales in Composite Section 5 to no younger than the uppermost part of the P21a planktonic zone (latest early Oligocene or earliest late Oligocene; approximately 30.0–30.5 Ma) (Figures 3, 8). Strontium isotopes suggest that the bottom of the section is 32.0–33.0 Ma (+/- 0.3 Ma). Therefore, the shale and thin limestones (large-foram wackestones/packstones) north of the main shelf limestone represent the slope to basinal part of sequence 2 (Figure 8). The sequence 2 shelf limestones at Jaan River can be divided into two main parts—lower transgressive systems tract and highstand systems tract (Figure 8).

Transgressive Systems Tract

The transgressive systems tract comprises the lower 130–140 m of the main Berai Limestone (Figure 8) at Jaan River. Carbonate sedimentation started on the slope and transgressed onto the shelf (Figure 8). The lower transgressive systems tract contains a relatively pure limestone at the shelf margin and more argillaceous carbonates farther back on the shelf. The relatively "clean" shelf-margin carbonates are dominantly coral wackestones, packstones, and boundstones. The more landward argillaceous carbonates include interbedded large-foram wackestones/packstones and coral wackestones/packstones. Deposition of the transgressive systems tract of sequence 2 occurred in normal marine water as indicated by an abundant, diverse assemblage of large, thin, benthic forams and abundant coral. High abundances of thin specimens of *Lepidocyclina (Eulepidina)*, and *Operculina*, together with low abundances of *Cycloclypeus* and planktonic foraminifera suggest that initial deposition of landward parts of sequence 2 occurred in moderately deep (20–30 m), slightly turbid, lagoonal environments. The greater abundance of planktonic specimens in shelf-margin and slope sections indicates more open-marine conditions.

Highstand Systems Tract

The highstand systems tract of sequence 2 includes 250 m of the upper Berai Limestone and contains several distinct facies belts (Figure 8). Shelf-edge facies aggrade (build upward) with only minor basinward progradation. In contrast, coralline-algae, large-foram packstone/grainstones prograde substantially in a lagoonward direction (Figure 8). A depositional model for the highstand systems tract is shown in Figure 13.

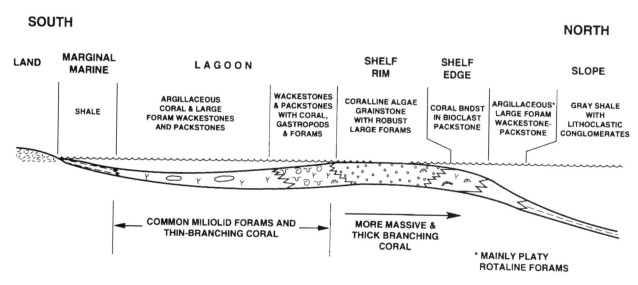

SOUTH   **NORTH**

| LAND | MARGINAL MARINE | LAGOON | SHELF RIM | SHELF EDGE | | SLOPE |

SHALE | ARGILLACEOUS CORAL & LARGE FORAM WACKESTONES AND PACKSTONES | WACKESTONES & PACKSTONES WITH CORAL, GASTROPODS & FORAMS | CORALLINE ALGAE GRAINSTONE WITH ROBUST LARGE FORAMS | CORAL BNDST IN BIOCLAST PACKSTONE | ARGILLACEOUS* LARGE FORAM WACKESTONE- PACKSTONE | GRAY SHALE WITH LITHOCLASTIC CONGLOMERATES

←—— COMMON MILIOLID FORAMS AND THIN-BRANCHING CORAL ——→

MORE MASSIVE & THICK BRANCHING CORAL ——→

* MAINLY PLATY ROTALINE FORAMS

Figure 13. Depositional model for the highstand systems tract of sequences 2 and 3 based on the Jaan River outcrops.

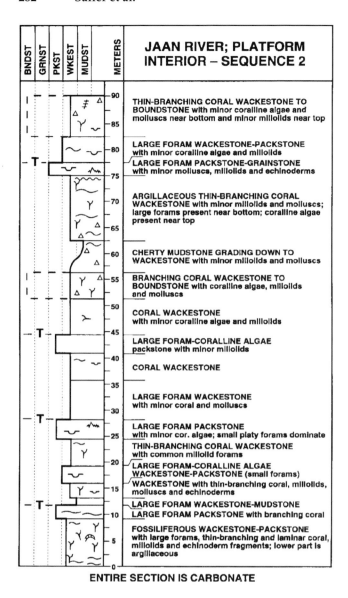

Figure 14. Detailed stratigraphic section in the southern, lagoonal part of the highstand systems tract of sequence 2 in the Jaan River area (approximately 100–190 m below the top of the shelf limestone in composite section 1; Figure 8). It is difficult to separate these strata into shallowing-upward cycles or parasequences. Possible cycle tops (T) are shown, but the confidence level on the location of cycle tops is low. See Figure 11 for legend.

Gray shale with thin interbeds of large-foram marl is the slope equivalent of the main carbonate shelf highstand systems tract (Figures 8, 13). The shale-rich slope deposits in basinward sections are only 150 m thick, in contrast to equivalent carbonate sections on the shelf which are 300–400 m thick (Figure 8). The shales and interbedded large-foram marls are rich in

planktonic forams but lack coralline algae, suggesting deposition on the slope below the photic zone (probably >70 m?). Argillaceous large-foram wackestones and packstones occur between the main carbonate shelf and slope. Planktonic forams, calcareous nannofossils, and very large platy forams in the argillaceous large-foram wackestones/packstones are typical of slope environments (30–90 m deep; Hallock and Glenn, 1986).

Bioclastic packstones and grainstones surrounding coral boundstones at the shelf edge suggest that the shelf edge contained scattered patch reefs surrounded by carbonate sand shoals and grainstone flats in the highstand systems tract (Figures 8, 13). Coralline-algae, large-foram packstones, and grainstones accumulated in shoal complexes near sea level along the shelf rim (Figure 13). Hallock and Glenn (1986) suggested that deposition of similar facies occurred on algal-stabilized open platforms in less than 5 m of water. Corals increase in abundance toward the basin in the grainstone complex (Figures 8, 13). The grainstone complex developed on the shelf rim and prograded substantially lagoonward (to the south), but only slightly basinward (to the north) (Figure 8).

Interbedded branching coral wackestone and bafflestone, large-foram wackestone/packstone, and coral-gastropod wackestone and packstone accumulated in lagoonal environments landward (south) of the shelf rim (Figures 8, 13). Thin-branching coral and miliolid forams (including *Borelis* and *Austrotrillina*) are common. The thin-branching coral are characteristic of quiet, open-lagoonal environments (S. Frost, personal communication). The miliolid forams also indicate deposition in protected lagoonal environments (Hallock and Glenn, 1986). A vertical succession of lithologies from the lagoonal assemblage is shown in a measured section in Figure 14.

Smaller-scale shallowing-upward cycles (parasequences) are not well developed in shelf carbonates in the Jaan River area. Coral and large-foram wackestones and packstones alternate in the transgressive systems tract, but it is not clear what these changing lithologies represent in terms of cyclicity. The shelf rim part of the highstand systems tract contains a fairly monotonous mixture of bioclastic packstones and grainstones in which shallowing-upward cycles were difficult to distinguish. This is similar to bioclastic packstones and grainstones which comprise most of the backreef sediments in the Pleistocene and Holocene of Enewetak Atoll (Couch et al., 1975; Goter and Friedman, 1988). Those Pleistocene and Holocene backreef sediments rarely show systematic vertical changes in depositional lithologies, even below subaerial exposure surfaces. Enewetak Atoll is characterized by relatively rapid Pleistocene subsidence (100 m in 1.7 m.y.; Wardlaw, 1989). Lagoonal strata in the sequence 2 highstand systems tract vary in depositional texture (mainly between wackestone and packstone) and fauna (Figure 14), but distinct shallowing-upward cycles are difficult to identify. Some grain-rich beds (packstones) may indicate the tops of shallowing-upward cycles, but deposition probably did not approach sea level in most cycles (Figure 14).

Sequence 2 at Lemo River

Strontium-isotope chronostratigraphy indicates that the main Berai Limestone in the Lemo River area is 30.5 to 28.0 Ma and can be divided into two units—an onlapping lower unit (30.5–29.7 Ma; sequence 2) and an upper shelf system (29.7–28.0 Ma; sequence 3) (Figure 9). The presence of *Heterostegina cf. H. borneensis* indicates a Te1–4 (late early to late Oligocene) age for the main Berai limestone at Lemo River and supports strontium isotope dating. This dating also suggests that a significant depositional hiatus exists between the end of sequence 1 (Td; 34.0 Ma) and deposition of the Berai Limestone (sequences 2 and 3) (Figure 3).

In the western part of the Lemo River, the lower Berai Limestone (30.5 to 29.7 Ma; sequence 2) is approximately 300 m thick and apparently onlaps a structural high in the eastern part of the Lemo River area (Figure 9). A meter-thick shale in the lower part of composite section 4 (Figure 9) contains the planktonic foraminiferal genus *Chiloguembelina*, which ranges no higher than the top of P21a. The shale also contains the calcareous nannofossil *Sphenolithus ciperoensis*, which constrains the sample to lowermost NP24, correlative to P21a (planktonic foraminiferal zone). Therefore, the shale is latest early Oligocene (ca. 30–30.5 Ma) based on concurrent biostratigraphic zones (Figure 3). A limestone below a meter-thick shale in the eastern outcrops (composite section 4) has a strontium-isotope age of 32.7 Ma (Figure 9). Composite section 4 (Figure 9) is in a more basinward area which apparently allowed deposition of a thin carbonate interval while the area to the west (landward) was subaerially exposed. Later (between 32.6–30.5 Ma), the location of composite section 4 was uplifted causing a depositional hiatus and erosional truncation at that location, while sequence 2 was onlapping the paleohigh. Composite section 4 is on the trend of a structural flexure along which subsequent carbonate shelf margins grew.

In the western parts of the outcrop area, deposition of onlapping limestones occurred in shallowing-upward cycles of bioclastic wackestones, packstones, and grainstones (Figure 15). Most corals in wackestones and boundstones are relatively delicate thin-branching forms characteristic of quiet lagoonal environments. Common miliolids (especially *Borelis*), widespread mollusks, and a scarcity of planktonic forams also suggest that many of the fossiliferous wackestones and packstones were deposited in lagoonal environments. A lack of micrite combined with the presence of robust forams and current-laminations in the grainstones indicate deposition in high-energy, shallow-marine and beach environments.

A thin package (60 m) of sequence 2 carbonates was deposited over the unconformity in the eastern part of the outcrop area (Figure 9). Those carbonates are mainly coral boundstones and bioclastic packstones and grainstones deposited in high-energy, shelf-margin environments.

Depositional Summary of Sequence 2

Sequence 2 accumulated during a period of rising relative sea level. A major transgression began at 32.0–33.0 Ma with coral wackestones and boundstones and large-foram wackestones/packstones deposited on the slope (Figure 8). Stratigraphic equivalents of these coral and large-foram limestones probably form reflectors onlapping the slope at the base of sequence 2 on seismic line KT86-25 (Figure 5). The shelf was flooded as subsidence and transgression continued, initiating carbonate deposition in a transgressive systems tract on the shelf (31–32.0 Ma). "Clean" carbonate sediments accumulated on the shelf margin where shallow, well-agitated waters allowed production of large volumes of carbonate sediment. As the rate of sea level rise slowed, the highstand systems tract developed, and shallow shelf-rim packstones and grainstones prograded over lagoonal wackestones and packstones with only minor basinward progradation of facies (Figure 8).

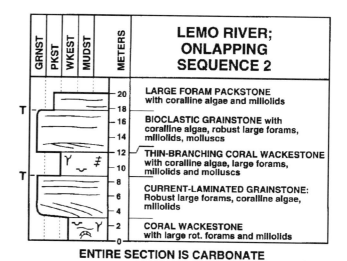

ENTIRE SECTION IS CARBONATE

Figure 15. Detailed stratigraphic section in onlapping sequence 2 in the Lemo River area (approximately 160 m above the top of sequence 1 in composite section 2; Figure 9). Tops of current-laminated grainstones (T) are probably the tops of shallowing-upward cycles (parasequences). This is one of the few places in the Lemo and Jaan River outcrops where well-defined, shallowing-upward cycles are present. See Figure 11 for legend.

SEQUENCE 3

Sequence 3 contains a second carbonate shelf complex in the southern part of the study area. The sequence 3 shelf margin either stepped back landward (southward) from the sequence 2 shelf margin or stacked on top of the sequence 2 shelf margin (Figures 4, 5). Outcrops of sequence 3 were studied along the Jaan and Lemo rivers.

Seismic Data

The boundary between sequences 2 and 3 on the shelf is commonly a surface separating relatively continuous, parallel reflectors at the top of sequence 2 from the more discontinuous to mounded reflectors in the lower part of sequence 3 (Figure 5). The sequence boundary is just below a zone of downlap (SP 1051-1151 on Figure 5). Younger strata of sequence 4 onlap and downlap onto the sequence 3 shelf margin and slope from the north (SP 840-935 on Figure 5 and SP 701-935 on Figure 16). Onlapping/downlapping strata in the basin in sequences 3 and 4 are mainly prograding deltaic strata which filled the basin and lapped up against the shelf margin (SP 551-851 in sequence 3 and SP 701-935 in sequence 4 on Figure 16). Landward-dipping clinoforms within the carbonate shelf start approximately 2.5 km landward of the shelf margin and extend approximately 5 km farther to the south (SP 1051-1151 on Figure 5). This suggests that, initially, the carbonate shelf margin built up rapidly, and then, as the rate of sea-level rise slowed, shallow-marine carbonates prograded 5 km from the shelf margin into the lagoon. Progradation of the shelf margin into the basin was minimal. Sequence 3 can be traced seismically to the surface and along depositional strike to the western Lemo River, where a carbonate section 400–500 m thick was dated with strontium isotopes as 29.9–28.0 Ma (+/- 0.3 Ma).

The Panran well (Figures 2, 7) penetrated 270 m of shelf interior limestones (approximately 8 km south of the shelf margin) correlated to seismic sequence 3. Those limestones are dated as Te1–3, on the basis of *Heterostegina borneensis*. Strontium isotope analyses of sidewall cores from the Panran well (Table 1) indicate that the top of sequence 3 has an age of 27.8 Ma (+/- 0.3 Ma). No suitable material for strontium isotopes was available from deeper in the well. Sidewall cores from sequence 3 in the Panran well are limestones dominated by lagoonal wackestones and micrite-rich packstones with common thin-branching corals, echinoderm fragments, and miliolid forams.

Sequence 3 at Jaan River

Sequence 3 at Jaan River includes a large-foram packstone/grainstone in the uppermost Berai Limestone overlain by a gray shale (Figure 8). These packstones/grainstones consist dominantly of thin, platy large forams with some fragments of coralline algae. Planktonic foraminifera and large thin

Cycloclypeus, Heterostegina, and *Eulepidina* (large forams) increase in abundance upward in the uppermost Berai Limestone. An increase in planktonic forams and a decrease in miliolids in the uppermost Berai suggest flooding to depths of perhaps 30–90 m (?) (Hallock and Glenn, 1986) prior to drowning. The contact of Berai shelf limestones with overlying shales is abrupt. No evidence of subaerial exposure or karstification was observed at the top of the limestone or in the upper 50 m of the Berai Limestone at Jaan River.

The gray shale above the main Berai Limestone contains thin limestone interbeds with abundant planktonic forams and glauconite. Samples from the upper shale contain the late Oligocene calcareous nannofossils *Sphenolithus ciperoensis* and *Helicosphaera recta*. The stratigraphic position of these shales suggest they are part of sequence 3. Abundant planktonic forams with minor platy large forams and no coralline algae suggest a relatively deep, open-marine environment of deposition (>100m ?). At Jaan River, the carbonate shelf deepened after sequence 2 and was drowned and covered by prodelta shales during sequence 3.

Sequence 3 at Lemo River

The carbonate shelf system of sequence 3 at Lemo River is 29.7–28.0 Ma, 400 m thick, and can be divided into three depositional areas—shelf interior, shelf margin, and slope (Figure 9).

Shelf Interior

Shelf interior strata can be divided into two parts—a lower interval of relatively pure limestones and an upper interval with a mixture of pure and argillaceous limestones (Figure 9). The lower interval of clean limestones contains low-energy and high-energy facies (Figure 17). Bioclastic grainstones are commonly interbedded with wackestones, packstones, and boundstones to form shallowing-upward cycles, 10–30 m thick (Figure 17). The upper parts of shelf interior strata are dominated by thin-branching coral wackestone and boundstone, many of which are argillaceous.

Shelf Margin

Shelf-margin carbonates are dominated by coral boundstones and coral-rich bioclastic packstones and grainstones (Figure 9). Corals include massive and thick-branching forms (>3 cm across). Coralline-algae fragments and robust large forams are common to very abundant, suggesting deposition in very shallow, high-energy, open-marine environments (Hallock and Glenn, 1986). No distinct shallowing-upward cycles were observed.

Slope

In the most basinward limestone outcrops along the Lemo River (most northeastward), dolomitized coral boundstones and other shallow-marine carbonates are overlain by interbedded globigerinid wacke-

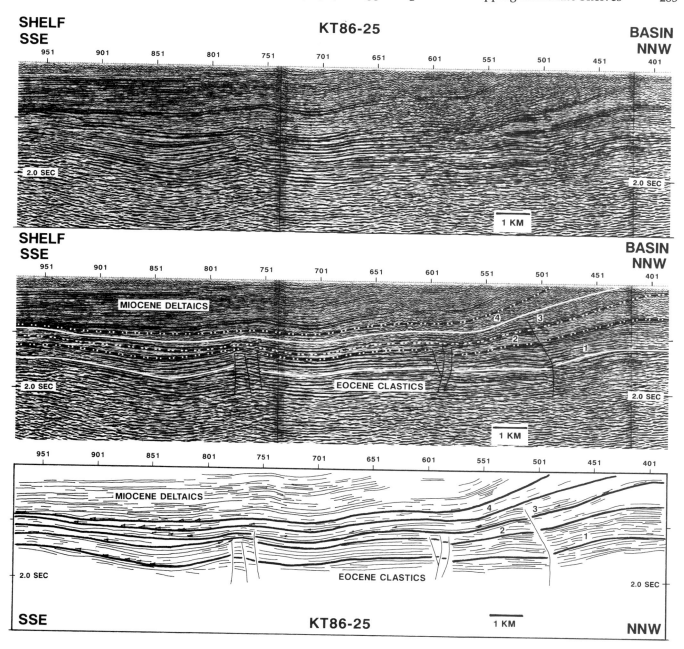

Figure 16. North part of seismic line KT86-25, uninterpreted, interpreted, and interpreted line drawing. This part of the line shows the basin north of the carbonate shelf. Downlapping strata in the basinal parts of sequences 3 and 4 prograde to the south, fill much of the basin, and lap up against the carbonate shelf. Where these reflector lap up against the shelf (SP 851-951), they locally appear to be onlapping, but the basin-filling strata are interpreted as deltaic deposits that prograded to the south and lapped up against the carbonate shelf. Also note truncation at the top of sequence 1 on the small anticlinal structure left of center (SP 760-780). This indicates that the anticlinal structure formed between deposition of sequences 1 and 2.

stones/packstones, graded packstones, and massive, lithoclastic conglomerates (Figures 9, 18). Lithoclasts include shelf-derived packstones, grainstones, and coral-algal boundstones with robust, shallow-water, larger foraminifera. The globigerinid-rich wacke-stones/packstones represent low-energy deposition on a slope between debris flows. Slope strata overlying shallow-marine carbonates (Figure 18) indicate

drowning of the shelf. The drowning and backstepping of the shelf occurred during the earliest part of sequence 3 deposition. Strontium isotope ratios indicate the underlying shelf strata are 29.8 Ma (+/- 0.3 Ma; sequence 2), and the overlying slope strata are 29.6 Ma (+/- 0.3 Ma), therefore drowning of the distal part of the shelf is constrained to approximately 29.7 Ma (+/- 0.3 Ma). This drowning surface correlates to

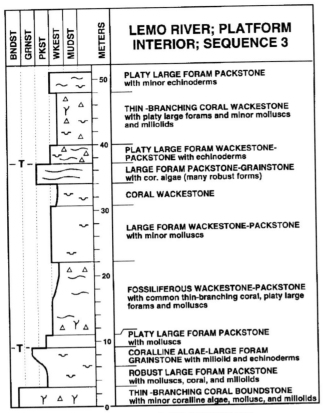

Figure 17. Detailed stratigraphic section of part of the platform interior of sequence 3 at Lemo River. These strata probably represent deposition in lagoonal environments. Shallowing upward cycles (parasequences) are difficult to identify, similar to lagoonal strata in the Jaan River area. Possible cycle tops (packstone/grainstone beds; "T") are marked, but the strata in between are a mixture of depositional lithologies which show no distinct patterns (lower part of sequence 3 in composite section 2; Figure 9). Transgressive (?) platy foram wackestones/packstones commonly occur at the bottom of depositional cycles, and shallow water packstones and grainstones occur at the top with deeper water coral wackestones often in between. See Figure 11 for legend.

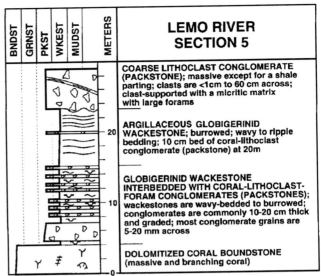

Figure 18. Most basinward section (most northeastern) in the Lemo River transect (section 5; Figure 9). Dolomitized coral boundstone (deposited in a shallow shelf environment) is overlain by interbedded globigerinid wackestone and lithoclast conglomerate deposited in a slope environment. This indicates a drowning of the carbonate shelf. Shelf strata immediately below this section have a strontium isotope age of 29.8 Ma (+/- 0.3 Ma). A large foram from the matrix of the upper lithoclastic conglomerate has a strontium isotope of 29.6 Ma (+/- 0.3 Ma), indicating drowning at approximately 29.7 Ma (+/- 0.3 Ma). See Figure 11 for legend.

the downlap surface observed between deposition of sequences 2 and 3 on seismic lines and also correlates temporally to the drowning of the Jaan River shelf.

Depositional Summary of Sequence 3

Sequence 3 was apparently deposited during another period of relative sea-level rise. The base of sequence 3 is characterized by rapid deepening, local drowning, and/or backstepping of the shelf margin (Figure 19). There is no evidence of widespread subaerial exposure in the upper part of sequence 2 or lower part of sequence 3. High sedimentation rates on

the shelf margin and lower carbonate accumulation rates in adjacent areas during the early parts of sequence 3 resulted in mounding and apparent downlap at the edges of the shelf margin as observed in seismic data (Figures 5, 19). This initial downlap is recognized as the boundary between sequences 2 and 3 on seismic lines.

Sequence 3 has well-developed carbonate facies tracts within the shelf margin and shelf interior. The carbonate shelf of sequence 3 on seismic line KT86-25 (Figure 5) has stratigraphic characteristics similar to sequence 2 at Jaan River (Figure 8). A transgressive systems tract apparently developed in the lower part of sequence 3, and the highstand systems tract developed as the rate of relative sea-level rise slowed. In more basinward parts of the sequence, seismically massive shelf-margin carbonates grade basinward into more continuous, inclined reflectors which probably represent slope strata (Figures 5, 8 and 19). The shelf-rim facies (horizontal reflectors representing coralline-algae, large-foram packstones/grainstones) apparently prograded 5 km landward and overlies lagoon and lagoon-margin facies (inclined reflectors; SP 1051-1151 on Figure 5).

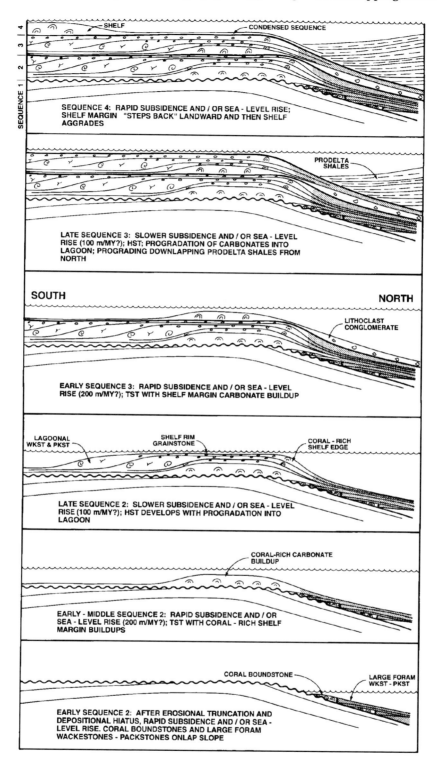

Figure 19. Schematic model for deposition of shelf carbonates in Oligocene sequences 2–4 in Central Kalimantan. Thick lines are sequence boundaries. Lagoonal wackestones and packstones contain thin-branching corals, miliolids, and mollusks and were deposited along the lagoonward-dipping clinoforms. Shelf-rim grainstones are composed dominantly of coralline algae and large forams, and were deposited in parallel beds at the shelf rim. Massive to mounded shelf-edge strata probably represent deposition of coral boundstones surrounded by bioclastic packstones and grainstones. Basinward-dipping clinoforms at the shelf margin probably represent upper slope deposition (large-foram wackestones and packstone). Sequence 1 is 34.0–36.5 Ma. Onlapping part of sequence 2 is 32.0–33.0 Ma. Shelf carbonates of sequence 2 are 29.7–32.0 Ma. Shelf carbonates of sequence 3 are 28.0–29.7 Ma. Shelf carbonates of sequence 4 are >24-28.0 Ma. Rates of subsidence/sea-level rise are approximations to give general ideas of possible rates.

SEQUENCE 4

The carbonate shelf margin of sequence 4 stepped back markedly from its location in sequence 3 (Figures 2, 4). Interbedded shales and carbonate in the Lemo River area were the only outcrops of sequence 4 studied. The carbonate shelf facies of sequence 4 was not observed in outcrop, so facies patterns were not systematically described, and shelf strata were not precisely dated.

Subsurface Data

Seismic lines show only the northernmost part of the carbonate shelf of sequence 4 (Figure 6). The carbonate shelf margin of sequence 4 "steps back" substantially to the south from the sequence 3 shelf margin (Figure 2). The relatively thick shelf deposits of sequence 4 thin substantially to the north, resulting in a condensed interval over the drowned shelf of sequence 3 (Figures 5, 6). The carbonate shelf in sequence 4 is composed largely of discontinuous to mounded reflectors, some of which have apparent downlap onto the top of sequence 3 (SP 840-940 on Figure 6). Strata above sequence 4 onlap and downlap the sequence 4 shelf margin and slope from the north (SP 840-1120 on Figure 5; SP 840-940 on Figure 6; SP 801-910 on Figure 16). Those onlapping/downlapping strata are mainly prograding deltaic shales that filled the basin from the north, lapped up against the shelf margin, and prograded over the shelf (Figures 5, 6, 16). Shales overlying sequence 4 are dated biostratigraphically as early Miocene (N4; approximately 24–? Ma).

Approximately 60–70 m of limestone in the upper part of sequence 4 were penetrated in the Montalat well (Figure 2). The limestones overlie a granite with a potassium-argon age of 63 Ma. The limestones in the upper part of seismic sequence 4 in the Montalat well were dated biostratigraphically as late Oligocene (N3; 28–24 Ma) (Figure 3). The top of "N3" is approximately 24 Ma, which is near the end of the Oligocene (Figure 3). Therefore, sequence 4 is apparently middle late to late late Oligocene in age (approximately 28.0 to >24 Ma). Unfortunately, no material suitable for strontium isotope analyses was obtained (i.e., large forams were diagenetically altered). Sidewall cores from the Montalat well indicate that the interior of the sequence 4 carbonate shelf is dominated by fossiliferous wackestones and packstones with relatively open-marine faunas including large rotaline forams.

Sequence 4 at Lemo River

Approximately 100–150 m of gray shale interbedded with coral wackestone, coral boundstone, and large-foram marls overlie the main Berai Limestone (sequence 3) in the Lemo River area. This interval contains *Eulepidina* and *Spiroclypeus* which indicate that the sections are no younger than Te5 (early Miocene). The upper part of Composite Section 3 contains *Lepidocyclina* (N.) *cf. N. isolepidinoides* indicating

a late Oligocene (Te1–4) age. An argillaceous coral boundstone occurring 10 m above the main Berai limestone (base of sequence 4) and interbedded with shale has a strontium isotope age of 28.0 Ma (+/- 0.3 Ma).

Carbonate interbeds thin and decrease in abundance upward as shale beds thicken and increase in abundance above the "clean" limestones of the main Berai Formation in the Lemo River area. This suggests that carbonate sedimentation gradually decreased as terrigenous sedimentation increased. Foraminiferal assemblages are dominated by small benthics with *Operculina* and juvenile planktonics, suggesting deposition in a deep inter-reef lagoon or inter-reef shelf with limited access to oceanic circulation. Shale with thin sandstones overlie shelf carbonates in the more basinward parts of the Lemo River area. Terrigenous sediments were apparently associated with progradation of prodelta environments over the carbonate shelf and slope.

Depositional Summary of Sequence 4

The shelf margin of sequence 3 drowned at approximately 28.0 Ma, and the locus of rapid carbonate accumulation shifted landward, forming the carbonate shelf of sequence 4 (Figures 4, 5, 19). No evidence of widespread subaerial exposure was observed in the upper part of sequence 3 or lower part of sequence 4 (Figure 9). Rapid carbonate deposition caused aggradation of the shelf, while a much thinner (condensed) sequence of shale and interbedded coral-rich wackestones and boundstones accumulated over the drowned shelf of sequence 3 to the north (Figures 6, 19). The boundary between sequences 3 and 4 is recognized on seismic lines on the shelf as the horizon of lowest downlap of mounded reflectors onto more parallel and continuous reflectors (Figure 6).

DEPOSITIONAL SYSTEM/ SEQUENCE MODEL

The depositional models and sequences in Oligocene carbonate systems of Central Kalimantan are different in some respects from carbonate models described by Sarg (1988). Many of the stratigraphic features observed in Oligocene carbonates of Central Kalimantan are similar to those described by Erlich et al. (1990) and Schlager (1989, 1991) as drowning unconformities. In Central Kalimantan, rapid subsidence and relatively low-amplitude eustatic sea-level fluctuations during the Oligocene apparently caused intervals of rapid carbonate aggradation to alternate with periods of slower aggradation and lagoonward progradation. Sea level did not drop below the shelf between deposition of several sequences (Figure 19). Early (lower) parts of sequences are characterized by rapid sedimentation at the shelf margin during rapid rises of relative sea level (rapid subsidence and/or eustatic sea-level rise). This caused the development of transgressive systems tracts with paleobathymetric highs at the shelf margin and open lagoons in landward areas (Figure 19). As sea-level rise slowed during the upper parts of sequences, highstand systems tracts developed with distinct facies belts including (1) coral-rich shelf-edge wackestones, packstones, grainstones, and boundstones; (2) shelf-rim grainstones; and (3) lagoon to lagoon-margin wackestones/packstones with thin-branching coral and milioid forams (Figures 13, 19). Shelf-rim bioclastic grainstones and lagoon-margin wackestones and packstones prograded into the lagoon as part of the highstand systems tracts (Figures 5, 6, 19). No evidence of subaerial exposure of the shelf was observed.

The most correlative features in the field and on seismic lines are indicative of rapid deepenings. Some shelf margins were drowned and "stepped back" landward (Figures 4, 5, 6, 8). In other cases, the shelf was not drowned, but a substantial deepening of depositional environments is observed. This rapid deepening caused rapid aggradation at the shelf margin, which also caused downlap in the vicinity of shelf margins on seismic lines (Figure 19). Surfaces correlating with the onset of deepening are the proposed sequence boundaries, because no erosional unconformities were observed on the shelf. The observed deepening events were probably the result of rapid subsidence rather than eustatic sea-level rise because the deepening events do not correspond to rises on the Haq et al. (1987) eustatic sea-level curve. Basin modeling indicates that the rapid subsidence observed in the Oligocene of Central Kalimantan could be flexural subsidence caused by sediment loading as thick packages of prograding deltaics filled basins adjacent to carbonate shelves and isolated platforms.

IMPLICATIONS FOR OTHER AREAS

(1) Sequence-stratigraphic studies require integration of regional geology. Oligocene carbonate shelf sequences in Central Kalimantan show baselap of basinal clastics at carbonate shelf margins (Figure 5). Without an understanding of the regional geology, an explorationist might interpret the baselapping reflectors as an onlapping lowstand systems tract and therefore might also predict subaerial exposure and development of porosity at the top of the carbonate shelf associated with the supposed lowstand. Instead of indicating lowstands of sea level, the baselapping reflectors are prodelta clastics deposited during highstands of sea level (Figures 5, 16, 19). Similarly, deltaic sediments were observed by Erlich et al. (1990) downlapping/onlapping (baselapping) drowned Miocene carbonate platforms in the South China Sea.

(2) Similar to the conclusions of Schlager (1991), Oligocene carbonate sequences in Central Kalimantan require a more flexible definition of a sequence boundary than that given in Van Wagoner et al. (1988). The "sequence boundary" of Van Wagoner et al. (1988) requires subaerial exposure with a significant hiatus or erosional truncation (even with

shelf-margin systems tracts and type 2 sequence boundaries). During periods of rapid subsidence and relatively low-amplitude eustatic sea-level fluctuations, sea level might not drop below the "flat-topped" shelf, not allowing subaerial exposure and/or a significant depositional hiatus. However, depositional cycles related to variable rates of subsidence and/or sea-level rise might still produce packages of carbonate rocks that are temporally distinct. Therefore, we prefer Schlager's (1989, 1991) definition that "a sequence boundary represents a geometrically manifest change in the pattern of sediment input and dispersal" in a basin.

(3) The mid-Oligocene (29.5–30 Ma) sea-level drop of Haq et al. (1987) was not observed in Central Kalimantan. Shallowing, subaerial exposure, and diagenesis expected from the 30 Ma sea-level drop were not recognized. Local subsidence in the Teweh region may have compensated for some sea-level drop, but it is difficult to imagine subsidence being severe and coincident enough to compensate for such a large and rapid sea-level fall (100 m in less than 0.1 m.y.) as depicted by Haq et al. (1987). Instead, deepening and apparent drowning of part of the Teweh shelf margin occurred between 29.5 and 30 Ma. These observations suggest that the mid-Oligocene sea-level drop was not as fast and/or as large in magnitude as depicted by Haq et al. (1987) or that the timing of the large sea-level fall is different than shown. (The time scales of Miller et al. [1988] and Haq et al. [1987] are based on the same Geomagnetic Polarity Time Scale.) Additional doubt has been cast over the magnitude of the mid-Oligocene eustatic sea-level drop by other workers. A large (100 m) mid-Oligocene sea-level drop was not recognized on the mid-Pacific Enewetak Atoll by Saller and Koepnick (1990). Based on oxygen-isotopic data on planktonic forams from deep sea cores, Williams (1988) suggests a much smaller sea-level drop (10 m ?) at 30 Ma.

(4) The Haq et al. (1987) curve needs to be tested to see what parts are truly eustatic (global), which are regional, and what are the true amplitudes and rates of sea-level fluctuation. This is not an academic question, because a great deal of porosity in carbonates is related to subaerial exposure during drops of relative sea level. Porosity in giant oil and gas fields like Bu Hasa, Arun, Yates, and Kelly–Snyder may have been enhanced by meteoric water during subaerial exposure. Our ability to predict the timing and magnitude of sea-level drops based on seismic and/or stratigraphic information should allow us to more accurately predict porosity in subsurface carbonates.

ACKNOWLEDGMENTS

We are grateful to G. Dixon, A. Fawthrop, G.A. Crawford, and A. Van de Weerd for setting up this project. Calcareous nannoplankton were analyzed by M.E. Hill, III. A number of people helped during the field work, and we especially thank E. Lumadyo, A. Prayoga, T. Rachwad, N. Arbi, S. Sutiyono, R. Syarif,

Y. Yusuf, Nenang, and W.G. Cutler. Discussions with S. Mahadi, G.D. Jones, Charles Stuart, R.C. Tjalsma, and S.H. Frost were very helpful. Insightful reviews were provided by J.F. Sarg, R.N. Erlich, C. Jordan, T.E. Elliott, and A. Van de Weerd. We thank Unocal, Katy Industries, Inpex, Pertamina, and BKKA (Indonesia) for permission to publish this. We also thank the Indonesian Petroleum Association (IPA) for permission to use figures and text that were previously copyrighted.

REFERENCES CITED

Adams, C.G., 1984, Neogene larger foraminifera, evolutionary and geological events in the context of datum planes, in N. Ikebe and R. Tsuchi, eds., Pacific Neogene Datum Planes: Tokyo, University of Tokyo Press, p. 44-67.

Berggren, W.A., D.V. Kent, and J.J. Flynn, 1985, Neogene chronology and chronostratigraphy, in N.J. Snelling, ed., The Chronology of the Geological Record: Geological Society Memoir 10, p. 141-195.

Berggren, W.A., and K.G. Miller, 1988, Paleogene tropical planktonic foraminiferal biostratigraphy and magnetobiochronology: Micropaleontology, v. 34, p. 362-380.

Blow, W.H., 1979, The Cainozoic Globigerinida: Leiden, E.J. Brill, 1413 p.

Bolli, H.M., and J.B. Saunders, 1985, Oligocene to Holocene low latitude planktonic foraminifera, in H.M. Bolli, J.B. Saunders, and K.D. Perch-Nielson, eds., Plankton Stratigraphy: Cambridge University Press, p. 155-262.

Burke, W.H., R.E. Denison, E.A. Hetherington, R.B. Koepnick, H.F. Nelson, and J.B. Otto, 1982, Variations of seawater $^{87}Sr/^{86}Sr$ throughout Phanerozoic time: Geology, v. 10, p. 516-519.

Couch, R.F., Jr., J.A. Fetzer, E.R. Goter, B.L. Ristvet, E.L. Tremba, D.R. Walter, and V.P. Wendland, 1975, Drilling operations on Enewetak Atoll during project EXPOE: Air Force Publication AFWL-TR-75-216, 270 p.

Erlich, R.N., S.F. Barrett, and B.J. Guo, 1990, Seismic and geologic characteristics of drowning events on carbonate platforms: AAPG Bulletin, v. 74, p. 1523-1537.

Goter, E.R., and G.M. Friedman, 1988, Deposition and diagenesis of the windward reef of Enewetak Atoll: Carbonates and Evaporites, v. 2, p. 157-180.

Hallock, P., and E.C. Glenn, 1986, Larger foraminifera: A tool for paleoenvironmental analysis of Cenozoic carbonate depositional facies: Palaios, v. 1, p. 55-64.

Hamilton, W., 1979, Tectonics of the Indonesian region: U.S. Geological Survey Professional Paper 1078, p. 1-345.

Haq, B.U., J. Hardenbol, and P.R. Vail, 1987, Chronology of fluctuating sea levels since the Triassic: Science, v. 235, p. 1156-1167.

Hardenbol, J., and W.A. Berggren, 1978, A new Paleogene numerical time scale, in G. Cohee, M. Glaessner, and H.D. Hieberg, eds., Contributions to the Geologic Time Scale: AAPG Studies in Geology No. 6, p. 213-234.

Miller, K.G., M.D. Feigenson, D.V. Kent, and R.K. Olsson, 1988, Oligocene stable isotope ($^{87}Sr/^{86}Sr$, $\delta^{18}O$, $\delta_{13}C$) standard section, Deep Sea Drilling Project Site 522: Paleoceanography, v. 3, p. 223-233.

Popp, B.N., F.A. Podosek, J.C. Brannon, T.F. Anderson, and J. Pier, 1986, $^{87}Sr/^{86}Sr$ ratios in Permo-Carboniferous sea water from analyses of well-preserved brachiopod shells: Geochimica et Cosmochimica Acta, v. 50, p. 1321-1328.

Saller, A.H., and R.B. Koepnick, 1990, Eocene to early Miocene growth of Enewetak Atoll: Insight from strontium isotope data: Geological Society of America Bulletin, v. 102, p. 381-390.

Sarg, J.F., 1988, Carbonate sequence stratigraphy, in C.K. Wilgus, B.S. Hastings, C.G.St.C. Kendall, H.W. Posamentier, C.A. Ross, and J.C. Van Wagoner, eds., Sea Level Changes: An Integrated Approach: SEPM Special Publication No. 42, p. 155-181.

Schlager, W., 1989, Drowning unconformities on carbonate platforms, in P.D. Crevello, J.L. Wilson, J.F. Sarg, and J.F. Read, eds., Controls on Carbonate Platform and Basin Development: SEPM Special Publication No. 44, p. 15-25.

Schlager, W., 1991, Depositional bias and environmental change—important factors in sequence stratigraphy: Sedimentary Geology, v. 70, p. 109-130.

Van de Weerd, A., R.A. Armin, S. Mahadi, and P. Ware, 1987, Geologic setting of the Kerendan gas and condensate discovery, Tertiary sedimentation and paleogeography of the northwestern part of the Kutai basin, Kalimantan, Indonesia: Proceedings of the Indonesian Petroleum Association, Sixteenth Annual Convention, p. 317-338.

Van Wagoner, J.C., H. W. Posamentier, R.M. Mitchum, P.R. Vail, J.F. Sarg, T.S. Loutit, and J. Hardenbol, 1988, An overview of the fundamentals of sequence stratigraphy and key definitions, in C.K. Wilgus, B.S. Hastings, C.G.St.C. Kendall, H.W. Posamentier, C.A. Ross, and J.C. Van Wagoner, eds., Sea Level Changes: An Integrated Approach: SEPM Special Publication No. 42, p. 39-45.

Wardlaw, B.R., 1989, Comment and reply on "Strontium-isotope stratigraphy of Enewetak Atoll"—Comment: Geology, v. 17, p. 190-191.

Williams, D.F., 1988, Evidence for and against sea level changes from the stable isotopic record of the Cenozoic, in C.K. Wilgus, B.S. Hastings, C.G.St.C. Kendall, H.W. Posamentier, C.A. Ross, and J.C. Van Wagoner, eds., Sea Level Changes: An Integrated Approach: SEPM Special Publication No. 42, p. 31-36.

Chapter 11

◆

Sequence Stratigraphy of a Miocene Carbonate Buildup, Java Sea

M. A. Cucci
M. H. Clark[1]
ARCO Exploration and Production Technology
Plano, Texas, U.S.A.

◆

ABSTRACT

Establishment of a sequence stratigraphic framework, based on an integration of seismic, well-log, core, and biostratigraphic data, indicates that the Tertiary-aged Gunung Putih carbonate complex in the East Java Sea comprises an asymmetric buildup. This asymmetry was inferred to occur in response to paleo-oceanographic circulation patterns that favored aggradation on the north face and progradation on the south face. The aggradational pattern was associated with carbonate reefal buildups, whereas the progradational patterns were associated with grain-rich forestepping deposits.

Widespread carbonate buildups were initiated on a beveled Cretaceous to early-Tertiary platform during the late Eocene. These buildups exhibit pronounced stratigraphic asymmetry; the northern side is inferred to lie on the paleowindward side and is characterized by bulwark-like framestone-rich buildups. Though the stacking pattern is predominantly aggradational, forestepping and backstepping can occur locally. The southern side is inferred to lie on the paleo-leeward side and is characterized by continuous forestepping buildups with clinoform reflection geometries. Mounded, toe-of-slope, lowstand deposits of several sequences form an apron around the entire Miocene complex.

From the late Eocene through the Miocene, differential subsidence progressively changed the style of carbonate buildup from widespread distribution across the entire platform to more restricted distribution associated with smaller structural highs that occurred on the platform. Eventually, as the platform continued to founder, carbonate deposition ended completely during the late Miocene. The carbonate complex subsequently was buried in terrigenous mudstones and siltstones.

Based on the evolution of the Gunung Putih carbonate complex, a general approach for the exploration for hydrocarbons within carbonate buildups has been developed. Initially, a structural high that caused the sea floor to be sufficiently raised so as to establish a carbonate factory must be identified.

---

[1]Present address: ARCO Oil and Gas Co., Midland, Texas, U.S.A.

Subsequently, evidence for deposition at the top and margins of the structure should be identified. The occurrence of backstepping (i.e., transgressive) deposits, in particular, is suggestive of an active carbonate factory on the platform. Further evidence for carbonate buildups is the re-establishment on the structural crests of shallow-water conditions through the development of stacked, mounded, and clinoforming seismic reflections. Shallow-water conditions also can be inferred through the identification of erosionally truncated or toplapping seismic reflections.

INTRODUCTION

The Tertiary Gunung Putih carbonate complex is situated on a Cretaceous platform that extended from Sulawesi in the northeast to Java in the southwest. The carbonate complex that is the subject of this chapter is a prominent structure on the platform and is located in the northeast quadrant of the Kangean Production Sharing Contract (Figure 1A). It covers an area of approximately 525 km^2 (202 mi^2) and is located approximately 70 km (45 mi) due east of Kangean Island and 140 km (87 mi) north of Bali. The entire study area is 125 km (77 mi) east–west and 30 km (19 mi) north–south (Figure 1B). The carbonate complex likely was deposited in a humid equatorial setting similar to that which exists today.

The Kangean Block in the East Java Sea is operated by Atlantic Richfield Bali North Inc. (55% participation) in partnership with British Petroleum. Although no known hydrocarbon production occurs from any Miocene buildups in this area, a noneconomic Eocene oil accumulation is associated with the early phase of the Gunung Putih buildup at the JS53A structure (Figure 1C).

The objective of this chapter is to present a case study illustrating the development and demise of carbonate buildups that occur in response to structurally associated bathymetric high areas. This chapter first will review the procedures followed in this study by reviewing the stratigraphic framework that was established. Subsequently, this case study will be used as a vehicle for a general discussion regarding hydrocarbon exploration in this type of setting.

Data Base

This study was based on an integration of multichannel reflection seismic, wireline log, and biostratigraphic data. The seismic data comprised an irregularly spaced grid of approximately 1000 km (600 mi) of data acquired between 1981 and 1989. The quality varied from poor to good and most of the data has been migrated. Meaningful information commonly occurs down to 4–5 sec (two-way travel time). Lithologic information was tied to the seismic data through synthetic seismograms generated for

nine wells. Well ties were facilitated by the use of velocity survey information that was available for seven of the nine wells.

The well data used in this study consisted of nine wells drilled by Cities Service Company and Atlantic Richfield Company during the late 1970s to late 1980s. These wells included the JS50-1, JS53A-1, JS53A-2, JS53B-1, Igangan-1, and the four straighthole Pagerungan wells (Figure 2 [Enclosure]). One well (JS50-1) reached total depth in the Miocene, the rest reached either the Eocene or the Cretaceous. The log curves that were most commonly used for correlation purposes in this study were the Gamma Ray, Deep Induction, and Sonic.

The biostratigraphic data was analyzed by Robertson Research (1986b). Sampling was done at 90 ft intervals taken from cuttings. Although palynologic work was done, the salient conclusions were drawn primarily from both planktonic and benthonic foraminiferal analyses.

Approach

Seismic data were interpreted according to the seismic sequence stratigraphic techniques of Vail et al. (1977) and Vail (1987). Seismic sequence stratigraphic terms used in this report follow the conventions of Mitchum (1977), Van Wagoner et al. (1987), and Posamentier et al. (1988).

Basic assumptions of seismic sequence analysis include:

1. primary seismic reflections parallel bedding planes and are inferred to represent time lines;
2. sequence boundaries are expressed as truncated and onlapping seismic reflections (i.e., unconformities) or by concordant reflections (i.e., correlative conformities);
3. seismic reflection discontinuities are used to subdivide sequences into systems tracts; and
4. seismic reflection terminations and geometries as well as reflection attributes (i.e., continuity and amplitude) are used to infer reservoir-prone and seal-prone intervals.

Seismic correlations were difficult in the more structurally complex parts of the study area. In these

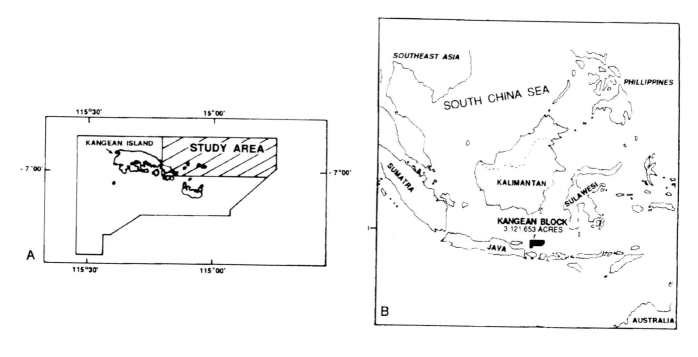

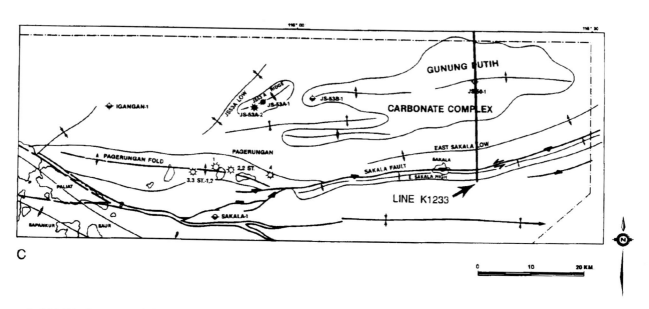

Figure 1. (A) Study area within the Kangean Block; (B) regional location map; (C) structural elements of the Northeast Kangean Block. The Gunung Putih carbonate complex lies on a broad east–west trending anticline. The Sakala fault zone and the Pagerungan fold are large inversion structures that mark the southern boundary of the study area. Position of seismic line used in this report is shown by the north–south black line through the JS-50-1 well.

areas, biostratigraphic and well-log data proved useful in bridging this gap in seismic data quality. The well log and biostratigraphic data were used iteratively with the seismic data in order to interpret across seismically difficult areas. The three data bases involved are characterized by inherently different resolution attributes. The resolution of the wireline log data commonly greatly exceeds that of the other two data bases. However, the detailed information avail-

able from wireline logs can be more difficult to cast into a regional temporal framework. In this area, the resolution of the biostratigraphic data commonly is equivalent to that of the seismic sequences for the Miocene, but somewhat less for the Eocene. In certain instances, where the paleontologic interpretations did not agree with the seismic and well-log sequence analyses, the seismic and well-log correlations were used and the paleontologic interpretations were discounted. This is because of problems commonly associated with biostratigraphic sample collection, preparation, and uphole cavings.

The establishment of a chronostratigraphic framework is complicated by the time-transgressive nature of the stratigraphic framework as developed by Robertson Research (1986a,b,c). Their approach involved the use of lithostratigraphy as an aid to correlation; their formation names, corresponding to lithostratigraphic units, are shown in Figure 3. These formation names are time-transgressive, commonly

encompassing intervals 10 m.y. or greater. Consequently, the Robertson Research stratigraphic framework does not lend itself well to correlation with seismically defined depositional sequences. Nonetheless, in spite of these shortcomings, this nomenclature has been adopted by Java Sea explorationists (e.g., Phillips et al., 1991).

Six key Miocene and four other older seismic sequence boundaries were correlated throughout the grid. These sequences were selected because they could be correlated with confidence throughout the

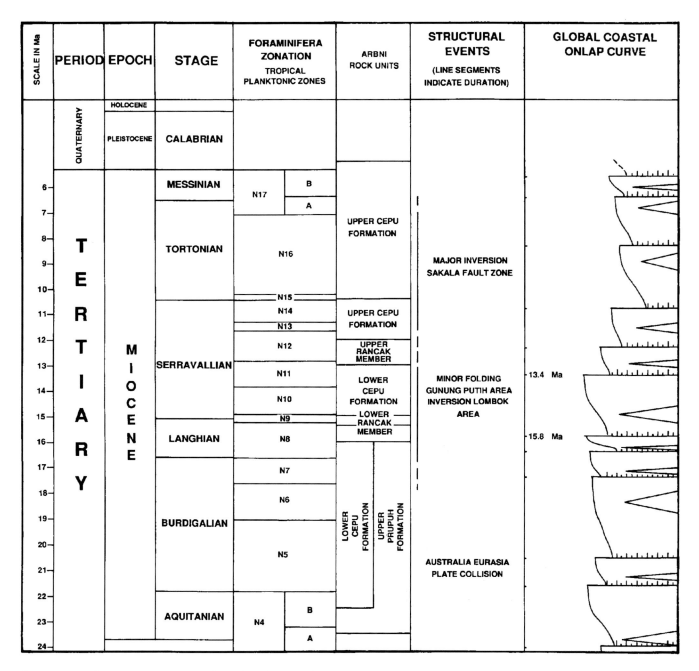

Figure 3. Exploration summary chart for the Northeast Kangean Block relating geologic time, paleontologic and palynologic zones, and the coastal onlap curve (after Haq et al., 1987) to sequences, formations, lithologies, and tectonic events of the study area.

study area. The bounding Miocene unconformities range from Burdigalian to Serravallian in age, and are related to the global sea level curve of Haq et al. (1987) as shown in Figure 3. Note that the geochronology shown in Figure 3 represents a modification of that which is shown in Haq et al. (1987) and is based on internal proprietary studies at ARCO Exploration and Production Technology (P. Thompson, personal communication, 1992).

STRATIGRAPHIC FRAMEWORK

Tectono-Stratigraphy

The initiation of Gunung Putih carbonate deposition occurred in late Eocene time with the development of widespread buildups on a beveled Cretaceous–early Tertiary platform. This platform comprised a tilted fault block of metamorphosed deep-sea deposits. Structural reactivation of pre-existing basement faults occurred during the early Tertiary in response to the collision of the Indian and Eurasian plates (Hutchison, 1989).

No Paleocene sediments are encountered on this platform suggesting that erosion and nondeposition characterized the early Tertiary. Deposition here began during the Eocene, and initially comprised thin siliciclastic sediments on the platform, followed by more substantial carbonate deposition during the subsequent late Eocene through Miocene (Figure 4A).

During Oligocene time, subsidence associated with local tectonism (e.g., basement-involved block faulting) began in the northeast part of the Kangean Block. This resulted in contraction in this area of the carbonate factory to smaller bathymetrically high areas. Small shelfal carbonate buildups were observed on the JS53B-1 structure (Figures 1C, 2) and other small nearby structures, whereas the remaining part of the platform had by then foundered sufficiently so that it was a site of upper bathyal marls and claystones deposition. Near the end of Oligocene time (29.7 Ma), relative sea-level fall, produced primarily by tectonic uplift, resulted in partial erosion of underlying Oligocene and Eocene deposits (Figure 4B).

During lower to middle Miocene time, the Gunung Putih complex was again affected by subsidence, this time, however, related to broad regional tectonism (e.g., folding). Carbonate deposition changed from Oligocene shelfal buildups related to local tectonic events, to high-relief, isolated platforms displaying vertical aggradation with relatively steep margins. The greater thickness of Miocene deposits relative to earlier deposits allows easier discernment of seismic facies within this part of the section (Figure 4C).

Miocene Sequence Stratigraphy

Six sequences can be recognized within the Miocene carbonates in the Kangean area (Figure 2 [Enclosure]). These sequences were defined primarily on the basis of seismic reflection geometries. Well data were used in conjunction with the seismic data

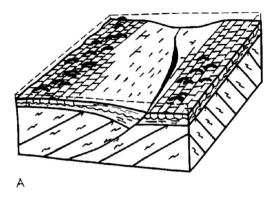

A

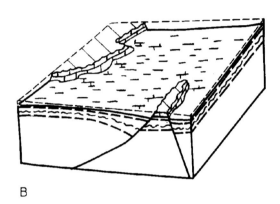

B

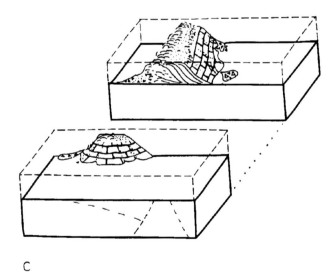

C

Figure 4. Block diagrams illustrating the evolution of the Gunung Putih from the (A) Eocene, to the (B) Oligocene, and the (C) Miocene. The widespread Eocene buildups (mounds) unconformably overlie Eocene siliciclastics. Cretaceous strata form an angular unconformity under the Eocene. Oligocene buildups were deposited on structural highs. They are truncated by the Chattian unconformity (parallel lines). Miocene buildups formed asymmetric buildups at the apices of broad structural folds. The dashed line represents sea level. North lies to the left side of the blocks.

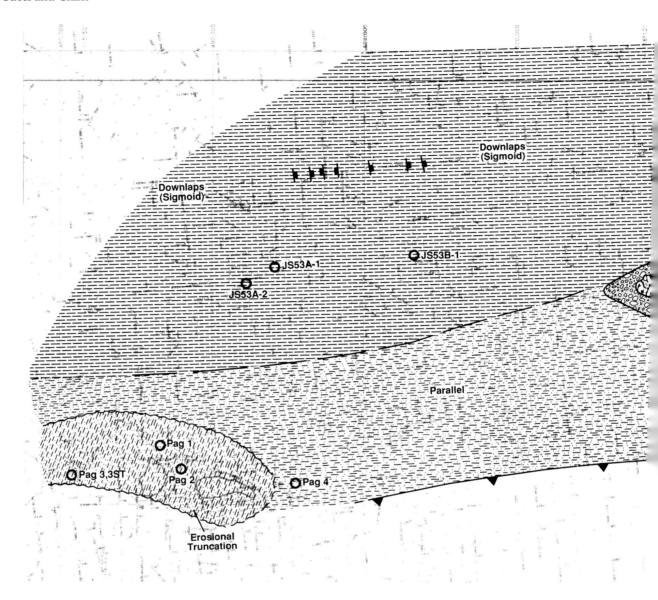

Figure 6. Seismic and geologic facies map of the 15.8–13.4 Ma, Miocene age interval. The strong asymmetry of the Miocene buildup is evident from the facies map showing the boundstone facies along the northern margins and the grain-rich facies along the southern margin. Stacked toe-of-slope, sediment gravity-flow deposits form an apron around the entire margin of the Gunung Putih carbonate complex.

to refine the sequence interpretation. Figure 5 illustrates the stratal geometries associated with one key Miocene sequence. Each sequence can be subdivided into lowstand, transgressive, and highstand systems tracts based on type of bounding surfaces, stratal geometries, and position within sequences (Figures 2 [Enclosure], 5).

For the purposes of this chapter, only one sequence will be analyzed in detail. Consequently, the following discussion will focus on the 15.8–13.4 Ma sequence as shown on Figure 3. This sequence includes excellent examples of each of the component systems tracts. The remaining sequences interpreted here also comprise these same systems tracts, though in some cases not as well expressed. The lowstand, transgres-

sive, and highstand systems tracts of the 15.8–13.4 Ma sequence will be described in detail in the following sections.

Lowstand Systems Tract

The lowstand systems tract represents those sediments deposited below the platform surface during an interval that the platform likely was subaerially exposed. The 15.8 Ma unconformity is the lower bounding surface of this systems tract and this sequence constitutes a Type 1 sequence in the sense of Sarg (1988); it is interpreted to have formed as a result of a major sea-level fall. Sarg (1988) and Rudolph and Lehmann (1989) interpret a similar sea-level fall in the nearby Natuna Basin. This unconformity is seismical-

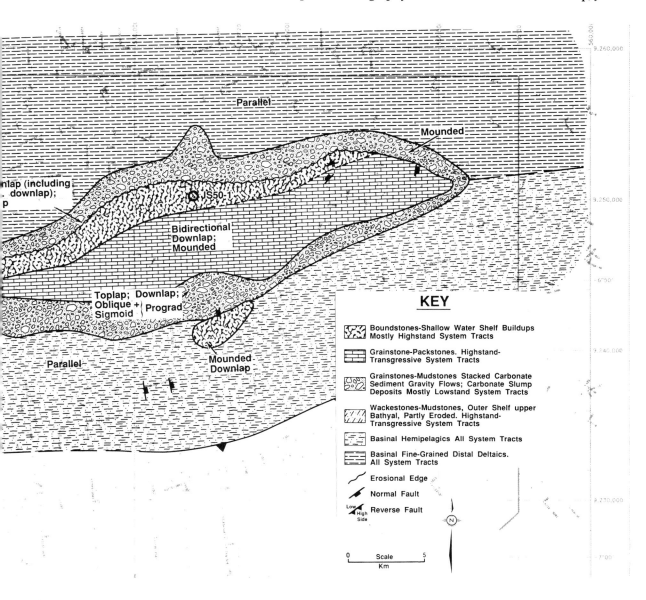

KEY

Boundstones-Shallow Water Shelf Buildups Mostly Highstand System Tracts

Grainstone-Packstones. Highstand-Transgressive System Tracts

Grainstones-Mudstones Stacked Carbonate Sediment Gravity Flows; Carbonate Slump Deposits Mostly Lowstand System Tracts

Wackestones-Mudstones, Outer Shelf upper Bathyal, Partly Eroded. Highstand-Transgressive System Tracts

Basinal Hemipelagics All System Tracts

Basinal Fine-Grained Distal Deltaics. All System Tracts

Erosional Edge

Normal Fault

Reverse Fault

Scale
Km

Figure 6. Continued

ly characterized by: (1) erosional truncation at the platform margin; and (2) reflections onlapping the truncation surface at the platform margin and downlapping the platform slope. These downlapping reflections represent the basinward movement of facies and the regression of the Miocene shoreline. This may be the carbonate expression of a "forced regression" in the sense of Posamentier et al. (1992) (Figure 5, shotpoints 600–750, at 1.6–1.9 sec).

The lowstand systems tract may be subdivided into two parts: (1) the early lowstand systems tract; and (2) the late lowstand systems tract. The two parts of this systems tract are separated by a downlap surface as shown on Figure 5 between shotpoints 700 and 800 at 1.9–2.0 sec. The early lowstand systems

tract is basinally restricted and surrounds the entire complex (Figure 6). The overlying late lowstand systems tract is less basinally restricted, though still below the highstand shelf edge. The early lowstand deposits occur during rapid relative sea-level fall. At this time, most of the carbonate complex is inferred to be subaerially exposed and the carbonate factory is limited to a narrow zone ringing the complex. Late lowstand deposits are formed during the ensuing slow relative sea-level rise (Sarg, 1988; Handford and Loucks, 1993).

The early lowstand deposits correlate to Foraminiferal Zones N8 (upper half) and N9 (see Figure 3). On seismic, these deposits are characterized by high-amplitude mounded reflections overlying the

15.8 Ma sequence boundary where it occurs in the basin. This reflection pattern is interpreted as detrital carbonate sediment gravity flows sourced from erosion and slumping of previous highstand platform carbonates. The mounded facies form an apron around the platform margin that formed during the previous highstand (Figures 6, 7A). The high-amplitude reflections may be due to the occurrence of porous carbonate material comprising the sediment gravity flows, encased in basinal shales. These shale-encased detrital carbonates may be prospective reservoirs and could form potential stratigraphic traps. The high-impedance interface between the carbonates and the shales are thought to be the cause of the high-amplitude reflections (Figure 5, shotpoints 320 to 420, at 1.6–1.5 sec).

The late lowstand systems tract corresponds to Foraminiferal Zone N9 and exhibits pronounced asymmetry in its geographic distribution. On the north side of the complex, these deposits comprise mounded, stacked, bidirectionally downlapping seismic reflections (Figures 5, 6, shotpoints 440 to 450, at 1.5–1.4 sec) that stack vertically, and are interpreted as buildups restricted to the platform margin. These deposits attain a maximum thickness of approximately 130 msec (c. 300 m, 1000 ft). On the south side of the complex, the late lowstand systems tract deposits form forestepping, sigmoidal clinoform, seismic reflection patterns that downlap underlying early lowstand deposits and onlap the older platform margin (Figure 5, shotpoints 700 to 770, at 1.9–2.0 sec). These deposits are interpreted as a prograding carbonate complex occurring below the platform surface. The maximum thickness attained by these deposits is approximately 100 msec (c. 230 m, 750 ft). Both the buildups and the upper portions of the clinoforms may contain porous facies and therefore may be prospective.

The lowstand systems tract is penetrated by the JS53A-1, JS53A-2, JS53B-1, and the Pagerungan-2 wells (Figure 2 [Enclosure]). All of these wells drilled into the distal facies of these depositional units and encountered scattered grainstones, claystones, and marl. The occurrence of scattered relatively coarse-grained detrital carbonates supports the interpretation that in more proximal locations, significant reservoir facies might be encountered. The units that encase these deposits were shown to be claystones and silts and would likely be good seal facies.

Transgressive Systems Tract

The transgressive systems tract is defined as those deposits that occur during an interval characterized by shoreline transgression. This occurs commonly during intervals of rapid relative sea-level rise (Sarg, 1988). In the Gunung Putih area, within the unit directly overlying the buildups and progradation of the late lowstand systems tract, a progressive back-stepping seismic reflection geometry is observed. These backstepping deposits occur in places as mounded seismic facies on the erosional surface atop

the earlier highstand platform (Figure 5, shotpoints 500 to 560, at 1.5 sec). The age of these transgressive deposits corresponds to the base of the N10 Foraminiferal Zone (Figures 2 [Enclosure], 3). The thickness of the transgressive systems tract reaches a maximum thickness of approximately 40 msec (c. 90 m, 300 ft) and is distributed as a widespread veneer on top of the platform. These sediments were penetrated by the JS50-1 well and reported as micritic limestone.

Highstand Systems Tract

Highstand systems tracts form during intervals of slowing relative sea-level rise and are deposited most thickly atop the platform (Sarg, 1988). In the Gunung Putih platform area, an interpreted highstand systems tract is observed to overlie the transgressive systems tract and is separated from it by a downlap surface associated with a possible condensed section. These interpreted highstand deposits are defined seismically by downlapping seismic reflection geometry below and toplapping and erosional truncation seismic reflection geometry above (Figure 5, shotpoints 520 to 540, at 1.3 sec and shotpoints 440 to 480, at 1.3–1.5 sec). The upper bounding surface is interpreted as the 13.4 Ma sequence boundary. Sarg (1988) and Rudolph and Lehmann (1989) interpret a similar sea-level fall as a Type 1 unconformity in the nearby Natuna Basin. The upper bounding surface is characterized by onlapping seismic reflections of subsequent deposits occurring at the platform margin (Figure 5, shotpoints 415 to 440, at 1.5–1.6 sec, and shotpoints 565 to 620, at 1.3–1.5 sec).

As with the late lowstand systems tract, the highstand tract exhibits a pronounced asymmetric stratal geometry (Figure 5). The north end of the platform is characterized by low-amplitude, stacked, and mounded seismic reflections. Reflections on the remaining two-thirds of the platform are parallel grading up into oblique reflections that forestep basinward (Figure 5, shotpoints 550–670, at 1.3–1.5 sec). The highstand systems tract is thickest (i.e., approximately 150 msec [c. 350 m, 1100 ft]) on top of the platform (Figure 7B) and thins into the surrounding basin.

Approximately 300 ft of highstand systems tract is penetrated by the JS50-1 well (Figure 2 [Enclosure]). Based on wireline-log data, these deposits are interpreted as microcrystalline dolomite with minor amounts of sparry calcite. Biostratigraphically, the sediments correspond to the mid-N10 to mid-N11 Foraminiferal zones (Figures 2 [Enclosure], 3).

This highstand systems tract is interpreted to be associated with deposits of both framework-building and grainstone-producing organisms. The framework-building organisms are inferred to be associated with reefs that occur as a rim around the northern edge of the Gunung Putih carbonate complex. The grainstone-producing organisms are inferred to be responsible for the progradational pattern occurring around the southern margin of the carbonate com-

plex. Both types of deposits may be characterized by porous facies locally and therefore may contain prospective reservoirs. Within this systems tract, porosity may be preserved within grainstone shoals associated with the upper parts of clinoforms and/or possibly as vuggy porosity within reefal buildups (Sarg, 1988).

DISCUSSION OF PLATFORM EVOLUTION

The architecture of the middle Miocene buildup was controlled by: (1) the amount and type of skeletal carbonate material produced on the platform; (2) subsidence; (3) eustatic change; and (4) dominant paleowind direction. Significant production of carbonate material is limited to shallow-water depths that can support abundant growth of photosynthetic organisms (Schlager, 1981). Therefore, during slow relative sea-level rises, isolated platforms are not as readily drowned and therefore can continue as "carbonate factories" for the production of skeletal material. The paleowindward side (i.e., present-day north) is where framework-building organisms nucleated (Figure 5, shotpoints 445–485, at 1.33–1.5 sec) and is characterized by low-amplitude, mounded seismic facies. These seismic facies are interpreted as keep-up reefs. The remaining two-thirds of the carbonate platform (Figure 5, shotpoints 485–670) is characterized by moderate- to high-amplitude, parallel to oblique seismic reflection geometry. The relatively thick deposits of the highstand systems tract on the platform, compared with the relatively thin deposits of the lowstand and transgressive systems tracts in the same area, suggests higher rates of deposition at that time (Wilber et al., 1990).

Both mounded and oblique seismic facies are indicative of "keep-up carbonate systems"; i.e., carbonate production was able to keep pace with the relative rise in sea level (Kendall and Schlager, 1981; Sarg, 1988). The influence of inferred dominant paleowind on current direction, from present-day north to south, accounted for the pronounced asymmetry of the platform. The framework builders presumably preferred the better-aerated and agitated oceanic waters on the windward side, whereas skeletal grains and mud generated on the platform would have been transported off the leeward side of the platform. Similar observations regarding platform asymmetry have been noted by Hine et al. (1981) in the Recent of the Bahamas; Wendte and Stoakes (1982) in the Devonian of the Western Canadian Sedimentary Basin; Eberli and Ginsburg (1989) in the Cenozoic of the Bahamas; Rudolph and Lehmann (1989) in the Miocene of the Natuna Platform, Indonesia; and by Handford and Loucks (1993) worldwide. Whereas Hine et al. (1981), Wendte and Stoakes (1982), Eberli and Ginsburg (1989), and Handford and Loucks (1993) explain the observed asymmetry by invoking a process similar to that invoked here, Rudolph and Lehmann (1989) ascribe this partitioning to differences in carbonate productivity across the platform top.

During relative sea-level fall or lowstand, the carbonate platform was subaerially exposed. During this time, the carbonate factory was significantly reduced in aerial extent. Those sediments deposited at this time were restricted to platform-fringing sediment gravity flow deposits sourced from the relatively steep platform margins (>30°). The evidence for extensive erosion and slumping in these areas is the occurrence of widespread seismic reflection truncations at the edge of the carbonate platform. The inference that these deposits constitute a part of the lowstand systems tract is based on their stratigraphic occurrence between highstand and transgressive deposits. The volume of sediments produced by erosion of the platform slopes may have been enhanced by periodic earthquake-generated collapse (scalloped edges) as has been proposed by Mullins and Hine (1989) for offshore Bahamas. Figure 6 shows a scalloped margin adjacent to a toe-of-slope "mounded downlap" seismic facies on the south side of the complex.

Carbonate production on the Gunung Putih platform eventually terminated when accelerated relative sea-level rise overwhelmed the rate of carbonate production producing what Schlager (1989) and Erlich et al. (1990) have referred to as a drowning unconformity. Apparently, the subsidence rates began increasing during the latter part of middle Miocene time (10.8 Ma) and by the late Miocene, the platform finally foundered, thereby ending carbonate production. Subsequently, later structural inversion of the portions of the platform area resulted in renewed carbonate deposition locally in the form of reefal buildups from the late Miocene to Holocene.

EXPLORATION IMPLICATIONS

Potential reservoir- and seal-prone sections were identified and mapped based on seismic reflection configuration and continuity, geometric relations of reflections to bounding surfaces, amplitude and frequency of reflections, and three-dimensional geometry of the seismic facies. These types of facies have been observed within each of the identified systems tracts and occur within specific and predictable areas within each depositional unit.

High-amplitude bidirectionally mounded reflections were identified in the lowstand systems tract of the 15.8–13.4 Ma sequence. These were interpreted as sediment gravity flow deposits, which form an apron around the relict (i.e., pre-15.8 Ma) platform margin (Figures 5, 6, 7A). A favorable reservoir-seal relationship is established by the onlapping basinal (or possibly prodelta) shales (Figure 5, shotpoints 300–350, at 1.50–1.65 sec). There is a risk that no lateral seal is present in a depositional updip position. Therefore, some component of tilting towards the platform needs to be established (e.g., a basinal high caused by faulting, regional tilting, draping, differential compaction, etc.) to provide a viable trap. Tilting of this type has been shown to be productive in the Golden

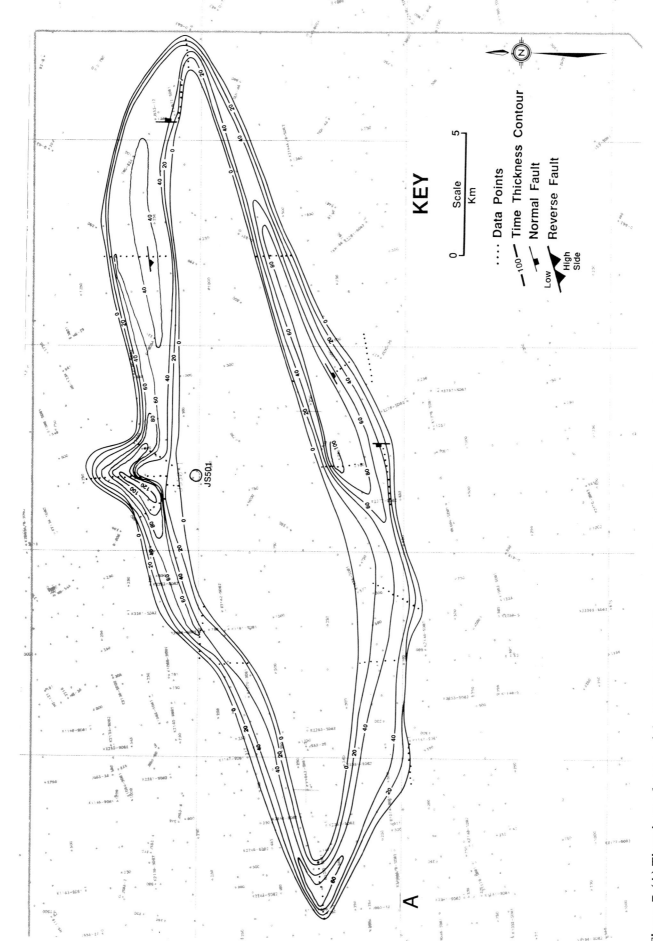

KEY

⋯ Data Points
—100— Time Thickness Contour
⊢ Normal Fault
Reverse Fault

Low ⊢
▼ High Side

Scale
0 ————— 5
Km

A

Figure 7. (A) Time-isopach map of the 15.8–13.4 Ma, Miocene age lowstand sediment gravity-flow deposits. The thickest deposits form downslope of where the buildup is most areally extensive.

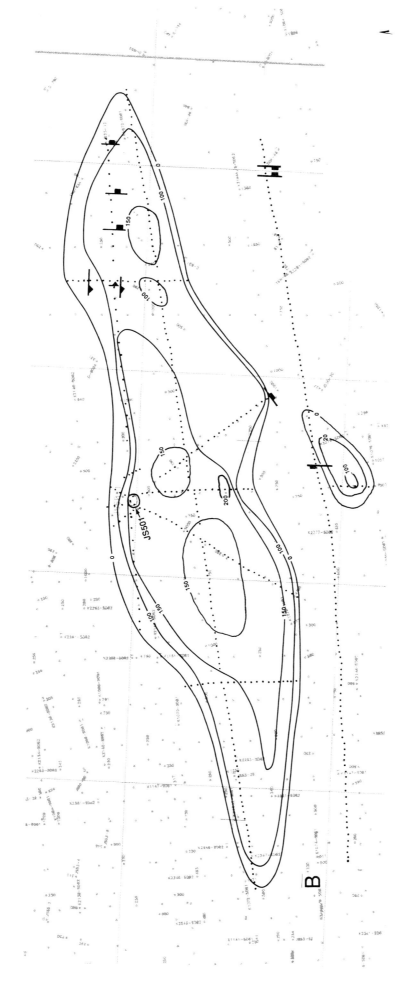

Figure 7. (B) Time-isopach map of the 15.8–13.4 Ma, Miocene age transgressive and highstand deposits. The transgressive-highstand buildups form locally thick deposits on the crest of the structure.

Lane of Mexico (Enos, 1983).

A potential seal-prone interval is established by the deposits of the transgressive systems tract. Within this systems tract, seismic reflections are characterized by progressive backstepping or onlap (Figure 5, shotpoints 500–600, at 1.5–1.6 sec) onto the relict platform margin (i.e., pre-15.8 Ma). This interval forms a thin veneer (<40 msec, 90 m, 300 ft) on top of the relict platform and is shown by well-bore cuttings to be shaly. Consequently, these deposits may provide a potential seal for the underlying highstand systems tract of the previous sequence (i.e., pre-15.8 Ma).

Two reservoir-prone facies, reef buildups, and prograding grainstone shoals were identified on the basis of seismic reflection geometries within the highstand systems tract of the 15.8–13.4 Ma sequence. Reef facies were defined by mounded seismic reflection geometries and fore-reef downlap; grainstone shoal facies were defined by oblique clinoform seismic reflections. Both facies were inferred to have been deposited under shallow-water conditions ("keep-up carbonate systems") conducive to good reservoir development.

CONCLUSIONS

Miocene carbonate reservoir and seal facies have been identified and mapped in the Gunung Putih area offshore Indonesia. This carbonate platform complex is shown to have evolved as a predictable succession of sequences and systems tracts. Six Miocene seismic sequences were identified and correlated on a grid of migrated seismic lines that covered the northeastern Kangean Block. A sequence stratigraphic analysis integrating seismic, wireline log, cuttings, and paleontologic data, provided criteria for the recognition of Miocene carbonate reservoirs and seals for this area.

Miocene carbonates of the Gunung Putih complex evolved in response to tectonic and eustatic events in a systematic manner. From the late Eocene through the Miocene, increased subsidence in the south progressively changed the style of carbonate buildup from widespread distribution across the entire platform to more restricted distribution associated with smaller intra-platform structural highs. In addition to relative sea-level effects, paleoclimatology may have played an important role in the depositional stratal architecture. Distinct depositional asymmetry observed during this interval, characterized by aggradation of carbonate deposits in the north and progradation of carbonate deposits in the south, is attributed to an inferred dominant paleowind direction oriented from north to south.

Within each sequence, the sedimentary responses to relative sea-level change are expressed by a predictable succession of systems tracts. Lowstand systems tracts are characterized by the basinward shift of sediment, subaerial exposure of the platform crest, and erosion and slumping of the platform margin. Lowstand carbonate reservoirs consist of sediment gravity flow deposits occurring during the early lowstand, and platform margin buildups and grainstones occurring during the later lowstand. Carbonate transgressive systems tracts are characterized by aggradationally stacked carbonate sediments occurring during platform flooding. These sediments are mud-prone in the Gunung Putih Complex. The condensed section at the top of the transgressive systems tract is expressed seismically as a downlap surface and acts as a potentially regional seal. Subsequently, highstand systems tract deposits are observed to overlie this regional downlap surface as deposition is reestablished in this area in response to slowing of relative sea-level rise. Carbonate reservoirs in this systems tract are associated with development of boundstone facies, and aggradation and progradation of grainstone facies.

ACKNOWLEDGMENTS

We thank T. Phillips, Geology Manager of Atlantic Richfield Bali North Inc.; D. Nicklin, Chief Geologist ARCO International; J. Pape, Exploration Manager of British Petroleum Exploration; and Pertamina for permission to publish this study. This study benefited from discussions with A. Brown, R. Handford, and R. Loucks, all of ARCO Exploration and Production Technology. We thank R. Bennett, who originally brought this data to our attention and recommended a detailed sequence stratigraphic study of the Gunung Putih sequences. We also thank Joanna Richardson for her help in typing earlier versions of the text, and J. Duke, L. Bradshaw, and G. Garrett for their contributions to the artwork which accompanies this text. This text was greatly improved by the reviews of H. W. Posamentier, M. Candelaria, R.G. Loucks, H.W. Mueller and colleagues, and J.F. Sarg.

REFERENCES CITED

Eberli, G.P., and R.N. Ginsburg, 1989, Cenozoic progradation of northwestern Great Bahama Bank, a record of lateral platform growth and sea-level fluctuations, in P.D. Crevello, J.L. Wilson, J.F. Sarg, and J.F. Read, eds., Controls on Carbonate Platform and Basin Development: SEPM Special Publication No. 44, p. 339-351.

Enos, P., 1983, Fore-reef slope environment, in P.A. Scholle, D.G. Bebout, and C.H. Moore, eds., Carbonate Depositional Environments: AAPG Memoir 33, p. 507-537.

Erlich, R.N., S.F. Barrett, and G.B. Ju, 1990, Seismic and geologic characteristics of drowning events on carbonate platforms: AAPG Bulletin, v. 74, p. 1523-1537.

Handford, C.R., and R.G. Loucks, 1993, Carbonate depositional sequences and systems tracts—response of carbonate platforms to relative sea level changes, this volume.

Haq, B., J. Hardenbol, and P.R. Vail, 1987, Chronology of fluctuating sea levels since the

Triassic: Science, v. 235, p. 1156-1167.

Hine, A.C., R.J. Wilber, and A.C. Neumann, 1981, Carbonate bodies along contrasting shallow bank margins facing open seaways in northern Bahamas: AAPG Bulletin, v. 65, p. 261-290.

Hutchison, C.S., 1989, Geologic Evolution of Southeast Asia: Clarendon Press, Oxford, 368 p.

Kendall, C.G.St.C., and W. Schlager, 1981, Carbonates and relative changes in sea level: Marine Geology, v. 44, p. 181-212.

Mitchum, R.M., Jr., 1977, Seismic stratigraphy and global changes of sea level, Part 11: Glossary of terms used in seismic stratigraphy, in C.W. Payton, ed., Seismic Stratigraphy—Applications to Hydrocarbon Exploration: AAPG Memoir 26, p. 205-212.

Mullins, H.T., and A.C. Hine, 1989, Scalloped bank margins: Beginning of the end for carbonate platforms?: Geology, v. 17, p. 30-33.

Phillips, T.L., R.A. Noble, and F.F. Sinartio, 1991, Origin of hydrocarbons, Kangean Block Northern Platform, Offshore N.E. Java Sea: Proceedings Indonesian Petroleum Association, 20th Annual Convention, October, IPA 91-11.12, p. 637-643.

Posamentier, H.W., G.P. Allen, D.P. James, and M. Tesson, 1992, Forced regressions in a sequence stratigraphic framework: Concepts, examples, and exploration significance: AAPG Bulletin, v. 76, p. 1687-1709.

Posamentier, H.W., M.T. Jervey, and P.R. Vail, 1988, Eustatic controls on eustatic deposition, I, conceptual framework, in C.K. Wilgus, B.S. Hastings, C.G.St.C. Kendall, H.W. Posamentier, C.A. Ross, and J.C. Van Wagoner, eds., Sea Level Changes: An Integrated Approach: SEPM Special Publication No. 42, p. 109-124.

Robertson Research, 1986a, Volume 1: Regional text and appendices East Java and Java Sea basinal area; stratigraphy, petroleum geochemistry and petroleum geology: Robertson Research Int. Ltd., Llandudno, Wales (jointly issued with Pertamina E & P, Jakarta).

Robertson Research, 1986b, Volume 2: Biostratigraphic and petroleum geochemistry well reports Alpha-1 to Terang-1 East Java and Java Sea basinal area; stratigraphy, petroleum geochemistry and petroleum geology: Robertson Research Int. Ltd., Llandudno, Wales (jointly issued with Pertamina E & P, Jakarta), component sections individually paged.

Robertson Research, 1986c, Volume 4: Biostratigraphic and petroleum geochemistry well reports East Java and Java Sea basinal area; stratigraphy, petroleum geochemistry and petroleum geology: Robertson Research Int. Ltd., Llandudno, Wales (jointly issued with Pertamina E & P, Jakarta), 68 plates.

Rudolph, K.W., and P.J. Lehmann, 1989, Platform evolution and sequence stratigraphy of the Natuna Platform, South China Sea, in P.D. Crevello, J.L. Wilson, J.F. Sarg, and J.F. Read, eds., Controls on Carbonate Platform and Basin Development: SEPM Special Publication No. 44, p. 353-361.

Sarg, J.F., 1988, Carbonate sequence stratigraphy, in C.K. Wilgus, B.S. Hastings, C.G.St.C. Kendall, H.W. Posamentier, C.A. Ross, and J.C. Van Wagoner, eds., Sea Level Changes: An Integrated Approach: SEPM Special Publication No. 42, p. 155-181.

Schlager, W., 1981, The paradox of drowned reefs and carbonate platforms: in P.D. Crevello, J.L. Wilson, J.F. Sarg, and J.F. Read, eds., Controls on Carbonate Platform and Basin Development: SEPM Special Publication No. 44, p. 16-25.

Schlager, W., 1989, Drowning unconformities on carbonate platforms: Geological Society of America Bulletin, v. 92, p. 197-211.

Vail, P.R., 1987, Seismic stratigraphy using sequence stratigraphy, Part 1: Seismic stratigraphy interpretation procedure, in A.W. Bally, ed., Atlas of Seismic Stratigraphy, v. 1: AAPG Studies in Geology 27, p. 1-10.

Vail, P.R., R.M. Mitchum, Jr., R.G. Todd, J.M. Widmier, S. Thompson, III, J.B. Sangree, J.N. Bubb, and W.G. Hatlelid, 1977, Seismic stratigraphy and global changes of sea level, in C.E. Payton, ed., Seismic Stratigraphy—Applications to Hydrocarbon Exploration: AAPG Memoir 26, p. 49-212.

Van Wagoner, J.C., R.M. Mitchum, Jr., H.W. Posamentier, and P.R. Vail, 1987, Seismic stratigraphy using sequence stratigraphy, part 2: Key definitions of sequence stratigraphy, in A. W. Bally, ed., Atlas of Seismic Stratigraphy, v. 1: AAPG Studies in Geology 27, p. 11-14.

Wendte, J.C., and F.A. Stoakes, 1982, Evolution and corresponding porosity distribution of the Judy Creek reef complex, Upper Devonian, Central Alberta, in W.G. Cutler, ed., Canada's Giant Hydrocarbon Reservoirs: Canadian Society of Petroleum Geologists 1982 Core Conference, p. 63-81.

Wilber, R.J., J.D. Milliman, Jr., and R.B. Halley, 1990, Accumulation of bank-top sediment on the western slope of Great Bahama Bank: Rapid progradation of a carbonate megabank: Geology, v. 18, p. 970-974.

Chapter 12

Parasequence Stacking Patterns, Third-Order Accommodation Events, and Sequence Stratigraphy of Middle to Upper Cambrian Platform Carbonates, Bonanza King Formation, Southern Great Basin

Isabel P. Montañez
David A. Osleger
Department of Earth Sciences
University of California
Riverside, California, U.S.A.

ABSTRACT

The Bonanza King Formation of the southern Great Basin is composed of 150 to 250 carbonate parasequences (0.5 to 7 m thick) that provide a remarkable "strip chart" of Middle to Late Cambrian third-order accommodation history. Six superb exposures of the Banded Mountain Member of the Bonanza King Formation were logged on a decimeter scale and correlated to generate a detailed platform-to-basin transect. These cyclic carbonates were deposited on a flat-topped, fully-aggraded platform that extended approximately 250 to 300 km across the early Paleozoic passive margin of southern Nevada and eastern California. The Banded Mountain Member ranges from 400 to 1330 m in thickness and forms a westward-thickening wedge from craton margin facies in the eastern Mojave to shelf-edge and base-of-slope facies in the Last Chance Range.

Thick successions of meter-scale carbonate parasequences show systematic changes in parasequence type, dominant lithofacies, and thickness vertically through the cyclic interval that were governed by long-term, third-order changes in accommodation. Fischer plots provide a graphic illustration of changes in accommodation space through time, and appear to be a valuable tool for correlating sequence boundaries and systems tracts. Similar patterns of positive and negative slopes on Fischer plots of the five platform sections of the Banded Mountain Member define four major sets of long-term increases in accommodation followed by long-term decreases in accommodation that correspond to four distinct depositional sequences.

Translating sequence stratigraphic principles originally defined for seismic-scale siliciclastic systems to outcrops of cyclic carbonates requires a de-emphasis of stratal geometries and an increased awareness of correlative vertical changes in stacking patterns of component parasequences. Sequence boundaries and transitions between systems tracts on flat-topped platforms, such as the Bonanza King, are believed to be *zones* rather than distinct horizons because of the effect of the higher frequency sea-level signal superimposed on the long-term, third-order event. Sequence boundary zones bracketing the four depositional sequences within the Banded Mountain Member are characterized by thin, tidal-flat dominated parasequences exhibiting abundant evidence for repeated episodes of exposure. Transgressive systems tracts in the Banded Mountain Member are characterized by a lower succession of parasequences composed of subequal amounts of tidal-flat and subtidal lithofacies passing upward into thicker parasequences with higher percentages of deeper subtidal facies. Highstand systems tracts in the Banded Mountain Member are characterized by thinning-upward stacks of parasequences showing progressively shallower peritidal conditions within the component lithofacies.

INTRODUCTION

Much of the current research in carbonate depositional systems has been oriented toward the application of sequence stratigraphic techniques to the carbonate rock record (e.g., Sarg, 1988; Franseen et al., 1989; Sonnenfeld, 1991). One major problem that has been encountered, however, is the translation of sequence stratigraphic concepts defined for siliciclastic systems into the carbonate realm. A primary difference is that carbonate sediment supply is not dependent upon an allochthonous source but is generated directly on the platform and is redistributed back onto tidal flats or out onto the deeper shelf. Because carbonate sediment production and the consequent stratal geometries are so sensitive to changes in the depth of the photic zone, shallow-water carbonates are probably better recorders of accommodation change than siliciclastics. The common and widespread occurrence of meter-scale shallowing-upward cycles (herein used synonymously and interchangeably with parasequences) throughout carbonate successions provides an ideal record of accommodation history. Systematic vertical changes in cycle type, dominant cycle composition, and cycle thickness (i.e., cycle-stacking patterns) provide an extremely valuable data set for identifying the various internal components of sequences and perhaps of the long-term accommodation events that may have generated them (Goldhammer et al., 1990; Mitchum and Van Wagoner, 1991; Osleger and Read, 1991; Montañez and Read, 1992).

Compared to the traditional methods of defining sequences and their component systems tracts by geometric relations of reflection contacts on seismic sections, distinct problems are encountered when translating sequence stratigraphic concepts to the outcrop. Ideally, sequence boundaries and systems tracts should be identified by lateral tracing of critical horizons suspected of being onlap or downlap surfaces. This is extremely difficult to do, however, within cyclic carbonates deposited on broad passive margins hundreds of kilometers wide, where outcrops may be isolated in mountain ranges tens of kilometers apart, and where biostratigraphic control may be lacking for accurate updip and downdip correlation. Individual surfaces may be traceable for kilometers along mountain fronts but, other than major regional unconformities, can rarely be tracked between ranges. Variations in parasequence stacking patterns provide a high-resolution data set for cross-platform correlation and systems tract identification on broad, fully aggraded carbonate platforms.

The Bonanza King Formation of the southern Great Basin is composed of from 150- to 250-m-scale carbonate parasequences that provide a remarkable strip record of Middle to Late Cambrian third-order accommodation history. The primary objective of this chapter is to use the excellent exposures, broad outcrop area, and extraordinary cyclicity of the Bonanza King Formation as a high-resolution data base to identify larger scale depositional sequences and their component systems tracts. The second objective is to illustrate how parasequence stacking patterns can be used to distinguish the long-term changes in accommodation that generated the depositional sequences in the Bonanza King.

STRATIGRAPHIC AND TECTONIC SETTING

Complete sections of the Banded Mountain Member of the Bonanza King Formation totaling 4200 m were logged on a decimeter scale in superb exposures in six locations spread out across the Middle to Late Cambrian passive margin of the southern Great Basin (Figure 1). The Banded Mountain Member was concentrated on because it is bounded below by a distinctive orange- to red-weathering silty dolomite and above by the Dunderberg Shale, two widespread marker units throughout the southern Great Basin. The Banded Mountain Member ranges from 400 to 1330 m in thickness and forms a westward-thickening wedge from craton margin facies near Las Vegas (total subsidence rates ~0.017 to 0.025 m/k.y.) to outer shelf and base-of-slope facies in the Last Chance Range (~0.08 to 0.11 m/k.y.) (Figure 2). In addition to standard lithologic description on the outcrop, five to ten samples of each lithofacies were collected for petrographic analysis to aid in the interpretation of paleoenvironments.

Biostratigraphic control has been established on the basis of trilobite zonation (Figure 3; Palmer and Hazzard, 1956; Palmer, 1965; Sundberg, 1990) but resolution in this area is poor because few trilobites lived in the generally restricted depositional environments. Areal distribution of the formation throughout the southern Great Basin has been established (Hazzard, 1937; Barnes and Palmer, 1961; Barnes et al., 1962; Gans, 1974), and excellent lithofacies work has been done on the Bonanza King and its equivalents (Kepper, 1972, 1976, 1981), but no systematic analysis of the cyclostratigraphy has been performed. The striking cyclicity visible along mountain front outcrops (Figure 4), however, suggests that a "strip chart" of superimposed orders of Middle to Late Cambrian relative sea-level oscillations are recorded in the cyclic strata of the Bonanza King Formation.

Platform Morphology

The Bonanza King Formation was deposited on the Cordilleran passive margin, which originated in response to breakup of a late Proterozoic super continent around 600 to 550 Ma (Stewart and Suczek, 1977; Bond et al., 1984; Levy and Christie-Blick, 1991). The continental platform edge during the Cambrian extended essentially east–west at about 10 to 15° N latitude (Scotese and McKerrow, 1990). The shelf morphology during Bonanza King time is believed to have been a flat-topped, fully aggraded platform that

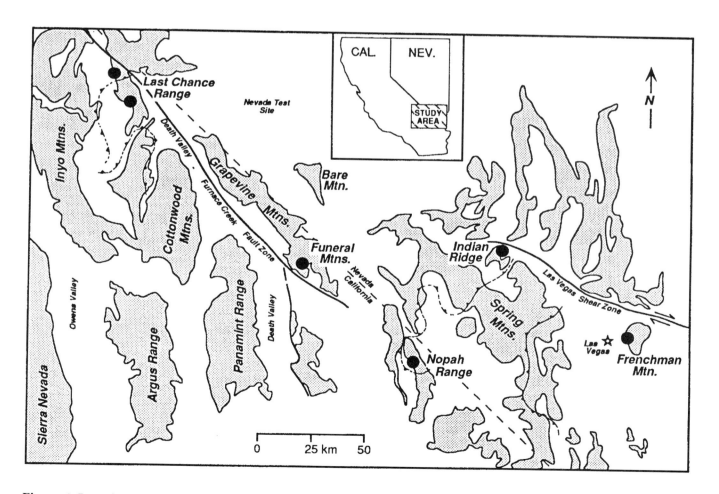

Figure 1. Location map of outcrop sections of the Bonanza King Formation in the southern Great Basin (adapted from Wernicke et al., 1988).

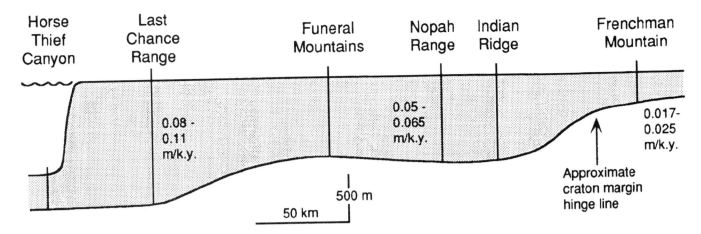

Figure 2. Palinspastically estimated distances between outcrop localities across the Middle to Late Cambrian passive margin of the southern Great Basin. Note the flat-topped, fully aggraded profile, the progressive basinward thickening, and the steep platform edge. Palinspastic distances estimated from Levy and Christie-Blick (1989, 1991), Snow and Wernicke (1989), Rowland et al. (1990), and Snow (1992). Total subsidence rates were calculated by dividing lithified thicknesses of individual sections by an estimated Middle to Late Cambrian time span of 15 (± 50%) m.y. (Palmer, 1983).

BIOMERE	TRILOBITE ZONE	FRENCHMAN MTN. - PROVIDENCE MTNS.		SPRING MTS. - NOPAH RANGE	INYO MTS.- LAST CHANCE RANGE
LATE CAMBRIAN Pterocephaliid	Aphelaspis	NOPAH FM. Dunderberg Mbr.		NOPAH FM. Dunderberg Mbr.	NOPAH FM
	Crepicephalus	BONANZA KING FORMATION		BANDED MOUNTAIN MEMBER	EMIGRANT FORMATION
MIDDLE CAMBRIAN Marjumiid	Cedaria				
	Bolaspidella				
	Ehmaniella			upper siltstone PAPOOSE LAKE MEMBER	
Corynexochid	Glossopleura	CADIZ FM	BRIGHT ANGEL SHALE	CARRARA FM	CARRARA FM

Figure 3. Biostratigraphic correlation chart for the Bonanza King Formation of the southern Great Basin.

extended approximately 250 to 300 km from the craton margin hinge line near Las Vegas to base-of-slope facies in the northern Last Chance Range (Figure 2). Restricted peritidal lithofacies extend across the entire platform indicating sedimentation rates in excess of subsidence rates and reflect a fully aggraded profile. Carbonates or clastics indicative of deeper water deposition (below the zone of storm wave reworking) are not found within the Banded Mountain Member. The "outer" platform was probably never more than 5 to 15 m deeper than the cratonward "inner" platform.

The shelf edge was probably a high-relief depositional, rather than a bypass, margin (Read, 1985) based on thick accumulations of peritidal lithofacies on the outer shelf (Last Chance Range) and a thin, starved, base-of-slope section 20 km to the north (Horse Thief Canyon). Off-platform lithofacies consist of monomict and polymict breccias, carbonate grain flow deposits, and lime mudstone-chert couplets. Palinspastic estimates of the distance between the outer shelf section and the off-platform section in the Last Chance Range suggest a minimum of 50 km (Snow, 1992). No exposures of a reefal rim exist but the composition of polymict breccias in base-of-slope facies suggest that "reefal" facies were extant during Bonanza King time. The regional shelf edge in the southern Great Basin may have been constructed of a diffuse mound and channel belt of thrombolite bioherms and lime sand shoals, common lithofacies on the outer platform. It has been postulated that syndepositional faulting may have created the high-relief shelf margin based on distinctive white algal boundstone clasts found in the base-of-slope breccias (Kepper, 1981). Detailed logging of outer platform sections shows that these clasts are not necessarily unique and do not require syndepositional tectonism to expose them on the outer shelf edge. This mechanism, however, cannot be discounted as a contributing factor on Bonanza King platform morphology.

PARASEQUENCE TYPES

Two main types of shallowing-upward parasequences are recognized in the Banded Mountain Member. Peritidal carbonate parasequences are the dominant component of the member and are characterized by shallow subtidal bases capped by tidal-flat

Figure 4. Typical outcrop expression of the Banded Mountain Member of the Bonanza King Formation, Nopah Range, eastern California. The interval is characterized by light–dark banding at the meter scale as well as thicker light–dark bands that collectively reflect superimposed scales of sea-level oscillations.

facies. Subtidal parasequences are composed entirely of subtidal lithofacies with no tidal-flat cap. Peritidal parasequences are typically thinner and range from 0.5 to 4 m, whereas subtidal parasequences range from 1 to 7 m. Approximately 150 to 250 parasequences can be recognized throughout the Banded Mountain Member with the number of parasequences preserved varying across the platform. The greatest number of parasequences occurs in the Nopah and Indian Ridge sections of the inner to mid-platform (± 180 to 250) with fewer parasequences being preserved on the craton margin section (± 150) and outer platform sections (± 160). By dividing the total duration of the Banded Mountain Member (~15 m.y. [± 50%]; Palmer 1983) by 150 to 250 parasequences, the duration of deposition for each parasequence ranges between 30 to 150 k.y.

The parasequences are extremely rhythmic vertically in outcrop with only minor variations in the arrangement of component lithofacies. Individual parasequences have been traced laterally for a few kilometers along the extensive mountain-front outcrops and show typical lateral facies transitions but surprisingly consistent lateral continuity. This is not

the case for correlation distances of tens of kilometers between isolated mountain ranges. Although individual parasequence correlation is difficult between the outcrop localities, distinct *intervals* of stacked parasequences can be correlated between ranges.

Peritidal Parasequences

Peritidal parasequences of the Banded Mountain Member (Figures 5, 6) typically begin with a lower subtidal unit of strongly burrowed peloidal wackestones and packstones often lateral to globose thrombolite and digitate algal bioherms. These lithofacies typically fine upward into less-burrowed ribbon rocks of alternating peloidal grainstones and dolomitic muds. Intraclastic grainstone lag deposits are relatively uncommon in the subtidal portion of most Banded Mountain parasequences, contrary to peritidal parasequences recognized in other Cambrian strata (Aitken, 1978; Demicco, 1985; Chow and James, 1987; Koerschner and Read, 1989; Osleger and Read, 1991). Tidal-flat caps overlying the subtidal base of these parasequences consist of low-relief laterally-linked hemispheroid stromatolites, centimeter-

Figure 5. Stacked peritidal cycles from the Banded Mountain Member at Indian Ridge, Spring Mountains, Nevada. Dark bands mark the subtidal bases of cycles (generally limestone) and light bands mark the dolomitized laminite cap of cycles.

scale mechanical laminites, and mud-cracked cryptalgal laminites. Evidence of prolonged subaerial exposure near the tops of some parasequences (Figure 7) is provided by disorganized breccias exhibiting abundant silicification, small tepees, and mud-cracked green silty marls interpreted to be sheet-flood deposits spread out across the tops of supratidal flats. Thin sections of some parasequence caps exhibit vadose diagenetic features such as pendant and meniscus cements, leached voids, and crystal silts geopetally filling solution-enhanced porosity. Evidence of exposure is greatest in the Frenchman Mountain craton margin section and decreases toward the outer platform where most peritidal parasequences are capped by nonmud-cracked, centimeter-scale mechanical laminites.

Lithofacies transitions within individual peritidal parasequences are typically gradational, whereas lithofacies transitions between adjacent parasequences are usually abrupt. This asymmetric succession of peritidal lithofacies records rapid flooding of

the tidal flat followed by progressive aggradation to sea level. Subsequent progradation is not recorded in individual parasequences but undoubtedly occurred as the entire platform responded to the decrease in accommodation space as infilling gradually proceeded. Progradation may not necessarily have migrated from the craton margin basinward but rather may have initiated from isolated islands or topographic highs across the platform. The direction of progradation may have been governed by the dominant orientation of storm tracks, similar to the modern tidal flats on Andros Island (Shinn, 1986). The possibility also exists that tidal flats may have migrated landward away from a shelf margin high and prograded over platform interior "lagoonal" carbonates (cf. Demicco, 1985). Indeed, mosaic shallowing (Laporte, 1967) probably occurred and contributes to the internal complexity of the platform, but the evidence for broad expansion of tidal-flat conditions across the entire 250 to 300 km platform is indisputable because similar intervals of stacked peritidal parasequences can be traced from craton margin locations all the way to the platform edge.

Subtidal Parasequences

Subtidal parasequences (e.g., Osleger, 1991) are a subordinate component of the Banded Mountain Member relative to peritidal parasequences. They are composed entirely of subtidal lithofacies with no tidal-flat laminite caps and show no evidence of episodic exposure (Figures 8, 9). Subtidal parasequences are only slightly asymmetric in that they exhibit progressive shallowing but, in contrast to the peritidal parasequences, they commonly have gradational boundaries between adjacent parasequences. The basal lithofacies is typically a massively bedded, strongly burrowed, peloidal packstone that gradually fines upward into wackestone textures with less-abundant burrowing. The parasequence cap is a wavy-bedded, slightly burrowed unit composed of centimeter-scale couplets of orange-weathering silty dolomite and gray lime mudstone. No mud cracks, fenestrae, or other evidence of deposition on tidal flats are exhibited. Transitions into overlying parasequences show gradually decreasing orange-weathering silty dolomites and a concomitant increase in burrowing intensity.

Subtidal parasequences of the Banded Mountain Member appear to manifest progressive aggradation within relatively shallow subtidal, restricted paleoenvironments. The burrow-homogenized peloidal muds of the basal lithofacies reflect an actively feeding infauna, perhaps in a lagoonal environment. The parasequence caps appear to be wave-reworked storm couplets, modified by discrete burrows in a firm substrate. Individual subtidal parasequences are not correlatable between mountain ranges but intervals of subtidal parasequences are traceable across the platform. Subtidal parasequences are more common toward outer platform localities of the Bonanza King. The controlling mechanism behind incomplete

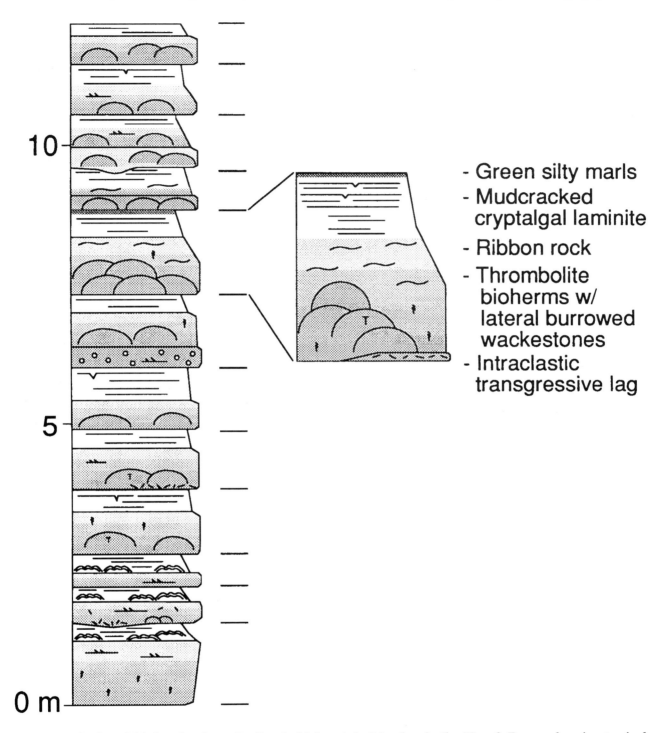

- Green silty marls
- Mudcracked cryptalgal laminite
- Ribbon rock
- Thrombolite bioherms w/ lateral burrowed wackestones
- Intraclastic transgressive lag

Figure 6. Stacked peritidal cycles from the Banded Mountain Member in the Nopah Range showing typical lithologies. Average cycle thickness is about 1 m.

shoaling in these subtidal parasequences may be related to their formation during long-term increases in accommodation when superimposed higher frequency oscillations were modulated by the long-term event (Osleger, 1991). The short-term falls may have been brief with renewed drowning occurring before aggradation to tidal flat depths was achieved.

PARASEQUENCE STACKING PATTERNS AND FISCHER PLOTS

Thick successions of meter-scale carbonate parasequences show systematic changes in parasequence type, dominant lithofacies, and thickness vertically through the cyclic interval. These stacks of genetically

Figure 7. Solution collapse breccia capping a meter-scale parasequence, Nopah Range, eastern California. Note the disoriented angular clasts and silicification of former evaporite(?) nodules.

related cycles are the parasequence sets of sequence stratigraphic terminology (Van Wagoner et al., 1987). Parasequence stacking patterns are most likely controlled by long-term, third-order changes in accommodation space and provide the crucial link between the individual meter-scale parasequences and the larger scale depositional sequences and their component systems tracts (Goldhammer et al., 1990; Osleger and Read, 1991; Montañez and Read, 1992). For ancient passive margin carbonate platforms such as the Bonanza King, where outcrops are limited to mountain ranges tens of kilometers apart and lateral tracing of individual horizons is difficult to recognize and conjectural in interpretation, parasequence stacking patterns may provide the primary means of deciphering the internal anatomy of depositional sequences. Before describing the distinctive characteristics of the parasequence stacking patterns that define various components of sequences within the Banded Mountain Member, a discussion of systematic variations in parasequence *thickness* will help to illustrate their control by long-term accommodation events.

Fischer plots provide a useful tool for graphically illustrating deviations from average cycle thickness throughout a cyclic succession (Fischer, 1964; Goldhammer et al., 1987; Read and Goldhammer, 1988; Sadler et al., 1993). Runs of thick cycles form positive slopes on the plots, whereas runs of thin cycles form negative slopes on the plots. The conceptual basis behind these changes in slope on the plots is that stacks of thick parasequences presumably form during times of increased accommodation space generated by long-term rises in relative sea level. Conversely, stacks of thin parasequences presumably form during times of reduced accommodation space generated by long-term falls in relative sea level. The method seems to be best suited for tidal-flat–capped parasequences that shallow to sea level although good results have also been attained for shallow subtidal parasequences (Osleger and Read, 1991; Goldhammer et al., 1992).

The Fischer plot generated from the Nopah Range section of the Bonanza King illustrates the relationship between parasequence stacking patterns and their effect on the form of Fischer plots (Figure 10). Negative slopes on the Nopah Fischer plot are created by stacks of thin parasequences dominated by tidal-flat lithofacies. Over an 18-m interval, 15 thin cycles ($\approx$1 m/cycle) were generated that are dominat-

Figure 8. Typical subtidal cycles, Indian Ridge, Spring Mountains, Nevada. Note the subtle light to dark variations that reflect changes in burrowing intensity. Cycle boundaries are slightly gradational. Person for scale is at base of cliff.

ed by mechanical laminites and mud-cracked cryptalgal laminites. Subtidal bases of these parasequences are characterized by low-relief thrombolite bioherms and burrowed ribbon rocks that show evidence of very shallow subtidal conditions. The interpretation of this interval is that the high-frequency sea-level oscillations that produced these stacked peritidal parasequences were modulated by a long-term decrease in accommodation space which inhibited episodic deepening. The succession of cycles in this parasequence set thin upward and exhibit an increase in diversity and abundance of exposure features (tepees and collapse breccias), suggesting progressively reduced accommodation space.

In contrast, positive slopes on the Nopah Fischer plot are created by stacks of thick subtidal parasequences. Over a 25-m interval, five subtidal cycles (5 m/cycle) were generated that are dominated by massively bedded, heavily burrowed peloidal wackestones and less-burrowed, centimeter-scale couplets

of silty dolomite and lime mudstone. The interpretation is that the high-frequency sea-level oscillations that produced these stacks of thick subtidal parasequences were modulated by a long-term increase in accommodation space that precluded episodic shallowing to near sea level.

The "noncyclic" intervals on the plot are composed of subtidal lithofacies (usually dark gray, burrowed, peloidal wackestone/packstone) showing inconsistent evidence for cyclic deposition. One or two recognizable parasequences may be bounded by tens of meters of noncyclic lithofacies. These noncyclic intervals tend to be correlatable across the Bonanza King platform and are commonly juxtaposed against subtidal parasequences both above and below. The absence of parasequences and the homogeneous lithofacies within these intervals is interpreted to reflect a subdued sedimentologic response to a long-term increase in accommodation. Sea level may have been oscillating too far above the sediment surface to cause a distinct sedimentologic change in facies. In essence, these "noncyclic" intervals manifest an indeterminate number of "missed beats" generated during a long-term third-order accommodation increase.

Depositional Sequences and Accommodation Events

Fischer plots provide a useful tool for cross-platform correlation (Read and Goldhammer, 1988; Montañez and Read, 1992) as well as for interbasinal correlation (Osleger and Read, 1993). Figure 11 shows Fischer plots generated for each of the five platform sections of the Banded Mountain Member. The lower datum (left side of diagram) is a very distinct interval of red siltstones at the base of the Banded Mountain Member and the upper datum is a major noncyclic oolitic grainstone. Both data are probably time transgressive but nevertheless provide good markers to constrain the cyclic intervals. The individual plots were correlated using similarity of patterns of rise and fall as well as interpreted zones of maximum exposure. Four major accommodation events can be recognized on the correlated Fischer plots (Figure 11). These events are defined by trends of long-term increases in accommodation space followed by long-term decreases in accommodation space. The good degree of correlation of accommodation events between individual Fischer plots reflects systematic variations in parasequence stacking patterns that occurred platform wide. Rising segments of all plots are defined by stacks of cycles that thicken upward and exhibit evidence of dominantly submergent conditions, and are thus interpreted to reflect long-term flooding events. Falling segments of all plots are composed of stacks of cycles that thin upward and exhibit evidence of prolonged exposure; these are interpreted to reflect decreasing accommodation space associated with a rate of sea-level fall in excess of tectonic subsidence at that platform location. Figure 11 provides a very useful visual aid for cross-platform correlation

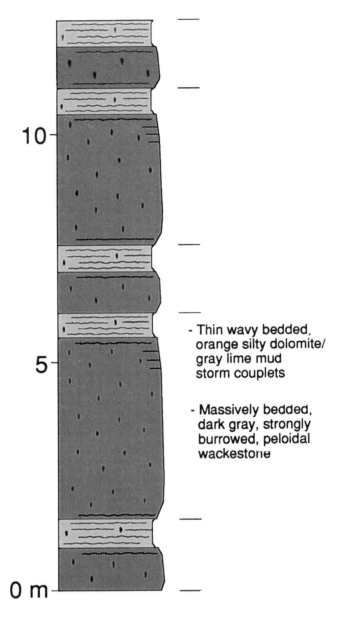

- Thin wavy bedded, orange silty dolomite/ gray lime mud storm couplets

- Massively bedded, dark gray, strongly burrowed, peloidal wackestone

Figure 9. Stacked subtidal cycles from the Spring Mountains, Nevada, showing typical lithologies. Average cycle thickness is about 2.6 m. The lower three cycles of this diagram correspond to the lower three cycles in the field photograph of Figure 8.

of stacking patterns within individual sections of the Banded Mountain Member and was instrumental in constructing the cross section shown in Figure 12.

The Banded Mountain Member exhibits several long-term variations in parasequence stacking patterns that have been used to define four depositional

sequences ranging in estimated duration from 0.8 to 5.0 m.y. (Figure 12). These sequences are hinted at visually along mountain-front exposures by conspicuous large-scale color banding tens to hundreds of meters in thickness and composed internally of several scales of cyclicity marked by light-dark alternations (Figure 13). In general, dark-weathering cliff-forming units correspond to intervals of thick subtidal parasequences or thick peritidal parasequences with thin laminite caps, whereas lighter weathering, ledge- and slope-forming units correspond to intervals of thinner peritidal parasequences dominated by tidal flat facies. The systematic changes in parasequence stacking patterns within each depositional sequence correspond to the four distinct accommodation events recognized on Figure 11. It appears that the correlated Fischer plots for the Banded Mountain Member define the *form* of the long-term accommodation events that generated the individual sequences during Middle to Late Cambrian time. It should be emphasized that interpretations of long-term third-order *eustatic* events should only be made from several Fischer plots correlated between separate basins.

Direct measurements of magnitudes and rates of the accommodation events illustrated on Fischer plots should not be interpreted as absolute because of the inherent assumptions in the technique (Osleger and Read, 1993; Sadler et al., 1993). (1) The plots are not corrected for isostatic response to sediment and water loading. (2) The assumption of constant cycle duration is made only for graphic reasons and is almost certainly erroneous in that the cycles probably formed over a range of durations. (3) "Missed beats" of sea level are highly likely, leading to condensation and amalgamation of cycles as sea level oscillates either above or below the platform (Goldhammer et al., 1990). (4) Age uncertainties in absolute time scales make calibration of the horizontal time axis unreliable.

SEQUENCE STRATIGRAPHY OF THE BONANZA KING

Translating sequence stratigraphic principles originally defined for seismic-scale siliciclastic systems to outcrops of cyclic carbonates deposited on broad passive margins requires a de-emphasis of stratal geometries and an increased awareness of correlative vertical changes in stacking patterns of component parasequences. The large-scale geometric relationships of stratal patterns recognizable on seismic sections have considerably improved our ability to identify stratigraphic sequences and their internal components. The limitations of seismic resolution may have inadvertently biased our understanding of stratigraphic systems, however, because of the seismic emphasis on individual stratal surfaces. Models based on seismic data alone may prove to be erroneous because the details evident on outcrop may reveal a distinctly different interpretation (Biddle et

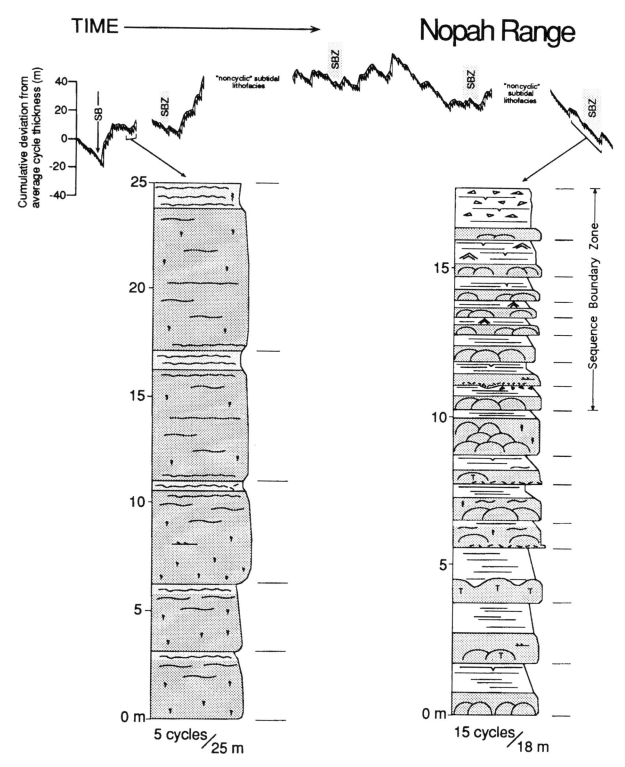

Figure 10. Fischer plot of the Nopah Range section of the Banded Mountain Member of the Bonanza King Formation illustrating characteristic stacking patterns and sequence boundary zones. SBZ = Sequence Boundary Zone. Note how considerably thicker the shallow subtidal parasequences are versus the thinner stacked peritidal parasequences. Parasequence thickness increases upward within sets that define the transgressive systems tract, whereas they thin upward within sets that define the transition from upper highstand systems tract into the sequence boundary zone. Noncyclic subtidal intervals are often bounded by subtidal parasequences and are interpreted to have occurred during flooding phases associated with the formation of the transgressive systems tract.

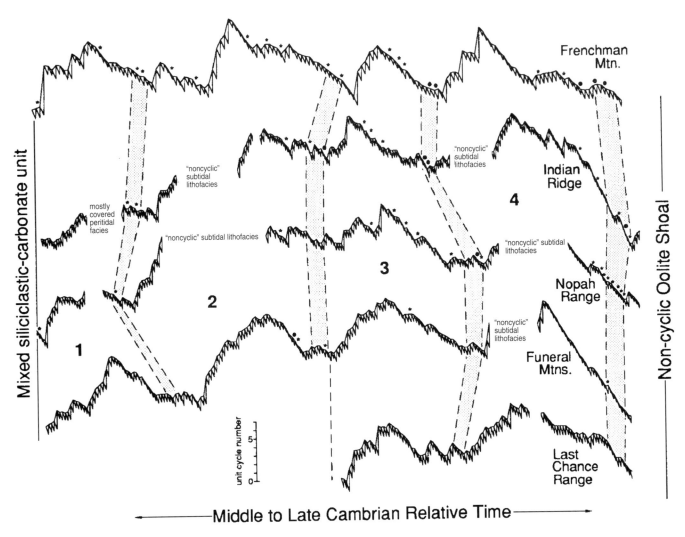

Figure 11. Fischer plot correlation diagram of the Banded Mountain Member, Bonanza King Formation. Vertical stippled bands correlate interpreted sequence boundary zones that were defined by stacks of thinning-upward parasequences and evidence for repeated episodes of exposure. The vertical scale is in mean cycle thickness units—all plots are normalized to this unit thickness. Asterisks mark intervals of cycles with evidence for prolonged subaerial exposure. Black dots mark intervals of cycles containing quartz silt laminae. The Fischer plot for the Last Chance section is abbreviated due to a covered interval below. Note the broad overall similarity in the form of the five plots. Each plot defines a wave train composed of four "cycles." The four "cycles" are interpreted to reflect roughly synchronous long-term changes in accommodation that generated the four major depositional sequences within the Banded Mountain Member.

al., 1992). Rather than conduct a difficult and perhaps misleading search for individual horizons across the Bonanza King platform, systematic variations in parasequence stacking patterns and cross-platform trends were used to identify the component systems tracts and sequence boundaries.

The following discussion on the internal components of depositional sequences within the Banded Mountain Member of the Bonanza King Formation will focus on (1) diagnostic lithofacies and parasequence stacking patterns of systems tracts and their bounding transitions in cyclic carbonates; (2) relative position of sequence components on correlated

Fischer plots; and (3) differences between seismically based definitions of systems tracts, sequence boundaries, and maximum flooding surfaces and the reality of the high-resolution data base exposed in outcrop.

Sequence Boundaries

No evidence is recognized anywhere on the Bonanza King platform for major penetrative karstification horizons that may manifest long-term subaerial erosion and a basinward shift of facies. Therefore, sequence boundaries bracketing the four depositional sequences in the Banded Mountain Member (Figure

12) are interpreted to be type 2 sequence boundaries generated when the rate of third-order eustatic sea-level fall is less than tectonic subsidence rates across the passive margin. Type 2 sequence boundaries are defined as "a regional surface marked by subaerial exposure and a downward shift in coastal onlap landward of the depositional-shoreline break" (Van Wagoner et al., 1987). This definition implies that a distinct *surface* should separate depositional sequences and that a shelf-margin systems tract should be expected above that surface.

Sequence boundaries within the Banded Mountain Member are interpreted to be transitional *zones* rather than discrete surfaces. Figure 14 illustrates the internal anatomy of a typical sequence boundary zone extending across the platform separating sequences 3 and 4. The sequence boundary zones are characterized by thin, tidal-flat dominated parasequences exhibiting abundant evidence for multiple episodes of exposure. On the inner to mid-platform, parasequence tops contain deeply mud-cracked cryptalgal laminites, highly disrupted breccias, tepees, quartz silt laminae, and erosionally truncated laminites. Toward the outer platform, the peritidal parasequences become capped by mechanical laminites (indicative of intertidal conditions) and show much less evidence for episodic exposure on a supratidal flat. The increasingly conformable nature of the sequence boundary zone toward the outer platform suggests that the examples from the Funeral Mountains and Last Chance Range manifest the "correlative conformities" inherent in the definition of a sequence (Mitchum, 1977). It is difficult to define an individual parasequence top to be the unequivocal sequence boundary horizon because of the difficulty of correlating any single surface across the platform and the lack of any distinguishing characteristics between the multiple exposure horizons within each interval (Figure 14).

On Fischer plots, the interpreted sequence boundary zones tend to be concentrated along the negative slopes (often near the troughs) because of the thinness of the component parasequences within the sequence boundary zones (Figure 11). The thinning-upward trend of parasequences toward the sequence boundary zone is well illustrated for the top of sequence 4 in the Nopah Range section on Figure 10 and is also displayed in outcrop on Figure 13. Within individual sections, it is difficult to determine whether the shoal-water facies near a sequence boundary zone represents the platform-wide upper highstand systems tract or the shelf-margin wedge with no updip facies equivalents. Onlap and downlap onto the sequence boundary zone cannot be recognized in the outer platform sections although it is suspected that shelf-margin wedge formation accounts for a large percentage of the drastic increase in thickness of the Last Chance section (Figure 12).

Our outcrop-based definition of type 2 sequence boundaries differs significantly from that of Van Wagoner et al. (1987) in that we see a diffuse transition recording multiple episodes of exposure at the tops of individual parasequences separating major depositional sequences, rather than a discrete surface. It is our contention that type 2 sequence boundaries are complicated by meter-scale parasequences generated in response to repeated pulses of high-frequency sea-level oscillations superimposed on the long-term sea-level event (Christie-Blick, 1991; Read et al., 1991; Goldhammer et al., 1992). Figure 15 shows the conceptual differences between changing rates of eustatic rise and fall on a simple "model" sea-level curve and a more realistic composite eustatic curve composed of complexly interfering high-frequency events superimposed on the long-term trend. An increase in the rate of sea-level fall on the model curve causes the development of a single horizon marking a discrete sequence boundary and an associated basinward shift in onlap (Vail, 1987). The composite curve suggests, in contrast, the development of a concentrated zone of thin parasequences characterized by evidence for multiple episodes of short-term subaerial exposure. The set of stacked parasequences from the Frenchman Mountain craton margin section (as well as Figure 14) illustrates the difficulty of defining an individual parasequence top as the "true" sequence boundary. The interval is marked by several parasequences exhibiting evidence for subaerial exposure with none of the parasequences being very distinctive from any of the others. Even in this craton margin location, where tectonic subsidence rates were probably <0.01 m/k.y., refining the sequence boundary to anything other than a transitional zone would be arbitrary. Indeed, the search for a single sequence-bounding surface on flat-topped platforms such as the Bonanza King is not only futile, but may also be misleading because the complex history of superimposed orders of sea-level oscillations suggests that a single significant surface does not even exist.

Fischer Plots and Sequence Components

The internal components of sequences are commonly interpreted to be deposited during a specific increment on a eustatic curve (Vail, 1987). It is important to note that we are not trying to suggest that Fischer plots can be interpreted as proxies for a eustatic curve. Fischer plots are clearly "noisy" and correlated plots (Figure 11) show similar patterns but certainly not a one-to-one correspondence of trends. Therefore, we resist forcing arbitrary boundaries onto the Fischer plots that separate the various components of depositional sequences in the Banded Mountain Member primarily because transitions between systems tracts and boundary zones are gradational in nature and are not easily defined. For example, our interpreted sequence boundary zones tend to occur along the base of the falling limbs of the plots (Figures 10, 11), but this does not mean that the following sets of cycles near the troughs should be interpreted as lowstand or shelf-margin deposits. We are only illustrating where the cycles that show the highest degree

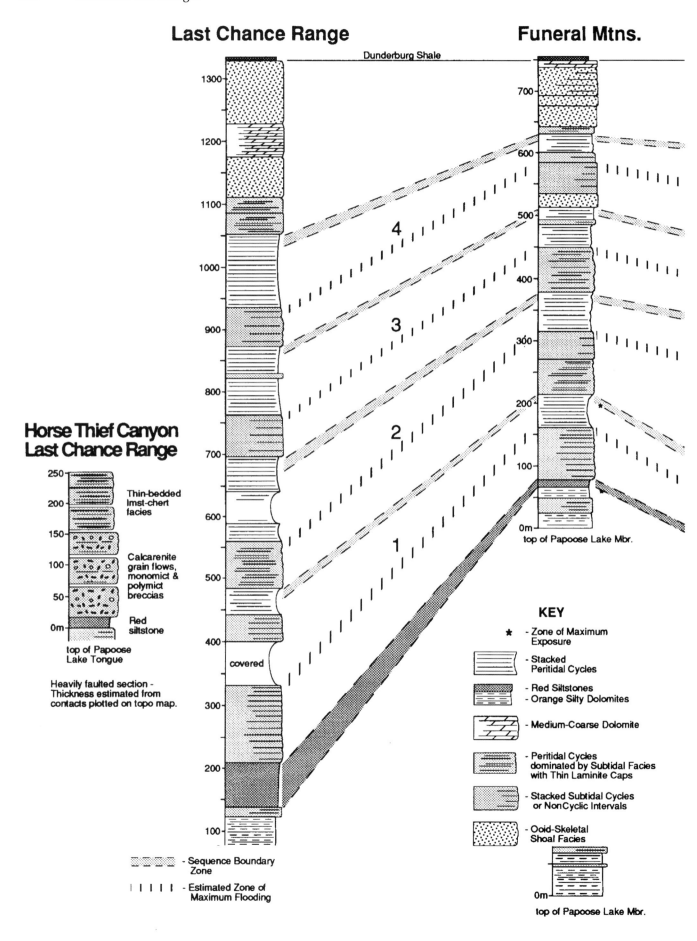

Last Chance Range

Funeral Mtns.

Dunderburg Shale

**Horse Thief Canyon
Last Chance Range**

Thin-bedded lmst-chert facies

Calcarenite grain flows, monomict & polymict breccias

Red siltstone

top of Papoose Lake Tongue

Heavily faulted section - Thickness estimated from contacts plotted on topo map.

covered

top of Papoose Lake Mbr.

KEY

★ - Zone of Maximum Exposure

- Stacked Peritidal Cycles

- Red Siltstones
- Orange Silty Dolomites

- Medium-Coarse Dolomite

- Peritidal Cycles dominated by Subtidal Facies with Thin Laminite Caps

- Stacked Subtidal Cycles or Non Cyclic Intervals

- Ooid-Skeletal Shoal Facies

top of Papoose Lake Mbr.

- Sequence Boundary Zone

| | | | | - Estimated Zone of Maximum Flooding

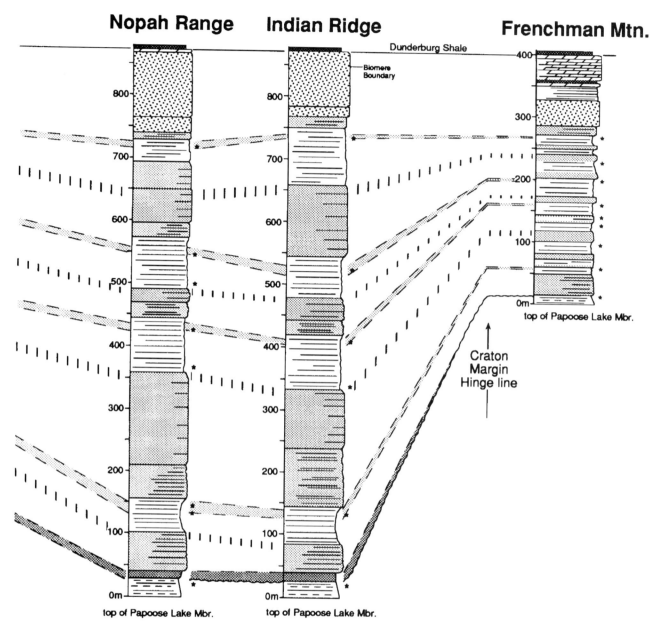

Figure 12 (opposite and this page). Cross-platform transect with interpreted sequence boundary zones (stippled) and maximum flooding zones (vertical lines). Upper datum is the Dunderberg Shale. Vertical scale in meters. Palinspastic distances between sections are shown in Figure 2. Four depositional sequences are recognized within the Banded Mountain Member below a major platform-wide oolitic shoal complex. Sequence correlations were determined on the basis of parasequence stacking patterns in conjunction with Fischer plots. These four depositional sequences correspond to the four "accommodation cycles" illustrated on Figure 11.

of exposure are located on the plots. It is best to view the Fischer plots as simply graphic illustrations of thickening and thinning stacks of parasequences and use them as a guide for identifying systems tracts and bounding transitions within the rocks themselves. We feel that only when several Fischer plots are correlated, both across a single platform and between basins, do they collectively approximate the general *form* of a eustatic curve that may have controlled the deposition of third-order sequences.

Transgressive Systems Tracts

Transgressive systems tracts are characterized by "retrogradational parasequence sets" that "onlap onto the sequence boundary in a landward direction and downlap onto the sequence boundary in a basinward direction" (Van Wagoner et al., 1987). Because onlapping and downlapping stratal geometries are extremely difficult to recognize on the very low paleoslopes of the Bonanza King platform, parasequence

Figure 13. Stacking patterns of parasequences within three of the four depositional sequences within the Banded Mountain Member of the Bonanza King Formation, Bare Mountain, Nevada. The largest scale of light to dark color bands aid in identifying third-order depositional sequences and are internally composed of an intermediate scale of light–dark bands and a meter-scale alternation of light–dark bands. The large arrows represent thickening-upward parasequence sets, intermediate-size arrows mark transitional intervals, and small arrows represent thinning-upward parasequence sets. SBZ = Sequence Boundary Zone; MFZ = Maximum Flooding Zone.

stacking patterns help to identify the transgressive systems tracts within each sequence. Transgressive systems tracts in the Banded Mountain Member are characterized by a lower succession of parasequences composed of subequal amounts of tidal-flat and subtidal lithofacies that show an overall upward deepening and upward thickening into parasequences with higher percentages of subtidal facies. Thickening-upward parasequence sets within transgressive systems tracts from a mid-platform location are illustrated in outcrop on Figure 13. The thick dark bands (located between the SBZ and MFZ) are dominantly composed of subtidal parasequences. Transgressive systems tracts in inner platform and craton margin locations are typically composed of thick parasequences with some thin laminite caps. Dominant lithofacies within transgressive systems tracts across the platform are burrowed peloidal wackestones, oolitic grainstones, and stacked thrombolitic and digitate algal bioherms.

On Fischer plots, transgressive systems tracts are located along the positive slopes, created by the stacked thick parasequences generated during long-term increases in accommodation (Figures 10, 11). The possibility of cycle amalgamation exists during the development of the transgressive systems tract, especially on the mid- to outer platform, due to the modulating effects of long-term third-order sea-level rise on the higher frequency events. This is most likely the reason for the development of subtidal *noncyclic* intervals within transgressive systems tracts across the platform. During the long-term flooding phase, sea-level oscillations may have occurred too far above the sediment surface to cause a distinct change in lithofacies, and therefore the high-frequency sea-level signal is not recorded in the rocks.

The transition from the transgressive to highstand systems tract (i.e., maximum flooding surface) is defined as the break between retrogradational to aggradational parasequence sets (Van Wagoner et al.,

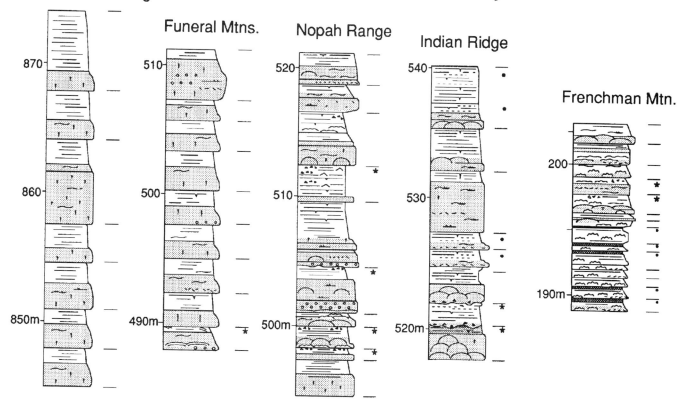

Figure 14. Correlative sequence boundary zones across the Bonanza King platform separating sequences 3 and 4. Sequence boundary zones were identified on the basis of thinning-upward parasequence sets (as revealed on Fischer plots) that show evidence of superimposed episodes of prolonged exposure. In the Funeral Mountains and Last Chance Range sections, decreasing parasequence thickness alone was used to define the zone because of the relative conformity of the type 2 sequence boundary on the outer platform. These sequence boundary zones can be correlated with their position on Fischer plots illustrated in Figure 11. Note how parasequence number, thickness, and composition vary across the platform. Dashes to right of columns indicate parasequence boundaries. Asterisks mark cycle tops that show evidence for prolonged subaerial exposure; black dots mark cycles containing quartz silt or green silty marls. Standard lithologic symbols are the same as in Figures 6 and 9. Shaded intervals are shallow subtidal lithofacies and unshaded intervals are tidal-flat caps.

1987). On platforms with outer shelf or basinal hemipelagic or pelagic sedimentation, the development of a condensed section marking very slow rates of deposition aids in the identification of the maximum flooding surface. On flat-topped, fully aggraded platforms like the Bonanza King, however, where no condensed section exists and other criteria for the recognition of maximum flooding (organic matter-rich sediments, glauconite or phosphatic hardgrounds) are lacking, parasequence stacking patterns provide the best data set for identifying the transition. Similar to sequence boundaries, the transition between the transgressive systems tract and highstand systems tract is complicated by high-frequency sea-level events (Figure 15) and is considered to be a *zone* of maximum flooding. Zones of maximum flooding in the Banded Mountain Member are char-

acterized by vertical transitions from very thick parasequences to successively thinner parasequences with a concomitant increase in tidal-flat facies. This shift in stacking patterns may reflect the progressive decrease in accommodation space as the rate of sea-level rise slows toward the eustatic peak. Correlated Fischer plots (Figure 11) frequently show the transition from thicker cycles to thinner cycles to be abrupt and "peaky" rather than rounded and gradational. This rapid transition can be seen in outcrop (Figure 13) by the abrupt change from thick dark bands of subtidal cycles (transgressive systems tract) into lighter bands of peritidal cycles (highstand systems tract).

The difficulty of assigning an individual horizon as the maximum flooding surface is illustrated in the stacked subtidal parasequences from the Funeral

322   Montañez and Osleger

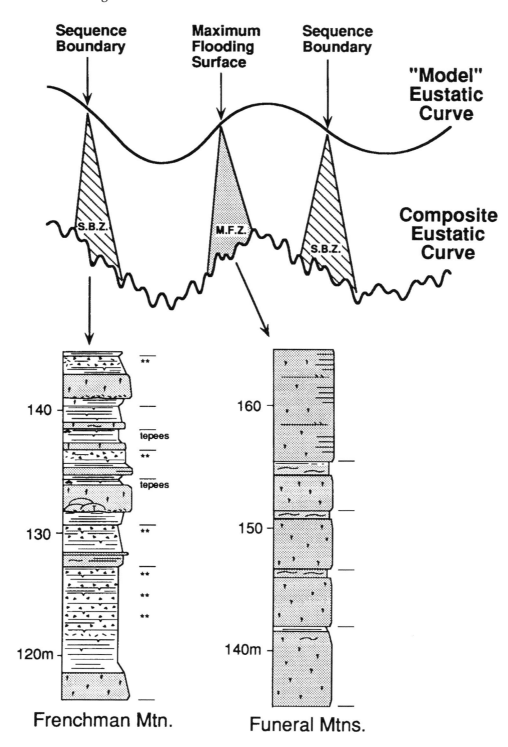

Figure 15. Comparison of a "model" sea-level curve (Vail, 1987) and a hypothetical composite sea-level curve. S.B.Z. = Sequence Boundary Zone; M.F.Z. = Maximum Flooding Zone. Asterisks indicate exposure horizons. Shaded intervals denote subtidal lithofacies. The lithologic columns are a representative sequence boundary zone (Frenchman Mountain) and a maximum flooding zone (Funeral Mountains). The problem is determining which cycle cap in the stack manifests a single sequence boundary horizon or which cycle base manifests the maximum flooding surface. The superimposed scales of sea-level oscillations suggest that we should not expect to find a single unconformity surface or maximum flooding surface. The composite sea-level history recorded within cyclic carbonates deposited on flat-topped platforms suggests that we should search for a transitional *zone* that records the complex sea-level history of maximum flooding or sequence boundary formation.

Mountains (Figure 15). The interval is marked by several parasequences exhibiting evidence for consistently submergent conditions with none of the parasequences being very distinctive from any of the others. Because of the composite sea-level history, an individual maximum flooding surface should not be expected to even exist on flat-topped platforms like the Bonanza King, but is better distinguished as a transitional zone into the lower highstand systems tract.

Highstand Systems Tracts

Highstand systems tracts are characterized by "one or more aggradational parasequence sets" overlain by "progradational parasequence sets with prograding clinoform geometries" (Van Wagoner et al., 1987). As has been emphasized previously, these strata geometries are difficult to recognize in outcrop and prograding clinoforms may not even be present on low-sloping platforms with high-relief margins. Highstand systems tracts in the Banded Mountain Member are characterized by stacks of parasequences composed of subequal amounts of subtidal and peritidal lithofacies that thin upward into parasequences showing progressively higher tidal-flat to subtidal ratios. Thinning-upward parasequence sets within highstand systems tracts from a mid-platform location are illustrated in outcrop on Figure 13. The thick light bands (located between the maximum flooding zone and sequence boundary zone) are dominantly composed of dolomitized peritidal parasequences that thin upward, merging imperceptibly into the sequence boundary zone. Tidal-flat caps of parasequences in the highstand systems tracts of Banded Mountain sequences tend to thin basinward, perhaps reflecting progradation direction and decreased durations of supratidal conditions on the outer platform (cf. Montañez and Read, 1992). Dominant lithofacies within highstand systems tracts across the platform are laterally-linked hemispheroid stromatolites, mechanical laminites, and cryptalgal laminites with random cycle caps showing exposure features. The transitional break between the upper highstand systems tract and the sequence boundary zone is interpreted to occur by the appearance of stacks of multiple cycle caps with solution collapse breccias, microkarsting, tepees, and erosional truncation, although we acknowledge that the exact transition may be arbitrary.

On Fischer plots, highstand systems tracts are located near the crest and along the negative slopes created by the stacks of thinning-upward parasequences generated during long-term decreases in accommodation (Figures 10, 11). During the development of the highstand systems tract, parasequence "condensation" and missed sea-level beats probably occurred, especially on the inner platform and craton margin, as the high-frequency sea-level events oscillated below the platform surface.

SUMMARY AND CONCLUSIONS

(1) The Middle to Upper Cambrian Banded Mountain Member of the Bonanza King Formation of the southern Great Basin is composed of a stacked succession of carbonate parasequences that record an extensive history of accommodation change. Systematic variations in parasequence stacking patterns can be used to recognize sequence boundaries and systems tracts that distinguish four depositional sequences within the Banded Mountain Member. Correlated Fischer plots graphically display deviations from average parasequence thickness and can be used for cross-platform correlation of depositional sequences and their internal components.

(2) A few limitations emerge when translating sequence stratigraphic concepts originally defined for seismic-scale siliciclastic systems to outcrops of cyclic carbonates deposited on broad passive margins. First, onlapping and downlapping strata geometries are extremely difficult to recognize on very low-sloping platforms such as the Bonanza King. Parasequence stacking patterns provide an alternative, high-resolution data set to identify components of sequences. Second, standardized definitions of the internal components of depositional sequences consistently refer to the necessity to find *surfaces* that separate systems tracts and sequences. The cyclic carbonates of the Bonanza King platform suggest, however, that type 2 sequence boundaries and transitions between systems tracts should be recognized as *zones* rather than distinct horizons because of the complicating effects of repeated pulses of high-frequency sea-level oscillations superimposed on the long-term sea-level event.

The search for individual horizons, as espoused in the sequence stratigraphic lexicon, may be an artifact of the "seismic bias" originating with the advent of the seismic stratigraphic method (Vail et al., 1977). An "outcrop unconformity" may not be visible as a "seismic unconformity" because of the limited resolution of seismic techniques (Schlager, 1991). An individual type 2 sequence boundary zone in the Banded Mountain Member, for instance, may be restricted to a thickness of a few tens of meters, thicknesses below the average limits of seismic resolution. In addition, similarities in lithologic composition and velocity across the sequence boundary zone may not generate a significant seismic response due to reduced impedance contrasts, although detailed seismic modeling would best determine this (e.g., Biddle et al., 1992). This suggests that outcrop sections of cyclic carbonates such as the Bonanza King Formation provide higher resolution data bases for sequence stratigraphic analysis, and also may provide better records of long-term sea-level change than lower resolution seismic sections.

(3) To establish a possible eustatic control on sequence development in the Bonanza King Formation, interbasinal correlation of coeval strata is essential. The excellent exposures of the Bonanza

King Formation across the southern Great Basin, the sensitive response of its carbonate lithofacies to changes in base level, and the systematic changes in stacking patterns of component parasequences suggest that the largely uninvestigated Bonanza King Formation provides a superb record of accommodation change throughout Middle to Late Cambrian time. If future correlations can be made interbasinally with a reasonable degree of synchroneity, the establishment of a eustatic sea-level history will greatly refine our chronostratigraphic resolution of this time interval.

ACKNOWLEDGMENTS

Constructive reviews by Art Saller, Rick Sarg, and Bob Goldhammer considerably improved the manuscript. Several of the ideas in this chapter were generated with J. Fred Read (VPI&SU) and we thank him for his input. Scott Edwards provided able field assistance. Funding for this research was provided by NSF grant EAR-9205839 (to IPM and DAO) and The Petroleum Research Fund, administered by the ACS (to IPM).

REFERENCES CITED

Aitken, J. D., 1978, Revised models for depositional grand cycles, Cambrian of the southern Rocky Mountains, Canada: Bulletin Canadian Petroleum Geology, v. 26, p. 515-542.

Barnes, H., R. L. Christiansen, and F. M. Byers, Jr., 1962, Cambrian Carrara Formation, Bonanza King Formation, and Dunderberg shale east of Yucca Flat, Nye County, Nevada: U.S. Geological Survey Professional Paper 450-D, D1-D27.

Barnes, H., and A. R. Palmer, 1961, Revision of stratigraphic nomenclature of Cambrian rocks, Nevada Test Site and vicinity, Nevada: U.S. Geological Survey Professional Paper 424, 100 p.

Biddle, K. T., W. Schlager, K. W. Rudolph, and T. L. Bush, 1992, Seismic model of a progradational carbonate platform, Picco di Vallandro, the Dolomites, northern Italy: AAPG Bulletin, v. 76, p. 14-30.

Bond, G. C., N. Christie-Blick, M. A. Kominz, and W. J. Devlin, 1984, An early Cambrian rift to post-rift transition in the Cordillera of western North America: Nature, v. 316, p. 742-745.

Chow, N., and N. P. James, 1987, Cambrian Grand Cycles: A northern Appalachian perspective: Geological Society of America Bulletin, v. 98, p. 418-429.

Christie-Blick, N., 1991, Onlap, offlap, and the origin of unconformity-bounded depositional sequences: Marine Geology, v. 97, p. 35-56.

Demicco, R. V., 1985, Patterns of platform and off-platform carbonates of the Upper Cambrian of western Maryland: Sedimentology, v. 32, p. 1-22.

Fischer, A. G., 1964, The Lofer cyclothems of the Alpine Triassic, in D. F. Merriam, ed., Symposium of Cyclic Sedimentation: State Geological Survey of Kansas Bulletin 169, p. 107-150.

Franseen, E. K., T. E. Fekete, and L. C. Pray, 1989, Evolution and destruction of a carbonate bank at the shelf margin: Grayburg Formation (Permian), western escarpment, Guadalupe Mountains, Texas, in P. D. Crevello, J. L. Wilson, J. F. Sarg, and J. F. Read, eds., Controls on Carbonate Platform and Basin Development: SEPM Special Publication No. 44, p. 289-304.

Gans, W. T., 1974, Correlation and redefinition of the Goodsprings Dolomite, southern Nevada and eastern California: Geological Society of America Bulletin, v. 85, p. 189-200.

Goldhammer, R. K., P. A. Dunn, and L. A. Hardie, 1987, High frequency glacio-eustatic sea-level oscillations with Milankovitch characteristics recorded in Middle Triassic platform carbonates in northern Italy: American Journal of Science, v. 287, p. 853-892.

Goldhammer, R. K., P. A. Dunn, and L. A. Hardie, 1990, Depositional cycles, composite sea-level changes, cycle stacking patterns, and the hierarchy of stratigraphic forcing: Examples from Alpine Triassic platform carbonates: Geological Society of America Bulletin, v. 102, p. 535-562.

Goldhammer, R. K., P. J. Lehmann, and P. A. Dunn, 1992, Third-order sequence stratigraphy and high-frequency cycle stacking patterns of Lower Ordovician platform carbonates, El Paso Group, Franklin Mountains west Texas, in M. P. Candelaria, and C. L. Reed, eds., Paleokarst, Karst Related Diagenesis and Reservoir Development: Examples from Ordovician-Devonian Age Strata of West Texas and the Mid-continent: SEPM, Permian Basin Section, Publication No. 92-93, p. 59-92.

Hazzard, J. C., 1937, Paleozoic section in the Nopah and Resting Springs mountains, Inyo County, California: California Journal of Mines and Geology, v. 33, p. 273.

Kepper, J. A., 1972, Paleoenvironmental pattern in Middle to lower Upper Cambrian interval in the eastern Great Basin: AAPG Bulletin, v. 56, p. 503-527.

Kepper, J. A., 1976, Stratigraphic relationships and depositional facies in a portion of the Middle Cambrian of the Basin and Range province, in R. A. Robinson, ed. Paleontology and Depositional Environments: Cambrian of Western North America: Brigham Young University Studies in Geology, v. 21, p. 75-91.

Kepper, J. A., 1981, Sedimentology of a Middle Cambrian outer shelf margin with evidence for syndepositional faulting, eastern California and western Nevada: Journal of Sedimentary Petrology, v. 51, no. 3, p. 807-821.

Koerschner, W. F., III, and J. F. Read, 1989, Field and modelling studies of Cambrian carbonate cycles, Virginia Appalachians: Journal of Sedimentary Petrology, v. 59, no. 5, p. 654-687.

Laporte, L. F., 1967, Carbonate deposition near mean sea level and resultant facies mosaic; Manlius Formation (Lower Devonian) of New York State: AAPG Bulletin, v. 51, p. 73-101.

Levy, M., and N. Christie-Blick, 1989, Pre-Mesozoic palinspastic reconstruction of the eastern Great Basin (western United States): Science, v. 245, p. 1454-1462.

Levy, M., and N. Christie-Blick, 1991, Tectonic subsidence of the early Paleozoic passive continental margin in eastern California and southern Nevada: Geological Society of America Bulletin, v. 103, p. 1590-1606.

Mitchum, R. M., 1977, Glossary of terms used in seismic stratigraphy, in C. E. Payton, ed., Seismic Stratigraphy—Applications to Hydrocarbon Exploration: AAPG Memoir 26, p. 205-212.

Mitchum, R. M., and J. C. Van Wagoner, 1991, High-frequency sequences and their stacking patterns: Sequence-stratigraphic evidence of high-frequency eustatic cycles: Sedimentary Geology, v. 70, p. 131-160.

Montañez, I. P., and J. F. Read, 1992, Eustatic sea level control on early dolomitization of peritidal carbonates: Evidence from the Early Ordovician, Upper Knox Group, Appalachians: Geological Society of America Bulletin, v. 104, p. 872-886.

Osleger, D. A., 1991, Subtidal carbonate cycles: Implications for allocyclic versus autocyclic controls: Geology, v. 19, p. 917-920.

Osleger, D. A., and J. F. Read, 1991, Relation of eustasy to stacking patterns of Late Cambrian cyclic carbonates: A field and computer modelling study: Journal of Sedimentary Petrology, v. 61, p. 1225-1252.

Osleger, D. A., and J. F. Read, 1993, Comparative methods of determining eustatic sea level history from cyclic carbonate sequences: An interbasinal approach, Late Cambrian, U.S.A.: American Journal of Science, v. 293, p. 157–216.

Palmer, A. R., 1965, Trilobites of the Late Cambrian Pterocephaliid biomere in the Great Basin, United States: U.S. Geological Survey Professional Paper 493, 105 p.

Palmer, A. R., 1983, The decade of North American geology 1983 geologic time scale: Geology, v. 11, p. 503-504.

Palmer, A. R., and J. C. Hazzard, 1956, Age and correlation of Cornfield Springs and Bonanza King Formations in southeastern California and southern Nevada: AAPG Bulletin, v. 40, p. 2494-2513.

Read, J. F., 1985, Carbonate platform facies models: AAPG Bulletin, v. 69, p. 1-21.

Read, J. F., and R. K. Goldhammer, 1988, Use of Fischer plots to define 3rd order sea level curves in peritidal cyclic carbonates, Early Ordovician, Appalachians: Geology, v. 16, p. 895-899.

Read, J. F., D. A. Osleger, and M. Elrick, 1991, Two-dimensional modeling of carbonate ramp sequences and component cycles, in E. K. Franseen, ed., Sedimentary Modeling: Computer Simulations and Methods for Improved Parameter Definition: Kansas Geological Survey Bulletin 233, p. 473-488.

Rowland, S. M., J. R. Parolini, E. Eschne, A. J. McAllister, and J. A. Rice, 1990, Sedimentologic and stratigraphic constraints on the Neogene trans-lation and rotation of the Frenchman Mountain structural block, Clark County, Nevada, in B. P. Wernicke, ed., Basin and Range Extensional Tectonics Near the Latitude of Las Vegas, Nevada: Boulder, Colorado, Geological Society of America Memoir 176.

Sadler, P. M., D. A. Osleger, and I. P. Montañez, 1993, On labelling, interpretation, and the length of Fischer plots: Journal of Sedimentary Petrology, v. 63, p. 360–368.

Sarg, J. F., 1988, Carbonate sequence stratigraphy, in C. Wilgus, H. Posamentier, C. Ross, and C. G. St. C. Kendall, eds., Sea Level Changes: An Integrated Approach: SEPM Special Publication No. 42, p. 156-181.

Schlager, W., 1991, Depositional bias and environmental change—important factors in sequence stratigraphy, in K. T. Biddle and W. Schlager, eds., The Record of Sea Level Fluctuations: Sedimentary Geology, v. 70, p. 109-130.

Scotese, C. R., and W. S. McKerrow, 1990, Revised world maps and introduction, in W. S. McKerrow, and C. R. Scotese, eds., Palaeozoic Palaeogeography and Biogeography: Geological Society of London Memoir 12, p. 1-24.

Shinn, E. A., 1986, Modern carbonate tidal flats: Their diagnostic features: Colorado School of Mines Quarterly, v. 81, p. 7-35.

Snow, J. K., 1992, Large-magnitude Permian shortening and continental-margin tectonics in the southern Cordillera: Geological Society of America Bulletin, v. 104, p. 80-105.

Snow, J. K., and B. Wernicke, 1989, Uniqueness of geological correlations: An example from the Death Valley extended terrain: Geological Society of America Bulletin, v. 101, p. 1351-1362.

Sonnenfeld, M. D., 1991, High-frequency cyclicity within shelf-margin and slope strata of the upper San Andres sequence, Last Chance Canyon, in S. Meader-Roberts, M. P. Candelaria, and G. E. Moore, eds., Sequence Stratigraphy, Facies, and Reservoir Geometries of the San Andres, Grayburg, and Queen Formations, Guadalupe Mountains, New Mexico and Texas: SEPM, Permian Basin Section, Special Publication No. 91-32, p. 11-51.

Stewart, J. H., and C. A. Suczek, 1977, Cambrian and latest Precambrian paleogeography and tectonics in the western United States, in J. H. Stewart, C. H. Stevens, and A. E. Fritsche, eds., Paleogeography of Western United States: SEPM, Pacific Section, Pacific Coast Paleogeography Symposium, p. 1-18.

Sundberg, F. A., 1990, Morphological diversification of the Ptychopariid Trilobites in the Marjumiid Biomere (Middle to Upper Cambrian): Virginia Polytechnic Institute and State University, Blacksburg, Virginia, unpublished Ph.D. dissertation, 425 p.

Vail, P. R., 1987, Seismic stratigraphy interpretation procedure, in A. W. Bally, ed., Atlas of Seismic Stratigraphy, Volume 1: AAPG Studies in Geology 27, p. 1-10.

Vail, P. R., R. M. Mitchum, and S. Thompson, III, 1977, Seismic stratigraphy and global changes of sea level, Part 4: Global cycles of relative changes of sea level, *in* C. E. Payton, ed., Seismic Stratigraphy—Applications to Hydrocarbon Exploration: AAPG Memoir 26, p. 83-97.

Van Wagoner, J. C., R. M. Mitchum, Jr., H. W. Posamentier, and P. R. Vail, 1987, Key definitions of sequence stratigraphy, *in* A. W. Bally, ed., Atlas of Seismic Stratigraphy, Volume 1: AAPG Studies in Geology 27, p. 11-14.

Wernicke, B., G. J. Axen, and J. K. Snow, 1988, Basin and range extensional tectonics at the latitude of Las Vegas, Nevada: Geological Society of America Bulletin, v. 100, p. 1738-1757.

Chapter 13

Sequence Stratigraphy and Evolution of a Progradational, Foreland Carbonate Ramp, Lower Mississippian Mission Canyon Formation and Stratigraphic Equivalents, Montana and Idaho

S. K. Reid
Department of Physical Sciences
Morehead State University
Morehead, Kentucky, U.S.A.

S. L. Dorobek
Department of Geology
Texas A&M University
College Station, Texas, U.S.A.

ABSTRACT

The Lower Mississippian Mission Canyon Formation and stratigraphic equivalents in Montana and Idaho were deposited on a progradational carbonate ramp that developed on the foreland side of the Antler foredeep. Shallow subtidal and peritidal lithofacies were deposited in ramp-interior settings across most of Montana. The ramp to basin transition in westernmost Montana was a relatively narrow belt of stacked skeletal grainstone banks. Farther west, skeletal grainstone banks prograded over and interfingered with outer ramp/slope cherty limestones. In east-central Idaho, coeval lower slope and basinal strata consisted of silty to argillaceous, spicular limestones, spiculites, and spicular calcareous siltstones/fine-grained sandstones.

In individual outcrops, stacked parasequence sets are the most prominent sequence stratigraphic units. Lateral lithofacies relationships across the deformed ramp to basin transition were reconstructed using regional biostratigraphic and lithostratigraphic correlations of measured sections. Depositional sequences and system tracts were identified from characteristics of bounding surfaces, stacking patterns of parasequences and parasequence sets, and lateral lithofacies relationships.

The reconstructed ramp to basin transect illustrates a progressive upward change in third-order sequence boundary type during evolution of the Mission Canyon platform. Type 2 sequence boundaries formed early in platform development, whereas type 1 sequence boundaries dominated later platform development. Associated ramp-margin wedges thicken and are composed of progressively larger proportions of peritidal lithofacies upward. Compared to correlative surfaces on global onlap-offlap curves, third-order sequence boundaries that formed early in Mission Canyon platform development appear subdued, whereas those that formed later appear enhanced. Combined with subsidence analysis, these relationships suggest that gradually waning flexural subsidence and falling second-order eustatic sea level permitted higher order eustatic sea levels to fall progressively farther basinward as the Mission Canyon platform evolved. This long-term decrease in accommodation profoundly influenced progradation of the Mission Canyon platform and was a major factor in maintaining a ramplike profile across the platform to basin transition.

INTRODUCTION

Modern sequence stratigraphy has continued to evolve since Sloss et al. (1949) outlined its basic concepts (see historical perspective of Ross, 1991). The advent of high-quality seismic data changed our perspective of regional stratigraphy by revealing chronostratigraphic relationships and stratal geometries (Vail et al., 1977a,b). Concepts of seismic stratigraphy subsequently have been refined and applied to well logs, cores, and outcrops, and detailed integrated studies now are common (e.g., Wilgus et al., 1988). More recently, workers have begun to integrate models of cyclic carbonate deposition with sequence stratigraphy (e.g., Goldhammer et al., 1990, 1991; Read et al., 1991). However, it is still necessary to apply sequence stratigraphy to a wider variety of carbonate platform types, preferably from different tectonic settings.

Paleozoic strata preserve a wide assortment of platform types deposited in various tectonic settings. However, the most critical portion of these Paleozoic platforms, the platform to basin transition, typically is strongly deformed during orogenesis. In addition, seismic profiles are few and of limited use for sequence stratigraphy in these areas. Therefore, studies of Paleozoic platforms rely heavily on outcrop and/or well data and require good biostratigraphic control. Studies that have successfully applied sequence stratigraphy to outcrops across carbonate platform to basin transitions (e.g., Sarg, 1988; Franseen et al., 1989) largely have been limited to well-exposed, undeformed rocks where stratal boundaries can be traced physically through mapping or on photo mosaics. In structurally complex regions, such as the North American Cordillera, undisturbed, laterally continuous outcrops are rare and stratal boundaries cannot be traced physically over great distances.

This chapter discusses the sequence stratigraphy and evolution of an Early Mississippian carbonate ramp that developed in a foreland tectonic setting in the northern Rocky Mountains of the United States. Sequence stratigraphic units were reconstructed across the tectonically shortened ramp to basin transition using six key stratigraphic sections in southwestern Montana (Figure 1). These measured sections are from a larger data set that covers western Montana and east-central Idaho (figure 1 in Dorobek et al., 1991b; Reid, 1991; S. Reid, S. Dorobek, and T. Smith, unpublished data). A transect across the ramp to basin transition was chosen because shallow-water, ramp-interior areas often are exposed during relative sea-level falls and some strata are eroded or not deposited, while a more complete record of platform evolution is preserved in coeval, deeper water deposits.

The sequence stratigraphic analysis presented in this chapter allows interpretation of the long-term evolution of a carbonate platform that appears to be transitional between a high-energy, distally steepened ramp (after Read, 1985) and a rimmed shelf. The relative influences of foreland tectonism and eustasy on development of this ramplike platform are discussed. In addition, this chapter illustrates that parasequence sets can be used as primary units for regional sequence stratigraphic analysis of thick stratigraphic intervals (~200–500 m) in areas where seismic data are absent (or unavailable) and outcrops or wells are widely separated. Lastly, this chapter provides a comparative framework for future sequence stratigraphic studies of other Lower Mississippian strata.

REGIONAL GEOLOGY

Paleogeography

During Early Mississippian time, a broad, shallow-marine carbonate platform occupied most of the central and western United States (Figure 2) (Sando, 1976; Gutschick et al., 1980; Gutschick and Sandberg,

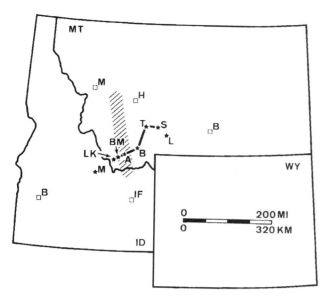

1983; Sandberg et al., 1983). The platform surrounded isolated remnants of the Transcontinental Arch. In North Dakota and Montana, this platform extended westward from the Williston basin to the rapidly subsiding Antler foredeep along the western margin of North America (Figure 2). The axis of the foredeep, whose northern extension is preserved in east-central Idaho, was approximately 1500–1800 km west of the Transcontinental Arch (Gutschick and Sandberg, 1983). Both the axis of the foredeep and the transition from shallow-water platform to slope environments trended approximately north–south relative to Mississippian plate reconstructions (Figure 2). These Early Mississippian paleogeographic features are preserved in the Mission Canyon Formation and stratigraphic equivalents in Montana, Idaho, and North Dakota.

Stratigraphic Framework

Foraminifera, conodonts, and corals have been used to define biozones within Mississippian strata of the northern Rocky Mountains (Figure 3) (Sando et al., 1969; Sandberg et al., 1983; Sando, 1985; Sando and Bamber, 1985). Because all three zonations have been used to correlate these strata, this study follows the composite biozonation of Sando (1985) as a matter of convenience. Composite biochronozones range from 0.75 to 3.4 m.y., depending on the absolute time scale used to calculate durations (Sandberg and Poole, 1977; Sandberg et al., 1983; Sando, 1985). The biostratigraphic zonations shown in Figure 3 have

Figure 1. Locations of measured sections in southwestern Montana (stars) (A = Ashbough Canyon; B = Baldy Mountain; BM = Bell/McKenzie Canyons; L = Livingston Canyon; LK = Lake Canyon; M = McGowan Creek; S = Sacajawea Peak; T = Trident). Hatched area is approximate position of ramp-margin skeletal banks. Bold line of section is for Figure 10. Boxes are major cities in the region [Boise (B) and Idaho Falls (IF), Idaho (ID); Missoula (M), Helena (H), and Billings (B), Montana (MT)]. Access to measured sections is summarized in Reid (1991, Appendix A).

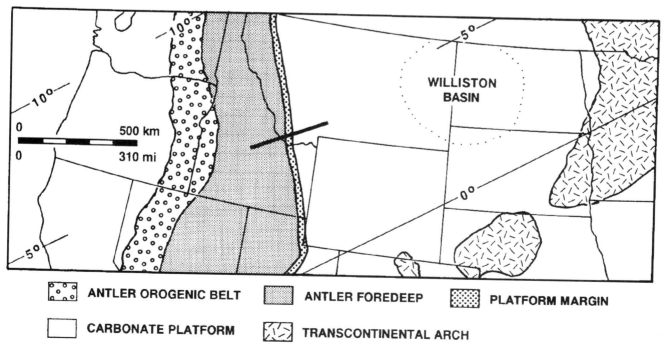

Figure 2. Early Mississippian paleogeography of the north-central and northwestern United States (after Gutschick and Sandberg, 1983). Bold line is approximate line of section for Figure 4.

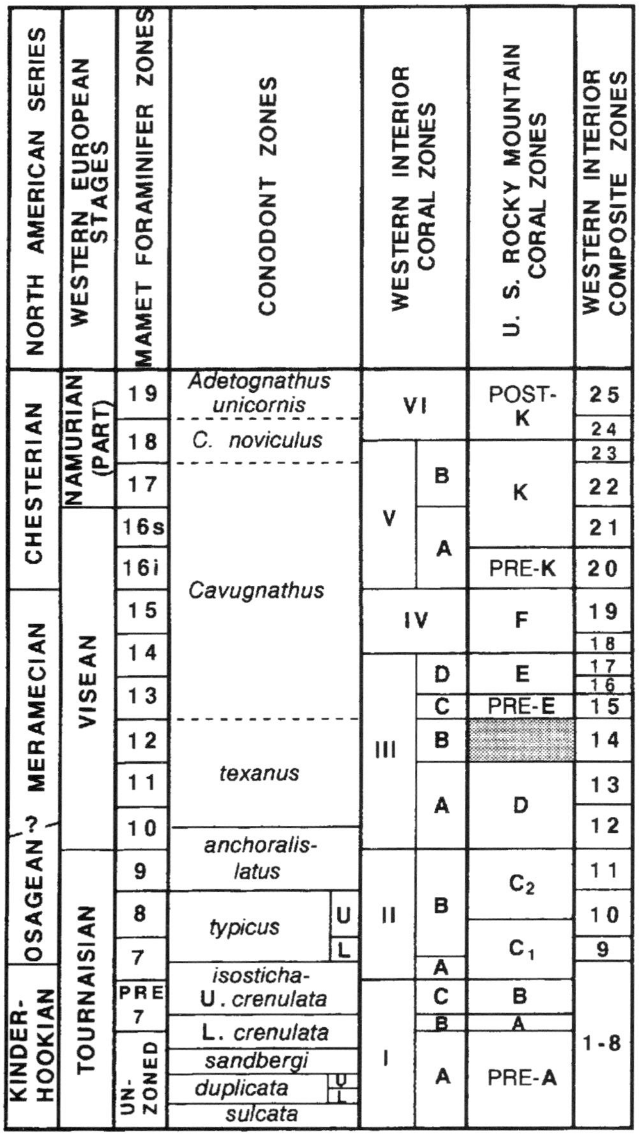

NORTH AMERICAN SERIES	WESTERN EUROPEAN STAGES	MAMET FORAMINIFER ZONES	CONODONT ZONES	WESTERN INTERIOR CORAL ZONES		U.S. ROCKY MOUNTAIN CORAL ZONES	WESTERN INTERIOR COMPOSITE ZONES
CHESTERIAN	NAMURIAN (PART)	19	Adetognathus unicornis	VI		POST-K	25
CHESTERIAN	NAMURIAN (PART)	18	C. noviculus	V	B	K	24
CHESTERIAN	NAMURIAN (PART)	17	Cavugnathus	V	B	K	23 / 22
MERAMECIAN	VISEAN	16s	Cavugnathus	V	A	K	21
MERAMECIAN	VISEAN	16i	Cavugnathus	V	A	PRE-K	20
MERAMECIAN	VISEAN	15	Cavugnathus	IV	F	19
MERAMECIAN	VISEAN	14	Cavugnathus	IV		F	18 / 17
MERAMECIAN	VISEAN	13	Cavugnathus	IV	D	E	16
MERAMECIAN	VISEAN	13			C	PRE-E	15
MERAMECIAN	VISEAN	12	texanus	III	B	(shaded)	14
MERAMECIAN	VISEAN	11	texanus	III	A	D	13
OSAGEAN?	VISEAN	10	texanus	III	A	D	12
OSAGEAN?	TOURNAISIAN	9	anchoralis-latus	II	B	C2	11
OSAGEAN?	TOURNAISIAN	8	typicus U	II	B	C2	10
OSAGEAN?	TOURNAISIAN	7	typicus L	II	A	C1	9
KINDER-HOOKIAN	TOURNAISIAN	PRE 7	isosticha-U. crenulata	I	C	B	1-8
KINDER-HOOKIAN	TOURNAISIAN	7 (UN-ZONED)	L. crenulata	I	B	A	1-8
KINDER-HOOKIAN	TOURNAISIAN	UN-ZONED	sandbergi	I	A	PRE-A	1-8
KINDER-HOOKIAN	TOURNAISIAN	UN-ZONED	duplicata U/L				
KINDER-HOOKIAN	TOURNAISIAN	UN-ZONED	sulcata				

Figure 3. Biostratigraphic zonation used in the region. Compiled from Sando et al. (1969), Sando (1985), and Sando and Bamber (1985).

been used to establish regional, formation-scale correlations in Mississippian strata throughout the western interior of the United States. Published biostratigraphic data, paleontologic samples collected by the senior author, and formation-scale correlations between stratigraphic sections examined in this study are summarized in Reid (1991).

The Mission Canyon Formation (200–400 m thick) and stratigraphic equivalents conformably overlie the McGowan Creek and Lodgepole formations and predominantly consist of shallow subtidal to peritidal platform facies throughout most of Montana (Figure 4). These facies grade westward into a relatively nar-

row belt of thick skeletal grainstone in westernmost Montana, which also is included in the Mission Canyon Formation (Huh, 1967, 1968; Rose, 1976; Nichols, 1980; Peterson, 1986; Reid, 1991). Near the Idaho–Montana border, grainstones of the Mission Canyon Formation interfinger with cherty limestones of the Middle Canyon Formation (Sando et al., 1985; Reid, 1991). Farther west, the upper member of the McGowan Creek Formation and the Middle Canyon Formation are correlative with the Mission Canyon Formation and consist of silty, spicular limestones, spiculites, calcareous siltstones, and fine-grained sandstones (Figure 4; Huh, 1967, 1968; Sandberg, 1975; Nilsen, 1977; Skipp et al., 1979; Gutschick and Sandberg, 1983).

The top of the Mission Canyon Formation is a regional unconformity that represents from 9 to 14 m.y. of subaerial exposure in central Montana (Figure 4) (Sando, 1976; Skipp et al., 1979; Gutschick et al., 1980; Sandberg et al., 1983; Sando, 1988). The Mission Canyon Formation was extensively karstified across most of Montana during this time (Middleton, 1961; Roberts, 1966; Sando, 1974, 1988). In westernmost Montana, however, deposition was continuous and shallow subtidal to peritidal facies of the McKenzie Canyon Formation (Figure 4) and equivalent deepwater facies of the upper Middle Canyon Formation were deposited in middle to late Meramecian time (Sando et al., 1985).

Platform Model

The Mission Canyon platform cannot be assigned to a single platform type (e.g., after Read, 1985). Previous studies have characterized the Mission Canyon platform as a ramp (sensu Ahr, 1973; Gutschick et al., 1980) and as a rimmed shelf with 200–400 m relief at its margin (Rose, 1976). A lack of balanced regional cross sections in the study area prevents accurate calculation of depositional slopes. However, the detailed stratigraphic cross section presented later in this chapter incorporates estimated palinspastic distances between measured sections (after Peterson, 1986) and suggests that depositional slopes were ramplike.

Although depositional slopes across the Mission Canyon platform appear ramplike, these slopes may not have been homoclinal. Thin, carbonate gravity-flow deposits and soft sediment deformation in slope facies (folding and sedimentary boudinage structures; Reid, 1991; W. Perry, personal communication, 1989) indicate that somewhat greater depositional slopes existed west of the platform to basin transition. In addition, the Antler foredeep was starved during deposition of the Mission Canyon platform (Sandberg, 1975; Nilsen, 1977; Skipp et al., 1979; Gutschick and Sandberg, 1983), which suggests that the difference between platform and basinal sedimentation rates may have been sufficient to generate a break in slope at the platform margin.

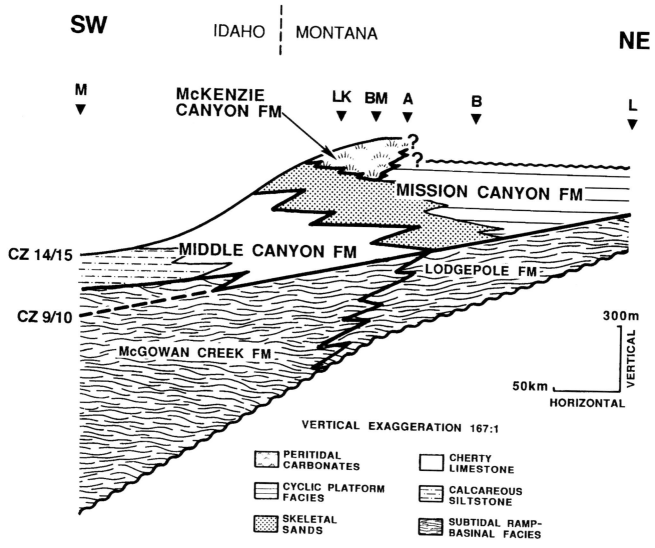

Figure 4. Highly schematic, formation-scale cross section of the Mission Canyon Formation and stratigraphic equivalents. CZ 9/10 and CZ 14/15 are composite biozone boundaries (Figure 3). Arrows indicate points of control (see Figure 1 for locations).

A reefal rim did not develop along the Mission Canyon platform margin because diverse assemblages of reef-building organisms apparently did not exist during Early Mississippian time (Heckel, 1974; James, 1984). Instead, wave-agitated skeletal banks formed a relatively narrow facies belt along the outer platform while much of the platform interior was covered by skeletal-ooid grainstone (Huh, 1968; Rose, 1976; Nichols, 1980; Peterson, 1986; Reid, 1991). The combination of ramplike slopes, widespread skeletal-ooid grainstones across the platform interior, a (subtle?) break in slope at the platform margin, and a relatively narrow skeletal grainstone belt along the platform margin suggests that the Mission Canyon platform was transitional between a high-energy, distally steeped ramp (Read, 1985) and

a rimmed shelf. In this chapter, we refer to this type of carbonate platform as a "progradational carbonate ramp."

LITHOFACIES AND DEPOSITIONAL ENVIRONMENTS

Peritidal Facies

Peritidal facies occur throughout the Mission Canyon Formation across most of Montana. Near the Montana–Idaho border, peritidal facies are confined to the McKenzie Canyon Formation. These facies form 0.5–10 m thick, shallowing-upward cycles that consist of, from bottom to top: skeletal-peloid grainstone/packstone; fenestral limestone; and cryptalgalaminite, evaporitic carbonates, and/or solution collapse breccia (Figure 5).

Skeletal-Peloid Grainstone/Packstone

Dark brown-gray to light tan-gray, color-mottled, skeletal-peloid packstone/grainstone occurs as thick to massive beds that overlie skeletal-ooid grainstones. Current-generated sedimentary structures are absent and thin to medium interbeds of intraclast-peloid packstone/grainstone and ooid-skeletal grainstone are rare. Grain types include peloids, intraclasts, ooids, calcispheres, foraminifera, gastropods, and crinoid fragments (Figure 5A).

Interpretation

Thick, skeletal-peloid packstone/grainstone formed in shallow, relatively low-energy subtidal settings. Color mottling and the lack of current-generated sedimentary structures suggest these sediments were subjected to intense bioturbation.

Fenestral Limestone

Medium gray to light tan-gray fenestral limestone overlies skeletal-peloid grainstone/packstone. Thin interbeds of bioturbated to laminated skeletal-peloid grainstone/packstone and cryptalgalaminite are common. Fenestral limestone fabrics range from mudstone to packstone with peloids, micritized grains, calcispheres, ostracods, gastropods, calcareous algae, and coated grains as primary constituents.

Fenestral fabrics are variable. In color-mottled fenestral limestone, tubular and irregular fenestrae are common. Fenestral limestone that is not color mottled usually contains irregular to laminoid fenestrae with inclusion-rich, isopachous calcite cement and geopetal sediment (Figure 5B). Fenestral limestones with abundant micritized/coated grains typically contain irregular and sheetlike fenestrae that are filled with laminated dolomite silt. This type of fenestral limestone also has scattered evaporite pseudomorphs and may cap parasequences.

Interpretation

Most fenestral limestone was deposited in intertidal to very shallow subtidal environments (Logan, 1974; Shinn, 1983). Color mottling and some tubular fenestrae probably were produced by burrowers in very shallow subtidal environments. Inclusion-rich, isopachous cements in some fenestrae are interpreted as marine phreatic cements that also formed in shallow subtidal environments. Micritized/coated grain fenestral limestones with sheetlike fenestrae and scattered evaporite pseudomorphs may represent intertidal to supratidal facies that have been overprinted by vadose meteoric diagenesis (Fischer, 1964; Goldhammer et al., 1987; Mazzullo and Birdwell, 1989).

Restricted Peritidal Facies

Cryptalgalaminite facies caps most parasequences. This facies consists of light gray to tan, fine-crystalline dolostone, calcareous dolostone, or dolomitic lime mudstone with crinkly to irregular, millimeter-scale laminations (Figure 5C). Shallow mud cracks and centimeter-scale tepee structures are common; evaporite pseudomorphs are rare. Thin layers of laminated intraclast-peloid grainstone/packstone and fenestral limestone commonly are interbedded with cryptalgalaminite facies.

Some parasequences are capped by very thin to thin, irregularly bedded, brown-gray to yellow-tan, peloidal mudstone/packstone. Dolomitization varies from slight (scattered dolomite rhombs) to complete. Calcite pseudomorphs after gypsum and anhydrite are common and occur as 1–10 cm nodules or 0.1–1-cm-long crystal pseudomorphs that commonly form irregular rosettes (Figure 5D). Coated grains (pisoids) with circumgranular and intragranular cracks are rare (Figure 5E). Where present, evaporitic facies overlie cryptalgalaminite or laminated, peloid grainstone/packstone. However, thin beds of dolomitic sandstone (10–30 cm thick) with evaporite pseudomorphs rarely overlie evaporitic facies and may form parasequence caps.

Interpretation

Restricted peritidal facies formed in supratidal and high intertidal areas of semiarid to arid tidal flats that developed when the Mission Canyon platform aggraded to sea level (cf. modern examples from Kinsman, 1964; Kendall and Skipwith, 1969; Purser, 1973; Logan et al., 1974). Storms intermittently deposited thin sheets of laminated intraclast-peloid grainstone/packstone on the flats. Evaporitic conditions prevented burrowing organisms from destroying mechanical and algal laminations. Pisoids in evaporitic facies formed where supratidal flats were subjected to vadose marine and/or vadose meteoric diagenesis (Read, 1974; Scholle and Kinsman, 1974; Esteban, 1976; Esteban and Klappa, 1983).

Solution-Collapse Breccia

Brecciated horizons from 0.5–75 m thick occur in the Mission Canyon Formation across most of Montana and in the McKenzie Canyon Formation in westernmost Montana. Lateral continuity and thickness of breccia horizons is highly variable. Contacts with unbrecciated rocks are irregular; fitted fabrics with surrounding, unbrecciated rocks are common around the margins of some breccia horizons. Breccias contain clasts and tabular blocks (<1 cm–2 m long) of all peritidal lithofacies (Figure 5F); clasts of skeletal-ooid grainstone are rare. Approximate original stratigraphic position of various clast lithologies still is preserved within many breccias. For example, fenestral limestone clasts typically grade upward into breccia dominated by cryptalgalaminite clasts, which in turn grade upward into brecciated evaporitic carbonates.

Interpretation

Most solution collapse breccias formed through dissolution of evaporite horizons and carbonate lithologies during karstification of the Mission Canyon platform, especially in middle to late Mera-

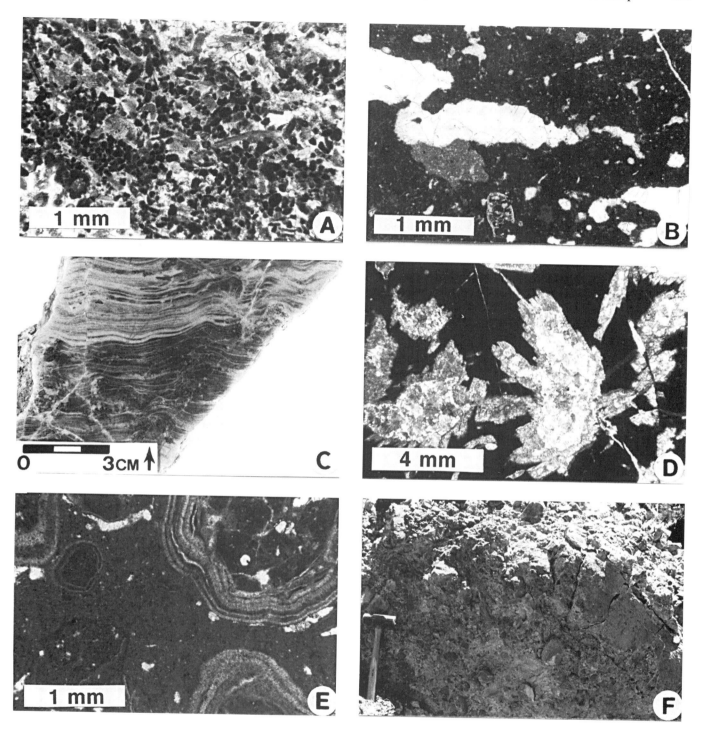

Figure 5. Peritidal facies. (A) Skeletal-peloid grainstone/packstone; (B) fenestral limestone with irregular to laminoid fenestrae; (C) slab of cryptalgalaminated dolostone, microporosity stained gray to enhance laminations; (D) evaporite pseudomorphs in restricted peritidal facies; (E) coated grains (pisoids) with circumgranular and intragranular cracks; and (F) solution collapse breccia.

mecian time when the major unconformity at the top of the Mission Canyon Formation developed (Sando, 1972, 1976, 1988). Some dissolution also probably occurred following Pennsylvanian(?) to Early Tertiary uplift of the Mission Canyon Formation.

Examination of clast lithologies, however, allows interpretation of original depositional environments. Most brecciated lithofacies originally were deposited in peritidal environments, except for rare clasts of subtidal grainstone.

Subtidal Ramp-Interior and Ramp-Margin Facies

Skeletal-ooid grainstone is the dominant subtidal lithofacies across most of Montana and occurs as thick to massive beds in the McKenzie Canyon and Mission Canyon formations. In westernmost Montana, massive skeletal grainstones dominate the upper Middle Canyon and Mission Canyon formations.

Skeletal-Ooid Grainstone

Skeletal-ooid grainstone is medium gray to light tan-gray and contains abundant ooids and skeletal grains (Figure 6A). Peloids also are common locally. Skeletal grains are abraded, bored, and partially micritized, and include crinoids, foraminifera, calcispheres, brachiopods, ostracods, and bryozoans. Cross-bedding and ripple cross-lamination are common. Beds commonly have irregular basal contacts (up to 10 cm of local relief) that are overlain by intraclast/skeletal lags. Skeletal-ooid grainstone commonly overlies peloid-skeletal packstone/wackestone.

Interpretation

Common erosional basal contacts, abundant ooids, cross-bedding, and ripple cross-lamination suggest that skeletal-ooid grainstones formed in mobile sand-shoal environments in shallow-water, ramp-interior settings (Ball, 1967; Hine, 1977). Shoals probably formed as laterally discontinuous barriers or isolated shoal complexes that partially protected peritidal environments. Peritidal facies accumulated adjacent to grainstone shoal complexes in quieter water and/or more restricted environments.

Massive Skeletal Grainstone

Skeletal grainstone is massively bedded and dominated by pelmatozoan grains, in contrast to skeletal-ooid grainstones across most of the rest of Montana. Low-angle, tabular cross-bedding and horizontal laminations are rare to common. Moderately sorted, fragmented to articulated pelmatozoans are the dominant grain type (Figure 6B). Ramose and fenestrate bryozoan debris, slightly abraded rugose corals and brachiopods, and large rugose and syringoporid coral colonies (0.5–1 m across) also are common locally. Massive skeletal grainstone typically overlies cherty, bioturbated, peloid-skeletal wackestone/packstone.

Interpretation

Massive skeletal grainstone in the Mission Canyon and upper Middle Canyon formations of western Montana were deposited in wave-agitated, skeletal bank environments along the ramp to basin transition in western Montana (cf. Laporte, 1969; Dorobek and Read, 1986). The less-fragmented, diverse fossil assemblage (including large coral colonies and articulated crinoid columnals) within these grainstones suggest approximately in situ sediment accumulation. Skeletal banks probably accumulated where dense thickets of pelmatozoans dominated benthic communities. Lime mud was not produced or was winnowed by fair-weather waves.

Outer Ramp, Slope, and Basinal Facies

Outer ramp, slope, and basinal facies become progressively richer in siliciclastics toward the axis of the Antler foredeep in east-central Idaho (Figure 4). Bioturbated cherty limestone is confined to the Middle Canyon and Mission Canyon formations of westernmost Montana and east-central Idaho. Laminated cherty limestone is confined to the Middle Canyon Formation of westernmost Montana and east-central Idaho. Mixed carbonate/siliciclastic and calcareous siltstone/sandstone facies are confined to the Middle Canyon and upper McGowan Creek formations in east-central Idaho.

Peloid-Skeletal Wackestone/Packstone

Dark brown-gray to medium gray, bioturbated, peloid-skeletal wackestone/packstone (Figure 6C) gradationally overlies bioturbated, cherty limestone and underlies massive skeletal grainstone facies and, less commonly, skeletal-ooid grainstone. Poorly sorted, whole to slightly abraded brachiopods, fenestrate bryozoans, rugose corals, and articulated crinoid stems are common. Bioturbation is moderate to intense. Irregular, bedding-parallel chert nodules and stringers (2–5 cm thick) are rare to common. Thin (3–10 cm thick), discontinuous beds of peloid-skeletal packstone/grainstone with irregular basal contacts (up to 2 cm local relief) also are common. These laterally discontinuous interbeds have coarse-grained skeletal debris at their bases, commonly fine upward, and have gently undulatory to horizontal laminations. Upper parts of these thin interbeds may be bioturbated and grade into surrounding wackestone/packstone facies.

Interpretation

Abundant lime mud, a diverse benthic fauna, and extensive bioturbation suggest that peloid-skeletal wackestone/packstone formed below fair-weather wave base in open-marine environments (Wilson, 1975). Thin, discontinuous packstone/grainstone interbeds probably represent storm deposits that were not destroyed by burrowing organisms (cf. Kreisa, 1981).

Bioturbated Cherty Limestone

Dark gray to dark brown-gray, bioturbated, cherty, peloid-skeletal wackestone/packstone gradationally overlies laminated cherty limestone facies and grades upward into peloid-skeletal wackestone/packstone facies. Skeletal grains include abundant sponge spicules that are mixed with fine sand-size echinoderm debris and peloids (Figure 6D). Whole rugose corals, brachiopods, and articulated crinoid stems are rare to common. Bedding in this facies is defined by very thin to medium, irregular limestone layers which alternate with thin (1–10 cm thick), discontinuous, black to dark gray chert beds and irregular, bedding-parallel chert nodules (Figure 7A). Thin- to medium-bedded, structureless layers of coarse-grained crinoid grainstone/packstone are rare. Bioturbation is slight to intense. Farther west in

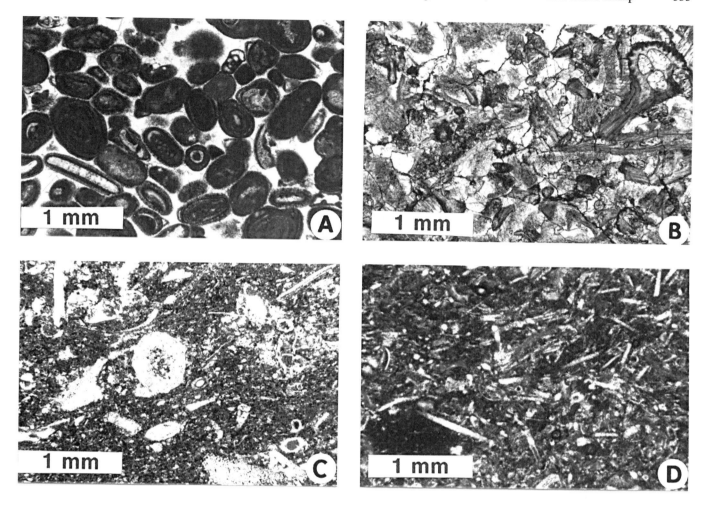

Figure 6. Subtidal platform and outer ramp–upper slope facies. (A) Skeletal-ooid grainstone; (B) skeletal grainstone; (C) peloid-skeletal wackestone/packstone; and (D) spicular peloid-skeletal packstone/wackestone.

Idaho, this facies becomes argillaceous, coarse-grained crinoid grainstone/packstone beds are absent, and black chert is rare. Instead, chert occurs as abundant, bedding-parallel incipient nodules or as silicified argillaceous seams and laminae.

Interpretation

Abundant chert and sponge spicules, a relatively low-diversity benthic fauna, and lack of preserved current-generated sedimentary structures suggest that bioturbated cherty limestone formed in outer ramp to upper slope environments below fair-weather wave base and perhaps below storm wave base (cf. Wilson, 1969; Cook and Enos, 1977). However, thin-bedded storm deposits may have been destroyed by abundant burrowing infauna. Rare structureless grainstone beds may represent sediment gravity flows (debris flows? or coarse-grained turbidites) derived from updip skeletal banks (Middleton and Hampton, 1976; Davies, 1977; Lowe, 1979).

Laminated Cherty Limestone

Black to dark gray, laminated to thin, even-bedded, cherty limestone and bedded chert with very thin (1–3 cm) limestone interbeds typically overlies mixed carbonate/siliciclastic facies. Centimeter-thick layers to very thin beds of black, spicule-rich chert alternate with spicular, peloid-skeletal wackestone/packstone beds (<10 cm thick). Skeletal-rich layers commonly fine upward and consist of fine-grained sand- to silt-sized echinoderm fragments and unidentifiable skeletal debris. Macrofossils are very rare. Horizontal and wavy laminations are abundant, but ripple cross-laminations are common (Figure 7B). Slight bioturbation is rare. Some soft sediment deformation occurs locally as contorted to gently folded horizons up to 1.5 m thick. Layers of structureless crinoid grainstone (10–30 cm thick) also are rare.

Interpretation

General absence of macrofossils and bioturbation suggests that laminated cherty limestone was deposited in middle- to lower-slope environments, below storm-wave base, and perhaps below the photic zone. Dysaerobic conditions probably limited burrowing infauna, thus preserving primary sedimentary structures. Alternating layers of chert and upward-fining, laminated limestone probably were deposited by turbidity currents generated updip; sediments were derived from shallower slope and ramp-margin areas to the east (cf. Davies, 1977). Structureless crinoid grainstone beds, similar to those in bioturbated cherty limestones, were deposited as rare sediment gravity flows (debris flows? or coarse-grained turbidites) that were derived from skeletal bank facies to the east. Contorted to gently folded horizons were produced by incipient failure and downslope movement of semicohesive slope deposits (cf. Cook and Taylor, 1977).

Mixed Carbonate/Siliciclastic Facies

Mixed carbonate/siliciclastic facies gradationally overlies calcareous siltstone/fine-grained sandstone facies. Bedding is even and very thin to thin. Internally, beds consist of graded laminae (3 mm to 1 cm thick) of black to dark tan-gray, silty to argillaceous, pyritic, skeletal-spicule wackestone/packstone that are intercalated with millimeter-scale, horizontal to wavy laminae of siliciclastic silt and fine sand. Interlayering appears to be random. Skeletal grains consist of unidentifiable, silt-size debris and rare radiolarian tests (Figure 7C); macrofossils are very rare to absent. In westernmost measured sections, this facies primarily consists of argillaceous spiculite. *Chondrites* and *Planolites* (0.5–5 mm diameter) disrupt laminations locally.

Interpretation

Mixed carbonate/siliciclastic rocks formed in lower slope to basin floor environments (cf. Wilson, 1969; Cook and Enos, 1977). Graded, alternating carbonate and siliciclastic laminae may represent thin, distal turbidites that were intermittently shed from the eastern Mission Canyon platform and the western Antler highlands, respectively. Abundant pyrite and lack of macrofossils and significant bioturbation suggest that deposition occurred below the photic zone in dysaerobic conditions (Seilacher, 1967; Rhoads, 1975).

Calcareous Siltstone/Sandstone

Calcareous siltstone/sandstone consists of black to dark gray, pyritic, laminated, calcareous siltstone and less-common fine-grained sandstone. This facies underlies and grades upward into mixed carbonate/siliciclastic facies. Very thin to thin even beds contain horizontal planar laminations (Figure 7D), but wavy laminations and slight to moderate bioturbation occur locally. Individual laminae fine upward. Locally, some laminae exhibit boudinage structures or are fractured perpendicular to bedding. Sponge spicules are common; unidentifiable, very fine-grained skeletal debris is rare.

Interpretation

Calcareous siltstone/sandstone probably was deposited in basin floor and lowermost slope environments. Deposition probably occurred in anaerobic(?) to dysaerobic conditions below the photic zone as indicated by abundant pyrite and a general lack of bioturbation or endemic fauna. Upward-fining laminae probably represent distal, low-density, siliciclastic turbidites that were derived from the western side of the basin (cf. Nilsen, 1977). Boudinage structures and fractured laminae probably are synsedimentary deformation features produced by downslope movement of partially lithified to well-lithified sediment. Very little platform-derived carbonate sediment reached this part of the basin during deposition of this facies.

SEQUENCE STRATIGRAPHIC NOMENCLATURE

Before discussing the vertical and lateral arrangement of the lithofacies described above, a very brief summary of basic sequence stratigraphic concepts and nomenclature is warranted. A full review of sequence stratigraphy is beyond the scope of this chapter.

It is well known that smaller scales of depositional cyclicity are superimposed on larger scales of depositional cyclicity (e.g., Goldhammer et al., 1990, 1991). Therefore, some nomenclatural hierarchy is required to distinguish the relative magnitudes (orders) of sequence stratigraphic units. In this chapter, we rank sequence stratigraphic units according to their respective distinguishing characteristics (after Sarg, 1988; Van Wagoner et al., 1988) and according to their durations (Vail et al., 1977a; Haq et al., 1988).

The smallest sequence stratigraphic unit is the *parasequence*, which is equivalent to the ubiquitous fifth-order (10^4–10^5 year duration), shallowing-upward, depositional cycles seen throughout the sedimentary record (Sarg, 1988; Van Wagoner et al., 1988). *Parasequence sets* are groups or bundles of stacked parasequences bounded by major marine flooding surfaces (Van Wagoner et al., 1988). The fundamental sequence stratigraphic unit, the *sequence*, ranges in duration from fourth-order (10^5–10^6 years) to third-order (10^6–10^7 years). Sequences are relatively conformable successions of genetically related strata bounded by unconformities and their correlative conformities (Mitchum, 1977; Vail et al., 1977a; Van Wagoner et al., 1988). Sequence bounding unconformities show evidence of subaerial erosional truncation (and, in some areas, correlative submarine erosion) or subaerial exposure and represent significant hiatuses (Van Wagoner et al., 1988). Internally, sequences are comprised of parasequences and parasequence sets that are arranged into systems tracts (see Sarg, 1988, and Van Wagoner et al., 1988, for explanations of systems tracts). A *supersequence* (second-order, 10^7–10^8 year duration) is a stacked set of sequences bounded by major unconformities (Vail et al., 1977a; Haq et al., 1988). Stacked sets of supersequences form *megasequences* (first order, >10^8 year duration; Vail et al., 1977a; Haq et al., 1988). Sequences identified by Sloss (1963) are examples of supersequences and megasequences (Mitchum et al., 1977; Haq et al., 1988).

Figure 7. Slope and basinal facies. (A) Heavily bioturbated cherty limestone. Chert stands in relief. Note near-ly complete destruction of layering in chert. Instead, chert follows burrows. (B) Horizontal and ripple cross-laminations in spicular peloid-skeletal packstone (turbidite). Lighter layers are weathered chert horizons. (C) Radiolarian tests and silt-size skeletal debris in mixed carbonate/siliciclastic facies. (D) Horizontal laminated calcareous siltstone/sandstone.

VERTICAL LITHOFACIES RELATIONSHIPS

Two scales of sequence stratigraphic units can be recognized based on vertical lithofacies relationships. In individual outcrops, the Mission Canyon Formation and its stratigraphic equivalents consist of prominent, 18–62-m-thick, shallowing-upward units composed of the lithofacies described previously. In shallow-water ramp-interior areas, these relatively thick intervals are comprised of stacked, smaller scale cycles; however, smaller scale cycles have not been recognized in outer ramp/slope and basinal strata. The set of lithofacies that defines larger scale, shal-lowing-upward units exhibits stratigraphic and pale-ogeographic variations.

Ramp-Interior Settings

In shallow-water, ramp-interior settings, large-scale shallowing-upward units consist of, in ascend-ing order (1) peloid-skeletal wackestone/packstone; (2) skeletal-ooid grainstone; (3) skeletal-peloid grain-stone/packstone; and (4) typically upward-thinning, stacked cycles of peritidal facies (Figure 8; Figure 9, upper left enlarged inset column). Thick-bedded to massive skeletal-ooid grainstone, the most common lithofacies at the bases of ramp-interior parasequence sets, appears noncyclic. All facies within this general-ized set of lithofacies may not be present within indi-vidual units. Facies that cap smaller scale cycles in ramp-interior settings often show evidence of subaer-ial exposure such as desiccation features, evaporite pseudomorphs, and vadose diagenetic fabrics. Larger scale unit boundaries in shallow-water platform set-tings are placed along the highest peritidal cap that is overlain by relatively thick (~3–10 m), subtidal facies (Figure 8). Careful observation of clast lithologies within solution collapse breccias allows some unit boundaries to be placed within brecciated horizons because overall shallowing-upward characteristics frequently are preserved within the breccias.

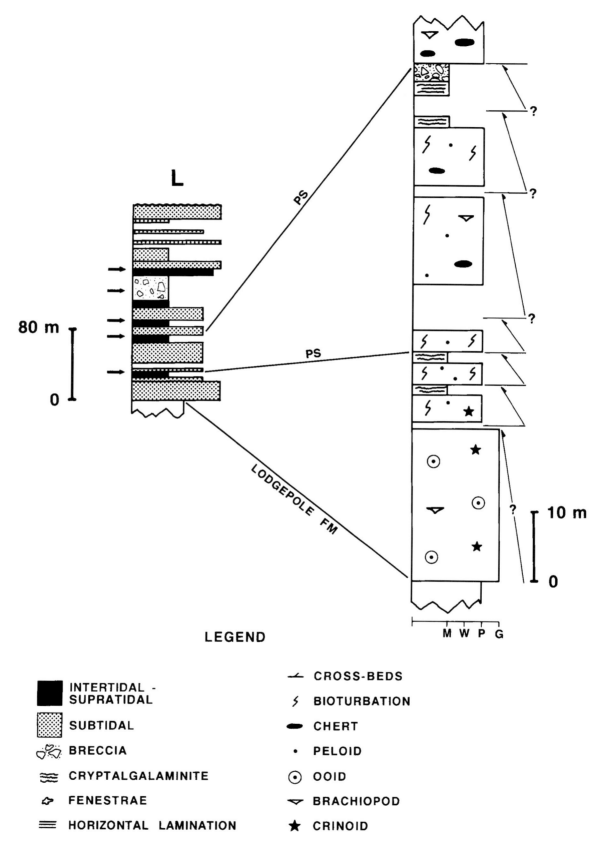

LEGEND

⬛ INTERTIDAL - SUPRATIDAL	⌐ CROSS-BEDS
▒ SUBTIDAL	∫ BIOTURBATION
◌ BRECCIA	⬤ CHERT
≋ CRYPTALGALAMINITE	• PELOID
↻ FENESTRAE	⊙ OOID
≡ HORIZONTAL LAMINATION	▽ BRACHIOPOD
	★ CRINOID

Figure 8. Typical ramp interior parasequence sets. Left column is the entire thickness of the Mission Canyon Formation at location L (Figure 1). Arrows beside left column indicate parasequence set boundaries. Enlargement of basal two parasequence sets and internal parasequences shown in right column. Lithofacies within parasequence sets are similar throughout the section (cf. Figure 10).

Outer Ramp/Slope Settings

Larger scale, shallowing-upward units can be identified in outer ramp/slope settings but meter-scale cycles have not been recognized. Outer ramp/slope units are characterized by repeated thick intervals of bioturbated cherty limestone overlain by thick-bedded to massive skeletal grainstone (Figure 9). Units in deeper slope settings consist of thick intervals of laminated cherty limestone overlain by bioturbated cherty limestone (Figure 10, locations BM, LK). Larger scale, shallowing-upward units from outer ramp/slope environments are bounded by surfaces that separate shallower-water deposits below from thick, deeper water facies above (Figure 9).

Sequence Stratigraphic Interpretation

Approximately 1–10 m thick, shallowing-upward cycles recognized in measured sections from ramp-interior settings are parasequences. The 18–62 m thick depositional cycles, originally interpreted by Reid (1991) and Reid and Dorobek (1991) as sequences, probably represent parasequence sets (J. Markello, personal communication, 1992). Defining boundaries are major marine flooding surfaces (sensu Van Wagoner et al., 1988). In ramp-interior settings, flooding surfaces are located at the bases of ~3–10-m-thick deeper water lithofacies that abruptly overlie peritidal carbonates (Figure 8). Parasequence set boundaries are placed at the bases of 5–50-m-thick cherty limestones that overlie massive skeletal grainstones in outer ramp/slope settings (Figure 9).

Approximate Durations of Parasequence Sets

Approximate durations of parasequence sets were calculated using the biostratigraphically best constrained measured section that likely represents essentially continuous deposition throughout Mission Canyon platform development (Figure 1, location BM; Figure 9). Biostratigraphic boundaries near the base and top of location BM (Figures 3, 9) bracket the time during which the Mission Canyon Formation and stratigraphic equivalents were deposited (Sando et al., 1985; Reid, 1991). Absolute ages of biozone boundaries were determined using the Mississippian time scales of Sandberg et al. (1983), Sando (1985), and Ross and Ross (1987). Stratigraphic equivalents of the Mission Canyon Formation at location BM were deposited over a time span of 10.9 to 12.4 m.y., depending on the time scale used.

Average durations of parasequence sets were calculated by dividing the amount of time represented by the Mission Canyon Formation and stratigraphic equivalents at location BM by the number of parasequence sets recognized in the measured section. Location BM contains at least 13 parasequence sets (Figure 9). The covered interval near the top of the section may conceal 1 or 2 additional parasequence sets. Given 13 to 15 parasequence sets deposited over a 10.9 to 12.4 m.y. period, average durations range

from 727 k.y. (10.9 m.y. divided by 15) to 954 k.y. (12.4 m.y. divided by 13).

Approximate durations of parasequence sets also were calculated based on measured thicknesses and average accumulation rates (again using location BM). A minimum average accumulation rate of 4.2 cm/k.y. was calculated by dividing measured thickness (525 m) by 12.4 m.y. A maximum average accumulation rate of 6.3 cm/k.y. was calculated by dividing a decompacted thickness of 683 m by 10.9 m.y. (assumes 30% decompaction throughout the section). Durations were calculated by dividing measured parasequence set thickness by minimum average accumulation rate (minimum durations) or by dividing decompacted unit thickness by maximum average accumulation rate (maximum durations). Parasequence sets range from 429 k.y. to 1.3 m.y. in duration based on these calculations. Therefore, parasequence sets appear to be fourth order (10^5–10^6 years duration). Parasequences within parasequence sets probably are fifth order (10^4–10^5 years duration) and sequences (discussed below) comprised of parasequences and parasequence sets probably are third order (10^6–10^7 years duration).

RECONSTRUCTION OF SEQUENCES AND SYSTEMS TRACTS

Sequence stratigraphic analysis across the Mission Canyon ramp to basin transition is difficult. Lateral facies relationships used to define sequences and systems tracts cannot be directly observed across the Mission Canyon platform because well-exposed, undeformed outcrops of Lower Mississippian strata in the Northern Rockies are widely separated and typically located in different thrust sheets.

In our earlier attempts at sequence stratigraphic analysis of the Mission Canyon ramp to basin transition (Reid, 1991; Reid and Dorobek, 1991), we overemphasized vertical lithofacies relationships and interpreted the previously described 18–62-m-thick parasequence sets as sequences. We interpreted the highest peritidal cap at the top of each unit as a sequence boundary (i.e., an unconformity representing a "significant hiatus") because these horizons typically show evidence of subaerial exposure in ramp-interior settings. However, our regional correlations of these "sequences" (figure 6 in Reid and Dorobek, 1991) did not illustrate important internal facies geometries (i.e., systems tracts) within individual "sequences." In this chapter, we reevaluate our earlier attempts to apply sequence stratigraphic analysis to the Mission Canyon ramp to basin transition.

Correlation Methods

Regional biostratigraphic and lithostratigraphic correlations were used to reconstruct facies relationships across the deformed ramp to basin transition. Then depositional sequences and system tracts within sequences were interpreted from characteristics of

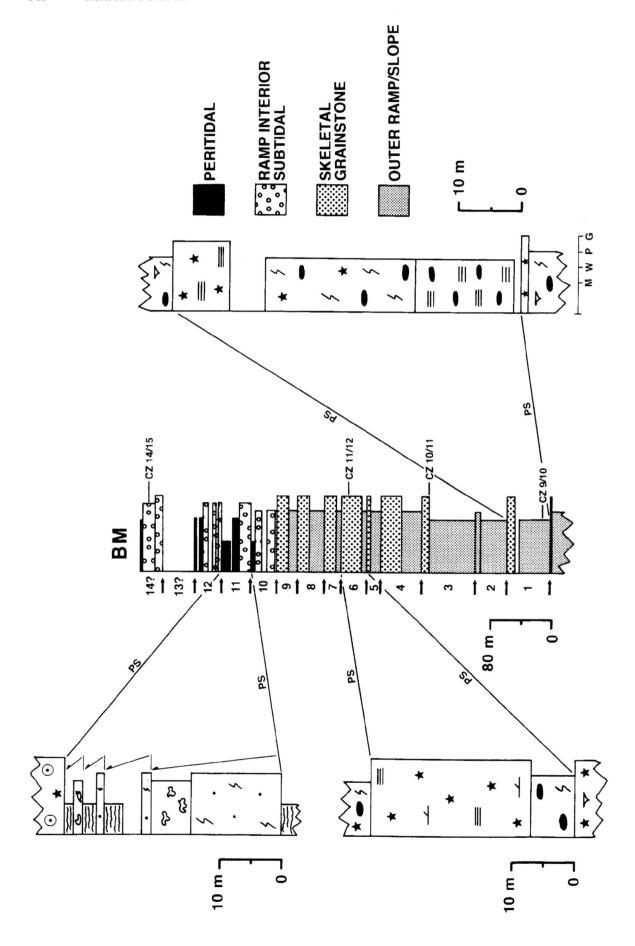

apparent bounding surfaces, stacking patterns of parasequence sets, and lateral lithofacies relationships.

Regionally extensive biozone boundaries were used to correlate six key measured sections that form a transect across the Mission Canyon ramp to basin transition (Figures 1, 3, 10). Datums are represented as basinward sloping surfaces because biostratigraphic control generally is contained in subtidal lithofacies (Gutschick et al., 1980; Sando, personal communication, 1988; Sando and Dutro, 1960, 1980; Sando et al., 1969; Sando et al., 1985; Reid, 1991) that probably dipped in a basinward direction originally.

Due to a lack of biostratigraphic control, key biozone boundaries were not recognized at two of the measured stratigraphic sections along the study transect (Figure 10, locations T, A). Location T was correlated using the thick grainstone at the base of the Mission Canyon Formation as a datum. This interval is an easily recognized and reliable regional lithologic marker that immediately overlies a known biozone boundary at other locations (Sando and Dutro, 1974; J. L. Wilson, personal communication, 1990; Reid, 1991). However, the likelihood of facies changes at and basinward of the ramp-margin precluded the use of this grainstone as a datum in more basinward settings.

Location A was included in the key transect because it is critically located near the ramp margin; however, the section is incomplete and lacks biostratigraphic control. The datum used to correlate this section is a prominent major marine flooding surface that can be recognized across southwestern Montana. At location LK, this surface is expressed as the contact between thick (30 m) skeletal grainstones, just above the center of the measured section, and overlying thick (28 m) cherty limestones (Figure 10). At location A, this major marine flooding surface probably coincides with the base of thick skeletal grainstones that overlie thick peritidal lithofacies just above the center of the measured section (Figure 10). The exceptionally thick (75 m) breccia at the top of location A contains clasts of mixed lithology that prevented the identification of sequence stratigraphic units in this interval.

Identification of Sequences

Figure 10 is an attempt at sequence stratigraphic analysis across the reconstructed Mission Canyon ramp to basin transition. Across the ramp-interior, sequence boundaries should be expressed as uncon-

formities and therefore are placed at horizons that appear to have experienced longer term subaerial exposure than that indicated at the tops of most parasequence sets. The major unconformity at the top of the Mission Canyon Formation and two stratigraphically lower breccia horizons are regionally correlatable (Figure 10; S. Dorobek, T. Smith, S. Reid, unpublished data; Sando, 1972, 1974, 1976, 1988) and are interpreted as sequence boundaries in ramp-interior settings. More localized breccias probably formed later during development of the major unconformity above the Mission Canyon Formation and during Pennsylvanian(?) to early Tertiary uplift of Mississippian strata (Sando, 1972, 1974, 1976, 1988).

Sequence boundaries should become conformable in more basinward settings. Therefore, sequence boundaries and transgressive flooding surfaces in outer ramp/slope settings are identified based on anomalous facies changes where Walther's Law of adjacent versus vertical facies relations appears to be violated. Sequence boundaries are placed at the bases of thick skeletal and skeletal-ooid grainstones that abruptly overlie cherty limestones (outer ramp/slope lithofacies) in the most basinward locations (Figure 10, locations BM, LK). Similarly, sequence boundaries in ramp-margin settings are placed at the bases of shallow subtidal to peritidal lithofacies that abruptly overlie outer ramp/slope or ramp-margin lithofacies (Figure 10, location A).

The uppermost sequence boundary can be recognized only basinward of the inferred ramp-margin (Figure 10, locations A, BM). This sequence boundary tentatively is placed at the base of the largely siliciclastic Kibbey Formation (not examined in this study), which conformably (Sando et al., 1985) overlies the McKenzie Canyon Formation (tops of locations A, BM, LK). The contact between the Mission Canyon and Kibbey formations defines the major regional unconformity across the rest of Montana (Figure 4). Cratonward increase in the duration of this major unconformity suggests that the Kibbey Formation onlapped the exposed Mission Canyon platform in middle to late Meramecian time (Sando, 1976; Skipp et al., 1979; Gutschick et al., 1980; Sandberg et al., 1983; Sando, 1988).

Identification of Systems Tracts

Systems tracts are identified based on position within sequences, relations to bounding surfaces, and

Figure 9. Parasequence sets from outer ramp/slope setting. Center column is entire thickness of Middle Canyon, Mission Canyon, and McKenzie Canyon formations at location BM (Figure 1). Arrows to left of column indicate boundaries of parasequence sets. Enlargements of specific parasequence sets shown on left and right. Parasequences are not recognized in outer ramp or slope lithofacies. Parasequence sets are numbered for comparison with Williston basin subintervals (Figures 13, 14). CZ 9/10, CZ 10/11, CZ 11/12, and CZ 14/15 are composite biozone boundaries (Figure 3). Note appearance of progressively shallower water lithofacies in stratigraphically higher parasequence sets suggesting long-term progradation. Lithologic symbols as in Figure 8.

lithofacies within constituent parasequences and parasequence sets (Sarg, 1988; Van Wagoner et al., 1988). However, systems tracts are difficult to recognize using widely spaced outcrops, particularly in ramp-interior settings where regional correlation of parasequences and parasequence sets may be complicated by erosion at sequence and parasequence boundaries and by possible shingling of depositional cycles (Read et al., 1991).

Identification of systems tracts across the Mission Canyon ramp interior is interpretive and largely based on vertical stacking patterns. Vertical parasequence stacking patterns within ramp-interior parasequence sets are progradational (Figures 8, 9, upper left enlarged inset column) while vertical stacking patterns of ramp-interior parasequence sets appear largely aggradational (Figure 10). Therefore, sequences in ramp-interior settings appear to consist primarily of highstand systems tracts dominated by skeletal-ooid grainstone ("keep-up" highstand systems tracts of Sarg, 1988). However, basal parts of sequences may be transgressive systems tracts that display progradational parasequence stacking within parasequence sets ("keep-up" transgressive systems tracts of Sarg, 1988).

In ramp-margin and outer ramp/slope settings, identification of systems tracts is more straightforward. Measured sections are more closely spaced, stacking patterns of parasequence sets are more obvious, and bounding surfaces are better expressed (Figure 10). Identification and deposition of deeper water systems tracts is discussed in more detail later.

CAUSAL MECHANISMS

Our basic approach to determine possible causal mechanisms for depositional cyclicity in the Mission Canyon Formation and stratigraphic equivalents has been to compare sequence stratigraphic analysis with quantitative subsidence analyses (Dorobek et al., 1991a,b; Reid, 1991), published Carboniferous onlap-offlap curves (Ross and Ross, 1987; Reid, 1991; Reid and Dorobek, 1991), and correlative strata in the Williston basin (Reid, 1991; Reid and Dorobek, 1991). The following brief summary of this previous work incorporates our revised sequence stratigraphic analysis (Figure 10); however, our primary conclusions about causal mechanisms still seem valid.

Biostratigraphic Dating of Sequence Boundaries

Sequence boundaries are dated using existing biostratigraphic control (summarized by Reid, 1991). Ages are assigned based on proximity of sequence boundaries to biozone boundaries, preferably using outer ramp/slope sections where sequence boundaries are more conformable. SB$_2$(A) is approximately equivalent to the composite zone 10/11 boundary (Figures 3, 10, locations BM, S). SB$_2$(B) is approximately equivalent in age to the composite zone 11/12

boundary (Figures 3, 10, locations LK, BM, S). SB$_1$(A) cannot be dated in outer ramp/slope sections because of poor biostratigraphic control. However, the age of this surface may be approximately equivalent to the composite zone 13/14 boundary, the youngest biozone boundary recognized below the major regional unconformity overlying the Mission Canyon Formation (Figures 3, 10) (Sando, 1976, 1985; Gutschick et al., 1980). SB$_1$(B) approximately coincides with the composite zone 14/15 boundary (Figures 3, 9, 10, location BM).

Subsidence History

Figure 11 is a subsidence analysis of section S (Figures 1, 10), one of six measured sections we have analyzed across the Antler foreland in Idaho and Montana (Dorobek et al., 1991a,b; Reid, 1991). These curves have been corrected to reflect only tectonic subsidence and eustasy (i.e., relative sea-level changes or accommodation; see methods in Dorobek et al., 1991a; Reid, 1991). All curves are similar in form and contain three Mississippian segments (Figure 11).

Tectonic subsidence probably was the dominant component of Mississippian accommodation events (Dorobek et al., 1991a,b; Reid, 1991). The steep Kinderhookian to early Osagean segment, represented by onlapping strata of the Lodgepole and lower McGowan Creek formations (Figure 4), is interpreted as a period of rapid tectonic subsidence concurrent with eustatic sea-level rise (Figures 11, 12). Subsidence slowed dramatically during Osagean to middle Meramecian time when the Mission Canyon Formation and stratigraphic equivalents were deposited (Figure 11). Tectonic subsidence decreased during this time but must have slightly outpaced coeval second-order eustatic sea-level fall until middle to late Meramecian time (cf. Figures 11, 12). From middle to late Meramecian time to at least the end of the Mississippian period, tectonic subsidence increased and was augmented by eustatic sea-level rise. Onlap of the major regional unconformity at the top of the Mission Canyon Formation occurred during this final Mississippian accommodation event (Figures 11, 12).

Comparison with Global Onlap-Offlap Curve

All sequence boundaries recognized in this study correlate with third-order sequence boundaries on the global onlap-offlap curve of Ross and Ross (1987) (Figure 12). Our interpretations of sequence boundary types do not always agree with those of Ross and Ross (1987) (cf. Figures 10, 12). In addition, we could not identify a sequence boundary in our study area that corresponds with the sequence boundary near the composite zone 12/13 biozone boundary (Figure 12). We attribute these discrepancies to local variations in tectonic subsidence (discussed in more detail later).

SACAJAWEA PEAK, MT

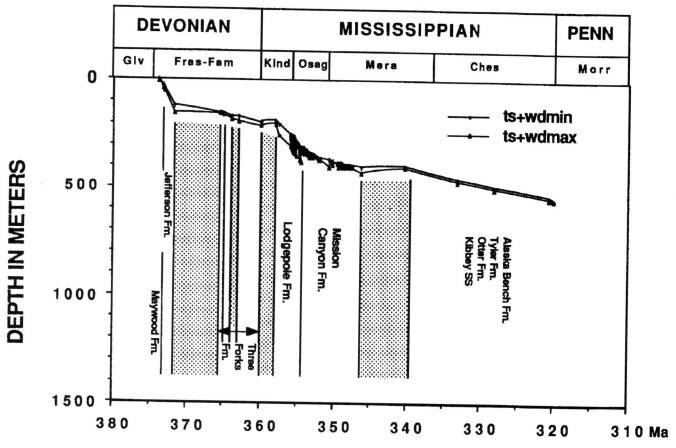

Figure 11. Subsidence analysis of section S (from Dorobek et al., 1991a,b; Reid, 1991). Stippled areas are unconformities. Note three Mississippian segments of the curve. The Mission Canyon Formation and stratigraphic equivalents were deposited during slow Osagean to middle Meramecian subsidence.

Correlations with Williston Basin Strata

In previous papers, we attempted to correlate parasequence sets in our most basinward measured sections (Figures 9, 10, locations LK, BM) with informal subintervals of the Mission Canyon and Charles formations in the central Williston basin (Figure 13) (Reid, 1991; Reid and Dorobek, 1991). Subintervals are ~10–40-m-thick, shallowing-upward units comprised of smaller scale cycles that generally contain, in ascending order (1) skeletal-ooid grainstone; (2) ooid-pisoid grainstone; (3) peloidal, pisoid grainstone/packstone; (4) cryptalgalaminite and evaporitic dolostone; (5) anhydrite and cryptalgalaminite; and (6) siliciclastics and mixed dolostone, siliciclastics, and anhydrite (Harrison and Flood, 1956; Harris et al., 1966; Hendricks, 1988; Petty, 1988). The datum used for these lithostratigraphic correlations was the composite zone 11/12 boundary, the only reliable biostratigraphic marker that can be used to correlate

between southwestern Montana and the Williston basin (Figures 3, 10, 13) (Waters and Sando, 1987; published and new biostratigraphic data summarized in Reid, 1991; Reid and Dorobek, 1991).

Correlations of parasequence sets in southwestern Montana with subintervals in the Williston basin illustrate several similarities between the two areas. The number of parasequence sets (>13) is similar to the number of subintervals (15) (Figures 9, 13, 14; numbers may be equivalent but covered intervals at locations LK and BM may conceal 2 parasequence sets). General vertical stacking patterns of parasequence sets and subintervals also are similar (Figures 9, 13, 14). In addition, the lowest maximum flooding surface and the maximum flooding surface above $SB_2(B)$ appear to correlate with "transgressive discontinuities" that separate the Tilston and Frobisher-Alida intervals and the Rival and Midale subintervals, respectively (Figures 10, 13, 14) (Petty, 1988; Reid, 1991; Reid and Dorobek, 1991). These sim-

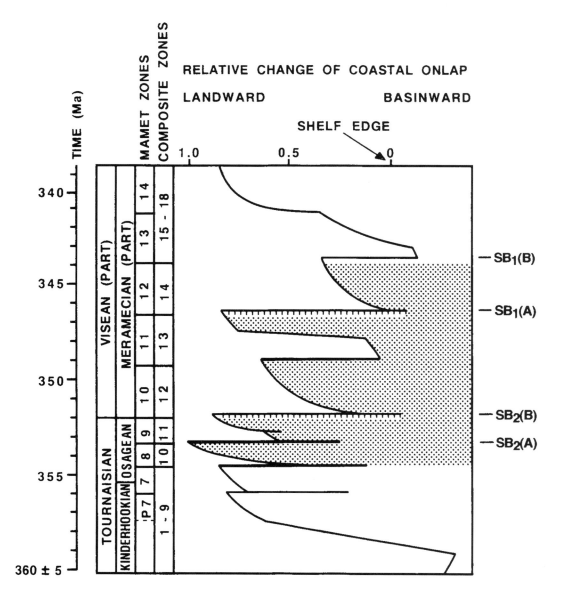

Figure 12. Coastal onlap-offlap curve of Ross and Ross (1987) modified to fit time scale of Sando (1985). Shaded portion corresponds with time of deposition of the Mission Canyon Formation and stratigraphic equivalents. Thin lines at sequence boundaries are "medium" sequence boundaries, heavy lines are "major" sequence boundaries, and hatched lines are "exposed lowstand surfaces" of Ross and Ross (1987). Correlative surfaces from this study are shown on the right (labels are the same as in Figure 10).

ilarities, while not conclusive, support our contention that parasequence sets formed in response to eustatic sea-level fluctuations.

Correlation of third-order sequence boundaries between southwestern Montana and the Williston basin is difficult because of poor biostratigraphic control and because a sequence stratigraphic analysis of the complete Mission Canyon and Charles formations has not been published. We can confidently correlate one third-order sequence boundary between these areas. In the central Williston basin, the composite zone 11/12 boundary is located within 6 m of the K-1 (Fryburg) gamma-ray marker which separates the Sherwood and Mohall subintervals (Figure 13)

(Harris et al., 1966; Waters and Sando, 1987; Petty, 1988). Interestingly, the top of the Mohall subinterval is incised by channels that are backfilled by the Kisbey Sandstone (Witter, 1988). Therefore, SB$_2$(B) probably correlates with the top of the Mohall subinterval in the Williston basin.

Summary of Controls on Platform Development

The Mission Canyon platform was deposited during relatively slow Osagean to middle Meramecian subsidence, which we interpret as a period of relative quiescence in the Antler thrust belt (cf. Figures 11, 12) (Dorobek et al., 1991a,b; Reid, 1991). Tectonic subsi-

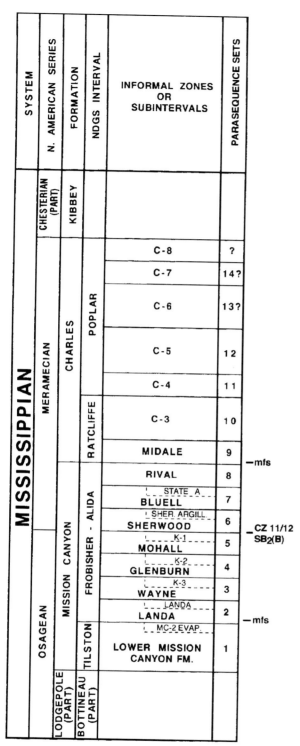

SYSTEM	N. AMERICAN SERIES	FORMATION	NDGS INTERVAL	INFORMAL ZONES OR SUBINTERVALS	PARASEQUENCE SETS
MISSISSIPPIAN	MERAMECIAN	KIBBEY	CHESTERIAN (PART)		
		CHARLES	POPLAR	C-8	?
				C-7	14?
				C-6	13?
				C-5	12
				C-4	11
			RATCLIFFE	C-3	10
				MIDALE	9
	OSAGEAN	MISSION CANYON	FROBISHER - ALIDA	RIVAL	8
				STATE A / BLUELL	7
				SHER. ARGILL. / SHERWOOD	6
				K-1 / MOHALL	5
				K-2 / GLENBURN	4
				K-3 / WAYNE	3
			TILSTON	LANDA / LANDA	2
				MC-2 EVAP. / LOWER MISSION CANYON FM.	1
		LODGEPOLE (PART)	BOTTINEAU (PART)		

Right-side markers: —mfs (at MIDALE/9), —CZ 11/12 SB₂(B) (at SHERWOOD/6), —mfs (at LANDA/2)

$SB_2(B)$

Figure 13. Correlation of parasequence sets ("sequences" of Reid, 1991; Reid and Dorobek, 1991) at locations BM and LK with informal subintervals of Harrison and Flood (1956) and Harris et al. (1966) from the central Williston basin. Numbers correspond to numbered parasequence sets on Figure 9 (also see figure 6 of Reid and Dorobek, 1991). CZ 11/12 boundary (composite zone 11/12 boundary, Figure 3) is biostratigraphic datum. Correlative surfaces from Figure 10 are shown on the right.

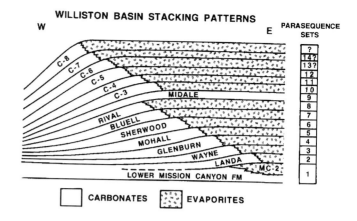

Figure 14. Schematic stratigraphic cross section illustrating large-scale stacking patterns of informal subintervals in the Williston basin (based on well log and core analyses of Harrison and Flood, 1956; Harris et al., 1966). Correlative parasequence sets (Figures 9, 13) are shown to the right of the cross section. Not to scale.

dence waned but generally outpaced gradual, second-order eustatic sea-level fall until middle to late Meramecian time. Relative tectonic quiescence during this time is suggested by a lack of coeval, coarse-grained, synorogenic siliciclastics in proximal Antler foredeep strata (Skipp and Mamet, 1970; Nilsen, 1977; Skipp et al., 1979; Dorobek et al., 1991a; Reid, 1991). Later, we discuss connotations of our sequence stratigraphic analysis on Antler foredeep subsidence history.

Third-order sequences within the Mission Canyon Formation and stratigraphic equivalents probably formed in response to third-order eustatic sea-level fluctuations (Figure 12) superimposed on the second-order accommodation trend. Parasequence sets and parasequences may have formed in response to fourth- and fifth-order sea-level oscillations superimposed on third-order eustatic cycles.

Third- to fifth-order sea-level fluctuations may have been glacioeustatic. Late Paleozoic (Gondwanan) glaciation apparently began in the Early Carboniferous (Kinderhookian to Osagean) and culminated in the Permian–Carboniferous (Frakes and Crowell, 1969; Crowell, 1978; Caputo and Crowell, 1985). Caputo and Crowell (1985) suggested that Gondwana was never covered by a single large ice cap. Instead, smaller ice centers migrated across Gondwana from South America (Early Carboniferous) to eastern Australia and Antarctica (early Late Permian). Volume changes in this relatively small Early Carboniferous ice cap would have produced relatively low-amplitude sea-level fluctuations, perhaps on the order of 10 to, at most, a few tens of meters.

Relatively low-amplitude sea-level oscillations are consistent with a lack of evidence for deep submergence or drowning of the Mission Canyon ramp-interior. Fourth-order sea-level rises flooded regional(?) tidal flat and supratidal areas (i.e., tops of parasequence sets) to produce major marine flooding surfaces. However, the predominance of skeletal-ooid grainstones at the bases of ramp-interior parasequence sets suggests that the rates and magnitudes of sea-level rises generally were not great enough to cause drowning of the Mission Canyon platform nor to submerge the ramp-interior below fair-weather wave base.

DEPOSITIONAL HISTORY

Long-term evolution of the Mission Canyon platform is interpreted based on our regional sequence stratigraphic analysis and our present understanding of eustasy, regional tectonism, and relative sedimentation rates (inferred after Schlager, 1981). We cannot directly assess the influence of all types and scales of variables that appear to control development of carbonate ramps (e.g., Read et al., 1991). However, major processes described below may have operated across other broad, progradational carbonate ramps that developed adjacent to foreland basins.

Tectonic Overprint on Sequence Development

The reconstructed ramp to basin transect illustrates a progressive upward change in sequence boundary type during evolution of the Mission Canyon platform. Type 2 sequence boundaries (sensu Sarg, 1988) formed early in platform development, whereas type 1 sequence boundaries (sensu Sarg, 1988) dominated later platform development (Figure 10). Second-order, eustatic sea-level fall must have outpaced tectonic subsidence during the later stages of platform deposition. This may reflect an increased rate of second-order eustatic sea-level fall during middle to late Meramecian time. However, we prefer to attribute this upward change in sequence boundary type to a gradual decrease in the rate of Antler foredeep subsidence. Slower flexural subsidence would have allowed third-order eustatic sea levels to fall well below the ramp margin, producing type 1 sequence boundaries. The Mission Canyon ramp margin is at or within typical flexural wavelengths for many foreland basins (cf. Jordan, 1981; Karner and Watts, 1983; Stockmal and Beaumont, 1987) based on existing palinspastic and paleogeographic reconstructions (Sando, 1976; Nilsen, 1977; Dover, 1980; Gutschick et al., 1980; Gutschick and Sandberg, 1983; Peterson, 1986).

Tectonic overprint of third-order eustatic sea-level cycles may explain differences between our interpretations of sequence boundary types and those of correlative surfaces on the Ross and Ross (1987) onlap-offlap curve (Figure 12). Type 2 sequence boundaries in the lower Mission Canyon platform correlate with sequence boundaries characterized as "major" and "exposed lowstand surfaces" on the

Ross and Ross (1987) curve. Conversely, the uppermost sequence boundary, which we interpret as type 1, correlates with a "medium" sequence boundary on the Ross and Ross (1987) curve. This is consistent with a progressive decrease in the rate of flexural subsidence near the Mission Canyon ramp margin. Local tectonic influence also is suggested by the fact that SB$_2$(B) has characteristics of a type 1 sequence boundary in the Williston basin (incised channels at the top of the Mohall subinterval; Witter, 1988) but appears to be a type 2 sequence boundary in southwestern Montana.

Ramp Progradation and Development of Ramp-Margin Wedges

The second-order decrease in accommodation forced the Mission Canyon platform to prograde into the distal Antler foredeep (Figure 10). In ramp-interior settings, the style of progradation is difficult to determine because of our inability to adequately define ramp-interior systems tracts. However, the predominance of progradational vertical stacking patterns of parasequences and largely aggradational vertical stacking of parasequence sets (Figures 8, 9, 10) suggests that ramp-interior sedimentation rates consistently kept pace with or exceeded rates of relative sea-level rises (characteristic of "keep-up" highstand and transgressive systems tracts of Sarg, 1988). Deposition of thick, apparently noncyclic grainstones at the bases of most ramp-interior parasequence sets is difficult to explain, but may reflect "extra" accommodation space generated by erosion or dominance of longer period (fourth-order?) and/or moderate amplitude (20–40 m ?) eustatic cycles (cf. Read et al., 1991).

The ability of the Mission Canyon ramp interior to accommodate large amounts of subtidal sediment may have affected progradation in outer ramp/slope settings. Conditions for optimal sediment production appear to have been largely confined to the ramp interior during most of third-order sequence development. Progradation into outer ramp/slope environments may not have begun until accommodation space across the ramp interior decreased during late highstands. In situ skeletal grainstone banks and/or thin sediment gravity flows derived from these banks may have prograded over outer ramp/slope lithofacies during late highstands and ensuing sea-level falls (Figure 10, note grainstones at the tops of outer ramp/slope parasequence sets). Major pulses of progradation into outer ramp/slope environments occurred when ramp-margin wedges (bounded below by sequence boundaries and above by transgressive surfaces) formed during third-order lowstands and early stages of succeeding sea-level rises (cf. Sarg, 1988).

Consistent upward variations in lithofacies and thicknesses of ramp-margin wedges support our interpretation that waning tectonic subsidence permitted third-order eustatic sea levels to fall

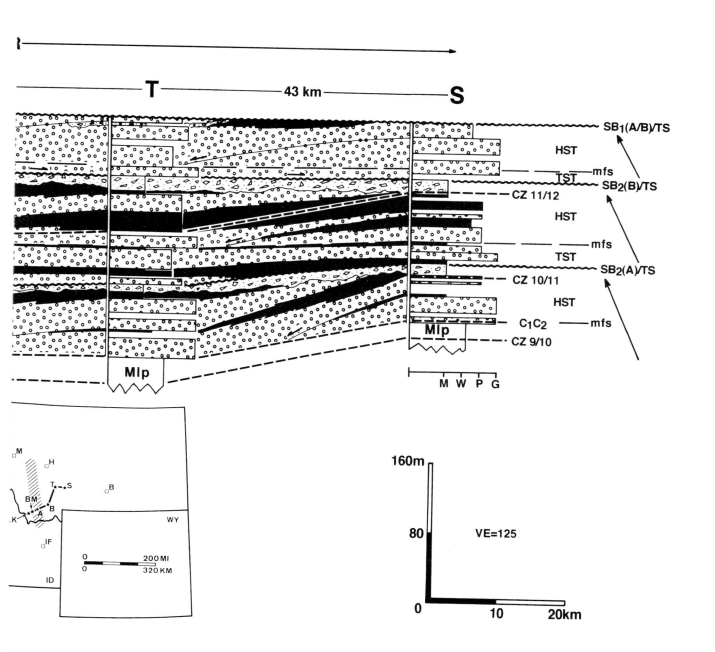

progressively farther basinward as the Mission Canyon platform evolved (Figure 10). Both $SB_2(A)$ and $SB_2(B)$ are type 2 sequence boundaries. However, the ramp-margin wedge above $SB_2(A)$ is thinner than that deposited above $SB_2(B)$, and contains only minor peritidal lithofacies that prograded into ramp-margin to outer ramp areas (Figure 10, location A). The thicker ramp-margin wedge above $SB_2(B)$ represents a more significant basinward shift of peritidal lithofacies over the ramp margin (Figure 10, location A) and also marks the appearance of well-developed, stacked skeletal banks in former outer ramp/slope settings (~30 m of massive skeletal grainstone, Figure 10, location LK).

The uppermost ramp-margin wedge is interpreted as a type 1 autochthonous carbonate wedge (sensu Sarg, 1988) that developed during major karstification of the Mission Canyon platform. This wedge is dominated by ramp-interior lithofacies that extend well beyond the former ramp margin (Figure 10). Third- to fifth-order sea-level fluctuations that controlled deposition of this wedge (and the overlying, areally limited highstand(?) systems tract) probably only affected former outer ramp/slope and ramp-margin areas. However, onlapping strata above $SB_1(A)$ may have been eroded when $SB_1(B)$ formed. Exceptionally thick breccias at location A suggest that part of this type 1 autochthonous wedge and much of the overlying highstand(?) systems tract were heavily karstified during development of $SB_1(B)$. The Mission Canyon platform probably remained emergent until the middle to late Meramecian (when the Kibbey Formation began to onlap the major karst surface) because it did not subside rapidly enough to allow flooding by relatively low-amplitude sea-level fluctuations.

Maintenance of Ramp Profile

The Mission Canyon platform appears to have maintained a ramplike profile throughout its development (Figure 10). Long-term maintenance of this ramplike profile may reflect a gradual basinward decrease in sedimentation rates across the platform to basin transition. However, the Antler foredeep was starved during deposition of the Mission Canyon platform (Sandberg, 1975; Nilsen, 1977; Skipp et al., 1979; Gutschick and Sandberg, 1983), which suggests that the difference between ramp-interior and basinal sedimentation rates may have been sufficient to generate a more pronounced break in slope across the ramp margin. We suggest that the upward change in sequence boundary type (and related changes in ramp-margin wedges) caused long-term maintenance of a ramplike profile. Basinward shifts in the locus of optimum sediment production and accumulation during each subaerial exposure of the ramp interior tended to "even-out" ramp-margin slopes. Had these shifts not occurred, the Mission Canyon platform probable would have developed characteristics more like those of a rimmed shelf.

SUMMARY AND CONCLUSIONS

The Lower Mississippian Mission Canyon Formation and stratigraphic equivalents in Montana and Idaho were deposited on a progradational carbonate ramp that developed on the foreland side of the Antler foredeep. In individual outcrops, these strata consist of prominent, stacked parasequence sets (18–62 m thick). These shallowing-upward units are comprised of smaller scale (1–10 m thick) parasequences in ramp-interior settings; however, parasequences have not been recognized in outer ramp/slope parasequence sets. Parasequence sets are our primary units for regional sequence stratigraphic analysis.

Lateral lithofacies relationships across the deformed ramp to basin transition were reconstructed by biostratigraphically and lithostratigraphically correlating six key measured sections from a larger regional database. Depositional sequences and systems tracts within sequences were identified from characteristics of bounding surfaces, stacking patterns of parasequences and parasequence sets, and lateral lithofacies relationships.

Third-order sequence boundaries show a progressive upward change in type; type 2 sequence boundaries formed early in ramp development, and type 1 sequence boundaries formed during the later stages of ramp deposition. Associated ramp-margin wedges become thicker and consist of progressively larger proportions of peritidal lithofacies toward the top of the Mission Canyon Formation and stratigraphic equivalents. Based primarily on quantitative subsidence analysis and comparisons with published onlap-offlap curves, we interpret this upward change in sequence boundary type (and related changes in ramp-margin wedges) as a depositional response to gradually decreasing Antler foredeep subsidence concurrent with second-order sea-level fall. Gradually waning flexural subsidence probably allowed higher order eustatic sea levels (superimposed on second-order eustasy) to fall progressively farther basinward as the Mission Canyon platform evolved.

The documented long-term decrease in accommodation caused progradation of the Mission Canyon platform and helped maintain a ramplike profile across the platform to basin transition. The Mission Canyon ramp interior was able to accommodate large amounts of subtidal sediment, which may have affected progradation into outer ramp/slope settings. Conditions for optimal sediment production may have been primarily confined to the ramp interior during most of third-order sequence development. Progradation into outer ramp/slope environments may not have begun until accommodation space decreased during late highstands. Major pulses of progradation into outer ramp/slope environments occurred during third-order lowstands and early stages of succeeding sea-level rises when ramp-margin wedges formed. Ramp-margin wedges tended to "even out" ramp-margin slopes and maintained a ramplike profile across the platform to basin transition. The long-term decrease in accommodation prevented the Mission Canyon platform from developing characteristics more like those of a rimmed shelf.

ACKNOWLEDGMENTS

This chapter is based on the senior author's dissertation research conducted at Texas A&M University. Research was supported by U.S. Department of Energy Grant DE–FG05–87ER13767, Basic Energy Sciences to S. Dorobek. Such support does not constitute an endorsement by DOE of the views expressed in this paper. Acknowledgment also is made to the Donors of The Petroleum Research Fund, administered by the American Chemical Society, for partial support of this research (Grant #19519–G2 to S. Dorobek). Additional support was provided by Geological Society of America Grants-in-Aid to S. Reid. Field assistance was provided by T. Smith and J. Harris. Field data from location T was provided by T. Smith. W. Sando identified corals collected by the senior author, allowed examination of unpublished data, and provided comments about stratigraphy that helped constrain interpretations. A. Harris kindly identified conodonts in samples collected at location LK by the senior author. Correlations with the Williston basin were better constrained by discussions with D. Witter. This chapter was improved by insightful reviews by R. Loucks and S. Greenlee. Special acknowledgment is made to J. Markello for his painstaking review of an earlier version of this paper. His suggestions regarding identification of third-order sequences on our regional cross section were invaluable. However, omissions or unintentional misrepresentations of reviewers' or contributors' suggestions is our responsibility.

REFERENCES CITED

Ahr, W.M., 1973, The carbonate ramp: An alternative to the shelf model: Transactions, Gulf Coast Association of Geological Societies, v. 23, p. 221-225.

Ball, M.M., 1967, Carbonate sand bodies of Florida and the Bahamas: Journal of Sedimentary Petrology, v. 37, p. 556-591.

Caputo, M.V., and J.C. Crowell, 1985, Migration of glacial centers across Gondwana during Paleozoic Era: Geological Society of America Bulletin, v. 96, p. 1020-1036.

Cook, H.E., and P. Enos, 1977, Deep-Water Carbonate Environments: SEPM Special Publication No. 25, 336 p.

Cook, H.E., and M.E. Taylor, 1977, Comparison of continental slope and shelf environments in the Upper Cambrian and lowest Ordovician of Nevada, in Cook, H.E., and P. Enos, 1977, eds., Deep-Water Carbonate Environments: SEPM Special Publication No. 25, p. 51-81.

Crowell, J.C., 1978, Gondwanan glaciation, cyclothems, continental positioning, and climate change: American Journal of Science, v. 278, p. 1345-1372.

Davies, G.R., 1977, Turbidites, debris sheets, and truncation structures in Upper Paleozoic deep-water carbonates of the Sverdrup Basin, Arctic Archipelago, in Cook, H.E., and P. Enos, eds., Deep-Water Carbonate Environments: SEPM Special Publication No. 25, p. 221-247.

Dorobek, S.L., and J.F. Read, 1986, Sedimentology and basin evolution of the Siluro-Devonian Helderberg Group, central Appalachians: Journal of Sedimentary Petrology, v. 56, p. 601-613.

Dorobek, S.L., S.K. Reid, and M. Elrick, 1991a, Antler foreland stratigraphy of Montana and Idaho: The stratigraphic record of eustatic fluctuations and episodic tectonic events, in Cooper, J.D., and C.H. Stevens, eds., Paleozoic Paleogeography of the Western United States—II: SEPM, Pacific Section, v. 67, p. 487-507.

Dorobek, S.L., S.K. Reid, M. Elrick, G.C. Bond, and M.A. Kominz, 1991b, Subsidence across the Antler foreland of Montana and Idaho: Tectonic versus eustatic effects, in Franseen, E.K., W.L. Watney, C.G.St.C. Kendall, and W. Ross, eds., Sedimentary Modeling: Computer Simulation and Methods for Improved Parameter Definition: Kansas Geological Society Bulletin 233, p. 231-251.

Dover, J.H., 1980, Status of the Antler orogeny in central Idaho—Clarification and constraints from the Pioneer Mountains, in Fouch, T.D., and E.R. Magathan, eds., Paleozoic Paleogeography of the West-central United States: SEPM, Rocky Mountain Section, Rocky Mountains Paleogeography Symposium 1, p. 371-386.

Esteban, M., 1976, Vadose pisolite and caliche: AAPG Bulletin, v. 60, p. 2048-2057.

Esteban, M., and C.F. Klappa, 1983, Subaerial exposure, in Scholle, P.A., D.G. Bebout, and C.H. Moore, eds., Carbonate Depositional Environments: AAPG Memoir 33, p. 1-54.

Fischer, A.G., 1964, The Lofer cyclothems of the Alpine Triassic: Kansas Geological Survey Bulletin, v. 169, p. 107-149.

Frakes, L.A., and J.C. Crowell, 1969, Late Paleozoic glaciation—Part I, South America: Geological Society of America Bulletin, v. 80, p. 1007-1042.

Franseen, E.K., T.E. Fekete, and L.C. Pray, 1989, Evolution and destruction of a carbonate bank at the shelf margin: Grayburg Formation (Permian), Western Escarpment, Guadalupe Mountains, Texas, in Crevello, P.D., J.L. Wilson, J.F. Sarg, and J.F. Read, eds., Controls on Carbonate Platform and Basin Development: SEPM Special Publication No. 44, p. 289-304.

Goldhammer, R.K., P.A. Dunn, and L.A. Hardie, 1987, High frequency glacio-eustatic sealevel oscillations with Milankovitch characteristics recorded in Middle Triassic platform carbonates in northern Italy: American Journal of Science, v. 287, p. 853-892.

Goldhammer, R.K., P.A. Dunn, and L.A. Hardie, 1990, Depositional cycles, composite sea level changes, cycle stacking patterns, and the hierarchy of stratigraphic forcing—examples from platform carbonates of the Alpine Triassic: Geological Society of America Bulletin, v. 102, p. 535-562.

Goldhammer, R.K., E.J. Oswald, and P.A. Dunn, 1991, Hierarchy of stratigraphic forcing: Example from Middle Pennsylvanian shelf carbonates of the Paradox basin, in Franseen, E.K., W.L. Watney, C.G.St.C. Kendall, and W. Ross, eds., Sedimentary Modeling: Computer Simulation and Methods for Improved Parameter Definition: Kansas Geological Society Bulletin 233, p. 362-413.

Gutschick, R.C., and C.A. Sandberg, 1983, Mississippian continental margins of the conterminous United States, in Stanley, D.J., and G.T. Moore, eds., The Shelfbreak: Critical Interface on Continental Margins: SEPM Special Publication No. 33, p. 79-96.

Gutschick, R.C., C.A. Sandberg, and W.J. Sando, 1980, Mississippian shelf margin and carbonate platform from Montana to Nevada, in Fouch, T.D., and E.R. Magathan, eds., Paleozoic Paleogeography of the West-central United States: SEPM, Rocky Mountain Section, Rocky Mountain Paleogeography Symposium 1, p. 111-128.

Haq, B.U., J. Hardenbol, and P.R. Vail, 1988, Mesozoic and Cenozoic chronostratigraphy and eustatic cycles, in Wilgus, C.K., B.S. Hastings, C.G.St.C. Kendall, H.W. Posamentier, C.A. Ross, and J.C. Van Wagoner, eds., Sea Level Changes: An Integrated Approach: SEPM Special Publication No. 42, p. 71-108.

Harris, S.H., C.B. Land, and J.H. McKeever, 1966, Relation of Mission Canyon stratigraphy to oil production in north-central North Dakota: AAPG Bulletin, v. 50, p. 2269-2276.

Harrison, R.L., and A.L. Flood, 1956, Mississippian correlation in the international boundary areas, in First International Williston Basin Symposium: North Dakota and Saskatchewan Geological Societies, Conrad Publishing, Bismark, North Dakota, p. 36-51.

Heckel, P.H., 1974, Carbonate buildups in the geologic record: A review, in Laporte, L.F., ed., Reefs in Time and Space: SEPM Special Publication No. 18, p. 90-155.

Hendricks, M.L., 1988, Shallowing-upward cyclic carbonate reservoirs in the lower Ratcliff interval (Mississippian), Williams and McKenzie Counties, North Dakota, in Goolsby, S.M., and M.W. Longman, eds., Occurrence and Petrophysical Properties of Carbonate Reservoirs in the Rocky Mountain Region: Rocky Mountain Association of Geologists, p. 371-380.

Hine, A.C., 1977, Lily Bank, Bahamas: History of an active oolite sand shoal: Journal of Sedimentary Petrology, v. 47, p. 1554-1581.

Huh, O.K., 1967, The Mississippian System across the Wasatch Line east central Idaho, extreme south-western Montana, in Centennial Basin of Southwest Montana—Guidebook, 18th Annual Field Conference: Montana Geological Society, Billings, Montana, p. 31-62.

Huh, O.K., 1968, Mississippian stratigraphy and sedimentology, across the Wasatch Line, east-central Idaho and extreme southwestern Montana, Ph.D. dissertation: The Pennsylvania State University, University Park, 176 p.

James, N.P., 1984, Reefs, in Walker, R.G., ed., Facies Models: Geoscience Canada, Reprint Series 1, p. 229-244.

Jordan, T.E., 1981, Thrust loads and foreland basin development, Cretaceous, western United States: AAPG Bulletin, v. 65, p. 2506-2520.

Karner, G.D., and A.B. Watts, 1983, Gravity anomalies and flexure of the lithosphere at mountain ranges: Journal of Geophysical Research, v. 88, p. 10449-10477.

Kendall, C.G.St.C., and P.A.D.E. Skipwith, 1969, Holocene shallow-water carbonate and evaporite sediments of Khor al Bazam, Abu Dhabi, southwest Persian Gulf: AAPG Bulletin, v. 53, p. 841-869.

Kinsman, D.J.J., 1964, The recent carbonate sediments near Halat el Bahrani, Trucial Coast, Persian Gulf, in Deltaic and Shallow Marine Deposits—Developments in Sedimentology, v. 1: Elsevier, Amsterdam, p. 185-192.

Kreisa, R.D., 1981, Storm generated sedimentary structures in subtidal marine facies with examples from the Middle and Upper Ordovician of southwestern Virginia: Journal of Sedimentary Petrology, v. 51, p. 823-848.

Laporte, L.F., 1969, Recognition of a transgressive carbonate sequence within an epeiric sea; Helderberg Group (Lower Devonian) of New York State, in Friedman, G.M., ed., Depositional Environments in Carbonate Rocks: SEPM Special Publication No. 14, p. 98-119.

Logan, B.W., 1974, Inventory of diagenesis in Holocene-Recent carbonate sediments, Shark Bay, Western Australia, in Logan, B.W., J.F. Read, G.M. Hagen, P. Hoffman, R.G. Brown, P.J. Woods, and C.D. Gebelein, eds., Evolution and Diagenesis of Quaternary Carbonate Sequences, Shark Bay, Western Australia: AAPG Memoir 22, p. 195-249.

Logan, B.W., P. Hoffman, and C.D. Gebelein, 1974, Algal structures, cryptalgal fabrics, and structures, Hamblin Pool, Western Australia, in Logan, B.W., J.F. Read, G.M. Hagen, P. Hoffman, R.G. Brown, P.J. Woods, and C.D. Gebelein, eds., Evolution and Diagenesis of Quaternary Carbonate Sequences, Shark Bay, Western Australia: AAPG Memoir 22, p. 140-194.

Lowe, D.R., 1979, Sediment gravity flows: Their classification and some problems of application to natural flows and deposits, in Doyle, L.J., and O.H. Pilkey, eds., Geology of Continental Slopes: SEPM Special Publication No. 27, p. 75-82.

Mazzullo, S.J., and B.A. Birdwell, 1989, Syngenetic formation of grainstones and pisolites from fenestral carbonates in peritidal settings: Journal of Sedimentary Petrology, v. 59, p. 605-611.

Middleton, G.V., 1961, Evaporite solution breccias from the Mississippian of southwest Montana: Journal of Sedimentary Petrology, v. 31, p. 189-195.

Middleton, G.V., and M.A. Hampton, 1976, Subaqueous sediment transport and deposition by sediment gravity flows, in Stanley, D.J., and D.J.P. Swift, eds., Marine Sediment Transport and Environmental Management: John Wiley & Sons, New York, p. 197-218.

Mitchum, R.M., Jr., 1977, Seismic stratigraphy and global changes of sea level, Part 2: Glossary of terms used in seismic stratigraphy, in Payton, C.E., ed., Seismic Stratigraphy—Application to Hydrocarbon Exploration: AAPG Memoir 26, p. 205-212.

Mitchum, R.M., Jr., P.R. Vail, and S. Thompson III, 1977, Seismic stratigraphy and global changes of sea level, Part 2: The depositional sequence as a basic unit for stratigraphic analysis, in Payton, C.E., ed., Seismic Stratigraphy—Application to Hydrocarbon Exploration: AAPG Memoir 26, p. 53-62.

Nichols, K.M., 1980, Depositional and diagenetic history of porous dolomitized grainstones at the top of the Madison Group, Disturbed Belt, Montana, in Fouch, T.D., and E.R. Magathan, eds., Paleozoic Paleogeography of the West-central United States: SEPM, Rocky Mountain Section, Rocky Mountain Paleogeography Symposium 1, p. 163-173.

Nilsen, T.H., 1977, Paleogeography of Mississippian turbidites in south-central Idaho, in Stewart, J.H., C.H. Stevens, and A.E. Fritsche, eds., Paleozoic Paleogeography of the West-central United States: SEPM, Pacific Section, Pacific Coast Paleogeography Symposium 1, p. 275-299.

Peterson, J.A., 1986, General stratigraphy and regional paleotectonics of the western Montana overthrust belt, in Peterson, J.A., ed., Paleotectonics and Sedimentation in the Rocky Mountain Region, United States: AAPG Memoir 41, p. 57-86.

Petty, D.M., 1988, Depositional facies, textural characteristics, and reservoir properties of dolomites in Frobisher-Alida interval in southwest North Dakota: AAPG Bulletin, v. 72, p. 1229-1253.

Purser, B.H., 1973, The Persian Gulf—Holocene carbonate sedimentation and diagenesis in a shallow water epicontinental sea: Springer-Verlag, Berlin, 471 p.

Read, J.F., 1974, Calcrete deposits and Quaternary sediments, Edel Province, Shark Bay, Western Australia, in Logan, B.W., J.F. Read, G.M. Hagen, P. Hoffman, R.G. Brown, P.J. Woods, and C.D. Gebelein, eds., Evolution and Diagenesis of Quaternary Carbonate Sequences, Shark Bay, Western Australia: AAPG Memoir 22, p. 250-282.

Read, J.F., 1985, Carbonate platform facies models: AAPG Bulletin, v. 69, p. 1-21.

Read, J.F., D. Osleger, and M. Elrick, 1991, Two-dimensional modeling of carbonate ramp sequences and component cycles, in Franseen, E.K., W.L. Watney, C.G.St.C. Kendall, and W. Ross, eds., Sedimentary Modeling: Computer Simulation and Methods for Improved Parameter Definition: Kansas Geological Society Bulletin 233, p. 473-488.

Reid, S.K., 1991, Evolution of the Early Mississippian Mission Canyon platform and Antler foredeep, Montana and Idaho, Ph.D. dissertation: Texas A&M University, College Station, Texas, 105 p.

Reid, S.K., and S.L. Dorobek, 1991, Controls on development of third- and fourth-order depositional sequences in the Lower Mississippian Mission Canyon Formation and stratigraphic equivalents, Idaho and Montana, in Cooper, J.D. and C.H. Stevens, eds., Paleozoic Paleogeography of the Western United States—II: SEPM, Pacific Section, v. 67, p. 527-541.

Rhoads, D.C., 1975, The paleoecological and environmental significance of trace fossils, in Frey, R.W., ed., The Study of Trace Fossils: Springer-Verlag, New York, p. 147-160.

Roberts, A.E., 1966, Stratigraphy of Madison Group near Livingston, Montana, and discussion of karst and solution-breccia features: U.S. Geological Survey Professional Paper 526-B, p. B1-B23.

Rose, P.R., 1976, Mississippian carbonate shelf margins, western United States: U.S. Geological Survey Journal of Research, v. 4, p. 449-466.

Ross, C.A., and J.P. Ross, 1987, Late Paleozoic sea levels and depositional sequences, in Ross, C.A., and D. Haman, eds., Timing and Depositional History of Eustatic Sequences: Constraints on Seismic Stratigraphy: Cushman Foundation for Foraminiferal Research, Special Publication No. 24, p. 137-149.

Ross, W.C., 1991, Cyclic stratigraphy, sequence stratigraphy, and stratigraphic modeling from 1964 to 1989: Twenty-five years of progress?, in Franseen, E.K., W.L. Watney, C.G.St.C. Kendall, and W. Ross, eds., Sedimentary Modeling: Computer Simulation and Methods for Improved Parameter Definition: Kansas Geological Society Bulletin 233, p. 3-8.

Sandberg, C.A., 1975, McGowan Creek Formation, new name for Lower Mississippian flysch sequence in east-central Idaho: U.S. Geological Survey Bulletin, v. 1405-E, 11 p.

Sandberg, C.A., R.C. Gutschick, J.G. Johnson, F.G. Poole, and W.J. Sando, 1983, Middle Devonian to Late Mississippian geologic history of the Overthrust Belt region, western United States: Rocky Mountain Association of Geologists, Geologic Studies of the Cordilleran Thrust Belt, v. 2, p. 691-719.

Sandberg, C.A., and Poole, F.G., 1977, Conodont biostratigraphy and depositional complexes of Upper Devonian cratonic-platform and continental shelf rocks in the western United States, in Murphy,

M.A., W.B.N. Berry, and C.A. Sandberg, eds., Western North America; Devonian: Campus Museum Contributions 4, California University, Riverside, p. 144-182.

Sando, W.J., 1972, Madison Group (Mississippian) and Amsden Formation (Mississippian and Pennsylvanian) in the Beartooth Mountains, northern Wyoming and southern Montana: Montana Geological Society, 21st Annual Field Conference, p. 57-63.

Sando, W.J., 1974, Ancient solution phenomena in the Madison Limestone (Mississippian) in north-central Wyoming: U.S. Geological Survey Journal of Research, v. 2, p. 133-141.

Sando, W.J., 1976, Mississippian history of the northern Rocky Mountains Region: U.S. Geological Survey Journal of Research, v. 4, p. 317-338.

Sando, W.J., 1985, Revised Mississippian time scale, western interior region, conterminous United States: U.S. Geological Survey Bulletin, v. 1605-A, p. A15-A26.

Sando, W.J., 1988, Madison Limestone (Mississippian) paleokarst: A geologic synthesis, in James, N.P., and P.W. Choquette, eds., Paleokarst: Springer-Verlag, New York, p. 256-277.

Sando, W.J., and E.W. Bamber, 1985, Coral zonation of the Mississippian System in the western interior province of North America: U.S. Geological Survey Professional Paper 1334, 61 p.

Sando, W.J., and J.T. Dutro, Jr., 1960, Stratigraphy and coral zonation of the Madison Group and Brazer Dolomite in northeastern Utah, western Wyoming, and southwestern Montana: Wyoming Geological Association Guidebook, Fifteenth Annual Field Conference, p. 117-126.

Sando, W.J., and J.T. Dutro, Jr., 1974, Type sections of the Madison Group (Mississippian) and its subdivisions in Montana: U.S. Geological Survey Professional Paper 842, 22 p.

Sando, W.J., and J.T. Dutro, Jr., 1980, Paleontology and correlation of the Madison Group at Baldy Mountain, in Hadley, J.B., Geology of the Varney and Cameron Quadrangles, Madison County, Montana: U.S. Geological Survey Bulletin 1459, p. 33-46.

Sando, W.J., B.L. Mamet, and J.T. Dutro, Jr., 1969, Carboniferous megafaunal and microfaunal zonation in the Northern Cordillera of the United States: U.S. Geological Survey Professional Paper 613-E, 29 p.

Sando, W.J., C.A. Sandberg, and W.J. Perry, Jr., 1985, Revision of Mississippian stratigraphy, Northern Tendoy Mountains, southwest Montana, in Sando, W.J., ed., Mississippian and Pennsylvanian Stratigraphy in Southwest Montana and Adjacent Idaho: U.S. Geological Survey Bulletin, v. 1656, p. A1-A10.

Sarg, J.F., 1988, Carbonate sequence stratigraphy, in Wilgus, C.K., B.S. Hastings, C.G.St.C. Kendall, H.W. Posamentier, C.A. Ross, and J.C. Van Wagoner, eds., Sea Level Changes: An Integrated

Approach: SEPM Special Publication No. 42, p. 155-181.

Schlager, W., 1981, The paradox of drowned reefs and carbonate platforms: Geological Society of America Bulletin, v. 92, p. 197-211.

Scholle, P.A., and Kinsman, D.J.J., 1974, Aragonitic and high-Mg calcite caliche from the Persian Gulf—a modern analog for the Permian of Texas and New Mexico: Journal of Sedimentary Petrology, v. 44, p. 904-916.

Seilacher, A., 1967, Bathymetry of trace fossils: Marine Geology, v. 5, p. 413-428.

Shinn, E.A., 1983, Tidal flat environment, in Scholle, P.A., D.G. Bebout, and C.H. Moore, eds., Carbonate Depositional Environments: AAPG Memoir 33, p. 171-210.

Skipp, B., and B.L. Mamet, 1970, Stratigraphic micropaleontology of the type locality of the White Knob Limestone (Mississippian), Custer County, Idaho: U.S. Geological Survey Professional Paper 700-B, p. B118-B123.

Skipp, B., W.J. Sando, and W.E. Hall, 1979, The Mississippian and Pennsylvanian (Carboniferous) Systems in the United States—Idaho: U.S. Geological Survey Professional Paper 1110-AA, 42 p.

Sloss, L.L., 1963, Sequences in the cratonic interior of North America: Geological Society of America Bulletin, v. 74, p. 93-113.

Sloss, L.L., W.C. Krumbein, and E.C. Dapples, 1949, Integrated facies analysis, in Longwell, C.R., chair, Sedimentary Facies in Geologic History: Geological Society of America Memoir 39, p. 91-124.

Stockmal, G.S., and C. Beaumont, 1987, Geodynamic models of convergent margin tectonics: The southern Canadian Cordillera and the Swiss Alps, in Beaumont, C., and A.J. Tankard, eds., Sedimentary Basins and Basin Forming Mechanisms: Canadian Society of Petroleum Geologists, Memoir 12, p. 393-411.

Vail, P.R., R.M. Mitchum, Jr., S. Thompson III, 1977a, Seismic stratigraphy and global changes of sea level, Part 4: Global cycles and relative changes of sea level, in Payton, C.E., ed., Seismic Stratigraphy—Application to Hydrocarbon Exploration: AAPG Memoir 26, p. 49-212.

Vail, P.R., R.G. Todd, and J.B. Sangree, 1977b, Seismic stratigraphy and global changes of sea level, Part 5: Chronostratigraphic significance of seismic reflections, in Payton, C.E., ed., Seismic Stratigraphy—Application to Hydrocarbon Exploration: AAPG Memoir 26, p. 99-116.

Van Wagoner, J.C., H.W. Posamentier, R.M. Mitchum, P.R. Vail, J.F. Sarg, T.S. Loutit, and J. Hardenbol, 1988, An overview of the fundamentals of sequence stratigraphy and key definitions, in Wilgus, C.K., B.S. Hastings, C.G.St.C. Kendall, H.W. Posamentier, C.A. Ross, and J.C. Van Wagoner, eds., Sea Level Changes: An Integrated Approach: SEPM Special Publication No. 42, p. 39-45.

Waters, D.L., and Sando, W.J., 1987, Coral zonules: New tools for petroleum exploration in the

Mission Canyon Limestone and Charles Formation, Williston Basin, North Dakota: Rocky Mountain Association of Geologists, 1987 Symposium, p. 193-208.

Wilgus, C.K., Hastings, B.S., Kendall, C.G.St.C., Posamentier, H.W., Ross, C.A., and Van Wagoner, J.C., 1988, Sea Level Changes: An Integrated Approach: SEPM Special Publication No. 42, 407 p.

Wilson, J.L., 1969, Microfacies and sedimentary structures in "deeper water" lime mudstones: SEPM Special Publication No. 14, p. 4-19.

Wilson, J.L., 1975, Carbonate facies in geologic history: Springer-Verlag, New York, 471 p.

Witter, D.N., 1988, Stratal architecture and volumetric distribution of facies tracts, Upper Mission Canyon Formation (Mississippian), Williston Basin, North Dakota, M.S. thesis: Colorado School of Mines, Golden, Colorado, 180 p.

Chapter 14

Sequence Stratigraphy and Systems Tract Development of the Latemar Platform, Middle Triassic of the Dolomites (Northern Italy): Outcrop Calibration Keyed by Cycle Stacking Patterns

R. K. Goldhammer
Exxon Production Research Co.
Houston, Texas, U.S.A.

M. T. Harris
University of Wisconsin at Milwaukee
Milwaukee, Wisconsin, U.S.A.

P. A. Dunn
Exxon Production Research Co.
Houston, Texas, U.S.A.

L. A. Hardie
Johns Hopkins University
Baltimore, Maryland, U.S.A.

ABSTRACT

The Middle Triassic Latemar platform (740 m thick, 5–6 km wide) provides a seismic-scale outcrop example of an intact carbonate shelf-to-basin transition, ideal for integrating sequence stratigraphy with facies and cyclic stratigraphy. This subcircular, high-relief buildup records two third-order (1–10 m.y.) accommodation sequences within the platform interior, the Lower Ladinian sequence (8 m.y.; 400 m thick) and the Upper Ladinian sequence (6 m.y.; 340 m thick). The Lower Ladinian sequence developed atop a widespread, low-relief Middle Anisian carbonate bank (60 m thick). Underlying subtidal cycles of the Middle Anisian bank thin upward into the basal, subaerial sequence boundary of the Lower Ladinian sequence reflecting decreasing third-order accommodation. Above this sequence boundary, platform-interior facies of the Lower Ladinian sequence retrograde. This retrogradation results in superimposition of Ladinian basinal and foreslope facies atop the underlying, horizontal, shallow-water bank along its periphery. The transgressive and highstand systems tract of the Lower Ladinian sequence and transgressive systems tract of the Upper Ladinian sequence are marked by long-term, systematic vertical facies changes (subtidal vs. subaerial exposure-dominated facies) and variation in stacking patterns of aggradational high-frequency, 20 k.y. cycles within the platform interior. The maximum flooding surface in the Lower Ladinian sequence is a prominent surface in a platform interior position that loses its identity laterally into reef margin, foreslope, and basinal facies. A stratigraphically transitional sequence boundary caps the Lower Ladinian sequence, marked by an interval of vertically superimposed thin subaerial tepees. Beneath this interval, high-frequency cycles are thinning-upward, and above they are thickening-upward. At this sequence boundary there is no downward shift in overlying facies, no lowstand wedge in the downdip position, particularly in the case of isolated buildups such as the Latemar, and no erosional hiatus. Only the transgressive systems tract of sequence L2 is preserved at the Latemar owing to Late Ladinian–Early Carnian volcanism and tectonism, which terminated carbonate platform deposition.

INTRODUCTION

Sequence stratigraphy developed from consideration of stratal geometries expressed on seismic sections, with a vertical resolution of several tens to hundreds of meters (e.g., Vail et al., 1984). Idealized and empirical models for **carbonate** sequence stratigraphy and systems tract development have been proposed and expanded upon (e.g., Sarg, 1988; Crevello et al., 1989; Tucker et al., 1990; Handford and Loucks, 1991; Hunt and Tucker, 1991; Tucker, 1991). Early models were based, in part, on a comparison of carbonate stratal architecture with siliciclastic counterparts, invoking third-order eustatic fluctuations in sea level as the primary control on sequence and systems tract architecture (Vail, 1987; Sarg, 1988).

Many studies applying carbonate sequence stratigraphic techniques to outcrop (Sarg and Lehmann, 1986; Sarg, 1988; Crevello et al., 1989; Tucker et al., 1990) have stressed third-order, seismic-scale geometries and recognition of sequence boundaries and systems tracts at the seismic scale. Attention has focused on criteria for recognizing third-order sequence boundaries and systems tracts boundaries as **physical surfaces** and tying these surfaces to the global cycle chart of Haq et al. (1987) to infer a eustatic origin.

Most recently, application of carbonate sequence stratigraphic techniques to outcrop has focused on the internal architecture of depositional sequences, with particular emphasis on the lateral and vertical **stacking patterns** of m-scale, high-frequency cycles (equivalent to parasequences of Van Wagoner et al., 1987, 1990) within a systems tract framework (e.g., Goldhammer et al., 1990, 1991a,b, 1992; Osleger, 1990,

1991; Read et al., 1990; Dunn, 1991; Elrick and Read, 1991; Montanez, 1992). The key to unraveling the details of depositional sequences relies on an appreciation of the concept of the hierarchy of stratigraphic cyclicity (Goldhammer et al., 1990, 1991b). The "stacking pattern" approach set forth in these studies deemphasizes the traditional siliciclastic-oriented approach to sequence stratigraphy (e.g., Vail, 1987), which focuses more on seismic-scale techniques of picking unconformities on either the basis of erosional truncation, stratal onlap, etc. ("surface picking") or on the basis of abrupt vertical facies changes in apparent discord with Walther's Law ("downward shifting").

Often, experience has demonstrated that such techniques will be of limited value in interpreting thick stratigraphic sections of shelfal carbonates where seismic-scale geometries are lacking, commonly the case in updip shelfal carbonate sections located landward of the shelf edge. One of the lessons learned from the Latemar is that despite the luxury of intact seismic-scale geometries and lateral continuity to guide our interpretations, we needed to turn our attention to more one-dimensional oriented analyses of stratal stacking patterns to ascertain multiple hierarchies of accommodation changes. This "stacking pattern" approach enabled us to deduce third-order accommodation changes "hidden" within the thick, monotonous, flat-lying stack of high-frequency platform cycles. Admittedly, in many settings, such as the mixed siliciclastic–carbonate regime in the Permian of west Texas (Sarg and Lehmann, 1986), sequence boundaries are often obvious, associated with erosional truncation, etc. But in many cases, particularly shallow-dipping ramps or flat-topped platforms developed on passive margins, third-order sequence boundary expression is less obvious, particularly in deformed terrains which lack two-dimensional dip continuity. Such examples include the Late Proterozoic Katakturuk Dolomite of the northeastern Brooks Range of Alaska (Clough, 1989; Clough and Goldhammer, 1992), the Cambro-Ordovician of the United States (Hardie, 1989; Koerschner and Read, 1989; Read, 1989; Osleger, 1990; Dunn et al., 1991; Goldhammer et al., 1992; Montanez, 1992), the Devonian of Nevada (Elrick, 1992), the Lower Mississippian of Wyoming and Montana (Elrick and Read, 1991), and the Upper Jurassic and Lower Cretaceous of northeast Mexico (Goldhammer et al., 1991a; Johnson et al., 1991). Even in ideal terrains, such as the Dolomites with their spectacular three-dimensional, seismic-scale exposures and exemplary geometries, liberal and noncritical utilization of sequence stratigraphic techniques may lead to misinterpretation of the stratigraphic record (Yose, 1991).

In this chapter we present a detailed analysis of the Middle Triassic Latemar Platform (Dolomites, northern Italy) that may bridge the gap from m-scale cyclic stratigraphy to seismic-scale sequence stratigraphy. Our study integrates well-preserved seismic-scale geometries and stratal architecture viewed from a distance (e.g., progradational downlap), with a detailed analysis of facies and high-frequency, fifth-order cycles, which are the fundamental building blocks of third-order sequences (Van Wagoner et al., 1990). While crossing this bridge, we are confronted with the nature of sequence and systems tract boundaries. In some cases, these boundaries are "physical surfaces" that "fit" the carbonate sequence models rather well. In other cases, they are difficult, if not impossible, to define in the field, although one can "force" a boundary through certain stratigraphic horizons. This study is not so much a check on existing models as it is an inventory of the stratigraphic realities of a well-exposed, nonstructured, and small, yet complex, system.

GEOLOGIC SETTING

The Dolomites are located in northern Italy on the southern margin of the Alpine chain within the major autochthonous structural element of the little deformed Southern Alps. During the Permian–Triassic, the Southern Alps region was part of a coherent continental block (consisting of Eurasia, Africa, and the Americas; Bernoulli and Lemoine, 1980), which underwent rifting (Bechstadt et al., 1978; Winterer and Bosellini, 1981). In the Dolomites, Middle Triassic (Late Anisian–Ladinian) rifting was marked by sinistral strike-slip tectonics (Doglioni, 1984a,b), with both extensional and compressional structures, followed by Late Ladinian volcanism and magmatic intrusions (Figure 1). Thick, Upper Anisian–Ladinian carbonate buildups (700–1300 m thick; Figures 1–3) grew from structurally induced topographic highs, as vertical aggradation and lateral progradation matched accommodation generated by combined eustasy and regional subsidence (Bosellini and Rossi, 1974; Bosellini, 1984). Deep-water starved basins developed adjacent to the buildups, which prograded and partly infilled basinal areas (Figures 1, 2). The differential topography was short-lived because volcanics and genic clastics partially to totally filled the depositional basins, onlapping and, in some volcanoe cases, burying pre-existing carbonate buildups (e.g., the Mount Agnello buildup south of the Latemar; Bosellini and Rossi, 1974; Goldhammer and Harris, 1989). In early to Middle Carnian time, shallow-marine carbonate platforms (as thick as 1 km) nucleated on submerged topographic highs of Ladinian buildups and prograded out into remaining basinal areas (Bosellini, 1984; Yose, 1991).

Upper Anisian–Ladinian paleogeography of the Dolomites is illustrated in Figure 1. The Latemar buildup is Late Anisian–Ladinian in age (Figure 3) and was an isolated, "atoll-like" entity (Figures 1, 2) based on the occurrence of basinal strata (Livinallongo Formation) and foreslope deposits (Marmolada Limestone of Gaetani et al., 1981), which essentially surrounded the platform (Figure 4; Goldhammer and Harris, 1989). The depositional geometry of the Latemar and other Middle Triassic buildups is spectacularly exhumed as a result of Quaternary erosion that has preferentially removed surrounding volcanogenic basinal sediments (Figure 2). Another isolated coeval buildup, the Mount Agnello buildup, occurs immediately southwest of the Latemar (Figure 4). This buildup is positioned structurally on the southern side of the master syndepositional strike-slip fault which crosscuts the region (Figures 1, 4; Doglioni, 1984a,b), and experienced enhanced overall subsidence as compared to the Latemar (Goldhammer, 1987).

LATEMAR BUILDUP

The stratigraphy and sedimentology of the Upper Anisian–Ladinian platforms of the Dolomites have received significant study in recent years (Cros and Lagny, 1969; Bosellini and Rossi, 1974; Cros, 1974; Fois and Gaetani, 1981; Gaetani et al., 1981; Fois, 1982; Blendinger, 1986; Goldhammer et al., 1987; Goldhammer and Harris, 1989; Hardie et al., 1991). The following synopsis of the Latemar is summarized from the studies of Hardie et al. (1986), Goldhammer (1987), Harris (1988), Goldhammer and Harris (1989), Dunn (1991), and Hardie et al. (1991). The Latemar buildup (Figures 4–6) is 5–6 km across with a central core of flat-lying strata 760 m thick, rimmed by an apron of steeply dipping foreslopes (Figures 7, 8). The platform core consists of a 420 m thick Ladinian unit (termed the Latemar Limestone by Gaetani et al.,

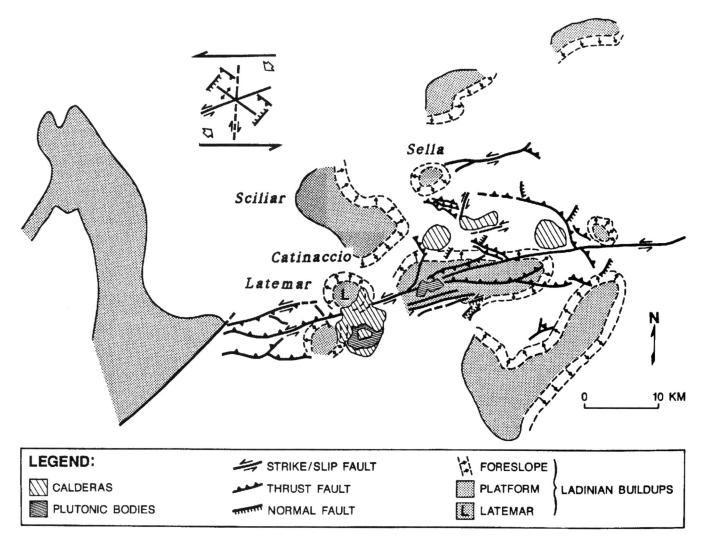

Figure 1. Middle Triassic paleogeography of the Central Dolomites superimposed on the paleotectonic map of Doglioni (1984a). Mount Agnello buildup is the platform immediately to the south–southwest of the Latemar. From Goldhammer and Harris (1989).

1981), which conformably overlies a 250 m thick Upper Anisian unit that Goldhammer (1987) informally designated the Lower Platform Facies (essentially the "Lower Edifice" of Gaetani et al., 1981). The Lower Platform Facies, in turn, unconformably overlies the 90 m thick Anisian Contrin Formation (Figures 4, 6). Gaetani et al. (1981) designated the reef margin and surrounding wedge of coarse foreslope breccias and grainstones as the Marmolada Limestone. Basinward, the foreslope deposits intertongue with, and prograde over, the flat-lying basinal Livinallongo Formation (Figures 5, 6; Bosselini and Rossi, 1974; Harris, 1988).

A narrow reef margin facies separates the Latemar foreslope deposits from the platform strata (Figure 5; Harris, 1988; Goldhammer and Harris, 1989). The margin facies consists of massive to thickly bedded boundstones and grainstones. The boundstones are

primarily a framework of encrusting **Tubiphytes** with minor sponges and corals cemented by submarine radial fibrous cements that line or occlude boundstone-supported cavities (Harris, 1988). The **Tubiphytes**-algal boundstone facies, which completely surrounds the platform interior as a narrow (approximately 20 m wide) massive platform edge, is interpreted as a shallow-water reef (Harris, 1988). The thickly bedded (20–30° basinward dip) diverse-biota grainstone facies (with only scattered boundstone patches) marks the transition from the margin to the steeply dipping foreslope clinoforms. The diagenesis of both margin facies is dominated by a variety of syndepositional submarine cements. Evidence of subaerial exposure or vadose diagenesis is lacking, in contrast to the platform interior.

The upper foreslope deposits (Figures 7, 8) which encircle the shallow-water interior constitute approxi-

Figure 2. Aerial panoramic view of exhumed Triassic buildups in the Dolomites showing the Latemar platform (foreground), Sciliar-Catinaccio complex (midground), and Sella buildup (background). Photograph taken southwest of the Latemar, looking northeast. Scale bar = approximately 650 m.

mately 40–50% of the buildup volume and are principally grain-supported breccias and megabreccias composed of blocks of reef margin boundstone and redeposited composite foreslope clasts ranging in diameter from 1 cm to over 1 m (Harris, 1988). Depositional matrix is entirely absent and syndepositional radial fibrous cements (20–40% by volume) fill the interparticle space. Tongues of megabreccia 2–5 m thick overlie one another with erosional contacts or are separated by grainstone lenses (Figure 5). The foreslope clinoforms that are so readily observed from a distance are defined by these erosional surfaces (Figures 7, 8). Upper foreslope clinoforms maintain steep primary depositional dips of 30–35°, but near the foreslope toes, the dips decrease rather abruptly to 5–10° before merging into the flat-lying Livinallongo Formation (Figures 5, 6). This change in slope dip is paralleled by a notable increase in grainstones and concomitant decrease in megabreccias (Harris, 1988).

At the toe-of-slope, the grainstones give way to graded peloidal–skeletal grainstones, which fine upward into wackestones and mudstones. Each very

thinly bedded to thickly bedded graded grainstone bed consists of a single graded unit of redeposited platform and margin sediment without any internal sedimentary structures, and is interpreted as a proximal turbidite. Basinward, these graded peloidal–skeletal grainstones become finer grained and intertongue with the dark shaly limestones of the deep water Livinallongo Formation (Bosellini and Ferri, 1980; Bosellini, 1984; Harris, 1988). The grain-supported breccias and megabreccias of the foreslope wedges are talus piles derived from erosion of the reef margin boundstones and resedimentation of updip foreslope material. The grainstone lenses and beds intercalated within the megabreccias are composed of shelf–lagoon sediment washed off the platform top and transported downslope by grain flows and turbidity currents (Bosellini and Rossi, 1974; Harris, 1988).

Throughout Anisian–Ladinian time, the vertical growth of the Latemar kept pace with subsidence (about 6 cm/k.y.) as necessitated by the nearly 800 m thick central core of Anisian–Ladinian shallow water platform facies. For most of its development, the Latemar platform core aggraded vertically rather

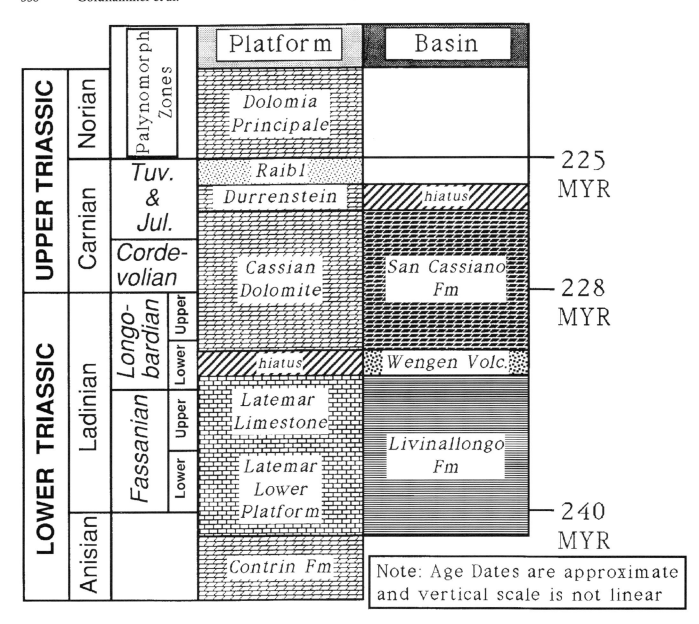

Figure 3. Stratigraphic column of the Middle and Upper Triassic rocks of the Dolomites. The positions of the top of the Latemar Limestone, the bottom of the overlying Wengen volcanics, and the bottom of the San Cassiano basinal sediments, are based on macrofossil data of Ulrichs (1977) and pollen data of Van Der Eem (1983) and Peter Litmann and Lyndon Yose (1991, unpublished study) and Yose (1991). Time scale based on Palmer (1983). Diagram from Hardie et al. (1991).

Figure 4. Geologic map of the Latemar and Mount Agnello buildups showing aerial distribution of major facies. Arrows in the foreslope facies indicate clinoform dip directions. Triassic age faults are shown as lines with "U" on the upthrown side and "D" on the downthrown side. Line of sight of photographs in Figures 7, 8, 20, and 25 are also illustrated.

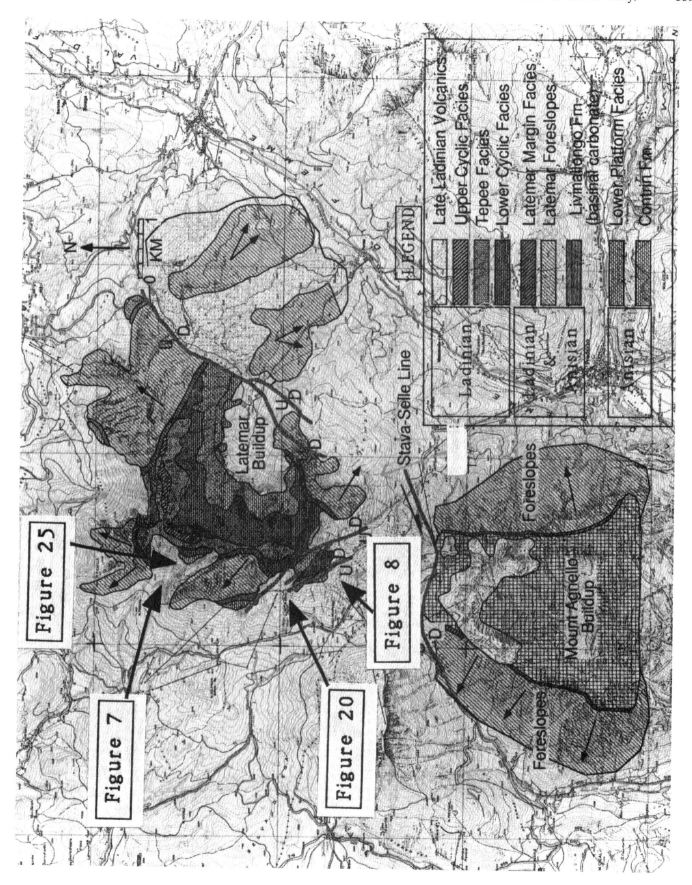

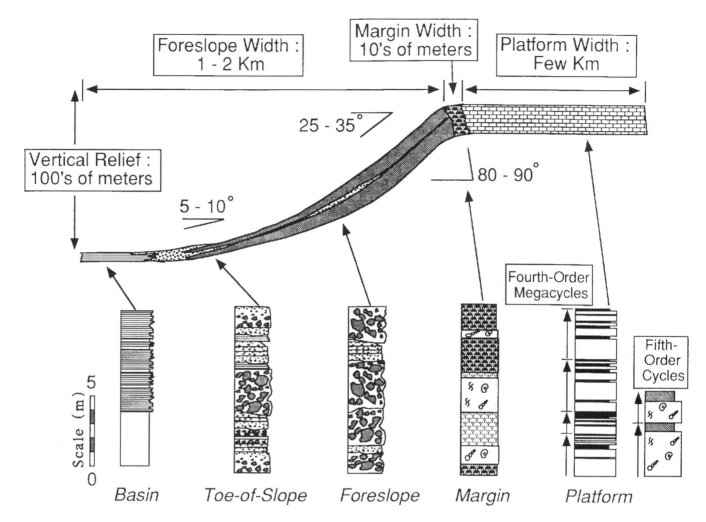

Figure 5. Schematic cross section showing platform-to-basin geometry and lateral succession of coeval facies of the Latemar buildup. From Goldhammer and Harris (1989).

than prograded outward (Harris, 1988). This is recorded by the near vertical contact between the margin facies separating flat-lying platform interior deposits from the steeply inclined foreslope deposits (Figure 6; Harris, 1988; Goldhammer and Harris, 1989). The Livinallongo basinal deposits are less than 200 m thick. During the final stages of buildup evolution, shelf to basin vertical relief was on the order of 500–600 m (Harris, 1988). During the final stage of development, the Latemar platform prograded at a low angle, similar to other Ladinian platforms (Catinaccio, Mount Agnello; Bosellini, 1984; Harris, 1988). This change from vertical to lateral growth resulted from a combination of a decrease in the rate of tectonic subsidence accompanied by the onset of rapid infilling of basinal areas by volcanics and volcaniclastics (Harris, 1988). In the Late Ladinian (Longobardian, Figure 3), the Latemar and coeval buildups were subject to a major volcanic–tectonic event that terminated the Lower-Middle Ladinian phase of carbonate deposition in the Dolomites (Bosellini, 1984). One of the main plutonic centers

was located essentially beneath the Latemar (Figure 1). The southeastern portion of the Latemar and part of the adjacent Mount Agnello buildup were down-dropped along a circular caldera fault linked with the intrusive event. Additionally, numerous narrow dikes and explosive diatremes intruded the Latemar platform and acted as conduits for subaqueous lava flows and volcaniclastic debris, which eventually buried Ladinian platforms near the Latemar, such as the Mount Agnello and Sciliar platforms (Figure 1), and filled up much of the surrounding basinal areas (Bosellini and Rossi, 1974; Bosellini, 1984).

HIGH-FREQUENCY DEPOSITIONAL CYCLES: THE FUNDAMENTAL BUILDING BLOCKS OF THE LATEMAR PLATFORM

The most striking aspect of the central platform core of the Latemar are the monotonous layer-cake stacks of shallow-water carbonates that lack any

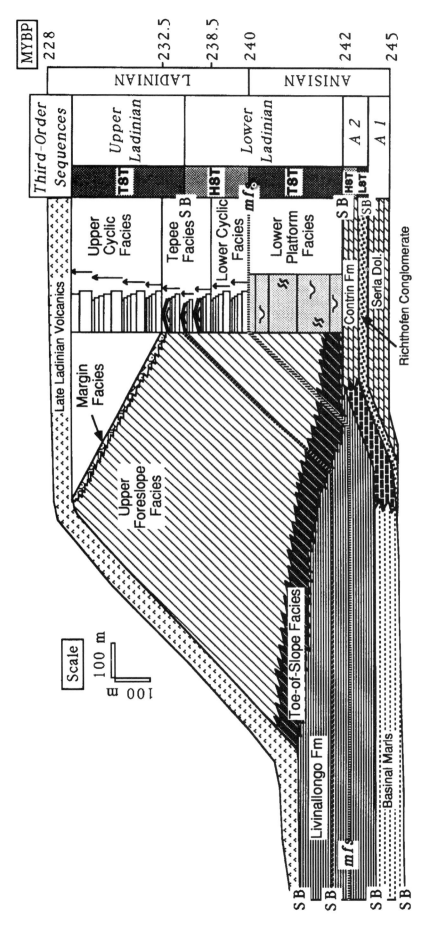

Figure 6. Restored true scale cross section of the Latemar buildup illustrating sequence stratigraphy of the Upper Anisian–Ladinian Latemar sequences and the underlying Anisian Contrin and Serla platforms. Diagram displays four third-order depositional sequences and sequence and system tract boundaries identified on the platform and their inferred basinal extensions. Late Ladinian volcanics that cap the buildup are inferred from the coeval Mount Agnello buildup, as the top of the Latemar is eroded. SB = sequence boundary; MFS = maximum flooding surface; LST = lowstand systems tract; TST = transgressive systems tract; HST = highstand systems tract. Age dates are approximate as large errors are involved and published chronologies differ from author to author.

Figure 7. Ground view from the west–northwest (looking east–southeast) of the flat-lying central platform core of the Latemar flanked on both sides by steeply dipping foreslopes. Note position of top of the Contrin bank (sequence A2) in dark gray. Vertical relief on the order of several hundred meters. Scale bar = 750 m.

internal geometry (Figures 2, 7, 8). The 420 m thick Ladinian portion of the platform consists of over 500 m-scale fifth-order cycles with periods in the 10^4-year range, which are the fundamental building blocks of the Latemar third-order sequences (Goldhammer et al., 1990). Goldhammer (1987) subdivided this cyclic Ladinian portion of the Latemar platform into three facies—the Lower Cyclic Facies, the Tepee Facies, and the Upper Cyclic Facies (Figure 6).

Each fifth-order cycle is a couplet consisting of a basal subtidal grainstone (0.6–0.7 m average thickness) overlain by a 1–15 cm thick subaerial exposure cap (Figures 9, 10; Goldhammer et al., 1987, 1990). Each cycle records alternating submergence and emergence of the Latemar platform. Individual fifth-order cycles are packaged into five-part **fourth-order megacycles** (up to 5 m thick) characterized by progressive upward thinning of successive cycles within each megacycle (Figures 9, 10). Despite large errors in radiometric dates of the Ladinian stage boundaries (e.g., Palmer, 1983), the average duration of an individual cycle must have been on the order of 10^4 years (8000–20,000 years), and that of a megacycle must have been on the order of 10^5 years.

Goldhammer et al. (1987) argued that the five-part thinning-upwards megacycles of the Ladinian Latemar platform were the result of high frequency (10^4–10^5 year) composite sea-level oscillations in tune to Milankovitch astronomical rhythms. The only mechanism known at this time that can produce a 5:1 bundling at these time scales is composite eustasy driven by Milankovitch 20 kyr precession rhythms superimposed on 100 kyr eccentricity rhythms. Their arguments were based on comparison with the Pleistocene sea-level record and on computer simula-

tion of the Latemar cyclic stratigraphy using Milankovitchian sea-level oscillations. Such a composite eustasy with fifth-order oscillations superimposed upon fourth-order oscillations has been demonstrated to have operated in the latest Pleistocene in response to Milankovitch forcing of growth and melting of polar ice caps (Goldhammer et al., 1987). Using the Pleistocene eustatic sea-level curves as an analogy, Goldhammer et al. (1987) were able to quantitatively simulate the Latemar cyclic stratigraphy of five-part megacycles composed of upward-thinning cycles (submeter scale) with thin soil caps using the "Mr. Sediment" program (Figure 11). Based on the Pleistocene comparison and the quantitative simulation, and in the absence of any other known mechanism that could produce five-part composite sea-level oscillations in the 10^4–10^5 year range, Goldhammer et al. (1987) called on Milankovitch-driven eustasy to explain the cyclic stratigraphy of the Triassic Latemar platform carbonates.

LATEMAR PLATFORM FACIES STRATIGRAPHY AND CYCLE STACKING PATTERNS: RESPONSE TO THIRD-ORDER EUSTASY

The platform interior of the Latemar can be subdivided into four facies on the basis of systematic vertical changes in the characteristics of the fifth- and fourth-order cycle stacking patterns (Figure 6; Goldhammer, 1987; Goldhammer et al., 1990; Hardie et al., 1991).

1.   Lower platform facies (Figures 12, 13)—250

Top Lower Platform
Facies (MFS)

Foreslope

Figure 8. Aerial view from the south-southwest (looking northeast) of the Latemar platform and Sciliar (platform) and Catinaccio (foreslope) complexes. Note the position of the Contrin bank (sequence A2, dark gray line), the maximum flooding surface and upper sequence boundary of the Lower Ladinian sequence (labeled Top Lower Platform Facies and Mid-Ladinian Sequence Boundary, respectively). Note also foreslope aprons flanking the central platform core of the Latemar. From the top of the Contrin bank (A2) to the maximum flooding surface is approximately 250 m. Note also the position of the Stava Line, the master sinistral strike slip fault in Doglioni's 1984a tectonic scheme (see Figure 1). Scale bar = 250 m.

m of subtidal, shelf–lagoon, skeletal–lithoclastic packstone–grainstone, full of a variety of early marine diagenetic features (pervasive early marine cements, submarine cemented sheet cracks, subaqueous tepee antiforms, hardgrounds). This massively bedded limestone is punctuated by thin, dolomitic, subaerial exposure horizons (< 1–3 cm thick) roughly every 10–15 m on average (Figure 12). This facies depicts monotonous conditions of subtidal sedimentation and submarine diagenesis, with rare subaerial exposure breaks. Repetitive m-scale, fifth-order cycles are not abundant.

2. Lower cyclic facies—90 m of fifth-order cycles (73 cycles; average 1.24 m/cycle) each composed of bioturbated, subtidal shelf–lagoon, skeletal–peloidal packstone–grain-

stone overlain by a thin (5–15 cm thick) vadose, dolomitic caliche crust. Fifth-order cycles are grouped into asymmetric, thinning-upwards fourth-order megacycles.

3. Tepee facies (Figures 14, 15)—120 m of m-scale, fifth-order cycles (295 cycles; average 0.4 m/cycle) periodically interrupted by tepee antiforms (1–5 m thick), composed of a few pre-existing, disrupted thin fifth-order cycles. These tepee antiforms indicate extended subaerial exposure and more intensive vadose diagenesis as compared to the typical thin caliche crust associated with each cycle. Fifth-order cycles are grouped in packages of five into thinning-upwards megacycles capped by tepees. The tepee facies, with numerous tepee intervals (35 tepee zones vertically) intercalated within the normal,

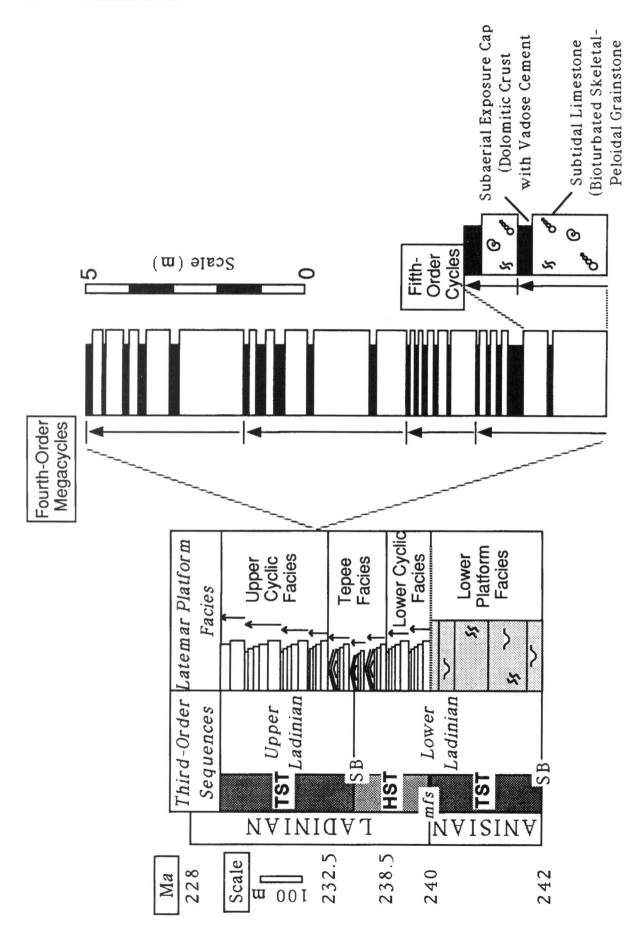

Figure 9. Typical example of the stratigraphic characteristics of both the Lower and Upper cyclic facies of the Latemar buildup. The part of the measured section from which the example was taken is shown by converging dashed lines. TST = transgressive systems tract; HST = highstand systems tract; SB = sequence boundary; mfs = maximum flooding surface.

Figure 10. Outcrop photographs of depositional cycles of the Upper Cyclic Facies of the Latemar platform. View at left shows a succession of thin beds of limestone separated by vadose diagenetic caps (dark, recessive bands), with arrows marking the tops of thinning-upward, fourth-order megacycles. Photograph at right gives detailed view of the megacycles, with individual cycle caps marked and the total number of cycles in each megacycle indicated. Scale bar = 2 m.

rhythmic cyclic succession, depicts depositional conditions dominated by subaerial exposure. The thickness and degree of diagenetic disruption is symmetrically distributed around the middle of the tepee facies. The lowermost and uppermost tepees are generally the thinnest and display the least amount of diagenetic overprint. The thickest tepees occur at the stratigraphic midpoint of the facies, coincident with the thinnest fifth-order cycles.

4.   Upper cyclic facies (Figures 9, 10)—210 m of fifth-order cycles (230 cycles, average 0.96 m/cycle) similar to the lower cyclic facies.

Goldhammer et al. (1990) have explained this vertical facies succession as a response to changes in accommodation space produced by a third-order sea-level oscillation with a period approximating 11 m.y. and an amplitude of about 60 m (derivation discussed by Goldhammer and Harris, 1989). In a sequence stratigraphic context, this facies succession records one complete third-order sequence (**Lower Ladinian sequence**) from the unconformity above the Contrin

Formation (basal sequence boundary of the Lower Ladinian sequence) to the middle of the tepee facies, and one partial third-order sequence (**Upper Ladinian sequence**) from the middle of the tepee facies to the eroded top of the upper cyclic facies (Figure 6). In the Lower Ladinian sequence, the lower platform facies constitutes the transgressive systems tract and the lower cyclic facies plus the lower part of the tepee facies makes up the highstand systems tract. The sequence boundary separating the Lower from the Upper Ladinian sequence (labeled mid-Ladinian sequence boundary in Figure 8) is a transitional sequence boundary rather than an unconformity, and is located within the thickest tepee zone in the middle of the tepee facies. The upper part of the tepee facies and upper cyclic facies constitute initial stages of the transgressive systems tract of the Upper Ladinian sequence, which is erosionally truncated at its top (Figure 6).

The first Latemar sequence records qualitative systematic changes in stratigraphic and lithologic character which reflect systematic changes in third-order **accommodation potential**; that is, the sum of long-

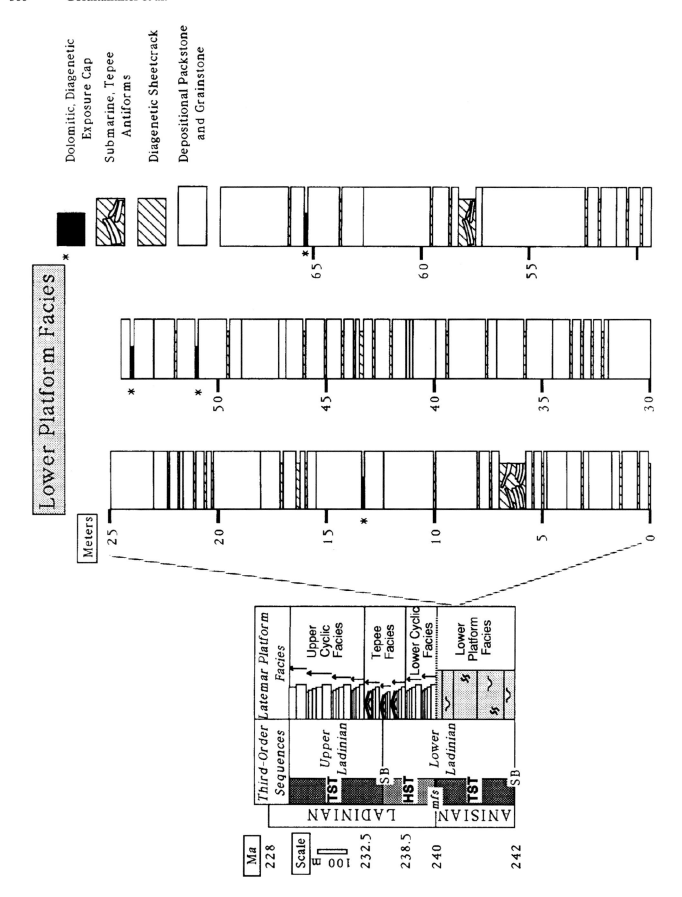

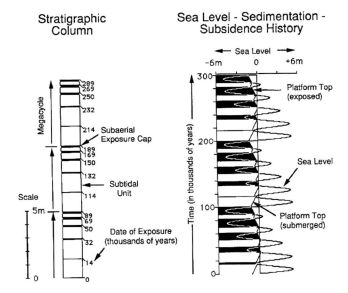

Stratigraphic Column

Sea Level - Sedimentation - Subsidence History

Figure 11. Computer simulation of Latemar platform megacycles. Stratigraphic column (left) shows three complete fourth-order megacycles consisting of five meter-scale fifth-order cycles composed of subtidal limestone (white) and dolomitic exposure cap (black). Chart (right) illustrates the interaction of composite sea level (asymmetric, 100,000 yr sawtooth wave and superimposed 20,000 yr sinusoidal oscillation) with platform sedimentation and subsidence during the run. From Goldhammer et al. (1990).

term third-order eustatic sea-level changes plus background (tectonic) platform subsidence. First of all, the lower platform facies differs from the other facies in that it lacks repetitive m-scale fifth-order cycles. Instead, subtidal deposits are punctuated by subaerial exposure caps on average roughly every 10 m (nonrhythmically, as confirmed by autocorrelation tests). In contrast, within the overlying lower cyclic facies and tepee facies, fifth-order cycles occur with regularity and decrease in thickness overall from the base of the lower cyclic facies (individual cycles up to 4 m thick) to the top of the tepee facies (individual cycles < 0.5 m thick). This progression from a noncyclic interval to cyclic intervals marked by decreasing cycle thickness upwards indicates that third-order accommodation potential for each depositional pulse was reduced.

Secondly, the proportion of marine submergence (and concomitant early marine diagenesis) to subaerial exposure (and concomitant subaerial diagenesis) varies systematically upwards through the sequence. The lower platform facies is dominated by subtidal deposits full of early marine syndepositional diagenetic features (hardgrounds, sheet cracks, early marine cements; Goldhammer, 1987). In contrast, evidence for such intensive early marine diagenesis is lacking in the overlying lower cyclic facies and tepee facies, which contain vadose diagenetic features (caliche fabrics, vadose cements; Goldhammer, 1987; Goldhammer et al., 1987). The progressive increase in subaerial exposure features clearly records a shift from generally submergent conditions during lower platform facies deposition to conditions whereby subtidal deposition was frequently interrupted by intervals of subaerial exposure (lower cyclic facies) to conditions dominated by subaerial exposure (tepee facies). The upper cyclic facies that overlies the tepee facies records a return to more submergent conditions. Thus, the stratigraphic variation of early diagenesis parallels the changes in cycle thickness, both being direct responses to systematic changes in third-order accommodation potential.

The style of vertical changes in fifth-order cycle stacking patterns depicted by the systematic succession of submergence-dominated (lower cyclic facies) to emergence-dominated (tepee facies) and then back to submergence-dominated (upper cyclic facies) conditions can be viewed semiquantitatively by constructing a Fischer diagram (Fischer, 1964) of a continuous measured section that displays the Latemar cyclostratigraphy (Figures 16, 17). Assuming that the cycles are stratigraphic time units equal to 20,000 years, this technique graphically portrays long-term deviations from mean subsidence (approximately 6 cm/k.y.). The background subsidence is assumed to be constant, or at least changing at a very slow rate over the time span of the Latemar buildup. Deviations in third-order accommodation are generated by the cumulative effect of progressive changes in fifth-order cycle thicknesses, which are plotted successively against time. The Fischer diagram in Figure 17 is based on a continuous record of 488 fifth-order cycles (individual fifth-order cycles plot as discrete triangles) and illustrates deviations from mean subsidence. Positive deviations (intervals rising to the upper right of the diagram) depict increased third-order accommodation. Negative

Figure 12. Typical example of the stratigraphic characteristics of the Lower Platform Facies of the Latemar buildup approximately 50 m beneath the base of the Lower Cyclic Facies. The part of the measured section from which the example was taken is shown by converging dashed lines. TST = transgressive systems tract; HST = highstand systems tract; SB = sequence boundary; mfs = maximum flooding surface.

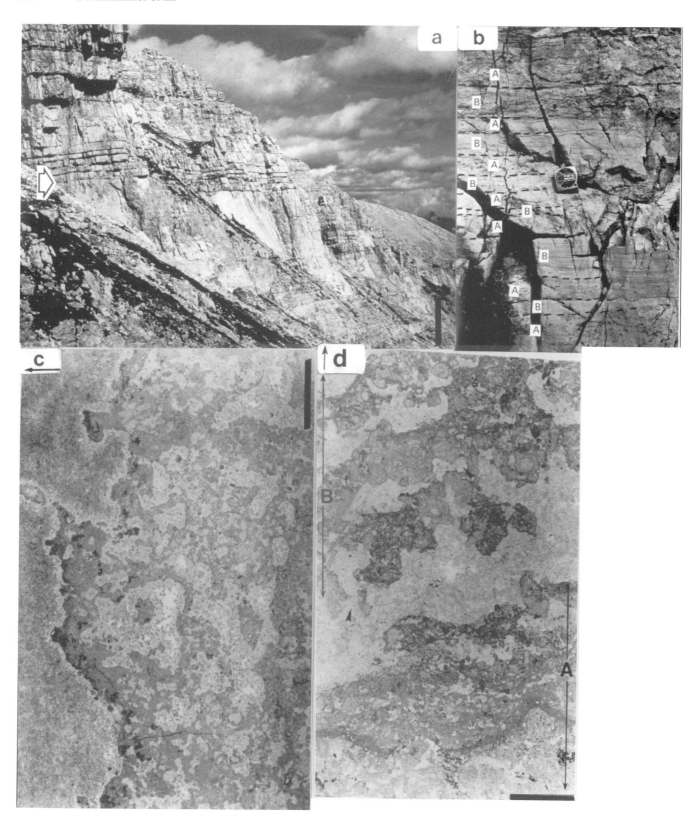

deviations (intervals falling to the lower right) mark a decrease in third-order accommodation. This curve for the lower cyclic facies through upper cyclic facies stratigraphy, which forms a third-order accommodation cycle of positive deviations (thick accumulation) giving way to negative deviations (thin accumulation), passing in turn back to positive deviations, substantiates the qualitative stratigraphic trends outlined above. Note that the position of the sequence boundary which caps the Lower Ladinian sequence (termed the mid-Ladinian sequence boundary in Figure 17) plots along the inflection point of the overall trend. When data from the lower platform facies is added (figure 16 in Goldhammer and Harris, 1989), the third-order accommodation trend for the entire Upper Anisian–Ladinian section forms a cycle of relative rise and fall, which Goldhammer and Harris (1969) interpreted as a eustatic sea-level change with an amplitude of approximately 60 m (after correction for isostatic loading) and a period of about 11 m.y.

The nature of the vertical changes in cycle stacking patterns depicted by the actual Latemar succession can be clearly illuminated using a "Mr. Sediment" computer simulation in which a fifth-/fourth-order composite wave has been superimposed on a lower-frequency third-order wave under conditions of constant platform sedimentation and long-term subsidence. The simulated sea-level subsidence-sedimentation history is shown in Figure 18. For this run, the rate of third-order eustatic fall is less than the rate of background platform subsidence (run data are summarized in the caption to Figure 18). The resultant simulated stratigraphy, displayed in Figure 18, is

remarkably similar to the cyclic stratigraphy of the Latemar (Figure 16), although on a reduced time scale of 3 m.y. instead of the actual 11 m.y. (selected for clarity of graphic presentation).

The simulation demonstrates the manner in which high-frequency cycles (fourth- and fifth-order) systematically change their stacking patterns (i.e., their thicknesses, subfacies characteristics, and early diagenetic attributes) across a third-order depositional sequence. This "generic" third-order sequence is thus characterized by a systematic vertical change in cycle and megacycle stacking patterns (i.e., thinning and thickening) accompanied by a systematic change in submergence-dominated versus emergence-dominated cycles and megacycles, precisely analogous to the pattern observed at the Latemar (Figure 18). It is these systematic changes in fourth- and fifth-order cycle thicknesses that generate the relative rises and falls in the long-term Fischer plot (Figure 17). These simulated cyclostratigraphic patterns are replicas, at least in qualitative terms, of the patterns observed at the Latemar.

Summary: A Composite Eustasy Model for the Systematic Succession of Latemar Facies

Based on the characteristics of each of the Latemar platform facies, the overall facies stratigraphy, and on our computer simulations (Goldhammer, 1987; Goldhammer et al., 1987, 1990; Dunn, 1991), a eustatic model for the development of the Latemar vertical facies succession and cycle stacking patterns can be formulated as follows:

1. Lower Platform Facies (Figure 12)—Prolonged, relatively rapid third-order sea-

Figure 13. Lower Platform Facies. (a) Outcrop photograph showing contact (arrow) between the well-bedded Lower Platform Facies and the massive-weathering Lower Cyclic Facies at the south–southwestern end of the Latemar platform. Note how the Lower Cyclic Facies cycles stand out as a result of the recessive weathering of the dolomitic caps. The faint thin "bedding" of the Lower Platform Facies is really a diagenetic layering resulting from closely spaced sheet cracks (see b). The arrow marks the top of the transgressive systems tract. Scale bar = 5 m. (b) Outcrop photograph of apparent bedding in the Lower Platform Facies made up of cement-filled sheet cracks (B) in massive grainstones (A). Scale = 5 cm. (c) Thin-section photomicrograph of intensely burrowed and/or bored peloidal packstone with a capping hardground surface. The convolute, irregular, and scalloped hardground morphology suggests bioerosion of early lithified sediment. Note 20–25 micron diameter dark dense micritic clots and tubes along the periphery, interpreted to result from micro borers. Sample taken from the maximum flooding surface marked by an arrow in (a) above. Scale bar = 4 mm. (d) Thin-section photomicrograph of bored, early cemented (submarine) peloidal packstone (A) and skeletal grainstone (B) separated by an elongate mm-scale diagenetic sheet crack filled with spherulitic-botryoidal calcite cement. Note closely spaced hardground surfaces in A and inverted borings (arrows) in layer B. Scale bar = 3 mm.

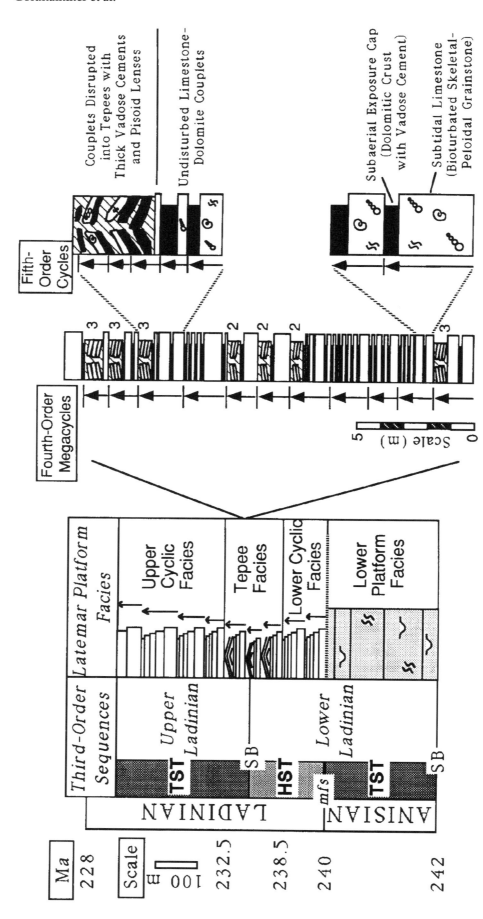

Figure 14. Typical example of the stratigraphic characteristics of the Tepee Facies of the Latemar buildup. Numbers to the right of individual tepees indicate number of fifth-order cycles disrupted into tepee zone. The part of the measured section from which the example was taken is shown by converging dashed lines. TST = transgressive systems tract; HST = highstand systems tract; SB = sequence boundary; mfs = maximum flooding surface.

level rise dominated the system maximizing third-order accommodation potential, maintaining subtidal conditions over the entire platform. Subtidal sedimentation was unable to quite keep up and thus not every beat of fifth-order, high-frequency sea level can "touch down" on the platform top and subaerially expose the top of the sediment column. These "missed beats" of subaerial exposure result in the formation of thick (average approximately 10 m), fourth-order **amalgamated megacycles**. During this phase of subtidal carbonate aggradation, subaerial exposure occurs infrequently, and lengthy periods of marine submergence promote abundant syndepositional marine diagenesis.

2. Lower Cyclic Facies—During development of the lower cyclic facies, the rate of third-order rise progressively declines as the crest of the third-order wave is approached, resulting in reduced third-order accommodation potential. Now, subtidal sedimentation can keep up with net sea-level change which, near the third-order crest, is due almost entirely to subsidence. Under these conditions, every fifth-order, high-frequency oscillation can subaerially expose the top of the sediment column and form a legible cycle boundary. Thus, complete five-part, fourth-order, **rhythmic megacycles** with roughly equivalent durations of submergence and exposure were formed. The progressive slowing of third-order rise results in sequentially thinner cycles upward from the base to the top of the lower cyclic facies.

3. Tepee Facies (Figure 14)—During the evolution of the tepee facies, third-order sea level was falling and third-order accommodation potential was minimized, but in a realm of continuous subsidence the net effect is close to, but not quite that of, a stillstand. Now sedimentation easily kept up and very thin cycles form (thinner than those of the lower cyclic facies), which are grouped into fourth-order **condensed megacycles** with tepee tops representing more lengthy periods of subaerial exposure. The tepees formed as a result of prolonged subaerial exposure due to one or more "missed beats" of submergence at the lowstand stage of a fourth-order oscillation. The Lower Ladinian Latemar third-order sequence (Figure 6) is capped by a platform-wide complex interval (15 m thick) of numerous thin, superimposed tepee zones in the middle of the tepee facies, which we interpret as the third-order sequence boundary to the Lower Ladinian sequence (Figure 19). With a decline in the rate of third-order fall, the rate of decreasing accommodation declines and

thicker, tepee-capped, condensed megacycles form the upper half of the tepee facies and the start of another third-order sequence.

4. Upper Cyclic Facies (Figure 9)—As the third-order sea-level fall that deposited the tepee facies bottomed out, the accommodation potential increased again. This resulted from a near zero rate of change of third-order sea level in the trough of the third-order wave so that subsidence now controls the accumulation space. The upper cyclic facies therefore is a replica of the lower cyclic facies formed under similar conditions that produces **rhythmic megacycles**, but this time in the trough rather than the crest of the third-order sea-level oscillation. In keeping with our eustatic model, the upper cyclic facies succession exhibits thicker cycles and upward-thickening of rhythmic megacycles as the third-order sea level begins a new oscillation near the end of the Ladinian. The upper boundary of the upper cyclic facies has been removed by Quaternary erosion of the top of the Latemar buildup, but other coeval buildups are capped by a regionally widespread unconformity and overlain by volcanics and volcaniclastics related to a late Ladinian–early Carnian volcano-tectonic event that catastrophically terminated the Anisian–Ladinian phase of carbonate buildups in the Southern Alps (Bosellini, 1984). This sequence boundary which caps the Upper Ladinian sequence (Figure 6) displays erosional truncation and lacks the systematic cycle stacking patterns associated with the Lower Ladinian sequence.

SEQUENCE STRATIGRAPHY AND SYSTEMS TRACT DEVELOPMENT OF THE LATEMAR BUILDUP: THE LINK BETWEEN CYCLIC STRATIGRAPHY, FACIES STRATIGRAPHY, AND SEQUENCE STRATIGRAPHY

The Latemar sequences (Figure 6) provide an optimum stratigraphic framework for integrating m-scale cyclostratigraphy and facies stratigraphy with seismic-scale sequence stratigraphy (Vail et al., 1984; Vail, 1987; Van Wagoner et al., 1987; Sarg, 1988). This is significant because most analyses of basin stratigraphy utilizing sequence stratigraphy concepts focus primarily on the third-order scale sequences (1–10 m.y. duration; a few to several hundred meters in thickness) and typically seismic data cannot resolve the higher frequency stratigraphic building blocks (i.e., the fourth- and fifth-order stratigraphic cycles). Thus, the detail of the internal architecture of sequences and systems tracts is often unresolved

(e.g., Sarg, 1988; Eberli and Ginsburg, 1989). As we have demonstrated (Figures 16–19), the stacking patterns of high-frequency cycles (fourth- and fifth-order) aid in determining changes in long-term accommodation, and therefore we can use the high-frequency cycle stacking patterns of third-order sequences to define the systems tracts of the sequence (Vail, 1987; Van Wagoner et al., 1987).

Anisian Sequences at the Latemar

Within the central Dolomites, two Anisian sequences are recognized (Figure 6), a Lower Anisian sequence (sequence A1) and an Upper Anisian sequence (sequence A2), which consist of blanket carbonates which formed low-relief carbonate banks (Bosellini, 1984). They are divided by a regional unconformity surface recognized in shallow-water platform sections by the occurrence of a widespread exposure surface or a coarse-grained siliciclastic unit (the Richthofen Conglomerate).

The Lower Anisian sequence (sequence A1; Serla Formation) is locally missing in the central Dolomites due to erosion over topographically elevated regions. Here, the Richthofen Conglomerate overlies the Lower Triassic Werfen Formation (for example, at the Catinaccio). The Contrin Formation is the platform facies of the Upper Anisian sequence (sequence A2) in the central Dolomites, while the basinal facies have a variety of names (Harris, 1988). The intervening slope facies (observed at the Latemar and Catinaccio buildups) contains large slump and channel features dramatically recording the presence of some syndepositional relief. These low-relief carbonate banks nucleated on localized upthrown basement blocks, formed by small normal faults (Doglioni 1984a,b). The Anisian age of the structural features is indicated by the fault geometry, the distribution and clast composition of conglomerates around the highs, and the off-structure thickening of the more basinal facies. These fault-bounded uplifts appear related to development of a sinistral strike-slip fault system (Doglioni, 1984a,b).

Within the Latemar buildup, the Lower Ladinian sequence developed atop a widespread, low-relief Anisian carbonate bank (the Contrin Formation =

depositional sequence A2) and is bounded at its base by a subaerial unconformity. This top Contrin sequence boundary (sequence boundary A2) can be physically walked out around the periphery of the Latemar (Figures 7, 8). As noted above, the Contrin bank (sequence A2) underlies many of the Ladinian buildups (e.g., the Sciliar/Catinaccio complex; Figure 8). Figure 20 provides a spectacular cross sectional panorama of the southwestern face of the Latemar viewed from a distance of approximately 3 km. This platform-to-basin transect can be compared to our reconstructed true-scale cross section of Figure 6. At the Latemar, the nature of this surface (i.e., sequence boundary A2) is well exposed.

Along the platform-to-basin transect illustrated in Figure 20, we can observe the contact separating sequence A2 from the Lower Ladinian sequence in three distinct depositional settings. Figure 21 illustrates a basinal position where the surface is a flooding surface, marked by toe-of-slope carbonates and shales of the Livinallongo Formation resting unconformably atop the massive Contrin bank. There is no evidence for subaerial exposure at that surface at this locality. Rather, this boundary is characterized by m-scale channels, infilled with slumped carbonate debris and megabreccias, incised into the slope facies of the Contrin bank. From a distance, this stratal relationship marks a "downlap surface" (Figure 20). Figure 22 portrays proximal coarse megabreccia foreslope clinoforms resting in angular discordance atop shallow-water bank cycles of the Contrin. At this locality, flat-lying, m-scale bank interior cycles thin upward into a master exposure boundary with evidence for subaerial exposure (calichification). Thus the A2 surface here is a subaerial sequence boundary downlapped by megabreccia clinoforms. Figure 23 displays the lower platform facies of the Lower Ladinian sequence directly resting on flat-lying interior bank cycles that thin upward into the master sequence boundary (sequence boundary A2), once again marked by a thin cm-scale caliche. Above the surface, the basal lower platform facies is markedly noncyclic, represented by amalgamated deeper subtidal deposits.

The shallow-water, interior bank cycles which thin upward into the A2 surface reflect decreasing third-

Figure 15. Tepee Facies. (a) Outcrop photograph of Tepee Facies in the Latemar buildup showing tepee zones interrupting the evenly bedded, flat-lying, meter-scale, fifth-order cycles. Note the thinner tepee zones (small arrows) which cap individual fourth-order megacycles, and the thicker more complex zones (large arrows), which incorporate more than one fourth-order megacycle. Scale bar = 10 m. (b) Outcrop photograph of a Latemar tepee zone which consists of a few stacked, offset tepees. Tepees have symmetrical chevronlike cross sections and a flat base, and consist of uparched layers inclined about a central, cement-filled tepee core. Scale bar = 1 m.

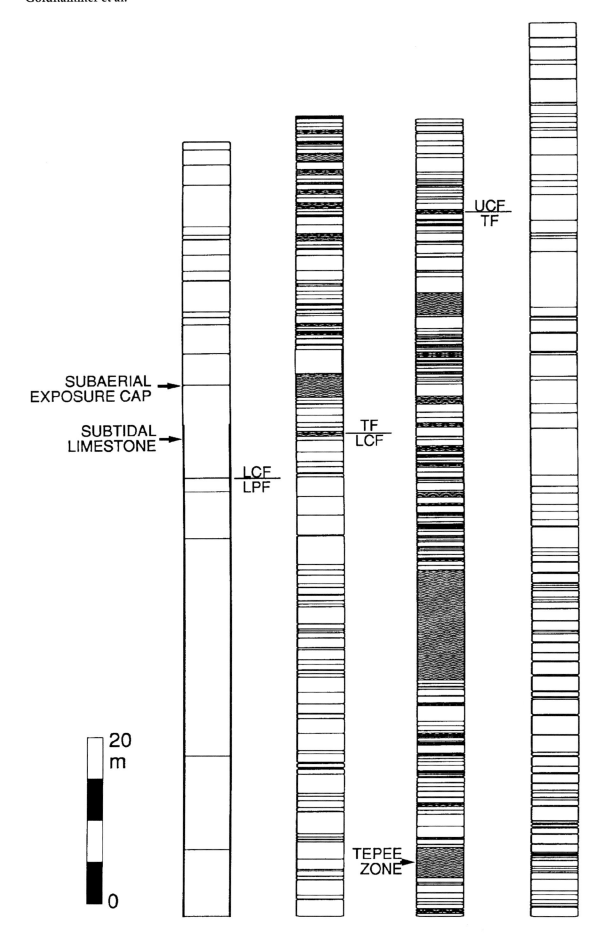

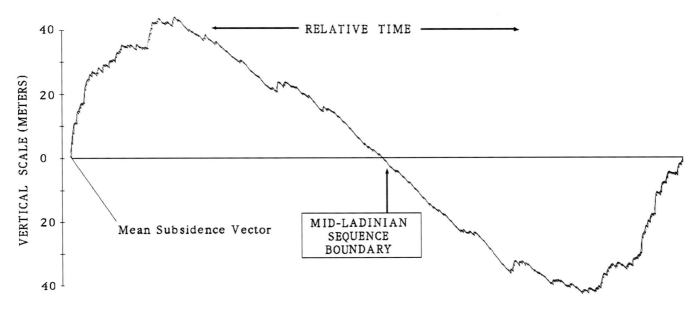

Figure 17. Fischer diagram based on a continuous record of 488 fifth-order cycles, with individual fifth-order cycles plotting as discrete triangles, illustrating deviations from mean subsidence. Note position of mid-Ladinian sequence boundary at inflection point on the falling limb of the third-order accommodation trend.

order accommodation of the underlying A2 sequence. However, above the A2 surface platform–interior facies significantly retrograde such that in a downdip setting, basinal and foreslope facies are superimposed atop the underlying shallow-water bank at the Latemar exterior. In an updip position, platform interior deposits of the lower platform facies abruptly deepen. These observations indicate a significant third-order relative rise in sea level initiating the Lower Ladinian sequence, with rates of rise that outpaced normal banktop accumulation, resulting in abrupt back-stepping of shallow-water platform facies. This contrasts markedly with the Lower Ladinian sequence boundary in the middle of the tepee facies (Figure 19), where a symmetrical turnaround in the high-frequency stacking patterns is observed. Thus, depending on paleogeographic position, the nature of the A2 surface varies from a subaerial sequence boundary to a subaqueous flooding surface. Viewed from a distance (i.e., at seismic scale), the A2 surface appears to be a "downlap surface" beneath prograding toe-of-slope and proximal fore-

slope facies. However, beneath the flat-lying lower platform facies it has no perceptible geometric stratal patterns, and rather than an overlying downward shift in facies, there is an "upward" shift. At the Latemar, lowstand deposits in a more basinal setting (lowstand wedge, basin-floor fan, etc.) associated with the A2 sequence boundary, are notably absent, perhaps due to the lack of a clastic source and the isolated nature of the Latemar.

Ladinian Sequences at the Latemar

The transgressive systems tract of the Lower Ladinian sequence is delineated by the lower platform facies (Figure 6). At the Latemar, as well as at the coeval Catinaccio buildup, in downdip positions, the Contrin Formation (sequence A2) is overlain by dark, laminated siliceous mudstone and pelagic carbonates of the Livinallongo Formation, recording the increase in accommodation associated with a rapid relative sea-level rise (Figures 8, 20). Within the transgressive systems tract of the Latemar platform interi-

Figure 16. High-frequency cyclostratigraphy of the Latemar platform carbonates based on several measured sections correlated by lateral tracing of key cycles. This composite section includes the uppermost part of the Lower Platform Facies (lower platform facies), the complete Lower Cyclic Facies (lower cyclic facies), the complete Tepee Facies (tepee facies), and the lower part of the Upper Cyclic Facies (upper cyclic facies). The stacking progression from amalgamated megacycles at the base (lower platform facies) through rhythmic megacycles (lower cyclic facies) to condensed megacycles (tepee facies) and back to rhythmic megacycles (upper cyclic facies) is well displayed in this approximately 200-m thick section. Cycles and packages of cycles disrupted by tepees (tepee zones) are shown in zig-zag pattern. The thick tepee zone in the middle of the Tepee Facies is a 15 m thick complex of tepee-capped cycles highly disrupted by multiple antiformal buckling and vadose cementation.

Stratigraphic Column

**Sea Level - Sedimentation -
Subsidence History**

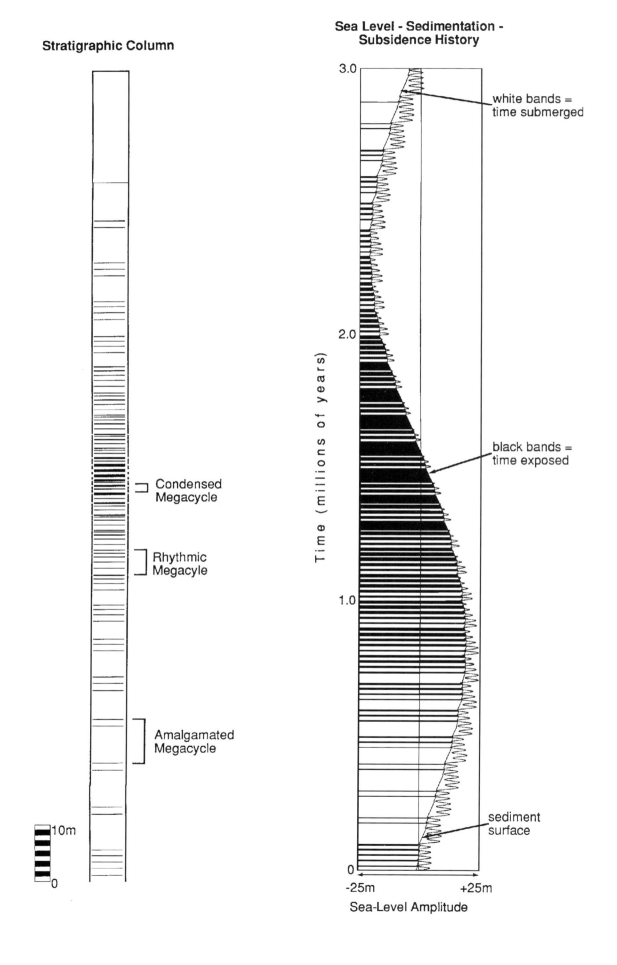

white bands =
time submerged

black bands =
time exposed

sediment
surface

Condensed
Megacycle

Rhythmic
Megacyle

Amalgamated
Megacycle

10m

0

3.0

2.0

Time (millions of years)

1.0

0

-25m   +25m

Sea-Level Amplitude

or, amalgamated megacycles vertically comprise an aggradational upward-deepening stacking pattern, as reflected by a general thinning of beds towards the top of the lower platform facies, as well as an increase in the intensity of early marine diagenesis marking a progressive decline in sedimentation rates related to increased paleowater depths. During deposition of the transgressive systems tract, the platform margin reef aggraded vertically. At the same time, the talus foreslopes downlapped the starved basin deposits of the lower Livinallongo Formation around the flanks of the buildup (Bosellini and Rossi, 1974; Bosellini, 1984; Harris, 1988).

The changeover from the transgressive systems tract to the highstand systems tract is marked by a dramatic change in cycle and megacycle stacking patterns in the platform interior of the Latemar buildup. Notably, fourth-order amalgamated megacycles of the lower platform facies give way to rhythmic fourth-order megacycles of the lower cyclic facies (Figures 8, 13a, 20). Thus, as demonstrated by our "Mr. Sediment" simulation of the Latemar cyclostratigraphy (Figure 18), this changeover from amalgamated to rhythmic megacycles at the top of the lower platform facies approximates the maximum flooding surface. This maximum flooding surface picked in a platform interior position cannot be traced laterally with confidence through margin and foreslope facies and must be located somewhere within the basinal Livinallongo Formation (Figure 6). In Figure 6, the maximum flooding surface is schematically "forced through" the foreslope and basinal facies because we surmise that the equivalent stratigraphic surface must exist somewhere within these facies, but we cannot pinpoint it precisely, and certainly there is no abnormal "marker" bed within these facies that we can identify. Most significantly, the maximum flooding surface as picked in a platform position does not equate to the downlap surface recognized downdip, as this downlap surface is simply a lateral facies change, which in detail comprises a narrow zone of interfingering grainstones (proximal turbidites) and breccia tongues (downdip ends of talus falls; Harris, 1988). The interfingering occurs over a scale of a few meters to perhaps several tens of meters. Additionally, the downlap surface is marked-

ly diachronous as the downlap geometry extends through two sequences (Lower Ladinian and Upper Ladinian) from the onset of platform growth to its demise (Figure 6).

Traditional sequence models (Vail, 1987; Sarg, 1988) link the geometrically defined downlap surface with a condensed interval. At the Latemar, the Livinallongo Formation is essentially the "condensed interval" (in the sense of Vail, 1987) and is a starved basin section (200 m thick) equivalent to both the Lower and Upper Ladinian sequences (700–1000 m thick on the platform). It consists of a slowly deposited mixture of distal carbonate turbidites, pelagic lime muds, thin shales, and volcanic fragments (Bosellini and Ferri, 1980; Harris, 1988). The base of this condensed interval is Upper Anisian (Cros, 1974), equivalent to the lower platform facies.

The lower Livinallongo Formation (Plattenkalk Member; Bosellini and Ferri, 1980; Bosellini, 1984) is dominated by shales and argillaceous carbonates. The carbonate content increases upward through the nodular limestones of the middle Livinallongo (Knollenkalk member; Bosellini and Ferri, 1980; Bosellini, 1984) into the thin bedded, redeposited carbonates of the upper Livinallongo (Banderkalk member; Bosellini and Ferri, 1980; Bosellini, 1984), the top of which is equivalent to the graded grainstone facies of the Latemar foreslope (Harris, 1988). The entire unit is capped by the toes of the prograding foreslopes of the Ladinian sequences or Upper Ladinian volcanics (Figure 6; Bosellini, 1984; Harris, 1988). The three Livinallongo members are, at least in part, lateral facies equivalents with increasing carbonate content adjacent to the prograding foreslopes. Thus, the condensed interval is a diachronous stratigraphic marker, which contains the maximum flooding surface.

The highstand systems tract of the Lower Ladinian sequence may be subdivided into an early highstand phase (the lower cyclic facies with rhythmic fourth-order megacycles) and a late highstand phase (the lower half of the tepee facies with condensed fourth-order megacycles). Within the highstand systems tract, the style and abundance of early diagenesis changes dramatically from that of the transgressive systems tract as early marine diagenetic features are virtually absent and instead subaerial exposure fea-

Figure 18. Simulation of the submergence–emergence record on a carbonate platform of composite high-frequency sea-level oscillations (20,000 yr > 100,000 yr) when superimposed on a third-order (3 m.y.) sea-level change using the "Mr. Sediment" computer program. Resulting cyclostratigraphy is shown on the left of the figure. On the third-order rise, some of the high-frequency "beats" fail to expose the platform top resulting in the deposition of a "missed beat" stratigraphy of amalgamated megacycles. On the third-order fall, "missed beats" result from failure of some of the high-frequency oscillations to flood the platform, producing condensed megacycles. At the crests and troughs of the third-order curve, the conditions are just right for submergence and emergence of the platform with every successive high-frequency oscillation, resulting in rhythmic megacycles ("Goldilocks" megacycles). Input data: sedimentation rate = 0.1 m/k.y.; subsidence rate = 0.05 m/k.y.; sinusoidal 20 kyr wave with 3 m amplitude; asymmetrical 100 kyr wave with 2 m amplitude; third-order wave with 40 m amplitude and 3 m.y. duration. From Goldhammer et al. (1990).

Figure 19. Outcrop photograph of fifth-order cycle and fourth-order cycle stacking patterns from the top of the Upper Cyclic Facies into the lower half of the Tepee Facies. Note the progressive thinning-upward stratal patterns of cycles that pass into a 15 m thick tepee complex, and the thickening-upward patterns above the complex. The mid-Ladinian sequence boundary is placed within the middle of the complex, in the absence of any one particular significant physical surface with evidence for erosion, etc.

tures are the rule. Higher frequency cycles form an aggradational, upward-shallowing, stacking pattern as reflected by an overall thinning of fifth-order cycles and increase in the intensity of early subaerial diagenesis (i.e., thicker caliche caps to cycles and subaerial tepees) from the base of the lower cyclic facies on up

into the tepee facies. The reef margin of the highstand systems tract reef vertically aggraded but supplied increasing amounts of eroded talus debris to the foreslope (Harris, 1988; Goldhammer and Harris, 1989).

The first Latemar sequence (the Lower Ladinian sequence) is capped by a complex 15 m thick strati-

Figure 21. Outcrop photograph of Contrin bank–Livinallongo contact. The top Contrin bank (sequence boundary A2) is marked by toe-of-slope carbonates and shales of the Livinallongo Formation resting unconformably atop the massive Contrin bank. From a distance, this stratal relationship marks a "downlap surface." See Figure 20 for location of photograph. Height of Contrin bank is approximately 60 m.

graphic interval which occurs in the middle of the tepee facies, consisting of numerous very thin tepee complexes (< 0.5-1.5 m thick) vertically superimposed one atop the other, giving the appearance of one very thick tepee zone (Figure 19). Beneath this interval, fifth-order cycles and fourth-order megacycles are overall thinning upward; above this interval, they are overall thickening upwards into the overlying upper half of the tepee facies and the overlying upper cyclic facies. Thus this interval, which records the shift from continuously decreasing to continuously increasing third-order accommodation (e.g., Figure 18), must contain the upper sequence boundary to the first Latemar sequence (Figure 6). This upper sequence boundary is characterized by pronounced subaerial exposure, which in detail is the superimposition of numerous higher frequency subaerial exposure surfaces. The boundary lacks subaerial erosion and cannot be demonstrably linked downdip with any lowstand deposit in the foreslope environment. It is interpreted to have formed where the rate of third order eustatic sea-level fall was less than the rate of subsidence of the platform. This interpretation is sub-

stantiated by inspection of our third-order Fischer diagram (Figure 17; note the rate of fall on the deviation curve—interpreted as eustasy—is less than the rate of mean subsidence). This capping boundary is thus a conformable boundary (on the scale of biostratigraphic resolution) composed of numerous closely spaced (in time, as well as in stratigraphic space) disconformities or diastems, lacking a correlative unconformity. J.F. Sarg (personal communication) has suggested that this boundary is a type 2 sequence boundary (in the sense of Vail, 1987; Sarg, 1988) and that the overlying remaining portion of the tepee facies is a shelf margin wedge systems tract.

Recalling our "generic" third-order Latemar stratigraphic simulation (Figure 18), the third-order sequence boundary formed along the third-order inflection point on the eustatic fall occurs within the interval of thinnest condensed megacycles and is placed at the time of maximum rate of long-term accommodation decrease, which is marked stratigraphically by the turnaround from fourth-order megacycles which progessively thin upwards to fourth-order megacycles which progressively thicken

upwards (Figure 18). Thus, the simulated sequence boundary, similar in nature to that which caps the platform, is **stratigraphically transitional** without a basinward facies shift and is located on the basis of stacking patterns of the higher frequency fifth- and fourth-order megacycles. There is no singular, long-term (i.e., greater in duration than the period of a few fifth-order oscillations) erosional unconformity generated in the simulation because the rate of third-order sea-level fall is less than the rate of background platform subsidence.

Goldhammer et al. (1990) have suggested that this Latemar-type sequence boundary with transitional instead of unconformable boundaries will be a particular attribute of shallow water carbonates because of the characteristic flat-topped geometry of most ancient and existing carbonate buildups (see also Hardie, 1989). It is reasoned that with such flat-topped platforms, any slow third-order sea-level fall (i.e., rate of fall less than the rate of subsidence) will be transmitted with essentially equal effect across the entire, or almost entire, buildup. This, in turn, would allow the deposition of "conformity-bounded" sequences across the entire, or almost entire, buildup. Such conformity-bounded sequences would not be identified in seismic sections because of the lack of seismic discontinuities at their boundaries (i.e., lack of erosion in a pure carbonate system), and the lack of pronounced stratal geometries (termed the mid-Ladinian sequence boundary in Figures 19 and 24). This sequence boundary capping the Lower Ladinian sequence contrasts with the basal sequence boundary (sequence boundary A2; Figures 20 and 23) in terms of the symmetry of the stratal architecture. The base of the lower platform facies underwent a deepening with renewed third-order rise, much more rapid than that which occurred above the upper mid-Ladinian sequence boundary.

Along the margin of the platform, localized erosion at the sequence boundary removed the boundstone facies so that foreslope breccias abut directly against the platform interior grainstones (Figure 25). The scour is several tens of meters deep and cuts laterally through 10–20 m of section. It occurs along the west edge of the platform at the horizon of the thick tepee zone, which marks the sequence boundary in the platform section. This interruption sharply contrasts with the consistent 20–40 m wide reef zone observed throughout the rest of the margin facies. In the foreslope breccia facies, lithified platform clasts persistently occur in low abundance at horizons equivalent to this boundary, indicating localized scour through

the reef margin. However, evidence of subaerial exposure of the reef margin is lacking, supporting the interpretation that the sequence boundary was generated without sea level falling significantly below the platform edge. Erosion of the reef margin occurred throughout the buildup evolution and supplied large quantities of sediment to the foreslope talus piles. The reduced accommodation which marks the sequence boundary probably resulted in increased erosion of the margin and platform edge due to increased wave and current effects.

Above the sequence boundary to the Lower Ladinian sequence (Figure 6), the Upper Ladinian sequence is incomplete owing to Alpine erosion of the top of the platform. In other coeval buildups throughout the Dolomites, Late Ladinian volcanism and tectonism terminated carbonate platform deposition. This partial sequence consists of the upper half of the tepee facies and the overlying upper cyclic facies. Only the transgressive systems tract of this Upper Ladinian sequence is preserved at the Latemar. The rate of third-order accommodation increase associated with this next sequence was apparently much less than that of the Lower Ladinian sequence, and the transgressive systems tract is characterized by the systematic vertical shift from condensed fourth-order megacycles of the upper portion of the tepee facies to rhythmic megacycles of the upper cyclic facies. The rather abrupt interruption of carbonate deposition by the Late Ladinian tectono-volcanic event just as third-order sea level was on the rise has considerable significance for the interpretation of the unconformity most workers place at the Ladinian–Carnian boundary. Clearly, if our cycle stacking pattern analysis is correct, then this unconformity cannot be a type 1 sequence boundary due to a eustatic fall of third-order sea level, as proposed by Haq et al. (1987) and Sarg (1988, his figure 23), and furthermore should not be used in reconstruction of global sea-level changes during the Triassic (e.g., Haq et al., 1987). Rather, it is the more subtle high-frequency sea-level changes within the conformable Upper Anisian–Lower Ladinian platform carbonates of the Latemar that reveal a third-order eustatic cycle, which is notably absent in the global sea-level curves of Haq et al. (1987) and Sarg (1988, his figure 23).

The buildup margins of the Upper Ladinian sequence are highly progradational (Figure 6; also observed at the Mount Agnello buildup and the Catinaccio), in contrast to the vertical geometry of the Lower Ladinian sequence. The progradation of plat-

Figure 22. Outcrop photograph of proximal coarse megabreccia foreslope clinoforms (inclined white arrows) resting in angular discordance atop shallow-water bank cycles of the Contrin bank (gray line; top sequence A2). At this locality, flat-lying, m-scale bank interior cycles thin upward into a master exposure boundary with evidence for subaerial exposure (vertical white arrow). Viewed from a distance, the A2 surface here appears as a "downlap surface." See Figure 20 for location of photograph. Scale bar = 5 m.

Figure 23. Outcrop photograph of the basal Lower Platform Facies of the Latemar platform interior resting on flat-lying interior bank cycles that thin upward into the master sequence boundary (dark gray line; sequence boundary A2). Above the surface, the basal lower platform facies is markedly noncyclic, represented by amalgamated deeper subtidal deposits. See Figure 20 for location of photograph. Scale bar = 10 m.

form facies over the foreslopes required infilling of the basin margin. Talus deposits which comprise the foreslopes of both sequences show no discernible change in character (Harris, 1988). This geometry of enhanced late progradation required a pronounced increase in the volume of foreslope material. We speculate that two factors may have contributed to the change in margin geometry: (1) the volcaniclastics within the Livinallongo significantly infilled the basin; and/or (2) large volumes of talus were added to the foreslope during the Upper Ladinian sequence. Both factors would enhance platform progradation, but which, if either, occurred cannot be demonstrated from the available outcrops.

Figure 24. Outcrop photograph of the Latemar interior platform, viewed from the south looking north. The "mid-Ladinian" sequence boundary that caps the Lower Ladinian sequence shows no pronounced stratal geometries and is essentially imperceptible within the thick stack of Latemar platform carbonates. The stratigraphic elevation from the top of the Lower Platform Facies to the "mid-Ladinian" sequence boundary is approximately 160 m. Upper white line marks the facies change from platform (foreground) to foreslope facies (background), which are dipping into the plane of the photograph. MFS = maximum flooding surface.

CONCLUSIONS

This analysis of the Latemar sequence stratigraphy reveals several significant aspects about carbonate platform responses to third-order sea-level oscillations that are different from the standard sequence stratigraphic model developed from siliciclastic shelf-to-basin transitions (e.g., Haq et al., 1987; Vail, 1987).

1. Third-order sequence boundaries may be transitional zones rather than sharp unconformities and recognizable only by cycle stacking patterns, a feature that could cause sequences to be overlooked. For example, in his analysis of the Triassic sequence stratigraphy of the Dolomites, Sarg (1988, his figure 23) did not recognize the two Upper Anisian–Ladinian sequences we have documented at the Latemar. Specifically, Sarg (1988), relying on seismic-scale techniques, failed to identify the upper sequence boundary of the Lower Ladinian sequence, as it is a less obvious "stacking pattern" boundary.

2. At these transitional boundaries (e.g., the upper boundary of the Lower Ladinian sequence), there will be (a) no downward shift in overlying facies, (b) no lowstand wedge in the downdip position, particularly in the case of isolated buildups such as the Latemar, and (c) no erosional hiatus.

3. In a platform position, the maximum flooding surface separating the transgressive systems tract from the highstand systems tract will not correlate downdip to a downlap surface per se but will simply record a cyclofacies change as the rate of third-order sea-level rise passes through its maximum value. The downlap surface in a downdip position is a diachronous surface, marked by a lateral facies change, which in detail comprises a narrow zone of interfingering grainstones (proximal turbidites) and breccia tongues (downdip ends of talus falls).

4. Superposition of composite fifth-/fourth-order ($10^4/10^5$ year) sea-level oscillations on a third-order (10^6 year) sea-level rise–fall cycle can produce a systematic succession of amalgamated —> rhythmic —> condensed megacycle facies and systematic changes in cycle stacking patterns, as is clearly demonstrated by the Latemar example. This succession allows a genetic connection to be made between cyclic stratigraphy, facies stratigraphy, and sequence stratigraphy.

ACKNOWLEDGMENTS

We would like to acknowledge Drs. Carlo Doglioni and Alfonso Bosellini (University of Ferrarra, Italy) for logistical support. In particular, Carlo Doglioni arranged the aerial overflight. The majority of this research was funded by Exxon Production Research Co. in the form of graduate student grants to R.K.

Figure 25. Outcrop photograph of the margin of the Latemar platform along the west edge of the platform at the horizon of the thick tepee zone which marks the sequence boundary in the platform section. View is to the south and clinoforms dip to the northwest. Scale bar = 50 m. White line marks the boundary between platform and foreslope facies.

Goldhammer, M.T. Harris, and P.A. Dunn. Ursula Hammes, Dave Osleger, and J.F. Sarg provided thorough and thoughtful reviews of the manuscript, and their efforts are appreciated.

REFERENCES CITED

Bechstadt, T., Brandner, R., Mostler, H., and Schmidt, K., 1978, Aborted rifting in the Triassic of the Eastern and Southern Alps: Neues Jarbuch fur Geologie und Palaontologie Abhandlungen, v. 156, p. 157-178.

Bernoulli, D., and Lemoine, M., 1980, Birth and early evolution of the Tethys: The overall situation: Memoire du Bureau de Recherches Geologiques et Minieres, No. 115, p. 168-179.

Blendinger, W., 1986, Isolated stationary carbonate platforms: The Middle Triassic (Ladinian) of the Marmolada area, Dolomites, Italy: Sedimentology, v. 33, p. 159-183.

Bosellini, A., 1984, Progradation geometries of carbonate platforms: Examples from the Triassic of the Dolomites, northern Italy: Sedimentology, v. 31, p. 1-24.

Bosellini, A., and Ferri, R., 1980, La Formazione di Livinallongo (Buchenstein) nella Valle di Pale di San Lucano (Ladinico Inferiore, Dolomite Bellunesi): Annali dell'Universita di Ferrara, nuova serie, Sez. IX, v. 6, p. 63-89.

Bosellini, A., and Rossi, D., 1974, Triassic carbonate buildups of the Dolomites, northern Italy: SEPM Special Publication No. 18, p. 209-233.

Clough, J. G., 1989, General stratigraphy of the Katakturuk Dolomite in the Sadlerochit and Shublik Mountains, Arctic National Wildlife Refuge, Alaska: State of Alaska Department of Natural Resources, Division of Geological and Geophysical Surveys, public data File 89-4a, 9 p.

Clough, J. G., and Goldhammer, R. K., 1992, Third-order vertical variations in parasequence character of the Lower Craggy Member, Katakturuk Dolomite (Proterozoic), northeastern Brooks Range, Alaska [abstract]: AAPG Bulletin, v. 76, p. 775.

Crevello, P., Sarg, J. F., Read, J. F., and Wilson, J.L., eds., 1989, Controls on Carbonate Platform to Basin Development: SEPM Special Publication No. 44, 405 p.

Cros, P., 1974, Evolution sedimentologique et paleo-ostructural de quelques platformes carbonates biogenes (Trias des Dolomites Italiennes): Sciences de la Terre, v. 14, p. 299-379.

Cros, P., and Lagny, P., 1969, Paleokarsts dans le Trias moyen et superieur des Dolomites et des Alpes Carniques occidentales, importance stratigraphique et paleogeographique: Sciences de la Terre, v. 14, p. 139-195.

Doglioni, C., 1984a, Tettonica Triassica transpressiva nelle Dolimiti: Giornale di Geologia (Bologna), v. 46, p. 47-60.

Doglioni, C., 1984b, Triassic diapiric structures in the Central Dolomites: Ecologae Geologicae Helvetiae, v. 77, p. 261-285.

Dunn, P. A., 1991, Diagenesis and cyclostratigraphy: An example from the Middle Triassic Latemar platform, Dolomites Mountains, northern Italy: unpublished Ph.D. dissertation, The Johns Hopkins University, Baltimore, Maryland, 865 p.

Dunn, P. A., Goldhammer, R. K., Hardie, L. A., and Nguyen, C. T., 1991, Two-dimensional forward modeling of Lower Ordovician platform carbonate sequences (Beekmantown Gp, central Appalachians): The search for high-frequency autocycles (abstract): AAPG Bulletin, v. 75, p. 565.

Eberli, G. P., and Ginsburg, R. N., 1989, Cenozoic progradation of northwestern Great Bahama Bank, a record of lateral platform growth and sea level fluctuations, in Crevello, P., Sarg, J. F., Read, J. F., and Wilson, J. L., eds., Controls on Carbonate Platform to Basin Development: SEPM Special Publication No. 44, p. 339-352.

Elrick, M., 1992, Cyclostratigraphy of dolomitized ramp deposits in the Middle Devonian Simonson Formation, Eastern Great Basin (abstract): AAPG Bulletin, v. 76, p. 775.

Elrick, M., and Read, J. F., 1991, Cyclic ramp-to-basin carbonate deposits, Lower Mississippian, Wyoming and Montana: A combined field and computer modeling study: Journal of Sedimentary Petrology, v. 61, p. 1194-1224.

Fischer, A. G., 1964, The Lofer cyclothems of the Alpine Triassic: Kansas Geological Survey Bulletin No. 169, p. 107–149.

Fois, E., 1982, The Sass da Putia carbonate buildup (western Dolomites): Biofacies succession and margin development during the Ladinian: Rivista Italiana di Paleontolgia e Stratigrafia, v. 87, p. 565-598.

Fois, E., and Gaetani, M., 1981, The northern margin of the Civetta buildup, evolution during the Ladinian and the Carnian: Rivista Italiana di Paleontolgia e Stratigrafia, v. 86, p. 469-542.

Gaetani, M., Fois, E., Jadoul, F., and Nicora, A., 1981, Nature and evolution of Middle Triassic carbonate buildups in the Dolomites (Italy): Marine Geology, v. 44, p. 25-57.

Goldhammer, R. K., 1987, Platform carbonate cycles, Middle Triassic of Northern Italy: The interplay of local tectonics and global eustasy: unpublished Ph.D. dissertation, The Johns Hopkins University, Baltimore, Maryland, 468 p.

Goldhammer, R. K., Dunn, P. A., and Hardie, L. A., 1987, High frequency glacio-eustatic sea level oscillations with Milankovitch characteristics recorded in Middle Triassic platform carbonates in northern Italy: American Journal of Science, v. 287, p. 853-892.

Goldhammer, R. K., Dunn, P.A., and L. A. Hardie, 1990, Depositional cycles, composite sea level changes, cycle stacking patterns, and the hierarchy of stratigraphic forcing: Examples from platform carbonates of the Alpine Triassic: Geological Society of America Bulletin, v. 102, p. 535-562.

Goldhammer, R. K., and Harris, M. T., 1989, Eustatic controls on the stratigraphy and geometry of the Latemar buildup (Middle Triassic), the Dolomites of northern Italy, in Crevello, P., Sarg, J. F., Read, J. F., and Wilson, J.L., eds., Controls on Carbonate Platform to Basin Development: SEPM Special Publication No. 44, p. 323-338.

Goldhammer, R. K., Lehmann, P. J., and Dunn, P. A., 1992, Third-order sequence boundaries and high-frequency cycle stacking patterns in Lower Ordovician platform carbonates, El Paso Group (Texas): Implications for carbonate sequence stratigraphy, in Candelaria, M. P., and Reed, C. L., eds., Paleokarst, Karst-Related Diagenesis, and Reservoir Development: SEPM, Permian Basin Section Publication No. 92-33, p. 59-92.

Goldhammer, R. K., Lehmann, P. J., Todd, R. G., Wilson, J. L., Ward, W. C., and Johnson, C. R., 1991a, Sequence stratigraphy and cyclostratigraphy of the Mesozoic of the Sierra Madre Oriental, northeast Mexico, a field guidebook: Gulf Coast

Section, SEPM Foundation, 85 p.

Goldhammer, R. K., Oswald, E. J., and Dunn, P. A., 1991b, The hierarchy of stratigraphic forcing: An example from Middle Pennsylvanian shelf carbonates of the Paradox Basin, in Franseen, E. K., Watney, W. L., Kendall, G. C. St. C., Ross, W., eds., Sedimentary modeling: Computer simulations and methods for improved parameter definition: Kansas Geological Survey Special Publication, p. 361-414.

Handford, C. R., and Loucks, R. G., 1991, Unique signature of carbonate strata and the development of depositional sequence and systems tract models for ramps, rimmed shelves, and detached platforms [abstract]: AAPG Bulletin, v. 75, p. 588-589.

Haq, B. U., Hardenbol, J., and Vail, P.R., 1987, Chronology of fluctuating sea levels since the Triassic: Science, v. 235, p. 1156-1166.

Hardie, L. A., 1989, Cyclic platform carbonates in the Cambro-Ordovician of the central Appalachians, in Walker, K. R., Read, J. F., and Hardie, L. A., (leaders), Cambro-Ordovician Carbonate Banks and Siliciclastic Basins of the United States Appalachians: 28th International Geologic Congress Field Trip Guidebook T161, p. 51-81.

Hardie, L. A., Bosellini, A., and Goldhammer, R. K., 1986, Repeated subaerial exposure of subtidal carbonate platforms, Triassic, northern Italy: Evidence for high frequency sea level oscillations on a 10,000 year scale: Paleoceanography, v. 1, p. 447-457.

Hardie, L. A., Wilson, E. N., and Goldhammer, R. K., 1991, Cyclostratigraphy and dolomitization of the Middle Triassic Latemar buildup, The Dolomites, northern Italy: Dolomieu Conference on Carbonate Platforms and Dolomitization, Guidebook Excursion F, Ortisei, Val Gardena, p. 1-37.

Harris, M. T., 1988, Margin and foreslope deposits of the Latemar carbonate buildup (Middle Triassic), the Dolomites, northern Italy: Ph.D. dissertation, The Johns Hopkins University, Baltimore, Maryland, 433 p.

Hunt, D., and Tucker, M., 1991, Responses of rimmed shelves to relative sea level rises; a proposed sequence stratigraphic classification (abstract): Dolomieu Conference on Carbonate Platforms and Dolomitization, Abstracts, p. 114-115.

Johnson, C. R., Ward, W. C., and Goldhammer, R. K., 1991, Mechanisms for high-frequency cyclicity in the Upper Jurassic limestone of northeastern Mexico (abstract): AAPG Bulletin, v. 75, p. 603.

Koerschner, W. F., III, and Read, J. F., 1989, Field and modelling studies of Cambrian carbonate cycles, Virginia Appalachians: Journal of Sedimentary Petrology, v. 59, p. 654-687.

Montanez, I. P., 1992, Controls of eustasy and associated diagenesis on reservoir heterogeneity in Lower Ordovician, Upper Knox carbonates, Appalachians, in Candelaria, M. P., and Reed, C. L., eds., Paleokarst, Karst-Related Diagenesis, and Reservoir Development: SEPM, Permian Basin Section Publication No. 92-33, p. 165-181.

Osleger, D. A., 1990, Cyclostratigraphy of Late Cambrian cyclic carbonates: An interbasinal field and modeling study, U. S. A.: unpublished Ph.D. dissertation, Virginia Polytechnic Institute and State University, Blacksburg, Virginia, 303 p.

Osleger, D. A., 1991, Subtidal carbonate cycles: Implications for allocyclic versus autocyclic controls: Geology, v. 19, p. 917-920.

Palmer, A. R., 1983, The decade of North American geology, 1983 geologic time scale: Geology, v. 11, p. 503-504.

Read, J. F., 1989, Controls on evolution of Cambrian-Ordovician passive margin, U. S. Appalachians, in Crevello, P., Sarg, J. F., Read, J. F., and Wilson, J. L., eds., Controls on Carbonate Platform to Basin Development: SEPM Special Publication No. 44, p. 147-166.

Read, J. F., Osleger, D. A., and Elrick, M., 1990, Computer modeling of cyclic carbonate sequences: Geological Society of America Short Course Manual, Dallas, Texas.

Sarg, J. F., 1988, Carbonate sequence stratigraphy, in Wilgus, C. K., Hastings, B. S., Kendall, C. G. St. C., Posamentier, H. W., Ross, C. A., and Van Wagoner, J. C., eds., Sea Level Changes: An Integrated Approach: SEPM Special Publication No. 43, p. 155-181.

Sarg, J. F., and Lehmann, P. J., 1986, Lower-Middle Guadalupian facies and stratigraphy San Andres/Grayburg formations, Permian Basin, Guadalupe Mountains, New Mexico, in Moore, G. E., and Wilde, G. L., eds., Field Trip Guidebook, San Andres/Grayburg Formations, Guadalupe Mountains, New Mexico and Texas: Permian Basin Section, SEPM Special Publication 86-25, p. 1-36.

Tucker, M. E., 1991, Sequence stratigraphy of carbonate-evaporite basins: Models and application to the Upper Permian (Zechstein) of northeast England and adjoining North Sea: Journal of the Geological Society, London, v. 148, p. 1019-1036.

Tucker, M. E., Wilson, J. L., Crevello, P. D., Sarg, J. F., and Read, J. F., 1990, Carbonate platforms, facies, sequences and evolution: International Association of Sedimentologists Special Publication No. 9, 328 p.

Ulrichs, V. M., 1977, Zur Altersstellung der Pachycardientuffe und der Unteren Cassianer Schichten in den Dolomiten (Italien): Mitt. Bayer Staatssig. Palaont. hist. Geol., v. 17, p. 15-25.

Vail, P. R., 1987, Seismic stratigraphy interpretation procedure, in Bally, A. W., ed., Atlas of Seismic Stratigraphy, volume 1: AAPG Studies in Geology 27, p. 1-11.

Vail, P. R., Hardenbol, J., and Todd, R. G., 1984, Jurassic unconformities, chronostratigraphy, and sea-level changes from seismic stratigraphy and biostratigraphy, in Schlee, J. S., ed., 1984, Interregional Unconformities and Hydrocarbon Accumulation: AAPG Memoir 36, p. 129-144.

Van Der Eem, J. G. L. A., 1983, Aspects of middle and late Triassic palynology. 6. Palynological investigations in the Ladinian and lower Karnian of the western Dolomites, Italy: Review of Paleobotany and Palynology, v. 39, p. 189-300.

Van Wagoner, J. C., Mitchum, R. M., Campion, K. M., and Rahmanian, V. D., 1990, Siliciclastic sequence stratigraphy in well logs, cores, and outcrops: AAPG Methods in Exploration Series, No. 7, 55 p.

Van Wagoner, J. C., Mitchum, R. M., Jr., Posamentier, H.W., and Vail, P.R., 1987, The key definitions of stratigraphy, *in* Bally, A. W., ed., Atlas of Seismic Stratigraphy, volume 1: AAPG Studies in Geology 27, p. 11-14.

Winterer, E. L., and Bosellini, A., 1981, Subsidence and sedimentation on a Jurassic passive continental margin, Southern Alps, Italy: AAPG Bulletin, v. 65, p. 394-421.

Yose, L. A., 1991, Sequence stratigraphy of mixed carbonate/volcaniclastic slope deposits flanking the Sciliar (Schlern)–Catinaccio buildup, Dolomites, Italy: Dolomieu Conference on Carbonate Platforms and Dolomitization, Guidebook excursion A, Ortisei, Val Gardena, p. 17-39.

Chapter 15

High-Resolution Sequence Stratigraphy in Prograding Miocene Carbonates: Application to Seismic Interpretation

L. Pomar
Departament de Ciencies de la Terra
Universitat de les Illes Balears
Palma de Mallorca, Spain

ABSTRACT

Complete exposure of the Upper Miocene Reef Complex in the sea cliffs of the Island of Mallorca (Spain) allows for a high-resolution sequence-stratigraphic analysis. A 6-km-long cross section parallel to the direction of reef progradation displays four hierarchical orders of accretionary units. These accretionary units are composed of lagoonal horizontal beds, reefal facies with sigmoidal bedding, and gently dipping slope deposits. They are bounded at the top by erosion surfaces and basinward by correlative conformities. Thus, each of the four orders of accretionary units have many characteristics of depositional sequences. Due to the close relationship between coral reefs and the sea surface, vertical shifts of the reef accretionary units record four orders of sea-level fluctuations, longer and larger fluctuations being modulated by lesser ones of higher frequency. Additionally, the position of accretionary units in lower-order progradations allows the definition of: (1) progradational low-stillstand units; (2) aggradational units (sea-level rise); (3) progradational high-stillstand units; and (4) offlapping units (sea-level fall). Erosion truncates the previous progradational high-stillstand and offlapping units during low positions of sea level.

These architectural patterns are similar to progradational sequences seen on some seismic lines. The progradational pattern generally shows the following geometric details: (1) discontinuous climbing high-amplitude reflectors; (2) truncation of clinoforms by these high-amplitude reflectors; (3) discontinuous high-amplitude reflectors with basinward dips; and (4) transparent areas intercalated between clinoforms and horizontal low-amplitude reflectors. Based on facies architecture in the outcrops of Mallorca, the discontinuous climbing high-amplitude reflectors are interpreted as erosion surfaces truncating aggradational reef-core units (sequence boundaries), and are overlain (onlapped) by lagoonal sediments deposited during the next sea-level rise. The discontinuous high-energy reflectors with basinward dips are interpreted as downlap surfaces. The transparent zones correspond to aggradational reef-core facies.

INTRODUCTION

Excellent examples of seismic lines across carbonate platforms have been published (Jansa, 1981; Gamboa et al., 1985; Fontaine et al., 1987; Ladd and Sheridan, 1987; Eberli and Ginsburg, 1987, 1988, 1989; Sarg, 1988, 1989; Epting, 1989; Tyrrell and Davis, 1989; Haan et al., 1990; Harris and Walker, 1990; Davies et al., 1991). These seismic lines show the seaward migration of the platform margin with sigmoid and sigmoid-oblique internal reflection patterns, downlap and onlap surfaces, and sequence boundaries. Often, however, facies interpretation only displays the classic zig-zag distribution on the prograding platform margin. One reason for this is that seismic lines do not provide sufficient resolution for accurate facies interpretations. An expression of this limited resolution is the fact that only third-order sequences are widely interpreted and clearly accepted. Below the resolution of third-order units, higher-order units ("parasequences," "subsequences," "high-frequency sequences," "simple sequences") are still under discussion and revision.

This chapter aims to provide an approach to the high-resolution sequence stratigraphic analysis of prograding carbonates. This approach has been developed from the exceptional outcrops of the Upper Miocene Reef Complex unit in the southern sea cliffs of Mallorca (Balearic Islands, Spain). These vertical cliffs allow a detailed analysis of the reef and associated lithofacies, the bedding geometries and their stacking patterns, and offer an unsurpassed opportunity to test the signatures of high-frequency sea-level fluctuations in the architecture of a prograding carbonate platform margin. Study of the facies packages in these outcrops is simplified because of (1) the absence of terrigenous influx; (2) the absence of significant compaction (preservation of primary porosity); and (3) the absence of significant tectonic or load-induced subsidence (Pomar, 1991).

The approach developed in Mallorca is applicable to prograding carbonates elsewhere and at different scales. Published seismic examples by Eberli and Ginsburg (1987, 1988, 1989) and Tyrrell and Davis (1989) are reviewed here. These lines show seismic facies and reflection patterns which are similar to the facies architecture, bedding geometries, and scale of progradation in the Mallorca example. It is remarkable that these three examples, which occur in quite different geographic settings, present such strong similarities. This fact reinforces the argument that the approach developed here is of general applicability for refined seismic stratigraphic analysis, and for prediction of lithologies and sequence-stratigraphic packages in prograding carbonate complexes.

THE UPPER MIOCENE REEF PLATFORM OF CAP BLANC, MALLORCA

An exceptional example of a progradational reef-platform complex crops out along the sea cliffs of the Cap Blanc area in southern Mallorca (Figure 1). The vertical cliffs are up to 90 m high and about 10 km long, and display the internal architecture of the Upper Miocene Reef Complex (Figure 2). Pomar (1991) presents a detailed description of the lithofacies and their architectural patterns.

The Reef Complex Unit is one of three upper Miocene depositional sequences (Pomar, 1979, 1991; Pomar et al., 1983a,b, 1985; Barón and Pomar, 1985). It is considered to correspond to the Late Tortonian TB 3.2 cycle of Haq et al. (1987) and Haq (1991). The upper Miocene sequences, as well as Pliocene and Pleistocene units, extend in horizontal platforms and infill localized basins. Basins are related to normal faulting, which occurred in the late Miocene and principally during the Pliocene and lower Pleistocene times. The upper Miocene coral reefs rimmed some of these basins (Palma, Alcudia) and prograded across the shallow platforms. In the Llucmajor platform (Marina de Llucmajor, Figure 1), extensive southwesterly progradation of up to 20 km is well documented from borehole data (Fuster, 1973; Pascual and Barón, 1973; Barón, 1977).

Four groups of lithofacies are distinguished in the Reef Complex: (1) lagoon; (2) reef core; (3) reef slope; and (4) open shelf. These lithofacies are described in detail in Pomar et al. (1983a, 1985) and Pomar (1991).

The lagoonal facies are composed of horizontal beds bounded by flat erosion surfaces. Outer-lagoonal facies consist mainly of coral patch reefs, 5–10 m in diameter, and bioturbated skeletal grainstones-packstones. Inner-lagoon lithofacies are composed of bioturbated grainstone, packstone and mudstone, stromatolites and thrombolites, mangrove-swamp packstones, subaerial crusts, and paleosols. Cross-bedded skeletal grainstones also are present in some beds. The lagoonal facies interfinger basinward with the reef-core facies.

The reef-core lithofacies is mainly composed of coral boundstone with sigmoidal bedding. Most of the corals are in life position. Coral colony morphology shows a vertical zonation which reflects the reef-core paleobathymetry. Dish and platy corals always appear at the base and grade upward to branching

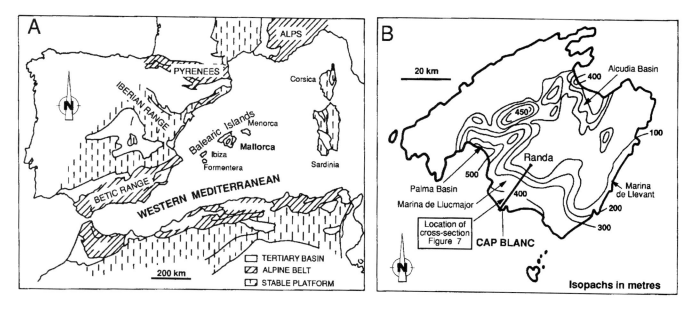

Figure 1. (A) Location of the Balearic Islands in the western Mediterranean (from Pomar et al., 1990b). (B) Isopach map of Tortonian to Pleistocene units of Mallorca (modified from Pomar et al., 1983b).

and stick colonies. Massive headlike or columnar colonies with encrusting biota characterize the upper part of the reef-core lithofacies. Allochemical components within the coral framework are coarse- to fine-grained skeletal packstone and grainstone.

Proximal reef-slope deposits underlie and interfinger with the reef-core lithofacies. They are composed of seaward-dipping clinobeds (up to 20°) of skeletal and intraclastic grainstone, packstone, rudstone, and floatstone. Lenticular coral rubble and coarse-grained skeletal grainstone layers are common. Proximal-slope facies change basinward to gently inclined layers (distal reef-slope lithofacies) of burrowed packstones and grainstones. Red-algae biostromes are locally present. The basal facies (open-shelf lithofacies) are poorly defined flat-lying, extensively bioturbated beds composed of fine-grained skeletal packstones and wackestones, rich in planktonic foraminifers.

Cap Blanc Reef Architecture

The clean and vertical sea cliff outcrops permit detailed correlation between the different lithofacies and related boundaries and thus, in turn, allow the progradational architecture to be established (Pomar, 1991; Pomar et al., in press). Hierarchies of sigmoidal depositional units is the characteristic architectural pattern. These units comprise four lithofacies tracts—horizontal lagoonal beds passing basinward to reef-core, and thence to reef-slope and open-shelf clinobeds. Boundaries are erosion surfaces and their basinward correlative conformities. These erosion surfaces are hierarchically ranked, with larger-scale surfaces truncating smaller-scale. The stacking pattern of the sigmoidal units and the hierarchy of ero-

sion surfaces define the hierarchy of accretional units (sigmoid, sigmoid-set, sigmoid-coset, and set of sigmoid-cosets) (Table 1).

The "sigmoid" (Figure 3 [1]) is the basic accretionary unit (fourth magnitude: 4-M).[1] It comprises a tract of the four lithofacies. The boundaries are erosion surfaces (C surfaces) on top of sigmoids that connect basinward to correlative conformities. The upper reef-slope and the lower reef-core facies are usually preserved (Figure 3 [2]), whereas the upper reef-core and the lagoon facies are truncated in some sigmoids. A wedge-shaped unit results where this truncation is intense (Figure 3 [3]).

The sigmoids are stacked in progradational and/or aggradational bundles.[2] The aggradational bundles usually have well-developed lagoonal beds (Figure 4 [1]), in contrast to the progradational bundles (Figure 4 [2]). The stacking of bundles of sigmoids, with a wavy prograding configuration (Figure 4 [3]), allows the definition of a "set of sigmoids" (third-magnitude accretionary units: 3-M). A set of sigmoids is composed of four bundles of sigmoids: (1) a lower prograding bundle; (2) an aggrading bundle in the middle; (3) an upper prograding bundle; and (4) an offlapping[3] bundle. The development of these bundles is not constant and they show size variations. The offlapping reef downlaps onto the open shelf and

---

[1] Magnitude refers to the rank of sigmoidal depositional units, as well as the order of the inferred relative sea-level cycles (high frequency). Table 1 illustrates the plausible correlation between the magnitudes of the Mallorcan units (and cycles) and depositional sequences (and global sea-level cycles).

[2] The term "bundle" is used as an assemblage of several depositional units and not as a formal unit.

[3] The term "offlap" is used sensu Swain (1949) in Bates and Jackson (1987).

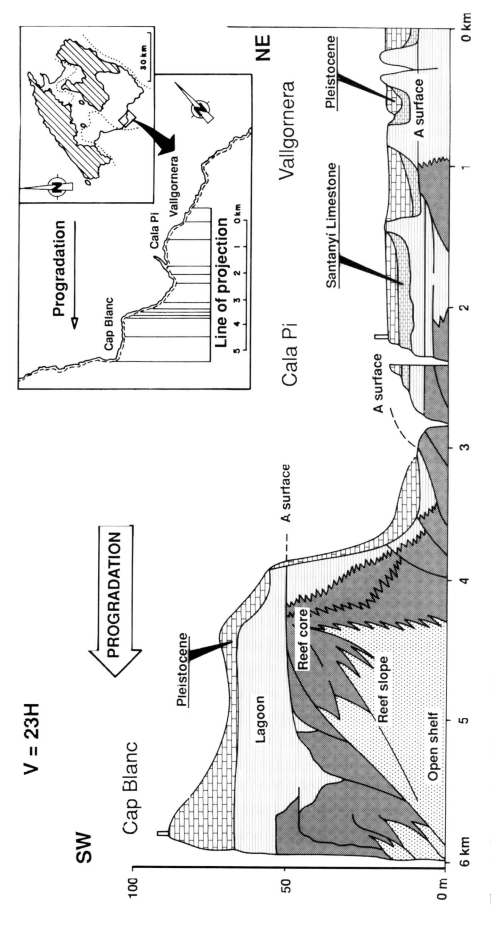

Figure 2. Cross section of the sea cliffs in the Cap Blanc area. This cross section is made by projecting the field data on a line parallel to the direction of reef progradation (modified from Pomar, 1991).

Table 1. Relationships between the Cap Blanc rock units, their boundaries (erosion surfaces), and the magnitude of the inferred sea-level cycles. The periods for these cycles and the correlation with global cycles are based on progradation rates (Pomar, 1991).

Rock Units	Erosion Surfaces	Magnitude of Sea-Level Change	Estimated Period	Global Cycles Haq et al. (1987)
Llucmajor Platform		—	1.9 Ma	3rd order
Sets of Sigmoid-Cosets		1-M	400 ka ?	4th order
Sigmoid-Cosets	A	2-M	100 ka	5th order
Sigmoid-Sets	B	3-M	?	6th order
Sigmoids	C	4-M	?	7th order

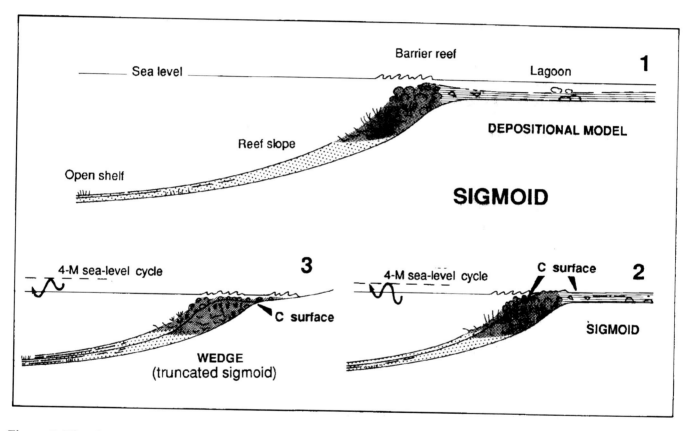

Figure 3. The sigmoid is the basic accretionary unit in the Cap Blanc reefal platform. It is interpreted to be the result of the highest frequency sea-level cycles (4-M).

correlates basinward with a condensed section (Figure 5). The sigmoid-set boundaries are erosion surfaces (B surfaces) connecting basinward to correlative conformities. These B surfaces usually truncate the upper prograding bundle of sigmoids and the sigmoid boundaries (C erosion surfaces).

The sigmoid-sets stack in a "coset of sigmoids" (set of sigmoid-sets) also with a characteristic wavy configuration (second-magnitude accretionary units: 2-M; Figure 6). Four bundles of sigmoid-sets can also be recognized: (1) lower-progradational at the base; (2) aggradational in the middle; (3) high-progradational;

and (4) an offlapping bundle. The boundaries of a coset of sigmoids are major erosion surfaces (A surfaces) and their basinward correlative conformities. This erosion surface truncates the top of the high-progradational and offlapping bundles of sigmoid-sets, as well as the sigmoid-set boundaries (B surfaces). The offlapping reef in a sigmoid-coset also downlaps on the open shelf and correlates basinward with a condensed section.

Upward and downward shifts of the reef-core lithofacies can be seen across the whole Llucmajor platform from water-well core data (Fuster, 1973;

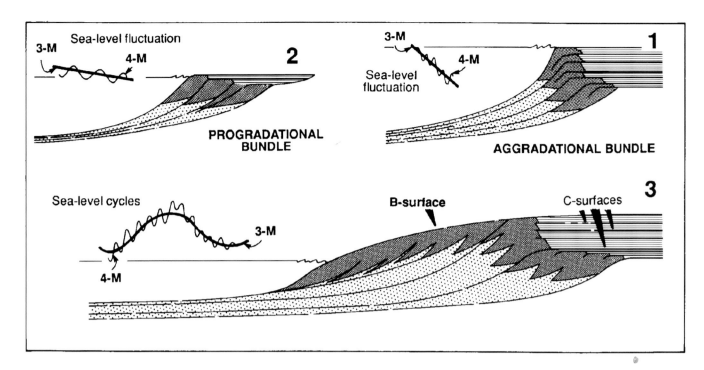

Figure 4. A set of sigmoids is composed of bundles of sigmoids with a wavy prograding configuration. It is bounded by B erosional surface and its correlative basinward conformity. The sigmoid sets are interpreted as the product of lower-order (3-M) sea-level cycles.

Pascual and Barón, 1973; Barón, 1977). Using these data and the outcropping facies geometries, it is possible to determine the architecture of the Llucmajor platform (Figure 7A). The progradational stacking of the sigmoid-cosets also shows a characteristic wavy configuration, which allows the identification of three lower-order accretionary units (Figure 7B) (sets of sigmoid-cosets). They are interpreted as first-magnitude (1-M) accretionary units (Pomar, 1991).

All these different accretionary units (1-M to 4-M) display similar characteristics in terms of facies and boundaries: (1) lagoonal facies in the upper part (inner belt); (2) reef-front facies in the core of the units (middle belt) containing clinoforms of slope facies; (3) open-platform facies in the lower part (basinward belt); and (4) the boundaries are erosional surfaces that become conformities basinward.

SEA-LEVEL FLUCTUATIONS AND REEF-PLATFORM ARCHITECTURE

Accretionary Units and High-Frequency Sea-Level Fluctuations

The upward and downward shifts of the reef-core lithofacies, which characterize the progradation in the Cap Blanc reefs, have been interpreted as products of sea-level fluctuation (Pomar, 1988, 1991). This conclusion is based upon the close relationship between the growth of coral reefs and the sea surface. Additionally, high rates of sediment production,

early cementation, and the susceptibility of reefal deposits to erosion and rapid karstic dissolution, determine the strong tendency of coral reef systems to record sea-level fluctuations. Thus, the hierarchy of accretionary-unit stacking implies a hierarchy of relative sea-level fluctuations of different amplitudes and frequencies (Figures 3, 4, 6, 7). Tectonic subsidence is considered to have been insignificant in the direction of reef progradation in the Cap Blanc area. Primary framework porosity is still open and original cavities up to 6 m in diameter are preserved within the coral framework.

Measuring these vertical shifts, it is possible to define a reef-crest curve (Pomar, 1991) by the successive positions (measured or inferred) of the reef-crest facies[4] (Figure 6 [2]). The curve reflects relative sea-level fluctuations as a function of progradation. The relationship between water paleodepth and coral morphology (based on Cap Blanc model) constitutes an essential tool for paleobathymetric reconstructions.

The periods of the four magnitudes of relative sea-level cycles are uncertain, but there is a good correlation between the 2-M cycles and the 100-k.y. Pleistocene cycles, and 1-M cycles could correlate with the 400-k.y. cycles (Pomar, 1991). Additionally, computer forward modeling using the inferred 2-M

[4] Note that the reef-crest curve is not a sea-level curve. It is an expression of the relative sea level changes in relation to progradation, a function of time, sediment production, and accommodation. It does not take into account subsidence or tectonic uplift. Nevertheless, this reef-crest curve, must be close to the actual sea-level curve.

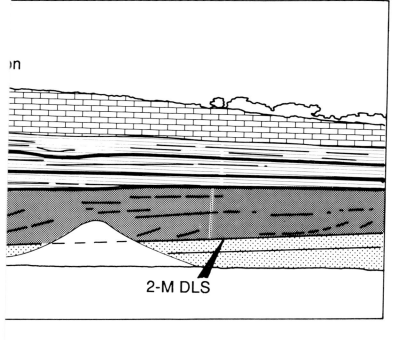

on

2-M DLS

Two-way travel time (milliseconds)

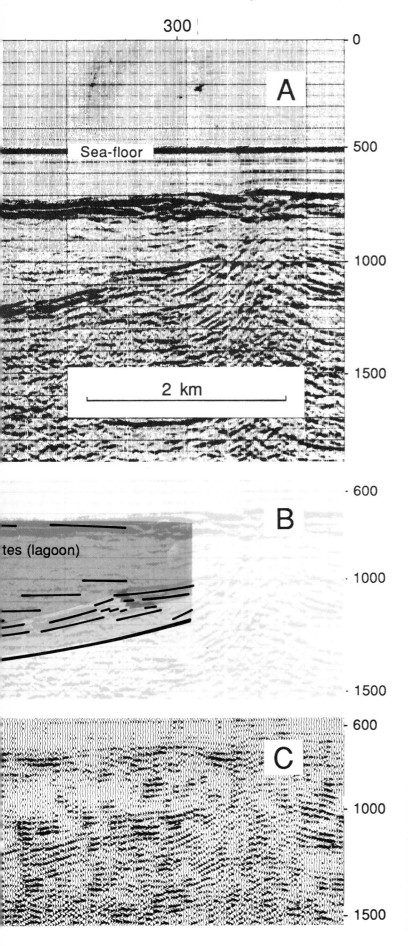

Figure 14. (A) Migrated time
section Amoco Line Y across a
Miocene carbonate shelf margin, Bali-Flores Sea, Indonesia
(from Tyrrell and Davis, 1989).
(B) Interpretation of "sequence
2" displaying the slant-tau
reflectors. (C) Band-pass filter
(30–60 Hz) image of the same
part of this line as B.

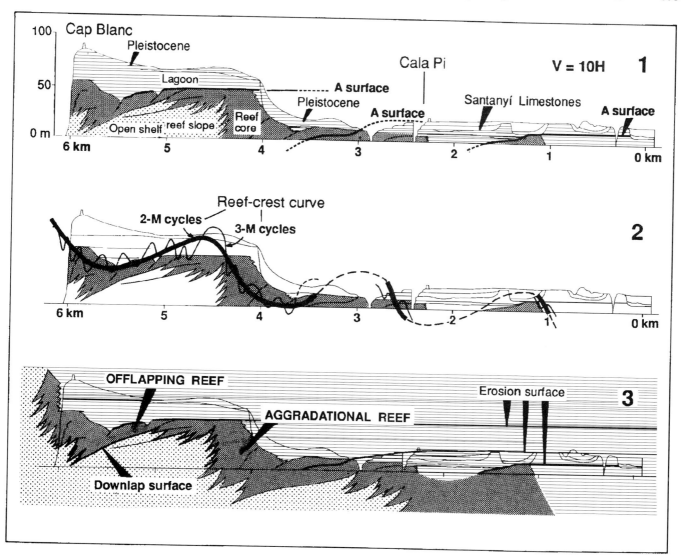

Figure 6. The coset of sigmoids is formed by stacking of sigmoid sets and it is bounded by A erosion surfaces and its basinward correlative conformities. The sigmoid-cosets are thought to record second magnitude (2-M) sea-level cycles. (1) Three cosets of sigmoids have been identified in the Cap Blanc area sea cliffs. (2) The reef-crest curve is defined by the successive positions of the reef-crest facies and reflects relative sea-level changes in relation to progradation. (3) Reconstruction of the three sigmoid cosets in the Cap Blanc area, before late Miocene erosional truncation.

sea-level curve deduced from the outcrops and pre-sent-day production and erosion rates (Pomar et al., 1990a,b; Bosence et al., 1991) is consistent with the 100-k.y. period. In any case, the four magnitudes of accretionary units seen on the Llucmajor platform are hierarchical subdivisions of the Reef Complex unit, which is considered to correlate with the late Miocene TB 3.2 third-order cycle of Haq et al. (1987), a third-order depositional sequence (sensu Vail et al., 1977) (Table 1).

The sea cliffs at Cap Blanc document a high-resolu-tion reefal architecture as a record of different fre-quencies of sea-level fluctuations. Although of different magnitudes, each accretionary unit has pat-terns similar to the third-order depositional sequences of Vail et al. (1977). This similarity is a con-sequence of like genetic processes (modulation of rel-ative sea-level changes) and is independent of the period of the sea-level cycles (Pomar, 1991).

The facies "package" on this prograding platform appears to be controlled by: (1) the amplitude of sea-level fluctuation; (2) the paleodepth of coral reef growth; (3) the period of the sea-level fluctuation; (4) carbonate production and erosion rates; and (5) the depth and profile of the basin. The Llucmajor plat-form had a critical accommodation profile, with a shallow open-shelf (basin).

Key to the Architecture

The 2-M, 3-M, and 4-M accretionary units present similar architectural patterns linked to sea-level fluc-

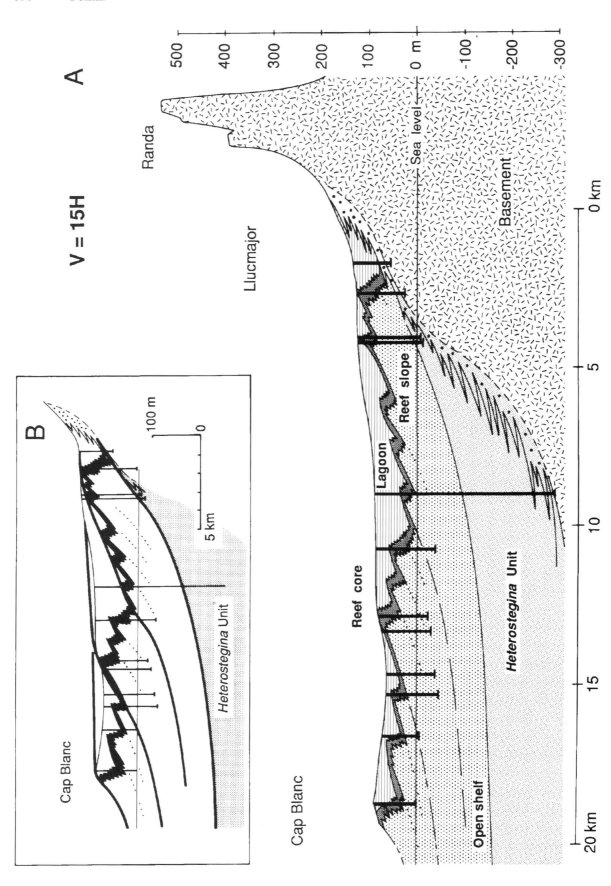

Figure 7. (A) Cross section of the Llucmajor platform (Marina de Llucmajor) showing 2-M accretional units, based on core data (vertical bars) and on the Cap Blanc facies architecture. (B) The sigmoid cosets (2-M units) stack, with a wavy configuration, in three lower order accretional units (1-M = sets of sigmoid cosets). They are considered to be the products of 1-M sea-level cycles. The whole Reef Complex in the Llucmajor platform is interpreted as a third-order sequence.

tuations, implying that carbonate accumulation was not constant during a sea-level cycle (Figure 8). Barrier reefs developed during rises of sea level (2-M, 3-M, and 4-M cycles). Reefs grew upward and aggradation also was important in the other systems—lagoon, reef slope, and open shelf.

The volume of carbonate sediments related to sea-level stillstands was smaller. During low stillstands, there was no aggradation of the lagoonal facies though the reef-core facies prograded with some aggradation of the open-shelf facies. During high stillstands, the reef core prograded with some aggradation on the open-shelf facies, but the lagoonal beds—if present—were truncated. Fringing reefs existed during relative falls in sea level (2-M, 3-M, and 4-M cycles) producing an offlapping-reef configuration. This was due to downward-shifting progradation over the open shelf and consequent erosion to landward of the now-emerged reef. The volume of sediment related to sea-level falls was much smaller: (1) lagoon deposits are absent or very reduced; (2) the reefs, with short reef-slopes, downlap onto the open shelf; and (3) each reef, in general, correlates basinward with a condensed section.

Based on the Cap Blanc example, it is possible to predict the architecture of a prograding-reef platform that possesses similar accommodation space for different configurations of relative sea-level curve. Thus, in a general relative sea-level rise (or in a subsiding basin) (Figure 9A), the high-frequency accretionary units will be stacked in a climbing configuration with a seaward migration of the platform margin. The reefs offlap during sea-level falls and aggrade during subsequent sea-level rises. The lagoon onlaps both the older offlapping reef and the previously truncated

accretionary unit (reef-core and lagoon lithofacies) towards the land.

In a general relative sea-level stillstand, the higher-frequency accretionary units will be stacked in a horizontally prograding configuration (Figure 9B). Similarly, during a general relative sea-level fall (or tectonic uplift) (Figure 9C), the higher-order accretionary units will be stacked in a downward-shifting configuration. Erosional truncation during the high-frequency sea-level falls can remove part of the previous lagoon and reef-core lithofacies, as well as part of the reef slope. During the sea-level rises, the reefs aggrade and the lagoon might onlap landward over the previous lagoon, reef-core, and reef-slope lithofacies.

Expected Reflector Patterns

Using the Cap Blanc example, it is possible to predict the seismic reflection patterns that may be expected for the three examples shown in Figure 9. Prediction is based on the assumption that because of contrast in acoustic impedance across physical surfaces in the rocks (mainly bedding and unconformities with contrasting lithologies), these are likely to act as high-amplitude reflectors (Vail and Mitchum, 1977). Likewise, boundaries without significant lithologic contrast (and, by inference, low acoustic impedance) would act as weak seismic reflectors.

Each magnitude of accretionary unit has similar architecture but differences will exist in facies contrast. The sigmoid and set of sigmoid boundaries (C and B surfaces) would not show important facies contrasts and should act as weak reflectors. Such reflectors could be expected from lagoonal, reef-core, and

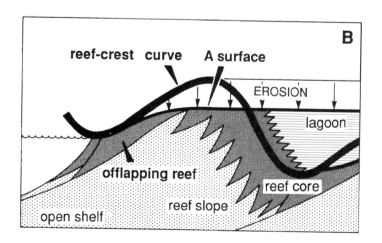
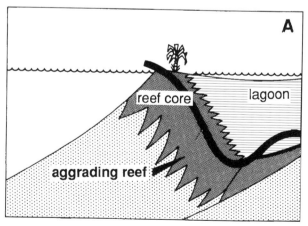

Figure 8. Cap Blanc platform architecture in relation to high-frequency sea-level cycles. (A) During relative sea-level rises (2-M, 3-M, and 4-M), reef, lagoon, and open-shelf deposits aggraded. (B) During relative sea-level falls (2-M, 3-M, and 4-M), offlapping reefs configuration was produced.

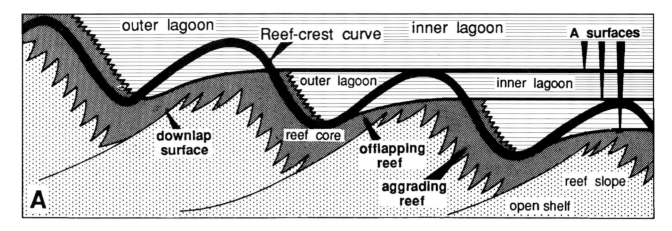

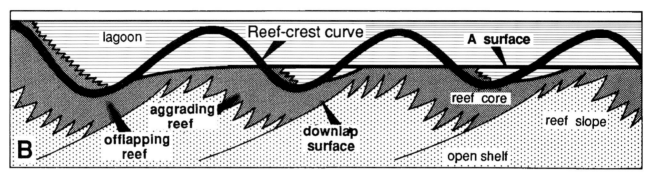

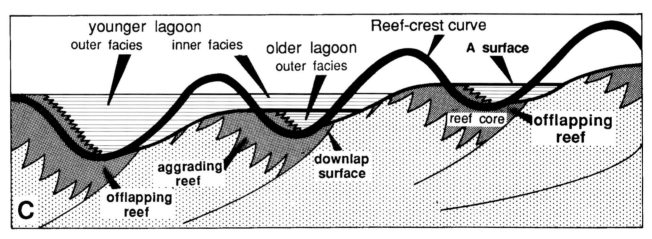

Figure 9. Prediction of the architecture of a prograding reef-rimmed platform in (A) a more general (lower-order) relative sea-level rise (or tectonic/loading subsidence); (B) stillstand of relative sea level; and (C) relative sea-level fall (or tectonic uplift).

reef-slope beds and bed sets. For the coset of sigmoids, facies contrasts exist (Figure 10A) across the erosion surfaces on top of the aggradational bundles of sigmoid-sets, truncating the reef-core lithofacies. This surface is overlain by lagoon deposits, usually fine-grained (inner facies). Contrasts of facies and acoustic impedance also exist along the downlap surface at the base of the offlapping sigmoid-sets (proximal reef-slope and reef-core lithofacies overlies fine-grained open-shelf facies) and its correlative condensed section.

Thus, the erosion surfaces on top of the aggradational reef-core lithofacies appear as horizontal and discontinuous high-amplitude reflectors (Figure 10B). Similarly, the downlap surfaces appear as high-amplitude clinoforms ending updip below the discontinuous horizontal reflectors. This reflection pattern can be termed "slant-tau" (τ) configuration (the horizontal reflector as the top of the tau and the basinward-inclined reflector as the stem). A reflection-free zone, corresponding to the aggradational bundle of sigmoid-sets, can be expected below the

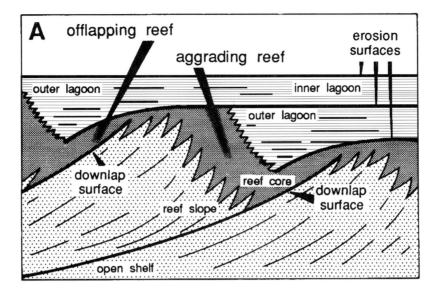

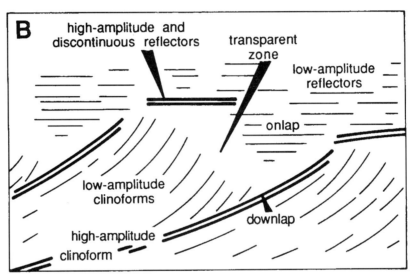

Figure 10. Expected seismic reflection patterns of the Cap Blanc high-frequency accretional units. This prediction is based on the assumption that physical surfaces with lithology contrast are likely to act as high-amplitude reflectors. These reflection patterns can be defined as "slant-tau" reflectors (see text for explanation).

horizontal high-amplitude reflectors. This transparent zone also could be due to equalization shadows. Horizontally, this transparent zone will occur between low-amplitude horizontal reflectors (lagoonal beds) and weak clinoforms (reef-slope clinobeds). The low-amplitude horizontal reflectors (lagoonal beds) apparently onlap the downlap surface.

Figure 11 illustrates the expected distribution of high- and low-amplitude seismic reflectors for the facies architecture shown on Figure 9. In a general rise of relative sea level (or in a subsiding basin) (Figure 9A), the discontinuous horizontal high-amplitude reflectors appear in basinward-climbing patterns (climbing slant-tau reflectors) (Figure 11A). This results in a complex oblique-sigmoid reflection pat-

tern. In a general stillstand of relative sea level (Figure 9B), the erosion surfaces truncating the higher-frequency aggradational reef-core segments appear as discontinuous horizontal high-amplitude reflectors. These discontinuous reflectors can be aligned horizontally, depending on the amount of erosion, and will result in horizontally stacked, slant-tau reflectors (Figure 11B). This results in an apparent unconformity reflection pattern. Similarly, in a general relative sea-level fall (or tectonic uplift) (Figure 9C), the discontinuous horizontal high-amplitude reflectors show a downstepping configuration (downstepping slant-tau reflectors) (Figure 11C), resulting in an apparent unconformity and onlap reflection pattern.

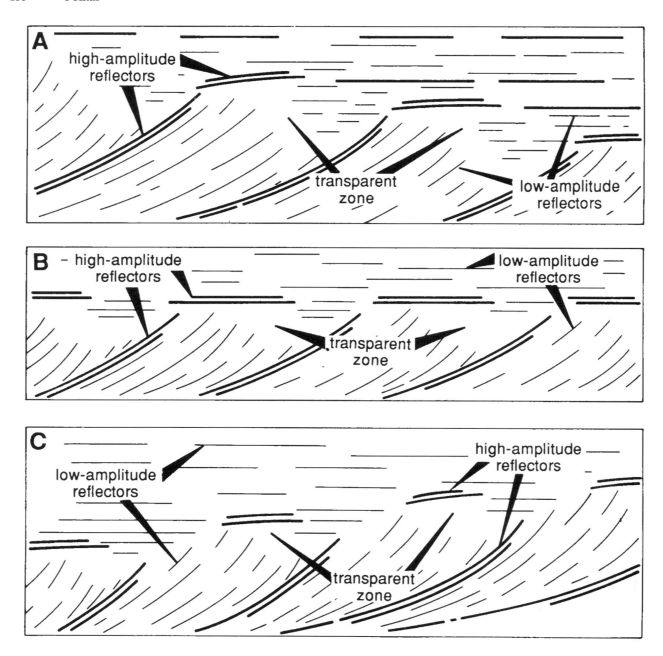

Figure 11. Expected reflection patterns for the three examples shown in Figure 9.

APPLICATION TO SEISMIC INTERPRETATION

Two examples of seismic lines displaying internal reflection patterns similar to those predicted from the Upper Miocene Reef Complex on Mallorca are presented here: (1) the Western line across the Straits of Andros, Great Bahama Bank (Eberli and Ginsburg, 1987, 1988, 1989); and (2) the Amoco Line Y from the Bali-Flores Sea, Indonesia (Tyrrell and Davis, 1989).

Straits of Andros, Western Line, Great Bahama Bank

Seismic lines across the northwestern Great Bahama Bank document an important progradation event of up to 25 km. In the Straits of Andros, an intraplatform seaway, the progradation probably took place during Miocene time (Eberli and Ginsburg, 1989). It is punctuated by horizontally stacked sequences of sigmoidal clinoforms that are interpreted to be a result of repeated fluctuations of sea level (Eberli and Ginsburg, 1987, 1989).

These authors distinguish two types of sigmoids in the progradational intraplatform deposits—simple and complex-oblique. The complex-oblique sigmoids are well developed, showing a complicated internal reflection pattern. From the configuration of this internal reflection pattern, Eberli and Ginsburg (1987, 1989) identify and interpret the main seismic facies as: (1) a "chaotic to reflection-free seismic signal" at the edge of the platform (reefal buildups or rim of skeletal sediments), which changes backward to (2) subhorizontal reflections (lagoonal sediments) and basinward to (3) steep reflections (forereef deposits) and to (4) inclined reflections (periplatform carbonate and downslope deposits). A high-amplitude reflector at the base acts as a downlap surface. The aggrading part of a sequence climbs progressively higher and onlaps the underlying sequence. These authors propose that these sequences were formed during periods of sea-level rise and highstand, and the boundaries correspond to the periods of lowstand sea level. Although not clearly visualized in the seismic data, high-frequency and lower amplitude sea-level fluctuations are interpreted to be a basic mechanism controlling reef growth on the prograding margin.

The outcropping example of prograding platform at Cap Blanc corroborates and illustrates the Eberli and Ginsburg (1987, 1989) interpretation. In the Straits of Andros, the internal reflection pattern, as well as the shape of the prograding sequences (complex-oblique sigmoids), are similar to the expected distribution of high-amplitude and low-amplitude seismic reflectors from the Cap Blanc outcrops. It corresponds to the reflections expected from the facies architecture predicted in this chapter for a general rise in relative sea level punctuated by high-frequency sea-level cycles (Figure 9A), i.e., the climbing slant-tau reflector configuration.

However, although there are similarities in shape and internal patterns, differences exist in the size of the accretional units. The amount of progradation in the Straits of Andros sequences is about 2 km, which is similar to the amount of progradation in Mallorca's 2-M accretional units (cosets of sigmoids). However, the Bahamas section is about ten times thicker than the Mallorca section and the Straits of Andros sequences are two to three times thicker than the Mallorcan 2-M units. This may be the result of differences in the periods of time represented by these examples; in Mallorca, the 2-M units seem to represent a 100-k.y. period, while the individual sequences in the Straits of Andros may be third-order sequences (Eberli and Ginsburg, 1989). Although the time periods estimated for these two examples are imprecisely known, this fact demonstrates that similarities in architectural patterns are a consequence of similar genetic processes (modulation of relative sea-level changes, sediment production rates, and accommodation) independent of the period of the sea-level cycles (Pomar, 1991).

Thus, in the Straits of Andros (Figure 12), the chaotic to reflection-free area at the edge of the sigmoids can be interpreted as aggradational bundles of reef-core sigmoids or sigmoid-sets. The low-amplitude horizontal reflectors, shelfward of the transparent zone, correspond to the back-reef lagoonal beds, and the low-amplitude clinoforms to the fore-reef slope. The sequence boundaries have discontinuous high-amplitude reflectors on top of the reflection-free zone. These reflectors can be interpreted as erosion surfaces (like A surfaces) on top of the aggradational

reef-core lithofacies overlain by lagoonal deposits, which correspond to a later sequence. Landward, these reflectors become weaker, probably because of the absence of important facies contrasts across the erosion surfaces that bound the lagoonal beds. Similarly, the high-amplitude clinoforms ending updip below the discontinuous horizontal reflectors correspond to downlap surfaces and can be interpreted as the bases of the offlapping reefs. Although it is difficult to recognize on the seismic lines, an offlapping reef would be expected over the downlap surface. At Cap Blanc, the offlapping reef is a thin unit downstepping onto the open-shelf beds and shows erosional truncation at the top. Seismic lines still do not provide enough resolution to differentiate these thin rock units. Nevertheless, clearly visible on the seismic line are the downward and basinward shift of the chaotic/reflection-free area (reef-core facies), as well as the low-amplitude horizontal reflectors (lagoonal beds) progressively onlapping the previous sequence.

Seismic facies are often characterized by "apparent frequency" as well as "amplitude" (Dumay and Fournier, 1988). Poststack processing tests on a Landmark interactive workstation proved that a high-frequency seismic section (frequencies restricted to 30–60 Hz instead of 10–60 Hz) discriminated best between the transparent low- and high-dip facies zones. Figure 13 shows the interpretation of the high-frequency section of the Straits of Andros. On the filtered section (Figure 13), more high-amplitude reflectors (horizontal and downlap) appear than on the seismic section shown on Figure 12A. The internal reflection pattern in the individual sequences also becomes more complex on the filtered section (Figure 13). Nevertheless, it is possible to see reflection patterns which are similar to the facies architecture characteristic of higher-order sigmoidal units. Climbing and downstepping "slant-tau" reflectors become more apparent and can be interpreted as higher-order accretional units.

Amoco Line Y, Central Lombok Block, Bali-Flores Sea, Indonesia

The Amoco Line Y (Tyrrell and Davis, 1989) is a "textbook" example of a Miocene carbonate complex prograding about 9 km seaward (Figure 14A). The authors differentiate four sequences in this line. "Sequence 2," which is attributed to the Miocene section and possibly part of the lower Pliocene section, shows a well-developed prograding carbonate platform. The southwest-dipping prograding reflectors pass basinward into nearly flat bottomset reflectors and landward into flat low-amplitude or null topset reflectors. From the reflection patterns, velocity relationships, regional geology, and well data, these authors interpret the flat low-amplitude reflectors as shallow-water bedded strata which grade basinward into unbedded reef or shelf-margin carbonates. The shelf-margin progrades basinward over sloping foresets or talus deposits (southwest-dipping reflectors),

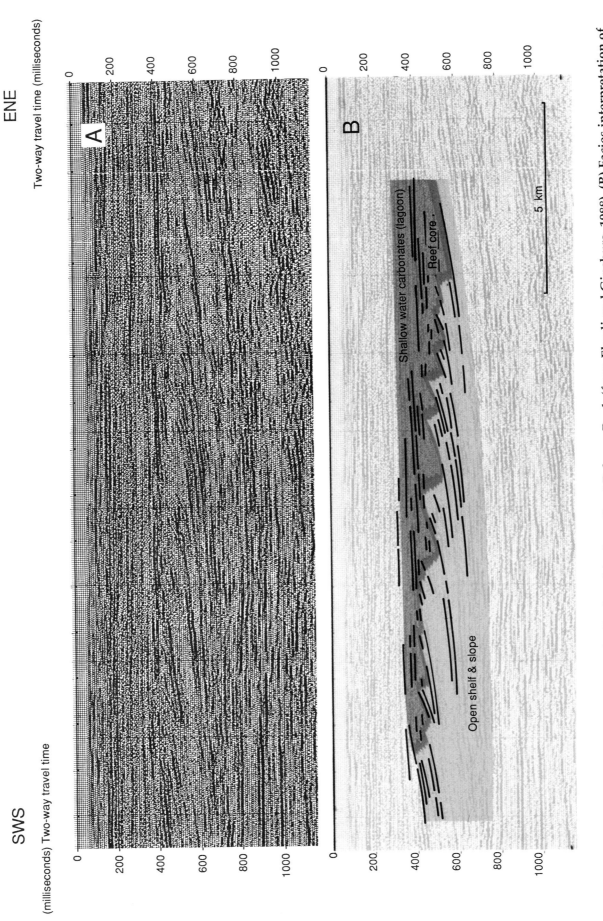

Figure 12. (A) Seismic Western line across the Straits of Andros, Great Bahama Bank (from Eberli and Ginsburg, 1988). (B) Facies interpretation of this line, based on the Cap Blanc reef-architecture model, displaying climbing slant-tau reflector configuration.

Two-way travel time (milliseconds)

(milliseconds) Two-way travel time

Figure 13. Band-pass filter (30–60 Hz) image of the Straits of Andros seismic line and facies interpretation. This post-stack processing test discriminated best between the transparent, low- and high-dip facies zones.

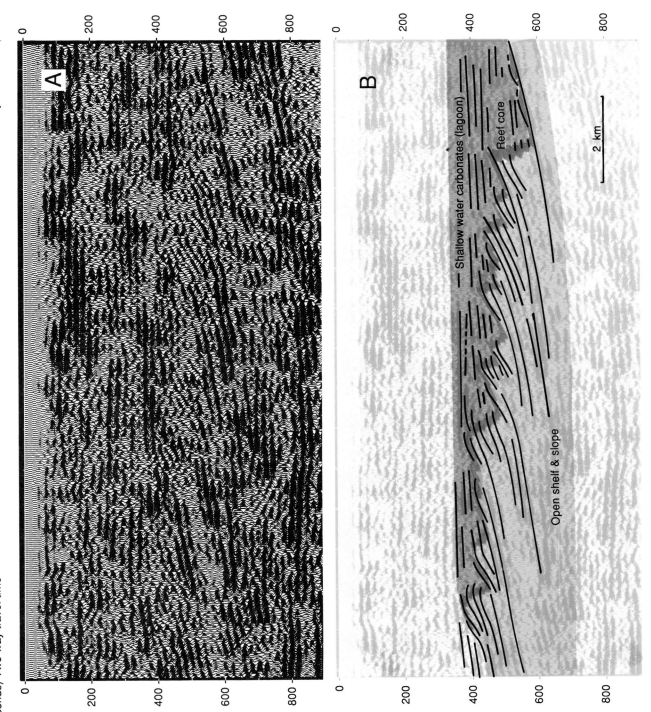

which in turn grade to basinal deposits (nearly flat bottomset reflectors).

Aside from the anomalous reflections between shot-points 220 and 300 (that "possibly represent a shelf-margin carbonate buildup"; Tyrrell and Davis, 1989), several segments can be differentiated within this progradational sequence. Between shotpoints 300 and 440 (Figure 14A), three units with weak reflectors can be seen in downstepping slant-tau configuration. Between shotpoints 440 and 500, a second part is characterized by climbing progradational clinoforms, followed by progradational and slightly aggradational clinoforms between shotpoints 500 and 600. The younger part, between shotpoints 600 and 700, clearly displays two accretionary sigmoids with climbing slant-tau–reflector configuration. The prograding shelf-margin ends with downstepping high-amplitude clinoforms below the high-amplitude reflector boundary at the top of sequence 2 (shotpoints 700–760).

Based on the Cap Blanc facies architecture, a more detailed and specific interpretation is proposed for the progradational shelf margin shown in sequence 2 of Amoco Line Y (Figure 14B).

1. Three downstepping reefal (or rimmed shelf) sigmoid-shaped units can be suggested for the first part (shotpoints 300 to 440) of this prograding shelf margin. The weak reflections are likely to be due to the absence of significant acoustic impedance (lithologic contrast).

2. The reflection patterns on the second part of Line Y (shotpoints 440–600) suggest aggradation and progradation of a nonreef-rimmed shelf; the presence of a weak shelf-slope break suggests a grainstone-rimmed margin. In the migrated time section of this line, it is possible to pick out three accretionary "subsequences" (shotpoints 500–600) whose base is formed by a high-amplitude reflection. The updip clinoforms fade out landward into a reflection-free area. Band-pass filtering (30–60 Hz) in the Landmark interactive workstation (Figure 14C), discriminated better between the transparent and dip-facies zones. A downward shift on the reflection-free area over the high-amplitude reflection suggests that these high-amplitude reflectors are downlap surfaces.

3. The climbing slant-tau–reflector configuration shown in the third part of this prograding shelf margin (shotpoints 600–700), defines two reef-rimmed sigmoid-shaped units. A last reef-rimmed accretional unit can be suggested at the shelf-margin end, just below the downstepping high-amplitude clinoforms (shotpoints 700–710).

The Mallorcan example also permits a refinement of the interpretation of the Indonesian seismic line. In this line, the shape as well as the internal reflection pattern of the prograding "subsequences" display similar geometries to the architecture predicted for different configurations of lower-order sea-level changes (lower-order sea-level rise, stillstand, or fall; see Figure 9). However, despite the similarities in shape and internal patterns, a difference exists in the size of the accretional units. In Line Y, the amount of progradation in the "subsequences" is from 1–2 km, which is similar to the Mallorcan sigmoid-cosets (2-M accretional units). However, the thickness of the Indonesian subsequences is about triple that of the Mallorcan 2-M units. The difference in the periods of time represented by these units could explain the differences in thickness; the Indonesian subsequences may correspond, like those of the Straits of Andros, to third-order sequences as opposed to the 100-k.y. period estimated for the Mallorcan sigmoid-cosets.

CONCLUSIONS

1. Facies architecture analysis of a Miocene prograding reef and associated carbonates (Cap Blanc, Mallorca) provides a high-resolution sequence-stratigraphy outcrop model. This model can be used to refine interpretation of seismic lines in terms of facies distribution and relative sea-level history.

2. The facies architecture can be compared with the reflector patterns on seismic lines by assuming that contrasts in lithologies (and inferred acoustic impedance) are likely to act as good seismic reflectors.

3. Comparison between the Cap Blanc example and two published seismic lines shows a number of similarities: (1) geometry of the accretionary units; (2) scale of progradation; and (3) facies architecture vs. reflection patterns.

4. These similarities allow a more detailed interpretation of the seismic lines by permitting: (1) interpretation of the architectural patterns; (2) prediction of facies distribution; and (3) identification of high-frequency sea-level fluctuations.

5. Although the three examples discussed here occur in different geographic settings, each developed during the Miocene, when the amplitude of sea-level fluctuations related to glacioeustasy was considerable (Shackleton and Kennett, 1975; Wise, 1981; Galloway, 1989).

6. Because the evidence for high-frequency sea-level fluctuations in the geologic record is well established (e.g., Fischer, 1964; Wilson, 1975; Kendall and Schlager, 1981; Berger et al., 1981; Grotzinger, 1986; Hardie et al., 1986; Goldhammer et al., 1987, 1989; Borer and Harris, 1991; Mitchum and Van Wagoner, 1991; Plint, 1991), these architectural patterns must be expected in progradational carbonate platforms when: (1) the relation between the amplitude of high-frequency sea-level fluctuations, carbonate production, and accommo-

dation space was similar to the Mallorcan example; (2) the amount of subsidence did not exceed the growth potential of the reef and associated carbonate systems; and (3) terrigenous influx was absent. The amplitude of the high-frequency sea-level fluctuation determines the upward and downward shifts of the reef and associated lithofacies, while the production rates and the accommodation space determine the package density of the high-frequency accretionary units and, as a consequence, the limits of resolution for the outcrop and seismic data. Any aggrading, prograding, or offlapping carbonate platform must be punctuated by higher-frequency accretionary units genetically related to high-frequency sea-level cycles.

ACKNOWLEDGMENTS

The continuing development and discussion with Mateo Esteban (Erico P.I.) and William C. Ward of many of the ideas set down in this chapter is gratefully acknowledged. The author also appreciates the scanning and vectorizing of the published seismic lines by Charles Hewlett (Links Information Systems Ltd.) and the contributions of John D. Kerr and Andrew J. Reader (Landmark UK) who loaded and processed the seismic data and provided technical support throughout the interactive interpretation. The author also thanks A. Barón and J.A. Fayas (Junta d'Aigües del Govern Balear) for the information about core data. Much appreciated are the comments and suggestions of Dominic Emery, Jonathan M. Henton, and Peter J. Unstead (BP Exploration, London), P.M. Harris, J. García-Mondéjar, R. Ginsburg, and G. Eberli, which improved this chapter. The author also thanks P. Choquette, R. Handford, and R. Loucks for their useful reviewers' comments, as well as help with improving the English by Bryan and Val Lynas, and with drafting by Arantxa Gállego. Financial support for this research has been provided by Spanish DGICYT Project 87-0812.

REFERENCES CITED

Barón, A., 1977, Estudio estratigráfico y paleontológico del Mioceno medio y superior postorogénico de la Isla de Mallorca: Premio Ciudad de Palma, unpublished, 180 p.

Barón, A., and L. Pomar, 1985, Stratigraphic correlation tables: Area 2c Balearic Depression, in F. F. Steininger, J. Senes, K. Kleemann, and F. Rögl, eds., Neogene of the Mediterranean, Tethys and Paratethys: Institute of Paleontology, University of Vienna, p. 17.

Bates, R. L., and J. A. Jackson, 1987, Glossary of Geology: American Geological Institute, Alexandria, Virginia, 788 p.

Berger, W. H., E. Vincent, and H. R. Thierstein, 1981, The deep-sea record: Major steps in Cenozoic ocean evolution, in J. E. Warme, R. G. Douglas, and E. L. Winterer, eds., The Deep Sea Drilling Project: A Decade of Progress: SEPM Special Publication No. 32, p. 489-504.

Borer, J. M., and P. M. Harris, 1991, Lithofacies and cyclicity of the Yates Formation, Permian Basin: Implications for reservoir heterogeneity: AAPG Bulletin, v. 75, p. 726-779.

Bosence, D., L. Pomar, and D. Waltham, 1991, Computer modelling late Miocene carbonate platforms, Spain: Dolomieu Conference on Carbonate Platforms and Dolomitization Abstracts, p. 30-31.

Davies, P. J., P. A. Symonds, D. A. Feary, and C. J. Pigram, 1991, The evolution of the carbonate platforms of Northeast Australia, in P. D. Crevello, J. L. Wilson, J. F. Sarg, and J. F. Read, eds., Controls on Carbonate Platform and Basin Development: SEPM Special Publication No. 44, p. 233-258.

Dumay, J., and F. Fournier, 1988, Multivariate statistical analyses applied to seismic facies recognition: Geophysics, v. 53, p. 1151-1159.

Eberli, G. P., and R. N. Ginsburg, 1987, Segmentation and coalescence of Cenozoic seaways, Northwestern Great Bahama Bank: Geology, v. 15, p. 75-79.

Eberli, G. P., and R. N. Ginsburg, 1988, Aggrading and prograding infill of buried Cenozoic seaways, northwestern Great Bahama Bank, in A. W. Bally, ed., Atlas of Seismic Stratigraphy: AAPG Studies in Geology, v. 27, p. 97-103.

Eberli, G. P., and R. N. Ginsburg, 1989, Cenozoic progradation of northwestern Great Bahama Bank, a record of lateral platform growth and sea-level fluctuations, in P. D. Crevello, J. L. Wilson, J. F. Sarg, and J. F. Read, eds., Controls on Carbonate Platform and Basin Development: SEPM Special Publication No. 44, p. 339-351.

Epting, M., 1989, The Miocene carbonate buildups of Central Luconia, offshore Sarawak, in A. W. Bally, ed., Atlas of Seismic Stratigraphy: AAPG Studies in Geology, v. 27, p. 168-173.

Fischer, A. G., 1964, The Lofer cyclothems of the Alpine Triassic: Bulletin Kansas Geological Survey, v. 169, p. 107-149.

Fontaine, J. M., R. Cussey, J. Lacaze, R. Lanaud, and L. Yapaudjian, 1987, Seismic interpretation of carbonate depositional environments: AAPG Bulletin, v. 71, p. 281-297.

Fuster, J., 1973, Estudio de los recursos hidráulicos totales de las Baleares: Informe de síntesis general, Ministerio de Obras Públicas, Ministerio de Industria, Ministerio de Agricultura, Comité de Coordinación. 2 vol., Marzo 1983.

Galloway, W. E., 1989, Genetic stratigraphic sequences in basin analysis II: Application to northwest Gulf of Mexico Cenozoic basin: AAPG Bulletin, v. 73, p. 143-154.

Gamboa, L. A., M. Truchan, and P. L. Stoffa, 1985, Middle and Upper Jurassic depositional environments at outer shelf and Slope of Baltimore Trough: AAPG Bulletin, v.69, p. 610-621.

Goldhammer, R. K., P. A. Dunn, and L. A. Hardie, 1987, High frequency glacio-eustatic sea level oscillations with Milankovitch characteristics recorded in Middle Triassic platform carbonates in northern Italy: American Journal of Science, v. 287, p. 853-892.

Goldhammer, R. K., E. J. Oswald, and P. A. Dunn, 1989, The hierarchy of stratigraphic forcing—an example from Middle Pennsylvanian (Desmoinesian) shelf carbonates of the southwestern Paradox basin, in E. K. Franseen and W. L. Watney, eds., Sedimentary Modeling: Computer Simulation of Depositional Sequences: Subsurface Geology Series, Kansas Geological Survey, v. 12, p. 27-30.

Grotzinger, J. P., 1986, Upward shallowing platform cycles: A response to 2.2 billion years of low-amplitude, high-frequency (Milankovitch band) sea level oscillations: Paleoceanography, v. 1, p. 403-416.

Haan, E. A., S. G. Corbin, M. W. Hughes Clarke, and J. E. Mabillard, 1990, The Lower Kahmah Group of Oman: The carbonate fill of a marginal shelf basin, in A. H. F. Robertson, M. P. Searle, and A. C. Ries, eds., The Geology and Tectonics of the Oman Region: Geological Society Special Publication, v. 49, p. 109-125.

Haq, B. U., 1991, Sequence stratigraphy, sea level change, and significance for the deep sea, in D. I. M. Macdonald, ed., Sedimentation, Tectonics and Eustasy: Special Publication of the International Association of Sedimentologists, v. 12, p. 3-39.

Haq, B. U., J. Hardenbol, and P. V. Vail, 1987, Chronology of fluctuating sea levels since the Triassic: Science, v. 235, p. 1156-1167.

Hardie, L. A., A. Bosellini, and R. K. Goldhammer, 1986, Repeated subaerial exposure of subtidal carbonate platforms, Triassic, Northern Italy: Evidence for high frequency sea level oscillations on a 10^4 year scale: Paleoceanography, v. 1, p. 447-457.

Harris, P. M., and S. D. Walker, 1990, McElroy Field—U.S.A. Central Basin Platform Permian Basin, Texas, in E. A. Beaumont and N. H. Foster, eds.: AAPG Atlas of Oil and Gas Fields—Stratigraphic Traps I, 296 p.

Jansa, L. F., 1981, Mesozoic carbonate platforms and banks of the Eastern North American margin: Marine Geology, v. 44, p. 97-117.

Kendall, C. G. St. C., and W. Schlager, 1981, Carbonates and relative changes in sea level: Marine Geology, v. 44, p. 181-212.

Ladd, J. W., and R. E. Sheridan, 1987, Seismic stratigraphy of the Bahamas: AAPG Bulletin, v. 71, p. 719-736.

Mitchum, R. M. Jr., and J. C. Van Wagoner, 1991, High-frequency sequences and their stacking patterns: Sequence-stratigraphic evidence of high-frequency eustatic cycles: Sedimentary Geology, v. 70, p. 131-160.

Pascual, D., and A. Barón, 1973, Estudio hidrogeológico de la Unidad de Llucmajor-Campos: Informe interno, Servicio Hidráulico de Baleares, Ministerio de Obras Públicas.

Plint, A. G., 1991, High-frequency relative sea level oscillations in Upper Cretaceous shelf clastics of the Alberta foreland basin: Possible evidence for glacio-eustatic control?, in M. Macdonald, ed., Sedimentation, Tectonics and Eustasy: Special Publication of the International Association of Sedimentologists, v. 12, p. 409-428.

Pomar, L., 1979, La evolución tectonosedimentaria de las Baleares: Análisis crítico: Acta Geológica Hispánica, v. 14, p. 293-310.

Pomar, L., 1988, Reef architecture and high-frequency relative sea level oscillations, Upper Miocene, Mallorca, Spain, in R. Swennen, ed., Abstracts 9th European Regional Meeting of Sedimentology: International Association of Sedimentologists, Leuven, Belgium, p. 174-175.

Pomar, L., 1991, Reef geometries, erosion surfaces and high-frequency sea level changes, Upper Miocene Reef Complex, Mallorca, Spain: Sedimentology, v. 38, p. 243-269.

Pomar, L., M. Esteban, F. Calvet, and A. Barón, 1983a, La unidad arrecifal del Mioceno superior de Mallorca, in L. Pomar, A. Obrador, J. Fornós, and A. Rodriguez-Perea, eds., El Terciario de las Baleares (Mallorca-Menorca), Guia de las Excursiones del X Congreso Nacional de Sedimentologia: Institut d'Estudis Baleàrics and Universidad de Palma de Mallorca, p. 139-175.

Pomar, L., M. Marzo, and A. Barón, 1983b, El Terciario de Mallorca, in L. Pomar, A. Obrador, J. Fornós, and A. Rodriguez-Perea, eds., El Terciario de las Baleares (Mallorca-Menorca), Guia de las Excursiones del X Congreso Nacional de Sedimentologia: Institut d'Estudis Baleàrics and Universidad de Palma de Mallorca, p. 21-44.

Pomar, L., J. J. Fornós, and A. Rodriguez-Perea, 1985, Reef and shallow carbonate facies of the Upper Miocene of Mallorca, in M. D. Milà and J. Rosell, eds., 6th European Regional Meeting Excursion Guidebook: International Association of Sedimentologists and Universitat Autònoma de Barcelona, p. 495-518.

Pomar, L., D. W. J. Bosence, and D. A. Waltham, 1990a, Computer modelling of a Miocene carbonate platform, Mallerca, Spain: Abstracts 13th International Sedimentological Congress, Nottingham, England, p. 431.

Pomar, L., A. Rodriguez-Perea, F. Sabat, and J. J. Fornós, 1990b, Neogene stratigraphy of Mallorca island, in J. Agustí, R. Domènech, R. Julià, and J. Martinell, eds., Iberian Neogene Basins: Paleontologia i Evolució, Memòria especial 2, Institut Paleontològic Dr. M. Crusafont, Sabadell, p. 269-320.

Sarg, J. F., 1988, Carbonate sequence stratigraphy, in C. K. Wilgus, B. S. Hastings, C. G. St. C. Kendall, H. W. Posamentier, C. A. Ross, and J. C. Van

Wagoner, eds., Sea Level Changes: An Integrated Approach: SEPM Special Publication No. 42, p.152-181.

Sarg, J. F., 1989, Middle-Late Permian depositional sequences, Permian basin, West Texas and New Mexico, *in* A. W. Bally, ed., Atlas of Seismic Stratigraphy: AAPG Studies in Geology, vol. 27, p. 140-154.

Shackleton, N. J., and J. P. Kennett, 1975, Paleo-temperature history of the Cenozoic and the initiation of Antarctic glaciation: Oxygen and carbon isotope analyses in DSDP Sites 277, 279 and 281: Initial Reports of the Deep Sea Drilling Project, v. 29, p. 743-755.

Swain, F. M., 1949, Onlap, offlap, overstep, and overlap: AAPG Bulletin, v. 33, p. 634-636.

Tyrrell, W. W., and R. G. Davis, 1989, Miocene carbonate shelf margin, Bali-Flores Sea, Indonesia, *in* A. W. Bally, Atlas of Seismic Stratigraphy: AAPG Studies in Geology, vol. 27, p. 174-179.

Vail, P. R., and R. M. Mitchum, Jr., 1977, Seismic stratigraphy and global changes of sea level, Part 1, *in* C. E. Payton, ed., Seismic Stratigraphy—Applications to Hydrocarbon Exploration: AAPG Memoir 26, p. 51-52.

Vail, P. R., R. M. Mitchum Jr., R. G. Todd, J. M. Widmier, S. Thompson III, J. B. Sangree, J. N. Bubb, and W. G. Hatlelid, 1977, Seismic stratigraphy and global changes of sea level, *in* C. E. Payton, ed., Seismic Stratigraphy—Applications to Hydrocarbon Exploration: AAPG Memoir 26, p. 49-212.

Wilson, J. L., 1975, Carbonate Facies in Geologic History: Springer-Verlag, New York, 471 p.

Wise, S. W., Jr, 1981, Deep sea drilling in the Antarctic: Focus on late Miocene glaciation and implications of smear-slide biostratigraphy, *in* J. E. Warme, R. G. Douglas, and E. L. Winterer, eds., The Deep-Sea Drilling Project: A Decade of Progress: SEPM Special Publication No. 32, p. 471-487.

Chapter 16

Sequence Stratigraphy of Miocene Carbonate Complexes, Las Negras Area, Southeastern Spain: Implications for Quantification of Changes in Relative Sea Level

Evan K. Franseen
Kansas Geological Survey
University of Kansas
Lawrence, Kansas, U.S.A.

Robert H. Goldstein
Tracey E. Whitesell
Department of Geology
University of Kansas
Lawrence, Kansas, U.S.A.

ABSTRACT

Exceptional exposures near the coastal village of Las Negras in southeastern Spain contain undeformed late Miocene carbonate complexes deposited on the slopes of isolated, earlier Neogene volcanic basement highs that were part of an archipelago. Detailed field studies identified five carbonate depositional sequences, locally admixed with volcaniclastic debris, with onlapping and downlapping geometries that reflect a complex history of relative sea-level fluctuations. The first four sequences record an overall evolution from an open-marine platform (ramp) to a fringing reef complex. The open-marine platform (depositional sequences DS1A, DS1B, DS2) is predominantly composed of red algal, bryozoan, mollusk, benthic and planktonic foraminiferal, echinodermal wackestones and packstones. Locally, DS2 contains megabreccia composed predominantly of *Tarbellastraea* and *Porites* coral reef blocks. The fourth depositional sequence (DS3) is composed of fringing reef complex strata characterized by *Porites* reefs. The fifth and youngest depositional sequence (TCC) is composed of normal to restricted marine cyclic carbonates consisting of red algal and mollusk packstones, oolite grainstones, and stromatolites that were deposited upon reflooding of shelf areas after a salinity crisis. The sequence boundaries are characterized by evidence of relative falls in sea level.

These depositional sequences preserve over 200 m of Miocene paleotopography. Such preservation allows for quantification of the relative sea-level curve through the development of "pinning points," which we define as a position of ancient sea level relative to a geologically useful starting elevation. A pinning point curve is constructed by plotting ancient relative elevations of sea level (pinning points) versus relative time. This is accomplished by determining relative elevations of facies deposited at or just below sea level, tracing downslope subaerial exposure on marine rocks, or from an upward transition of marine to nonmarine deposits. Relative elevations for pinning points are determined from preserved paleotopography. Pinning point curves can be used as templates to construct more interpretive relative sea-level curves that incorporate stratal geometries, facies relationships, and stacking patterns. The pinning point curve developed at Las Negras illustrates a complex history with large-scale relative sea-level changes of more than 150 m combined with smaller scale fluctuations. Comparison of relative sea-level histories for equivalent strata from different areas can provide for the quantification of the effect of local, regional, and eustatic variables on carbonate sequence development. The pinning point method illustrated here can be used for carbonate sequences of all ages.

INTRODUCTION

Development of appropriate sequence stratigraphy models for carbonate strata are dependent on field models where paleotopography is well preserved and where the outcrop is sufficient to allow for lateral tracing of important bounding surfaces. Such exposures are also significant in that they may allow for quantification of the history of relative sea-level change, which in turn can lead to a better understanding of the causes of sequence development and sequence character.

This study examines outcrops of Miocene carbonate strata from southeastern Spain, describes the sequence stratigraphic framework, and develops the concept of "pinning points" to quantify the history of relative sea-level fluctuations for these strata. Rather than concentrating on current sequence stratigraphic models, this chapter deals with description of the sequences, and what can be shown quantitatively about forcing mechanisms that affected sequence development. It can serve as a case study by which current models can be refined and modified. Also, quantitative "pinning-point" relative sea-level curves constructed for age-equivalent strata from separate areas can be compared to determine the roles of local, regional, and eustatic variables on sequence development. Moreover, the pinning point method developed herein can serve as a methodology for understanding the quantitative constraints on sea-level history that should be applicable to understanding the variables that affect carbonate sequences from many areas and of many ages.

GEOLOGIC SETTING

According to Esteban (1979) and Esteban et al. (in press), the Betic and Balearic provinces offer the highest variety of types and ages of late Miocene reefs in the western Mediterranean. During late Miocene time, the Neogene basins of the Betic Ranges occurred as a series of interconnected corridors, passageways, and basins, resulting in a complex and changing paleogeography around an emergent archipelago of islands (Esteban et al., in press) (Figure 1). Main tectonic events occurred from the Oligocene to middle Miocene time in different zones of the Betics and Rif with phases of regional uplift occurring in late Tortonian and early Messinian times through the Pliocene and later times (Rehault et al., 1985), contributing to the present-day basin-and-range topography. A major northeast-trending sinistral strike-slip fault (Carboneras fault) separates the Cabo de Gata volcanic complex from the Mesozoic–Paleozoic basement of the Betic Range to the northwest (Figure 1). This fault was active throughout Tortonian, Messinian, and Pliocene times (Montenat et al., 1987). De Larouzière et al. (1988) show through seismic studies that the crustal thickness associated with the Cabo de Gata volcanic complex ranges from about 20 to 27 km compared to a thickness of about 40 km for the more "continental" type crust of the Betic basement rocks to the northwest of the fault zone.

According to Esteban (1979) and Esteban et al. (in press), the progressive severing and restriction of the Betic Straits and Rif Straits in Morocco during the late Miocene was the key factor leading to the isolation of

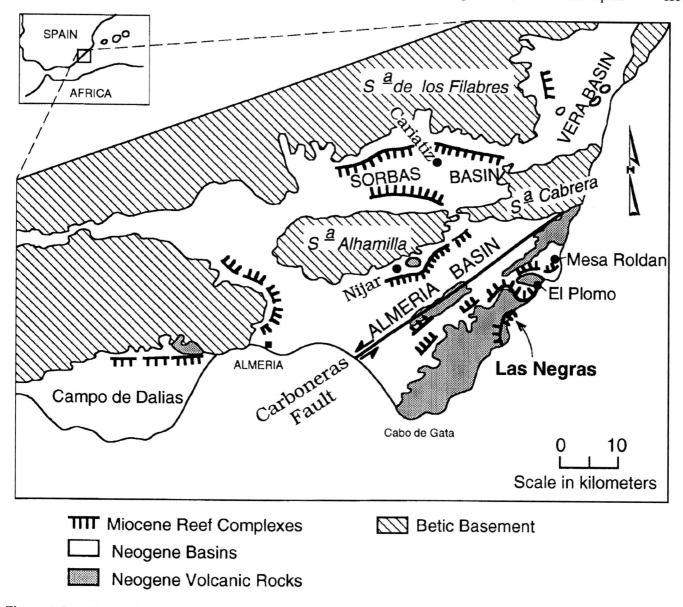

Figure 1. Location and general geologic map showing distribution of late Miocene reefs in southeastern Spain. The Betic basement consists predominantly of Paleozoic and Mesozoic metamorphic and sedimentary rocks. Modified from Dabrio et al. (1985).

the Mediterranean basin and onset of the "Messinian salinity crises" of Hsü (1973) and Hsü et al. (1977). The late Miocene carbonate strata exposed on the margins of the western Mediterranean basin provide an important record of sea-level history prior to, during, and after the major sea-level drop that culminated in the Messinian salinity crisis.

The area around the small coastal village of Las Negras is the site of this study (Figure 1). The studied units consist of carbonate rocks, locally mixed with volcaniclastic debris, that developed on the slopes of volcanic basement highs. During deposition, basement consisted of isolated volcanic highs that formed

in association with development of the approximately 11-m.y.-old Rodalquilar caldera (Rytuba et al., 1990). Volcanic rocks associated with the caldera range between andesitic and rhyolitic composition. The youngest known volcanic rocks of the sequence may be as young as 7.5 m.y. (Di Battistini et al., 1987). Deposition of marine carbonates and sandstones occurred after deformation of the caldera because marine sediments within the caldera are largely undeformed (Franseen, 1989; Rytuba et al., 1990). These carbonate rocks can be divided into five depositional sequences based on stratal geometries, facies patterns, and sequence boundaries. These sequences, the facies stacking patterns within the sequences, and the sequence boundaries record several scales of relative sea-level fluctuation (Franseen and Mankiewicz, 1991).

PREVIOUS STUDIES

Many workers have been involved in various sedimentologic and paleontologic/paleoecologic studies of Miocene reefs around the western Mediterranean, particularly those in southeastern Spain, for many years. Some of those important studies in southeastern Spain include Permanyer and Esteban, 1973; Dabrío, 1974, 1975; Montenat, 1975, 1977; Alvarez et al., 1977; Dronkert and Pagnier, 1977; Pagnier, 1977; Addicott et al., 1978; Dabrío and Martín, 1978; Esteban, 1979, 1981, 1988; Armstrong et al., 1980; Esteban and Gíner, 1980; Santisteban and Taberner, 1980; Dabrío et al., 1981; Esteban and Pray, 1981; Poore and Stone, 1981; Veeken, 1983; Dabrío et al., 1985; Megías, 1985; Müller and Hsü, 1987; Braga and Martín, 1988; Martín et al., 1989; Braga et al., 1990; Martín and Braga, 1990; Serrano, 1990; and Esteban et al., in press. Franseen (1989) and Franseen and Mankiewicz (1991) were the first to apply sequence stratigraphic concepts to Miocene carbonate complexes in southeastern Spain.

DEPOSITIONAL SEQUENCES

The Las Negras carbonate complex can be divided into five depositional sequences (DS) of Miocene age that are labeled from bottom (oldest) to top (youngest) DS1A, DS1B, DS2, DS3, and TCC (Terminal Carbonate Complex of Esteban, 1979; Esteban and Gíner, 1980). For the most part, the depositional sequences are composed of nine facies in the Las Negras area: volcaniclastic/carbonate conglomerate (VCC), red algal-rich packstone/grainstone (RAPG), fine-grained wackestone/packstone (FGWP), coarse-grained packstone/grainstone (CGPG), megabreccia 1 (MB1), megabreccia 2 (MB2), volcaniclastic sandstone/conglomerate (VSC), reef core/talus (RCT), and *Halimeda*-rich facies (HR). Generalized depositional sequence characteristics are shown in the stratigraphic column of Figure 2 and in the cross section of La Molata in Figure 3. Facies characteristics are summarized in Table 1. Figures 4, 5, and 6 are La Molata hillside photo mosaics and line drawings showing representative stratal geometries and facies patterns for the depositional sequences. Additional details of depositional sequences DS1B, DS2, and DS3 and their facies relationships have been described previously (Franseen, 1989; Franseen and Mankiewicz, 1991). We now divide DS1 into DS1A and DS1B because of a newly discovered sequence (DS1A) and sequence boundary (SB1b) in the area (Franseen and Goldstein, 1992). Herein, we further define the TCC strata as a depositional sequence and subdivide it into three to four separate units.

Depositional Sequence 1A (DS1A)

DS1A is the lowest depositional sequence and ranges in thickness from 0 to approximately 20 m. In general, DS1A contains the following succession from base to top.
1. Subaerial exposure of volcanic basement to form sequence boundary 1a (SB1a).
2. Mixed marine carbonate packstones/grainstones and volcaniclastic sandstones/conglomerates. Massive to planar bedded strata (some with low-angle cross-bedding) that onlap SB1a.
3. Local deposition of volcanic mass flow deposit and its erosion.
4. Mixed marine carbonate packstones/grainstones and volcaniclastic sandstones/conglomerates that onlap SB1a.
5. Subaerial exposure of marine carbonates along sloping surface with 25–30 m relief to form sequence boundary 1b (SB1b).

The lower sequence boundary (SB1a) occurs at the contact between volcanic basement and marine carbonate (Figure 2). In some places, this contact is marked by a polymictic volcanic conglomerate. The top of the volcanic basement rocks contain subvertical fissures several centimeters wide, vugs 10–20 cm wide, autoclastic breccia, and concentric cracks that wrap around large volcanic clasts (Figure 7A). Some of the clasts exhibit reddened rims in which volcanic clasts have been altered (Figure 7B). Some of the fissures are lined with crusts and highly altered zones rich in iron oxides. Most of the remaining pore space is filled with the pinkish molluskan wackestone and packstone that just overlies the volcanic conglomerate. The infill of this marine sediment constrains the age of fissure formation, concentric cracking, autoclasts, vug formation, and precipitation of iron oxide crust as occurring before deposition of this molluskan lithology. Although no well-developed paleosols have been found along this surface, the formation of the concentric cracks is best explained by the process of spheroidal weathering in a subaerial environment. Moreover, any confusion of these structures with pillows from submarine extrusion of the volcanic substrate is unlikely because the rocks lack the "space-filling" habit and vesicle pattern of typical pillow structures, air-fall volcanics are well known in the complex, and some of the volcanic rocks are interbedded with lacustrine deposits (Rytuba et al., 1990). Precipitation of iron oxide crusts and formation of fissures and other pore systems are consistent with an event of subaerial exposure predating deposition of the molluskan carbonate. Also, the overall morphology of SB1a appears terraced in this area (Franseen and Mankiewicz, 1991) and in other areas (Esteban and Gíner, 1980) as if differential wave planation was important in shaping the sequence boundary during subsequent transgression. In some areas, the evidence for subaerial exposure along the basal

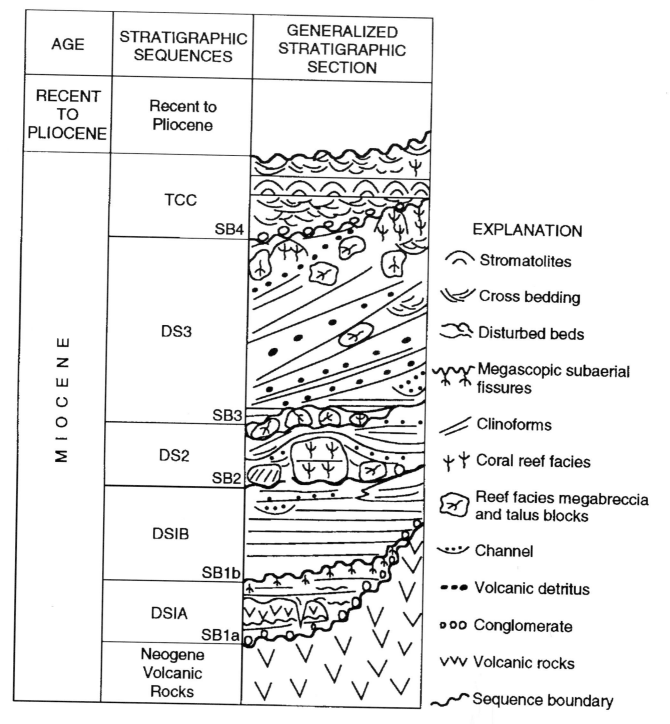

Figure 2. Generalized stratigraphic section of the depositional sequences in the Las Negras area.

volcanic contact can be traced from 140 m above sea level down to present-day sea level.

The strata in DS1A are mostly shallow-dipping, normal-marine strata that were deposited on a shallow-water ramp. Locally, strata dip up to 10°. For the most part, DS1A strata onlap volcanic basement and are massive to planar bedded with low-angle cross-bedding locally developed. Basal DS1A strata are pre-dominantly a packstone/grainstone lithology consisting mostly of fragments of bivalves, gastropods, bryozoans, benthic foraminifera, echinoderms, and various amounts of admixed volcaniclastic grains. A 10–15-m-thick interbedded volcanic mass flow unit was locally deposited, apparently only at the Cerro del Cuervo Beach locality (Figure 8). The volcanic mass flow deposit consists of large (cm to several m

Table 1. Facies characteristics.

Reef Core or Talus (RCT)	
Bedding (dips)	Massive; locally tabular (0–30°)
Skeletal & nonskeletal components	*Porites* dominant framework, locally serpulids; encrusting, fragmented, and branching red algae, foraminifers, bryozoans, bivalves, gastropods, echinoderms, peloids, coated grains, intraclasts, volcaniclastics
Dominant grain sizes	Mud to boulder
Porosity (visual estimates)	Less than 5–25%; avg. less than 10% MO, BP, GF, VUG, FR
Cements	Dolomite-equant or spherulitic/polyhedral crystals, micritic and fibrous to bladed calcite, coarse (poikilitic) calcite
Other	Coral framework is minor compared to matrix; local vadose fabrics

Halimeda-Rich Beds (HR)	
Bedding (dips)	Massive; very thinly to thickly bedded (0–30°)
Skeletal & nonskeletal components	*Halimeda;* coralline algae (fragments, rhodoliths, crusts), serpulids, bivalves, *Porites*, bryozoans, gastropods, volcaniclastics
Dominant grain sizes	Coarse sand to pebble
Porosity (visual estimates)	5–50%; avg. 15–25% MO, BP, VUG
Cements	Dolomite, fibrous to bladed isopachous calcite, equant calcite, poikilitic calcite
Other	Occurs in reef core to distal foreslope positions

Coarse-Grained Packstones/Grainstones (CGPG)	
Bedding (dips)	Well bedded, thinly to very thickly bedded, local cross-bedding and cross-lamination (less than 10–30°)
Skeletal & nonskeletal components	Bivalves, coralline algae (fragments, rhodoliths, crusts, branching), echinoderms, bryozoans, gastropods, serpulids, *Porites* fragments, foraminifers, ostracods, peloids, coated grains, volcaniclastics, intraclasts
Dominant grain sizes	Medium sand to pebble
Porosity (visual estimates)	5–40%; Avg. 10–15% MO, BP, VUG
Cements	Equant dolomite, fibrous to bladed, isopachous calcite, equant calcite
Other	Normally graded units locally; common as channel-fill deposits

Fine-Grained Wackestones/Packstones (FGWP)	
Bedding (dips)	Massive; locally thinly to thickly bedded, locally laminated and cross-laminated (less than 10°)
Skeletal & nonskeletal components	Planktonic foraminifers, coralline algae (fragments) bivalves, echinoderms, gastropods, peloids, volcaniclastics
Dominant grain sizes	Mud to fine sand; locally pebble
Porosity (visual estimates)	5–40%; avg. 10–15% MO, BP, VUG
Cements	Rare equant dolomite, equant calcite locally
Other	Extensive bioturbation characteristic

Table 1. Continued.

Red Algal-Rich Packstone/Grainstones (RAPG)	
Bedding (dips)	Massive, thinly to thickly bedded (less than 10 degrees)
Skeletal & nonskeletal components	Coralline algae (fragments, rhodoliths), bryozoans, bivalves, gastropods, benthic and planktonic foraminifers, solitary corals, echinoderms, serpulids, peloids, intraclasts, volcaniclastics
Dominant grain sizes	Silt to granule
Porosity (visual estimates)	5–35%; avg. 15–20% MO, BP, VUG
Cements	Equant dolomite, equant calcite
Other	Common as channel-fill deposits or wedge-shaped deposits

Volcaniclastic/Carbonate Conglomerate (VCC) and Volcaniclastic Ss/Cgl (VSC)		
	VCC	**VSC**
Bedding (dips)	Massive	Thinly to thickly bedded (0–30°)
Skeletal & nonskeletal components	Volcanic clasts, coralline algae (fragments), bryozoans, bivalves, solitary corals, gastropods, serpulids, benthic and planktonic foraminifers, echinoderms	Volcanic clasts, coral, serpulids, coralline algae (fragments) bivalves, gastopods, bryozoans, peloids
Dominant grain sizes	Mud to boulder	Mud to cobble
Porosity (visual estimates)	5–45%; Avg. 15–20% MO, BP, WP, VUG	10–35%; Avg. 20% MO, BP, WP, VUG
Cements	Equant dolomite, equant calcite	Rare: locally equant dolomite
Other	Mostly matrix-support texture	Common as channel-fill deposits or wedge-shaped deposits
Megabraccia 1 (MB1) and Megabreccia 2 (MB2)		
	MB1	**MB2**
Bedding (dips)	Massive, isolated blocks; encased in strata dipping less than 10°	Layers 1–3 m thick; some isolated blocks encased in strata dipping 0–10°
Skeletal & nonskeletal components	*Tarbellastraea, Porites,* coralline algae (crusts, fragments), bryozoans, peloids, planktonic and benthic foraminifers, serpulids, bivalves, echinoderms, gastropods	*Porites,* coralline algae (crusts, fragments), serpulids, bryozoans, foraminifers, gastropods, bivalves, echinoderms, peloids, intraclasts, composite grains, volcaniclastics
Dominant grain sizes	Mud to boulder	Coarse sand to boulder
Porosity (visual estimates)	0–25%; Avg. 5–15% MO, BP, GF, VUG	5–40%; Avg. 10–20% MO, BP, GF, VUG
Cements	Fibrous to bladed isopachous calcite; equant calcite cement; bladed-equant dolomite	Fibrous to bladed isopachous calcite; equant calcite, equant dolomite
Other	Clast matrix is more abundant than coral framework	Matrix in clasts is more abundant than coral framework

Figure 7. Sequence Boundary 1a (SB1a) characteristics. (A) Concentric, spheroidal fractures in volcanic clasts. These concentric fractures form at the contact between volcanic basement and overlying carbonate. For the most part, fractures are filled with marine carbonate debris. The concentric fracturing is likely indicative of spheroidal weathering of volcanic clasts in a subaerial environment before influx of the marine carbonate. (B) Volcanic conglomerate at SB1a with marine carbonate matrix. A thin reddish weathering rind on the rounded volcanic clasts is likely indicative of alteration during subaerial exposure.

Figure 8. Depositional Sequence 1A (DS1A), Cerro Del Cuervo Beach location. Deep reentrants were eroded into a volcanic mass flow deposit before deposition of later carbonates indicating the volcanic mass flow deposit formed coastal headlands that were later filled by carbonate deposits. On the left, marine carbonate characterized by deformed bedding and large clasts consisting of volcanic mass flow deposit fill in a paleoerosional reentrant in the volcanic mass flow unit (exposed on the right) and lap out against its erosional margin.

in diameter) dark gray to black, jointed volcanic clasts in a lighter gray matrix. The mass flow deposit deformed and brecciated some underlying carbonates. However, the volcanic mass flow unit apparently did not thermally alter the underlying carbonate beds. The mass flow unit was extensively eroded into coastal headlands shortly after deposition, likely by wave erosion in very shallow-marine environments. This erosion formed reentrants at least 10 m deep in the volcanic deposit that were filled by later carbonates and clasts eroded off of the mass flow unit itself (Figure 8). Upper DS1A deposits consist of carbonate packstones/grainstones and volcaniclastic sandstones/conglomerates. These strata contain large bivalves, benthic foraminifera, bryozoans, corals, echinoderms, red algae, and local red algal rhodolith beds.

The upper sequence boundary (SB1b) can be traced along 25–30 m of relief over a lateral distance of 150–700 m, from present-day sea level upslope to where it onlaps volcanic basement and merges with SB1a. The upper unit of the sequence is extensively fissured and this fissured unit forms SB1b (Figure 9A). Most fissures extend subvertically about 1 m below the sequence boundary and are only centimeters wide. Some of the fissures form three-dimensional networks that break sediment in the unit into autoclasts (Figure 9A, B). Some fissures widen upward toward the top of the sequence boundary. Many fissures are lined with a crust of iron oxide and are filled with multiple generations of sediment, some of which appear to be marine in origin and others that appear to be nonmarine in origin (Figure 9C). Generations of fissure fillings tend to crosscut one another indicating multiple events of fracturing and filling. Fissures are filled with tan and pink marine carbonate and volcaniclastic sediments. In samples where marine sediment fills fissures, the volcanic fragments in the host rock matrix are extremely altered and stained with iron oxides, whereas the volcanic grains in the fissure fillings are unaltered and lack the iron oxide staining (Figure 9D, E). Similar alteration is concentrated at the sequence boundary (SB1b) and decreases downward (Figure 9B). Thus, alteration preceded influx of marine sediment. Some fillings are autoclastically brecciated micrite and microspar (Figure 9F). Others are grainy and contain iron oxide-rich internal sediment and grain coatings, some of which have meniscus fabrics or form bridges between grains (Figure 9G). Some of the fillings contain sinuous cylindrical pores filled with porous microspar that appear to be rhizoliths. The above described features are indicative of alteration by subaerial exposure and karst formation before influx of overlying marine sediments.

Depositional Sequence 1B (DS1B)

DS1B is the second depositional sequence in the Las Negras area and ranges in thickness from about 20 m to over 150 m (Figure 3). DS1B strata occur in the following general vertical succession from bottom to top.

1. Initial development of a surface of subaerial exposure (SB1b).
2. Volcaniclastic sandstone/conglomerate mixed with carbonate matrix immediately overlies SB1b.
3. Fining-upward cycles composed of coarse-grained carbonates at the base and burrowed fine-grained carbonates at the tops. For the most part, cycles onlap paleotopography. Several wedges of coarse-grained carbonates locally display basinward progradational geometries.
4. Development of the upper sequence boundary, SB2.

The basal sequence boundary (SB1b) is the subaerial exposure surface described at the end of the DS1A section. This sequence boundary is overlain by a volcaniclastic/carbonate sandstone and conglomerate or a molluskan packstone.

The rest of DS1B is composed mostly of normal-marine strata that were deposited on an open-marine ramp. Most strata dip less than 12° and onlap SB1b and volcanic basement rocks (SB1a) (Figures 3, 4, 10A). Locally, beds downlap just above SB1b. Also locally, in proximal positions at the top of DS1B, several wedges of coarse-grained packstones/grainstones or red algal-rich packstones/grainstones appear to have basinward progradational geometries. DS1B strata are characterized by red algal-rich packstones/grainstones or coarse-grained packstones/grainstones interbedded with burrowed fine-grained (foraminiferal?) wackestones/packstones (Figures 4–6, 10B). Locally, these interbedded strata are arranged in five to six fining-upward cycles (approximately 2–8 m thick) with red algal-rich packstones/

Figure 9 (pages 418 and 419). Sequence Boundary 1b (SB1b) characteristics. (A) Three-dimensional networks of fissures are characteristic of this sequence boundary. Such fissures are not consistent with simple submarine dilational opening of fractures and instead are more consistent with alteration during subaerial exposure at SB1b. (B) Arrow points to the upper surface of SB1b. The vertical fissures that extend downward from the surface are filled with sediment from the overlying sequence. The strata immediately underlying the upper surface are stained red and there is a gradational change downward to a greenish unaltered color. This gradational change downward is indicative of alteration during subaerial exposure and paleosol formation. (C) Oblique cut through fissure associated with SB1b. Arrows point to margins of a fissure first lined with iron oxide crust and then later filled with marine carbonate sediment from the overlying sequence. Iron oxide crusts support an origin of subaerial exposure for the sequence boundary. (D) Arrows point to the margins of a fissure. The volcaniclastic-rich carbonate host material surrounding the fissure is highly altered, stained with iron oxide, whereas the fissure-filling is unaltered even though it is also rich in volcaniclastic debris. The fissure is filled with later mixed carbonate/volcaniclastic sediment. The observation that the fissure filling is unaltered indicates an event of oxidative alteration (probably subaerial exposure) before fissure filling. (E) Photomicrograph of fissure margin. Note the iron oxide crust lining the fissure and later filling patterns of carbonate debris along irregularities of the margin. The iron oxide crust is associated with subaerial exposure alteration during fissure formation. Scale bar is 0.25 mm. (F) Photomicrograph of autoclastic brecciation in a fissure filling. Autoclastic brecciation may be associated with desiccation and wetting. Scale bar is 0.5 mm. (G) Photomicrograph fissure fill just below SB1b surface. Arrow points to iron oxide–rich micritic cement that forms meniscus fabric indicating vadose conditions likely associated with subaerial exposure along the SB1b surface. Scale bar is 0.25 mm.

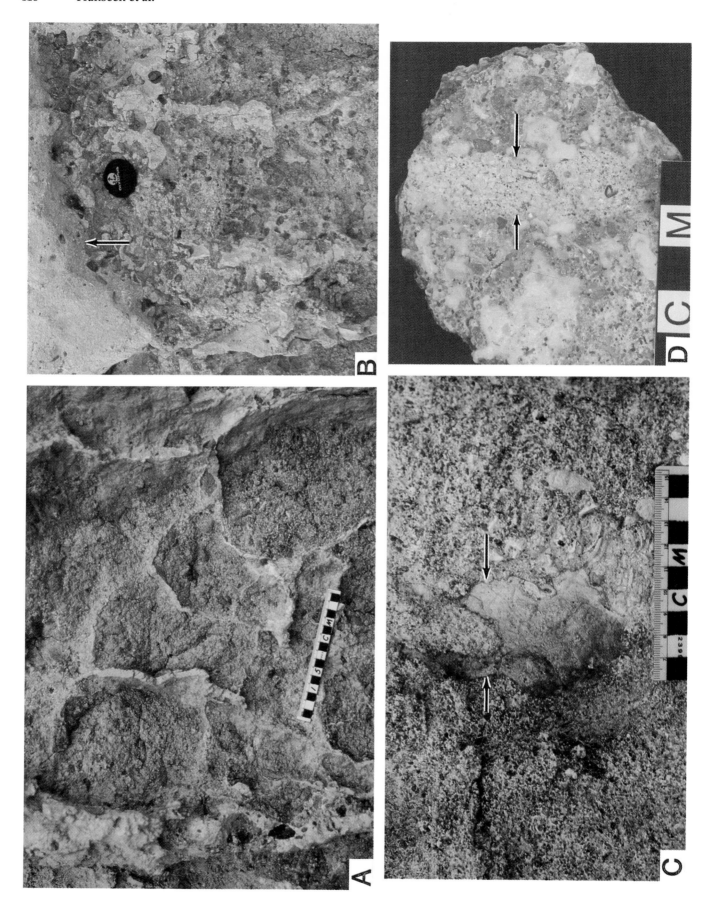

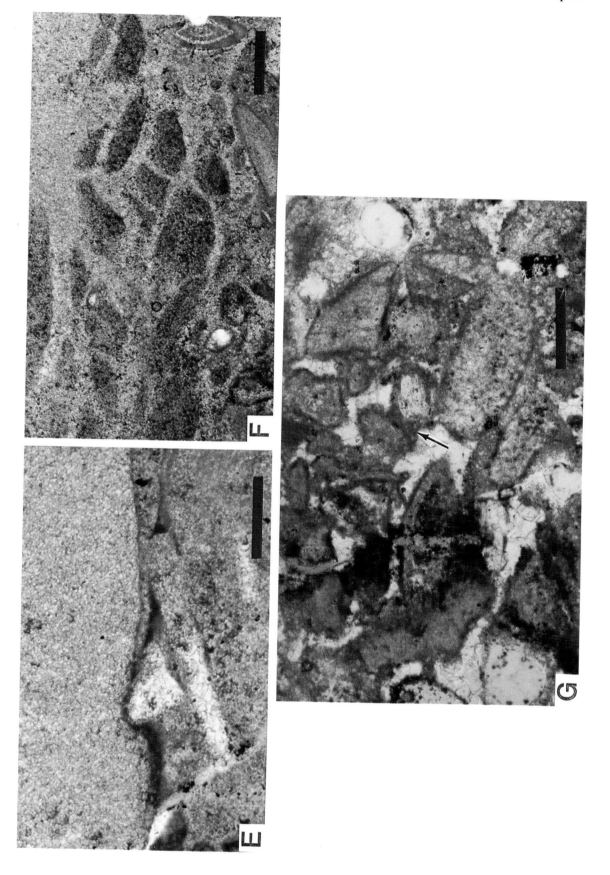

Figure 10. Depositional Sequence 1B (DS1B) characteristics. (A) DS1B strata are characterized by shallow dips and onlapping geometry. Here, at the base of Cerro El Romeral at the Big Foot location in the Las Negras area, DS1B beds clearly onlap volcanic basement. The volcanic basement contact is shown by the arrows. (B) Coastal exposure of DS1B strata here consists of fine-grained wackestones/packstones, characteristically forming recessives, interbedded with some coarse-grained, more resistant carbonate layers.

grainstones or coarse-grained packstones/grain-stones at the base of cycles and fine-grained wacke-stones/packstones at the tops.

The upper sequence boundary, SB2, occurs at the base of a laterally traceable megabreccia horizon (megabreccia 1), which will be described in the next section.

Depositional Sequence 2 (DS2)

DS2 is the third depositional sequence in the Las Negras area and ranges in thickness from 1 to 30 m (Figure 3). From base to top, DS2 strata generally occur in the following vertical lithofacies succession.

1. Megabreccia 1 occurs locally at the base and consists primarily of meter- to tens of meters-sized reef blocks of *Tarbellastraea* and *Porites* framework. Steeply dipping (up to 25°) flank beds may develop on the sides of blocks. Locally, volcanic, grainstone, and wackestone clasts are admixed within the megabreccia 1 unit.
2. Local development of coarse-grained carbonate beds on the tops of megabreccia 1 blocks.
3. Burrowed fine-grained carbonates (locally draping megabreccia 1 blocks; locally interbedded with normally graded coarse-grained carbonates).
4. Local coarse-grained carbonates (channels, cross-bedding, current-ripple lamination, silt-clay drapes, rip-up clasts, and scoured bases filled with shell lag deposits occur locally).
5. Formation of upper, sharp, and erosional sequence boundary, SB3.

The lower DS2 sequence boundary (SB2) occurs at a stratigraphic horizon marked by the base of megabrec-cia 1 (Figures 3–6, 11). Megabreccia 1 consists mostly of allochthonous reef facies blocks, but also consists of

fine-grained wackestone/packstone blocks, red algal-rich packstone/grainstone blocks, coarse-grained packstone/grainstone blocks, and volcaniclastic blocks (Figure 11). Although in most places megabreccia 1 is not a continuous, traceable layer, mapping of the Las Negras area suggests separated megabreccia 1 out-crops occur at a correlatable horizon. Locally, where megabreccia 1 is in contact with underlying red algal-rich facies, the contact is sharp and erosional. No evi-dence of subaerial exposure occurs along this contact. In some distal positions, where no megabreccia 1 blocks occur, the DS1-DS2 contact is commonly grada-tional (conformable) (Figure 5). Correlation of the megabreccia 1 unit indicates a minimum shelf-to-basin relief of 160 m over a lateral distance of 1.5–2.0 km during initial DS2 deposition.

The basal DS2 sequence boundary (SB2), at the base of megabreccia 1, is interpreted as a sequence boundary resulting from a lowering of base level for the following reasons.

1. The polymictic texture of megabreccia 1 strongly suggests erosion of preexisting materi-al and not just shedding of reef materi-al during highstand reef growth.
2. Megabreccia 1 was deposited onto lower energy carbonates of DS1B and occurs as a debris-flow bed, channelized deposits, and isolated blocks or groups of several isolated blocks (some with flanking beds that locally downlap onto previously deposited megabreccia blocks). It appears that this unit was deposited over a significant amount of time and was not just simply a single storm or earthquake deposit.
3. Megabreccia 1 is areally extensive and can be correlated throughout the area.
4. Shallower-water conditions are suggested by possible downslope reef growth after initial

Figure 11. Sequence Boundary 2 (SB2) and Megabreccia 1 (MB1). (A) MB1 reef blocks (arrows) overlie fine-grained wackestones/packstones and form the basal deposits of DS2. Here the MB1 unit forms a traceable layer. Note the person on top of the block on the far left side of the photograph for scale. (B) MB1 unit scoured into underlying fine-grained carbonate beds. Here, MB1 consists of volcaniclastic pebbles and cobbles, reef blocks, and blocks of fine-grained wackestones/packstones and coarse-grained packstones/grainstones (note upturned bedding). MB2 forms the top of the hill in the photograph and is in sharp contact with the underlying flat-bedded fine-grained carbonates that overlie MB1. (C) Reef block consisting of robust *Tarbellastraea* coral sticks. These partly moldic sticks, about 6 cm in diameter and 10–75 cm long, are oriented on their sides in an allochthonous block. (D) Transported red algal facies block in channelized MB1 deposit. Note vugs filled in with greenish volcaniclastic material indicating the vugs formed prior to transport, likely in an upslope setting subjected to subaerial exposure.

transport, the development of coarse-grained carbonate grainstone beds as flank beds and capping layers on debris blocks, and the transition from fine-grained wackestones/packstones directly below (in DS1B) into reef blocks and grainstone beds.

5. Local erosion of a red algal-rich grainstone flank bed which is overlain by an echinoderm, fossil fragment grainstone bed that laps out onto the erosion surface as traced into a reef block. This likely indicates high-energy conditions associated with shallow-water facies.

6. Vugs in some transported red algal-rich and *Porites*-rich blocks are filled with green volcaniclastic debris indicating the vugs likely formed prior to transport from possible sub-

aerial exposure in an upslope area (Figure 11D).

DS2 beds overlying megabreccia 1 have shallow primary dips and locally onlap underlying strata (Figures 4–6). Some beds drape underlying strata, mostly megabreccia 1 blocks (Figure 5). Some lenticular-shaped beds are interpreted as channel-fill deposits. The fine-grained wackestone/packstone facies is volumetrically the most abundant facies in DS2. The abundance of fine-grained wackestones/packstones and the shallow dips suggest deposition in a deeper ramp/slope setting. However, the presence of megabreccia 1 reefal blocks and grainstone beds indicates shallow-marine conditions. It is likely that transported megabreccia 1 reef blocks originally formed as (patch?) reefs in upslope (surf zone?) locations at the edges of grainstone wedges during upper

DS1B time, prior to their erosion and transportation to preserved downslope positions.

The upper sequence boundary (SB3) is a sharp, erosional surface marked by deposition of another megabreccia (megabreccia 2), or by other coarse-grained deposits which will be described in the next section.

Depositional Sequence 3 (DS3)

DS3, volumetrically, is the most significant of the depositional sequences in Las Negras and has a preserved thickness ranging from 20–70 m (Figure 3). The original thickness is unknown, but was probably higher, as the uppermost portions have been eroded and are unconformably overlain by the Terminal Carbonate Complex (TCC) sequence. DS3 is characterized by fringing reef and clinoform strata (Figures 3–6). DS3 strata show the following generalized vertical succession from base to top.

1.   Formation of the sharp, erosional basal sequence boundary (SB3).
2.   Deposition of megabreccia 2, coarse-grained carbonates or volcaniclastic sandstones/conglomerates. Megabreccia 2 occurs in medial and distal slope locations and coarse-grained carbonates, volcaniclastic sandstones/conglomerates occur in proximal to distal slope positions.
3.   Possible subaerial exposure of megabreccia 2.
4.   Deposition of fringing reef strata characterized by local preservation of in-place *Porites* framestone, reef talus of mostly slumped *Porites* framestone reef blocks (with rotated geopetals) deposited on high-angle foreslopes, and downlapping foreslope strata consisting of coarse-grained carbonates, fine-grained carbonates, or volcaniclastic sandstones/conglomerates.
5.   Deposition of *Halimeda*-rich facies forms distinct units in youngest DS3 reef core to distal slope strata.
6.   Subaerial exposure of DS3 to form the upper sequence boundary (SB4).

The trace of the lower DS3 sequence boundary, SB3, has a gently sloping profile with a minimum preserved shelf-to-basin relief of 120-130 m. SB3 is a sharp surface (locally erosional on a scale of meters) and is overlain by megabreccia 2, coarse-grained packstones/grainstones, or volcaniclastic sandstones/conglomerates (Figures 3–6, 12A,B). SB3 is a distinct, mappable horizon throughout the Las Negras area.

The base of DS3 is picked as a sequence boundary (SB3) for the following reasons.

1.   SB3 is a sharp, erosional, and scoured surface that is laterally extensive. Shallower water reef facies, coarse-grained packstones/grainstones (locally cross-bedded), or volcaniclastic sandstones/conglomerates overlie SB3 as compared to deeper water fine-grained wackestones/packstones directly below.

2.   Megabreccia 2 occurs throughout the area as debris flows and as isolated blocks or groups of isolated blocks that locally developed flank grainstone beds. Megabreccia 2 is composed of a variety of clast types including *Porites* reef-facies debris, volcanic clasts, coarse-grained packstone/grainstone blocks, and fine-grained wackestone/packstone blocks. Apparently, megabreccia 2 deposition occurred over some significant time and was not a single storm or earthquake deposit nor just simply shedding of reef material during highstand reef growth.
3.   The top of megabreccia 2 in medial and distal slope positions exhibits red, yellow, and orange coloration that is not present along other horizons in the field area (Figure 12C). The tops of some areas of megabreccia 2 contain some sinuous and cylindrical pores that appear to cut across rock fabric (Figure 12D). These pores are ancient features that have similar diagenetic histories to primary pores in the blocks with the exception of submarine cement. They cannot be attributed to dissolution of any known marine biotic component. They lack typical morphologies associated with borings. Most of these cylindrical pores are associated with minor autoclastic brecciation and circumgranular cracking. Although we have searched for evidence of marine hardground development at this surface, none has yet been found. Thus, we are left with a tentative interpretation that perhaps the alteration on the top of megabreccia 2 is from subaerial exposure of megabreccia 2 just after its deposition and that the cylindrical pores are rhizoliths. This interpretation should be considered tentative because evidence for subaerial exposure is subtle and not altogether convincing as observed in the field.
4.   Local onlap geometry of basal coarse-grained packstones/grainstones onto SB3 in distal and proximal slope positions on several hills in the area.

DS3 strata consist of low-angle (less than 12°) to high-angle (up to 30°) prograding fringing reef complex strata, composed predominantly of *Porites* framestone (Figure 13A), that downlap above SB3 (Figures 3–6). Some lowermost DS3 strata in distal positions and proximal positions are flat bedded and apparently onlap the lower DS3 sequence boundary. Locally, in-place massive to flat bedded reef-core strata (reef core/talus facies) are present. Some reef and fore-reef slope deposits alternate with wedges of siliciclastic debris (Figures 3, 6). There are several cycles of alternation and the latest two cycles display in-place reef deposits that developed on fore-reef slope beds of previous cycles representing "downstepping" of the last two cycles (Figure 6). Trough cross-bedded, coarse-grained carbonates occur locally (Figure 13B) and lenticular-shaped beds (channel-fill deposits) and wedge-shaped foreslopes (thinning and

Figure 12. Sequence Boundary 3 (SB3) and Megabreccia 2 (MB2). (A) Allochthonous *Porites* reef blocks of MB2 locally overlie the sharp, erosional surface that forms SB3 (arrow). (B) As traced into proximal positions on La Molata, SB3 is a sharp, scoured, and channeled surface underlain by fine-grained wackestones/pack-stones (upper DS2) and overlain by basal DS3 coarse-grained packstone/grainstone. (C) Sample taken from the top of MB2 in a medial slope position on La Molata containing red staining from iron oxide. This staining is characteristic of the top portion of MB2 and does not occur in any regular sense in the strata above or below. Some red-stained patches appear to have been brecciated autoclastically. These observations are consistent with but not diagnostic of subaerial exposure on the top of MB2. (D) Photomicrograph of sample from top of MB2 with voids that appear to be "lined up" as if they were a cross section through a single cylindrical and sinuous pore. Such pores appear to cut across the rock fabric and may represent small-scale rhizoliths. These features are lined with a dolomite cement, which indicates they did not form from late-stage processes and instead may have resulted from subaerial exposure of MB2 shortly after deposition. Scale bar is 1 mm.

pinching-out downslope) are common. For upper DS3 strata, the earliest traceable reef core-to-distal slope clinoforms indicate a relief of 40 m (Figure 3). Additional tracing of equivalent strata exposed in basinal positions indicates a minimum relief of 150–180 m. Youngest DS3 reef strata, characterized by abundant *Halimeda*, indicate a minimum reef core-to-distal foreslope relief of 90 m (Figures 3, 5, 6, 13C).

The sequence boundary at the top of DS3 (SB4) represents erosion and subaerial exposure of the underlying units and will be discussed in the next section.

Terminal Carbonate Complex Sequence (TCC)

The uppermost depositional sequence (TCC) is up to 30 m thick and caps the surface of subaerial exposure at the top of DS3 (SB4) in the Las Negras area

(Figure 3). The TCC deposits form the last Miocene sequence. The TCC sequence shows the following generalized vertical succession from base to top.

1. Subaerial exposure and erosion of DS3 to form SB4.
2. Local deposition of volcaniclastic/carbonate conglomerate.
3. Deposition of three to four cycles of marine carbonate dominated by oolite and minor stromatolites.
4. Subaerial exposure of Miocene strata before deposition of Pliocene deposits.

The basal TCC sequence boundary (SB4) locally contains evidence of subaerial exposure consisting of chalkification, autoclastic brecciation, micritization, rhizoliths, laminated crust, and fenestrae (Figure 14). Several meters to several 10s of meters of underlying

424   Franseen et al.

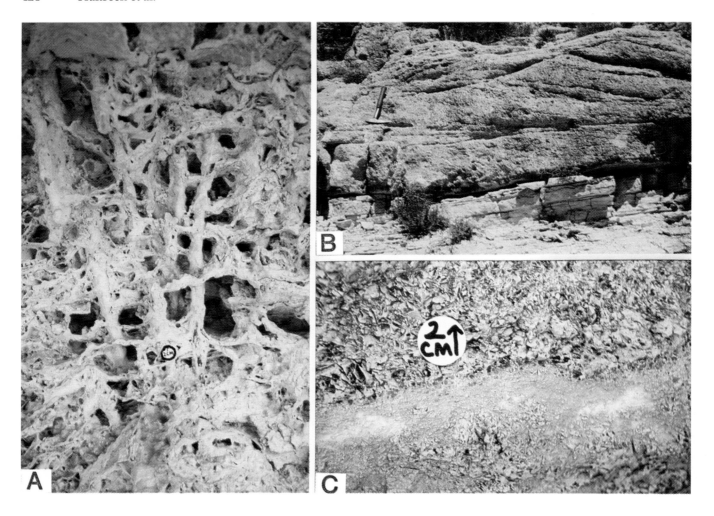

Figure 13. Depositional Sequence 3 (DS3). (A) Reef framework in upper DS3 fringing reef strata consisting of stick and horizontal *Porites* morphologies. Interframework porosity may be original primary porosity or may be from leached matrix. White disc (bottom center) is 2 cm in diameter. (B) Trough cross-bedded, coarse-grained carbonate strata locally occur in DS3. (C) *Halimeda*-rich facies. *Halimeda* occurs with abundance only in youngest DS3 strata. On La Molata, it forms several distinct layers traceable from reef crest to distal slope positions. In distal slope locations (this photograph), *Halimeda* is the exclusive carbonate grain type in some beds.

DS3 reef core/talus facies strata are estimated to have been eroded prior to TCC deposition. SB4 has a demonstrable 90 m of relief from proximal to basinal positions in Las Negras and shows local relief on the top of La Molata of approximately 35 m (Figure 3). Locally, the basal deposit of the TCC consists of volcaniclastic conglomerate with some admixture of carbonate clasts and oolite. TCC strata consist predominantly of topography-onlapping and topography-draping strata consisting of three to four cycles of cross-bedded, coated grain carbonates (including oolites), stromatolites, and restricted marine facies (Figure 15).

The subaerial exposure surface at the top of the TCC is represented by a well-developed caliche breccia similar to and presumably the same as what Esteban and Gíner (1980) termed the "Pre-Pliocene breccia." In some areas such as Mesa Roldan, just 10 km to the north, strata mapped as Pliocene overlie

Miocene strata in topographic positions near present-day sea level. This contact is marked by karstic fissures that have cut into the underlying Miocene units. Fissures are filled with calcite cement and multiple generations of internal sediment deposited in the vadose zone (Figure 16) overlain by the sediment mapped as Pliocene age strata. If the mapping is correct, the pre-Pliocene exposure surface is traceable to present-day sea level and has a demonstrable relief of 200 m.

QUANTITATIVE RELATIVE SEA-LEVEL CONTROLS ON SEQUENCE DEVELOPMENT

In this section we interpret relative sea level and the associated depositional history for each sequence using facies patterns, stratal geometries, nature of

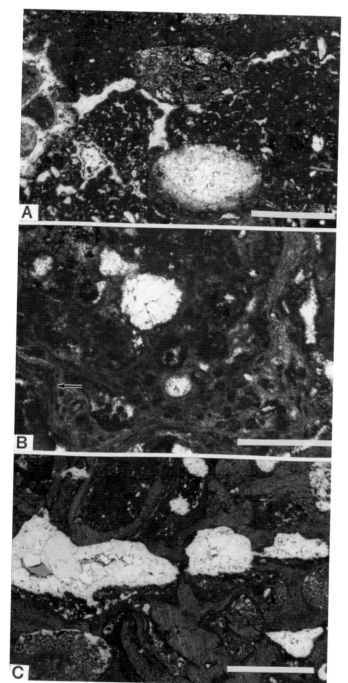

Figure 14. Sequence Boundary 4 (SB4). (A) Photomicrograph from sample immediately below SB4 showing well-developed fenestral fabric indicative of subaerial exposure along SB4. Scale bar is 1 mm. (B) Photomicrograph from top of DS3 (SB4). Note laminated crust (arrow) of alternating fine spar and micrite coatings of clots. Such crusts are well known from formation of paleosols and indicate SB4 was a surface of subaerial exposure. Scale bar is 0.5 mm. (C) Photomicrograph from top of DS3 (SB4). Notice aligned voids that cut across the rock fabric, which suggest a sinuous and cylindrical shape and appear to have formed as rhizoliths indicating subaerial exposure at SB4. Scale bar is 1 mm.

sequence boundaries, and pinning points. The basic principals used in this section were developed in earlier studies (e.g. Wilson, 1967). Some further application and methodologies appear in some recent works (e.g., Goldstein, 1988; Goldstein et al., 1990, 1991).

For construction of the quantitative relative sea-level curves (Figure 17), we introduce the term "pinning point." Pinning point is herein defined as a point of quantitative constraint on the topographic position of sea level relative to an arbitrarily defined geologically useful starting elevation. For example, in a marine sequence, the downslope extent of subaerial exposure along a surface represents a known elevation of relative sea level and therefore is a pinning point. If sea level rises and covers the slope and in-place reef-crest facies, which generally forms at or very near sea level, is deposited at an elevation 100 m above the initial pinning point, then another pinning point is provided and indicates a relative rise in sea level of 100 m. We point out, however, that in most instances, pinning points are useful only as minimum magnitudes of relative sea-level elevation change due to traceability limits imposed by modern outcrop position or preservation, because many marine lithologies are not precise depth indicators, and because many nonmarine sediments or surfaces cannot be used to determine elevation above paleosea level. The applicability of pinning points assumes that some quantitative measure of sea-level position can be determined. If a paleoslope is preserved without deformation or if paleoslope can be determined by onlap of overlying units or geopetal fabrics, pinning points of relative sea-level elevation can be easily applied. For some applications, stratigraphic thickness can serve to define pinning points. For all of these applications, compaction and lateral variation in amount of compaction must either be determined or assumed. Because pinning points define changes in relative sea-level elevation, they include a component of compaction, subsidence or uplift, and a component that is true sea-level variation.

Pinning points form the quantitative template on which more interpretive relative sea-level curves can be based. The curve in Figure 17A was constructed by simply connecting pinning points with straight lines, and is termed a pinning point curve. Where pinning points are absent and there is uncertainty in maximum or minimum sea-level position, dashed lines are used and curves are left open-ended. The horizontal axis on the curves represents relative sea-level position in meters; zero is present-day sea level. The vertical axis represents time. For the curves, time is represented in only a generalized sense due to the lack of precise age dating for the sequences. The interpretive relative sea-level curve in Figure 17B is constructed by using the pinning point curve as a quantitative template and incorporating facies patterns, stratal geometries, and stacking patterns to build a more interpretive, more detailed, and less quantitative version of the relative sea-level history.

In construction of the curves, it is assumed that Miocene paleotopography is preserved along sloping

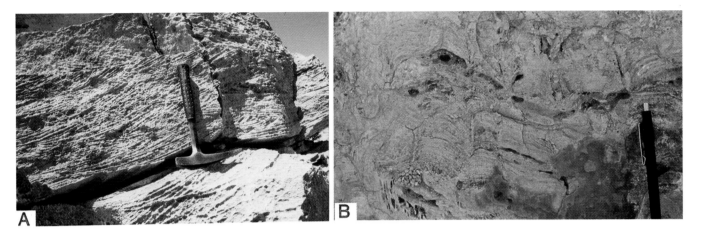

Figure 15. Terminal Carbonate Complex (TCC). (A) Cross-bedded oolites are a common facies in TCC strata. (B) Stromatolitic layers occur along several horizons in TCC strata in proximal locations on La Molata.

Figure 16. Karstic filling below possibly pre-Pliocene surface of subaerial exposure, Mesa Roldan. Notice flow stone overlain by iron-oxide-rich karstic internal sediment. These observations indicate that subaerial exposure of Miocene strata may have predated deposition of Pliocene strata. The exposure features at this upper TCC horizon are termed the "Pre-Pliocene Breccia" by Esteban and Gíner (1980).

strata. Although at least minor compaction is likely to have occurred, we made no correction for compaction because of the lack of stylolites, pressure-solution features, or other evidence of significant compaction, and because maximum burial depth was probably small (Addicott et al., 1978; Franseen, 1989). Similarly, no adjustments were made for tectonic tilting because geopetal fabrics suggest little if any tectonic tilting. In addition, Esteban and Gíner (1980) showed that

marine planation surfaces (pre-Miocene carbonates and pre-TCC) occur at essentially the same heights throughout southeastern Spain. However, we cannot rule out minor tectonic tilting of the strata. Therefore, for every 1° correction in dip of the strata, the vertical relief would change 1.7–1.9 m/100 m of horizontal distance (using data from La Molata geologic map and the basic formula in which the difference in elevation = horizontal distance × tangent of vertical angle). The magnitudes of relative sea-level fluctuations determined from pinning points would be adjusted similarly. Thus, although some minor adjustments could be made, it would appear that differences in Miocene paleotopographic elevation are preserved in the modern exposures of Miocene strata in the Las Negras area, and that interpreted changes in the elevation of sea level relative to the Miocene strata represent relative changes in the elevation of Miocene sea level. In total, over 200 m of relief is exposed along the Miocene outcrops of Las Negras, allowing for development of pinning point curves that show maximum relative sea-level variation of more than 200 m. Apparently, several scales of duration and amplitude of relative sea-level fluctuations are represented by the curves. However, it will remain unknown if the scales of fluctuations are third-, fourth-, fifth-, or some other order of sea level fluctuation until better age control is obtained for these strata.

DS1A

Pinning Point Curve

The basal contact with the volcanic basement, SB1a, is a surface of subaerial exposure that can be traced to present-day sea level. Therefore, the entire volcanic island on which the now-exposed overlying carbonates were deposited was originally entirely subaerially exposed. The lowermost exposure in which marine carbonates overlie SB1a represents a

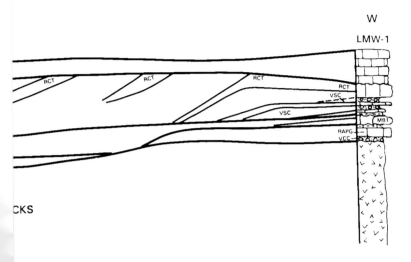

W

LMW-1

RCT

RCT

RCT

RCT

VSC

VSC

MBT

RAPG

VCC

CKS

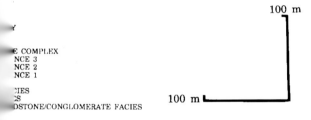

100 m

100 m

E COMPLEX
NCE 3
NCE 2
NCE 1

:IES
:S
0STONE/CONGLOMERATE FACIES

:STONE/GRAINSTONE FACIES
:ONATE CONGLOMERATE FACIES

SW

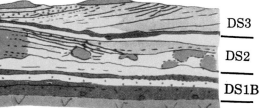

DS3

DS2

DS1B

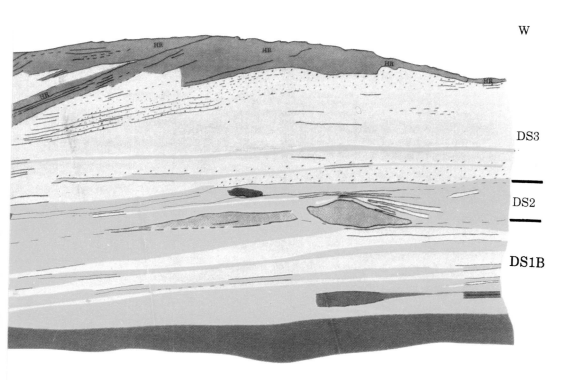

W

DS3

DS2

DS1B

E

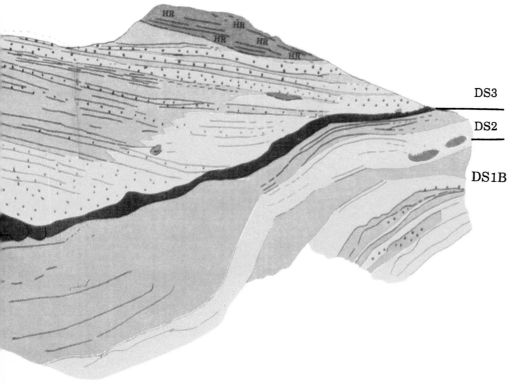

DS3

DS2

DS1B

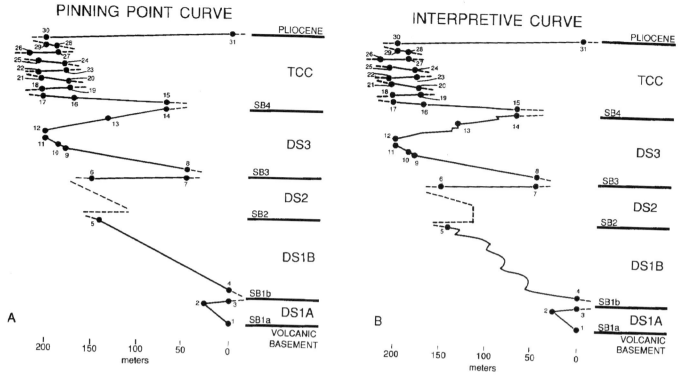

Figure 17. Relative sea-level curves for the Las Negras area. (A) Pinning point relative sea-level curve. Horizontal axis is relative sea-level position in meters; zero is present-day sea level. Vertical axis is time, here depicted in only a general sense. Pinning point numbers are explained in the text. (B) Interpretive relative sea-level curve using pinning points, facies relationships stratal geometries, and stacking patterns. Horizontal axis is relative sea-level position in meters; zero is present-day sea level. Vertical axis is time, here depicted in only a general sense.

pinning point (pinning point #1) to start deriving a quantitative relative sea-level curve for the region. Much of the volume of DS1A consists of coarse-grained, mud-poor carbonate facies, some with low-angle cross-beds. Moreover, the volcanic mass flow unit shows evidence of having had at least 10-m-deep reentrants and caves eroded into it before further deposition of DS1A carbonates within the reentrants (Figure 8). It appears that the volcanic mass flow deposit was eroded into coastal headlands during deposition of DS1A. Thus, much of DS1A was deposited in high-energy marine conditions in which water depth for the most part was not deeper than several meters. Because mapping of the top of DS1A (SB1b) shows that it onlaps volcanic basement at a consistent elevation of 25–30 m above present-day sea level, the onlap of DS1A strata up to that point represents a relative rise in sea level of approximately 25–30 m (pinning point #2).

The sequence boundary between DS1A and DS1B (SB1b) represents a surface of subaerial exposure that can be traced upslope from present-day sea level to where it onlaps against volcanic basement and merges with SB1a at a position approximately 25–30 m above present-day Mediterranean sea level. The subaerial exposure evidence on SB1b can be traced downslope

to present-day sea level. Thus, SB1b represents a minimum relative sea-level fall after DS1A deposition of approximately 25–30 m (pinning point #3).

Interpretive Curve

Because DS1A facies are all shallow-water marine carbonates and represent essentially the same water depth, the curve between pinning points #1 and #2 is smooth and lacks smaller scale deepening and shallowing trends to reflect that deposition essentially kept pace with sea level as the volcanic basement was transgressed. There was no sedimentary record deposited during the relative fall in sea level indicated between pinning points #2 and #3. Therefore, the curve is drawn as a smooth line for that segment. However, the actual sea-level history of this segment could have been much more complex.

DS1B

Pinning Point Curve

DS1B strata consist of marine deposits that, for the most part, onlap SB1a and SB1b from the basal SB1b exposure surface near present-day sea level (pinning point #4) to the top of DS1B, which can be traced to elevations of 150 m. Thus, the flooding of SB1b and deposition of mostly onlapping DS1B strata in normal-marine water in upslope positions required a major relative sea-level rise of at least 150 m (pinning point #5).

Interpretive Curve

Several locations in the area contain cycles within the onlapping DS1B sequence. As revealed on the Big Foot location of Cerro El Romeral, there are five to six of these onlapping cycles characterized by coarse-grained carbonates (locally cross-bedded and containing red algal rhodoliths) at the base of the cycle that pass gradationally upward into (foraminiferal?) fine-grained carbonates that are generally extensively bioturbated. Both the coarse- and fine-grained lithologies appear to have been subtidally deposited because no peritidal facies or subaerial exposure features were identified. However, the coarse-grained carbonates were likely deposited in somewhat shallower marine water, although exact water depths are not known. In this interpretation, each coarse- to fine-grained carbonate cycle would represent a relative deepening upward. Assuming that these cycles are not autogenic in origin (an assumption that has not as yet been effectively evaluated), the interpretive curve is drawn to show the five to six cycles as smaller scale fluctuations in an overall rising relative sea level between pinning points #4 and #5. If the cycles are allogenic, it is still not known if the base of each cycle represents a relative fall in sea level, a stillstand, or a slow down in rate of relative rise. Magnitudes of any fluctuations are also unknown but are presumed to be lower than fluctuations that caused the major sequence boundaries.

DS2

Pinning Point Curve

The basal DS2 sequence boundary (SB2) occurs at the basal contact with megabreccia 1. SB2 is interpreted as a sequence boundary caused by a relative fall in sea level for the reasons discussed earlier. However, even though the base of megabreccia 1 apparently represents a relative fall in sea level, there is no evidence in this study area of subaerial exposure at the base or at the top of megabreccia 1; any subaerial exposure would have occurred in an up-slope position and evidence for it must have been eroded away. Because of the lack of "in-place" subaerial exposure in the rocks, we are unable to place a known pinning point for relative sea level and, therefore, show the inferred drop of several tens of meters by a dashed line.

Upper DS2 deposition was dominated by fine-grained (foraminiferal?) carbonates suggestive of lower energy, more distal conditions and similar to the deposits in upper portions of DS1B. This return to fine-grained marine deposits in upper parts of DS2 likely represents a relative rise in sea level, but the magnitude of relative rise in sea level is difficult to determine from the rocks. Because of the similarity of the upper DS2 facies with upper DS1B facies, and because these strata can be traced to a similar upslope position, we indicate a relative rise in sea level of unknown amount with a dashed line.

Interpretive Curve

Due to the lack of in-situ subaerial exposure features, the amount of fall interpreted for the base of megabreccia 1 is only an estimate and is drawn on the curve as a dashed line. The interpretation that megabreccia 1 deposition occurred over some time and the presence of coarse-grained carbonate beds flanking and overlying some megabreccia 1 blocks are represented on the curve as a stillstand of relative sea level. The return to fine-grained carbonate deposition in upper DS2 is interpreted to represent a relative rise in sea level. Because there is no pinning point for the top of DS2, the relative rise is drawn with a dashed line to the same level as for the top of DS1B, plus the additional thickness of DS2 strata, due to similarity in facies.

DS3

Pinning Point Curve

The base of DS3 is marked by a sharp, locally erosional surface (SB3) that is overlain by megabreccia 2, coarse-grained carbonates, or volcaniclastic sandstones/conglomerates. The base of DS3 is picked as a sequence boundary caused by a relative sea-level drop for reasons discussed earlier.

The evidence for possible subaerial exposure at the top of megabreccia 2, which has been documented by tracing in the field, reveals important data about when megabreccias developed with respect to relative changes in sea level. Erosion and downslope deposition of megabreccia was likely induced by a relative fall in sea level. After megabreccia deposition began, sea level continued to fall and eventually the megabreccia unit may have been subaerially exposed. The basal surface of the megabreccia is the surface that most likely correlates with initial subaerial exposure upslope and the top surface of the megabreccia is a discontinuity surface.

Tracing of the possible subaerial exposure surface at the top of megabreccia 2 and correlating SB3 into proximal positions (pinning point #6) indicates paleotopographic relief of about 100 m on a possible subaerially exposed surface of marine carbonate. Thus, if evidence for exposure is correct, SB3 represents a fall in relative sea level of at least 100 m (pinning point #7). The amount of further relative fall is unknown because of limitations imposed by the downslope extent of the outcrop.

The remainder of DS3 deposits are composed of alternating cycles of marine carbonates and volcaniclastic strata that were deposited in distal to proximal foreslope and *Porites* reef environments. To deposit marine rocks on top of megabreccia 2 in the most basinward locations that contain possible subaerial exposure evidence (pinning point #8) to the point in which DS3 fringing reef strata are preserved with in-situ horizontally bedded reef-core facies (likely representing reef flats that developed at or very near sea level) required a relative rise in sea level of 130 m (pinning point #9).

The DS3 fringing reefs pass laterally into thick deposits of fore-reef slope material. An interval of aggradation and progradation of reef and fore reef slope deposits, some alternating with volcaniclastic debris, occurs in the earliest portion of reef development (Figure 3). Locally, in a later part of the early fringing reef unit in which pinning point #9 is identified, reef-crest facies (following Riding et al., 1991) occurs at a topographic position that is approximately 5 m higher than pinning point #9 and therefore represents a relative rise in sea level of 5 m (pinning point #10). As later fringing reef strata are traced farther basinward, in-situ reef-core facies with horizontal bedding (likely representing reef flats that developed at or very near sea level) occurs about 15 m higher topographically than pinning point #10 and, therefore, represents a relative sea-level rise of 15 m (pinning point #11).

The sequence boundary above the reef-flat deposits of pinning point #11 (base of TCC) is marked by significant erosion. Thus, it is not known if sea level continued to rise, was at a stillstand, or fell directly after pinning point #11. Therefore, a break is included in the pinning point curve between pinning points #11 and #12. Pinning point #12 is at the same elevation as pinning point #11, but drawn separately to illustrate that if relative sea level had continued to rise after pinning point #11, it would have had to fall past this elevation on its downward trace toward pinning point #13.

Several later reef/volcaniclastic cycles step downward toward the basin so that successively younger reefs sit on the fore-reef slope deposits of previously deposited reefs. At least one downstepped reef cycle preserves in-situ reef-crest facies (following Riding et al., 1991) that developed at or very near sea level and thus provides a pinning point for sea level position. This downstepped reef cycle with reef-crest facies (near the LMN-1 section, Figure 3) occurs approximately 65 m lower than pinning point #12 and represents a relative drop in sea level of 65 m (pinning point #13).

The surface or sequence boundary at the top of DS3 strata (SB4) displays alteration by subaerial exposure processes. This sequence boundary can be traced downslope for 110 m (from pinning point #12 to pinning point #14) and likely extends much farther as this particular sequence boundary could represent the significant drawdown of the Mediterranean Sea during the Messinian salinity crisis.

Interpretive Curve

The shape of the relative fall interpreted for the base of megabreccia 2 is unknown because stratigraphic data are lacking. Therefore, the drop between pinning points #6 and #7 is drawn in its simplest smooth form.

A relatively rapid smooth rise in sea level from pinning point #8 to pinning point #9 is interpreted from the stratigraphically abrupt landward shift of shallow-water facies (including reef crest) after deposition of megabreccia 2 and other basal DS3 shallow-water facies. A continued, possibly slower rise in relative sea level is interpreted from an interval of progradational and aggradational reef-crest and fore-reef slope strata, with the reef-crest facies of each successive unit occurring stratigraphically higher than the previous reef deposit (pinning points #10, #11). Although the significant erosion at the base of the TCC leaves the portion of the curve between pinning points #12, #13, and #14 extremely speculative, it appears that the overall rate of relative sea-level fall during the latter phases of DS3 deposition was slow enough to allow reefs to develop. The curve from pinning point #12 to 170 m is drawn as a relatively gradual drop in sea level to reflect that each successive reef cycle after pinning point #12 appears to have prograded basinward at essentially the same elevation or a slightly lower topographic position than the previous cycle.

An increase in rate of sea-level fall from 170 to 140 m is indicated on the interpretive relative sea-level curve. This is interpreted from field data and mapping which reveals in-situ reef-core facies developed on a volcaniclastic wedge (near the NL-3 section, Figure 3) about 30 m lower than in-situ reef-core facies in the previous cycle at 170 m (near the L2 section, Figure 3). Because no reef-crest facies was identified in this first downstepped cycle, exact water depth for deposition is unknown. However, if it is assumed that reef-core facies of different cycles were deposited in approximately the same water depths, then there likely was a relative drop of sea level of about 30 m from the previous occurrence of reef core at 170 m. The development of reef-core facies at 140 m is indicated on the curve by a slower rate of sea-level fall. A subsequent abrupt fall of about 5 m is indicated on the curve because of the development of another basinward downstepped volcaniclastic wedge/reef cycle at about 135 m. The stillstand in the curve at about 135 m represents the development of reef-crest facies at this position (pinning point #13). The volcaniclastic wedges (Figures 3, 6) are interpreted to have been deposited as fan-delta lobes in submarine conditions evidenced by admixed marine carbonate material and some volcanic pebbles that contain marine encrustations. The depth of water for deposition of the volcaniclastic wedges is unknown and may have been essentially the same depth as for the overlying carbonate foreslope/reef-core deposits. These cycles could have resulted from autocyclic processes such as basinward progradation of shifting fan-delta lobes or from storm events. Alternatively, each cycle can be interpreted to represent a higher frequency relative sea-level fluctuation or stillstand during an overall slow fall in relative sea level (Franseen, 1990). The interpretive sea-level curve represents this latter alternative.

Further relative fall in sea level is indicated by pinning point #14. A minor rise during the overall drop is interpreted and drawn on the curve prior to pinning point #14 because of the development of some finer grained carbonates basinward of and stratigraphically above pinning point #13, and immediately below the SB4 trace which extends to pinning point #14.

TCC

Pinning Point Curve

Marine deposition resumed after subaerial exposure of DS3 strata and resulted in deposition of TCC strata composed of a sequence of three to four shallow-marine carbonate cycles (Figure 18). Deposition of TCC strata on the topography expressed from the lowest exposure of SB4 (at 70 m above present-day sea level; pinning point #15) to initial marine deposition of TCC in proximal locations (at about 170 m above present-day sea level; pinning point #16) required an initial relative sea-level rise of about 100 m. Correlations show that cycles within the TCC drape significant paleotopography (Figure 18). If the oolites and stromatolites were deposited in no more than several meters of water depth, then position of cross-bedded oolite and stromatolites represents approximate pinning points for sea-level elevation.

Drape of such deposits over significant paleotopography requires significant fluctuation in relative sea level. Because some cycles contain packstone that may have been deposited in water deeper than several meters, calculated relative rises in sea level are minimum estimates based on tracing the shallow-water deposits over paleotopography. Deposition of shallow-water deposits in successive cycles requires relative falls in sea level. The calculated magnitudes of relative falls are minimum estimates because extent of outcrop precludes further tracing downslope. Thus, the paleotopographic drape of cycles in the TCC represents:

1. Cycle 1—a relative sea-level rise of at least 35 m (pinning point #17) for further shallow-water deposition of cycle 1 in upslope positions. Deposition of shallow-water facies in the base of cycle 2 required a relative fall in sea level of at least 30 m (pinning points #18 to #19).

2. Cycle 2—downslope initiation of shallow-water deposition on the cycle 1–2 boundary (pinning point #20) was followed by further upslope deposition of cycle 2 shallow-water deposits which required a relative rise in sea level of at least 30 m (pinning point #21). Downslope initiation of shallow-water depo-

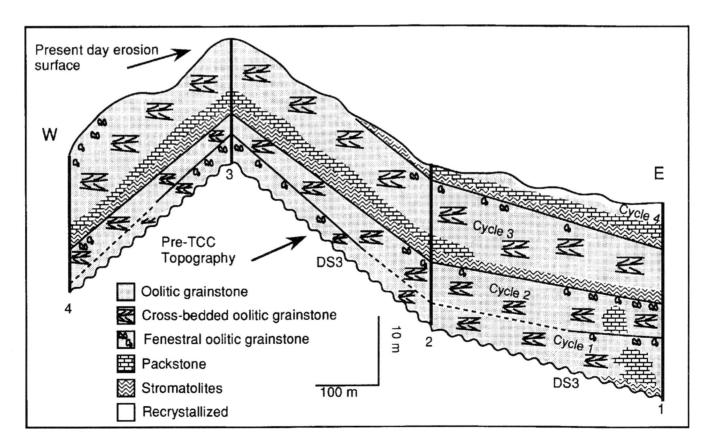

Figure 18. Schematic diagram illustrating pre-TCC topography and generalized TCC cycle patterns on highest part of La Molata. Numbered vertical lines are measured section locations.

sition on the cycle 2–3 boundary required a relative fall in sea level of at least 30 m (pinning points #22, #23).

3. Cycle 3—downslope initiation of shallow-water deposition on the cycle 2–3 boundary (pinning point #24) was followed by further upslope deposition of cycle 3 shallow-water deposits, which required a relative rise in sea level of at least 28 m (pinning point #25). Downslope deposition of shallow-water facies and development of fenestral fabrics in the top of cycle 3 required a relative fall in sea level of at least 33 m (pinning points #26, #27).

4. Cycle 4—downslope initiation of deposition on the cycle 3–4 boundary (pinning point #28) was followed by upslope deposition of shallow-water deposits (pinning point #29) which required a relative rise in sea level of at least 11 m.

The exposure surface at the top of the TCC is represented by a well-developed caliche breccia which appears similar to the "Pre-Pliocene breccia" of Esteban and Gíner (1980). The pre-Pliocene exposure surface can be traced to present-day sea level. Thus, it represents a relative fall of sea level of at least 202 m (pinning points #30, #31).

Interpretive Curve

The interpretive curve for the initial relative rise between pinning points #15 and #16 is drawn in its simplest smooth form because of the lack of additional data due to the equivocal relationships of lower elevation TCC strata to the other proximal TCC stratal units. The interpretive curve between pinning points #16 and #30, and pinning points #30 and #31 is the same as for the pinning point curve due to the lack of any additional data that could add more detail.

REGIONAL SIGNIFICANCE

It is likely that the sequences in the Las Negras area were controlled, at least in part, by regional-interregional processes (e.g., tectonism, eustasy) as opposed to local controls (Franseen and Mankiewicz, 1991). The overall similarity of sequences and ages in other southeastern Spain areas, Mallorca (some 600 km to the northeast) and Morocco, is also supportive of regional control on sequence development. Correlations of the different phases of carbonate/reef development with eustatic curves have been proposed by workers for several study areas in the western Mediterranean (e.g., Saint-Martin and Rouchy, 1990; Pomar, 1991; Braga and Martín, 1992; Esteban et al., in press). Even though there is much similarity in the sequences from these other areas to those in Las Negras, until we have better age control for the strata in the Las Negras area, correlations to other sequences and the role of local, regional, tectonic, and eustatic processes remain speculative.

However, it is worth noting that the facies and geometries observed in the upper Las Negras

sequences are similar to those in other better dated Miocene exposures of the Mediterranean. Based on this similarity, speculative correlation to a eustatic curve would place the DS3 fringing reef sequence and TCC sequence in Las Negras in the TB 3.2 (8.2 to 6.3 m.y.) and TB 3.3 (6.3 to 5.5 m.y.) cycles respectively of Haq et al. (1987) (as interpreted for similar Mallorca strata by Pomar, 1991, and for Algeria, Spain, Morocco by Saint-Martin and Rouchy, 1990) or in the TB 3.3 (6.3 to 5.5 m.y.) and TB 3.4 (5.5 to 4.2 m.y.) cycles respectively of Haq et al. (1987) (as interpreted for Sorbas basin strata by Braga and Martín, 1992, and for the western Mediterranean in general by Esteban et al., in press). There are not yet sufficient age data for the lower sequences of Las Negras to allow speculation about their comparison to other areas.

DISCUSSION

Our studies in the Las Negras area illustrate that scale of observation is important. For instance, DS1A through the end of DS3 (base of TCC) could be interpreted as one large-scale rise and fall of relative sea-level. Thus, DS1A and DS1B would be classified as lowstand to transgressive systems tracts, DS2 as a transgressive systems tract to early highstand systems tract, and DS3 as a highstand systems tract. However, there are interpreted sequence boundaries at the base of DS1B, DS2, and DS3, the strata of DS3 record falling sea level, and TCC strata record a complex history of relative sea-level change, all of which are beyond the resolution of the systems tract approach. In this sense, the Las Negras exposures are instructive in showing details of the internal architecture of relatively larger scale sequences and systems tracts, and illustrate how smaller scale sequences stack to form larger scale sequences in a setting where carbonate deposits developed as fringes on relatively steep-sided basement rocks.

Determining whether carbonate debris and megabreccias are shed during sea level highstands or lowstands is of considerable importance in carbonate sequence stratigraphic interpretations and placement of sequence boundaries (e.g., Sarg, 1988; Schlager, 1992). The Las Negras area serves as an excellent example to understand the nature of megabreccia deposition. There, megabreccias formed in at least two ways, one that is associated with falling sea level (exemplified by megabreccia 1 at the base of DS2 and megabreccia 2 at the base of DS3) and another that may not necessarily be associated with a relative fall in sea level, but is associated with shedding of reefal debris at a high position of relative sea level. Randomly distributed reef talus blocks in upper DS3 strata are examples of the latter process. Erosion and transport of megabreccia 2 was likely initiated by a relative sea-level fall and evidence suggests that this megabreccia eventually might have been exposed subaerially in a basinward slope position as sea level continued to drop down the paleoslope. Thus, the base of this megabreccia represents the initial fall of sea level, but its upper surface may contain the actual exposure evidence. In such cases, it may be better to consider the megabreccia as a "sequence boundary zone."

The development of a quantitative record of relative sea-level fluctuations, such as for our example on carbonate sequences in the Las Negras area, has many important implications for improving our understanding of the controls on carbonate stratigraphic sequences. For example, by comparing the magnitudes of relative sea-level fluctuations for separate areas that contain age-equivalent sequences, the role of local and regional processes can be determined as well as other important depositional parameters such as sedimentation rates. If the magnitudes of relative sea-level fluctuations in a number of areas are similar, then the role of regional and possibly global processes (e.g., eustasy, tectonism, climate) would be the dominant controlling factor. If the magnitudes differ significantly in separate areas, then local processes (e.g., local tectonism, basin-wide sea-level changes, storms, autocyclic processes) would also be important in sequence development. The quantification of relative sea-level fluctuations also can provide important data to be incorporated into the increasingly utilized area of sedimentary simulation modeling. As exemplified by papers in Franseen et al. (1991), quantitative rock-derived data will lead to improved computer models, which in turn can aid in interpretations and understanding of controlling processes on carbonate stratigraphic sequences.

CONCLUSIONS

1. Detailed field studies of exceptional exposures of undeformed late Miocene carbonate strata near the coastal village of Las Negras, southeastern Spain, reveal five depositional sequences deposited on earlier Neogene volcanic rocks in an archipelago setting. These five depositional sequences record several scales of relative sea-level fluctuations during an overall evolution from open, normal-marine carbonate platform to fringing reef complex to normal and restricted marine carbonate platform. The sequence boundaries that separate the five sequences are marked by surfaces or megabreccia units that contain evidence of subaerial exposure due to relative falls in sea level.

2. The five depositional sequences were deposited over paleotopography of more than 200 m relief. Because that paleotopography is preserved today, these rocks allow for quantification of a significant part of the relative sea-level curve through development of "pinning points." Pinning points are defined as a position of ancient sea level relative to a geologically useful starting elevation.

3. Pinning points can be used to construct a pinning point curve by plotting ancient relative elevations of sea level versus relative time. This can be accomplished by identification of

facies deposited at or just below sea level (e.g., reef crest, tidal flat, shallow-water oolite, beach deposits), from identification of subaerial exposure of marine rocks, or from an upward transition of marine to nonmarine deposits. Relative elevations of pinning points can be determined if the paleoslope is preserved without deformation, if slope can be reconstructed, and if overall compaction and differential compaction can be either determined or assumed. Pinning point curves can be used as templates to construct more interpretive relative sea-level curves that incorporate stratal geometries, facies relationships, and stacking patterns.

4. The Las Negras study area serves as an excellent example to demonstrate the usefulness of pinning point curves. The pinning point curve developed there illustrates a complex history with large-scale relative sea-level changes of greater than 200 m combined with smaller scale relative sea-level changes. Comparing this quantitative record to the strata of each sequence allows a better determination of the interaction between sedimentation and sea-level change, and improvement of sequence stratigraphy models. Furthermore, with adequate age control, comparison of the Las Negras area relative sea-level record to those of other areas, and to global eustatic sea-level curves can provide for the quantification of the effect of local, regional, and eustatic variables that constitute relative sea-level change. The pinning point method offered in this study can be used for understanding the quantitative constraints on relative sea-level history of carbonate sequences of all ages.

ACKNOWLEDGMENTS

We are especially grateful to Exxon Production Research Company for financial support of our research in the Las Negras area. British Petroleum Oil Company, Texaco Research Company, and Amoco Research Company also provided valuable financial support. Our research in the Las Negras area has benefited from discussions with A. Arribas, Jr., J.C. Braga, M. Esteban, T.E. Fekete, C. Mankiewicz, J.M. Martín, E.J. Oswald, L.C. Pray, J.F. Sarg, and A. Símo. We thank J.F. Sarg for his encouragement to prepare this chapter. R.L. Loucks, J.R. Markello, and J.F. Sarg critically reviewed the manuscript and provided useful comments that improved the final product. We thank L. Davidson for help in manuscript preparation and R. Hensiek for preparing illustrations.

REFERENCES CITED

Addicott, W.D., P.D. Snavely, Jr., D. Bukry, and R.Z. Poore, 1978, Neogene stratigraphy and paleontology of southern Almeria Province, Spain: An overview: U.S. Geological Survey Bulletin, v. 1454, 49 p.

Alvarez, G., S.P. Busquet, A. Permanyer, and M. Vilaplana, 1977, Growth dynamic and stratigraphy of the Sant Pau D'Ordal Miocene patch-reef (Province of Barcelona, Catalonia): 2nd International Symposium on Corals and Fossil Coral Reefs, Paris, 1975, Memoir B.R.G.M., v. 89, p. 367-377.

Armstrong, A.K., P.D. Snavely, Jr., and W.O. Addicott, 1980, Porosity evolution of upper Miocene reefs, Almeria Province, southern Spain: AAPG Bulletin, v. 64, p. 188-208.

Braga, J.C., and J.M. Martín, 1988, Neogene coralline-algal growth-forms and their palaeoenvironments in the Almanzora river valley (Almeria, S.E. Spain): Palaeogeography, Palaeoclimatology, Palaeoecology, v. 67, p. 285-303.

Braga, J.C., and J.M. Martín, 1992, Messinian carbonates of the Sorbas basin: Sequence stratigraphy, cyclicity, and facies, in E.K. Franseen, R.H. Goldstein, J. C. Braga, and J.M. Martín, eds., Late Miocene Carbonate Sequences of Southeastern Spain: A Guidebook for the Las Negras and Sorbas Areas, in conjunction with the SEPM/IAS Research Conference on Carbonate Stratigraphic Sequences: Sequence Boundaries and Associated Facies, August 30-September 3, La Seu, Spain.

Braga, J.C., J.M. Martín, and B. Alcala, 1990, Coral reefs in coarse-terrigenous sedimentary environments (upper Tortonian, Granada basin, southern Spain): Sedimentary Geology, v. 66, p. 135-150.

Dabrío, C.J., 1974, Los niveles arrecifes del Neogene de Purchena (SE Cordilleras Beticas): Cuadernos Geologia, v. 5, p. 79-88.

Dabrío, C.J. 1975, La sedimentacion arrecifal Neogena en la region del Rio Almanzora: Estudios Geologicos, v. 31, p. 285-296.

Dabrío, C.J., M. Esteban, and J.M. Martín, 1981, The coral reef of Nijar, Messinian (uppermost Miocene), Almeria Province, SE Spain: Journal of Sedimentary Petrology, v. 51, p. 521-539.

Dabrío, C.J., and J.M. Martín, 1978, Las arrecifes Messiniense de Almeria (SE de Espana): Cuadernos Geologia, v. 8-9, p. 85-100.

Dabrío, C.J., J.M. Martín, and A.G. Megías, 1985, The tectosedimentary evolution of Mio-Pliocene reefs in the Province of Almeria (SE Spain), in M.D. Mila and J. Rosell, eds., Excursion Guidebook: International Association of Sedimentologists 6th European Regional Meeting, p. 271-305.

De Larouzière, F.D., J. Bolze, P. Bordet, J. Hernandez, C. Montenat, and P. Ott D'Estevou, 1988, The Betic segment of a lithospheric Trans-Alboran shear zone during upper Miocene: Tectonophysics, v. 152, p. 41-52.

Di Battistini, G., L. Toscani, S. Iaccarino, and J.M. Villa, 1987, K/Ar ages and the geological setting of calc-alkaline volcanic rocks from Sierra de Gata, SE Spain: N. Jb. Miner. Mh., v. H8, p. 369-383.

Dronkert, H., and J. Pagnier, 1977, Introduction to

Mio-Pliocene of the Sorbas Basin: Messinian Seminar 3, Field Trip 2 University of Malaga, p. 1-21.

Esteban, M., 1979, Significance of the upper Miocene coral reefs of the western Mediterranean: Palaeogeography, Palaeoclimatology, Palaeoecology, v. 29, p. 169-188.

Esteban, M., 1981, Miocene carbonate models: Internal Report, Erico Company, Inc., London, 46 p.

Esteban, M., 1988, Miocene reefs in western Mediterranean (abstract): AAPG Bulletin, v. 72, p. 182.

Esteban, M., J.C. Braga, J.M. Martín, and C. De Santisteban, (in press), Western Mediterranean.

Esteban, M., and J. Gíner, 1980, Messinian coral reefs and erosion surfaces in Cabo de Gata (Almeria, southeastern Spain): Acta Geologico Hispanica, v. 15, p. 97-104.

Esteban, M., and L.C. Pray, 1981, A guidebook to the Tertiary reef carbonate and associated facies, southeastern Spain and the Balearic Platform: Internal Report, Erico Company, Inc., London, 74 p.

Franseen, E.K., 1989, Depositional sequences and correlation of Middle to Upper Miocene carbonate complexes, Las Negras Area, Southeastern Spain: unpublished Ph.D. dissertation, University of Wisconsin-Madison, 374 p.

Franseen, E.K., 1990, Middle to upper Miocene mixed carbonate and volcaniclastic slope deposits, Las Negras and Rodalquilar area, southeastern Spain (abstract): 13th International Sedimentological Congress, Nottingham, England, p. 80-81.

Franseen, E.K., and Goldstein, R.H., 1992, Sequence stratigraphy of Miocene strata, Las Negras area, southeastern Spain: Implications for quantification of relative change in sea level, in E.K. Franseen, R.H. Goldstein, J.C. Braga, and J.M. Martin, eds., Late Miocene Carbonate Sequences of Southeastern Spain: A Guidebook for the Las Negras and Sorbas Areas, in conjunction with the SEPM/IAS Research Conference on Carbonate Stratigraphic Sequences: Sequence Boundaries and Associated Facies, August 30-September 3, La Seu, Spain, p. 1-77.

Franseen, E.K., and C. Mankiewicz, 1991, Depositional sequences and correlation of middle(?) to late Miocene carbonate complexes, Las Negras and Nijar areas, southeastern Spain: Sedimentology, v. 38, p. 871-898.

Franseen, E.K., W.L. Watney, C.G.St.C. Kendall, and W. Ross, eds., 1991, Sedimentary modeling: Computer simulations and methods for improved parameter definition: Kansas Geological Survey Bulletin, v. 233, 524 p.

Goldstein, R.H., 1988, Paleosols of Late Pennsylvanian cyclic strata, New Mexico: Sedimentology, v. 35, p. 777-803.

Goldstein, R.H., J.E. Anderson, and M.W. Bowman, 1991, Diagenetic responses to sea level change: Integration of field, stable isotope, paleosol, paleokarst, fluid inclusion, and cement stratigraphy research to determine history and magnitude of sea level fluctuation, *in* E.K. Franseen, W.L. Watney, C.G.St.C. Kendall, and W. Ross, eds., Sedimentary Modeling: Computer Simulations and Methods for Improved Parameter Definition: Kansas Geological Survey Bulletin, v. 233, p. 139-162.

Goldstein, R.H., E.K. Franseen, and M.S. Mills, 1990, Diagenesis associated with subaerial exposure of Miocene strata, southeastern Spain: Implications for sea level change and preservation of low-temperature fluid inclusions in calcite cement: Geochimica et Cosmochimica Acta, v. 54, p. 699-704.

Haq, B.U., J. Hardenbol, and P.R. Vail, 1987, Chronology of fluctuating sea levels since the Triassic: Science, v. 235, p. 1156-1166.

Hsü, K.J., 1973, The desiccated deep-basin model for the Messinian events, *in* C.W. Drooger, ed., Messinian Events in the Mediterranean: Geodynamiques Scientific Report No. 7, North-Holland Publishing Co., Amsterdam, p. 60-67.

Hsü, K.J., L. Montedart, D. Bernoulli, M.B. Cita, A. Erickson, R.E. Garrison, R.B. Kidd, F. Melieres, C. Muller, and R. Wright, 1977, History of the Mediterranean salinity crisis: Nature, v. 267, p. 399-403.

Martín, J.M., and J.C. Braga, 1990, Arrecifes Messinienses de Almeria: Tipologias de crecimiento, posicion estratigrafica y relacion con las evaporitas: Geogaceta, v. 7, p. 66-68.

Martín, J.M., Braga, J.C., and Rivas, P., 1989, Coral successions in upper Tortonian reefs in SE Spain: Lethaia, v. 22, p. 271-286.

Megías, A.G., 1985, Tectosedimentary relationships between Mio-Pliocene reefs and evaporites in Almeria and Sorbas Basins, SE Iberian Peninsula (abstract): International Association of Sedimentologists 6th Annual Meeting, Lleida, Spain, p. 292-295.

Montenat, C., 1975, Le Neogene des Cordilleres Betiques, essay de synthese stratigraphique et paleogeographique: Internal Report, Paris, C.N.R.S., 187 p.

Montenat, C., 1977, Les basins Neogenes du levant d'Alicante et de Murcia (Cordilleres Betiques orientales-Espangne); stratigraphie, paleogeographie et evolution dynamique: Doc. Lab. Geol. Fac. Sci., Lyon, v. 69, 345 p.

Montenat, C., P. Ott D'Estevou, and P. Masse, 1987, Tectonic-sedimentary characters of the Betic Neogene basins evolving in a crustal transcurrent shear zone (SE Spain): Bull. Centres Rech. Explor.-Prod., Elf-Aquitaine, v. 11, p. 1-22.

Müller, D.W., and K.J. Hsü, 1987, Event stratigraphy and paleoceanography in the Fortuna Basin (southeast Spain): A scenario for the Messinian salinity crisis: Paleoceanography, v. 2, p. 679-696.

Pagnier, H., 1977, Excursion to Messinian reef deposits in the northern part of Sorbas Basin: An introduction: Messinian Seminar 3, Field Trip, University of Malaga, p. 44-54.

Permanyer, A., and M. Esteban, 1973, El arrecife

Mioceno de Sant Pau d'Ordal (Provincia de Barcelona): Rev. Inst. Inv. Geol., Dept. Prov., Barcelona, v. 28, p. 45-72.

Pomar, L., 1991, Reef geometries, erosion surfaces and high-frequency sea level changes, upper Miocene reef complex, Mallorca, Spain: Sedimentology, v. 38, p. 243-270.

Poore, R.Z., and S.M. Stone, 1981, Biostratigraphy and paleoecology of the upper Miocene (Messinian) and Lower Pliocene (?), Cerro de Almendral section, Almeria Basin, southern Spain: Geological Survey Professional Paper, v. 774-F, p. F1-F11.

Rehault, J.-P., G. Boillot, and A. Mauffret, 1985, The western Mediterranean basin, *in* D.J. Stanley and F.C. Wezel, eds., Geologic Evolution of the Mediterranean Basin: Springer-Verlag, New York, p. 101-130.

Riding, R., Martín, J.M., and Braga, J.C, 1991, Coral-stromatolite reef framework, Upper Miocene, Almeria, Spain: Sedimentology, v. 38, p. 799-819.

Rytuba, J.J., A. Arribas, Jr., C.G. Cunningham, E.H. McKee, M.H. Podwysocki, J.G. Smith, W.C. Kelly, and A. Arribas, 1990, Mineralized and unmineralized calderas in Spain; Part II, evolution of the Rodalquilar caldera complex and associated gold-alunite deposits: Mineralium Deposita, v. 25 (Suppl.), p. S29-S35.

Saint-Martin, J.P., and J.M. Rouchy, 1990, Les plates-formes carbonatees messiniennes en Mediterranee occidentale: Leur importance pour la reconstitution des variations du niveau marin au Miocene terminal: Bull. Soc. geol. France, v. 8, p. 83-94.

Santisteban, C., and C. Taberner, 1980, The siliciclastic environments as a dynamic control in the establishment and evolution of reefs: Sedimentary models (abstract): International Association of Sedimentologists 1st European Meeting, Bochum, p. 208-211.

Sarg, J.F., 1988, Carbonate sequence stratigraphy, *in* C.K. Wilgus, B.S. Hastings, C.G.St.C. Kendall, H.W. Posamentier, C.A. Ross, and J.C. Van Wagoner, eds., Sea Level Changes: An Integrated Approach: SEPM Special Publication No. 42, p. 155-181.

Schlager, W., 1992, Sedimentology and sequence stratigraphy of reefs and carbonate platforms: AAPG Continuing Education Course Note Series, v. 34, 71 p.

Serrano, F., 1990, Presencia de Serravalliense marino en la cuenca de Nijar (Cordillera Betica, Espana): Geogaceta, v. 7, p. 95-97.

Veeken, P.C.H., 1983, Stratigraphy of the Neogene-Quaternary Pulpi Basin, Provinces Murcia and Almeria (SE Spain): Geologie en Mijnbouw, v. 62, p. 255-265.

Wilson, J.L., 1967, Cyclic and reciprocal sedimentation in Virgilian strata of southern New Mexico: Geological Society of America Bulletin, v. 78, p. 805-818.

Chapter 17

Volumetric Partitioning and Facies Differentiation within the Permian Upper San Andres Formation of Last Chance Canyon, Guadalupe Mountains, New Mexico

Mark D. Sonnenfeld
Timothy A. Cross
Department of Geology and Geological Engineering
Colorado School of Mines
Golden, Colorado, U.S.A.

ABSTRACT

San Andres Formation (Permian) outcrops in Last Chance Canyon are interpreted to contain two large-scale, fourth-order depositional sequences, upper San Andres sequence 3 and upper San Andres sequence 4. Embedded within each of these sequences are numerous higher frequency sequences. Reciprocal siliciclastic/carbonate sedimentation patterns enhance the recognition of cyclicity within both large-scale sequences and within many of the higher frequency sequences.

Upper San Andres sequence 4 contains at least 12 smaller scale sequences that occur in landward-stepping, vertically stacked, stratigraphically rising seaward-stepping, and stratigraphically falling seaward-stepping geometric patterns. In outer-ramp/slope environments, landward-stepping, high-frequency sequences of the lowstand to transgressive systems tracts record a progressive decrease in sediment accumulation rate, depositional energy, and siliciclastic content. This reflects a long-term transition from siliciclastic-dominated slope sedimentation to increasingly carbonate-dominated slope sedimentation. Thick outer-shelf carbonate strata, highly aggradational to mounded fusulinid-rich shelf-margin deposits, and pronounced stratigraphic rise of the fusulinid facies tract all reflect long-term increases in accommodation. Concurrent with the highly aggradational and moderately progradational mode of outer-shelf to slope carbonates, siliciclastic shelf-feeder systems are inferred to have stepped progressively landward. These contrasting reflections of increasing accommodation during lowstand to transgressive systems tract deposition accentuate the dissimilar nature of carbonate versus siliciclastic sediment supply. The most carbonate-rich and bioherm-bearing interval of the entire sequence overlies a distinctive maximum flooding surface capping the transgressive systems tract. Ensuing seaward-stepping high-frequency sequences of the middle to late highstand systems tract show pronounced progradational offlap and record a progressive increase in the volume of siliciclastics accumulated in the basin with a concomitant decrease in the volume of siliciclastics accumulated on the outer shelf. A karsted toplap surface represents a subaerial unconformity and sequence boundary capping upper San Andres sequence 4. Peritidal cycles within the overlying Grayburg Formation exhibit a significant seaward shift in coastal onlap across this sequence boundary.

Depositional topography and time-significant surfaces within upper San Andres sequence 4 were traced across multiple geomorphic environments that extend from shelf to basin. As a result, the evolution of seismic-scale stratal geometries, depositional topography, facies associations, and volumetric proportions of carbonate and siliciclastic strata can be related with a high degree of temporal precision. By calibrating these changes to position within a stacking pattern of high-frequency sequences, we observe ordered, predictable changes in the volumetric partitioning of siliciclastics, grain-supported carbonates, and mud-supported carbonates. This volumetric partitioning is accompanied by differentiation of carbonate facies that accumulated in similar paleobathymetric settings, but in different positions of the large-scale stacking pattern. For example, siliciclastic-free, mud-supported, bioherm-prone slope carbonates attain their greatest seaward extent during the latest transgressive to earliest highstand systems tract period of long-term maximum shelfal accommodation. By contrast, grain-rich carbonate and mixed carbonate-siliciclastic facies predominate in lower-slope settings during periods of reduced shelfal accommodation such as the lowstand, early transgressive, and middle-late highstand systems tracts. If facies models are to become predictive at the scale of reservoirs, they must be calibrated to the systematically changing sediment supply and accommodation conditions that appear to drive volumetric partitioning and facies differentiation.

INTRODUCTION AND PURPOSE

As stratigraphic concepts and methods have evolved, geologists with different perspectives have repeatedly examined Permian outcrops of the Guadalupe Mountains of New Mexico and Texas. The attraction of the Guadalupe Mountains as a stratigraphic testing ground is attributed to a diversity of siliciclastic and carbonate facies and to excellent three-dimensional exposures that clearly display large-scale stratal geometries. Last Chance Canyon's outcrops permit physical examination of strata and bounding surfaces continuously from the outer shelf to the basin margin. This uninterrupted exposure enhances description and understanding of the relationships among facies, rock volumes, stratigraphic geometries, and time-significant stratigraphic bounding surfaces.

Early stratigraphic studies in Last Chance Canyon focused on the nature and distribution of rock types and fossils within the context of correlating formations and their boundaries (Darton and Reeside, 1926; King, 1942; Skinner, 1946; Boyd, 1958; Hayes, 1959). Other studies emphasized the interpretation of environments of deposition (Harrison, 1966; Jacka et al., 1968; Williams, 1969; Naiman, 1982; McDermott, 1983). More recently, Sarg and Lehmann (1986a,b) described depositional facies, large-scale stratal geometries, and sequence boundaries within the framework of seismic-scale sequence stratigraphy. We extend sequence stratigraphic concepts and methods into the domain of high-resolution stratigraphic correlation, with an emphasis on how the stratigraphic concepts of accommodation and sediment supply relate to volumetric partitioning and facies differentiation within prograding shelf-margin strata.

Within the threefold hierarchy of sequences discussed in this chapter, high frequency, relatively small-scale sequences are embedded within sequences of progressively larger physical scale and lower temporal frequency. In these sequences, regardless of scale, we observe that different volumes of sediment accumulated in similar geomorphic settings according to changes in the geographic location of accommodation and preservation potential. For example, we observe that during middle to late highstand systems tract deposition, detrital sediment accumulation is biased toward the slope and basin at the expense of the shelf. Conversely, during transgressive systems tract deposition, sediment accumulation increases on the shelf, and the slope and basin are progressively more starved of siliciclastics. The concept of volumetric partitioning is implicit to seismic-scale sequence stratigraphy (Vail et al., 1977b; Jervey, 1988), yet is equally important at smaller physical and temporal scales. Recognition of volumetric partitioning and how time is represented in the stratigraphic record establishes the basis for a high-resolution correlation methodology of rock-to-rocks, rocks-to-surfaces, and surfaces-to-surfaces.

In many geologic settings, clinoforms of offlapping strata progress from low-angle sigmoidal geometries to higher angle, nearly oblique geometries. Clinoform geometries have been used to infer long-term accommodation trends (e.g., Barrell, 1912; Cotton, 1918; Sarg, 1988; Bosellini, 1989), yet premises concerning the significance, genesis, and likely facies associated with different clinoform geometries have remained quite general (Brown and Fisher, 1977; Mitchum et al., 1977; Sangree and Widmier, 1977; Bosellini, 1984; Sarg, 1988). Relationships between seismic-scale clinoform geometries and coeval depositional facies are directly observable in Last Chance Canyon because depositional topography and time-significant surfaces directly reveal geomorphic environments. We observe that the volumetric partitioning of sediment responsible for evolving clinoform geometries is accompanied by "facies differentiation," that is, significant changes in the nature of facies that occur within comparable depositional environments and similar paleobathymetric positions during different phases of a depositional sequence. For example, facies preserved in the upper-slope facies tract of landward-stepping to vertically stacked cycles involving sigmoidal clinoforms are different in many respects from those preserved in the same bathymetric tract of seaward-stepping cycles involving complex sigmoid-oblique clinoforms. Facies differentiation within sequences of various scales appears orderly and regular when viewed in the context of time, clinoform stratal geometries, and stacking patterns.

An appreciation for volumetric partitioning and facies differentiation emphasizes the potential pitfall of becoming preoccupied with the identification of surfaces and the labeling of systems tracts while potentially overlooking important variations within intervening strata. Volumetric partitioning and facies differentiation highlight the need to incorporate a dynamic and changing sedimentology into sequence stratigraphic analysis and expose a need for revised facies models that incorporate dynamic stratigraphic controls on the accumulation and preservation of sediment.

In this chapter, we first present the geologic setting and regional sequence stratigraphic framework, followed by a description of facies, facies distributions, and large-scale stratigraphic relationships with an emphasis on upper San Andres sequence 4. The direct applicability of these observations to other shelf-margin settings may be limited by site-specific configurational aspects including pronounced reciprocal sedimentation, the clay-free siliciclastic sediment supply, the arid climate, and the lower Guadalupian biota. Attributes of facies and stratal geometries in Last Chance Canyon may be extrapolated to other situations by emphasizing relationships between stratigraphic *processes* such as accommodation and sediment supply and stratigraphic *responses* such as volumetric partitioning and facies differentiation. With this in mind, we focus on the specific stratigraphic responses of progradation/aggradation ratios, the variation over time in the lateral extent of grain-dominated versus mud-dominated carbonate facies, and the timing and nature of bioherm development.

STRATIGRAPHIC APPROACH

We employ the stratigraphic concepts of accommodation (Barrell, 1912, 1917; Cotton, 1918; Sloss, 1962; Allen, 1964; Jervey, 1988) and sediment supply to describe and explain the volumetric partitioning and facies differentiation that we observe within San Andres and Cherry Canyon strata. Accommodation is the cumulative sum of space available for sediment deposition and is the product of eustasy, subsidence, and compaction.

Strata in Last Chance Canyon are naturally divisible into depositional sequences of multiple physical and temporal scales. We consider a sequence of any scale as the stratigraphic record of a base-level transit cycle (Wheeler, 1964). An unconformity-bounded depositional sequence of Vail et al. (1977a) is one type of sequence in this usage. High-frequency sequences are arranged or "stacked" in geometric patterns that are describable by the terms "landward-stepping," "seaward-stepping," and "vertically stacked." Though similar to the retrogradational, progradational, and aggradational descriptors used by Frazier (1974) and Van Wagoner (1985), these terms accentuate the episodic nature of stratal accumulation and are connotative of stratigraphic geometries as opposed to depositional processes. In landward-stepping sequences, most facies tracts in one sequence occur in more landward positions than equivalent facies tracts of the preceding sequence of equivalent scale. In seaward-stepping sequences, most facies tracts in one sequence occur in more seaward positions than equivalent facies tracts of the preceding sequence of equivalent scale. In vertically stacked sequences, most facies tracts in one sequence are essentially superpose over equivalent facies tracts of the preceding sequence of equivalent scale. These geometric arrangements reflect the dynamic balance between accommodation and sediment supply through time and space.

In outer-shelf, ramp, or slope settings affected by storm-wave action, we use the term "wave-base razor" to describe a natural upward limit of accumulation or base level that is not sea level, but rather a zone characterized by an unsteady outer-shelf equilibrium between sediment supply and removal that occurs in the water-depth zone between storm- and fair-weather wave-base (Osleger, 1991). Below this threshold, shelf sediments accumulate, but above the threshold they are removed (either shelfward or seaward) at rates slightly slower to slightly faster than the rate of sediment influx. Repeated landward and seaward transits of the zone of impingement of the wave-base razor on the seafloor, linked to changes in water depth, produce a high-resolution record of base-level transit cycles in subtidal deposits both above and below storm wave base.

GEOLOGIC SETTING

Last Chance Canyon lies 20 km northwest of the upper Guadalupian Capitan escarpment that separates the Permian-aged Delaware basin province from the Northwestern Shelf province (Figures 1, 2). The Huapache fault zone obliquely crosses the eastern edge of the Last Chance Canyon area and is expressed on the surface as the Huapache Monocline of Tertiary age. This fault zone was also active in

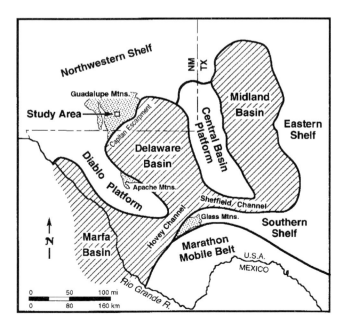

Figure 1. Regional paleogeographic setting during the Late Guadalupian (Permian). Modified from King (1948).

Pennsylvanian to earliest Permian (Wolfcampian) times and defines the northeastern margin of the Huapache paleohigh. (Figure 3a; Meyer, 1966; Galley, 1968). The lower Leonardian Abo carbonate shelf margin has a regional northeast trend along the northwest margin of the Delaware basin, however, in the Last Chance Canyon area it bends sharply northwest following the trend of the Huapache paleohigh (Figure 3a). Pennsylvanian and Permian (Wolfcampian) isopachs (Meyer, 1966) show a northwest–southeast trending area of increased subsidence northeast of the Huapache paleohigh and informally termed the Huapache subbasin. An underfilled Huapache subbasin is reflected by the east-southeast direction of upper San Andres shelf-margin progradation in the Last Chance Canyon area, further illustrating the persistent influence of the inactive Huapache fault zone on Leonardian and Guadalupian sedimentation patterns.

Both the Delaware basin and its probable clastic provenance area hundreds of kilometers to the north (Uncompahgre and Sierra Grande Uplifts: Newell et al., 1953) are thought to have been tectonically quiescent in post-Wolfcampian time (Oriel et al., 1967). The presence of polar ice caps in the Permian Period (Veevers and Powell, 1987), coupled with the apparent lack of tectonism, leads to the common assumption that eustasy is an important, if not dominant, forcing function for Guadalupian depositional sequences.

The northwestern margin of the Permian Delaware basin was an arid region ± 5° of the paleoequator (Scotese et al., 1979; Fischer and Sarnthein, 1988). The arid climate limited fluvial runoff while promoting the development of inferred aeolian dune fields during episodic subaerial exposure of carbonate-dominated shelves (Fischer and Sarnthein, 1988). The low paleolatitude, the arid climate, and the lack of turbid, terrestrial runoff promoted the growth of basin-rimming carbonate banks and reefs. Negligible volumes of hemipelagic clay, coupled with the absence of calcareous and siliceous planktonic organisms prior to the Jurassic, led to the development of a sediment-starved basin during relative highstands of sea level. By contrast, siliciclastics bypassed the shelves in a "reciprocal" fashion during relative lowstands of sea level, resulting in the accumulation of predominantly siliciclastic basinal deposits of the Delaware Mountain Group (Meissner, 1967, 1972).

SEQUENCE STRATIGRAPHIC FRAMEWORK

Studies of Delaware basin strata traditionally were approached within the context of lithostratigraphic units and nomenclature. As a result, temporal relationships were poorly established among diverse lithofacies, particularly those involving shelf-to-basin transitions across significant depositional topography (Figures 3b, 4). The early sequence stratigraphic studies of Meissner (1967, 1972) and Silver and Todd

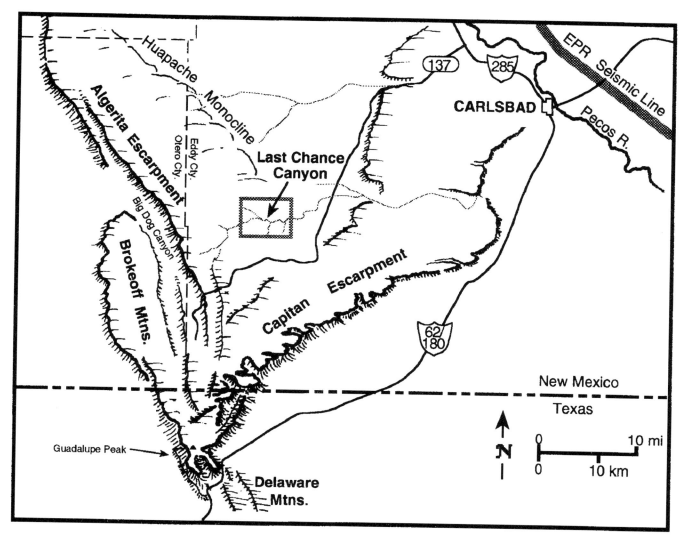

Figure 2. Location map of Guadalupe Mountains region and Last Chance Canyon study area. Modified from King (1948) and Babcock (1977).

(1969) refined temporal correlations by applying the concept of "reciprocal sedimentation" (Van Siclen, 1958; Wilson, 1967) to the Delaware basin. These workers recognized the cyclic alternation of basin-centered siliciclastics with carbonate-dominated shelf and slope deposits and presumed that eustasy was responsible for basinwide Late Wolfcampian through Guadalupian depositional cycles.

Meissner (1967, 1972) and Silver and Todd (1969) also interpreted the lower San Andres Formation as the product of a long-term, punctuated transgression commencing at or shortly before the end of the Leonardian Epoch. Relatively deep-water lower Guadalupian strata were deposited upon flooded Leonardian platforms, resulting in deposits at the toe of depositional slopes that represent 40–100+ m water depth, rather than the 200–500+ m water depths inferred for Delaware Mountain Group sandstones in the center of the Delaware basin proper (Figure 4; Garber et al., 1989). This transgression was followed

by long-term, punctuated regression during deposition of the middle to upper San Andres Formation. Multiple episodes of siliciclastic bypass across the upper San Andres shelf resulted in deposition of Brushy Canyon and lower Cherry Canyon sandstones in the Delaware Basin (Meissner, 1972).

Sarg and Lehmann's (1986a) identification and correlation of seismic-scale sequences within Guadalupian strata further refined the regional stratigraphic framework. They proposed a temporal correspondence between two seismically resolvable San Andres sequences in the subsurface (location of Exxon Production Research seismic section shown on Figures 2 and 3) and two sequences recognized in Last Chance Canyon. Sarg and Lehmann (1986a) demonstrated that the complex transition between shelf and slope carbonates of the San Andres Formation, and slope and basinal siliciclastics of the Cherry Canyon Formation is composed of two genetically distinct stratigraphic units. The lower unit is an

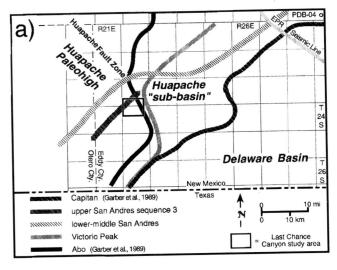

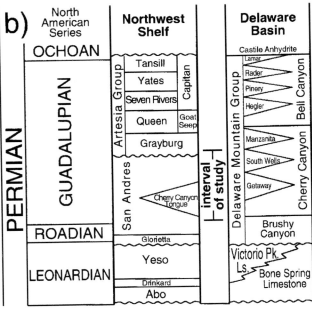

Figure 3. (a) Terminal shelf-margin trends for several Leonardian and Guadalupian units. Note the pronounced bend of the lower Leonardian Abo Reef trend in response to the remnant paleohigh formed by Pennsylvanian through Wolfcampian activity along the Huapache fault zone. The lower-middle San Andres margin's trend records a subdued yet evident expression of the Huapache subbasin, and was derived from Rossen and Sarg (1988) and Fitchen et al. (1992). (b) Generalized early to middle Permian stratigraphic columns. After Hills and Kottlowski, 1983.

onlapping, predominantly siliciclastic unit termed the "lower Cherry Canyon sandstone tongue." The upper unit consists of offlapping, mixed siliciclastic/carbonate strata of the uppermost San Andres Formation and "upper Cherry Canyon sandstone tongue." These strata downlap the upper surface of the lower Cherry Canyon sandstone tongue (Figures 4, 5).

Compilations of recent outcrop studies have subdivided the San Andres Formation into at least eight sequences of similar spatial scales (Figure 4; Fitchen et al., 1992; Kerans et al., 1992). Additionally, some sequences within the Brushy Canyon Formation may lack shelfal equivalents (Fitchen et al., 1992; M.H. Gardner, personal communication, 1992), thus *at least* ten large-scale sequences may be embedded within the two seismic-scale, third-order sequences of Sarg and Lehmann (1986a; Figure 4). Estimates of the duration of the entire San Andres Formation range from 2.1–4.0 m.y., depending on the absolute time scale used (Figure 5). Assuming equal durations, each of the ten presently recognized San Andres sequences of similar spatial scale could range from 210–400 k.y., well below the 1–5 m.y. duration of Leonardian and Guadalupian biostratigraphic zones (Ross and Ross, 1987; Harland et al., 1989; Wilde, 1990). Because our estimates of temporal duration are imprecise, we use a site-specific terminology. Probable fourth-order cycles (0.1–1.0 m.y.) are informally termed "large-scale" (rock) and "long-term" (time), and probable fifth-order cycles (<0.1 m.y.) embedded within the two large-scale San Andres sequences exposed in Last Chance Canyon are termed "high-frequency sequences" or simply "cycles."

SEQUENCE STRATIGRAPHY OF LAST CHANCE CANYON

In this section, we describe lithofacies distributions and large-scale stratigraphic relationships within sequences of the San Andres Formation and Cherry Canyon sandstone tongue. Photo mosaics were used in the field to record facies and stratal geometries along outer-shelf to toe-of-slope depositional profiles while physically tracing surfaces between measured sections. Most measured sections (21 of 28) are projected onto cross section A–A', which is oriented parallel to inferred depositional dip (Figures 6, 7). Although it will not be referenced repeatedly, all subsequent discussions are keyed to cross section A–A' (Figure 7). For most facies, we consider position along depositional profiles as a descriptive rather than interpretive attribute because this observation can be made directly due to the shelf-to-basin continuity and quality of exposure in Last Chance Canyon. Measured sections and extensive facies descriptions can be found in Sonnenfeld (1991a,b).

Only three of the eight to ten large-scale San Andres sequences are present within Last Chance Canyon's 120–145 m thick San Andres Formation outcrops. These are upper San Andres sequences 2, 3, and 4, denoted as "uSA$_2$," "uSA$_3$," and "uSA$_4$" (Figures 4, 5) (also referred to, in much of the rest of the literature, as uSA2, uSA3, and uSA4). The oldest large-scale sequence in Last Chance Canyon, uSA$_2$, is incomplete in outcrop but continues in the subsurface where an additional 255 m of San Andres strata were penetrated by the Panoil 1-33 Huapache well (Figure 6). This chapter emphasizes uSA$_4$ for two reasons.

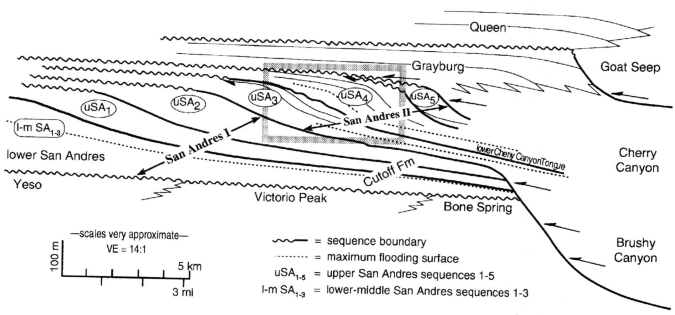

Figure 4. Schematic stratigraphic setting of Leonardian and Guadalupian strata along the northwestern shelf of the Delaware basin. Modified in part from Sarg and Lehmann (1986a), Pray (1988), and Kerans and Nance (1991). Box represents area of Last Chance Canyon cross section A–A' (Figure 7). San Andres "I" and "II" represent seismic-scale third-order sequences defined by Sarg and Lehmann (1986a).

First, cyclicity within uSA₄ slope strata is unusually well resolved due to high-frequency reciprocal siliciclastic/carbonate sedimentation. Second, stratigraphic architecture and facies relationships are well established for most phases of the uSA₄ sequence due to the accessibility and shelf-to-basin continuity of its exposures.

Upper San Andres Sequence 3 (uSA₃)

Upper San Andres sequence 3 (uSA₃) is highlighted by large-scale reciprocal siliciclastic/carbonate sedimentation. The basal sequence boundary of uSA₃ is interpreted to underlie a 20 m thick, very fine-grained hummocky cross-stratified sandstone deposited within storm wave base on a low-angle ramp or slope. The sandstone abruptly overlies a cherty bioturbated dolomudstone that is devoid of sedimentary structures and is interpreted to have been deposited below storm wave base in an outer-ramp or slope environment. An equivalent sandy interval that thickens eastward into the Huapache subbasin can be traced in geophysical well logs throughout the area. We interpret the surface below this sandstone as a submarine surface of siliciclastic bypass developed during a basinward shift of facies tracts and a relative fall of sea level. This 20 m sandstone is interpreted as a tongue of the Brushy Canyon sandstone because it marks the first occurrence of outer-ramp to slope siliciclastics within the San Andres (Figures 4, 5).

Five higher frequency, smaller scale cycles are resolved within uSA₃ (Figures 7, 8a). Pronounced sea-ward dislocations of facies across some of these high-frequency, bounding surfaces imply that they probably formed in response to high-frequency relative falls of sea level. Because cycles exposed in uSA₃ outcrops lack a common reference, such as a shoreface or offlap break, stacking patterns were inferred by comparing the seaward limits of facies deposited within storm wave-base (Figure 8a). The seaward limits of facies deposited within storm wave-base were defined by the presence of swaley or hummocky cross stratification and/or by the transition from winnowed to unwinnowed carbonate fabrics (e.g., the fusulinid grainstone/packstone to brachiopod-fusulinid wackestone facies transition). This approach reveals a landward-stepping stacking pattern below an interpreted maximum flooding surface capping cycle 2, followed by a seaward-stepping stacking pattern that culminates with an outer-shelf toplap surface (Figures 8a, 9).

The formation of the outer-shelf toplap surface, an ensuing episode of siliciclastic bypass that fed lower slope to basinal lower Cherry Canyon sandstones, and an observed seaward shift of facies deposited within storm wave base (Figure 8) are all interpreted as responses to a relative fall in sea level resulting in a sequence boundary between uSA₃ and uSA₄. Although the toplap surface lacks evidence for sub-aerial exposure in western Last Chance Canyon, it is tentatively correlated with the highest internal San Andres karst surface at Algerita Escarpment, 10–20 km shelfward (Fitchen et al., 1992; Kerans et al., 1992). The basis for this correlation is the observation that both surfaces immediately underlie the first occur-

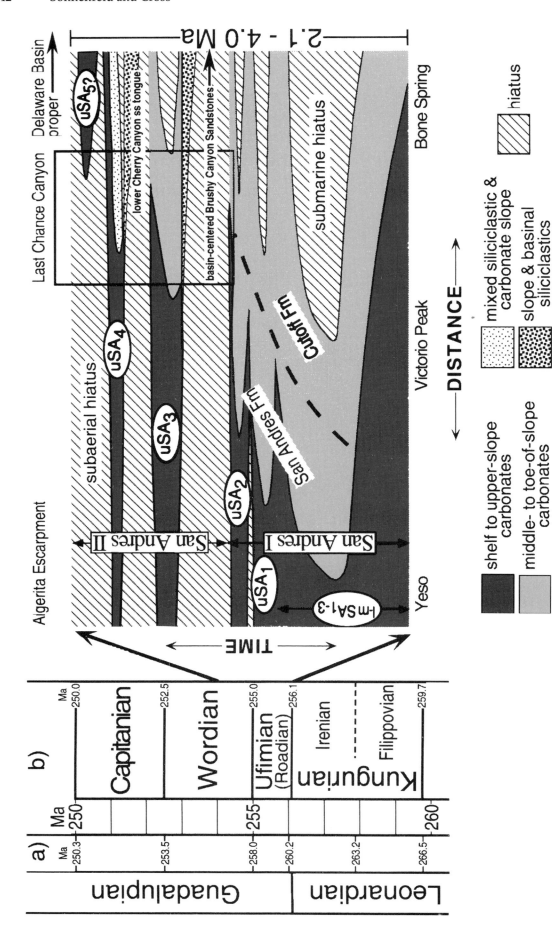

Figure 5. Absolute time scales, geochronologic units, and Wheeler (chronostratigraphic) diagram for the ten San Andres sequences recognized in the western Delaware basin and Guadalupe Mountains region. Time scale (a) of Ross and Ross (1987) and Wilde (1990) differs from time scale (b) of Harland et al. (1989), particularly with respect to the duration ascribed to the Roadian and Wordian stages. These discrepancies impart a poten-

.

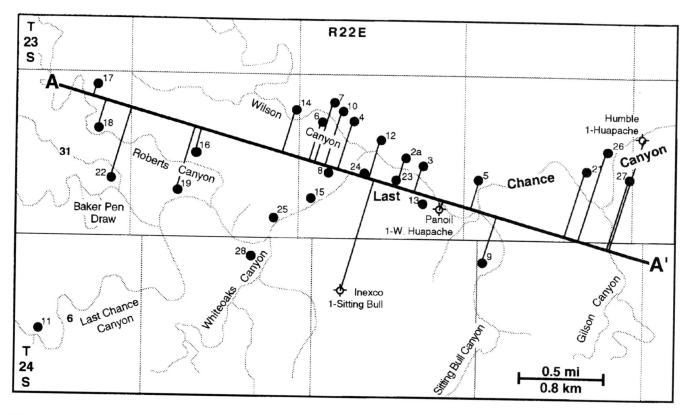

Figure 6. Geographic map of Last Chance Canyon area showing locations of measured sections and projection lines used to construct depositional-dip cross section A–A' (Figure 7).

rence of siliciclastic deposits within the upper San Andres.

Upper San Andres Sequence 4 (uSA₄)

Most high-frequency cycles within uSA_4 are composed of slope- and basin-restricted sandstone prisms that constitute transgressive hemicycles, overlain by outer-shelf and slope mixed carbonate and siliciclastic strata that constitute regressive hemicycles (Figures 10–12). Most high-frequency cycles in Last Chance Canyon also involve a lateral gradation from asymmetric regressive hemicycles commonly observed in shelf settings (e.g., "parasequences" and "PACs"), to less frequently documented symmetric transgressive and regressive cycles in slope settings, to asymmetric transgressive hemicycles in toe-of-slope settings.

Stacking patterns within uSA_4 are difficult to define by geometries and thickness patterns alone because all high-frequency cycles show some net offlap. Additionally, selection of the long-term maximum flooding surface is complicated by the presence of downlap surfaces within individual high-frequency cycles. Downward excursions of storm wave base (Figure 8b) facilitated cycle resolution within the predominant subtidal strata of uSA_4. We infer storm wave base from the seaward limit of the well-bedded, partially winnowed fusulinid packstone facies or by

the presence of the sigmoidal sand wave and trough cross-stratified fusulinid sandstone facies (Figure 13). We also use the volume, frequency of deposition, and hydrodynamic energy of slope siliciclastics as indirect criteria for stacking pattern analysis because changes in the location of shelfal accommodation are inferred to control whether siliciclastics are stored on the inner shelf, the outer shelf, or whether they tend to be bypassed to slope and basinal settings. Changes in siliciclastic influx to slope settings, coupled with the geometric pattern of facies tracts deposited within storm wave base, help define a landward-stepping to vertically stacked pattern for cycles 1–4, followed by a strongly seaward-stepping pattern for the remaining cycles in uSA_4 (Figure 8b).

Lowstand Systems Tract

In Last Chance Canyon, the lower Cherry Canyon sandstone tongue comprises toe-of-slope siliciclastic strata within cycles 1–4. At the start of uSA_4, siliciclastic sediments completely bypassed the Last Chance Canyon area and accumulated farther seaward in the Huapache subbasin and Delaware basin proper (Figures 4, 5). Cycles 1 and 2 are the oldest uSA_4 strata exposed in Last Chance Canyon and onlap the basal sequence boundary seaward of the terminal shelf break of uSA_3. On the basis of these geometric constraints, we assign cycles 1 and 2 to the lowstand systems tract.

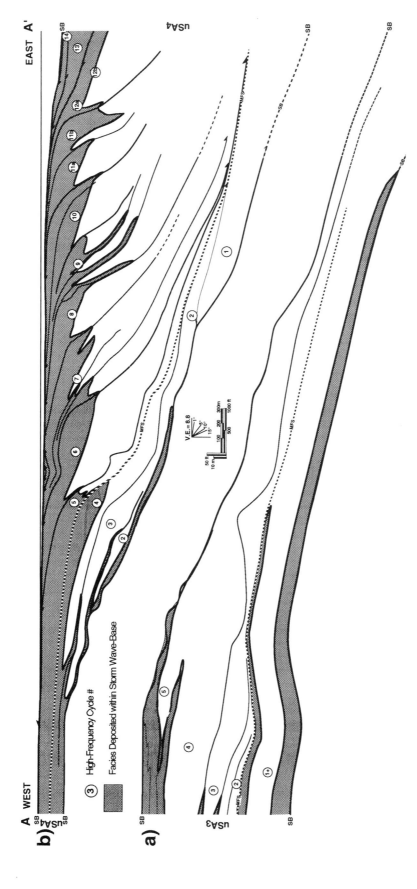

Figure 8. Simplified version of Figure 7 emphasizing facies deposited within storm wave base. uSA₃ and uSA₄ are spaced apart from each other to better distinguish their internal stacking patterns. (a) Storm-wave-base transits within uSA₃. Note the landward-stepping, to vertically stacked, to seaward-stepping long-term pattern within cycles 1–5. (b) Storm-wave-base transits within uSA₄. Note the land-ward-stepping to vertically stacked long-term pattern within cycles 2–5.

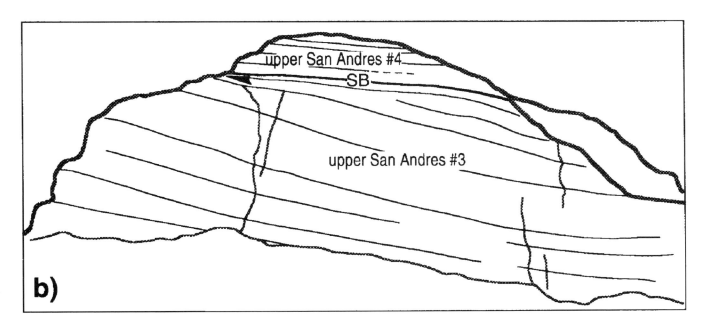

Figure 9. Photo (a) and line drawing (b) of a prominent toplap surface in Baker Pen Draw. This toplap surface overlies prograding fusulinid packstones of uSA$_3$ cycle 5 and is interpreted as the sequence boundary separating uSA$_3$ from uSA$_4$. Siliciclastics bypassed across this surface during an interpreted relative fall of sea level. A silty recess and flat-lying peloid wackestone/packstone outer-shelf topsets overlie the toplap surface and are correlated with uSA$_4$ cycles 3 and 4(?).

Cycle 1 is confined to toe-of-slope settings and is composed of channelized skeletal-peloid sandstone with coarse-tail normal grading and convolute lamination (Figure 14). This facies is interpreted as the product of rapid deposition by sediment gravity flows. Although complete Bouma sequences are absent, we interpret the graded beds as possible top-truncated turbidites. Channel-form scours, probably cut by bypassing sediment gravity flows, were back-filled by multiple 10- to 60-cm thick turbidites. Within cycle 1, skeletal size and diversity decrease upward from a coarse tail composed of brachiopod, gastropod, and echinoderm fragments up to 5 cm in maximum dimension, to a coarse tail exclusively dominated by fusulinids less than 1 cm in length. Concurrently, medium- and coarse-grained peloids in the basal beds

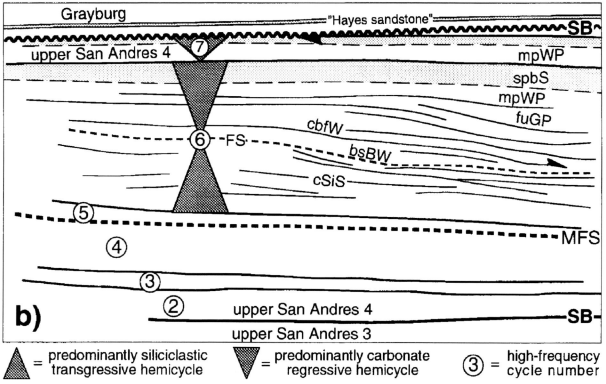

Figure 10. Photo (a) and line drawing (b) of expanded lower- to upper-slope deposits of high-frequency uSA$_4$ cycle 6 just east of measured section #4. Underlying cycles 2–5 are represented by thin toe-of-slope deposits and overlying cycles 7–8 are represented by thin outer-shelf beds. The transgressive hemicycle of cycle 6 is composed of a churned sandstone (cSiS) prism capped by a recessive condensed section and flooding surface. Silty sponge-brachiopod wackestone beds (bsBW) and prograding fusulinid packstone foresets (fuGP, cbfW) downlap the condensed section and form the regressive hemicycle. The fusulinid wackestone/packstone facies grades landward into more massive gray peloid packstone beds (mpWP) that are succeeded gradationally upward by outer-shelf sandstones (spbS).

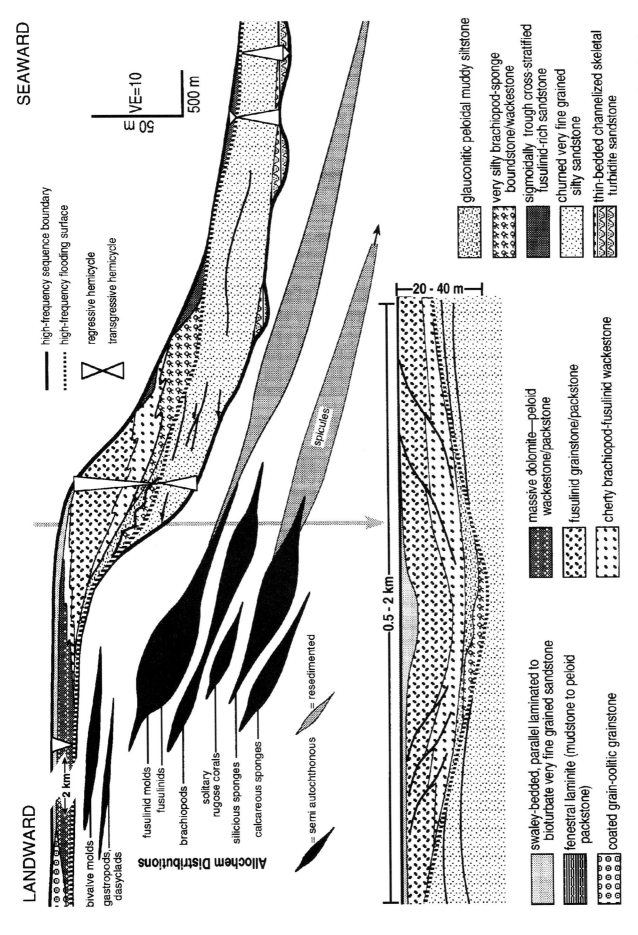

Figure 11. Schematic strike and dip cross sections through a high-frequency cycle typical of the uSA$_4$ middle highstand systems tract showing the distribution of facies and skeletal allochems.

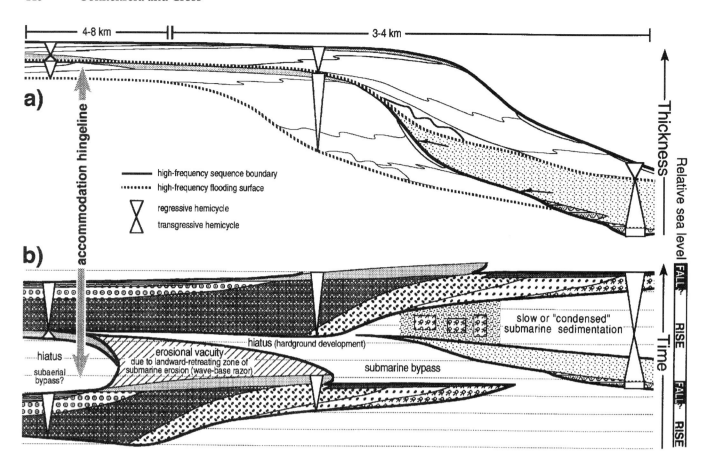

Figure 12. (a) Schematic cross section through a high-frequency cycle typical of the uSA$_4$ middle highstand systems tract with siliciclastics patterned and carbonates left white to emphasize reciprocal sedimentation patterns. (b) Wheeler (chronostratigraphic) diagram of (a). Key to facies patterns on Figure 11.

of cycle 1 grade upward to very fine and fine-grained peloids within the upper beds of cycle 1 and all of cycle 2. The siliciclastic fraction is uniformly very fine grained, probably due to presorting by aeolian or shallow-marine processes on the shelf.

Medium- to thick-bedded channelized turbidite sandstones of cycle 1 are separated from overlying more thinly bedded fusulinid- and peloid-rich sandstones of cycle 2 by a thin, recessive weathering, glauconitic silty sandstone. This silty sandstone represents slow rates of sedimentation and a pause in sediment gravity flow deposition (Figure 15). Thin-bedded fusulinid-peloid sandstones exhibit coarse-tail normal grading, an abundance of soft-sediment deformation structures, and rare flute casts and climbing ripple stratification. They are interpreted as top-truncated turbidites that are locally reworked by traction currents of uncertain origin. Toe-of-slope sandstones of cycle 2 continue the trend established in cycle 1 involving an upward decrease in depositional energy or flow competence, as inferred from diminishing flow unit thicknesses, maximum grain sizes, and basal scour depths. The toe-of-slope turbidite facies of cycles 1 and 2 form a fan-shaped lobe that emanates from an inferred point source (Figure 16).

The abundance of peloids, fusulinids, and other skeletal grains within sandstones of cycle 1 and 2 suggests that carbonate-producing communities were flourishing, at least intermittently, and that an autochthonous fusulinid grainstone/packstone facies may have existed upslope in the 10–30 m water depth range ascribed to fusulinids (Ross, 1983). The absence of any outer-shelf or slope deposits within cycle 1 or of any upper-slope deposits within cycle 2 suggests, however, that these facies tracts lacked preservation potential. Along strike preservation of cycle 1 and 2 slope carbonates remains an unresolved possibility. At the base of cycle 3, allodapic skeletal packstones and grainstones overlie a major truncation surface and are interpreted as compound sediment gravity flows (Figure 17). Vague indications of current winnowing within these skeletal grainstones suggest that a local subaqueous base level of accumulation, perhaps dictated by storm activity and the wave-base razor, may have been responsible for the incomplete preservation of upper-slope carbonate facies (Figures 8b, 17). We infer that upper-slope carbonate facies approximately coeval with the allodapic grainstones were removed by seaward transits of the wave-base razor during high-frequency relative falls of sea level.

Figure 13. Sigmoidal sand wave and trough cross-stratified fusulinid sandstone facies. (a) The cross-stratified bedset above the 15-cm pencil contains fusulinid-rich sigmoids that are draped by white, relatively recessive, silty micrites. These drapes are interpreted to have been deposited during slack periods when little or no traction transport occurred. Episodic offshore-oriented storm gradient currents (Aigner, 1985) focused down slope-conduits or gullies may have generated these bedforms. Alternatively, episodic saline density (Harms, 1974) or tidally influenced currents may have been responsible. (b) Trough cross-stratified fusulinid sandstone (above pencil) enclosed within churned sandstone and silty sponge-brachiopod wackestone facies and lacking the sigmoidal sand waves and heterolithic bundles of (a). Located just downslope of a fusulinid bank (LCC #5).

Transgressive Systems Tract

Cycles 3 and 4 are the first cycles within uSA$_4$ to preserve a full suite of autochthonous outer-shelf and slope carbonate facies. These cycles involve highly aggradational sigmoidal fusulinid banks with relatively muddy foreslopes predominantly composed of cherty microskeletal-spicule wackestone and cherty brachiopod-fusulinid wackestone facies. Autochthonous fusulinid wackestones and packstones of cycles 3 and 4 aggraded and produced an irregular mounded topographic surface with up to 5 m relief along ramp or slope profiles (Figures 7, 17). The top of cycle 4 occurs approximately midway within the transition from aggradational sigmoidal clinoforms to progressively steeper, more progradational clinoforms.

Rare burrows within lower- to toe-of-slope turbidites of lowstand cycles 1 and 2 contrast with abundant cherty burrows (Thalassinoides) within overlying bioturbated fusulinid-peloid sandstones that compose the lower- to middle-slope portions of cycles 2–4. A gradual upward increase in dolomicrite and glauconitic peloids within the bioturbated fusulinid-peloid sandstone facies suggests the resumption of downslope transport from an expanding carbonate factory on the outer shelf. Concurrently, an upward decrease in bed amalgamation and bed thickness (Figure 15) is associated with a diminished preservation of sedimentary structures, limiting our ability to attribute a precise depositional mechanism to these deposits. These vertical patterns record a progressive decrease in siliciclastic content, sediment accumulation frequency, and hydrodynamic energy, and are interpreted as toe-of-slope expressions of a landward-stepping siliciclastic feeder system on the shelf concurrent with episodic landward movement of the zone of impingement of the wave-base razor on the seafloor (Figure 8b).

As the siliciclastic content progressively diminishes within slope deposits of cycles 3 and 4, episodic omission surfaces are increasingly littered with bryozoa, brachiopods, echinoid spines, and articulated crinoid columnals and cirri. These nearly autochthonous life-assemblage accumulations are interpreted as bryozoan-echinoid-crinoid meadows. In lower-middle slope positions, a distinctive 0.2–1.0 m, cherty, burrowed interval with glauconitic peloids caps cycle 4 and records the onset of a sediment-starved period when the siliciclastic feeder system on the shelf is inferred to have retreated farthest landward (Figure 15). The top of this interval is interpreted as the transgressive maximum or maximum flooding surface of uSA$_4$. Cycles 3 and 4 are assigned to the transgressive systems tract because they extend landward of the uSA$_3$ terminal shelf margin and are capped by the interpreted maximum flooding surface (Vail, 1987). The progressive changes from lowstand systems tract cycles 1 and 2 through transgressive systems tract cycles 3 and 4 reflect an evolution from siliciclastic-dominated slope sedimentation to carbonate-dominated slope sedimentation in association with an interpreted long-term relative rise in sea level.

Highstand Systems Tract

Ten high-frequency cycles (cycles 5–12b) are identified within the uSA₄ highstand systems tract. These cycles have evolving facies associations and stratal geometries, coupled with changing siliciclastic:carbonate sediment ratios that invite division into the early, middle, and late portions of the highstand systems tract of uSA₄. Early highstand systems tract cycle 5 is highly aggradational and almost exclusively carbonate in outer-shelf and slope settings; by contrast, middle highstand systems tract cycles 6–10 exhibit increased rates of offlap and mark the resumption of siliciclastic input to the outer shelf and slope. In contrast to the apparently abrupt discharge and accumulation of Brushy Canyon and lower Cherry Canyon sandstones at the base of sequences uSA₃ and uSA₄, high-frequency reciprocal sedimentation within the uSA₄ highstand systems tract reflects a progressive increase in the volume of siliciclastics accumulated in the basin with a concomitant decrease in the volume of siliciclastics accumulated on the outer shelf. As a result, large-scale reciprocal siliciclastic/carbonate sedimentation within uSA₄ is more obscure than the straightforward pattern of large-scale reciprocal siliciclastic/carbonate sedimentation within uSA₃. The late highstand systems tract is similar in many respects to the middle highstand systems tract, however, late highstand systems tract cycles 11a–12b were deposited during a long-term relative fall in sea level as indicated by a downstepping shelf-margin geometry and stratigraphic fall of the bathymetrically sensitive fusulinid facies tract.

A possible karst surface caps late highstand systems tract cycle 12b, hence cycles 13–14 are tentatively assigned to a fifth large-scale upper San Andres sequence ("uSA₅") that is only partly exposed in outcrops and that is completely absent by erosion or nondeposition in the central to western portions of the study area. Complex sigmoid-oblique clinoforms with fusulinid dominated slopes change across the karsted sequence boundary separating uSA₄ from uSA₅, to high-angle oblique clinoforms with peloid grainstone dominated slopes. The San Andres/Grayburg sequence boundary is defined by a karsted subaerial unconformity, which is onlapped by peritidal cycles of the overlying Grayburg Formation, demonstrating that a significant seaward shift in onlap occurred (Sarg and Lehmann, 1986a,b).

Early Highstand Systems Tract

Early highstand systems tract cycle 5 has a slope facies association that is almost exclusively carbonate in lithology and is distinct from preceding cycles of the transgressive systems tract as well as from ensuing cycles of the middle highstand systems tract. Siliciclastic sediments were almost completely partitioned on the shelf during the maximum long-term transgression represented by the latter portion of transgressive systems tract cycle 4 and early highstand systems tract cycle 5. In lower-slope to toe-of-slope positions, a glauconitic micritic to shaly

siltstone overlies the uSA₄ maximum flooding surface at the base of cycle 5 (Figure 15). The volumetrically limited basinward transport of siliciclastics represented by this condensed deposit is interpreted as the product of aeolian dust storms that ultimately settled out of suspension in the water column. Shelf sandstones attain their seaward limit 1.7 km from the coeval shelf-slope break. In the western limits of Last Chance Canyon, these sandstones grade from bioturbated sandstone upward into trough cross-stratified and herringbone ripple-stratified sandstone, and finally into coated grain-oolitic grainstone capped by fenestral laminites.

Cycle 5 includes conspicuous lower-slope crinoid-bryozoan bafflestone bioherms (Figure 18a) and bryozoan-brachiopod-sponge reefs (Figure 18b) at

Figure 14. Channelized skeletal-peloid turbidite sandstone facies within uSA₄ cycle 1. (a) Outcrop close-up of a 0.6 m flow unit with coarse-tail (skeletal) normal grading. Fusulinid and disarticulated brachiopod molds are most abundant. Note large gastropod at the top of the skeletal-rich portion. Figure 14. Continued. (b) Despite the well-sorted

very fine-grained to silty sand and the absence of a clay fraction, diverse skeletal and peloid grain sizes define coarse-tail, normally graded beds interpreted as top-truncated A, AB, and ABC Bouma sequences. Fusulinid and brachiopod molds are most common; additionally, gastropods, bivalve and/or green algal molds are present, suggestive of an allochthonous outer-shelf to upper-slope faunal assemblage. (c) Large flame structures capping 0.2- to 1.0-m flow units interpreted as turbidites. Soft sediment deformation can fully obscure lamination over intervals of several meters. Bioturbation rarely extends more than 5–10 cm down from the top of individual flow units. Recessive pock marks are fusulinid molds that frequently define coarse-tail grading.

Figure 15. Seven high-frequency cycles within toe-of-slope strata of uSA$_3$ and uSA$_4$. The 1-m-thick glauconitic peloidal muddy siltstone (gpSi) separates cycles 1 and 2 where both are present; however, cycle 1 is absent in this photo due to onlap. Note the increasingly well-defined thin-to-medium bedding within uSA$_4$ cycles 3 and 4. In a lower- to toe-of-slope setting, such a thinning-upward pattern is indicative of *increasing* accommodation on the shelf. Toe-of-slope sandstones of uSA$_4$ cycle 6 downlap pure carbonate beds of cycle 5. fissM = fissile (wackestone)/mudstone; cmsW = cherty microskeletal spicule wackestone; tfpS = thin-bedded fusulinid-peloid turbidite sandstone; bpWS = bioturbated fusulinid-peloid dolopackstone-wackestone/sandstone; cSiS = churned very fine-grained sandstone. Photo located between measured sections #2a and #3.

inferred paleowater depths of 30–50 m. These autochthonous buildups are downlapped by aggradational to weakly progradational fusulinid wackestones (Figures 7, 18b). Finely laminated toe-of-slope dolomudstones with rare horizontal burrows (Figure 19) indicate that cycle 5 represents the only time during uSA$_4$ deposition when toe-of-slope bottom waters became dysaerobic, possibly due to landward encroachment of a stratified water column during long-term relative rise in sea level.

Middle Highstand Systems Tract

High-frequency cycles 6–10 have many common aspects that are summarized in a generalized depositional model for the middle highstand systems tract and that provide a basis for comparison with high-frequency cycles in other portions of the uSA$_4$ sequence (Figures 11, 12). Lower-slope to toe-of-slope

siliciclastic strata within high-frequency transgressive hemicycles typically overlie an erosive basal bounding surface and are composed of thin bedded, channelized, skeletal turbidite sandstones that grade upward into bioturbated, very fine grained to silty sandstones (Figure 20). Sandstones interpreted as turbidites show coarse-tail grading of carbonate allochems and display ABC and BC Bouma divisions (Figure 21a). Centimeter- to decimeter-scale turbidite beds fill channels that are 5–30 m wide and 0.5–3.5 m in relief (Figure 21b). The stratigraphic backfilling and sporadic distribution of these channels along strike imply that multiple small-scale point sources discharged siliciclastic turbidity currents. Erosive surfaces at the base of transgressive hemicycles are probably the product of higher energy turbidites that bypassed the Last Chance Canyon area during regressive maxima.

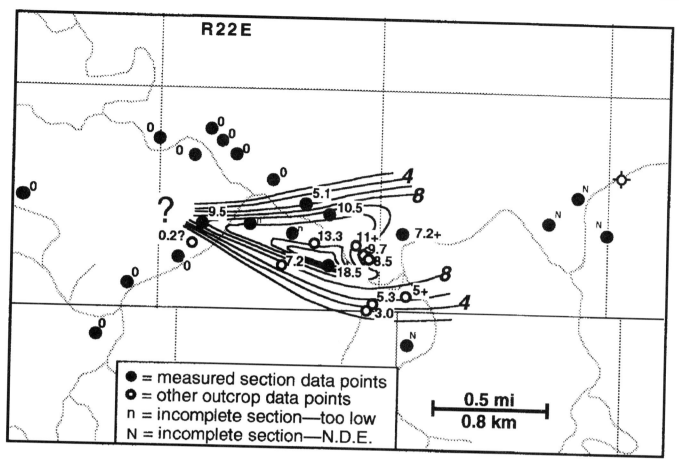

Figure 16. Isopach defining a small-scale point-source for turbidites of the lower Cherry Canyon sandstone tongue (cycle 1 and toe-of-slope of cycle 2). The axis of this lobe was probably a conduit for sand bypass to the east during prior high-frequency cycles of the lowstand systems tract.

The turbidite facies grade upward into increasingly burrowed, massive sandstones that probably were derived from airborne suspensions of sand and silt and/or from subaqueous shelf sands that were reworked and transported seaward by storm-induced low-density sediment gravity flows. The bioturbation and loss of bedding upward is a product of decreased frequency of deposition and hydrodynamic energy and reflects landward retreat of shelf sand sources and storm wave base during transgression. We infer that erosion and resuspension in storm-generated currents reworked outer-shelf sands deposited during the previous regressive hemicycle, sourcing much of the sediment in the bioturbated slope to toe-of-slope sandstone prisms of the ensuing transgressive hemicycle (Figure 12b). Thus, during inferred retreat of shoreline-attached sediment sources, sands were still delivered to the slope from outer-shelf sources, albeit with decreasing volume and frequency.

Slope sandstone prisms are often capped by cherty omission surfaces interpreted as high-frequency flooding surfaces. Peloid and micrite-rich siltstones, very silty brachiopod-sponge wackestones, and small autochthonous brachiopod-sponge bioherms overlie the omission surfaces. These rocks accumulated at anomalously slow rates and are interpreted as high-frequency condensed sections (Figures 11, 12b). During this sediment-starved transition period, the siliciclastic feeder system on the shelf retreated to its most landward position, siliciclastic sedimentation on the slope was minimal, and carbonate production and accumulation on the outer shelf and slope rejuvenated slowly. Note the stratigraphic similarity of these relationships to those already described for the siliciclastic-carbonate sedimentation patterns within the overall large-scale uSA_4 sequence.

Autochthonous and parautochthonous fusulinid wackestones and packstones at the depositional shelf-slope break mark the onset of active carbonate production on the shelf and slope. These facies volumetrically dominate regressive hemicycles. Fusulinid facies of cycles 6–12b apparently aggraded to water depths near the upper limit of fusulinid tolerances ($\approx$10 m) at a relatively early stage of each regressive hemicycle. This precluded development of hummocky mounded fusulinid foreslopes character-

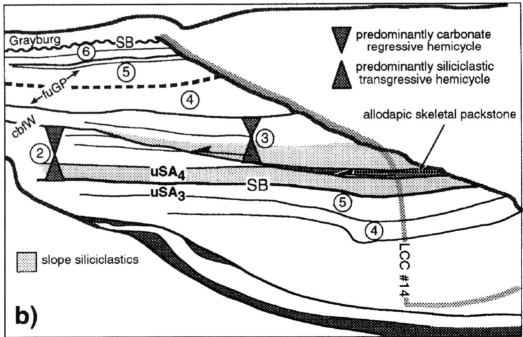

Figure 17. Photo (a) and line drawing (b) of the northwest wall of Wilson Canyon dominated by two symmetrical mid-slope cycles within uSA$_4$. A prominent truncation scar forms the base of uSA$_4$ cycle 3 and is overlain by a channelized allodapic skeletal packstone/grainstone at its lower right. This surface may be associated with a descent of storm wave base and the erosive action of storm gradient currents. The middle-slope, cherty brachiopod-fusulinid wackestones (cbfW) within cycles 2–3 and upper-slope, fusulinid packstones (fuGP) within cycles 4–5 are much more aggradational than comparable facies tracts within the strongly progradational middle highstand systems tract (Figure 10).

istic of earlier, higher accommodation cycles, and instead fostered the strongly progradational nature of middle to late highstand systems tract cycles. Prograding fusulinid wackestones and packstones downlap the siliciclastic slope prisms and sponge-brachiopod bioherms (Figures 10, 20). Upper slope fusulinid facies tracts grade landward into massive bioturbated ostracod-peloid dolopackstones interpreted as slightly restricted outer-shelf facies.

Outer-shelf hemicycles are composed of 1–10 m thick carbonate-sandstone couplets. Outer-shelf carbonates grade upward over decimeters from massive peloid wackestone/packstone to very fine-grained sandstone (Figures 10–13). These ≤6 m sheetlike sandstones exhibit minor thickening and thinning over depositional embayments and promontories within underlying fusulinid shelf margins (Figure 11). Vague bedding within the sandstones exhibits parallel laminated bases, bioturbated tops, and minor relict hummocky to swaley cross stratification, indicating intermittent seaward transport as combined tracted and suspended loads in storm-generated waves and currents. We interpret an outer-shelf marine environment of deposition for these sandstones ranging from just below storm wave base to just below fair-weather wave base, or within a water depth range of about 10–20 m as measured from positions on clinoforms that are bathymetrically calibrated by the fusulinid facies tract.

Outer-shelf sandstones are capped by submarine disconformities interpreted as firmgrounds or hardgrounds because they are penetrated by extensive networks of straight to branched burrows and/or borings (Figure 22). These surfaces formed during the period of sediment starvation following the transgressive erosion described above, but prior to reestablishment of a vigorous outer-shelf carbonate factory.

Outer-shelf carbonate-sandstone couplets in Last Chance Canyon are purely subtidal, in contrast to peritidal inner-shelf and shelf-crest settings landward of the Last Chance Canyon area. In peritidal inner-shelf cycles of the San Andres, Grayburg, and other Guadalupian formations, carbonates shoal to subaerial exposure and *sandstone-to-carbonate* cycles are the rule (Hurley, 1989; Borer and Harris, 1991) because stratigraphic preservation of siliciclastics was limited to the period of increasing accommodation and transgression following subaerial exposure. In subtidal outer-shelf settings, by contrast, carbonate facies do not shoal to subaerial exposure, and the remaining unfilled accommodation makes outer-shelf sandstone accumulation possible during times of either regression or transgression, explaining why *carbonate-to-sandstone* cycles are common in Last Chance Canyon. A diffuse, yet important zone within each high-frequency cycle, here termed the "accommodation hingeline" (Figure 12) separates outer-shelf *carbonate-to-sandstone* couplets typical of uSA₄ in Last Chance Canyon from *sandstone-to-carbonate* couplets of the inner shelf to oolite/pisolite shelf crest.

Late Highstand Systems Tract

The late highstand systems tract is characterized by a progressive increase in the volumetric partitioning of siliciclastic sediment from the shelf to the slope and basin, stratigraphic fall of the bathymetrically sensitive fusulinid facies tract, and a possible karst surface separating cycle 12b and cycle 13 that may represent a large-scale sequence boundary between uSA₄ and uSA₅. The late highstand systems tract also contains prominent crinoid-brachiopod-sponge-algal reefs that are absent in the early to middle highstand systems tract. These were rigid lithified topographic structures (Figure 23) and have fabrics ranging from boundstones to massive intraclastic skeletal-peloid floatstone/rudstone (Figure 24). Calcareous sponges, brachiopods, well-articulated crinoids, and subordinate *Archaeolithoporella* are the most common skeletal components of these reefs (Figure 24). Allodapic skeletal, coated-grain, peloid grainstones compose the fore reef slopes and are interpreted as sediment gravity flows.

Despite the increase in volume of siliciclastic sediment reaching the slope and basin during the late highstand systems tract, fusulinids and crinoid-brachiopod-sponge-algal reefs could apparently prosper during high-frequency highstands when siliciclastic influx was minimal. Although there is negligible preservation of outer-shelf sandstone within cycles 11a–12b, sand-filled cracks and cavities associated with discrete surfaces along the reef crest attest to sand bypass, synsedimentary reef lithification, and the possible first occurrence of subaerially exposed high-frequency cycle boundaries.

VOLUMETRIC PARTITIONING AND FACIES DIFFERENTIATION

In this section, we compare stratal geometries, sediment volume distributions, and differences in carbonate facies associations in different parts of the uSA₄ sequence. There is a pronounced volumetric partitioning of siliciclastics onto either the shelf or the slope that operates in response to both long- and short-term changes in accommodation. The presence or absence of siliciclastics in outer-shelf and slope settings has important consequences on the facies and geometries of carbonate strata with which they alternate in high-frequency cycles. Not only is siliciclastic "poisoning" an influence on carbonate productivity and diversity, but the volume of siliciclastics discharged to the slope affects carbonate progradation patterns by providing the essential platform across which the depth-sensitive carbonate sediments prograde. We stress the following stratigraphic responses to changes in accommodation because they are sufficiently generalized to be applied to other shelf-margin settings: (1) progradation/aggradation ratios; (2) the variation over time in the lateral extent of grain-dominated versus mud-dominated facies; and (3) the timing and nature of bioherm development.

Progradation:Aggradation Ratios of Outer-Shelf to Slope Strata

Within sequence uSA$_4$, clinoforms evolved from low-angle sigmoidal, to high-angle sigmoidal, to complex sigmoid-oblique, and finally to high-angle oblique. This evolution is a product of long-term changes in accommodation and volumetric partitioning, as are numerous other attributes of stratal geometries and facies. Volumetric partitioning of carbonate strata in Last Chance Canyon is measurable by comparing the volumes of sediment accumulated in outer-shelf topset, slope foreset, and toe-of-slope bottomset strata within different high-frequency cycles. We quantify volumetric partitioning by determining progradation:aggradation ratios and offlap angles (Figure 25). Any procedure attempting to quantify geometric parameters such as these is dependent upon knowing what was an originally horizontal or very near horizontal datum. Cross section A–A' (Figure 7) utilizes as a datum an intertidal to supratidal high-frequency cycle cap within the Grayburg Formation that is overlain by a prominent white sandstone marker bed, locally termed the "Hayes sandstone."

Long-term variations of the progradation:aggradation ratio were evaluated after normalizing the data for high-frequency fluctuations in accommodation. The first step in this procedure is to select comparable bathymetric tie points within each high-frequency cycle. These tie points are defined as the middle of the upper-slope fusulinid packstone facies tract, interpreted as representing approximately 20 m (±5 m) water depth based on the 10–30 m depth range for autochthonous fusulinid growth (Ross, 1983). This facies is the most depth-sensitive and depth-restricted facies, and occurs in all but a few cycles. The second step is selecting a comparable phase within each cycle in order to limit potential biases imposed by minor stratigraphic rises and falls resulting from high-frequency relative changes in sea level. Because fusulinid grainstone/packstone facies accumulated principally during high-frequency regressive hemicycles, tie points were selected at the midpoints of each regressive hemicycle. When such tie points are connected, long-term stratigraphic rises and falls become quite evident (Figure 25).

Unfortunately, the selection of comparable phases and bathymetric tie points within each high-frequency cycle remains somewhat subjective. Additionally, subtle along-strike variations in sediment supply and progradation directions may bias the representation of stratigraphic rise and fall when projected or "collapsed" onto the depositional dip cross section. For example, although cycle 6 shows minor stratigraphic fall, stratal geometries within its regressive hemicycle suggest that this may be an artifact of a more southerly progradation direction than other cycles of the middle highstand systems tract. As a result, the depicted depositional dip profile may be distorted and represent a partially strike-oriented view. The

unusual erosional pattern below the San Andres/Grayburg unconformity at measured section #7 also suggests that some differential compaction may have occurred across the transgressive to early highstand carbonate bank margin (Figure 7). In order to minimize such problems, it is best to average the observed stratigraphic rise or fall over a number of cycles.

The offlap angle is the arctangent of the aggradation:progradation ratio and is zero for clinoform surfaces that prograde without rising such that topsets of clinoforms merge into a single horizontal surface. This angle is positive when there is stratigraphic rise and negative when there is stratigraphic fall.

The progradation:aggradation ratio and offlap angle were not calculated between cycles 1 and 2 because they lack the common tie point discussed above. Preserved facies that were deposited in water depths as shallow as the fusulinid tie point (~20 m

a)

Figure 18. Selected facies from early highstand systems tract cycle 5. (a) Delicate 3-D bryozoan network in the core of crinoid-bryozoan bafflestone bioherm overlying the maximum flooding surface at measured section #10.

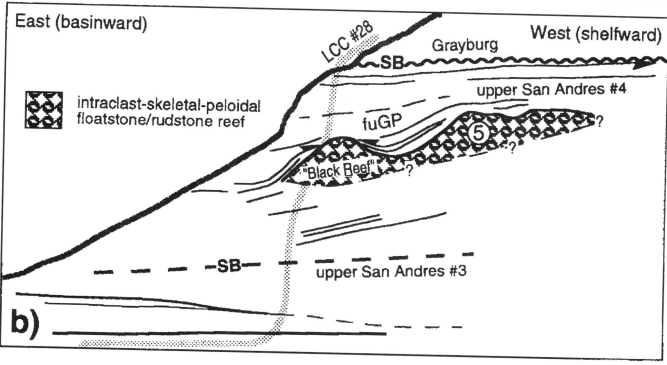

Figure 18. Continued. (b) Outcrop photo and line drawing of massive intraclast-skeletal-peloidal floatstone/rudstone facies primarily composed of bryozoans, brachiopods, and calcareous sponges forming a reef with over 25 m depositional relief. Note the oversteepened (>35°) fusulinid foreslope beds on the down-slope (east) side of the reef and the upturned fusulinid wackestone beds just right of the upper reef knob. These relationships, coupled with multiple internally brecciated horizons, attest to early cementation and subsequent differential compaction. Located at mouth of White Oaks Canyon.

Figure 19. Toe-of-slope laminated dolomudstone with ≤1 cm flow units with parallel laminated caps. Note the local erosive surface accentuated by soft-sediment uplift in response to an underlying burrow. Some gradational waxing and waning of parallel lamination implies that deposition was not necessarily from discrete bottom-hugging sediment gravity flows. Instead, variable rates of suspension-fallout from higher in the water column are likely.

water depth) first occur within the upper part of cycle 3. Geometric relationships between cycles 3 and 4 demonstrate significant progradation with a progradation:aggradation ratio of +281:1 and a low offlap angle of +0.2°. Previously reviewed sedimentologic evidence in coeval siliciclastic slope strata led to the inference that siliciclastic shelf feeder systems were stepping landward at this time. Nevertheless, the transgressive systems tract is characterized by net offlap due to the high indigenous carbonate sediment supply within the peloidal outer-shelf and fusulinid-rich upper-slope facies tracts.

Late transgressive systems tract cycle 4 and early highstand systems tract cycle 5 were deposited when the geographic position of new accommodation space had shifted the farthest landward. Cycles 4 and 5 are nearly vertically stacked (pronounced stratigraphic rise) and occur at the transition from a landward-stepping to a stratigraphically rising seaward-stepping stacking pattern. Compared to the cycle 3-4 pair,

cycles 4 and 5 have a much lower progradation: aggradation ratio (+17:1) and a relatively high positive offlap angle of 3.4° (Figure 25).

During deposition of highstand systems tract cycles 5–12b, high-angle sigmoidal clinoforms evolved transitionally into complex sigmoid-oblique clinoforms of the middle to late highstand systems tract. This evolution in geometry reflects the changing balance between sediment supply and accommodation, and is associated with long-term decrease in shelfal accommodation due to decreasing rates of relative sea-level rise (Sarg, 1988). The sigmoid to oblique transition is a product of the progressive decrease in carbonate sediment volumes accumulated in outer-shelf topset strata in conjunction with a progressive increase in carbonate sediment volumes accumulated in slope foresets. Volumetric partitioning from the shelf to the slope is expressed in Figure 25 by the dramatic increase in the progradation: aggradation ratio for cycles 6–10 to an average of

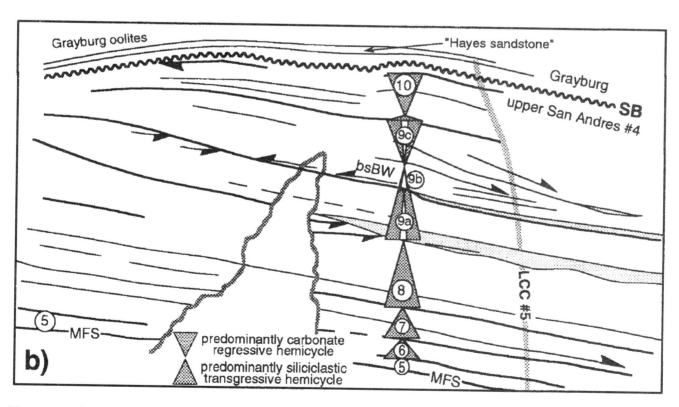

Figure 20. Thin-bedded channelized skeletal turbidite sandstone facies. Photo (a) and line drawing (b) of a depositional-dip view of uSA$_4$ highstand systems tract showing condensed lower slope portions of cycles 5–8, expanded slope portion of cycle 9, and thin outer-shelf portions of cycles 10 and 11. A 3-m turbidite sandstone channel overlies the basal boundary of cycle 9 at the right. Truncation surfaces and channels define bases of subcycles 9b and 9c and reflect an even higher frequency of cyclicity. Also note the onlap of darker colored sponge-brachiopod silty wackestones (bsBW) onto an erosional surface at the base of subcycle 9b. Overlying light-colored fusulinid wackestones and packstones show prominent offlapping geometries.

+209:1, as compared to +17:1 for cycles 4 and 5. Concomitantly, the offlap angle for cycles 6–10 decreases to +0.3°, as compared to +3.4° for cycles 4 and 5, reflecting pronounced progradation and diminished stratigraphic rise. Late highstand systems tract cycles 11a–12b represent the onset of stratigraphic fall in conjunction with the onset of a long-term relative fall in sea level. These cycles record a downstepping or topographically descending shelf margin with a negative progradation:aggradation ratio of –47:1 and a negative offlap angle of –1.2°.

The evolution in clinoform geometries, the reduction in outer-shelf topset thicknesses, the progressive decrease in stratigraphic rise, and ultimately the onset of stratigraphic fall all reflect a long-term decrease in shelfal accommodation. It is important to note that this interpretation is supported by other topset stratal patterns, discussed below, which demonstrate that these changes are neither the exclusive product of differential compaction within the underlying Cherry Canyon sandstone tongue, nor an artifact of cross-section construction.

The San Andres/Grayburg sequence boundary is well defined by toplap along the entire outcrop belt. However, topset stratal patterns indicate different origins for toplap stratal discordances to the west (shelfward) versus those to the east (basinward). Topset strata to the west (early to middle highstand systems tract cycles 5–8) show little or no landward convergence prior to erosional truncation at the sequence boundary (Figure 25a). This implies erosion of the landward reaches of the peloid wackestone/packstone and other more shelfward facies tracts. These facies tracts probably extended 0.5–1.5 km farther west prior to erosion and karsting at the San Andres/Grayburg sequence boundary. By contrast, topset strata to the east (middle to late highstand systems tract cycles 9–13) show considerable landward convergence before terminating by toplap at the sequence boundary (Figure 25b). The convergent geometric relationship suggests that nondeposition by subaerial exposure or by subaqueous bypass occurred during formation of high-frequency cycle boundaries prior to erosion associated with the large-scale sequence boundary. This change in topset geometry—unlikely to be discernible on conventional seismic data—indicates decreasing accommodation through time on the outer shelf.

The upper slope fusulinid facies tract used as a tie point in progradation:aggradation calculations was replaced in cycles 13–15 by an obliquely prograding peloid grainstone facies tract. Nevertheless, an estimation of the progradation:aggradation ratio (+6.1:1) and offlap angle (+9°) reveals a dramatic resumption of stratigraphic rise interpreted as the product of a relative rise in sea level. The existence of predominantly subsurface uSA_5 is supported by the presence of a possible karst surface capping cycle 12b, coupled with the stratigraphic rise exhibited by overlying uSA_5 cycles 13–14.

Spatial and Temporal Differentiation of Grain- versus Mud-Dominated Facies

To facilitate a comparison of carbonate facies with clinoform geometries, carbonate and mixed carbonate/siliciclastic facies were grouped into three fabric-defined categories: (1) grain-supported facies; (2) matrix-supported facies with a skeletal fraction; and (3) carbonate mudstones (Figure 26). Both grain-supported and matrix-supported carbonate facies exhibit an accommodation-controlled volumetric partitioning. To the extent that carbonates and siliciclastic sediments are subjected to similar mechanical transport processes, the interpretation of long-term accommodation patterns in carbonates and siliciclastics employs a similar logic. However, siliciclastics and carbonates can have dissimilar responses to long-term accommodation patterns. These differences reflect the largely autochthonous nature of carbonate sediment supply.

Toe-of-slope turbidites of uSA_4 cycle 1 are predominantly siliciclastic and lack carbonate micrite yet contain a significant "grainy" carbonate component, including an allochthonous outer-shelf and upper-slope fauna and very fine to medium-grained peloids. Cycles 2–4 record the first preservation within the uSA_4 sequence of relatively micritic lower- to middle-slope carbonates. We interpret that the seaward limit of the wave-base razor stepped progressively landward within these cycles causing less erosion to occur at the top of each high-frequency cycle boundary and allowing increased preservation of unwinnowed carbonate slope facies (Figure 8b). In conjunction with a long-term relative rise in sea level and an increase in shelfal accommodation, these middle slope

Figure 21. Thin-bedded channelized skeletal turbidite sandstone facies. (a) Center turbidite shows a basal silicified skeletal lag (Bouma A), parallel lamination (Bouma B), and climbing ripple drift to convolute lamination (Bouma C). Siliciclastic grain size is limited to very fine grained, but the coarse skeletal material attests to significant flow competence. Flow units are highlighted by vertical white lines. (b) Depositional-dip view showing steep (10–15°) erosive channel base that is backfilled by vertical accretion of individual turbidite flow units. Surrounding massive to flaggy weathering strata are churned very fine grained silty sandstone facies.

microskeletal dolowackestone facies tracts expand progressively in width from 650 m in cycle 2 to 1200 m in cycle 5. As this occurred, the toe-of-slope carbonate mudstone facies tract expanded progressively downslope as well culminating with cycle 5 where laminated dolomudstones extend over 3 km seaward from the coeval fusulinid packstone shelf-slope break. Note that these conspicuous seaward-extending micrite-rich carbonate tongues at the toe-of-slope are not the result of bypass due to limited accommodation on the shelf or slope; rather, they are expressions of an active carbonate mud factory during cycles 4 and 5, coupled with effective sediment dispersal mechanisms probably related to storm waves affecting the outer shelf and upper slope to depths of 20–40 m.

Hummocky mounded clinoforms and the onset of the fusulinid grainstone/packstone facies in cycles 3 and 4 record the aggradation of carbonate wackestones from below storm wave base into a zone of episodic current winnowing. From cycles 3 to 5, the maximum width of the winnowed fusulinid grainstone/packstone facies tract increases from 600 to 1200 m. The expanded suite of autochthonous slope and outer-shelf carbonate facies during cycles 4 and 5 may be viewed as the product of the bidirectional (landward and seaward) expansion of the outer shelf to slope carbonate factory coupled with increased preservation potential during long-term maximum accommodation.

Long-term constant to decreasing shelfal accommodation during highstand systems tract cycles 6–12b caused high-angle sigmoidal clinoforms of the early highstand systems tract to evolve transitionally into high-angle oblique clinoforms of the late highstand systems tract. Depositional slopes of 1–3° within aggradational fusulinid deposits of the transgressive to early highstand systems tract steepen progressively to 5–15° within cycles of the middle to late highstand systems tract. Increased slope gradients caused bathymetrically constrained grain-rich carbonate facies tracts on the slope to contract in width. For example, the width of the upper-slope fusulinid packstone facies tract, representing water depths of 10–30 m, contracted from 600 m in cycle 5 to 300 m in cycle 6 and finally to 100 m in cycle 11a.

The previous analysis of topset stratal patterns below the San Andres/Grayburg sequence boundary leads to the inference that the aerial extent of outer-shelf carbonate sediment production must have progressively diminished in response to long-term constant to decreasing accommodation (Figure 25b). This trend reduced the micrite volume available for redeposition in slope and toe-of-slope settings. Consequently, matrix-supported lower-slope carbonates facies tracts (cherty brachiopod-fusulinid wackestone and very silty brachiopod-sponge boundstone/wackestone facies) contracted from over 2 km in width within cycle 5 to less than 150 m in width in cycle 11a.

A common tendency in siliciclastic volumetric partitioning is for sediment accumulation either on the shelf or on the slope. By contrast, matrix-supported carbonates in Last Chance Canyon extended farther basinward into slope and toe-of-slope settings during times of long-term increasing accommodation, concurrent with an inferred landward and seaward *bidirectional expansion* of outer-shelf carbonate facies tracts. In mixed carbonate-siliciclastic systems such as this, the farthest basinward extent of micritic carbonate drape deposits of platform origin can be expected to correspond with long-term maximum transgression (also note on Figure 26 the basinward extent of toe-of-slope micrites within uSA_3 cycles 2–3). When long-term accommodation decreases, outer-shelf facies tracts progressively contract while slope facies become progressively more grain-rich and less micritic.

Timing and Nature of Bioherm Development

The occurrence of large bioherms in cycle 5 is augured by the occurrence of very small bioherms and skeletal concentrations on multiple very high-frequency omission surfaces within cycles 3 and 4. These bioherms are restricted to times when influx of allochthonous sediment (both carbonate and siliciclastic) is anomalously low or temporarily suspended, and occur on condensed sections and flooding surfaces of various temporal scales. The development of several lens-shaped crinoid-bryozoan wackestone to bafflestone bioherms up to 5 m thick and 25 m in diameter (Figure 18a), and of other more rigid intraclast-rich bryozoan-brachiopod-sponge reefs (Figure 18b) on the medial flooding surface of cycle 5, culminates the trend of increasingly autochthonous skeletal accumulations on sediment-starved surfaces and within condensed sections.

◄

Figure 22. (a) Subtidal outer-shelf sandstone disconformably overlain by massive peloid dolowackestone/packstone facies. This submarine disconformity is interpreted as transgressive in origin and represents a bored hardground or open-burrowed firmground developed on outer-shelf sandstones that accumulated during the maximum regression. Borings often extend half a meter and are infilled with siliciclastic-free skeletal peloid dolopackstone. Hammer head (10 cm) at base of photo for scale. (b) Close-up of a sandstone capping an outer-shelf high-frequency cycle. Note angularity of boring walls that are later infilled by peloidal carbonate with a micro skeletal lag.

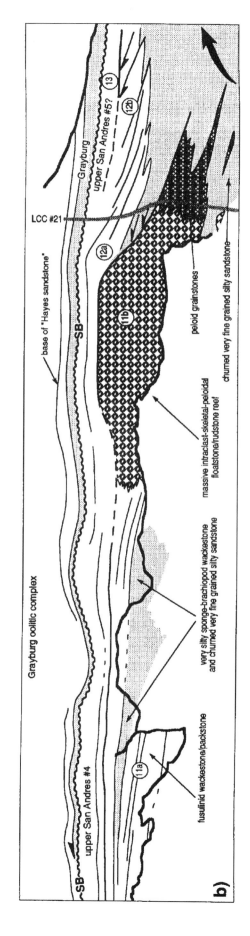

Figure 23. Photo mosaic (a) and line drawing (b) of cycles 11–13. Cycle 11b includes the massive reef with very steep foreslopes dipping away from the picture as well as to the right. Note the sandstone wedge (transgressive hemicycle of cycle 12a) onlapping the margin of the first reef phase. These siliciclastic incursions temporarily suspended reef growth during high-frequency lowstands when sands bypassing the shelf apparently utilized natural conduits (heavy arrow) between reefs or fusulinid shoals (also see Figure 11).

Cycle 5 is the most carbonate-dominated unit within the uSA$_4$ sequence and represents a period of reduced to suspended siliciclastic sediment influx just above the condensed section of large-scale sequence uSA$_4$. Additionally, the crinoid-bryozoan bafflestone bioherms developed prior to active shedding of fusulinids, peloids, and other shelf-derived detritus. The overlying fusulinid wackestones that downlap the bioherms were shed during high-frequency regressive hemicycle of cycle 5 and apparently terminated both reef and biohermal growth (Figure 18b). Analogous to an influx of siliciclastic sediment, the discharge of carbonate detritus may limit the growth of filter-feeding communities (Newell et al., 1953). Downslope shedding of allochthonous siliciclastic and carbonate detritus during the greater part of each high-frequency cycle may, therefore, have restricted the time of bioherm growth to the maximum transgressive phases of each high-frequency cycle (Figures 11, 12).

Fusulinid wackestones and packstones are the exclusive middle- to upper-slope carbonate facies throughout most of the uSA$_4$ sequence. However, in late highstand systems tract cycles 11b and 12a, the fusulinid facies tract was replaced by large conspicuous algal-sponge-crinoid-brachiopod reefs (Figures 23, 24). In contrast to the crinoid-bryozoan bafflestone bioherms of earlier cycles, which developed on the lower-middle slope during times of rapidly increasing or maximum accommodation space on the outer shelf, the algal-sponge-crinoid-brachiopod reefs developed on the middle-upper slope during a long-term decrease in shelfal accommodation. Are there attributes of relative sea-level fall and decreasing shelfal accommodation that would contribute to development of these relatively diverse reefs, just as attributes of relative sea-level rise are inferred to have contributed to development of lower-slope bioherms?

During long-term relative fall in sea level, the geographic zone of alternating emergence and submergence of the seafloor produced by high-frequency sea-level fluctuation was progressively encroaching toward the shelf margin. Concomitant with seaward migration of the emergent zone, the width of the submergent zone and, therefore, the area of the carbonate factory on the shelf were reduced, thereby limiting the cumulative duration of carbonate production. This, in turn, reduced the volume of peloidal and micritic sediment available for offshelf shedding. Thus, during the highstand systems tract there was an early period of carbonate detritus shedding and poisoning of reefal communities, but, as the manufacture and discharge of carbonate detritus waned, a diverse reefal community proliferated and formed topographic structures. Later, with an increase in the rate of relative sea-level fall, reefal communities were again poisoned, possibly by increased discharge of siliciclastics.

Other plausible hypotheses exist for the causes of reef development during the late highstand systems tract. The reefs occur in a position coincident with the crest of the Huapache Monocline. Subsurface well control defines a pronounced thickening of Delaware Mountain Group sandstones coincident with the trace of the Huapache Monocline. Thus, the trace of the present Huapache Monocline corresponds to the trace of the ancestral Huapache wrench zone and probably to the approximate trend of the pre-San Andres physiographic bank-margin of Leonardian age (Figures 2–4). Enhanced upwelling and nutrient flux related to antecedent relief at the underlying physiographic bank-margin could also have contributed to reef initiation.

The development of bioherms on the flooding surface of cycle 5 and coincident with long-term maximum transgression is a predictable response to long-term accommodation trends within the uSA$_4$ sequence. A remarkably similar pattern was also observed within each high-frequency cycle where sponge-brachiopod "mini-bioherms" develop above and within high-frequency condensed sections during transgressive maxima (Figures 10–12). By contrast, the development of reefs within cycles 11b and 12a seems counterintuitive at first, and may represent a less predictable response to the late highstand systems tract decrease in shelfal accommodation. Alternatively, initiation of the late highstand systems tract reefs may be linked to extrinsic configurational factors related to the underlying Leonardian bank-margin.

DISCUSSION

We have shown that volumetric partitioning and facies differentiation are systematic within strata of the large-scale upper San Andres sequence 4 and its component high-frequency sequences. Changes in the amount and locus of accommodation are inferred to control the degree of volumetric partitioning and the styles of facies differentiation. In this section, we discuss some implications of these observations for stratigraphic analysis in general. First, we stress the importance of stacking pattern analysis in providing a framework for the recognition and interpretation of volumetric partitioning and facies differentiation. Second, we propose a high-resolution correlation methodology based upon the consequences of volumetric partitioning. Finally, we consider the implications of facies differentiation on the use and development of facies models.

Stacking Pattern Analysis in Offlapping Strata

When there is little or no depositional topography, strata accumulate vertically upon horizontal or subhorizontal depositional surfaces and stacking pattern analysis may be confidently performed from one-dimensional stratigraphic sections such as well logs, cores, or outcrops of limited lateral extent. In carbonate shelf strata, cycle thicknesses are relatively straightforward indicators of stacking patterns if cycle thicknesses are constrained by subaerial exposure. By contrast, potential pitfalls exist in the analysis of stacking patterns within offlapping subtidal strata because individual cycle thicknesses are not direct reflections of accommodation potential. Cycles intersected by vertical stratigraphic sections will yield a thin (toe-of-slope strata), to thick (slope strata), to thin (shelf strata) apparent stacking pattern that does not necessarily reflect an increase followed by a decrease in accommodation—it merely reflects progradation of a clinoform-bounded rock body or "clinothem" (Rich, 1951; Figure 20).

In Last Chance Canyon, all carbonate clinoforms show some net offlap, irrespective of systems tract position, primarily due to high autochthonous carbonate sediment supply within outer-shelf and upper-slope settings. During transgressive systems tract deposition, equivalent paleobathymetric tie points within siliciclastic portions of clinoforms are likely to retreat in concert with landward stepping

shorelines (Vail, 1987); on the other hand, tie points along carbonate clinoforms are more likely to show net offlap—*but at an increased angle of climb*—due to the maintenance or expansion of an autochthonous outer-shelf sediment supply (Figure 25).

From uSA$_4$ cycles 1 to 4, the waning siliciclastic influx to slope environments suggests that the shoreline and associated siliciclastic feeder systems stepped landward in response to landward increases in accommodation. By contrast, net offlapping geometries for outer-shelf to slope carbonates within cycles 3 and 4 reflect continued shelf-margin carbonate sediment productivity during transgressive systems tract deposition. The geometries of uSA$_4$ cycles 1–4, the offsets of facies tracts across these cycle boundaries, and the facies successions along contiguous offlapping clinoforms within these cycles have

Figure 24. Massive intraclastic skeletal-peloid floatstone/rudstone facies within reefs of the late highstand systems tract. (a) Exceptionally well-articulated crinoid stems over 40 cm long and 2 cm in diameter are surrounded by an intraclastic microskeletal peloid packstone matrix. (b) Primary reefal cavity formed by an encrusted arched network of *Lepdotus* brachiopods. White conical brachiopod at center is probably a *Richtofenid*, a common Guadalupian reef dweller (Bowsher, 1985). Brachiopods above cavity at right have in-situ geopetal fills. (c) Local sponge-algal boundstone fabric. Note calcareous sponge at upper-center of photo. The bulbous sparry to cream-colored micritic laminae encrusting the sponge are probably *Archaeolithoporella*. *Archaeolithoporella* also encrusts angular intraclasts at right, suggestive of early marine lithification. Black pen for scale. Figure continued on page 467.

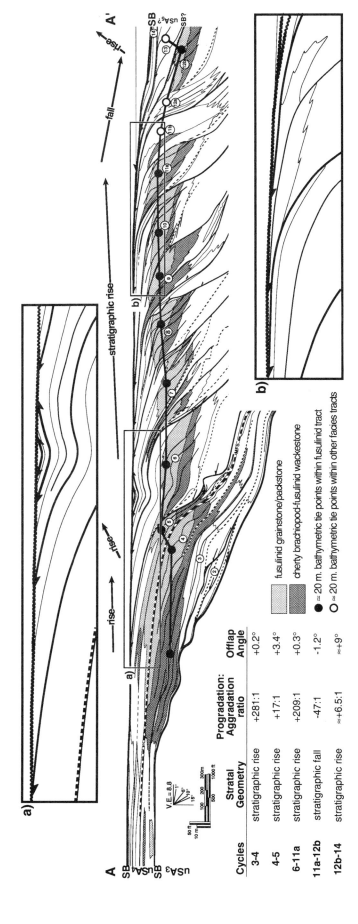

Figure 25. Simplified version of depositional dip cross section A–A′ (Figure 7) emphasizing progradation:aggradation ratios, offlap angles, the stratigraphic rise and fall of the fusulinid facies tracts, and topset stratal patterns within upper San Andres sequence 4. (a) Subparallel topset stratal patterns denote toplap by erosional truncation during large-scale sequence boundary formation. (b) Convergent topset stratal patterns denote toplap by nondeposition or bypass during high-frequency sequence boundary formation.

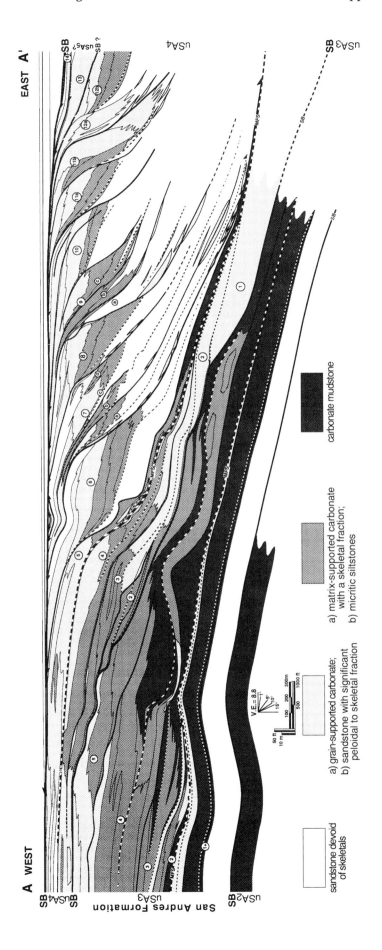

Figure 26. Simplified version of depositional dip cross section A–A' (Figure 7) emphasizing matrix-supported versus grain-supported carbonate facies. Note that carbonate clinoform toes attain their maximum seaward extent at the transgressive to highstand systems tract "turnaround"; that is, cycles 3–4 for uSA3 and cycles 4–5 for uSA4.

characteristics of *both* landward-stepping and vertically stacked cycles. This is interpreted as a product of potentially differing responses of carbonate and siliciclastic systems to long-term accommodation trends. In contrast to the landward shifts of siliciclastic facies within the lowstand to transgressive systems tracts, rates and magnitudes of high-frequency relative rises of sea level were apparently insufficient to shift the locus of carbonate sediment production appreciably landward. Landward movement of shelfal siliciclastics concurrent with bidirectional landward expansion and seaward progradation of outer-shelf and slope carbonates illustrates why stacking patterns must sometimes be evaluated from the perspective of ratioed or normalized rather than absolute changes.

Progradation:aggradation ratios and offlap angles are useful geometric criteria for the interpretation of stacking patterns in strongly offlapping strata because they emphasize relative rather than the absolute offsets of normalized stratigraphic piercing points (Figure 25). These criteria are a function of the volumetric partitioning of sediment in outer-shelf topsets versus slope foresets. Recognition of changes in the offlap angle assists in the precise determination of long-term accommodation patterns and provides a measure of predictive insight into coeval stacking patterns and facies differentiation on the shelf.

Volumetric Partitioning and Stratigraphic Correlation

In sequences spanning at least two temporal and spatial scales, we observe that the volume of sediment accumulated and preserved within each facies tract varies as a function of position within interpreted depositional sequences. This volumetric partitioning of sediment reflects changes in geographic locations of accommodation and sediment preservation through time, and establishes the basis for a high-resolution correlation methodology of rock-to-rocks, rocks-to-surfaces, and surfaces-to-surfaces.

High-frequency cycles in Last Chance Canyon record a seaward transition from asymmetric regressive hemicycles commonly observed in shelf settings, to less frequently documented symmetric cycles composed of both transgressive *and* regressive components in slope settings, to asymmetric transgressive hemicycles in toe-of-slope settings (Figures 11, 12). Variations in cycle symmetry can also be viewed in terms of the spatial arrangement of surfaces bounding each hemicycle (Figure 12a). Most previously recognized high-frequency cycles occur in shallow shelf to very broad low-angle shelf or ramp environments. In most shelf or inner ramp settings, maximum transgressive surfaces (high-frequency flooding surfaces) tend to be spatially coincident with maximum regressive surfaces (high-frequency cycle boundaries), rendering cycles whose boundaries are arbitrarily defined by flooding surfaces (e.g., parasequences, PACs) identical to those arbitrarily bounded by maximum regressive surfaces. In slope settings, by contrast, transgressive surfaces tend to be spatially separated from regressive surfaces, thus cycles defined by one or the other type of surface are distinct (Figure 12a). Correlations of asymmetric regressive hemicycles in shelf settings, to symmetric transgressive and regressive cycles in slope settings, to asymmetric transgressive hemicycles in toe-of-slope settings should incorporate the fundamental genetic distinction between these surfaces. Anticipated well log correlations using a model that incorporates variable cycle symmetries would be drastically different than if one tried to "force" a shallowing-upward asymmetric cycle model from shelf to slope to toe-of-slope settings.

Symmetric high-frequency cycles may be characteristic of mixed carbonate and siliciclastic slope strata in general, but are not necessarily limited to these settings. High-frequency cycles 2–4 of the uSA$_3$ sequence involve slope or distal-ramp facies that are devoid of siliciclastics, yet still record both transgressive and regressive hemicycles. Although the preservation of symmetric cycles is commonly recognized for seismic-scale third-order depositional sequences, few examples in the literature document these physical stratigraphic relationships from shelf to basin for high-frequency fourth- and fifth-order cycles (Aigner, 1985; Brett and Baird, 1986). We propose that preservation of symmetric cycles and asymmetric transgressive hemicycles may be more common than presently recognized in deeper water, distal ramp, slope, and basinal settings. One component that may be critical to the accumulation of deeper water transgressive hemicycles is the net seaward, rather than landward, transport of sediment during individual storm events that occur during the landward movement of storm wave base.

Facies Differentiation and the Need for a Dynamic Sedimentology

Many sequence stratigraphic studies seem preoccupied with the labeling of systems tracts and bounding surfaces. A common inference within and to a lesser extent between systems tracts is that cycles simply step landward or seaward without significant differentiation in facies associations. We find that volumetric partitioning of sediment is accompanied by significant changes in facies types and associations despite evidence that the paleobathymetry and many characteristics of the geomorphic environments that they represent remained unchanged. Facies differentiation appears orderly and regular in the context of stacking patterns and a hierarchy of cycles, however, systematic facies differentiation may be overlooked unless placed in a stratigraphic context. Facies differentiation apparently reflects changes in accommodation, preservation potential, and ratios of allochthonous versus autochthonous sediment supply within paleobathymetrically identical facies tracts during different phases of a depositional sequence.

The existence of deterministic facies differentiation has potential implications for reservoir characterization and for the development and use of facies models. Due to pronounced depositional topography coupled with rapid offlap rates in Last Chance Canyon, long-term facies differentiation occurs over the relatively short lateral distance of 2–4 km (Figure 7). Because this scale corresponds to the dimensions of many hydrocarbon accumulations, facies differentiation presents an additional challenge for reservoir characterization, distinct from, and more subtle than the potential reservoir compartmentalization caused by the discrete clinoform-bounded high-frequency cycles. In Last Chance Canyon, lateral reservoir heterogeneity reflective of facies differentiation would be influenced by the changing spatial arrangement and relative proportions of potential reservoir facies tracts such as the outer-shelf sandstones, massive peloid dolowackestones/packstones, and fusulinid dolopackstones. For example, lateral bed continuity within both the fusulinid and the outer-shelf sandstone facies tracts decreases progressively during the highstand systems tract. In geologic settings with significantly less depositional topography and lower offlap rates than Last Chance Canyon, analogous aspects of facies differentiation might become pertinent to reservoir prediction at an exploration scale because they could occur over distances of 10–100 km.

Volumetric partitioning and facies differentiation within depositional sequences also add an important caveat to the use and construction of facies models. The derivation of both local and general facies models typically involves comparison, abstraction, and synthesis of multiple occurrences of facies that are representative of similar or identical depositional environments (Walker, 1984). The procedures used to erect a facies model implicitly assume that the deposits of a particular environment (e.g., tidal channel belt, ooid or skeletal shoal, lagoon) appear basically similar, regardless of stratigraphic context and that most descriptive attributes may be generalized from multiple specific examples. The possibility that entirely different facies may occur and accumulate in identical paleobathymetric positions during different phases of a depositional sequence has not been explicitly considered in existing facies models. As a consequence, the likelihood increases that depositional system or depositional environment scale facies models have been abstracted from occurrences drawn from multiple stacking pattern positions. While such models are useful for environmental interpretation, their predictive capabilities are compromised because potentially deterministic facies differentiation has been "abstracted away" by mixing facies successions accumulated during different accommodation/sediment supply conditions. If facies models are to become predictive at the scale of reservoirs, they must be calibrated to the systematically changing accommodation conditions that drive volumetric partitioning and facies differentiation. Facies models must therefore incorporate a more dynamic sedimentology that is tied to position within stacking patterns and depositional sequences.

CONCLUSIONS

San Andres Formation (Permian) outcrops in Last Chance Canyon are interpreted to contain two large-scale fourth-order depositional sequences, upper San Andres sequence 3 and upper San Andres sequence 4. Embedded within each of these two sequences are numerous higher-frequency sequences. High-frequency cycles in Last Chance Canyon contain a complete spatial gradation from asymmetric regressive hemicycles commonly observed in shelf settings (e.g., parasequences and PACs), to less frequently documented symmetric cycles in slope settings, to asymmetric transgressive hemicycles in toe-of-slope settings.

Last Chance Canyon's outcrops of the San Andres and Cherry Canyon formations provide an example of dynamic outer-shelf to slope facies associations that vary over sequences of several temporal scales (approximately fourth- to fifth-order). We observe that the volume of sediment accumulated and preserved within each facies tract varies systematically as a function of position within a depositional sequence. Progradation:aggradation ratios and offlap angles help quantify volumetric partitioning in outer-shelf and slope settings. Volumetric partitioning is accompanied by significant changes in facies associations that occur at comparable paleobathymetric positions of clinoform profiles that vary transitionally from sigmoidal to oblique. In the late transgressive to early highstand systems tract of upper San Andres sequence 4, for example, upper-slope fusulinid packstone facies form laterally widespread (0.6–1.2 km), mounded deposits with sigmoidal clinoform profiles. Within the middle to late highstand systems tract, by contrast, fusulinid facies deposited in similar water depths form laterally contracted (0.1 km) deposits with high-angle sigmoid to sigmoid-oblique profiles.

Other generalized stratigraphic responses to changes in accommodation coupled with variations in the ratio of allochthonous to autochthonous sediment supply include the variation over time in the lateral extent of grain-dominated versus mud-dominated carbonate slope facies and the timing and nature of bioherm development. For example, siliciclastic-free, mud-supported, bioherm-prone slope carbonates attain their greatest seaward extent during the latest transgressive to earliest highstand systems tract period of maximum shelfal accommodation. By contrast, grain-rich carbonate and mixed carbonate-siliciclastic facies predominate in lower-slope settings during periods of reduced shelfal accommodation such as the lowstand, early transgressive, and middle-late highstand systems tracts. The existence of deterministic facies differentiation has implications for the development and use of facies models, particularly at the scale of reservoirs, where they must be calibrated to the systematically changing accommodation conditions that drive volumetric partitioning and facies differentiation.

ACKNOWLEDGMENTS

Financial support for this Master's thesis research was provided by AAPG, GSA, and Sigma Xi grants; an Amoco Fellowship; Marathon Oil Company; Mobil Research and Development Corporation; Shell Oil Company; the Texaco Foundation; and the Department of Geology and Geological Engineering at Colorado School of Mines. This study would have been less fruitful, and probably would not have been undertaken, without the sequence stratigraphic framework established by Rick Sarg and Pat Lehmann. Numerous discussions with Bill Fitchen, Mike Gardner, Charlie Kerans, Fred Meissner, and the Genetic Stratigraphy Research Group at CSM were of great benefit. The Permian Basin Section of SEPM is thanked for permission to reproduce certain figures. We thank AAPG reviewers Bob Loucks, Rick Sarg, and Terry Twyman for their constructive comments on the manuscript.

REFERENCES CITED

Aigner, T., 1985, Storm depositional systems: Dynamic stratigraphy in modern and ancient shallow-marine sequences: Lecture Notes in Earth Sciences, v. 3, Springer-Verlag, Berlin, 174 p.

Allen, P., 1964, Sedimentologic models: Journal of Sedimentary Petrology, v. 34, p. 289-293.

Babcock, J.A., 1977, Calcareous algae, organic boundstones, and the genesis of the Upper Capitan Limestone (Permian, Guadalupian), Guadalupe Mountains, West Texas and New Mexico, in M.E. Hileman and S.J. Mazzullo, eds., Upper Guadalupian Facies, Permian Reef Complex, Guadalupe Mountains, New Mexico and West Texas: Permian Basin Section, SEPM 77-16, p. 3-44.

Barrell, J., 1912, Criteria for the recognition of ancient delta deposits: Geological Society of America Bulletin, v. 23, p. 377-446.

Barrell, J., 1917, Rhythms and the measurement of geologic time: Geological Society of America Bulletin, v. 28, p. 75-904.

Borer, J.M., and P.M. Harris, 1991, Lithofacies and cyclicity of the Yates Formation, Permian Basin: Implications for reservoir heterogeneity: AAPG Bulletin, v. 75, p. 726-779.

Bosellini, A., 1984, Progradational geometries of carbonate platforms: Examples from the Triassic of the Dolomites, northern Italy: Sedimentology, v. 31, p. 1-24.

Bosellini, A., 1989, Dynamics of Tethyan carbonate platforms, in P.D. Crevello, J.L. Wilson, J.F. Sarg, and J.F. Read, eds., Controls on Carbonate Platform Development: SEPM Special Publication No. 44, p. 3-13.

Bowsher, A.L., 1985, Geology of the backreef sediments equivalent to the Capitan reef complex through Dark Canyon to Sitting Bull Falls, Eddy County, New Mexico: Roswell Geological Society Guidebook, p. 1-24.

Boyd, D.W., 1958, Permian sedimentary facies, central Guadalupe Mountains, New Mexico: New Mexico Bureau of Mines and Mineral Resources Bulletin 49, 100 p.

Brett, C.E., and G.C. Baird, 1986, Symmetrical and upward shallowing cycles in the Middle Devonian of New York State and their implications for the punctuated aggradational cycle hypothesis: Paleooceanography, v. 1, p. 431-445.

Brown, L.F., and W.L. Fisher, 1977, Seismic-stratigraphic interpretation of depositional systems: Examples from Brazilian rift and pull-apart basins, in C.E. Payton, ed., Seismic Stratigraphy: Applications to Hydrocarbon Exploration: AAPG Memoir 26, p. 213-248.

Cotton, C.A., 1918, Conditions of deposition on the continental shelf and slope: Journal of Geology, v. 26, p. 135-160.

Darton, N.H., and J.B. Reeside Jr., 1926, Guadalupe group: Geological Society of America Bulletin, v. 37, p. 413-428.

Fischer, A.G., and M. Sarnthein, 1988, Airborne silts and dune-derived sands in the Permian of the Delaware Basin: Journal of Sedimentary Petrology, v. 58, p. 637-643.

Fitchen, W.M., M.H. Gardner, C. Kerans, M.D. Sonnenfeld, S. Tinker, and N.L. Wardlaw, 1992, Evolution of platform and basin architecture in mixed carbonate-siliciclastic sequences: Latest Leonardian through Guadalupian, Delaware Basin (abstract): 1992 AAPG Annual Convention Program, Abstracts, p. 41.

Frazier, D.E., 1974, Depositional sequences: Their relationship to the Quaternary stratigraphic framework in the northwestern part of the Gulf Basin: University of Texas at Austin Bureau of Economic Geology, Geological Circular 74-1, 28 p.

Galley, J.E., 1968, Some principles of tectonics in the Permian Basin, in W. Stewart, ed., Basins of the Southwest (vol. 1): West Texas Geological Society Publication, p. 5-20.

Garber, R.A., G.A. Grover, and P.M. Harris, 1989, Geology of the Capitan shelf margin: Subsurface data from the northern Delaware Basin, in P.M. Harris, and G.A. Grover, eds., Subsurface and Outcrop Examination of the Capitan Shelf Margin Northern Delaware Basin: SEPM Core Workshop No. 13, p. 3-278.

Harland, W.B., R.L. Armstrong, A.V. Cox, L.E. Craig, A.G. Smith, and D.G. Smith, 1989, A Geologic Time Scale 1989: Cambridge University Press, Cambridge, 263 p.

Harrison, S.C., 1966, Depositional mechanics of Cherry Canyon Sandstone Tongue: unpublished M.Sc. thesis, Texas Tech. College, Lubbock, 114 p.

Harris, J.C., 1974, Brushy Canyon Formation, Texas: A deep-water density current deposit: Geological Society of America Bulletin, v. 85, p. 1763–1784.

Hayes, P.T., 1959, San Andres Limestone and related Permian rocks in Last Chance Canyon and vicinity, southeastern New Mexico: AAPG Bulletin, v. 43, p. 2197-2213.

Hills, J.M., and F.E. Kottlowski, 1983, Southwest/Southern Mid-Continent Region, Correlation of Stratigraphic Units of North America (COSUNA) Project, AAPG, Tulsa.

Hurley, N.F., 1989, Facies mosaic of the lower Seven Rivers Formation, McKittrick Canyon, New Mexico: in P.M. Harris, and G.A. Grover, eds., Subsurface and Outcrop Examination of the Capitan Shelf Margin, Northern Delaware Basin: SEPM Core Workshop No. 13, p. 325-346.

Jacka, A.D., R.H. Beck, L.C. St. Germain, and S.C. Harrison, 1968, Permian deep-sea fans of the Delaware Mountain Group (Guadalupian), Delaware Basin: Guadalupian facies, Apache Mountains area, West Texas: Permian Basin Section, SEPM Publication 68-11, p. 49-90.

Jervey, M.T., 1988, Quantitative geological modeling of siliciclastic rock sequences and their seismic expression, in C.K. Wilgus, B.S. Hastings, C.G.St.C. Kendall, H.W. Posamentier, C.A. Ross, and J.C. Van Wagoner, eds., Sea Level Changes: An Integrated Approach: SEPM Special Publication No. 42, p. 47-69.

Kerans, C., and H.S. Nance, 1991, High-frequency cyclicity and regional depositional patterns of the Grayburg Formation, Guadalupe Mountains, New Mexico, in S. Meader-Roberts, M.P. Candelaria, and G.E. Moore, eds., Sequence Stratigraphy, Facies, and Reservoir Geometries of the San Andres, Grayburg, and Queen Formations, Guadalupe Mountains, New Mexico and Texas: Permian Basin Section, SEPM Publication 91-32, p. 53-96.

Kerans, C., W.M. Fitchen, M.H. Gardner, M.D. Sonnenfeld, S.W. Tinker, and B.R. Wardlaw, 1992, Styles of sequence development within latest Leonardian through Guadalupian strata of the Guadalupe Mountains (abstract): 1992 AAPG Annual Convention Program, Abstracts, p. 65.

King, P.B., 1942, Permian of west Texas and southeastern New Mexico: AAPG Bulletin, v. 26, p. 535-763.

King, P.B., 1948, Geology of the southern Guadalupe Mountains, Texas: U.S. Geological Survey Professional Paper 215, 183 p.

McDermott, R.W., 1983, Depositional processes and environments of the Permian sandstone tongue of the Cherry Canyon Formation and the upper San Andres Formation, Last Chance Canyon, southeastern New Mexico: unpublished M.A. thesis, University of Texas at Austin, 172 p.

Meissner, F.F., 1967, Cyclic sedimentation in mid-Permian strata of the Permian Basin (abstract), in J.G. Elam, and S. Chuber, eds., Cyclic Sedimentation in the Permian Basin (1st edition): West Texas Geological Society Symposium, p. 37-38.

Meissner, F.F., 1972, Cyclic sedimentation in middle Permian strata of the Permian Basin, in J.G. Elam, and S. Chuber, eds., Cyclic Sedimentation in the Permian Basin (2nd edition): West Texas Geological Society Publication 72-60, p. 203-232.

Meyer, R.F., 1966, Geology of Pennsylvanian and Wolfcampian Rocks in Southeast New Mexico: New Mexico Bureau of Mines and Mineral Resources Memoir 17, 123 p.

Mitchum, R.M., P.R. Vail, and J.B. Sangree, 1977, Seismic stratigraphy and global changes of sea level, Part 6: Stratigraphic interpretation of seismic reflection patterns in depositional sequences, in C.E. Payton, ed., Seismic Stratigraphy: Applications to Hydrocarbon Exploration: AAPG Memoir 26, p. 117-133.

Naiman, E.R., 1982, Sedimentation and diagenesis of a shallow marine carbonate and siliciclastic shelf sequence: The Permian (Guadalupian) Grayburg Formation, southeastern New Mexico: unpublished M.A. thesis, University of Texas at Austin, 197 p.

Newell, N.D., J.K. Rigby, A.G. Fischer, A.J. Whiteman, J.E. Hickox, and J.S. Bradley, 1953, The Permian Reef Complex of the Guadalupe Mountains Region, Texas and New Mexico: W.H. Freeman & Co., San Francisco, 236 p.

Oriel, S.S., D.A. Myers, and E.J. Crosby, 1967, West Texas Permian Basin Region, in Paleotectonic Investigations of the Permian System in the United States: U.S. Geological Survey Professional Paper 515C, p. C17-C60.

Osleger, D., 1991, Subtidal carbonate cycles: Implications for allocyclic vs. autocyclic controls: Geology, v. 19, p. 917–920.

Pray, L.C., 1988, The western escarpment of the Guadalupe Mountains, Texas and day two of the field seminar, in S. Tomlinson Reid, R.O. Bass, and P. Welch, eds., Guadalupe Mountains Revisited—Texas and New Mexico: West Texas Geological Society Publication 88-84, p. 23-31.

Rich, J.L., 1951, Three critical environments of deposition and criteria for recognition of rocks deposited in each of them: Geological Society of America Bulletin, v. 62, p. 1-20.

Ross, C.A., 1983, Late Paleozoic foraminifera as depth indicators: AAPG Bulletin, v. 67, p. 542.

Ross, C.A., and J.R.P. Ross, 1987, Late Paleozoic sea levels and depositional sequences, in C.A. Ross and D. Haman, eds., Timing and Depositional History of Eustatic Sequences: Constraints on Seismic Stratigraphy: Cushman Foundation for Foraminiferal Research, Special Publication 24, p. 137-149.

Rossen, C., and J.F. Sarg, 1988, Sedimentology and regional correlation of a basinally restricted deep-water siliciclastic wedge: Brushy Canyon Formation—Cherry Canyon Tongue (Lower Guadalupian), Delaware Basin, in S.T. Reid, R.O. Bass, and P. Welch, eds., Guadalupe Mountains Revisited, Texas and New Mexico: West Texas Geological Society Publication 88-84, p. 127-132.

Sangree, J.B., and J.M. Widmier, 1977, Seismic stratigraphy and global changes of sea level, Part 9:

Seismic interpretation of clastic depositional facies, *in* C.E. Payton, ed., Seismic Stratigraphy: Applications to Hydrocarbon Exploration: AAPG Memoir 26, p. 165-184.

Sarg, J.F., 1988, Carbonate sequence stratigraphy, *in* C.K. Wilgus, B.S. Hastings, C.G.St.C. Kendall, H.W. Posamentier, C.A. Ross, and J.C. Van Wagoner, eds., Sea Level Changes: An Integrated Approach: SEPM Special Publication No. 42, p. 155-182.

Sarg, J.F., and P.J. Lehmann, 1986a, Lower-Middle Guadalupian facies and stratigraphy, San Andres/Grayburg formations, Permian Basin, Guadalupe Mountains, New Mexico, and Texas, *in* G.E. Moore, and G.L. Wilde, eds., Lower and Middle Guadalupian Facies, Stratigraphy, and Reservoir Geometries, San Andres/Grayburg Formations, Guadalupe Mountains, New Mexico and Texas: Permian Basin Section, SEPM Publication No. 86-25, p. 1-8.

Sarg, J.F., and P.J. Lehmann, 1986b, Facies and stratigraphy of lower-upper San Andres shelf-crest and outer shelf and lower Grayburg inner shelf, *in* G.E. Moore, and G.L. Wilde, eds., Lower and Middle Guadalupian Facies, Stratigraphy, and Reservoir Geometries, San Andres/Grayburg Formations, Guadalupe Mountains, New Mexico and Texas: Permian Basin Section, SEPM Publication No. 86-25, p. 9-35.

Scotese, C.R., R.K. Bambach, C. Barton, R. Van Der Voo, and A.M. Ziegler, 1979, Paleozoic base maps: Journal of Geology, v. 87, p. 217-277.

Silver, B.A., and R.G. Todd, 1969, Permian cyclic strata, northern Midland and Delaware basins, west Texas and southeastern New Mexico: AAPG Bulletin, v. 53, p. 2223-2251.

Skinner, J.W., 1946, Correlation of Permian of west Texas and southeast New Mexico: AAPG Bulletin, v. 30, p. 1858-1874.

Sloss, L.L., 1962, Stratigraphic models in exploration: AAPG Bulletin, v. 46, p. 1050-1057.

Sonnenfeld, M.D., 1991a, Anatomy of offlap in a shelf-margin depositional sequence: Upper San Andres Formation (Permian, Guadalupian), Last Chance Canyon, Guadalupe Mountains, New Mexico: unpublished M.Sc. thesis T-3915, Colorado School of Mines, Golden, Colorado, 297 p.

Sonnenfeld, M.D., 1991b, High-frequency cyclicity within shelf-margin and slope strata of the upper San Andres sequence, Last Chance Canyon, *in* S. Meader-Roberts, M.P. Candelaria, and G.E. Moore, eds., Sequence Stratigraphy, Facies, and Reservoir Geometries of the San Andres, Grayburg, and Queen Formations, Guadalupe Mountains, New Mexico and Texas: Permian Basin Section, SEPM Publication 91-32, p. 11-51.

Vail, P.R., 1987, Seismic stratigraphy interpretation procedure, *in* A.W. Bally, ed., Atlas of seismic stratigraphy: AAPG Studies in Geology 27, v. 1, p. 1-10.

Vail, P.R., R.M. Mitchum, and S. Thompson III, 1977a, Seismic stratigraphy and global changes of sea level, Part 2: The depositional sequence as a basic unit for stratigraphic analysis, *in* C.E. Payton, ed., Seismic Stratigraphy: Applications to Hydrocarbon Exploration, AAPG Memoir 26, p. 53-62.

Vail, P.R., R.M. Mitchum, and S. Thompson III, 1977b, Seismic stratigraphy and global changes of sea level, Part 3: Relative changes of sea level from coastal onlap, *in* C.E. Payton, ed., Seismic Stratigraphy: Applications to Hydrocarbon Exploration, AAPG Memoir 26, p. 63-82.

Van Siclen, D.C., 1958, Depositional topography: Examples and theory: AAPG Bulletin, v. 42, p. 1897-1913.

Van Wagoner, J.C., 1985, Reservoir facies distribution as controlled by sea-level change (abstract), SEPM Mid-Year Meeting, Golden, Colorado, p. 91-92.

Veevers, J.J., and C. McA. Powell, 1987, Late Paleozoic glacial episodes in Gondwanaland reflected in transgressive-regressive depositional sequences in Euramerica: Geological Society of America Bulletin, v. 98, p. 475-487.

Walker, R.G., 1984, General introduction: Facies, facies sequences, and facies models, *in* R.G. Walker, ed., Facies Models, Second Edition: Geoscience Canada Reprint Series 1, p. 1-9.

Wheeler, H.E., 1964, Baselevel, lithosphere surface, and time-stratigraphy: Geological Society of America Bulletin, v. 75, p. 599-610.

Wilde, G.L., 1990, Practical fusulinid zonation: The species concept; with Permian Basin emphasis: West Texas Geological Society Bulletin, v. 29, p. 5-34.

Williams, K., 1969, Principles of cementation, environmental framework and diagenesis of the Grayburg and Queen formations, New Mexico and Texas: unpublished Ph.D. thesis, Texas Technical College, Lubbock, 198 p.

Wilson, J.L., 1967, Cyclic and reciprocal sedimentation in Virgilian strata of southern New Mexico: Geological Society of America Bulletin, v. 78, p. 805-818.

Chapter 18

Ancient Outcrop and Modern Examples of Platform Carbonate Cycles—Implications for Subsurface Correlation and Understanding Reservoir Heterogeneity

P. M. Harris
Chevron Petroleum Technology Company
La Habra, California, U.S.A.

Charles Kerans
D. G. Bebout
Bureau of Economic Geology
University of Texas
Austin, Texas, U.S.A.

ABSTRACT

Detailed geologic studies of hydrocarbon reservoirs in platform carbonates commonly show that reservoir zones occur within 1–15-m-thick, upward-coarsening successions of lithofacies, i.e., within upward-shallowing cycles. Our understanding of the depositional history and reservoir characteristics of such cycles and their component facies is enhanced by observations of ancient outcrop examples and models derived from modern analogs.

Outcrops of the Permian San Andres Formation along the Algerita Escarpment of the Guadalupe Mountains contain cycles 3–12 m thick, with thin mudstone/wackestone bases, overlain by burrowed wackestones and packstones, and capped by thick massive to planar or cross-bedded packstones and grainstones. These facies formed during relative rise and/or stabilization of sea level during which carbonate sand shoals developed. The outcrops also display lateral facies relationships within the cycles on the scale of hundreds of meters that are representative of those commonly observed in analogous hydrocarbon reservoirs of the Permian basin of Texas and New Mexico.

Core and surface sediment mapping in the Holocene Joulters Cays ooid-shoal complex of Great Bahama Bank reveals the three-dimensional complexity of an upward-coarsening and shallowing cycle. This facies mosaic is like that observed in two dimensions at Algerita Escarpment or in one dimension in a core from a reservoir. This modern example points out difficulties in interpretation and correlation of grainstone cycles in subsurface studies of platform carbonate reservoirs. The modern shoal complex, which extends over 400 km², varies greatly in thickness but averages 4 m thick. Shoal growth, largely in a response to a relative rise of sea level, records rapid expansion of ooid sands, island formation and associated meteoric diagenesis, local shoal stabilization and reworking by burrowing, and generation of hardground layers.

Sand generation and topography varied greatly in the Joulters Cays area during flooding of the platform and development of the shoal. Such variation should be expected in ancient examples, as was observed at Algerita Escarpment. Within the upper grain-dominated part of the cycle at Joulters, depositional facies geometries and early diagenetic alteration contribute to fine-scale heterogeneities. This is at a scale equivalent to documented hydrocarbon reservoir heterogeneities.

INTRODUCTION

Upward-coarsening, 1–15-m-thick successions of lithofacies, i.e., upward-shallowing cycles (James, 1979), parasequences (Van Wagoner et al., 1988), or fundamental depositional sequences (Wanless, 1991), are the basic building blocks of carbonate platforms. By analyzing the distribution of reservoir-quality facies within a framework of upward-shallowing cycles, several studies have also shown the cycles to be the basic architectural element of carbonate-platform reservoirs (Goldhammer et al., 1991; Kerans and Nance, 1991; Lindsay, 1991).

The cycle framework therefore has direct applications for reservoir characterization and flow-modeling studies in carbonate reservoirs. The systematic stacking of high- and low-permeability layers serves to stratify a reservoir vertically while lateral facies changes establish restrictions to simple layer-parallel flow. The upward change from fine- to coarse-grained textural fabrics can be translated into predictable, petrophysically significant relationships that aid log interpretation of permeability (Lucia, 1983). Finally, dividing a reservoir interval into smaller time-bounded units facilitates the accurate assessment of depositional and diagenetic facies, which can form the basis of stochastic reservoir modeling.

In this chapter, an ancient outcrop example is first used to illustrate the lateral facies variation documented within an upward-shallowing cycle that is analogous to that found in platform carbonate reservoirs in the Permian basin of Texas and New Mexico. A modern example from the Bahamas is then presented to further stress the potential difficulties of interpretation and correlation of upward-shallowing cycles in ancient subsurface examples.

UPWARD-SHALLOWING CYCLES OF THE ALGERITA ESCARPMENT

The Algerita Escarpment of the northern Guadalupe Mountains, southeastern New Mexico (Figure 1), exposes the complete section of the Permian (Leonardian–Guadalupian) San Andres Formation. Here a 366-m-thick San Andres section spanning 27 km of an oblique-dip carbonate-ramp profile includes a diverse array of dolomitized carbonate-ramp facies.

Stratigraphic Framework

Kerans et al. (1991) refined the sequence stratigraphic framework previously developed for the San Andres and related units in the Guadalupe Mountains. Specifically, the upper San Andres third-order sequence of Sarg and Lehmann (1986) was shown to consist of four unconformity-bounded sequences of higher order (Figure 2, uSA1–4). Each of these sequences in turn is made up of upward-shallowing cycles. The measured sections and oblique aerial photo mosaic mapping of the escarpment by Kerans et al. (1991) demonstrate that the upper San Andres carbonate-ramp complex, like many thick carbonate-platform units, is composed of multiple depositional sequences that exhibit basinward shifts in facies tracts across sequence boundaries. Reservoir quality follows these shifts in facies. This chapter examines only the first (oldest) of the four upper San Andres sequences, i.e., uSA1, where it is exposed at the Lawyer Canyon locality of the Algerita Escarpment (Figure 3).

The cyclic platform carbonates of the first upper San Andres sequence at Lawyer Canyon formed in a ramp-crest facies tract (Figure 2), as demonstrated by the dominance of cross-stratified peloid-ooid grainstones that pass updip in 5 km into mud-rich dasyclad-peloid facies and downdip in 2 km into massive to burrowed fusulinid-peloid packstones. In the ramp-crest position, the sequence consists of seven upward-shallowing cycles that total 30 to 40 m in thickness (Figure 3), with individual cycles ranging from 3–12 m and averaging 4.5 m in thickness.

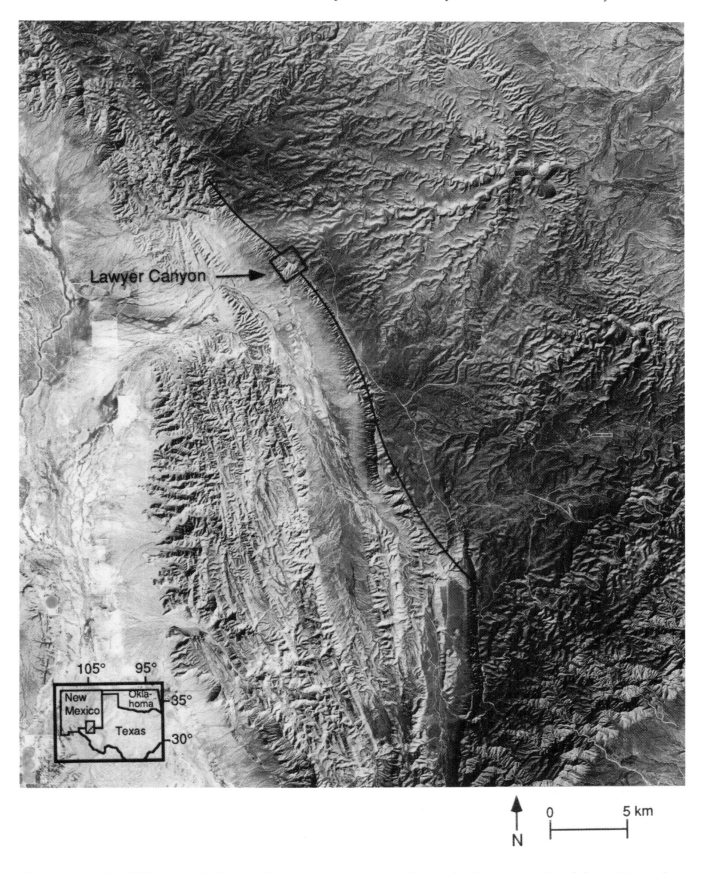

Figure 1. Landsat TM image of Algerita Escarpment outcrop study area in the western Guadalupe Mountains. The line of section for the regional cross section of Figure 2 is shown, and the Lawyer Canyon locality is indicated by a box. Depositional strike is north–northwest for the San Andres, making the cross section an approximate dip orientation.

Cycle detailed in figure 5

Cycle detailed in figure 4

20 m

0

—— Sequence boundary
—— Cycle boundary
—— Measured section

QA 20251

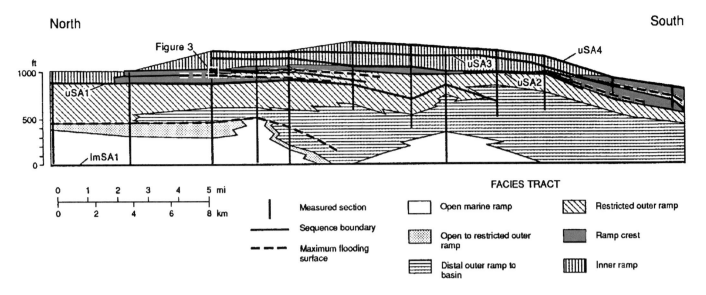

Figure 2. Dip-oriented cross section of San Andres Formation along Algerita Escarpment showing stratigraphic relationships and facies tracts (modified from Kerans et al., 1991). Cross section is based on measured sections (vertical black lines) and analysis of oblique aerial photos. Ramp deposits of the upper San Andres prograded basinward, with abrupt shifts of facies tracts occurring across sequence boundaries associated with four sequences, i.e., uSA1-4. Grainstone-rich upward-shallowing cycles typify the ramp crest as in the detailed study area at Lawyer Canyon (inset location).

Facies Within Cycles

Detailed mapping of facies within cycles was completed for a 50-m high by 800-m long study area (Figure 3). This area was selected to cover a lateral distance comparable to the spacings of several wells in an equivalent subsurface reservoir of the San Andres Formation, where the well spacing is generally 300–600 m.

The uppermost cycle examined in detail at Lawyer Canyon provides a striking example of the possible facies variation within a single cycle (Figure 4). The development of this cycle was initiated by a rapid sea-level rise recorded by mudstone deposition interpreted to represent slow, quiet-water sedimentation. Following this transgressive event, a burrowed to flaser/lenticular-bedded wackestone/packstone was deposited. As shoaling occurred, shallower water, bar-crest, and bar-flank grainstones developed. Bar-crest grainstones exhibit small-scale trough and planar-tabular cross stratification, indicating reworking within fair-weather wave-base and tidal influence. Bar-flank deposits contain parallel-laminated grainstones reworked by storms from the active bar crest. There reworked sands are intercalated with thin,

lower energy, wackestone/packstone layers deposited in a more protected environment.

In addition to the well-documented upward-coarsening and shallowing facies succession of a typical carbonate cycle, there is also an equally significant lateral succession (Figure 5; Kerans et al., 1991). Generally, the thicker the cycle, the greater the internal facies variability within that cycle. This greater facies diversity reflects the longer time interval and wider paleo-depth range across which local depositional environments are established.

Because grainstones are the most porous and permeable facies within the outcropping cycles at Lawyer Canyon (Kerans et al., 1991) and also in analogous San Andres reservoirs (Major et al., 1988, 1990; Garber and Harris, 1990; Ruppel, 1990), it is critical that their geometry be well understood. Grainstone bar complexes of individual cycles at Lawyer Canyon range from 3–5 km in a dip direction. In the seven cycles of the first upper San Andres sequence (Figure 3), individual grainstone bodies have maximum thicknesses ranging from 1.5–11.5 m (average of 5 m), and they are from less than 30 m to greater than 800 m long in a dip direction. Grainstones shown in the cycle of Figure 5 attain a thickness of 11.5 m, but thin

Figure 3. Oblique aerial photograph of Lawyer Canyon area outlining the seven upward-shallowing cycles contained in the first upper San Andres sequence (uSA1 of Figure 2). Upward change in sedimentary structures and facies is illustrated for a single cycle in Figure 4 and the lateral distribution of facies within the uppermost cycle is detailed in Figure 5. Substantial local relief defined by the upper surface of the uppermost cycle reflects depositional topography developed at the crest of the grainstone shoal.

Cross Bedded Grainstone

Massive To Planar Bedded Grainstone

Burrowed Packstone

Massive Mudstone

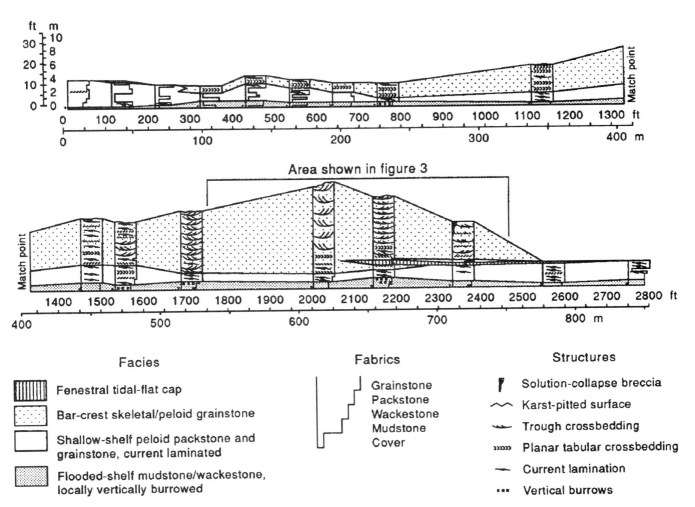

Facies

▯ Fenestral tidal-flat cap

▯ Bar-crest skeletal/peloid grainstone

▯ Shallow-shelf peloid packstone and grainstone, current laminated

▯ Flooded-shelf mudstone/wackestone, locally vertically burrowed

Fabrics

Grainstone
Packstone
Wackestone
Mudstone
Cover

Structures

Solution-collapse breccia

Karst-pitted surface

Trough crossbedding

Planar tabular crossbedding

Current lamination

Vertical burrows

Figure 5. Detailed cross section of 800- by 10-m area of the uppermost cycle in the uSA1 sequence shown in Figures 2 and 3. Marked changes in facies composition and thickness of facies are noted in this section, which covers an area comparable to or less than interwell spacings in typical hydrocarbon reservoirs. Datum for cross section is fenestral tidal-flat cap of underlying cycle, assumed to have negligible depositional topography.

laterally within less than 200 m, or well within typical well spacing. The orientation of the thicker grainstone accumulations is a strike-elongate trend using data from the somewhat irregular outcrop face and nearby shallow boreholes.

Abrupt variations in thickness and lateral extent of bar-crest grainstones within individual cycles (Figure 5) matches observations from modern grainstone depositional sites and corroborates observations from San Andres and Grayburg reservoirs. These studies suggest limited interwell connectivity of grainstone facies and associated permeability pathways. Mapping of permeability patterns within bar-crest

grainstones demonstrates a highly varied but generally uncorrelated permeability structure on the scale of centimeters and meters that justifies averaging of petrophysical properties within depositional facies with uniform fabrics (Lucia et al., 1992).

Recognition of grainstone bodies that are laterally discontinuous on the interwell scale (hundreds of meters) and that possess uniform or homogeneously heterogeneous permeability structure has important implications for accurate portrayal of reservoir structure for fluid flow modeling. Deterministic mapping of these facies through high-resolution geophysics (improved surface three-dimensional seismic data or

Figure 4. The recessive to resistant outcrop weathering pattern reflects upward coarsening from mudstone through wackestone to ooid-peloid grainstone. This change in depositional textures is also reflected in petrophysical properties, with mudstones having approximately 5% porosity and less than 0.01 md permeability and grainstones having 20% porosity and 16 md permeability.

downhole geophysical data) coupled with stochastic modeling of grainstone distribution, will aid in construction of more realistic models in these settings.

Continuous lateral mapping of carbonate cycles like those of Algerita Escarpment is essential to evaluate the continuity of depositional, and ultimately petrophysical, facies for comparison to subsurface reservoir equivalents. Although the vertical succession of facies and their changes in time (stacking pattern) are commonly documented in subsurface studies, it is the combined effect of vertical and lateral changes that will best constrain an understanding of depositional facies continuity and its control on such derivatives as fluid flow behavior and reservoir performance.

MODERN DEVELOPMENT OF UPWARD-SHALLOWING CYCLES

The interpretation of depositional environments in subsurface studies of ancient carbonates depends largely on comparisons with modern carbonate analogs that emphasize depositional textures, sedimentary structures, stratigraphic successions, and diagenetic overprint. The Bahamas have several ooid shoal complexes that were examined by coring and seismic profiling (Ball, 1967; Hine, 1977; Harris, 1979, 1983; Hine et al., 1981; Bebout et al., 1991) and can be used as modern analogs to the San Andres platform carbonate cycles.

The Joulters Cays area (Figure 6), one of the most intensely studied ooid shoal complexes, is important because it displays a variety of environments in which ooid sands can accumulate, some of which are quite different from the environment in which the ooids are formed. The shoal is a 400 km² sand flat, partially cut by numerous tidal channels and fringed on the ocean-facing borders by mobile sands. The active border of ooid sands in the Joulters area covers 0.5–2 km in a dip direction, comparable to the width of the grainstones at Lawyer Canyon. At Joulters, the active ooid sands extend approximately 25 km parallel to the shelf break and 5 km bankward of it.

The relationship between relative sea level and rate of sedimentation in the Joulters Cays area was such that a 4-m-thick upward-coarsening and shallowing succession (cycle) of nonskeletal sands accumulated. The Holocene deposits in the Joulters Cays area and elsewhere in the Bahamas are separated from the underlying cemented Pleistocene limestones by a pronounced exposure surface representing in excess of 100 k.y. of subaerial weathering during the last major lowstand of sea level. The sea-level curve of Fairbanks (1989) and those of others show the most recent rise of sea level began some 18–20 Ka. Early during this rise, deposition occurred along the steep sides of the platform (Ginsburg et al., 1991; Grammer

et al., 1993), while the platform top remained subaerially exposed. Deposition in the Joulters Cays area commenced only when the platform top was recently reflooded, approximately 6 Ka. The nature of the upper boundary of the Holocene succession in the Joulters Cays area is still evolving. Locally, however, the boundary is already one of subaerial exposure where islands are developed, and much of the shoal top is presently near mean sea level. Whether sea level abruptly rises or falls will determine the type of boundary that ultimately caps the Holocene succession.

Regional Facies Relations

Harris (1979) employed an extensive coring program to document the facies relations in the Joulters Cays area. Sixty cores were taken with an average spacing between core locations of 1.5 km. This work provides valuable information on facies variability at the cycle scale across distances of tens of kilometers and is considered regional in nature.

The relief of the Joulters Cays shoal above the surrounding sea floor is primarily a result of ooid sands in one of three facies (Harris, 1979, 1983) (Figure 7). Ooid sands are present as a narrow belt along the active ocean-facing borders of the shoal where ooid accumulation coincides with ooid formation. Ooid and fine-grained peloid muddy sands, the more widespread facies exposed on the sand flat and platform interior west of the shoal, respectively, are the result of mixing of ooids with other grain types and carbonate mud. Collectively, these modern sands are more than 3 m thick and extend for 22 km in a dip direction in a 260-km² irregularly-shaped portion of the shoal. They exceed 7 m in thickness in an area coinciding with the Joulters Cays islands.

The basic facies pattern within the shoal is a fringe of ooid sand bordering opposing wedges of muddy ooid sand and muddy fine-grained peloid sand. Ooid sand directly overlies Pleistocene limestone along the seaward margin of the shoal and interfingers with muddier sediments in a bankward direction. Throughout most of the shoal, the vertical succession is of lithoclast sand and/or pellet mud at the base, muddy fine-peloid sand in the middle, and muddy ooid sand at the top. This succession shows distinct upward trends of increasing grain size, sorting, ooid content, stratification, and grain-supported fabric. The succession thins to the south because of a shallowing Pleistocene limestone surface and to the north and west because sediment thickness decreases.

Reservoir-Scale Facies Variability

Interwell-scale heterogeneities in hydrocarbon reservoirs of the San Andres Formation have been documented to be on the scale of hundreds of meters

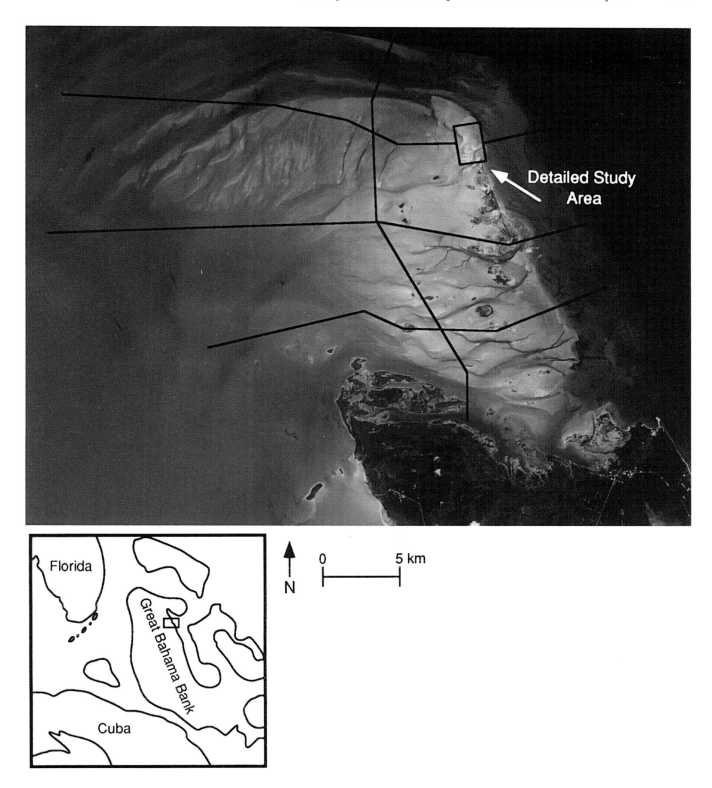

Figure 6. Landsat TM image of Joulters Cays ooid shoal complex of Great Bahama Bank showing location of cross sections for the fence diagram of Figure 7 and detailed study area of Figure 8. Inset shows location of Joulters area in the Bahamas.

484   Harris et al.

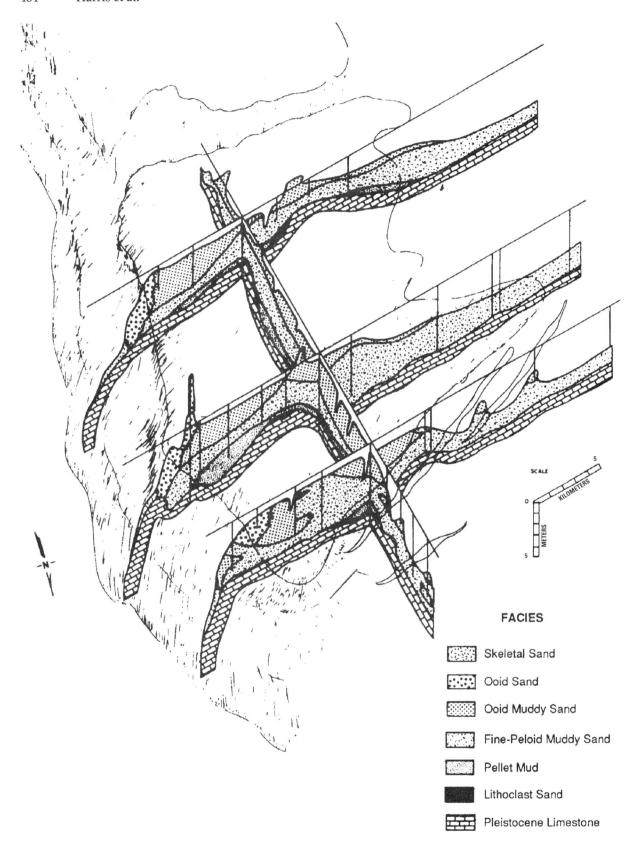

FACIES

Skeletal Sand

Ooid Sand

Ooid Muddy Sand

Fine-Peloid Muddy Sand

Pellet Mud

Lithoclast Sand

Pleistocene Limestone

Figure 7. Fence diagram modified after Harris (1983) showing facies relationships revealed by extensive coring of the Joulters Cays area. Vertical lines show core locations and depth of penetration, datum for cores is mean sea level, and lines of section are located in Figure 6. Figure 9 is an interpretation of the growth history of the shoal based on this fence diagram.

or less (Bebout et al., 1987; Ruppel and Cander, 1988; Bebout and Harris, 1990; Major et al., 1990). Depositional facies variability and early diagenetic alteration both contribute to fine-scale heterogeneities in the grain-rich upper portions of the succession of the Joulters Cays area that appear to be at an equivalent scale. To capture more information from the modern analog at the reservoir scale, Bebout et al. (1991) built upon the regional framework in the Joulters Cays area with additional coring in a subarea of the active portion of the shoal (Figure 6). Thirty-nine cores were taken in a 2.3 km² area, with an average core spacing of 330 m.

Three depositional facies, identified from cores by Bebout et al. (1991), all occur within the ooid sand of Harris (1979, 1983) discussed above (Figure 8). Well-sorted, cross-bedded ooid sand occurs on the active high-energy bar crest in the area of the shoal investigated in detail. Burrowed ooid sand accumulated along the edges of the shoal. A poorly sorted ooid sand, which is stabilized by seagrass and algae, occurs just bankward of the bar crest and represents a transition into the adjacent sand-flat environment. Differences between the three facies that commonly lead to heterogeneity are grain size, grain sorting, and sedimentary structures.

Marine-cemented hardground layers occur as thin seaward-offlapping layers near the base of the section (Figure 8). These hardgrounds have been correlated laterally as far as 600 m by sediment-probing transects. Porosity and permeability have not been greatly reduced in these modern lithified zones, but these zones could form barriers to fluid flow should they subsequently serve as preferential nucleation sites for additional cementation.

These subtle changes in depositional facies and early diagenetic overprint at Joulters upon burial would respond to compaction and cementation differently, and ultimately result in significant permeability variability within a single grainstone unit.

Growth of Shoal

The distribution of lithofacies within the Joulters shoal is a product of both today's depositional environments and changes in depositional patterns during development of the shoal. The changes were primarily a response to rising sea level, a corresponding increase of platform accommodation, and rapid sedimentation. Another reason that patterns of deposition changed is that there was topographic relief on the underlying Pleistocene limestone surface. During the Holocene rise of sea level, the depth of the Pleistocene limestone surface underlying the Joulters Cays area below present mean sea level determined the flooding history and the beginnings of marine sediment production and accumulation. Topographic highs and lows created hydrodynamic conditions influencing sediment distribution, until water depth or an evolving syndepositional topography masked these irregularities.

A precise sea-level curve for the Holocene has not been constructed for the northern Bahamas. The stratigraphy of Holocene sediments on the tidal flats of northwest Andros Island (Shinn et al., 1969; Hardie, 1977) and in the Joulters Cays area is consistent with a sea-level scenario like that portrayed on the curves for South Florida of Scholl et al. (1969) and for Barbados by Fairbanks (1989). Both of these curves indicate sea level was continuously rising during the Holocene but at a slower rate for the last few thousand years. If this interpretation of Holocene flooding is reasonably correct for the Joulters Cays area, then the succession of sediments would consist of basal facies indicating restricted water movement overlain by facies indicating more open circulation. Modifications of this succession would be the result of antecedent or syndepositional topography locally affecting sedimentation.

Holocene deposition in the Joulters Cays area occurred in three stages—bank flooding, shoal formation, and shoal (tidal sand bar and barrier) development (Harris, 1979, 1983). During bank flooding, rising sea level was still approximately 5 m below present sea level and the platform underlying today's shoal complex was only shallowly submerged (Figure 9A). Muddy sands of fine-grained peloids and pellets accumulated in protected lows on the underlying Pleistocene limestone floor. At a sea level 3 m below present, the shoal initially formed as the muddy sands extended seaward over the now-submerged Pleistocene surface, and ooid production began where bottom agitation was most pronounced (Figure 9B). During subsequent shoal development, the production and bankward dispersal of ooid sands through tidal sand bar and barrier environments established the present size and physiography of the shoal and changed the general nature of sediments throughout the area from muddy peloidal sands to ooid sands. During the tidal sand bar stage, ooid sands were transported farther bankward as a belt of active bars increased in width (Figure 9C). Eventually a barrier stage was achieved, in which the exchange of water between the seaward and bankward sides of the shoal was increasingly restricted by widespread sediment buildup approaching sea level, restriction of tidal-channel flow, and island formation along the shoal's ocean-facing margin (Figure 9D). What began as a series of subtidal bars and channels became the shallow subtidal to intertidal sand flat that now occupies most of the shoal.

Despite the lack of a local sea-level curve as previously discussed, timing for the onset of deposition in the Joulters Cays area, and therefore the entire development of the shoal, can be approximated from the average depth to the Pleistocene limestone surface in the area, the sea-level curves of Scholl et al. (1969) and Fairbanks (1989), and radiocarbon ages determined on core samples by Harris (1979). From this information (Figure 10), it is apparent that widespread deposition in the area began approximately 6 Ka. Thus, the entire complex Holocene succession,

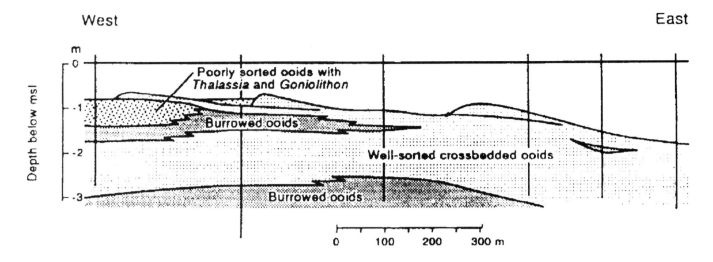

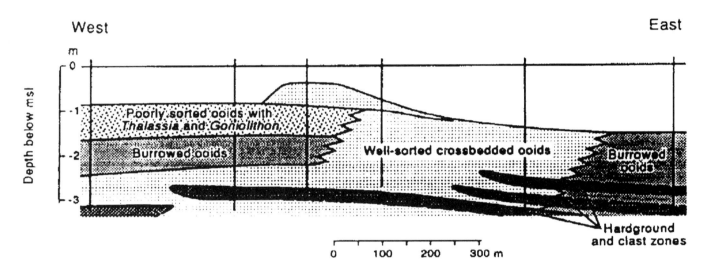

Figure 8. Cross sections modified after Bebout et al. (1991) based on coring within portion of active shoal of the Joulters Cays area. Vertical lines show core locations and depth of penetration, datum is mean sea level, and area of this detailed study is shown on Figure 6.

Figure 9. Block diagrams modified after Harris (1983) showing geologic development of Joulters Cays area. This growth history scenario is based on data discussed in detail by Harris (1979), including detailed mapping of the Pleistocene limestone surface, the flooding history during the Holocene rise of sea level, and lithofacies relationships revealed by extensive coring as shown on Figure 7. (A) Bank flooding stage; (B) shoal formation stage; (C) and (D) shoal development stages of tidal sand bar and barrier development are discussed in text. Orientation and lithology symbols are the same as Figure 7. The cross section within each block diagram shows the sequential development of the middle east–west cross section panel from Figure 7.

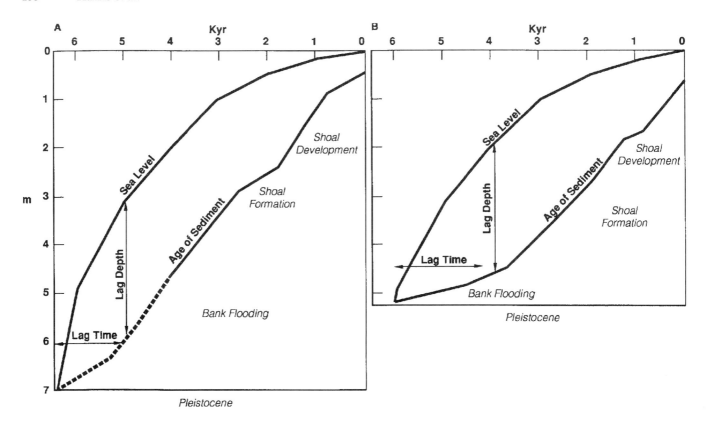

Figure 10. Rate of rise of Holocene sea level from Scholl et al. (1969) and Fairbanks (1989) compared with accumulation rates determined by radiocarbon dating by Harris (1979) for two sediment cores from the Joulters Cays area (A and B). Lag time, lag depth, and approximate timing of stages of shoal growth are shown.

with both appreciable vertical and lateral facies variation, formed very rapidly.

IMPLICATIONS OF MODERN EXAMPLE

Coring studies like that of the Joulters Cays area provide valuable calibration of depositional textures and sedimentary structures with depositional environments and give insight into the vertical and lateral patterns of facies that are similar to those encountered in ancient platform carbonate settings like that of the Algerita Escarpment. This modern example graphically shows that ooid sands may accumulate in varied environments that are quite different and perhaps remote from the site of formation of the ooids. Other implications equally important relative to the recognition and correlation of platform carbonate cycles concern the lateral extent and geometry of the deposits, facies preservation, lag time, and rapidity of deposition.

Lateral Extent and Geometry of Deposits

The Bahama Banks are long-lived, immense, flat-topped carbonate platforms. A veneer of Holocene sediments now covers the platforms in a complex fashion (Enos, 1974). These sediments of reef, sand-shoal, tidal-flat, and platform-interior origin do not form a widespread sheet of equal thickness but generally occur in localized depocenters or buildups of varying size, shape, and orientation. Their occurrence and geometry are due in large part to controls on deposition imparted by the preexisting topography as well as the hydrographic setting (Halley et al., 1983). It is not difficult to imagine the complexity that is likely to develop as subsequent sediments are deposited between the existing Holocene sediment thicks.

More striking perhaps is the variation that exists in depositional facies within the Joulters shoal itself. Coring has shown that in a few-meters-thick interval of sediment representing a brief period of accumulation, a complex succession of sediments can form largely as a result of localized topography and changes in relative sea level. The Joulters succession covers an area in excess of 25 km in a strike direction and 20 km in a dip direction. A thinner interval of sediment extends farther to the north and west as part of a sheet-like accumulation covering much of Great Bahama Bank. There are significant facies and thickness changes across the width of the shoal, and the surrounding bank equivalents are totally different facies. Grainstones are located in complex patterns toward the seaward edge of the shoal and in localized portions of the sand flat, beaches around small islands, and levees or lobe-shaped fans along tidal channels. Textural and diagenetic variation within grainstones impart a fine-scale heterogeneity that is also present in analogous reservoirs. Finally, the fine-scale time lines outlining the development of the Holocene in the Joulters Cays area (Figure 9) emphasize again the generalizations likely made when cycles are correlated in subsurface examples between widely spaced well locations.

Facies Preservation

The preservation potential of depositional facies in the modern example is also variable. Major changes in depositional environment have taken place in the Joulters Cays area within a brief span of time. Much of the area now occupied by a vast sand flat originally formed as a belt of active bars and tidal channels. Thus, the widespread belt of muddy ooid sands began as ooid sands that were subsequently modified during formation of the sand flat. Mud was mixed with the sediments during burrowing, and pellets along with various skeletal grain types were added by the sand-flat biota.

Certain depositional environments may stand a better chance for preservation due to the effects of early cementation. Island dune ridges up to 5 m above present sea level, for example, have become rapidly cemented during freshwater cementation (Halley and Harris, 1979) and are less likely to be reworked during storms or with change in sea level. Beaches are partially cemented by both marine and freshwater cements and similarly may not be eroded as readily during storms or by longshore currents. Marine-cemented hardground layers, as discussed earlier, punctuate the grain-rich portions of the shoal and may potentially represent early forming permeability barriers. The cemented islands may also play a role in the development of subsequent deposits in the area. If the next major sedimentation event follows a rise of relative sea level, the islands can form depositional highs around which bottom agitation would be centered and additional ooid shoals would develop.

Considering the degree of reworking of facies presented in the growth-history model for the Joulters Cays area and the variability of preservation potential imparted by diagenesis, one must question whether the facies patterns recognized in the Joulters area represent a common theme. Certainly the upward-coarsening and shallowing succession documented for the Holocene section in the Joulters area is common in ancient settings and is one of the several types of shallowing-upward sequences discussed by James (1979). The lateral facies variability (Figure 7) is also apparently not unique. Strikingly similar facies relations have been recognized in the Pleistocene limestones of South Florida (Halley et al., 1977; Halley and Evans, 1983; Evans, 1984) and the Bahamas (Beach and Ginsburg, 1980; Garrett and Gould, 1984;

Williams, 1985) and also in the Permian San Andres Formation of the Algerita Escarpment.

Facies variability in the Joulters area has served as an analog to guide interpretations of numerous subsurface examples of ancient carbonate sand deposits (e.g., Ottmann et al., 1973; Bebout and Schatzinger, 1978; Smosna, 1984). The more recent coring in the Joulters area by Bebout et al. (1991) illustrates textural and diagenetic variation occurring on a scale that is coincident with variations in reservoir quality in reservoirs from the San Andres Formation of the Permian basin and other platform carbonate examples.

Lag Time and Rapidity of Deposition

The radiocarbon ages reported by Harris (1979) provide approximate estimates of the lag time between initial platform flooding and onset of rapid sediment accumulation and also the sediment accumulation rates for the Holocene deposits of the Joulters Cays area (Figure 10). A comparison of the suspected position of sea level during the stages of shoal growth (Figure 9) and available sea-level curves with the apparent radiocarbon ages in the lower portions of two cores suggests a lag time of 1500–2000 yr (Figure 10). By this time, much of the platform was flooded to a depth of 2 m or more, likely a depth (lag depth of Enos, 1991) at which sufficient cross-bank tidal flow was first initiated and ooid-shoal formation began.

The overall rate of accumulation estimated from the cores paced the rise of sea level at approximately 1 m/k.y., or 1000 Bubnoffs (mm/10^3 yr) (Figure 10). However, the rate was initially somewhat lower during the bank-flooding stage. An increase in the rate of accumulation coincides with increased amounts of ooids produced during the shoal formation and development stages. The decreasing rate of rise of sea level and the high sediment accumulation rate resulted in the sediment surface approaching sea level through time as was indicated by the shoaling succession of facies documented in the cores. These results corroborate the findings of others, e.g., James (1979), Schlager (1981), Read et al. (1986), and Enos (1991), that (1) carbonate sedimentation on the platform top tends to lag behind the relative rate of sea-level rise, resulting in an initially deepening succession, but (2) sediments accumulate at overall rates greater than the relative rise of sea level and, thus, quickly build to near sea level.

The rapidity of deposition in the Joulters Cays area reinforces the notion that platform carbonates have high growth potential and shows that the facies relations can become complex at the cycle scale. Here, there was little problem keeping up and/or catching up to a rising sea level and sediments have accumulated essentially to sea level over 400 km^2 of the platform margin. Estimates of an average accumulation rate of approximately 1 m/k.y. are based on an understanding of the depth to the preexisting limestone surface, lithofacies descriptions of sediment cores, and radiocarbon ages determined on core samples. We can speculate that accommodation space may fill equally quickly and complexly during the development of cycles in ancient platform settings.

CONCLUSIONS

It is becoming increasingly apparent during the geologic and engineering evaluation of hydrocarbon reservoirs in platform carbonates that a detailed geologic framework must serve as the template for a reasonable understanding of porosity and permeability distribution and delineation of fluid-flow units. The ancient outcrop example and modern analog discussed here illustrate the lateral extent and complexity of facies relations that are likely to occur within an upward-shallowing cycle. The vertical and lateral variability of facies of the Joulters modern cycle is directly analogous to that of the Permian cycle at the Algerita Escarpment. Both are initiated by a mud-dominated subtidal section that grades upward into grain-dominated facies. Laterally, an increase in the extent of burrowing, amount of mud, and types of grains occurs away from active portions of the shoal.

Certainly this complexity cannot be fully addressed in normal subsurface situations with downhole log and limited core information, especially when some of the relations occur at distances less than normal well spacing even in extensively developed fields. Nevertheless an appreciation of the general nature of such facies variability is needed in reservoir studies to guide correlations of cycles and flow units between wells and constrain the input into reservoir models or forward-looking geologic models.

ACKNOWLEDGMENTS

Publication was permitted by Chevron and by the Director, Bureau of Economic Geology, The University of Texas of Austin. We thank coworkers at both institutions who worked with us on the Algerita Escarpment outcrops.

REFERENCES CITED

Ball, M. M., 1967, Carbonate sand bodies of Florida and the Bahamas: Journal of Sedimentary Petrology, v. 37, p. 556-591.

Beach, D. K., and Ginsburg, R. N., 1980, Facies succession, Plio-Pleistocene carbonates, Northwestern Great Bahama Bank: AAPG Bulletin, v. 64, p. 1634-1642.

Bebout, D. G., and Harris, P. M., eds., 1990, Geologic and Engineering Approaches in Evaluation of San Andres/Grayburg Hydrocarbon Reservoirs—Permian Basin: The University of Texas at Austin, Bureau of Economic Geology, 297 p.

Bebout, D. G., Lucia, F. J., Hocott, C. R., Fogg, G. E., and Vander Stoep, G. W., 1987, Characterization of the Grayburg reservoir, University Lands Dune field, Crane County, Texas: The University of Texas at Austin, Bureau of Economic Geology Report of Investigations No. 168, 98 p.

Bebout, D. G., Major, R. P., Tyler, N., Harris, P. M., and Kerans, C., 1991, Platform-edge, shallow-marine ooid grainstone shoals, Joulters Cays, Bahamas: A modern analog of carbonate hydrocarbon reservoirs, in Coastal Depositional Systems in the Gulf of Mexico, Quaternary Framework and Environmental Issues: Gulf Coast Section, SEPM, 12th Annual Research Conference, Program and Abstracts, p. 26-29.

Bebout, D. G., and Schatzinger, R. A., 1978, Distribution and geometry of an oolite-shoal complex—Lower Cretaceous Sligo formation, South Texas: Gulf Coast Association Geological Societies Transactions, v. 78, p. 33-45.

Enos, P., 1974, Surface sediment facies of the Florida-Bahamas Plateau: Geological Society of America Map, 5 p.

Enos, P., 1991, Sedimentary parameters for computer modeling, in Franseen, E. K., Watney, W. L., Kendall, C. G. St. C., and Ross, W., eds., Sedimentary Modeling: Computer Simulations and Methods for Improved Parameter Definition: Kansas Geological Survey Bulletin 233, p. 63-99.

Evans, C. C., 1984, Development of an ooid sand shoal complex: The importance of antecedent and syndepositional topography, in Harris, P. M., ed., Carbonate Sands: SEPM Core Workshop No. 5, p. 392-428.

Fairbanks, R. G., 1989, A 17,000-year glacio-eustatic sea level record: Influence of glacial melting rates on the Younger Dryas Event and deep-ocean circulation: Nature, v. 342, p. 637-642.

Garber, R. A., and Harris, P. M., 1990, Depositional facies of Grayburg/San Andres dolostone reservoirs, Central Basin Platform, Permian Basin, in Bebout, D. G., and Harris, P. M., eds., Geologic and Engineering Approaches in Evaluation of San Andres/Grayburg Hydrocarbon Reservoirs—Permian Basin: The University of Texas at Austin, Bureau of Economic Geology, p. 1-19.

Garrett, P., and Gould, S. J., 1984, Geology of New Providence Island, Bahamas: Geological Society of America Bulletin, v. 95, p. 209-220.

Ginsburg, R. N., Harris, P. M., Eberli, G. P., and Swart, P. K., 1991, The growth potential of a bypass margin, Great Bahama Bank: Journal of Sedimentary Petrology, v. 61, p. 976-987.

Goldhammer, R. K., Oswald, E. J., and Dunn, P. A., 1991, Hierarchy of stratigraphic forcing: Example from Middle Pennsylvanian shelf carbonates of the Paradox Basin, in Franseen, E. K., Watney, W. L., Kendall, C. G. St. C., and Ross, W., eds., Sedimentary Modeling: Computer Simulations and Methods for Improved Parameter Definition: Kansas Geological Survey Bulletin 233, p. 361-413.

Grammer, G. M., Ginsburg, R. N., and Harris, P. M., 1993, Timing of deposition, diagenesis, and failure of steep carbonate slopes in response to a high-amplitude/high-frequency fluctuation in sea level, Tongue of the Ocean, Bahamas, this volume.

Halley, R. B., and Evans, C. C., 1983, The Miami Limestone: A Guide to Selected Outcrops and Their Interpretation: Miami Geological Society, Miami, Florida, 67 p.

Halley, R. B., and Harris, P. M., 1979, Fresh-water cementation of a 1000-year-old oolite: Journal of Sedimentary Petrology, v. 49, p. 969-987.

Halley, R. B., Harris, P. M., and Hine, A. C., 1983, Bank margin environments, in Scholle, P. A., Bebout, D. G., and Moore, C. H., eds., Carbonate Depositional Environments: AAPG Memoir 33, p. 463-506.

Halley, R. B., Shinn, E. A., Hudson, J. H., and Lidz, B. H., 1977, Pleistocene barrier bar seaward of ooid shoal complex near Miami, Florida: AAPG Bulletin, v. 61, p. 519-526.

Hardie, L. A., 1977, ed., Sedimentation on the Modern Carbonate Tidal Flats of Northwest Andros Island, Bahamas: Johns Hopkins University Studies in Geology No. 22, Johns Hopkins Press, Baltimore, Maryland, 202 p.

Harris, P. M., 1979, Facies anatomy and diagenesis of a Bahamian ooid shoal: Sedimenta 7, Comparative Sedimentology Laboratory, University of Miami, Florida, 163 p.

Harris, P. M., 1983, The Joulters ooid shoal, Great Bahama Bank, in Peryt, T. M., ed., Coated Grains: Springer-Verlag, Berlin, p. 132-141.

Hine, A. C., 1977, Lily Bank, Bahamas: History of an active oolite sand shoal: Journal of Sedimentary Petrology, v. 47, p. 1554-1581.

Hine, A. C., Wilber, R. J., and Neumann, A. C., 1981, Carbonate sand bodies along contrasting shallow bank margins facing open seaways in Northern Bahamas: AAPG Bulletin, v. 65, p. 261-290.

James, N. P., 1979, Shallowing-upward sequences in carbonates, in Walker, R. G., ed., Facies Models: Geoscience Canada Reprint Series 1, Toronto, p. 109-119.

Kerans, C., Lucia, F. J., Senger, R. K., Fogg, G. E., Nance, H. S., Kasap, E., and Hovorka, S. D., 1991, Characterization of reservoir heterogeneity in carbonate-ramp systems, San Andres/Grayburg, Permian Basin: Reservoir Characterization Research Laboratory, The University of Texas at Austin, Bureau of Economic Geology Final Report, 245 p.

Kerans, C., and Nance, H. S., 1991, High frequency cyclicity and regional depositional patterns of the Grayburg Formation, Guadalupe Mountains, New Mexico, in Meader-Roberts, S., Candelaria, M. P., and Moore, G. E., eds., Sequence Stratigraphy, Facies, and Reservoir Geometries of the San Andres, Grayburg, and Queen Formations, Guadalupe Mountains, New Mexico and Texas: Permian Basin Section SEPM Society of Sedimentary Geology Publication 91-32, p. 53-69.

Lindsay, R. F., 1991, Grayburg Formation (Permian-Guadalupian): Comparison of reservoir character-

istics and sequence stratigraphy in the Northwest Central Basin Platform with outcrops in the Guadalupe Mountains, New Mexico, in Meader-Roberts, S., Candelaria, M. P., and Moore, G. E., eds., Sequence Stratigraphy, Facies, and Reservoir Geometries of the San Andres, Grayburg, and Queen Formations, Guadalupe Mountains, New Mexico and Texas: Permian Basin Section SEPM Society of Sedimentary Geology Publication 91-32, p. 111-118.

Lucia, F. J., 1983, Petrophysical parameters estimated from visual descriptions of carbonate rocks: A new field classification of carbonate pore space: Journal of Petroleum Technology, v. 35, p. 629-637.

Lucia, F. J., Kerans, C., and Senger, R. K., 1992, Defining flow units in dolomitized carbonate-ramp reservoirs: Society of Petroleum Engineers Annual Meeting, Washington, D.C., Publication SPE 24702, p. 399-406.

Major, R. P., Bebout, D. G., and Lucia, F. J., 1988, Depositional facies and porosity distribution, Permian (Guadalupian) San Andres and Grayburg Formations, P. J. W. D. M. Field Complex, Central Basin Platform, West Texas, in Lomando, A. J., and Harris, P. M., eds., Giant Oil and Gas Fields: SEPM Core Workshop No. 12, p. 615-648.

Major, R. P., Vander Stoep, G. W., and Holtz, M. H., 1990, Delineation of unrecovered mobile oil in a mature dolomite reservoir: East Penwell San Andres Unit, University Lands, West Texas: The University of Texas at Austin, Bureau of Economic Geology Report of Investigations No. 194, 52 p.

Ottmann, R. D., Keyes, P. L., and Ziegler, M. A., 1973, Jay field—A Jurassic stratigraphic trap: Gulf Coast Association Geological Societies Transactions, v. 23, p. 146-157.

Read, J. F., Grotzinger, J. P., Bone, J. A., and Koerschner, W. F., 1986, Models for generation of carbonate cycles: Geology, v. 14, p. 107-110.

Ruppel, S. C., 1990, Facies control of porosity and permeability: Emma San Andres Reservoir, Andrews County, Texas, in Bebout, D. G., and Harris, P. M., eds., Geologic and Engineering Approaches in Evaluation of San Andres/Grayburg Hydrocarbon Reservoirs—Permian Basin: The University of Texas at Austin, Bureau of Economic Geology, p. 145-173.

Ruppel, S. C., and Cander, H. S., 1988, Effects of facies and diagenesis on reservoir heterogeneity: Emma San Andres field, West Texas: The University of Texas at Austin, Bureau of Economic Geology Report of Investigations No. 178, 67 p.

Sarg, J. F., and Lehmann, P. J., 1986, Lower-Middle Guadalupian facies and stratigraphy, San Andres-Grayburg formations, Permian Basin, Guadalupe Mountains, New Mexico, in Moore, G. E., and Wilde, G. L., eds., Lower and Middle Guadalupian Facies, Stratigraphy, and Reservoir Geometries, San Andres-Grayburg Formations, Guadalupe Mountains, New Mexico and Texas: Permian Basin Section, SEPM Special Publication No. 86-25, p. 1-36.

Schlager, W., 1981, The paradox of drowned reefs and carbonate platforms: Geological Society of America Bulletin, v. 92, p. 197-211.

Scholl, D. W., Craighead, F. C., and Stuvier, M., 1969, Florida submergence curve revised: Its relation to coastal sedimentation rates: Science, v. 163, p. 562-564.

Shinn, E. A., Lloyd, R. M., and Ginsburg, R. N., 1969, Anatomy of a modern carbonate tidal flat, Andros Island, Bahamas: Journal of Sedimentary Petrology, v. 39, p. 1202-1228.

Smosna, R. A., 1984, Sedimentary facies and early diagenesis of an oolite complex from the Silurian of West Virginia, in Harris, P. M., ed., Carbonate Sands: SEPM Core Workshop No. 5, p. 20-57.

Van Wagoner, J. C., Posamentier, H. W., Mitchum, R. M., Jr., Vail, P. R., Sarg. J. F., Loutit, T. S., and Hardenbol, J., 1988, An overview of the fundamentals of sequence stratigraphy and key definitions, in Wilgus, C. K., Hastings, B. S., Kendall, C. G. St. C., Posamentier, H. W., Ross, C. A., and Van Wagoner, J. C., eds., Sea Level Changes: An Integrated Approach: SEPM Special Publication No. 42, p. 39-46.

Wanless, H. R., 1991, Observational foundation for sequence modeling, in Franseen, E. K., Watney, W. L., Kendall, C. G. St. C., and Ross, W., eds., Sedimentary modeling: Computer simulations and methods for improved parameter definition: Kansas Geological Survey Bulletin 233, p. 43-62.

Williams, S. C., 1985, Stratigraphy, facies evolution, and diagenesis of Late Cenozoic limestones and dolomites, Little Bahama Bank, Bahamas: Ph.D. dissertation, University of Miami, Miami, Florida, 217 p.

Chapter 19

Parasequence Geometry as a Control on Permeability Evolution: Examples from the San Andres and Grayburg Formations in the Guadalupe Mountains, New Mexico

Susan D. Hovorka
H. S. Nance
Charles Kerans
Bureau of Economic Geology
The University of Texas at Austin
Austin, Texas, U.S.A.

ABSTRACT

Depositional porosity and permeability distribution in Guadalupian-age grainstone-dominated parasequences of the central Guadalupe Mountains have been moderately to strongly modified by diagenesis. Early diagenetic features in the ramp crest and inner shelf crest can be related to inferred paleohydrology that developed during high-frequency (parasequence-scale) sea-level lowstands. In the San Andres example, leaching of carbonate grains and resulting precipitation of intergranular cement in a meteoric lens beneath an emergent bar crest produced a well-defined, highly porous but relatively low-permeability moldic zone in the thickest part of a grain-dominated parasequence. In the Grayburg example, preferential preservation of intergranular porosity and minor leaching along a paleo-water table produced a thin (less than 30 cm) stratiform zone of higher than average permeability. A zone of preferentially preserved intergranular porosity developed within an inferred meteoric lens beneath an emergent bar crest. Diagenetically influenced lateral and vertical changes in permeability distribution are predicted to be typical of intermittently exposed ramp crest and inner shelf crest environments. These examples demonstrate the efficacy of parasequence-scale mapping of depositional and diagenetic facies for improving prediction of permeability distribution in analogous strata.

INTRODUCTION

Relationships among depositional facies type, facies geometries, and permeability distribution in carbonates are typically complicated by diagenetic modification, which alters pore type and pore distribution. Diagenetic modification of depositional fabric is intrinsic to most shallow-water carbonate sediments because the sediment includes metastable aragonite and high-Mg calcite and the setting favors introduction of chemically reactive meteoric and evaporitic fluids. The resulting complex diagenetic patterns hinder efforts to use geologic facies models for predicting permeable reservoir zones. Models are needed that correlate between diagenetic processes that augment or limit permeability and mappable and predictable geologic parameters such as depositional facies, carbonate platform constructional histories, and burial histories. Techniques developed from understanding depositional geometries and platform evolution based on sequence stratigraphic concepts can be applied to defining the distribution of diagenetically modified permeability.

The Guadalupe Mountains provide some of the best exposures of Permian carbonate and mixed-siliciclastic-carbonate ramp and rimmed-margin systems in the world. The San Andres and Grayburg (lower to middle Guadalupian) outcrops have been used to define facies and systems tracts, facies geometries and their petrophysical parameters, and sequence-stratigraphic organization of carbonate and siliciclastic depositional facies (Fekete et al., 1986; Moore and Wilde, 1986; Sarg and Lehmann, 1986; Franseen et al., 1989; Kerans, 1991; Kerans and Nance, 1991).

Documentation of sequence-stratigraphic relationships in continuous outcrop has the advantage of allowing the lateral continuity of depositional units to be traced and their geometries determined. This study documents the relationship between depositional geometries and permeability distribution for two parasequences deposited in ramp crest and inner-shelf crest settings, one in the upper San Andres Formation and one in the Grayburg Formation. Based on parasequence mapping, we interpret the depositional topography and distribution of primary depositional facies that define the probable paleo-hydrologic patterns developed during the relative sea-level lowstand that terminated deposition of each parasequence. Permeability distribution was measured in 5- to 10-m-thick parasequences by sampling at closely spaced (typically at 0.5 or 1 ft) intervals along vertical measured sections spaced along 700- to 1500-m transects. Permeability was determined from 1-in. core plugs removed with a portable core plugger and analyzed using standard laboratory techniques. Petrographic observations define the diagenetic sequence in each part of the facies tract. Integration of interpreted paleohydrology with petrographic and permeability data allows for interpretation of the diagenetic processes active in bar-crest deposits during relative sea-level lowstands and for determining the permeability modification of each

facies at this time of earliest and, possibly, most significant diagenetic modification of depositional fabrics.

GUADALUPIAN SHELF—GEOLOGIC SETTING AND DEPOSITIONAL HISTORY

Sequence and Parasequence Description

The San Andres and Grayburg formations (lower to middle Guadalupian) are exposed in the western Guadalupe Mountains of New Mexico and west Texas (Figure 1). Three third-order depositional sequences (Figure 2) have been identified in the San Andres and Grayburg formations—the lower to middle San Andres, the upper San Andres, and the Grayburg sequences. In the Algerita Escarpment study area (Figure 1), the San Andres Formation unconformably overlies the Yeso Formation (Leonardian) and comprises approximately 365 m of cyclic dolostone and a few quartz sandstone beds that are concentrated mainly in the upper 30 m of the section. Four unconformity-bounded fourth-order sequences (uSA1 through uSA4) in the upper San Andres Formation are recognized by locally to regionally extensive karst surfaces and associated facies stacking patterns (Kerans, 1991). The boundary with basal sandstones of the overlying Grayburg Formation is unconformable.

In the Guadalupe Mountains, the Grayburg Formation comprises interbedded dolostones and siliciclastics and is 75–100 m thick on the inner shelf (Kerans and Nance, 1991) to 335 m thick in basin-margin exposures on Bush Mountain (Franseen et al., 1989). The Grayburg Formation unconformably underlies the siliciclastic-rich Queen Formation. Widespread conspicuous karst surfaces (type 1 sequence boundaries in the terminology of Van Wagoner et al., 1988) occur at the top and base of the Grayburg Formation in the southern and central Guadalupe Mountains (Fekete et al., 1986; Kerans and Nance, 1991). The Grayburg depositional sequence has been subdivided into a combined lowstand-transgressive system tract and highstand systems tract by Kerans and Nance (1991). Detailed parasequence windows (Figure 2) in which diagenetic studies were undertaken are in the basal upper San Andres fourth-order sequence (uSA1) on the Algerita Escarpment near Lawyer Canyon (Figure 3) and in the highstand systems tract of the Grayburg Formation at Stone Canyon, north of El Paso Gap (Figure 4).

Parasequence Geometry

Cyclic deposits in San Andres and Grayburg inner shelf and ramp-crest settings are identified as parasequences (terminology of Van Wagoner et al., 1988). San Andres/Grayburg parasequences are 1- to 10-m-thick, asymmetric, progradational, upward-shoaling cycles. The basal transgressive facies is composed of

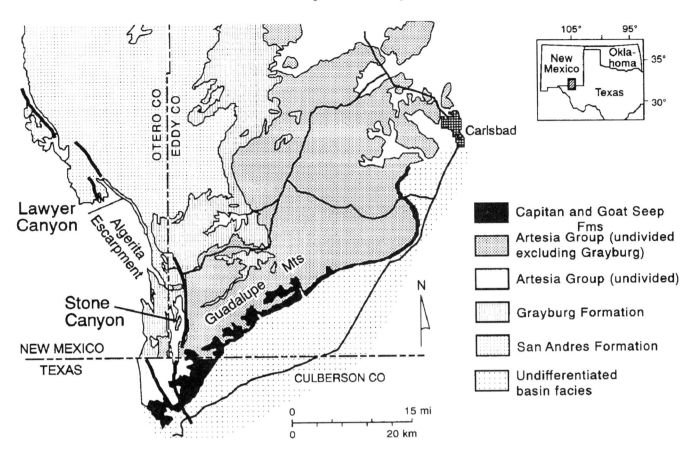

Figure 1. Location map showing outcrops at Lawyer Canyon and Stone Canyon.

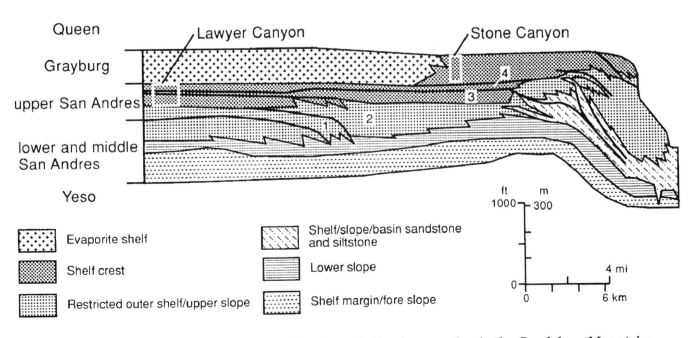

Figure 2. Schematic dip section of the San Andres through Grayburg section in the Guadalupe Mountains showing the geologic setting of the detailed study areas at Lawyer Canyon and Stone Canyon. Numbers denote four unconformity-bounded fourth-order sequences in the upper San Andres.

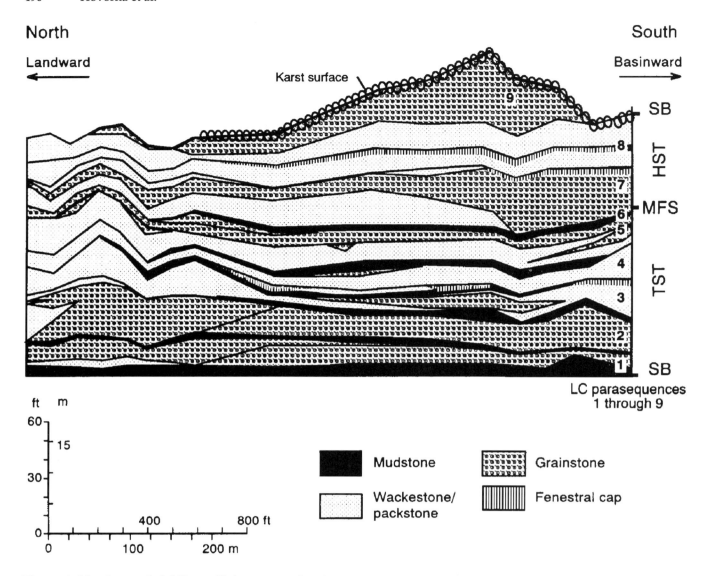

Figure 3. North–south (oblique dip) cross section in Lawyer Canyon Upper San Andres sequence 1 (uSA1). Parasequences LC 1–LC 6 comprise the transgressive systems tract (TST), LC 7 contains the maximum flooding surface (MFS), and LC 8 and LC 9 form the highstand systems tract (HST). SB = sequence boundary.

dolomitic quartz sandstone (Grayburg Formation) or a relatively nonporous and impermeable carbonate wackestone or packstone (San Andres and/or Grayburg Formations). This basal facies is overlain by coarser grained, ooid-peloid-rich carbonate facies deposited during a progradational phase. Some of the parasequences investigated in Lawyer Canyon (San Andres Formation) and many of the parasequences at the Stone Canyon (Grayburg Formation) study area are capped by laterally discontinuous tidal-flat deposits characterized by fenestral laminations and tepee structures, which correlate over a distance of 80–100 m with peloid packstones of probable shallow subtidal origin. Lateral discontinuity of tidal-flat deposits suggests development of islands or emergent shoals late in the evolution of each parasequence. Tidal-flat deposits that extend vertically through several parasequences are interpreted to

record the persistent location of islands during deposition of these cycles.

In the Lawyer Canyon study area, nine parasequences (LC 1 through LC 9) are recognized in 43–55 m of uSA1 (Figure 3). Parasequences LC 1 through LC 6 of the transgressive systems tract are aggradational to slightly retrogradational with more completely developed shallow-water, high-energy facies in the lowermost units (Kerans, 1991). The basal part of LC 7 contains the maximum flooding surface defined by a fusulinid-peloid packstone that marks the maximal updip transgression of outer ramp facies. The regressive hemicycle of LC 7 and the overlying parasequences LC 8 and LC 9 make up the highstand systems tract. Discontinuous fenestral caps at the tops of these units occur preferentially in the southern part of the transect and indicate stacking of topographic highs in this area. LC 9 is a peloid-echin-

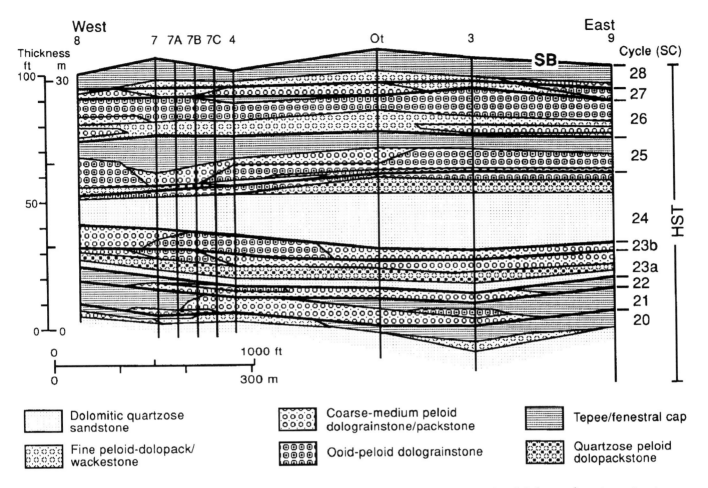

Figure 4. West–east stratigraphic cross section of uppermost Grayburg Formation highstand systems tract (HST), northern wall of Stone Canyon. Shown are locations of measured sections, generalized facies, parasequence boundaries (thin lines), and Grayburg–Queen sequence boundary (SB, thick line).

oderm grainstone-packstone with 11 m of depositional relief on the top. A karst surface was developed on top of the sequence. Basal uSA2 parasequences onlap the topographic high. LC 7 was selected for petrographic study of the origin of unusually abundant moldic pores. The porosity, permeability, and petrographic fabrics of this parasequence were sampled at 7 out of 18 measured sections. Samples were collected at 1-ft spacing (vertically) in sections H, A, Z, and X and at wider spacing at downdip sections R, O, and Q (Figure 5), resulting in collection of 123 core plugs from which thin sections were prepared for transmitted light microscopy. Samples strongly altered by leaching of large vugs or by calcite or travertine cement were eliminated from further consideration.

The Grayburg Formation at Stone Canyon comprises beds of dolomitic sandstone, quartzose dolostone, and a variety of relatively pure dolostone facies that can be divided into at least 28 parasequences (SC 1–SC 28) (Figure 4). Parasequences of the lowstand-transgressive system tract and highstand systems tract are separated from each other by the maximum flooding surface interpreted here to be at the base of a 2-m-thick fusulinid-peloid packstone bed (SC 7) located 25 m above the canyon floor. This is the only interval of fusulinid-peloid rock occurring in Stone Canyon, and it is the deepest water carbonate depositional facies in the canyon.

Stone Canyon parasequences comprise three lithologic associations: (1) dolomitic quartz sandstone throughout (SC 2 in the lowstand-transgressive system tract); (2) sandstone grading up to dolopackstone and dolograinstone (e.g., SC 23a and SC 24; Figure 4); and (3) relatively siliciclastic-free dolopackstone and dolograinstone throughout (e.g., SC 23b, SC 26, SC 27, and SC 28; Figure 4). Parasequence SC 23 was chosen for detailed sampling because of its representative facies composition, relatively easy access, and manageable size. The Stone Canyon data base consisted of thin sections and petrophysical analyses of 1-in. core plugs taken at 6-in. vertical spacing from eight measured sections spaced 100 to 900 ft apart on the northern wall of Stone Canyon.

San Andres and Grayburg Diagenesis

The subsurface San Andres and Grayburg formations have had a complex diagenetic history. Multiple

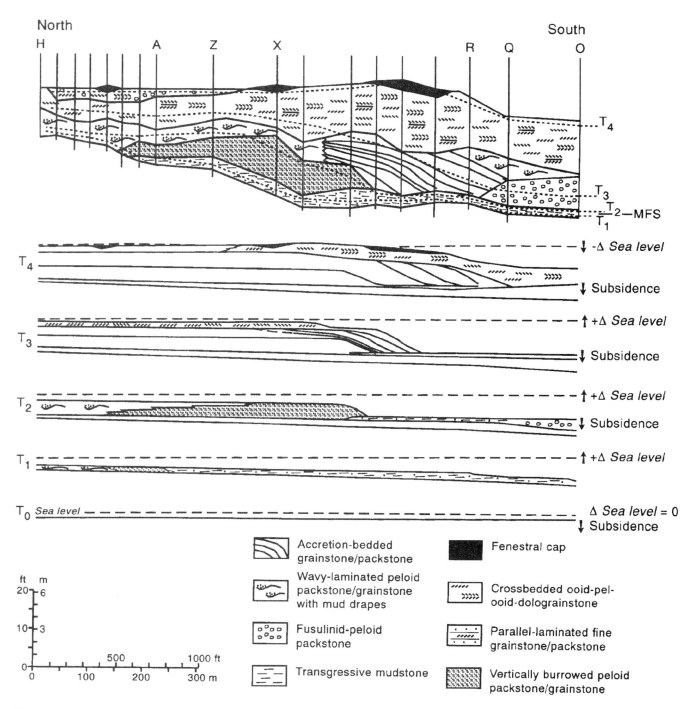

Figure 5. Sequential development of parasequence LC 7. Time lines are largely traceable in outcrop except T 2, wherein bioturbation obscures time-line-parallel bedding surfaces. See text for description of depositional events. MFS = maximum flooding surface.

episodes of dolomitization by hypersaline and meteoric or mixed marine-meteoric fluids are recognized (Ruppel and Cander, 1988; Leary and Vogt, 1990). Anhydrite was emplaced as cement and nodules at several different stages (Ruppel and Cander, 1988; Leary and Vogt, 1990; Major et al., 1990). Anhydrite dissolution, gypsum and calcite precipitation, and feldspar alteration are late events (Leary and Vogt, 1990; Major and Holtz, 1990). Outcrop samples examined in this study have undergone additional near-surface diagenesis, including sulfate dissolution, cavern formation, and precipitation of calcite spar and travertine (Scholle et al., 1992). Subsurface dolomitization and sulfate precipitation and outcrop diagenetic features probably have an impact on permeability distribution, but detailed description and interpretation of the entire spectrum of these features are outside the scope of this discussion. The effect of near-surface alteration on the results of this study have been minimized, in so far as possible, by removing samples from several centimeters below the outcrop surface and rejecting samples that contain large vugs or abundant calcite cements.

UPPER SAN ANDRES LAWYER CANYON PARASEQUENCE 7 (LC 7)

Facies

The depositional history of LC 7 as recorded at Lawyer Canyon is characteristic of fully developed ramp-crest parasequences. Parasequence development can be summarized in four time steps (Figure 5). At T_0, initial relative sea-level rise exceeded the ability of underlying LC 6 carbonates to keep pace, causing termination of LC 6 deposition. The proximal ramp-crest was flooded to depths of 5–10 m. During T_1, gradual reequilibration of carbonate depositional environments and productivity resulted in deposition of low-energy deeper subtidal (<10 m) mudstone to wackestone in a downdip position. Updip in slightly shallower subtidal environments burrowed peloid wackestone–packstone was deposited. T_2 shows aggradational deposition of current-laminated and vertically burrowed peloid packstones (updip) to starved mudstone and fusulinid wackestone (downdip) during conditions of slowing relative rise of sea level and depositional "catch-up." T_3 is characterized by development of grainstone shoal sedimentation and initiation of progradational bar-accretion deposits, and (T_4) continued shoaling and offlap of grainstone facies with localized development of tidal flats. Maximum flooding within this parasequence occurs during T_2 deposition, the parasequence-level maximum flooding surface (here coincident with the sequence-level maximum flooding surface) running beneath the bar-accretion downlap surface.

Initial deposits of LC 7 are a transgressive basal mudstone that pinches out in an updip (toward the north) direction within the study area. Peloid-fusulinid packstone defining the maximum flooding surface overlies the basal mudstone in the southern part of the transect. Grainstone in the upper part of the sequence is composed mainly of fine-grained, well-sorted ooids and peloids. Fusulinids occur scattered throughout the grainstones and as lag concentrations at the bases of cross-laminae. A thin fenestral bed caps the parasequence in the southern part of the transect (sections X to R, Figure 5).

Porosity and Permeability Distribution

Moldic porosity (quantified by point counting) is extensively developed in the middle of the LC 7 transect (sections Q through Z), whereas only minor amounts of moldic porosity exist in the northern and southern parts of the transect (Figure 6). Molds in the middle part of the study area compose 10–30% of the bulk volume, representing 20–80% of the original grains. At both the updip and more seaward parts of the cross section, moldic pores are less than 10% of the bulk volume and less than 10% of the original grains have been dissolved. The molds are dominantly leached peloids. No consistent vertical trend in abundance of molds is recognized, and no direct relationship can be identified between depositional facies and the distribution of molds (compare Figure 5 with Figure 6). However, a clear relationship is observed between the zone dominated by moldic porosity and the distribution of intertidal and emergent features at the top of parasequences LC 7, LC 8, and LC 9 (Figure 6). The thick part of LC 7 has a thin fenestral cap, documenting an episode of intertidal deposition and potential exposure that terminated deposition of that parasequence. LC 8, which overlies LC 7, contains fenestral beds overlying the area of moldic porosity in LC 7. The thick southern part of the next overlying parasequence, LC 9, is coincident with the thickest part of LC 7 and has karst pits at the top infilled with sediments of overlying cycles, documenting sea-level drop and prolonged exposure at the sequence boundary.

Highly moldic grainstones influence the porosity/permeability relationships (Figure 7). Mudstones have depositional-fabric-controlled low porosity and permeability; grainier rocks exhibit a wide scatter of porosities and permeabilities rather than a trend of increasing permeability in grainier rocks as would be expected if permeability was controlled by intergranular porosity (Figure 7a). However, grainstones with abundant moldic porosity (greater than 10% bulk volume) have lower permeability than less moldic grainstones with equally high but mainly intergranular porosity (Figure 7b). This difference demonstrates that, as expected (Lucia, 1983; Lucia and Conti, 1987), molds are isolated and are not as effective in transmitting fluids as intergranular porosity. The average and range of permeability values, however, are about the same in moldic and nonmoldic grainstones. The effect of moldic porosity in LC 7 is to reduce the permeability of thick, well-sorted grainstones to form an interval of average permeability rather than to create a barrier to flow.

Petrographic Relationships

Grain dissolution is a key element in porosity and permeability trends in LC 7. Peloids 150 µm in diameter are preferentially dissolved, as indicated by comparison of grain size and shape of molds to preserved grains (compare Figure 8a and 8b to 8c). The variability in mold abundance is caused by dissolution of these spherical grains. Mollusk fragments, identified on the basis of their elongate shape and rectangular ends, are moldic throughout the study area, and no correlation between the abundance of mollusk fragments and total abundance of moldic porosity was found. All grain types, including echinoderm plates and fusulinids, locally show some leaching.

Two types of dolomite cement are recognized in LC 7. A first generation of isopachous rim cement is

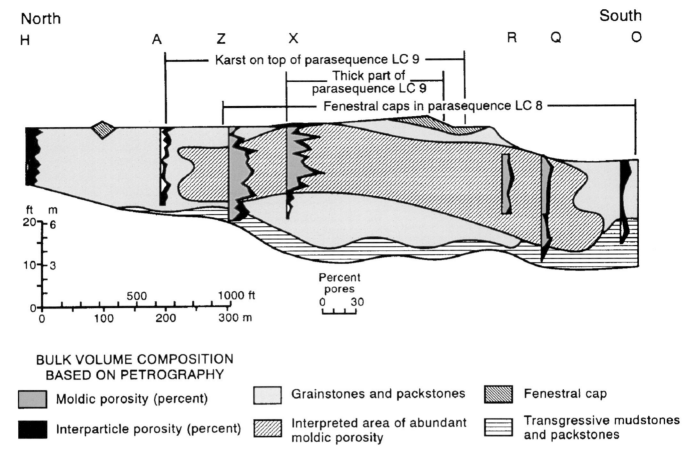

North   South

H   A   Z   X   R   Q   O

Figure 6. Facies distribution in LC 7. The geographic control of the distribution of total moldic porosity in parasequence 7 is apparent in the cross section along the Algerita Escarpment. Molds in LC 7 are abundant beneath the bar crest (defined by the thickest grainstone, fenestral caps, and karst) of LC 7, LC 8, and LC 9.

recognized in all grainstones. Variable amounts of finely crystalline (10–20 μm) equant intergranular dolomite cement postdate the rim cement in areas with preserved grainstone fabrics or molds. These two generations of cement cannot be consistently separated and therefore were qualitatively described but not separated during point counting. Neither generation of cement is syntaxial on echinoderms. Micrite rims or other features that would have prohibited syntaxial overgrowths were not identified.

A positive correlation is observed between leached grains and underpacking associated with abundant cement (Figure 9a). In intervals with the most abundant molds, only 40–60% of the bulk volume of the rock is composed of grains and grain molds. The remaining volume is dolomite spar (cement plus possible coarsely recrystallized grains and micrite), intergranular pores, and variable amounts of large vugs and calcite representing late diagenetic alteration. The increased cement is petrographically identified as equant intergranular dolomite spar precipitated after the rim cement. Increased cement in areas with abundant molds corresponds to decreased intergranular porosity (Figure 9b).

High cement volumes (more than 45%) observed in LC 7 grainstones may be partly the result of point-counting errors caused by inability to identify altered grains or carbonate-mud matrix. Porosity of wet well-sorted sand is experimentally determined as about 45% and varies only slightly with sorting and grain size (Beard and Weyl, 1973). Inspection of spacing between preserved grains in moldic rocks (Figure 8c), however, supports the observation based on point-counting that moldic rocks are characterized by abundant intergranular cement.

Deformation of peloids is common in LC 7 grainstones. Deformed concentric banding within ooids indicates that compactional strain has penetrated through the grain instead of being restricted to the grain margins. The 10- to 15-μm rim cement is not thinned or removed in areas where deformed grains contact each other, as might be expected in cases of simple chemical compaction (Figure 8d). The result is closely packed grains having little or no intergranular porosity, each grain conforming to the shapes of its neighbors but separated from them by a 10- to 15-μm rim of dolomite cement. Thin cement rims are rarely broken by compaction. Grains are typically flattened

(a)

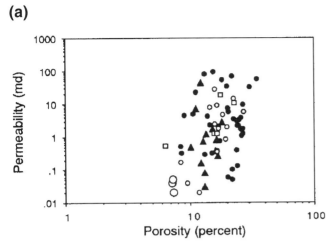

- ● Grainstone
- ▲ Compressed grainstone and packstone
- ○ Grain-dominated packstone
- □ Packstone and wackestone
- ◯ Mudstone

(b)

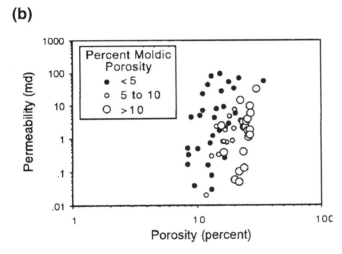

Figure 7. (a) Mudstones having lower porosity and permeability than those of other rocks, however, no clear relationship is seen between depositional fabric and porosity, indicating that depositional porosity and permeability distribution has been obscured by diagenetic overprints. (b) Porosity plotted against permeability shows offset trends for samples with abundant molds and those with few molds. Percent molds is in volume percent on the basis of point counting; porosity and permeability data are from plug analysis. Molds are identifiable leached grains, distinct from larger vugs of various origins.

to a third of their presumed original spherical shapes, but can be flattened to 10-μm-thick wisps. On a macroscopic scale, deformed layers are transitional to nondeformed layers, supporting the assumption that the deformed grains and spherical grains were initially similar. Micritic algal grains are most susceptible to deformation; skeletal grains such as mollusks are not deformed, and peloids and ooids exhibit an intermediate degree of deformation. Molds of spherical grains are not flattened, although some molds of elongated mollusk fragments are collapsed. In highly moldic rocks, neither molds nor grains are deformed. This style of compaction is generally associated with restricted-shelf or hypersaline grainstones rather than with open-marine intervals (Bebout et al., 1981; Bein and Land, 1982; Fracasso and Hovorka, 1986); however, its origin remains undetermined.

Interpretation of Diagenetic Processes

Porosity and permeability distribution in LC 7 has been altered by diagenesis. Several observations concerning the distribution of abundant molds and comparison of petrographic relationships between those areas and areas with few molds are summarized in Figure 10. Most grainstones in LC 7 have an early generation of 10-μm dolomite rim cement (Figure 10a). These rims are similar in thickness and abundance in deformed grainstones (Figure 10d), undeformed grainstones, and moldic grainstones (Figure 10b, c) and, therefore, are presumed to have formed contemporaneously throughout LC 7. The thickness and distribution of the 10-μm rim cement are unaffected by grain deformation; therefore, its precipitation predates compaction and grain dissolution.

Samples with abundant moldic porosity have the most open intergranular packing, reflected in the abundance of intergranular cement (Figure 8c). Leached grains correspond to loose packing and abundant intergranular dolomite cement, which suggests a genetic link between leaching and cementation. Intergranular cement could have been locally derived from dissolution of metastable aragonite and high-magnesium-calcite grains, now molds. Local derivation of intergranular cement might indicate dissolution and precipitation in a hydrologic system that was sluggish or partly closed, similar to the hydrology observed beneath modern ooid sand islands in the Bahamas (Budd, 1988; Vacher et al., 1990). The abundance of intergranular cement in highly moldic rocks compared with that of sparsely moldic rocks indicates cementation before compaction and dates grain leaching to the shallow-burial environment.

Grains of several different initial mineralogies were dissolved, indicating that pore water was significantly undersaturated with respect to aragonite, low-magnesium calcite, and high-magnesium calcite (Walter, 1985). This undersaturated water could have been formed under a variety of conditions, including influx of meteoric water accompanying sea-level drop

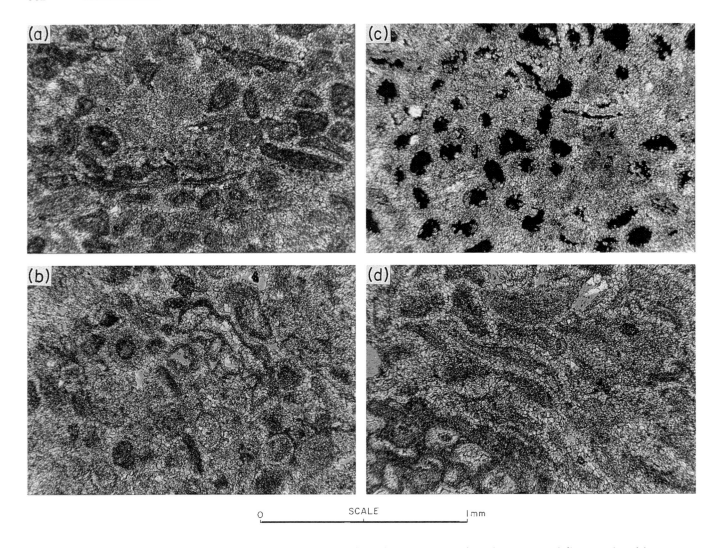

Figure 8. (a) Typical parasequence 7 grainstone with abundant preserved grains, normal fine-grained inter-granular dolomite cement, and some moldic pores. Sample O18, transmitted plane light. (b) Variable grain preservation; some grains are now 5- to 15-µm dolomite, others are 20- to 40-µm dolomite. Sample A93, trans-mitted plane light. (c) Abundant moldic porosity and associated abundant intergranular cement. Sample X 90, crossed nicols. (d) Distinctive deformed and overcompacted peloid grainstones are diagenetically formed low-permeability rocks. Note that the isopachous cement rims most grains even where they are compressed against one another. Photomicrograph of sample X 85, transmitted plane light.

and exposure and locally increased P_{CO_2}. Absence of syntaxial rims on echinoderms composed of calcite with no apparent micrite rims may indicate initial precipitation of noncalcitic carbonate (dolomite or aragonite) cement.

The zone dominated by moldic porosity is found in the thickest parts of LC 7 (Figures 5, 6). The distrib-ution of moldic porosity is coincident with the distri-bution of thin fenestral caps in LC 7 and LC 8, documenting episodes of intertidal deposition and potential exposure. The distribution of moldic porosi-ty in LC 7 also underlies the thick southern part of LC 9 where a karst surface documents sea-level drop and prolonged exposure. Evidence of potential exposure at the end of LC 7, LC 8, and LC 9 in the thick topo-graphically high bar crests suggests that freshwater

lenses may have developed. Coincidence between the distribution of abundant moldic pores with bar crests and the distribution of exposure surfaces capping parasequences LC 7, LC 8, and LC 9 suggests that moldic porosity may have originated in the fresh-water lens associated with one or more of these expo-sure events.

Combining these observations, we propose the fol-lowing diagenetic history (Figures 10, 11). (1) Precipitation of early 10-µm rim cements (mineralogy not determined) occurred in all parts of LC 7, proba-bly in the marine environment, because of their uni-form distribution. (2) The thickest and paleo-topographically highest parts of LC 7 were locally exposed. Exposure may have occurred after deposi-tion of LC 7 or LC 8 fenestral caps, or during karstifi-

(a)

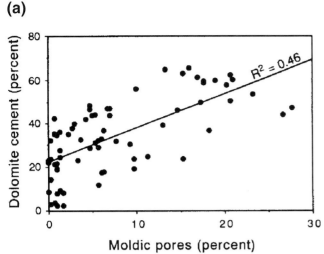

Altered samples (>10 percent calcite or large vugs)
and mud-dominated samples eliminated

(b)

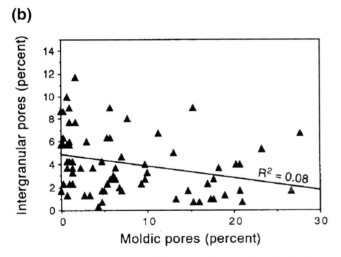

Altered samples and mud-dominated samples eliminated

Figure 9. (a) The percent dolomite cement (rim cement and intergranular cement not separated) increases directly in relation to the percent molds. Wide scatter of data reflects other variables that contribute to percent cement, such as compaction, amount of calcite and leached void space, and amount of rim cement. (b) Correspondingly, the percent intergranular pore space (based on point count) has a weak inverse relationship to percent of molds. Large scatter reflects other factors contributing to development of intergranular pore space, such as compaction and amount of leached void space and calcite.

cation after deposition of LC 9. (3) An undersaturated meteoric or mixed nonmarine-meteoric phreatic lens developed, causing local dissolution of grains and precipitation of intergranular cement in a sluggish or partly closed system. Analogous moldic porosity and early intergranular cement are observed in freshwater

lenses beneath modern ooid sand islands in the Bahamas (Budd, 1988; Vacher et al., 1990). Aragonite and high-magnesium calcite allochems dissolve, and equant calcite cement precipitates, in the freshwater lens in the very shallow burial environment.

GRAYBURG FORMATION, STONE CANYON PARASEQUENCE 23 (SC 23)

Facies

The Grayburg Formation at Stone Canyon was deposited in an inner shelf-crest setting located paleogeographically upslope of environments responsible for deposition of parasequences similar to LC 7. Stone Canyon parasequences, in general, and particularly SC 23 contain more siliciclastics and show greater tidal-flat development than Lawyer Canyon parasequences, including LC 7. Deposition of SC 22 terminated with tidal-flat development and a probable sea-level fall of several meters, enough to permit subarkosic sands transported from the northwest to bypass the Stone Canyon area to depositional locations basinward to the southeast. With a sea-level rise, siliciclastics were trapped and worked by marine processes progressively upslope until deposition commenced in the Stone Canyon area with sandstone deposition as the base of SC 23a. With continued relative sea-level rise, high-energy carbonate ooid-peloid grainstone bars developed, first including a siliciclastic component then becoming siliciclastic free. The maximum-flooding surface for SC 23 probably occurs in the vicinity of the transition from quartzose peloid packstone to ooid-peloid grainstone. Eventually a relative drop in sea level occurred, followed by a rise during which a thin transgressive skeletal grainstone/packstone was deposited as the base of SC 23b, prior to development of ooid-peloid bars. Tidal flats developed at the top of SC 23b prior to another relative sea-level drop, at which time an erosional unconformity (probable parasequence-set-scale sequence boundary) developed, creating a channel-like feature and removing most of the SC 23 tidal-flat deposits.

Parasequence SC 23 ranges from 2.75–4.9 m thick (Figure 12). The uneven thickness of SC 23 reflects depositional topography and erosion during sea-level lowstand preceding SC 24 deposition. The thickest sandstone interval found in the highstand systems tract at Stone Canyon (approximately 5.2 m) occurs in the base of SC 24 just above the most truncated part of SC 23 and records infilling of topography by SC 24 sand in a southeast-trending channel during SC 24 transgression. This southeast trend lies along approximate paleoslope.

Rock types identified from SC 23 compose marine depositional facies, all of which have been dolomitized. Basal massive to cross-bedded sandstone (well-sorted, fine-grained, dolomitic feldspar-moldic quartz sandstone) was deposited as a sheet during transgression (Figure 12a). SC 23 sandstone becomes more dolomitic at both the west and east ends of the study

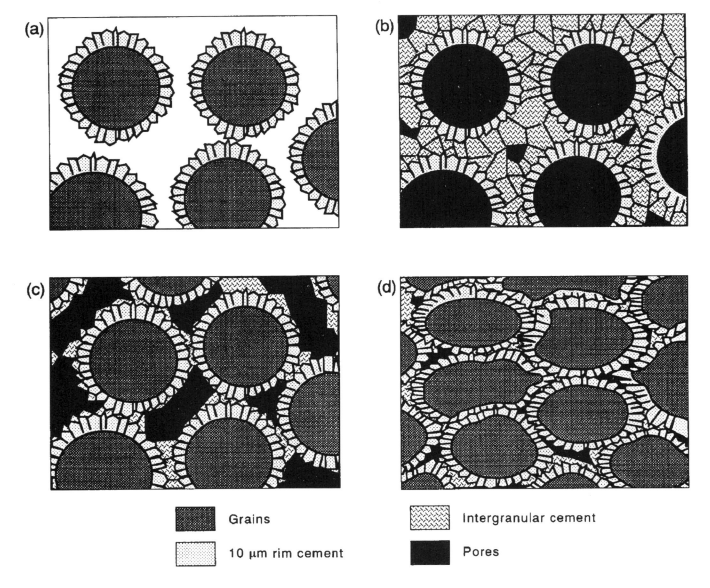

	Grains		Intergranular cement
	10 µm rim cement		Pores

Figure 10. Three end-member fabrics formed after (a) precipitation of initial 10-µm cement rims; (b) most grains were leached and abundant intergranular cement precipitated; (c) minor intergranular cement precipitated around preserved grains, allowing some compaction but preserving most of the intergranular porosity; and (d) compaction of grains, resulting in internal deformation and loss of most intergranular pore space.

area within a transition to more peloid-rich facies. SC 23 sandstone is overlain by massive moderately sorted fine- to medium-grained quartzose peloid dolopackstone. Sand content decreases upward in this facies and reflects a transition from siliciclastic- to carbonate-dominated deposition. Decrease of siliciclastic influx resulted from trapping of terrigenous sediments progressively farther up depositional slope during transgression.

Relatively pure dolostones deposited on the bar crest and flanks immediately overlie quartzose dolopackstone facies in SC 23. These dolostones comprise three depositional facies, one of which includes two diagenetic subfacies. First, massive to locally cross-bedded ooid dolograinstone (well-sorted medium- to coarse-grained ooid-peloid dolograinstone) deposited on or near a bar crest is divided into one diagenetic subfacies with relatively well-preserved intergranular porosity (Figure 14), and another with chemical compaction and wholesale pore occlusion by dolomite cement (Figure 14). Second, fine-grained dolopackstone/grainstone (fine-grained peloid dolopackstone/grainstone) represents bar-flank facies more distal than coarser grained dolograinstones. Third, skeletal dolograinstone/packstone (skeletal peloid dolograinstone/packstone) bears mollusk, brachiopod, and pelmatozoan fragments deposited with abundant peloids during marine

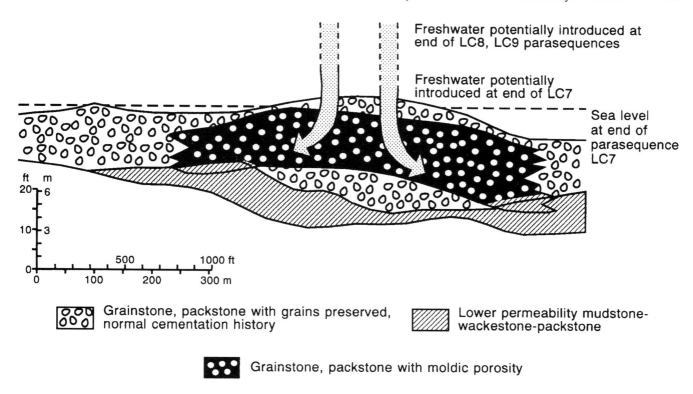

Figure 11. Moldic porosity formed in a freshwater lens during relative sea-level lowstand that occurred at the end of parasequence or sequence deposition.

transgression. The skeletal dolopackstone and the fine-grained dolopackstone/grainstone facies onlap bar crests that have relief of about 1 m each (Figure 12a). The occurrence within SC 23 of a transgressive (fossiliferous) facies suggests that SC 23 consists of two cycles. The lower cycle (SC 23a) has a sandstone base, whereas the upper cycle (SC 23b) has a carbonate (skeletal dolopackstone or dolograinstone) base.

Porosity and Permeability Distribution

Samples from SC 23 are characterized by a narrow range of low porosity values (less than 14%, according to core-plug analyses) and by low permeability values (with two exceptions, less than 2.3 md). Within the pure dolostones, the porosity values range from 1–6%, with a few exceptions. Quartz-bearing dolopackstones have up to 9% porosity. Three porosity groups are identified and generally correspond to three broadly defined facies (Figure 13)—dolomitic sandstone (greater than 9% porosity); medium- to coarse-grained, fossiliferous to nonfossiliferous, quartzose to nonquartzose ooid-peloid dolopackstone/grainstones (2–9% porosity); and fine dolopackstone/grainstone (less than 2% porosity).

Better permeabilities (k >0.1 md) are preferentially distributed in the sandstones and in the ooid dolograinstones, including fossil-bearing examples (Figure 12). Among the dolograinstones, the cross-laminated and fossiliferous examples have the higher

permeability values. Sandstones and dolograinstones show good correlation between permeability and porosity, whereas dolopackstones show little relationship between these parameters. However, of special interest are five samples with permeabilities >0.1 md that define a horizontal zone of above-average permeability (sections 7, 7A, 7B, and 7C; Figures 12b, 14c). These samples have permeabilities one to two orders of magnitude higher than the samples immediately above or below them. The horizontal zone of above-average permeability is confined to the paleotopographically high features and does not cross into the onlapping transgressive skeletal packstone, suggesting that its formation predated the transgression that led to deposition of SC 23b. A zone of relatively well-preserved intergranular porosity (average = 4.4% porosity) includes and underlies the interpreted stratiform high-permeability zone and is overlain and flanked by zones with prevalent chemical compaction fabrics and associated decreased porosity (average = 2.3% porosity).

Petrographic Relationships

Cementation history of Grayburg rocks in Stone Canyon includes well-developed isopachous rims that may have been syndepositionally precipitated in a marine setting or in a meteoric phreatic environment. Other cements are blocky and distributed irregularly around grains and at pore throats. Pen-

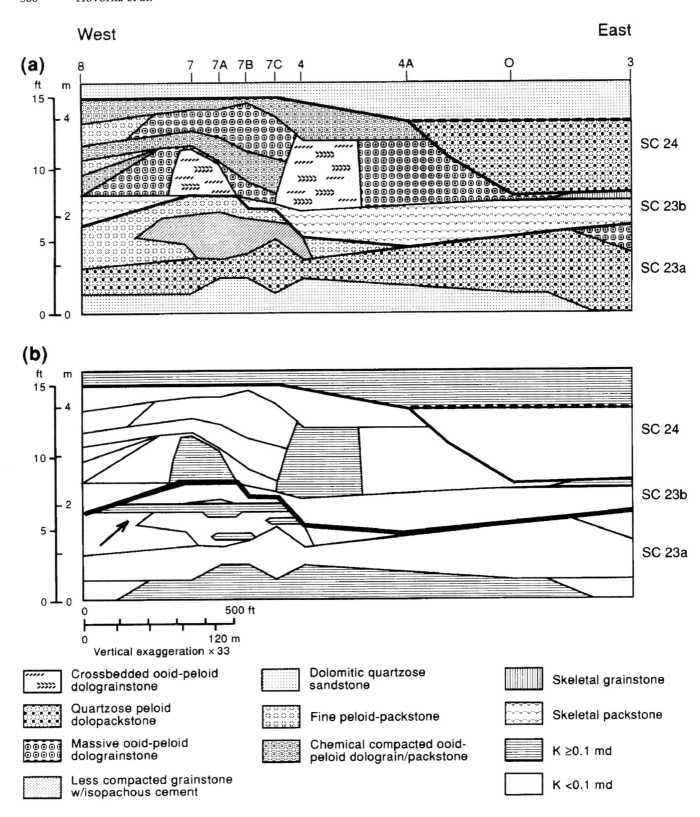

Figure 12. West–east cross section of SC 23, sections 8–3. Shown are (a) locations of sample sections, facies, and parasequence boundaries and (b) distribution of permeability values greater than 0.1 md. Stratiform permeability zone (arrow) in bar crest of cycle SC 23a is interpreted from samples recovered at sections 7, 7A, 7B, and 7C. Polygons in (b) are facies boundaries taken from (a). Section 4A used for total thickness measurement only. Heavy lines are parasequence boundaries, the upper of which may be an inter-Grayburg sequence boundary.

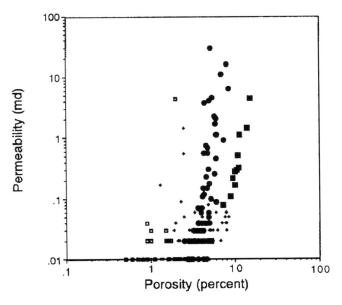

* Ooid-peloid dolopackstone
■ Dolomitic quartz sandstone
● Ooid-peloid and fossil dolograinstone
□ Fine-grained peloid dolopackstone

Figure 13. Porosity/permeability cross plot for samples recovered from SC 23. Three general facies types are reflected in three overlapping porosity ranges: fine-grained peloid dolopackstone (≤2% porosity); dolograinstone/packstone (2–9% porosity); and dolomitic quartz sandstone (>9% porosity). Sandstone and dolograinstone show good correlation between porosity and permeability; dolopackstones show poor correspondence between these parameters and fine-grained dolopackstones show, with one exception, permeabilities clustered near the lower limit of measurability (0.01 md).

dulous (microstalactitic) cements occur locally. Patchy poikilotopic blocky cements encompass numerous ooids and peloids and totally occlude intergranular pore space. Dolomitization provided additional cement as well as transformed calcitic species already present.

Partial molds after ooids (Figure 14e) within or a few inches above the stratiform permeability zone indicate selective dissolution of more soluble cortices of the ooids (ooid cores remain in geopetal positions). Generally, molds in Grayburg samples are filled with dolomite spar, although some are open pores. Pendulous (microstalactitic) dolomite cements (Figure 14f) are also present above the stratiform permeability zone, indicating precipitation in a vadose environment. Below the permeability zone, many of the ooids and peloids have isopachous rimming cements (Figure 14a). Significantly less isopachous cement is found in samples above this zone (Figure

14b); there, blocky dolomite cement or pore space is present instead. Patches of blocky cement that encompass numerous grains and occlude intergranular pores postdate the isopachous rim cements and are somewhat more abundant in rocks overlying the stratiform permeability zone than in those below it. Ooid and peloid grains are significantly more chemically compacted above the stratiform permeability zone and primary pore space that remained was filled with blocky cement; the cement was probably locally derived from pressure solution of grains. Some ooid cortices are well preserved (Figure 14b, d), indicating microscale (neomorphic) alteration of the original calcitic species. Neomorphic alteration is more common in SC 23 oolite above the permeability zone than below it.

Interpretation of Diagenetic Processes

This zone of above-average stratiform permeability is interpreted as the result of meteoric diagenesis within a floating meteoric (Ghyben–Herzberg) lens established on an island during emergence, perhaps associated with a parasequence-scale sea-level lowstand. A stable lowstand, rather than intermittent exposure of the bar crest, is suggested by the thin dimension of the stratiform permeability zone; it would have been fortuitous for intermittent exposure to have repeatedly established a freshwater table within centimeters of the same position. The interpreted permeability zone is positioned across the paleotopographically high bar crest in SC 23a in a plausible location for a paleometeoric lens. Petrographic fabrics in SC 23 are analogous to modern and ancient features interpreted as diagnostic of meteoric diagenesis within a floating freshwater lens (Figure 15). Diagenesis associated with a meteoric water table may have chemically stabilized parts of the facies tract and caused dissolution of relatively soluble carbonate components within a thin (~0.3-m thick) zone. Such diagenesis could explain the systematic fabric distribution around an interpreted thin zone of laterally continuous above-average permeability. Oomoldic fabrics occur within the zone, isopachous cements and preferentially preserved intergranular porosity occur within and below it, and neomorphically transformed grains, microstalactitic cements, and chemical compaction fabrics occur above it. Waters of the uppermost meteoric phreatic zone (beneath but proximal to the water table) are generally the most chemically aggressive because they are charged with carbon dioxide acquired from the atmosphere (which extends down through the vadose zone) and soils (Moore, 1989) and are likely to be undersaturated with respect to aragonite, Mg-calcite, and calcite. The SC 23 meteoric water table may have been established during a sea-level lowstand that occurred after SC 23a bar-crest deposition but prior to deposition of the immediately overlying SC 23b bar; alternatively, the sea-level lowstand may

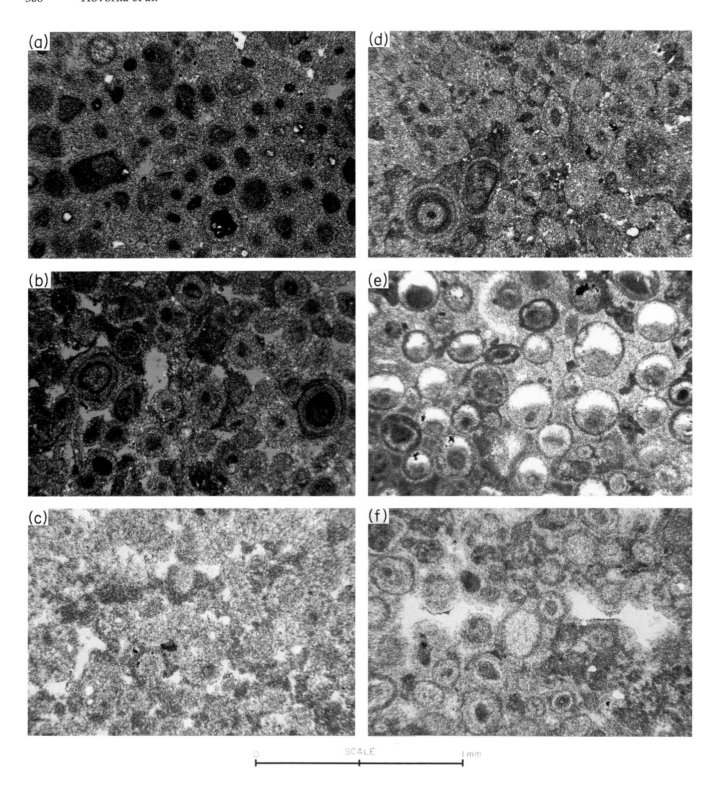

Figure 14. (a) Ooid-peloid dolograinstone with intergranular porosity reduced mainly by isopachous rimming cements. Sample shows relatively little chemical compaction (SC 23a, sample 7C-10, 4.5% porosity, 0.74 md). (b) Ooid-peloid dolograinstone with well-preserved intergranular porosity, blocky cements irregularly distributed on grain surfaces and concentrated at grain contacts, and neomorphic preservation of ooid cortices. Note penetrative intersection of grains caused by chemical compaction from cross-bedded bar crest (SC 23b, sample 4-18, 6.9% porosity, 11.22 md). (c) Ooid(?)-peloid dolograinstone from stratiform permeability zone SC 23, sample 7A-12, 8.0% porosity, 16.12 md). (d) Cement-occluded ooid-peloid dolograinstone/packstone that is

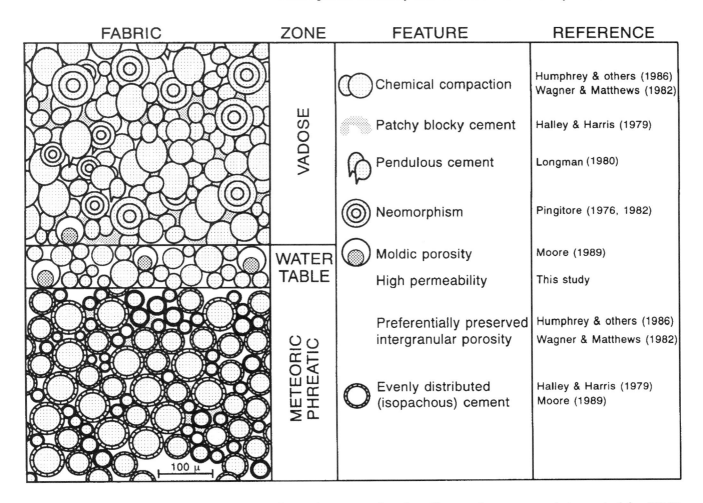

Figure 15. Schematic diagram showing pertinent features related to diagenetic processes interpreted for SC 23 and published modern or ancient analogs used for interpretations. White areas represent pores.

be related to the erosional surface at the top of SC 23 and textural or compositional properties of the onlapping skeletal dolograinstone/packstone precluded development of enhanced effective porosity.

In a modern Bahamian ooid shoal investigated by Halley and Harris (1979), meteoric diagenetic cements were found to be more evenly distributed around ooid grains below the meteoric water table (meteoric phreatic zone) than above it (vadose zone), and the overall appearance of cementation was more patchy in oolite residing above the meteoric water table than below it. These relationships reflect conditions in which grains in the meteoric phreatic zone are completely submerged in diagenetic fluids,

whereas grains in the generally unsaturated vadose zone are bathed in transient fluids that are nonuniformly distributed in the capillary fringe.

Neomorphic processes, such as those suggested for ooid cortical preservation above the permeability zone in SC 23, are more common in the vadose zone than in the meteoric or marine phreatic zone (Pingitore, 1976). In the vadose zone, chemical equilibrium is maintained between carbonate species and slow-moving or stationary capillary water (Pingitore, 1982). In the meteoric phreatic zone, chemically aggressive fluids tend to move more rapidly through sediment and are less likely to achieve chemical equilibrium. Macroscale dissolution results when soluble components are dissolved completely, destroying fine

◄

chemically compacted with neomorphic preservation of ooid (SC 23b, sample 7A-16, 3.3% porosity, 0.02 md). (e) Ooid-peloid dolograinstone, with partial molds after ooids and geopetal fabric, immediately overlies stratiform permeability zone (SC 23, sample 7–13, 2.1% porosity, 0.04 md). (f) Ooid-peloid grainstone with keystone voids hosting microstalactitic cements and geopetal fill (SC 23, sample 7-12, 2.2% porosity, 0.59 md).

details of the original grain and leaving a mold (Moore, 1989). The partial molds after ooids near and within the SC 23 permeability zone, some of which are filled with spar, illustrate this process.

Microstalatitic cements result from preferential precipitation of calcite on the undersides of particles in the vadose zone. These textures have been reported by Jacka and Brand (1977) from the Edwards Reef of Texas (Lower Cretaceous), by Purser (1969) from Middle Jurassic limestones of the Paris Basin, and summarized by Longman (1980).

Other parasequences identified in Stone Canyon have features suggestive of meteoric diagenesis similar to that interpreted for SC 23. Oomoldic grainstone was sampled from SC 26 just beneath tidal-flat facies at section 8, and skeletal-moldic grain/packstone was recovered from SC 25, also just beneath tidal-flat facies at section 3. The SC 25 sample was recovered from a 50-cm-thick zone that is laterally continuous for at least several meters along the outcrop before it disappears behind desert vegetation. This sample also has well-developed microstalactitic cements indicative of vadose cementation and has the highest permeability (59 md) of any sample recovered from Stone Canyon.

In Jurassic Smackover oolites of Louisiana (Humphrey et al., 1986) and Arkansas (Wagner and Matthews, 1982), porous relatively uncompacted oolites were overlain and underlain by chemically compacted oolites. Heterogeneous porosity preservation was interpreted to result from preferential early calcitic mineral stabilization by meteoric-phreatic diagenesis of the relatively uncompacted interval. Because pressure-solution proceeds more readily in metastable carbonate sediments, Smackover oolites in the vadose- and marine-phreatic zones were preferentially compacted and cemented.

It is important to consider whether the island that resulted from emergence of the SC 23a bar crest was sufficiently large to develop a meteoric lens thick enough to effect the distribution of fabrics observed. Budd and Vacher (1991) developed criteria from which the maximum thickness of a floating meteoric lens beneath an island could be calculated. Although facies types and diagenetically modified hydraulic paleo-conductivity affect the calculations, the authors established that maximum lens thickness varied mainly between 1 and 2% of the island width. Assuming an island width for SC 23a of about 183 m (the distance between points of interpreted intersection of the stratiform permeability zone with the bar flank; Figure 12b), meteoric lens thickness would have been a maximum of between 1.8 and 3.6 m beneath the bar crest. A meteoric lens this thick would extend at least into the sandstone at the base of SC 23 and, thus, could account for the preserved intergranular porosity in the grainstones below the permeability zone. According to the Ghyben–Herzberg relation, the ratio of lens thickness above sea level to lens thickness below sea level is about 1:40 (Ghyben [1888] and Herzberg [1901] both cited in Freeze and Cherry, 1979). Given a maximum lens

thickness of 1.8–3.6 m, the water table height would be 46–91 mm above sea level at the island center, or essentially flat and barely above sea level.

Given 0.5 m between the base of the stratiform permeability zone and the bar crest, a minimum of 0.5 m of sea-level fall is required at the end of SC 23a deposition. There is no tidal-flat facies on the SC 23a bar crest, however, so sea level was probably one or more meters above the bar crest prior to the fall preceding establishment of a meteoric lens. A sea-level fall of at least 1 m is probably more reasonable. Establishment of the meteoric lens at this position was probably rapid and short lived compared to rates of diagenetic alteration. This is suggested by the thin geometry of the anomalous permeability zone and by the lack of abundant grain dissolution. A stillstand at lowered sea level probably produced the zone of stratiform permeability because a much broader zone would have been affected if water-table related processes operated during the entire period of SC 23a emergence and resubmergence. For similar reasons, periodic exposure (at higher than parasequence-scale frequencies) of the bar crest would probably produce less well-defined diagenetic zones unless base level always returned to the same position relative to the stratiform permeability zone.

An alternatively less-favored process for preservation of the zone of stratiform permeability in SC 23 is leaching of anhydrite cement. A combination of pore filling and replacement by sulfate followed by sulfate dissolution would have both preserved and enhanced original depositional porosity. In this model, a back-barrier lagoon with an attendant hypersaline water table residing within the emergent bar crest would be established. Precipitation of evaporite minerals within primary depositional pores caused by evaporation and salt concentration along the water table could have protected them against the occlusion by carbonate cements experienced by overlying and underlying rocks. Dissolution of a few unstable ooid cortices could reflect dissolution caused by a thin ephemeral mixing zone of saline and meteoric water or replacement by evaporites. Dissolution of the evaporites would have occurred most likely during Tertiary basin-and-range uplift but would probably survive in the subsurface if not exposed to sulfate-undersaturated water. Arguing against this hypothesis, however, is the general lack of evidence for abundant previously occurring replacive sulfate (multigrain-size pores and truncated grain boundaries); such features were identified in only one of almost 275 samples analyzed from Stone Canyon, including samples collected from eight parasequences other than SC 23.

A comparison of SC 23 (Grayburg Formation) with LC 7 (San Andres Formation) demonstrates the variable effects of meteoric diagenesis on similar sediments (e.g., grainstone). In the San Andres example, a thick laterally restricted interval has been modified by dissolution of ooids and peloids. This resulted in a lens of mold-dominated porosity encased in rocks with mostly intergranular porosity. Effective porosity is intergranular in both the moldic and nonmoldic portions of the parasequence and there is no significant difference in permeability between the two.

In Grayburg SC 23, porosity is almost all intergranular; however, more primary porosity has been preserved in portions of the parasequence altered by diagenesis in a freshwater lens. Primary porosity was reduced in sediment above the freshwater lens. Minor moldic porosity only occurs in a thin zone of above-average permeability that developed proximal to the upper surface of the lens (water table). The result is a lens of rock with intergranular porosity and a thin, stratiform, above-average permeability zone at the top encased in chemically compacted rocks with reduced porosity and permeability.

Differences between these examples are probably related to different paleogeographic positions for the two parasequences. Grayburg SC 23 was deposited on the restricted inner shelf crest and, except for a thin transgressive bed, is faunally barren. It has well-developed tidal flats comprising abundant pisoids, fenestral laminations, and tepee-structures. The thin horizons of enhanced permeability and sparse molds were produced during a short-lived water table still-stand at an estimated relative sea-level reduction of 1 m. Dolomite crystal sizes are generally finer than in LC 7. San Andres LC 7, in contrast, was deposited on a ramp crest and has included marine fauna, but only minor tidal flat development. In LC 7, poorer development of supratidal fabrics may indicate lower salinity in the depositional environment. LC 7, therefore, may have been dolomitized later or, because of higher preserved porosity, been subjected to multiple episodes of dolomitization resulting in a coarser dolomite crystal size.

SUBSURFACE ANALOGS AND APPLICATIONS

Diagenetic features formed at the time of interpreted sea-level fall at the end of both parasequences described here modify the distribution of depositionally controlled porosity and permeability. The resulting distribution of permeable units and barriers to flow control the pathways available for flow of subsequent fluids. The moldic zone in LC 7 affects flow in reservoir simulations (Senger et al., 1991) by reducing permeability in what would otherwise have the potential to be a thick zone of high permeability. In reservoir simulations, the moldic part of LC 7 is poorly swept during waterflood.

The effect of variable cementation, compaction, and mold development on reservoir quality and geometry has been documented in carbonate reservoirs of a wide variety of ages and geologic settings. A few of the many reservoir examples in which relative sea-level–controlled diagenetic modifications are key elements in reservoir quality are the Ordovician of the Williston Basin (Reeckmann and Friedman, 1982), the Pennsylvanian of Kansas (Watney and French, 1988), the Miocene of the West Java Sea,

(Ardila and Kuswinda, 1984), and the Tertiary of Libya (Bebout and Pendexter, 1975). The impact of development of various types of pores on permeability distribution is closely tied to relative sea level (Choquette and Pray, 1970; Friedman, 1975; Lucia, 1983; Moore, 1989). Integrating sequence stratigraphic models with the patterns of diagenetic modification allows better understanding of the spatial and temporal controls on positive and negative influences on permeability, therefore improving the predictability of reservoir quality.

Lenses of intergranular porosity preserved in Smackover (Jurassic) grainstones by early meteoric phreatic diagenesis prior to burial have been interpreted for the producing zones in Oaks Field, Louisiana (Humphrey et al., 1986) and Walker Creek field, Arkansas (Wagner and Matthews, 1982). In both fields, individual producing zones reside within shoaling cycles consisting of thin siliciclastic-bearing grainstones and ooid-grainstones. These are similar to the siliciclastic-carbonate couplets that compose much of the Grayburg section at Stone Canyon. In Oaks and Walker Creek fields, reservoir seals are composed of tight grainstones that have been interpreted to be preferentially compacted upon burial because aragonite and Mg-calcite had not been altered to calcite by meteoric diagenesis.

The stratiform zones of diagenetically influenced permeability, with preferential preservation of intergranular porosity within and beneath, as observed in the Grayburg Formation at Stone Canyon, also has implications for production from hydrocarbon reservoirs. Such a zone, if it connects wells, could act as a "thief" zone allowing injected fluids to flow along one thin zone bypassing petroliferous compartments with lower permeability. An underlying compartment with well-preserved intergranular porosity but permeability an order of magnitude lower, such as is present in SC 23, may well contain significant hydrocarbon resources but would remain uncontacted by water or CO_2 floods. Thin zones of enhanced permeability would probably be missed in analyses of plugs taken from cores unless sampling was conducted at sufficiently close vertical spacing.

Intensity of sampling may be the key to identifying thin zones that exhibit distinct petrophysical characteristics resulting from early meteoric diagenesis. In ongoing reservoir analyses involving core analysis (e.g., North Foster Grayburg on the northern Central Basin Platform of Texas; Tyler et al., 1992), facies identifications are tested by thin section examination, but seldom at the close spacing used in these outcrop investigations. In North Foster, carbonate strata occur in shallow-water shoaling cycles similar to those studied at Lawyer Canyon and Stone Canyon. Various diagenetic fabrics, including moldic porosity, isopachous rimming cement, neomorphic preservation, and probable meniscus cements are nonuniformly distributed vertically; however, no individual cycle identified in core was sampled intensely enough to identify intracycle diagenetic patterns. Future reservoir analyses, when appropriate, may well include more intense sampling to detect such reservoir complications.

The absolute values of porosity and permeability observed in outcrop must be used with care in application to subsurface reservoirs because porosity and permeability of outcrop samples have been modified to an undetermined extent by outcrop alteration. In particular, the absence of the sulfates that characteristically occlude pores in many Permian reservoirs must be considered. However, the relationship between parasequences and porosity-modifying process observed in outcrop should be applicable to similar environments in the subsurface.

CONCLUSIONS

Distribution of porosity and permeability in carbonate rocks deposited in ramp-crest settings results from combined depositional and diagenetic processes. High-energy, shallow-water, ramp-crest deposits are dominantly grain-rich facies with abundant primary intergranular porosity. Early diagenesis associated with establishment of meteoric lenses during high-frequency sea-level lowstands modifies primary porosity. Grains in emergent bar crests and tidal-flat facies are altered by meteoric waters that nonuniformly percolate through the vadose zone. A complex nonuniform distribution of chemically reactive meteoric, mixed marine-meteoric, and hypersaline fluids causes variable grain preservation and neomorphism, grain dissolution, cementation, and mineralogic stabilization. Chemical and hydrologic variation between the uppermost meteoric phreatic zone (top of the water table) and lower meteoric phreatic zone results in variable grain preservation, stabilization, and cementation histories. The geometries of these diagenetic zones are predictable by mapping the distribution of emergent facies at the parasequence scale.

Early diagenetic alterations that redistribute porosity and permeability in inner shelf carbonate rocks are parasequence specific. Porosity and permeability characteristics change most rapidly where meteoric lenses have been established (i.e., on islands or coastal regions that emerge during high-frequency lowstands). Additional alterations may occur later in the geologic history of these rocks, but circulation of diagenetic fluids attendant with later events will be profoundly affected by the porosity and permeability distributions established during early diagenesis. Mapping geometry and depositional facies at a parasequence scale is a key to interpretation of permeability trends and has potential for prediction of permeability distribution in the subsurface.

Detailed parasequence-scale mapping of depositional and diagenetic facies patterns in bars in inner shelf-crest and ramp-crest settings allows development of a model that closely parallels that developed for present-day carbonate shelf-edge sand bodies (Halley and Harris, 1979; Budd and Vacher, 1991). The highly heterogeneous depositional and early diagenetic mosaics documented for these parasequences

associated with localized depositional topography and associated freshwater lenses should be considered the signature of the inner shelf-crest and ramp-crest facies tracts. Delineation of these high-energy facies tracts and recognition of this highly heterogeneous depositional and diagenetic style will be particularly significant when attempting to develop depositional and/or reservoir models for these strata.

The following specific observations are made based on examination of facies and diagenetic overprints in two outcrops.

1. Bar emergence occurred in response to high-frequency relative sea-level falls between deposition of parasequences; during emergence, meteoric diagenetic conditions were present within floating meteoric lenses (Ghyben–Herzberg lenses) established in the shallow subsurface of islands.
2. In San Andres LC 7, chemically aggressive diagenetic fluids within a meteoric lens produced a zone of moldic porosity. The dissolved carbonate components were reprecipitated locally and reduced permeability in the moldic rock.
3. Primary depositional porosity was preferentially preserved in Grayburg SC 23 and grainstones mineralogically stabilized by early meteoric phreatic diagenesis during bar emergence, perhaps associated with high-frequency sea-level lowstand episodes. Porosity and permeability in rocks not so stabilized prior to burial have been reduced by chemical compaction and reprecipitation in pores of carbonate mobilized by pressure solution.
4. Thin stratiform zones of above-average permeability were preserved (during burial) through emergent parts of parasequences SC 23 and SC 25 due to meteoric diagenetic grain stabilization and minor leaching along a water table associated with stillstands of lowered sea levels. Alternatively, occlusion of primary depositional porosity by evaporites could have protected pores from carbonate cementation seen in overlying and underlying rocks.
5. Diagenetic modification of primary porosity and permeability occur at the parasequence scale and results from paleohydraulic conditions created by relative sea-level changes coupled with depositional setting. Detection in core of parasequence-scale diagenetic patterns such as pore occlusion by anhydrite, nonuniform preservation of intergranular porosity, or development of thin laterally continuous zones of enhanced permeability may require closely spaced sampling intervals.

ACKNOWLEDGMENTS

This research was funded by sponsors of the San Andres Reservoir Characterization Research Laboratory of the Bureau of Economic Geology and by the U. S. Department of Energy. We thank R. P. Major for early discussion and helpful comments on previous versions of this chapter. Figures were drafted under the supervision of Richard Dillon. We thank Rick Sarg, Kwok-Choi S. Ng, Bob Loucks, and an anonymous reviewer for helpful review comments. Publication is authorized by the Director, Bureau of Economic Geology, The University of Texas at Austin.

REFERENCES CITED

Ardila, L. E., and Kuswinda, I., 1984, The Rama Field: An oil accumulation in Miocene carbonates, West Java Sea, in The Hydrocarbon Occurrence in Carbonate Rocks: Asian Council on Petroleum Technical Paper (ASCOPE) TP/2, p. 341-361.

Beard, D. C., and Weyl, P. K., 1973, Influence of texture on porosity and permeability of unconsolidated sand: AAPG Bulletin, v. 57, p. 349–369.

Bebout, D. G., Budd, D. A., and Schatzinger, R. A., 1981, Depositional and diagenetic history of the Sligo and Hosston Formations (Lower Cretaceous) in South Texas: The University of Texas at Austin, Bureau of Economic Geology Report of Investigations No. 109, 69 p.

Bebout, D. G., and Pendexter, C., 1975, Secondary carbonate porosity as related to early Tertiary depositional facies: AAPG Bulletin, v. 59, p. 665-693.

Bein, A., and Land, L. S., 1982, Carbonate sedimentation and diagenesis associated with Mg-Ca-chloride brines: The Permian San Andres Formation in the Texas Panhandle: Journal of Sedimentary Petrology, v. 53, p. 243–260.

Budd, D. A., 1988, Aragonite-to-calcite transformations during fresh-water diagenesis of carbonates: Insights from pore-water chemistry: Geological Society of America Bulletin, v. 100, p. 1260–1270.

Budd, D. A., and Vacher, H. L., 1991, Predicting the thickness of fresh-water lenses in carbonate paleoislands: Journal of Sedimentary Petrology, v. 61, p. 43–53.

Choquette, P. W., and Pray, L. C., 1970, Geological nomenclature and classification of porosity in secondary carbonates: AAPG Bulletin, v. 54, p. 207-250.

Fekete, T. E., Franseen, E. K., and Pray, L. C., 1986, Deposition and erosion of the Grayburg Formation (Guadalupian, Permian) at the shelf-to-basin margin, western escarpment, Guadalupe Mountains, Texas, in Moore, G. E., and Wilde, G. L., eds., Lower and Middle Guadalupian Facies, Stratigraphy, and Reservoir Geometries, San Andres-Grayburg Formations, Guadalupe Mountains, New Mexico and Texas: SEPM, Permian Basin Section, Publication No. 86-25, p. 69–81.

Fracasso, M. A., and Hovorka, S. D., 1986, Cyclicity in the middle Permian San Andres Formation, Palo Duro Basin, Texas Panhandle: The University of Texas at Austin, Bureau of Economic Geology Report of Investigations No. 156, 48 p.

Franseen, E. K., Fekete, T. E., and Pray, L. C., 1989, Evolution and destruction of a carbonate bank at the shelf margin: Grayburg Formation (Permian), Western Escarpment, Guadalupe Mountains, Texas, in Crevello, P. D., Wilson, J. L., Sarg, J. F., and Read, J. F., eds., Controls on Carbonate Platform Development: SEPM Special Publication No. 44, p. 289-304.

Freeze, R. A., and Cherry, J. A., 1979, Groundwater: Prentice-Hall, Inc., New Jersey, 604 p.

Friedman, G. M., 1975, The making and unmaking of limestones or the downs and ups of porosity: Journal of Sedimentary Petrology, v. 45, p. 379-398.

Halley, R. B., and Harris, P. M., 1979, Fresh-water cementation of a 1,000-year-old oolite: Journal of Sedimentary Petrology, v. 49, p. 969–988.

Humphrey, J. D., Ransom, K. L., and Matthews, R. K., 1986, Early meteoric diagenetic control of Upper Smackover production, Oaks Field, Louisiana: AAPG Bulletin, v. 70, p. 70–85.

Jacka, A. D., and Brand, J. P., 1977, Biofacies and development and differential occlusion of porosity in a Lower Cretaceous (Edwards) reef: Journal of Sedimentary Petrology, v. 47, p. 366-381.

Kerans, C., 1991, Geologic characterization of San Andres reservoirs: Outcrop-analog mapping, Algerita Escarpment, Guadalupe Mountains, and Seminole San Andres unit, Northern Central Basin Platform, in Kerans, C., Lucia, F. J., Senger, R. K., Fogg, G. E., Nance, H. S., Kasap, E., and Hovorka, S. D., Characterization of Reservoir Heterogeneity in Carbonate-ramp Systems, San Andres/Grayburg, Permian Basin: The University of Texas at Austin, Bureau of Economic Geology, Reservoir Characterization Laboratory Final Report, p. 3–46.

Kerans, C., and Nance, H. S., 1991, High-frequency cyclicity and regional depositional patterns of the Grayburg Formation, Guadalupe Mountains, New Mexico, in Meader-Roberts, S., Candezaria, M. P., and Moore, G. E., eds., Sequence Stratigraphy, Facies, and Reservoir Geometries of the San Andres, Grayburg, and Queen Formations, Guadalupe Mountains, New Mexico and Texas: SEPM, Permian Basin Section, Publication 91-32, p. 53–69.

Leary, D. A., and Vogt, J. N., 1990, Diagenesis in the San Andres Formation (Guadalupian) reservoirs, Central Basin Platform, Permian Basin, in Bebout, D. G., and Harris, P. M., eds., Geologic and Engineering Approaches in Evaluation of San Andres/Grayburg Hydrocarbon Reservoirs—Permian Basin: The University of Texas at Austin, Bureau of Economic Geology, p. 21–48.

Longman, M. W., 1980, Carbonate diagenetic textures from nearsurface diagenetic environments: AAPG Bulletin, v. 64, p. 461-487.

Lucia, F. J., 1983, Petrophysical parameters estimated from visual descriptions of carbonate rocks: A field classification of carbonate pore space: Journal of Petroleum Technology, v. 35, p. 629–637.

Lucia, F. J., and Conti, R. D., 1987, Rock fabric, permeability, and log relationships in an upward-shoaling vuggy carbonate sequence: The University of Texas at Austin, Bureau of Economic Geology Geological Circular 87-5, 22 p.

Major, R. P., and Holtz, M. H., 1990, Depositionally controlled reservoir heterogeneity at Jordan Field: Journal of Petroleum Technology, v. 2, p. 1304-1309.

Major, R. P., Vander Stoep, G. W., and Holtz, M. H., 1990, Delineation of unrecovered mobile oil in a mature dolomite reservoir: The east Penwell San Andres unit, University Lands, West Texas: The University of Texas at Austin, Bureau of Economic Geology Report of Investigations No. 194, 51 p.

Moore, C. H., 1989, Developments in Sedimentology 46: Carbonate Diagenesis and Porosity: Elsevier, Amsterdam, 338 p.

Moore, G. E., and Wilde, G. L., eds., 1986, Lower and middle Guadalupian facies, stratigraphy, and reservoir geometries, San Andres–Grayburg Formations, Guadalupe Mountains, New Mexico and Texas: SEPM, Permian Basin Section, Publication No. 86-25, 144 p.

Pingitore, N. E., Jr., 1976, Vadose and phreatic diagenesis: Processes, products and their recognition in corals: Journal of Sedimentary Petrology, v. 46, p. 985–1006.

Pingitore, N. E., Jr., 1982, The role of diffusion during carbonate diagenesis: Journal of Sedimentary Petrology, v. 52, p. 27–39.

Purser, B. H., 1969, Syn-sedimentary marine lithification of Middle Jurassic limestones in the Paris Basin: Sedimentology, v. 12, p. 205-230.

Reeckmann, A., and Friedman, G. M., 1982, Exploration for Carbonate Petroleum Reservoirs: John Wiley & Sons, New York, p. 100-107.

Ruppel, S. C., and Cander, H. S., 1988, Effects of facies and diagenesis on reservoir heterogeneity: Emma San Andres field, West Texas: The University of Texas at Austin, Bureau of Economic Geology Report of Investigations No. 178, 67 p.

Sarg, J. F., and Lehmann, P. J., 1986, Lower-Middle Guadalupian facies and stratigraphy, San Andres-Grayburg formations, Permian Basin, Guadalupe Mountains, New Mexico, in Moore, G. E., and Wilde, G. L., eds., Lower and Middle Guadalupian Facies, Stratigraphy, and Reservoir Geometries, San Andres-Grayburg Formations, Guadalupe Mountains, New Mexico and Texas: SEPM, Permian Basin Section, Publication 86-25, p. 1–36.

Scholle, P. A., Ulmer, D. S., and Melim, L. A., 1992, Late-stage calcites in the Permian Capitan Formation and its equivalents, Delaware Basin margin, west Texas and New Mexico: Evidence for replacement of the precursor evaporites: Sedimentology, v. 39, p. 207-234.

Senger, R. K., Lucia, F. J., Kerans, C., Fogg, G. E., and Ferris, M., 1991, Dominant controls on reservoir-flow behavior in carbonate reservoirs as determined from outcrop studies, in Burchfield, T. E., and Wesson, T. C., co-chairmen, Third International Reservoir Characterization Technical Conference, Conference ITT Research Institute: National Institute for Petroleum and Energy Research and U. S. Department of Energy, v. 1, Section 3RV-58, p. 1–44.

Tyler, N., Barton, M. D., Bebout, D. G., Fisher, R. S., Grigsby, J. D., Guevara, E., Holtz, M., Kerans, C., Nance, H. S., and Levey, R. A., 1992, Memorandum of understanding between the United States Department of Energy and the State of Texas: Characterization of oil and gas reservoir heterogeneity: University of Texas at Austin, Bureau of Economic Geology contract report, contract no. DE-FG22-89BC-14403.

Vacher, H. L., Bengtsson, T. O., and Plummer, N. L., 1990, Hydrology of meteoric diagenesis: Residence time of meteoric ground water in island freshwater lenses with application to aragonite-calcite stabilization rate in Bermuda: Geological Society of America Bulletin, v. 102, p. 223–232.

Van Wagoner, J. C., Posamentier, H. W., Mitchum, R. M., Vail, P. R., Sarg, J. F., Loutit, T. S., and Hardenbol, J., 1988, An overview of the fundamentals of sequence stratigraphy and key definitions, in Wilgus, C. K., Hastings, B. S., Kendall, C. G. St. C., Posamentier, H. W., Ross, C. A., and Van Wagoner, J. C., eds., Sea Level Changes: An Integrated Approach: SEPM Special Publication No. 42, p. 39–46.

Wagner, P. D., and Matthews, R. K., 1982, Porosity preservation in the upper Smackover (Jurassic) carbonate grainstone, Walker Creek field, Arkansas: Response of paleophreatic lenses to burial processes: Journal of Sedimentary Petrology, v. 52, p. 3–18.

Walter, L. M., 1985, Relative reactivity of skeletal carbonate during dissolution: Implications for diagenesis, in Schneidermann, N., and Harris, P. M., eds., Carbonate Cements: SEPM Special Publication No. 36, p. 3–16.

Watney, L. W., and French, J., 1988, Characterization of carbonate reservoirs in the Lansing-Kansas City groups, (Upper Pennsylvanian) in the Victory Field, Haskell County, Kansas, in Goolsby, S. M., and Longman, M. W., Occurrence and Petrophysical Properties of Carbonate Reservoirs in the Rocky Mountain Region: The Rocky Mountain Association of Geologists, Denver, Colorado, p. 27-45.

Chapter 20

Stratigraphic Patterns and Cycle-Related Diagenesis of Upper Yates Formation, Permian, Guadalupe Mountains

Maria Mutti[1]
J. Antonio (Toni) Simo
Department of Geology and Geophysics
University of Wisconsin-Madison
Madison, Wisconsin, U.S.A.

ABSTRACT

Capitan backreef strata of the upper Yates and lowermost Tansill formations are characterized by three cycles composed of a lower siliciclastic unit and an upper carbonate unit. The cycles are bounded by sharp and erosive surfaces on the shelf that become concordant and disappear within concordant outer-shelf margin strata. Breccias are occasionally associated with cycle boundaries. Composite eustatic variations (100-k.y. cycles superimposed on 400-k.y. cycles) control both the deposition and the internal organization of these siliciclastic/carbonate units. The sandstones were transported onto the shelf by eolian processes during sea-level lowstands and then reworked in the outer shelf by marine processes by the following transgression. The upper carbonates, transitional down into sandstones, are interpreted as subtidal to supratidal deposits. Relative sea-level drops subaerially exposed cycle tops, but marine reworking after exposure modified the cycle boundaries. Dolomitized and undolomitized clasts of the underlying rock are reworked locally at the base of cycles.

Syndepositional diagenetic events record several phases of marine cementation, neomorphism and dissolution, replacive dolomitization, and primary dolomite precipitation. The paragenetic constraints on the timing of syndepositional diagenetic events allow for the establishment of a relationship between the stratigraphic cyclic patterns and the early diagenetic features. Most of the marine cementation (Mg-calcite and aragonite) occurs during transgressive phases, as well as replacive dolomitization of the inner-shelf restricted facies. During relative sea-level falls, metastable phases are dissolved and sediments and early marine cements are replaced by dolomite in

[1]Present address: Geologisches Institut, ETH, Zürich, Switzerland.

the shelf-crest and outer-shelf facies, and dolomite cement precipitates in inner-shelf and shelf-crest fenestral pores. Integration of sedimentologic observations and timing of diagenetic events documents three relative sea-level falls. The association of dolomites with evaporite nodules and their enrichment in $\delta^{18}O$ (up to 4.1‰) with respect to normal marine values, suggest that precipitation occurred from evaporation-concentrated seawater.

Integration of sedimentologic and diagenetic studies of Guadalupe Mountains outcrops suggests that composite sea-level variations controlled the deposition of cyclic strata and distribution of syndepositional diagenetic features. Because syndepositional diagenesis records the effects of relative eustatic variations, its distribution can be predicted.

INTRODUCTION

Cyclic carbonate/siliciclastic shelf strata are very common on ancient platforms and recent studies of these have stimulated scientific interest (e.g., Budd and Harris, 1990; Lomando and Harris, 1991). Meter-scale siliciclastic-carbonate cycles are common throughout the geologic record, particularly in the Paleozoic (Van Siclen, 1958; Moore, 1959; Oliver and Cowper, 1963; Wilson, 1972, 1975). Vertical changes in rock types are abrupt and laterally persistent. The base of the cycles are conventionally considered to be limestones with sharp bases and overlain by siliciclastics. The concept of cyclic and reciprocal sedimentation (Van Siclen, 1958; Wilson, 1972) explains carbonate/siliciclastic cycles as a response to sea-level change. Carbonate shelf deposition occurs during sea-level highstands when the basin is sediment-starved, whereas siliciclastic deposition occurs during lowstands. Most of the sandstones bypass the shelf and are deposited in the basin, some sandstones are trapped on the shelf during the following transgression. Cyclic deposition in mixed systems by reciprocal sedimentation has become one of the foundation blocks for sequence stratigraphy (Van Siclen, 1958; Meissner, 1972; Sarg, 1988).

This study concentrates on the uppermost Yates Formation where three sandstone beds (Hairpin sandstone, and lower and upper Triplet sandstone) are interbedded with carbonate backreef strata (Hairpin and Triplet dolomites) (Neese and Schwartz, 1977; Candelaria, 1982). Borer and Harris (1991a,b) interpreted the stacking pattern of siliciclastic/carbonate depositional sequences in the Yates shelf as a response to three orders of orbitally forced, low-amplitude sea-level variations. They interpreted the deposition of the Yates shelf/siliciclastic deposition to have occurred during the lowstands and carbonate deposition during the highstands of asymmetric, 400-k.y. sea-level fluctuations.

This study has three objectives. (1) Further refine the depositional environments of the upper Yates car-

bonate and siliciclastic units, and describe the characteristics of the surfaces that bound the units. Although facies types present in the upper Yates Formation are well known from previous work, the lateral and vertical variations of facies tracts within each unit have yet to be described. (2) Study cement and pore types and cement chemistry in order to understand the diagenetic environments present both within the cyclic units and at cycle-bounding surfaces. This involves using the paragenetic sequence to document subaerial exposure at cycle tops, to document the hierarchy of the discontinuity surfaces, and changes in the chemistry of diagenetic fluids. (3) Relate the syndepositional diagenetic history to the stratigraphic evolution of siliciclastic/carbonate cycles. This will allow understanding of the response of early diagenetic syndepositional events to allocyclic controls on Permian cyclicity.

METHODS

Eleven stratigraphic sections were measured in detail. The sections follow a traverse parallel to the Permian shelf margin, extending from Rattlesnake Canyon to Dark Canyon, and two traverses perpendicular to the shelf margin, one in Rattlesnake Canyon and the other in Walnut Canyon (Figure 1). Stratigraphic sections were correlated throughout the area by physical tracing of carbonate and sandstone beds. A total of 200 samples were collected from each stratigraphic section, with detailed sampling at the top of the Hairpin dolomite. Diagenetic features were carefully described in the field where possible in order to have an outcrop-scale control on some of the microscopic features.

The samples were slabbed and polished. One hundred and twenty thin sections were made and stained with a dual Alizarin red S-potassium ferrocyanide solution (Evamy, 1963). Petrography of uncovered and polished thin sections was done with a standard petrographic and cathodoluminescence microscope.

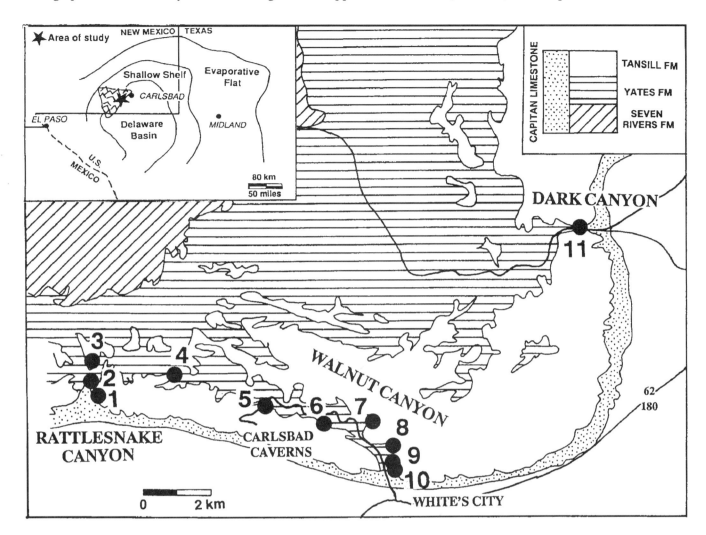

Figure 1. Geologic map of the study area with location of measured sections. Inset in the top left corner indicates the paleogeographic position of the study area during Guadalupian time (after Ward et al., 1986).

Trace element composition was measured with a nine spectrometer SEMQ electron microprobe used at the University of Wisconsin-Madison. Operating conditions for the microprobe were 15-KeV accelerating voltage, 15mA absorbed current, approximately 10-μm spot size and 80 s counting time for each element. Detection limits at the 95% level of significance for manganese, iron, strontium, and magnesium are approximately 110 ppm, 120 ppm, 100 ppm, and 280 ppm, respectively (E. Glover, personal communication).

Samples for stable carbon and oxygen isotopic analyses were drilled from polished 2-mm-thick thin sections using a dental drill attached to a binocular microscope. Samples were drilled with carbide burrs of 0.5 mm diameter. Carbon dioxide gas for the analyses was produced by reaction of 2 to 15 mg of powdered carbonate with 100% phosphoric acid at 25° C for eight hours. Analyses were made using a MAT 251 triple collector mass spectrometer at John Valley's stable isotope laboratory at the University of Wisconsin-Madison. Data were standardized using NBS-16. Precision of data ($\delta^{18}O$ = 0.02‰; $\delta^{13}C$ = 0.04‰) was monitored by multiple analyses of samples when sufficient powder was available. The reference composition of the average Permian Marine Calcite (PMC = $\delta^{18}O_{PDB}$ = −2.7‰ and $\delta^{13}C_{PDB}$ = +5.3‰) of Given and Lohmann (1985, 1986) is used.

REGIONAL SETTING

The study area is located in southeastern New Mexico (Eddy County) on the eastern margin of the Guadalupe Mountains (Figure 1) where Permian carbonates are exposed along a 50-mi northeast–southwest trend. The Delaware basin, a Permian intracratonic basin, has been subdivided by King (1934) into three main provinces, each characterized by different rock types and facies associations—shelf, shelf margin, and basin. Three major stratal complexes are distinguished (Pray, 1988) in Permian strata (Figure 2): (1) bank-ramp complex (Leonardian and early

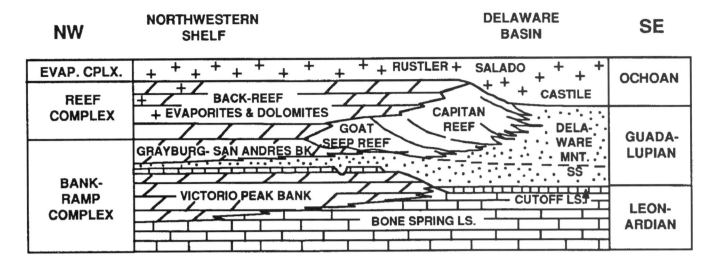

Figure 2. Three major stratal complexes distinguished in Permian strata (after Pray, 1988).

Guadalupian); (2) reef complex (late Guadalupian); and (3) evaporite complex (Ochoan), consisting of anhydrite, halite, and minor evaporitic dolomite. The reef complex is characterized by the abundance of depositional boundstones at the shelf margin, by steep basinward sloping foreslopes creating a relief of 400–600 m, and by abrupt shelfward changes in the facies caused by effective shelf-edge restrictions. The Capitan portion of the reef complex evolved during a two-phase growth history (Garber et al., 1989). The early phase occurred during the deposition of the Seven Rivers Formation, with maximum progradation on a low-slope margin. This is characterized by rare deposition on the shelf, emplacement of abundant foreslope detritus, and maximum deposition of siliciclastics in the basin. The late phase occurred during the deposition of the Yates and Tansill formations, with less progradation, and a steeper margin (30°) characterized by siliciclastic and carbonate deposition on the shelf.

Paleogeographic reconstructions locate the Delaware basin during the Permian time at a low latitude, slightly south of the equator (Smith et al., 1973; Scotese et al., 1979; Dott and Batten, 1981) or in the southern part of the northern trade-winds belt, analogous to the modern Sahara area (Fischer and Sarnthein, 1988). The widespread occurrence of evaporites in backreef strata indicate arid climatic conditions. Climatic conditions prevailing at time of deposition are an important controlling factor on diagenetic processes occurring during subaerial exposure. Kutzbach and Gallimore (1989) modeled the global climatic conditions for the Late Permian time and proposed the existence of a mega-monsoon regime. This would result in a strongly seasonal climate, with extended arid seasons and two wet periods a year.

STRATIGRAPHY AND SEDIMENTOLOGY

Different aspects of backreef units have been discussed in several papers (e.g., Sarg and Lehmann, 1986; Sarg, 1989); the Yates Formation received particular attention, as it is one of the major producing units in the Permian basin (Ward et al., 1986; Borer and Harris, 1991a).

Pray and Esteban (1977) suggested an informal terminology for outcrops of the upper Yates and lowermost Tansill formations, from bottom to top, as follows (Figure 3): (1) Hairpin dolomite, comprising the lower dolomite, Hairpin sandstone, and upper dolomite; (2) Triplet dolomite-sandstone unit, comprising the lower sandstone, middle dolomite, and upper sandstone; and (3) basal dolomite or lowermost Tansill. Facies of the upper Yates and the lowermost Tansill formations change significantly perpendicular to the Capitan reefal limestone trend and remain fairly constant parallel to the Capitan Limestone (Pray and Esteban, 1977; Neese, 1979; Schwartz, 1981; Neese and Schwartz, 1977). The Yates shelf can be divided into inner shelf, shelf crest, outer shelf, and reef, each characterized by a distinct association of carbonate and siliciclastic facies (Pray and Esteban, 1977; Pray, 1988). The distinction between inner shelf and shelf crest is well marked for Hairpin dolomite, which is characterized by rapid facies changes from mud-rich restricted facies to pisolitic grainstones and outer-shelf grainstone (Pray and Esteban, 1977; Neese, 1979; Schwartz, 1981). Lithofacies are generally richer in mud in Triplet dolomite, however, and changes are gradual, indicating a more continuous seaward sloping profile rather than a well-developed shelf crest (Pray and Esteban, 1977; Neese, 1979; Schwartz, 1981).

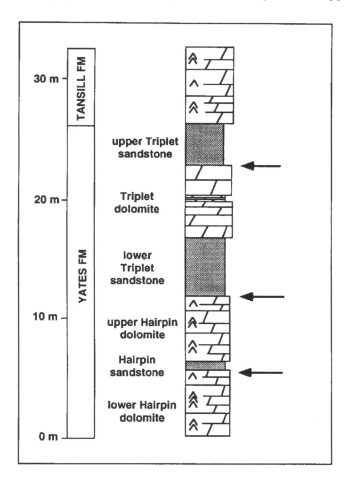

Figure 3. Stratigraphic terminology by Pray and Esteban (1977) for "classic outcrops" of the upper Yates and lower Tansill formations in Walnut Canyon. Arrows point to the base of siliciclastic/carbonate cycles.

Lithofacies and Depositional Environments

Sediments of the Hairpin dolomite, lower and upper Triplet sandstone, and Triplet dolomite have been subdivided into lithofacies (grain types and depositional textures), which have been related to depositional environments on the basis of their characteristics, occurrence, and bedding types (Table 1). Each lithofacies is also characterized by distinctive depositional porosity and cements, significant because of the distribution of syndepositional diagenetic features.

Sandstone

The sandstone lithofacies is characterized by (Candelaria, 1982; Mutti, 1990): (1) an extremely well-sorted grain size (coarse-grained silt to fine-grained sandstone); (2) a tabular geometry of the sand bodies; (3) sharp and erosive bases of sand bodies on the shelf; (4) thinning of beds toward the shelf margin (interbedding with marine carbonates; the pinch-out

occurs within 30–50 m of the Capitan Limestone); (5) association of massive bedding, fine flat and parallel lamination, and low- and high-angle cross beds and ripples close to the shelf margin; (6) presence of abundant anhydrite molds oriented parallel to lamination (Hairpin sandstone and lower Triplet sandstone); (7) juvenile tepee deformation (Hairpin sandstone, Dark Canyon); and (8) molds of marine fossils in sandstones closest to the shelf margin where they are interbedded with carbonates.

Three major models have been proposed for the environment of deposition of the sandstones: (1) on a subaerially exposed shelf where sands are transported by eolian processes and/or by wadi systems (Kendall, 1969; Silver and Todd, 1969; Smith, 1974; Meissner, 1972; Mazzullo et al., 1985; Fischer and Sarnthein, 1988); (2) transported in a shallow-marine environment (Pray, 1977; Sarg, 1977; Neese and Schwartz, 1977; Candelaria, 1982); and (3) deposited on a subaerially exposed shelf with siliciclastics transported by eolian processes and then reworked in the outer shelf by marine processes (Mutti and Simo, 1991a; Borer and Harris, 1991a,b). This interpretation is favored for the sandstones discussed in this chapter.

Mudstones and Wackestones with Evapo-Molds

Mudstones and wackestones with evapo-molds lithofacies is confined to the inner portion of the shelf. The lack of subaerial exposure features, and the abundance of mud and evapo-molds, associated with scarce and restricted biota (ostracods) indicate that deposition occurred in a geographically restricted evaporitic environment.

Peloidal Packstone/Wackestone

Peloidal packstone/wackestone facies occurs in the inner to middle portion of the shelf. The presence of restricted biota, algal mats, and fenestral fabrics indicate deposition in the intertidal to supratidal portion of a lagoon with semi-open circulation.

Pisoid Grainstone/Packstone

Pisoid grainstone/packstone facies occurs in the middle portion of the shelf and defines the depositional shelf-crest facies. Pisoids occur in lens-shaped beds that mainly pinch out against tepee cores. These pisolitic facies were originally considered to be caliche products (Dunham, 1972), but reinterpreted as primary precipitates of marine waters in subaqueous pools, which occasionally underwent modification in the vadose zone (Esteban and Pray, 1983).

Peloidal Skeletal Packstone/Wackestone

Peloidal skeletal packstone/wackestone facies is associated with the pisoid grainstone/packstone facies but covers a wider area on the shelf, extending basinward from the shelf crest. Bedding is mostly even and parallel. This facies is similar to the peloidal packstone/wackestone but is characterized by less mud and a generally more abundant and less restricted

Table 1. Rock types characteristics.

LITHOFACIES	OCCURRENCE AND BEDDING TYPE	DEPOSITIONAL ENVIRONMENT
(1) Sandstone -Fine-grained quartz and feldspar in a dolomicrite matrix	inner shelf to proximal outer shelf poorly defined parallel lamination; low- to high-angle cross beds near the sandstone pinch-out	subaerial to shallow marine nearshore
(2) Mudstone and wackestone with evapo-molds -peloids and restricted biota. Dolomicrite and siliciclastic silt as matrix; evapo-molds common -dolomitized	inner portion of shelf bedding is even and parallel, 5-20 cm rare wavy bedding, 5-10 cm	restricted evaporitic
(3) Peloidal packstone / wackestone -peloids, restricted biota, oncoids; occasionally bound by algal mats. Siliciclastic silt as matrix -dolomitized	inner shelf to shelf crest even and parallel beds, 0.5 to 1.5 m thick occasionally low-angle cross beds, 0.5 m thick	subtidal to intertidal semi-open circulation
(4) Pisoid grainstone and packstone -pisoids (1mm to 3 cm) are most abundant grain type; intraclasts are rare -dolomitized	shelf crest shoals beds are mostly lens-shaped and occur in intertepee depressions (0.5 m thick)	subtidal to supratidal
(5) Peloidal-skeletal packstone/grainstone -grain types are peloids, intraclasts, restricted biota (0.1-0.2 mm) -dolomitized	inner shelf to shelf crest even and parallel beds about 0.5 m thick	subtidal to supratidal lagoon with semi-open circulation
(6) Skeletal peloidal grainstone -biota are dasycladaceans, bryozoa, gastropods, bivalves, echinoids and forams; peloids and intraclasts decrease seaward. Micritic envelops and organic encrustations occur on grain surfaces. -mostly not dolomitized	outer shelf massive, thick parallel bedding or large-scale cross-beds	shallow marine ramp with high energy and open circulation

marine fauna. No features have been observed that can be related to supratidal conditions. The environment for the deposition of this facies is interpreted as being a subtidal to intertidal lagoon with relatively semi-open water circulation.

Skeletal Peloidal Grainstone

Skeletal peloidal grainstone facies occurs in the outer shelf and comprises the portion between the exposed shelf-crest and the submerged reefal Capitan Limestone. It is composed of normal marine fauna (dasycladaceans, gastropods, bryozoa, forams, echinoids, and bivalves) and minor peloids and intraclasts. Micritic envelopes and organic encrustation are commonly observed on grain surfaces. Bedding is either massive or large-scale cross-beds. The deposition of this facies occurred in a shallow-marine ramp, with relatively high-energy and open-water circulation.

Depositional Cycles and Cycle Boundaries

Figure 4 shows the vertical and lateral lithofacies changes and the correlation of stratigraphic sections.

The cyclic backreef succession is subdivided into three depositional cycles bounded by erosional surfaces at the base of the sandstones. The contact between carbonates and siliciclastics is sharp and erosive, indicative of a cycle boundary, whereas the passage from sandstones to the overlying carbonates is always transitional, suggesting a facies change within the cycle.

The cycle boundaries on the shelf crest are planar, sharp, and erosional with truncation up to 1 m and dip seaward. On the outer-shelf cycle, boundaries are erosional, irregular, and with relief of 5–10 cm (Mutti, 1990). The cycle boundaries truncate dolomitized and undolomitized (e.g., sections 8 and 9, Figure 4) lithofacies and disappear within the Capitan reef limestone. The truncated top of the Hairpin dolomite is fractured. The fractures are spoon shaped (Figure 5), a few centimeters wide, and are associated with the sequence boundary. Sedimentary fill of the fracture by the Triplet sandstone occurred in two stages, with an early deposition of fine-grained carbonate and later fine-grained sandstone (Figure 5). The early carbonate is composed of up to 90% carbonate mud and

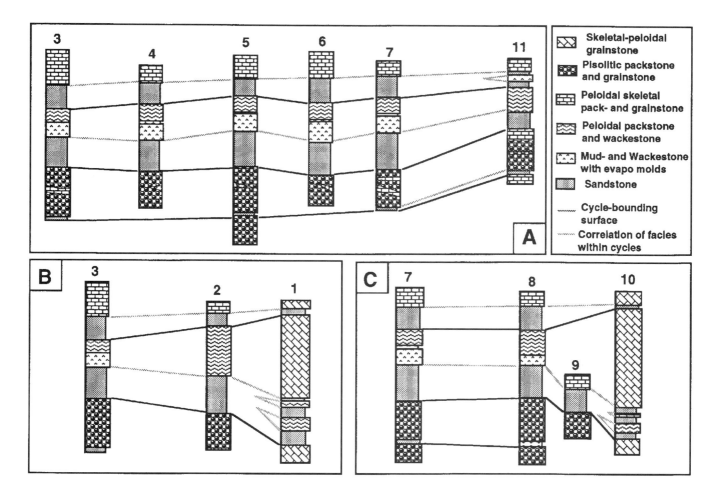

Figure 4. Cross section constructed with sections shown in Figure 1. (A) Cross section parallel to shelf margin, located about 2 km shelfward with respect to the Capitan Reef; (B) cross section perpendicular to the shelf margin (Rattlesnake Canyon); and (C) cross section perpendicular to the shelf margin (Walnut Canyon).

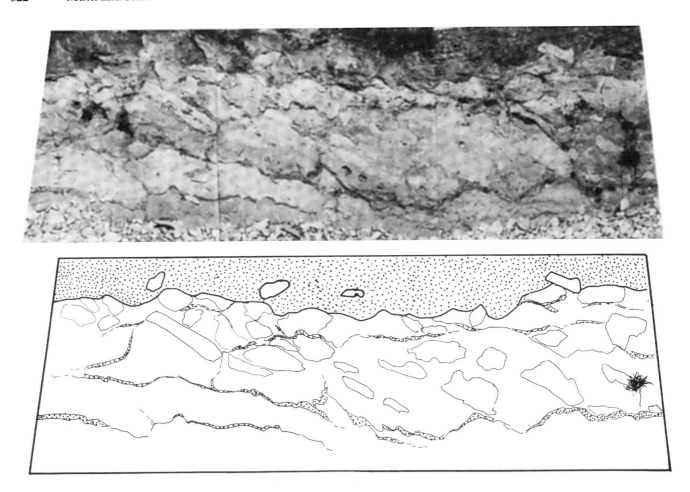

Figure 5. The contact between Hairpin dolomite and lower Triple sandstone at the "Elbow Locality" in Walnut Canyon is shown in the photograph and drawing. Note breccias at the top of Hairpin dolomite, postdated by fractures filled by siliciclastics from the lower Triplet sandstone. Some dolomitic breccia clasts are ripped up and incorporated in the lower Triplet sandstone.

minor quartz and feldspar grains, it is gray-yellow in color, and commonly occurs on fracture walls or geopetally within primary pores. It often shows a form which mimics prismatic to equant crystal morphologies (Figure 6). The fine-grained sandstone is brown-yellow and composed mostly of quartz grains (up to 90%) with a carbonate mud matrix. Fibrous and bladed cements postdate the emplacement of the fine sandstone. The origin of the fracture is problematic; a possible modern analog are the desiccation cracks in the Sahara that are filled with the overlying wind-blown sand (Breed et al., 1987). An alternative mechanism is dissolution, which would account for the irregular fracture walls but would not explain the fracture geometry. The overlying Triplet sandstone drapes the cycle boundary, and occasionally contains clasts of the underlying lithologies.

The cycle boundary between Hairpin dolomite and Triplet sandstone is locally overlain by a breccia (Figure 5). The breccia is variably both clast and matrix supported, and is composed of white dolomitized fenestral peloidal packstone/grainstone, rectan-

gular-shaped clasts, and gray-yellow undolomitized skeletal grainstone, subangular- and subrounded-shaped clasts. Clasts are generally imbricated and dip basinward. The source of the breccia clasts is believed to be landward where both dolomitized, tepee-deformed, fenestral peloidal packstone/grainstone, and undolomitized skeletal grainstone lithofacies occur. The breccia matrix is fractured and filled by Triplet sandstone. The base of the Triplet sandstone drapes the breccia and contains similar clasts to those in the breccia (Figure 5).

Brecciation and fracturing are related to the cycle boundary formation and require a polyphase origin. Brecciation may be related to tepee formation, but the presence of non-tepee clasts indicates additional erosion. Both lithoclast types are imbricated, reworked, and mixed with marine fauna, suggesting a high-energy marine depositional setting. Most probably, erosion of tepee and non-tepee clasts occurred during this time, although fracturing of the underlying carbonates and of the breccia matrix suggests exposure. We interpret the cycle boundaries as a combination of

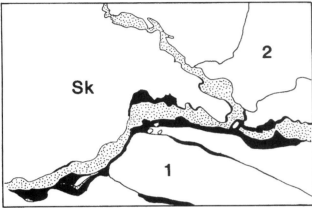

Figure 6. Detail of the top of Hairpin dolomite shown in Figure 5. Two types of breccia clasts [dolomitized skeletal-peloidal packstone (1) and calcitic skeletal grainstone (2)] are incorporated in a skeletal grainstone matrix (Sk). Fractures do not cross-cut breccia clasts and are filled in two stages by fine-grained carbonate and later by fine-grained sandstone.

marine reworking and subaerial exposure; breccias indicative of marine reworking preceded fracturing, and marine sandstone deposition postdated fracturing.

Depositional cycles are characterized by a lower homogeneous sandstone unit (sandstone lithofacies) and an upper carbonate unit composed of different lithofacies (Table 1). Cycles are capped by surfaces of subaerial exposure. The sandstone units thin basinward and split into two or three beds in that direction (Figure 4). These beds are separated by carbonate lithofacies that thicken basinward. The sandstones on the outer-shelf area contain some marine fauna and sedimentary structures suggesting marine reworking (Candelaria, 1982; Mutti, 1990). The carbonate units show distinctive facies belts parallel to the Permian shelf margin (Figure 4). Three depositional cycles are distinguished and are named from oldest to youngest, 1, 2, and 3, respectively.

Cycle 1 comprises Hairpin sandstone and upper Hairpin dolomite, and lithofacies within the cycle record a vertical transition on the shelf from a restricted-lagoonal environment to increased water circulation and high-energy conditions. Cycle 2 comprises upper Triplet sandstone and Triplet dolomite. Carbonate lithofacies within cycle 2 again record an overall transgressive trend, recorded by two small-scale transgressive cycles (Figure 4). Cycle 3 comprises upper Triplet sandstone and basal dolomite. The upper boundary of the cycle was not studied.

Origin of Cycles

Borer and Harris (1989, 1991a,b) analyzed both subsurface data and outcrops of the Yates Formation and recognized three orders of relatively low-amplitude, sea-level fluctuations that produced the observed stacking patterns of siliciclastic/carbonate cycles. They showed that the Yates Formation coincides with a third-order cycle that contains five 400-k.y. depositional cycles. The thickness, internal packaging, and dominant lithology of the 400-k.y. depositional cycle (fourth-order) is controlled by their position and a third-order, sea-level cycle and by the character of internal fifth-order (100-k.y.) cycles (Borer and Harris, 1991a,b). The Borer and Harris model suggests that widespread carbonate deposition occurs when 100-k.y. highstands are in phase with 400-k.y. highstands, and that siliciclastics will be trapped on the shelf when 100-k.y. highstands are superimposed on the 400-k.y. lowstands. When 100-k.y. lowstands are in phase with 400-k.y. lowstands, siliciclastics would bypass the shelf and be deposited in the basin.

According to the Borer and Harris model, each of the cycles recognized were deposited during a 400-k.y. relative sea-level fluctuation, including depositional hiatuses. The complex internal stratigraphy of the siliciclastics near the shelf margin, where they intertongue with carbonates, may represent a higher frequency signal, however. Similarly, the Triplet dolomite shows cycles of a higher frequency than the 100-k.y. cycles. Facies organization within cycles indicates that sedimentation kept pace with relative sea-level rise during cycle I but not during cycle II, thereby forcing facies belts to retrograde. Relative sea-level drops exposed cycle tops subaerially but marine reworking before and after exposure modified the cycle boundaries.

PETROGRAPHIC AND GEOCHEMICAL DATA

Several studies have dealt with various aspects of the Capitan platform diagenesis at different stratigraphic intervals. Dunham (1972), Neese (1979), Schwartz (1981), Candelaria (1982), Garber et al. (1989), and Borer and Harris (1991a,b) described the petrographic characteristics of carbonate grains, mineralogy, and cement types of the Yates and Tansill formations. Rosenblum (1984) studied the early cements associated with the tepees in the Yates and Tansill formations in Walnut Canyon, and Mruk (1985, 1989) and Given and Lohmann (1986) described the diagenesis of the Capitan Limestone along a single paleoslope in McKittrick Canyon. Hill (1987) studied Carlsbad Cavern and other caves in the area and established relationships between spar cements, cave formation, and pore-water evolution, and suggested that sulfuric acid and carbon dioxide migrated from the oil and gas fields in the basin to dissolve the Capitan Limestone.

In this section, we discuss the distribution of porosity types, cements, and dolomites present in the shelf and their temporal relationships with respect to depositional events. Three main stages are differentiated in the diagenetic history—early, middle, and late (Figure 7).

Pore Types

Several types of primary pores are recognized in the study area. Primary pores are related to lithofacies distribution (Figure 8). Interparticle porosity (micro and meso; terminology after Choquette and Pray, 1970) in the Hairpin and Triplet dolomites is common in pisoid grainstone on the shelf crest (Hairpin dolomite) and in outer-shelf skeletal grainstones (Hairpin and Triplet dolomite). Fenestral porosity (meso) is commonly associated with supratidal, algal-related, mud-dominated sediments or peloidal packstones. Intraparticle porosity (meso to mega) in the Hairpin and Triplet dolomites is mostly

located in dasycladacea and gastropods and its distribution is therefore concentrated in open-marine, outer-shelf grainstones. Shelter porosity (meso) is most commonly found in the Hairpin dolomite at the mouth of Walnut Canyon and is associated with bivalve shells, which act as a shelter for the siliciclastic matrix infiltrating into the rock from above.

Secondary pores occur in all rock types and throughout the shelf. Macroscopic fractures (up to 20 cm wide and 8 m long) are common and concentrated in the outer-shelf facies of Hairpin dolomite. The walls of the macrofractures show partial dissolution. The fractures are filled by siltstone infiltrated from the overlying Triplet sandstone. Microfractures (1 mm wide and 2 cm long) occur occasionally throughout the shelf and in both the Hairpin and Triplet dolomites. Skeletal and evapo-molds are found in different shelf positions, the skeletal molds being concentrated in outer-shelf skeletal grainstones and the evapo-molds in mud-rich sediments of the inner shelf. Skeletal molds are considered to have formed early since they are filled with syndepositional marine cements in Hairpin and Triplet dolomites. Evapo-molds form in the early postdepositional history as discussed in the paragenetic sequence. Vugs are related to limestone or dolostone dissolution and the cross-cutting relationships suggest that vug formation is related to a dissolution event occurring in a late diagenetic stage. The timing of vug formation with respect to other cement types suggests, a correlation with Hill's Pleistocene Stage III dissolution event (Hill, 1987). Evidence of dissolution and brecciation is also found throughout the PDB-04 core (Garber et al., 1989), but the pores are filled principally by anhydrite

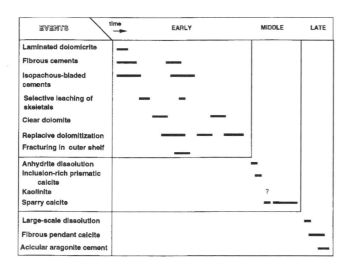

Figure 7. Paragenesis of events recognized in shelf strata, grouped into early, middle, and late stages. The early stage consists of synsedimentary cementation, dolomitization, dissolution, and fracturing. The middle diagenetic stage consists of precipitation of postdepositional cements in a shallow-burial environment. The last diagenetic stage corresponds to a Pleistocene large-scale dissolution event and precipitation of speleothems in a cave-related system.

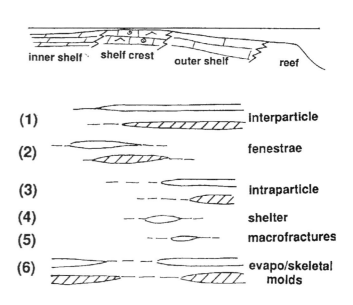

Figure 8. Distribution of porosity types across the shelf. The thickness of the horizontal bar is proportional to porosity abundance. Empty bars refer to Hairpin dolomite, bars with diagonal symbol refer to Triplet dolomite.

and gypsum and not by calcite cements. The timing and the diagenetic environment of dissolution cannot be determined (Garber et al., 1989).

Cement Types

The characteristics and position of syndepositional cements on the shelf are directly related to lithofacies and shelf location. Most syndepositional cements are specific to one of the lithostratigraphic units (Table 2). Postsedimentary cements cross-cut syndepositional cements and are present in both units and all lithofacies and therefore their precipitation postdates the deposition of both Hairpin and Triplet dolomites.

Syndepositional Cements

Three syndepositional groups of cements are recognized by their texture, crystal size, and crystal terminations (Figure 9). Syndepositional cements on the inner- to proximal outer-shelf have been replaced by

dolomite and have become calcitic toward the outer-shelf. The distribution of cements is summarized in Figure 10 and Table 2 and cement characteristics are summarized in Table 3. The geochemistry of cements that have been dolomitized is discussed in the section on dolomites.

Laminated or Pisoid-Enlarging Cement

Laminated or pisoid-enlarging cement is common in the shelf-crest facies; it is commonly dolomitized (Figure 9, Table 3). Nondolomitized pisoids have been found in a proximal outer-shelf location. Pray and Esteban (1977) refer to this cement type (laminated dolomicrite) as pisolite-enlarging cement. Calcitic samples of this cement from Hairpin dolomite have an isotopic composition averaging $\delta^{18}O = -3.5‰$ and $\delta^{13}C = +6‰$ (Figure 11). These are similar to Permian marine calcite values, which indicate that this cement precipitated from normal-marine waters. This interpretation is consistent with the interpretation of

Table 2. Occurrence of cement types by stratigraphic units and lithofacies.

CEMENT TYPES			STRATIGRAPHIC UNIT	LITHOFACIES ASSOCIATION AND SHELF LOCATION
SYNDEPOSITIONAL		Laminated dolomicrite	Hairpin dolomite	(4) shelf crest
	Fibrous cements	Fibrous dolomite	Hairpin dolomite	(4.5) shelf crest and inner shelf
		Radial-fibrous dolomite	Hairpin dolomite	(4.5) shelf crest and inner shelf
		Radial-fibrous calcite	Hairpin and Triplet dolomite	(6) outer shelf
	Bladed cements	Bladed dolomite	Hairpin dolomite	(4.5) shelf crest and inner shelf
		Bladed calcite I	Hairpin dolomite	(6) outer shelf
		Bladed calcite II	Triplet dolomite	(6) outer shelf
POSTDEPOSITIONAL		Inclusion-rich prismatic calcite	all	mostly outer shelf and margin
	Sparry calcite	Zone CLI	all	all
		Zone CLII		
		Zone CLIII		
	Kaolinite		all	mostly shelf crest and outer shelf
	Fibrous pendant calcite		all	all
	Acicular aragonite			

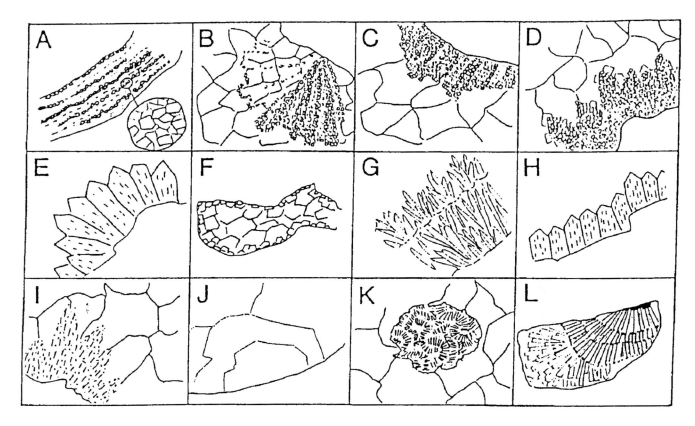

Figure 9. Sketches of cement types present in Capitan backreef strata. (A) laminated dolomicrite; (B) radial-fibrous dolomite; (C) fibrous dolomite; (D) bladed dolomite; (E) bladed calcite I; (F) clear dolomite; (G) radial-fibrous calcite; (H) bladed calcite II; (I) inclusion-rich prismatic calcite; (J) clear calcite; (K) kaolinite; and (L) fibrous pendant calcite and acicular aragonite.

Esteban and Pray (1983) who suggested, on the basis of field observations, that pisoid coatings form in a marine, subaqueous environment, rather than being related to soil processes (Dunham, 1972).

Fibrous Cements

Fibrous cements are abundant in shelf-crest facies (fibrous dolomite and radial-fibrous dolomite; Figure 11, Table 2) where they have been dolomitized, and in outer-shelf facies (radial-fibrous calcite; Figure 10, Table 3) where they are still composed of calcite (Table 2). Fibrous dolomite, radial-fibrous dolomite, and calcite are found in Hairpin dolomite; no fibrous cements have been observed in Triplet dolomite (Table 2). In both fibrous dolomite and radial-fibrous dolomite, square crystal terminations suggest a former aragonite mineralogy (Assereto and Folk, 1980). Radial-fibrous or botryoidal cement is well developed in certain Permian reefs and slope facies (Purser and Schroeder, 1986; Tucker and Hollingworth, 1986).

No radial-fibrous calcite has been analyzed for stable isotopes because it is too small for sampling. Data from Rosenblum (1984) on fibrous and radial-fibrous calcite cements from the lowermost Tansill Formation

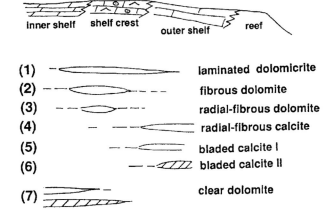

Figure 10. Distribution of syndepositional cement types across the shelf. The thickness of the horizontal bar is proportional to cement abundances. Empty bars refer to Hairpin dolomite, bars with diagonal symbol refer to Triplet dolomite. Postdepositional cements do not show any preferential distribution, except inclusion-rich prismatic calcite, which is concentrated in the margin/outer shelf.

Table 3. Cement types characteristics.

	CEMENT TYPES		DESCRIPTION	CRYSTAL TYPE	CL COLOR	TRACE ELEMENT	STABLE ISOTOPES
SYNDEPOSITIONAL	Fibrous cements	Laminated dolomicrite	laminae 50-200 µ thick form coatings (up to 0.4 mm thick) around pisoids and in primary pores; gravity features are common	turbid dolomite (1-10 µ) replaces original laminae; calcite spar (5-15 µ) in between laminae postdates dolomitization	dull-orange	Mg=13.30 Fe=358 ppm	δ18O 0.4‰ δ13C 5.8‰
		Fibrous Dolomite	isopachous crusts up to 0.4 mm thick	turbid crystals (10-20 µ) mimic fibrous crystals 80-150 µ wide and 300-600 µ long	dull-orange	Mg=12.47 Fe=285 ppm	
		Radial fibrous dolomite	fans up to 1 mm in length in intergranular pores; fans perpendicular to the substrate	turbid to clear dolomite crystals (5-20 µ)	dull-orange	Mg=13.23 Fe=220 ppm	
		Radial fibrous calcite	isopachous layers up to 2 mm thick in intergranular pores and early fractures; crystals are arranged in fibrous splays or radial cones perpendicular to the substrate	fibrous calcite crystals (50-200 µ long); sweeping extinction	patchy dull-brown	Mg=0.27 Fe=445 ppm	δ18O -2‰ δ13C 6.5‰ data from Rosenblum (1984)
	Isopachous-bladed cements	Bladed dolomite	isopachous pore-lining up to 0.5 mm thick in primary pores	turbid dolomite (5-50 µ) mimics original bladed crystals, 400-600 µ long and 100-200 µ wide	dull-red	Mg=13.36 Fe=728 ppm	
		Bladed calcite I	isopachous crusts up to 0.6 mm thick in intergranular and moldic pores; black inclusions impart a pseudo-fibrous texture	bladed calcite 200-600 µ long and 80-120 µ wide straight to sweeping extinction	patchy dull brown	Mg=0.42	δ18O -2.3‰ δ13C 4‰
		Bladed calcite II	isopachous lining up to 0.2 mm thick in intergranular pores numerous black inclusions	bladed calcite up to 200 µ long and 40-80 µ wide straight to sweeping extinction	patchy dull brown	Mg=0.50	
POSTDEPOSITIONAL	Inclusion-rich prismatic calcite		nucleates on pre-existing bladed cements and lines primary pores	prismatic calcite 50-70 m wide and 100-400 m long; dark inclusions outline crystal shape	patchy dull with bright inclusions	Mg=0.27	
	Clear calcite	CLI	occurs as pore-lining and -filling in primary and secondary pores	clear and equant calcite crystals, 30-250 µ	bright to dull orange alternating zones	Mg=0.13 Mn=183ppm	δ18O -11.8‰ δ13C 0.5‰
		CLII			bright orange-yellow	Mg=0.15 Mn=311ppm	
		CLIII			non-luminescent	Mg=0.10 Sr=183ppm	
	Kaolinite		patches 50-1000 µ wide replace calcite and cross-cut all CL zones	vermicular crystals poorly ordered	blue		
	Fibrous pendant calcite		several phases of cement bandings grow preferentially toward center of cavities	fibrous calcite up to 600 m long and 50 m wide; slightly sweeping extinction	non-luminescent	Mg=1.94 Sr=222 ppm	δ18O -6.8‰ δ13C -4.6‰
	Acicular aragonite		mesh of randomly oriented needles	aragonite needles 100-400 µ long and 20-50 µ wide; extinction parallel to C-axis	non-luminescent	Mg=0.39 Sr=271ppm	

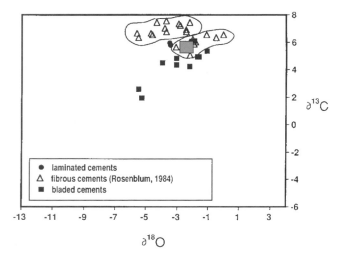

Figure 11. Stable isotope composition of syndepositional cements. The gray square indicates Permian marine calcite of Given and Lohmann (1986).

average $\delta^{13}C$ = +6.5‰ and $\delta^{18}O$ = –2.0‰ (Table 3). He interpreted these data to plot in two fields; field I centered around $\delta^{13}C$ = +7‰ and $\delta^{18}O$ = –3.5‰ and field II centered around $\delta^{13}C$ = +6.5‰ and $\delta^{18}O$ = –2‰ (Figure 11). Samples with more negative oxygen isotopic compositions are also characterized by a higher strontium content (over 2000 ppm), which has been interpreted by Rosenblum (1984) to reflect a former aragonitic composition for field I and a former high-Mg composition for field II.

Isopachous-Bladed Cements

Isopachous-bladed cements occur in the shelf crest (bladed dolomite; Figure 10, Table 2) where they have been dolomitized, and outer-shelf facies (bladed calcite I and II; Figure 10, Table 2) where they are composed of calcite. Bladed calcite I is restricted to Hairpin dolomite and bladed calcite II is restricted to Triplet dolomite (Table 2). No dolomitized bladed cements have been observed in Triplet dolomite.

Ten samples, with several degrees of mixing of radial-fibrous calcite, bladed calcite I, and postdepositional sparry calcite, have compositions ranging from $\delta^{18}O$ = –5.5‰ to –1.1‰ (average –2.3‰) and from $\delta^{13}C$ = +1.9 to +6.0‰ (average 4.0‰) (Figure 11, Table 3). The mixture of different cements is due to the small size that does not allow separate sampling using the drilling technique. Fractionation factors have been calculated with an average mixing of 70% dolomite and 30% calcite, estimated from petrographic observations. The covariant trend reflects different degrees of mixing between two end members, the Permian marine calcite and the composition of postdepositional sparry calcite (Figure 11). The heaviest values, centered around $\delta^{13}O$ = –2.3‰ and $\delta^{13}C$ = +4.9‰, are considered the most representative of the actual isotopic composition of the calcite cements, and reflect precipitation from isotopically normal- marine waters.

Postsedimentary Cements

Postsedimentary cements postdate dolomitization, are present in all lithofacies on the shelf, and are characterized by their texture, crystal habit, and termination (Table 3). Postsedimentary cements provide a record of middle and late diagenetic stages. The middle diagenetic stage consists of precipitation of inclusion-rich prismatic calcite in a mixed marine-meteoric environment and calcite cementation in a shallow-burial environment. Kaolinite is found in the proximity of the outer shelf and postdates calcite cementation. The late diagenetic stage corresponds to a large-scale dissolution event, and speleothems precipitate (pendant calcite and acicular aragonite) as a response to uplift in the late Pliocene–Pleistocene section.

Inclusion-Rich Prismatic Calcite Cement

Inclusion-rich prismatic calcite postdates all syndepositional cements, but predates precipitation of calcite cements. The best developed crystals occur in the outer-shelf area where they decrease in size shelfward and eventually disappear. This cement is texturally similar to inclusion-rich prismatic cements from Cambrian (James and Klappa, 1983) and Mississippian (Meyers and Lohmann, 1978) limestones. These were interpreted as former Mg-calcite precipitated in a zone of mixed meteoric and marine phreatic groundwater. Mruk (1985) and Mutti (1990) provide a similar interpretation for inclusion-rich cements in the Capitan Limestone, consistent with their distribution. Inclusion-rich prismatic calcite cements on the shelf are too small to be analyzed for their stable isotope content.

Sparry Calcite

Sparry calcite is composed of three CL zones, identified as CL-I, CL-II, and CL-III. These zones postdate dissolution of evaporite nodules in the inner-shelf facies. Crystal size in both CL-I and CL-II zones is too small for stable isotope sampling. Six samples of CL-III have a restricted isotope composition of typically $\delta^{18}O$ = –11.8‰ and $\delta^{13}C$ = +0.5‰ (Figure 12).

Sparry calcite cements are common in the Capitan reef complex and are interpreted to have precipitated from shallow-phreatic to shallow-burial meteoric lenses (Mruk, 1985; Given and Lohmann, 1986; Hill, 1987). Darke et al. (1990), on the basis of cross-cutting relationships and paleomagnetic data, relate this calcite cementation to uplift during Tertiary time. The correlation of cathodoluminescent zones from the shelf (Mutti, 1990), shelf edge, and slope (Mruk, 1985) based on cathodoluminescent color alone can be misleading, but stable isotopic data and petrographic observations suggest that CL-I and CL-II may be correlated with Mruk's Spar I. Although both CL-I and CL-II show color zonation, they are relatively homogeneous with respect to zonation observed in Spar I (Mruk, 1985). This is interpreted as indication of minor fluctuations in Eh in the recharge areas with respect to the variations observable downdip. Isotopic data suggest that Mutti's (1990) Cl-III zone

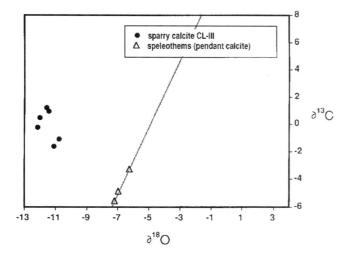

Figure 12. Stable isotopic composition for postdepositional cements. Line is CO_2-loss line by Hill (1987).

may be equivalent to Mruk's (1985) and Given and Lohmann's (1986) Spar II. CL-III zone is nonluminescent while Spar II is commonly brightly luminescent (Mruk, 1985). This change in luminescent color is interpreted as being related to a change in redox conditions downdip. The narrow oxygen isotopic range, however, suggests that CL-III sparry calcite cement precipitated from fluids within a uniform oxygen isotopic composition at a relatively constant temperature. Light carbon values indicate that fluids were in contact either with an organic carbon source or soil-derived CO_2. Mruk (1985), Given and Lohmann (1986), and Mutti (1990) interpreted Spar II and CL-III as precipitates from meteoric waters (initial $^{18}O/^{16}O$ isotopic ratio = –2.8) circulating at a burial depth of 1.0 to 1.4 km (Crysdale, 1986). Hill (1987) calculated the temperature of precipitation (55.4° C) and the isotopic ratios of single-phase fluid inclusions found in Spar II or CL-III ($\delta^{18}O_{SMOW}$ –46.5‰ and –41.9‰, $\delta^{13}C_{PDB}$ –19.6‰ and –19.9‰, and D_{SMOW} –86‰ and –68‰), which suggest a burial depth of less than 1 km. Our diagenetic work confirms this previous interpretation.

Kaolinite

Kaolinite occurrence is generally confined to a narrow belt close to the shelf margin, both in the Hairpin and Triplet dolomites. This observation on kaolinite distribution led Dunham (1972) to suggest that kaolinite is an early diagenetic feature, produced by meteoric alteration of feldspars during phases of subaerial exposure of the shelf. This hypothesis is invalidated by the presence of cross-cutting relationships with late-diagenetic sparry calcite cements. Furthermore, early meteoric alteration would be unlikely considering that arid conditions existed at the time of deposition. Borer and Harris (1991a) suggest that feldspar dissolution and kaolinite-growth are facies-specific, and are related to meteoric flushing of the shelf mar-

gin during uplift of the Guadalupe Mountains in the late Pliocene–Pleistocene time.

Speleothems

Fibrous pendant calcite cement was analyzed (three samples) for stable isotopes and the data points (Figure 12) define a line which coincides with the CO_2 line of Hill (1987). Hill's line was experimentally defined and was interpreted to be related to increasing CO_2 loss, which results in a carbon-isotope shift (enrichment of ^{13}C). Relatively constant oxygen values suggest precipitation from vadose meteoric water with a relatively uniform composition at a relatively constant temperature in a cave setting. The marked depletion in carbon suggests that this cement precipitated from waters that did not experience much interaction with the carbonate host rock and probably formed near the top of the water table. Uranium-series data on a genetically similar cement type indicate ages >350 y.b.p. and an Electron Spin Resonance age of 879,000+/–124,000 y.b.p., which indicates a Pleistocene age for these speleothems (Hill, 1987).

Fibrous pendant calcite cement is interpreted as a popcorn cement related to speleothems and cave formation. Popcorn precipitation requires thin films of water to lose carbon dioxide due to evaporation and to preferentially precipitate carbonate at the apex of irregularities. During precipitation of popcorn cement, the cave water remains saturated with respect to calcite and shifts toward relatively higher magnesium and strontium concentrations (Mg = 1.94 wt. % and Sr = 222 ppm). Acicular aragonite is also interpreted as a speleothem commonly referred to as anthodites. Precipitation of aragonite needles is the final stage of popcorn cement formation, probably occurring where slowly moving thin films of water sweep through pores in a meteoric environment (Hill, 1987). This stage is related to the uplift of the study area in the late Pliocene–Pleistocene, which brought the rocks into the vadose zone. Sulfuric acid and carbon dioxide migrated from the oil and gas accumulations in the basin and dissolved part of the Capitan Limestone (Hill, 1987). Speleothems precipitated from shallow-phreatic to vadose freshwater zones.

Dolomite Types

Shelf carbonates of the Capitan Reef complex have been pervasively replaced by fabric-conservative dolomite, as described in a number of papers (Dunham, 1972; Neese, 1979; Candelaria, 1982; Schwartz, 1981; Garber et al., 1989; Borer and Harris, 1991a,b). Dolomite occurs as both replacive and primary cement also in the Capitan massive (Garber et al., 1989) and in the foreslope (King, 1934; Mruk, 1985), but has a patchy distribution.

Replacive Dolomites

Hairpin and Triplet dolomite inner-shelf and shelf-crest strata are replaced by fabric-conservative cryptocrystalline to very fine crystalline dolomite. Replacive dolomite is by far the most common, replacing both depositional fabrics (grain type and matrix) and syndepositional cements (laminated, fibrous, and bladed cements) throughout the shelf. Replacive dolomite consists of micrite-sized dolomite crystals, commonly irregular, turbid, and yellow-brown in color. Original fabrics show different degrees of preservation, but are commonly recognizable. Replacive dolomites are commonly dull red in cathodoluminescence.

Different types of replacive dolomite have been analyzed for isotopic composition (Figure 13, Table 3). Dolomicrite samples are characterized by relatively constant $\delta^{13}C$ values (average 5.8‰) and by $\delta^{18}O$, ranging from –0.4 to 2.3‰, reflecting an enrichment of 2.3 to 5‰ with respect to Permian marine calcite (Figure 13). No significant differences are apparent in samples taken from different stratigraphic positions nor with different shelf locations along the shelf profile. Dolomite which replaced laminated cements show $\delta^{18}O$, ranging from +0.6‰ to +1.5‰ (average +0.4‰) and $\delta^{13}C$, ranging from +3.8‰ to +6.9‰ (average +5.8‰). Comparison with calcitic laminated cements (Figure 11) indicates that dolomitized cements are characterized by relatively constant oxygen values but show variable carbon values.

Data from a mixture of fibrous and radial-fibrous dolomitized cements have $\delta^{18}O$, ranging from +0.9‰ to +2.3‰ (average +1.4‰) and $\delta^{13}C$, ranging from +6.0‰ to +7.3‰ (average +7‰) (Figure 13). Oxygen shows an enrichment of +4.1‰ with respect to the Permian marine calcite. Mixtures of dolomitized fibrous and bladed cements and sparry calcite have $\delta^{18}O$, ranging from –6.5‰ to +1.5‰, and $\delta^{13}C$, ranging from +0.8‰ to +5.9‰. The heaviest values, typically $\delta^{13}C$ = +5.8‰ and $\delta^{18}O$ = +1.2‰, are considered the most representative of the actual composition of isopachous fibrous dolomite. Oxygen shows an

enrichment of +3.9‰ with respect to Permian marine calcite. The association of the upper Yates early dolomites with evaporite nodules and restricted peloidal mudstones/wackestones suggests that the initial dolomitization occurred in a sabkha environment, such as described in the Holocene Persian Gulf (McKenzie, 1981), or in other evaporite-related systems. Trace element and stable isotopic data are also consistent with a dolomitization model which involves evaporation of concentrated sea water with abundant iron and slightly reducing Eh character. These characteristics are similar to the semi-arid sabkha model in which the dolomitization process takes place during or immediately after deposition (McKenzie, 1981). Later partial reequilibration of initial dolomites by other type of waters cannot be excluded on the basis of the present data.

Primary Dolomite

Dolomite cement (or clear dolomite) is present both as a pore-lining and pore-filling cement in fenestral pores in the inner-shelf and shelf-crest strata of both Hairpin and Triplet dolomites (Figure 10, Table 2). Dolomite cement consists of 20- to 300-µm-wide clear and transparent dolomite crystals that are rhombohedric to irregular in shape (Table 3). Clear dolomite is bright-red to nonluminescent in cathodoluminescence. Several features suggest this is a primary dolomite cement rather than a dolomitized calcite cement: (1) dolomite crystals are clear, whereas replacive dolomite generally show abundant black solid inclusions; (2) intercrystalline boundaries of dolomite are planar; (3) clear dolomite lines the interiors of primary pores or grows out from dolomitized cements; (4) replacement of preexisting cements commonly results in a finer crystalline and more heterogeneous texture; (5) there is a different iron content with respect to replacive dolomite; (6) the association with fenestral pores never filled with other cement types; and (7) homogeneous nonluminescent cathodoluminescent color. Dolomite cement fills fenestral porosity, whereas equivalent rocks contain evapo-molds filled with sparry calcite. This relationship suggests that clear dolomite cement precipitated before evaporite dissolving fluids entered the system and it occurred in an environment where fluids were in equilibrium with evaporite minerals. Dolomite cement is rich in iron (up to 2600 ppm) (Table 3). Five samples from Hairpin dolomite shelf crest indicate $\delta^{18}O$, ranging from –3.1‰ to –1.0‰ (average –2.1‰), and $\delta^{13}C$, ranging from +4.1‰ to +5.2‰ (average +4.7‰). Samples are likely to have been contaminated with late pore-fill sparry calcite, which could account for $\delta^{18}O$ variations (Figure 13). Clear dolomite probably precipitated from reduced, evaporation-concentrated hypersaline waters below the sulfate reduction zone, possibly associated with methane production. This would explain the high carbon isotopic values and the incorporation of high amounts of iron in the dolomite lattice (Kastner, 1991, personal communication).

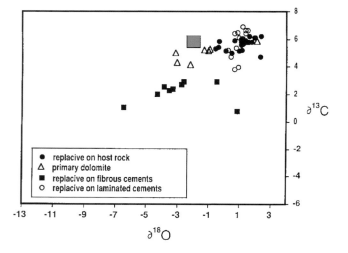

Figure 13. Stable isotopic composition for dolomites. The gray square indicates Permian marine calcite of Given and Lohmann (1986).

TIMING OF DIAGENETIC EVENTS

Timing of Dolomitization

Two dolomitization phases are documented on the basis of textural relationships in upper Hairpin dolomite and in the Triplet dolomite. The first dolomitization phase postdated deposition of upper Hairpin dolomite and is postdated by the truncation at the base of the lower Triplet sandstone. Replacive dolomite replaced inner-shelf and shelf-crest depositional facies and cements; dolomite cement precipitated in fenestral pores. The timing of dolomitization is also constrained by the occurrence of dolomitized clasts ripped from the underlying Hairpin dolomite and incorporated together with calcite clasts in the overlying lower Triplet sandstone.

The second dolomitization phase postdated deposition of Triplet dolomite and is responsible for replacive dolomitization of inner-shelf facies and precipitation of clear dolomite in fenestral pores. By analogy with Hairpin dolomite, dolomite cement is probably an early syndepositional cement, although scarcity of other inner-shelf syndepositional cements in Triplet dolomite precludes a better constraint on timing of its precipitation. The second dolomitization phase is relatively shallow and only affected Triplet dolomite. In a seaward direction, the Triplet Dolomite remains dolomitic, while the underlying Hairpin Dolomite outer-shelf facies is composed of calcite.

The dolomite-calcite interface shifts within each cycle following the backstepping of facies types (Figures 5, 14). Dolomitization occurs in the muddy supratidal facies that are shifted shelfward within cycles. Subaerial exposure at cycle tops allows dolomitization of former calcitic open-marine facies at cycle tops. These observations suggest that in a sabkha-type environment, the repetition of transgressive episodes capped by subaerial exposure forced by composite eustatic cycles, would result in multiple dolomitization phases (Mutti, 1990; Mutti and Simo, 1991b).

Timing of Syndepositional Diagenetic Features

The temporal relationships among syndepositional diagenetic features allow us to relate them to depositional cycles and their bounding surfaces. The main syndepositional diagenetic events consist of marine cementation, dissolution, and dolomitization.

Syndepositional to deposition of Hairpin dolomite marine cements (high-magnesium calcite and aragonite) precipitated in intergranular, intragranular, and shelter pores in the shelf-crest and outer-shelf facies. Anhydrite precipitated in muddy facies of the inner shelf. Selective leaching of skeletal grains occurs in proximal outer-shelf grainstones and predates the replacive dolomitization. Depositional facies and laminated fibrous and bladed (I) cements were mimically replaced by dolomite in shelf-crest facies, but remain calcitic in the outer shelf. Primary dolomite precipitated in fenestral pores in inner-shelf and shelf-crest

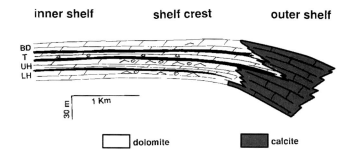

inner shelf   shelf crest   outer shelf

Figure 14. Distribution of dolomites as related to depositional cycles (LH = lower Hairpin dolomite; UH = upper Hairpin dolomite; T = Triplet dolomite; BD = basal dolomite). Thin black units are the siliciclastic sheets.

fenestral pores. Dissolution-enlarged, spoon-shaped, centimeter-scale fractures formed in the upper Hairpin dolomite outer-shelf facies and are filled by lower Triplet sandstone. Fracturing postdates the precipitation of bladed calcite I in Hairpin dolomite outer shelf. Precipitation of radial-fibrous calcite and calcite II postdate fracturing and sandstone fracture infilling.

Syndepositional to deposition of Triplet dolomite, bladed calcite II precipitated in outer-shelf facies. Anhydrite precipitated in muddy inner-shelf facies. No marine cements have been observed in Triplet dolomite inner and outer shelf. Dolomite replaced depositional fabrics in inner-shelf and shelf crest. Primary dolomite precipitated in fenestral pores in both inner shelf and the proximal shelf crest of the Triplet dolomite.

DIAGENETIC EVENTS IN A STRATIGRAPHIC FRAMEWORK

The integration of facies analysis with a detailed paragenesis of early syndepositional diagenetic features allows us to relate diagenetic events to depositional cycles. Our data show that syndepositional diagenesis is not a steady process starting immediately after deposition, but records, in analogy with depositional facies, the effects of relative eustatic variations. The main diagenetic processes occurring during development of the Yates depositional cycles are marine cementation, formation of new porosity, and dolomitization. Marine cementation (laminated, fibrous, and bladed cements) occurs during 100-k.y. cycles of transgressive phases of the 400-k.y. cycles. Extensive marine cementation occurs during deposition of cycle I, when sedimentation kept pace with a sea-level rise. During cycle II, facies belt retrogradation indicates that sedimentation did not keep pace with sea-level variations; this is suggested by the absence of marine cementation in the shelf crest.

Dolomitization occurs in muddy, evaporative, supratidal inner-shelf facies and is shifted shelfward within the deposition of each cycle. Exposure at cycle tops produces dolomitization environments over marine depositional facies, resulting in the replacement of primary calcitic fabrics and early syndepositional marine cements. In this context, the occurrence of replacive dolomitization of marine facies indicates a downward shift of facies and is related to sea-level falls at cycle tops. The repetition of transgressive-regressive cycles results in the dolomitization of shelf facies contrasted to nondolomitized outer shelf and slope.

Two relative sea-level drops (Top Hairpin dolomite and Top Triplet dolomite) are recognized in the paragenetic sequence characterized by subaerial exposure of the shelf. The relative drops have different effects on underlying sediments and this is interpreted to be related to the duration and the amplitude of subaerial exposure. The two sea-level drops are recorded by the following features. Sea level occurs at the top of Hairpin dolomite and is recorded by (1) moldic porosity formation and partial dissolution of fibrous dolomite; (2) replacive dolomitization and precipitation of clear dolomite in Hairpin dolomite shelf crest; and (3) truncation, deposition of dolomitized and nondolomitized breccia clasts, and formation of fractures in Hairpin dolomite outer shelf. Stages 1 and 2 record chemical evolution of fluids from meteoric waters to evaporation-concentrated brines; and sea-level drop occurs at the top of Triplet dolomite and is recorded by replacive dolomitization and precipitation of clear dolomite in fenestral porosity.

CONCLUSIONS

1. Three cyclic units can be distinguished in the uppermost Yates Formation, in the Capitan backreef strata, between Rattlesnake and Dark canyons. The units are bounded by sharp and erosive surfaces on the shelf that become concordant and disappear within concordant outer-shelf margin strata. The cycles are composed of a thin (0.5 to 6 m) sandstone sheet and a thicker (6 to 16 m) carbonate unit. The stacking of facies indicates that within each cyclic unit a transgressive systems tract of different thickness was deposited during a regional landward shift in the shoreline, capped by a surface of prolonged subaerial exposure.

2. Nine porosity types (four primary and five secondary) and 13 cement types (eight syndepositional and five postdepositional) are recognized. Dissolution of metastable carbonates, replacive dolomitization of shelf-crest and part of outer-shelf strata, precipitation of primary dolomite in shelf-crest fenestral pores, and formation fractures in outer-shelf strata is associated to the unconformity surfaces.

3. Replacive dolomite and primary dolomite cement are documented at the top of each cyclic unit. Replacive dolomite is found in inner-shelf and shelf-crest facies and primary dolomite cement precipitates in inner-shelf fenestral porosity. Repetition of subaerial exposure periods on the shelf related to repetitive relative sea-level oscillations leads to shelf dolomitization contrasted to non-dolomitized slope and basin.

4. The integration of facies analyses with distribution of porosity types and cement paragenetic sequence allows the assessment of relative sea-level changes (allocyclic model) and the mechanisms responsible for cyclic sandstone/carbonate deposition. Syndepositional diagenetic features are not a steady process, and the mechanisms are controlled by eustatic variations. The paragenetic sequence indicates two relative sea-level drops. Careful study of diagenetic events allows one to recognize high-frequency sea-level variations occurring within the formation of one unconformity. These features can be recognized only with a careful study of the diagenetic features.

ACKNOWLEDGMENTS

Financial support was provided by Exxon Production Research Company, The University of Wisconsin Graduate School, and GSA Grant-In-Aid. All financial support is gratefully acknowledged. We thank L.C. Pray (University of Wisconsin), R.H. Dott, Jr. (University of Wisconsin), and J.F. Sarg for their interest and technical support throughout the completion of this project. John Valley provided excellent advice, helped during the compilation of the stable isotope data, and opened the door at the University of Wisconsin Stable Isotope Laboratory. The field area is located within the boundaries of the Carlsbad Caverns National Peak, New Mexico. We are grateful for the help and cooperation of the Rangers and administrative staff of the Park. The manuscript was significantly improved by the review of P.M. Harris (Chevron), R.H. Goldstein (University of Kansas), and by AAPG reviewers E.K. Franseen (Kansas Geological Survey), R.G. Loucks (ARCO), and K.-C. Ng (ARCO). The English was improved by comments by Sara Spencer and Russell Sweeney (ETH).

REFERENCES CITED

Assereto, R.L.A.M., and R.L. Folk, 1980, Diagenetic fabrics of aragonite, calcite and dolomite in an ancient peritidal-spelean environment: Triassic Calcare Rosso, Lombardia, Italy: Journal of Sedimentary Petrology, v. 50, p. 371-395.

Borer, J.M., and P.M. Harris, 1989, Depositional facies and cycles in Yates Formation outcrops, Guadalupe Mountains, New Mexico, in P.M. Harris, and G.A. Grover, eds., Subsurface and Outcrop Examination of the Capitan Shelf Margin, Northern Delaware basin, New Mexico: SEPM Core Workshop No. 13, p. 319-324.

Borer, J.M., and P.M. Harris, 1991a, Lithofacies and cyclicity of the Yates Formation, Permian basin: Implications for reservoir heterogeneity: AAPG Bulletin, v. 75, p. 726-779.

Borer, J.M., and P.M. Harris, 1991b, Depositional facies and model for mixed siliciclastics and carbonates of the Yates Formation, Permian Basin, in A.J. Lomando, and P.M. Harris, eds., Mixed Carbonate-Siliciclastic Sequences, SEPM Core Workshop No. 15, p. 1-133.

Breed, C.S., J.F. McCauley, and P.A. Davies, 1987, Sand sheets of the eastern Sahara and ripple blankets on Mars, in L.E. Frostick, and I. Reid, eds., Desert Sediments: Ancient and Modern: Geological Society of London, Special Publication No. 35, p. 337-359.

Budd, D.A., and P.M. Harris, eds., 1990, Carbonate-Siliciclastic Mixtures: SEPM Reprint Series No. 14, 272 p.

Candelaria, M.P., 1982, Sedimentology and depositional environments of upper Yates Formation siliciclastics (Permian, Guadalupian), Guadalupe Mountains, southeast New Mexico: M.S. thesis, University of Wisconsin-Madison, 267 p.

Choquette, P.W., and P.C. Pray, 1970, Geologic nomenclature and classification of porosity in sedimentary carbonates: AAPG Bulletin, v. 54, p. 207-250.

Crysdale, B.L., 1986, Fluid inclusion evidence for the origin, diagenesis and thermal history of sparry calcite cement in the Capitan Limestone, McKittrick Canyon, West Texas: M.S. thesis, University of Colorado, Boulder, 78 p.

Darke, G., G. Harwood, and A. Kendall, 1990, Time-constraints on sulphate-related diagenesis, Capitan Reef Complex, West Texas and New Mexico, USA (abstract): 13th IAS Congress, p. 97-98.

Dott, R.H., and R.L. Batten, 1981, Evolution of the Earth, 4th ed.: McGraw-Hill, New York, 504 p.

Dunham, R.J., 1972, Capitan Reef, New Mexico and Texas: Facts and questions to aid interpretation and group discussion: Permian Basin Section, SEPM Publication 72-14, 294 p.

Esteban, M., and L.C. Pray, 1983, Pisoids and pisolite facies (Permian), Guadalupe Mountains, New Mexico and West Texas, in T.M. Peryt, ed., Coated Grains: Springer-Verlag, New York, p. 503-537.

Evamy, B.D., 1963, The application of a chemical staining technique to a study of dedolomitization: Sedimentology, v. 2, p. 164-170.

Fischer, A.G., and M. Sarnthein, 1988, Airborne silts and dune-derived sands in the Permian of the Delaware basin: Journal of Sedimentary Petrology, v. 58, p. 637-643.

Garber, R.A., G.A. Grover, and P.M. Harris, 1989, Geology of the Capitan shelf margin—Subsurface data from the Northern Delaware basin: SEPM Workshop No. 13, p. 3-269.

Given, R.K., and K.C. Lohmann, 1985, Derivation of the original isotopic composition of Permian marine cements: Journal of Sedimentary Petrology, v. 55, p. 430-439.

Given, R.K., and K.C. Lohmann, 1986, Isotopic evidence for the early meteoric diagenesis of the reef facies, Permian Reef Complex of West Texas and New Mexico: Journal of Sedimentary Petrology, v. 56, p. 183-193.

Hill, C.A., 1987, Geology of Carlsbad Caverns and other caves in the Guadalupe Mountains: New Mexico and Texas Bureau of Mines Bulletin 117, 150 p.

James, N.P., and C.F. Klappa, 1983, Petrogenesis of Early Cambrian reef limestones, Labrador, Canada: Journal of Sedimentary Petrology, v. 53, p. 1051-1096.

Kendall, C.S.St.C., 1969, An environmental reinterpretation of the Permian evaporite/carbonate shelf sediments of the Guadalupe Mountains: Geological Society of America Bulletin, v. 80, p. 2503-2526.

King, P.B., 1934, Permian stratigraphy of Trans-Pecos Texas: U.S. Geological Survey Professional Paper 187, 148 p.

Kutzbach, J.K., and R.G. Gallimore, 1989, Pangean climate: Megamonsoons of the megacontinent: Journal of Geophysical Research, v. 94, p. 3341-3357.

Lomando, A.J., and P.M. Harris, eds., 1991, Mixed Carbonate-Siliciclastic Sequences: SEPM Core Workshop No. 15.

Mazzullo, S.J., J. Mazzullo, and P.M. Harris, 1985, Eolian origin of quartzose sheet sands in Permian shelf facies, Guadalupe Mountains, Permian carbonate/clastic sedimentology, Guadalupe Mountains, Analogs for shelf and basin reservoirs: Symposium preceding the 1985 Spring Field Trip, Permian Basin Section, SEPM, Midland, Texas, program with abstract, 3 p.

McKenzie, J.A., 1981, Holocene dolomitization of calcium carbonate sediments from the coastal sabkha of Abu Dhabi, U.A.E.: A stable isotope study: Journal of Geology, v. 89, p. 185-198.

Meissner, F.F., 1972, Cyclic sedimentation in Middle Permian strata of the Permian Basin, in J.C. Elam, and S. Chuber, eds., Cyclic Sedimentation in the Permian Basin, 2nd ed.: West Texas Geological Society, Publication 72-60, p. 203-232.

Meyers, W.J., and K.C. Lohmann, 1978, Microdolomite-rich syntaxial cements: Proposed meteoric-marine mixing zone phreatic cements from Mississippi limestones, New Mexico: Journal of Sedimentary Petrology, v. 48, p. 475-488.

Moore, D., 1959, Role of deltas in the formation of some British Lower carboniferous cyclothems: Journal of Geology, v. 67, p. 522-539.

Mruk, D.H. 1985, Cementation and dolomitization of the Capitan Limestone (Permian), McKittrick Canyon, West Texas: M.S. thesis, University of Colorado, Boulder, 153 p.

Mruk, D.H., 1989, Diagenesis of the Capitan Limestone, Upper Permian, McKittrick Canyon, West Texas, in subsurface and outcrop examination of the Capitan shelf margin, Northern Delaware basin: SEPM Core Workshop No. 13, p. 387-406.

Mutti, M., 1990, Sedimentology and diagenesis of sili-ciclastic/carbonate cycles, Yates Formation, Permian, New Mexico: M.S. thesis, University of Wisconsin-Madison, 228 p.

Mutti, M., and J.A. Simo, 1991a, Stratigraphic patterns, sedimentology and diagenesis of Capitan Backreef strata, Guadalupe Mnts., New Mexico (abstract): AAPG Bulletin, v. 75, p. 643-644.

Mutti, M., and J.A. Simo, 1991b, Eustatic control of early dolomitization of cyclic shelf facies, Yates Formation (Guadalupian): Capitan Reef Complex (abstract): Dolomieu Conference, p. 187-188.

Neese, D.G., 1979, Facies mosaic of the Upper Yates and Lower Tansill formations (Upper Permian), Walnut Canyon, Guadalupe Mountains, New Mexico: M.S. thesis, University of Wisconsin-Madison, 110 p.

Neese, D.G., and A. Schwartz, 1977, Facies mosaic of upper Yates and lower Tansill formations, Walnut and Rattlesnake Canyons, Guadalupe Mountains, New Mexico, in M.E. Hileman, and S.J. Mazzullo, eds., Upper Guadalupian Facies, Permian Reef Complex: Permian Basin Section, SEPM Publication 77-16, p. 437-450.

Oliver, T.A., and N.W. Cowper, 1963, Depositional environments of the Ireton Formation, central Alberta: Canadian Petroleum Geology Bulletin, v. 2, p. 183-202.

Purser, B.H., and J.H. Schroeder, 1986, Conclusions, the diagenesis of reefs: A brief review of our present understanding, in P.O. Roehl, and R.W. Choquette, eds., Reef Diagenesis: Springer-Verlag, New York, p. 341-355.

Pray, L.C., 1977, The all wet constant sea level hypothesis of upper Guadalupian shelf and shelf edge strata, Guadalupe Mountains, New Mexico and Texas, in M.E. Hileman, and S.J. Mazzullo, eds., Upper Guadalupian Facies, Permian Reef Complex, Guadalupe Mountains, New Mexico and West Texas: Field Conference Guidebook, v. 1, Permian Basin Section, SEPM Publication 77-16, p. 437-450.

Pray, L.C., 1988, Geology of the western escarpment, Guadalupe Mountains, Texas, in J.F. Sarg, C. Rossen, P.J. Lehmann, and L.C. Pray, eds., Geologic Guide to the Western Escarpment Guadalupe Mountains, Texas: Permian Basin Section, SEPM Publication 88-30.

Pray, L.C., and M. Esteban, eds., 1977, Upper Guadalupian Facies, Permian Reef Complex, Guadalupe Mountains, New Mexico and West Texas: 1977 Field Conference Guidebook, v. 2, Permian Basin Section, SEPM Publication 77-16, 194 p.

Rosenblum, M.B., 1984, Early diagenetic sheet crack cements of the Guadalupian (Permian) shelf, Yates and Tansill formations, New Mexico, USA: M.S. thesis, University of Wisconsin-Madison, 95 p.

Sarg, J.F., 1977, Sedimentology of the carbonate-evaporate facies transition of the Seven Rivers Formation (Guadalupian, Permian) in southeast New Mexico, in M.E. Hileman, and S.J. Mazzullo,

eds., Upper Guadalupian Facies, Permian Reef Complex, Guadalupe Mountains, New Mexico and West Texas: 1977 Field Conference Guidebook, v. 1, SEPM Publication 77-16, p. 451-478.

Sarg, J.F., 1988, Carbonate sequence stratigraphy, in C.K. Wilgus, B.S. Hastings, C.G.C. St. Kendall, H.W. Posamentier, C.A. Ross, and J.C. Van Wagoner, eds., Sea Level Changes: An Integrated Approach: SEPM Special Publication No. 42, p. 155-181.

Sarg, J.F., 1989, Middle-Late Permian depositional sequences, Permian Basin, West Texas and New Mexico, in A.W. Bally, ed., Atlas of Seismic Stratigraphy: AAPG Studies in Geology, v. 3, p. 140-154.

Sarg, J.F., and P.J. Lehmann, 1986, Lower-Middle Guadalupian facies and stratigraphy, San Andres/Grayburg Formations, Permian Basin, Guadalupe Mountains, New Mexico: SEPM Annual Field Trip Guidebook, Permian Basin Section.

Schwartz, A.H., 1981, Facies mosaic of the Upper Yates and Lower Tansill formations (Upper Permian), Rattlesnake Canyon, Guadalupe Mountains, New Mexico, M.S. thesis, University of Wisconsin-Madison, 148 p.

Scotese, C.R., R.K. Bambach, C. Barton, R. Vandervoo, and A.M. Ziegler, 1979, Paleozoic base maps: Journal of Geology, v. 87, p. 217-277.

Silver, B.A., and R.G. Todd, 1969, Permian cyclic strata, northern Midland and Delaware basins, West Texas and Southeastern New Mexico: AAPG Bulletin, v. 53, p. 2223-2251.

Smith, D.B., 1974, Sedimentology of Upper Artesia (Guadalupian) cyclic shelf deposits of northern Guadalupe Mountains, New Mexico: AAPG Bulletin, v. 58, p. 1699-1730.

Smith, G.A., J.C. Bryden, and G.A. Drewry, 1973, Phanerozoic world maps, in N.F. Hughes, ed., Organisms and Continents Through Time: Paleontological Association, Special Papers in Paleontology, No. 12, p. 1-42.

Tucker, M., and H. Hollingworth, 1986, The Upper Permian reef complex (EZ1) of North East England: Diagenesis in a marine to evaporitic setting, in J.H. Schroeder, and B.H. Purser, eds., Reef Diagenesis: Springer-Verlag, New York, p. 270-290.

Van Siclen, D.C., 1958, Depositional topography-examples and theory: AAPG Bulletin, v. 42, p. 1896-1913.

Ward, R.F., C.G.St.C. Kendall, and P.M. Harris, 1986, Upper Permian (Guadalupian) facies and their association with hydrocarbons—Permian Basin, West Texas and New Mexico: AAPG Bulletin, v. 70, p. 239-262.

Wilson, J.L., 1972, Cyclic and reciprocal sedimentation in Virgilian strata of southern New Mexico, in J.G. Elam, and S. Chuber, eds., Cyclic Sedimentation in the Permian Basin, 2nd ed.: West Texas Geological Society, Publication 72-60, p. 82-99.

Wilson, J.L., 1975, Carbonate Facies in Geologic History: Springer-Verlag, New York, 471 p.

Index